Probability and
Conditional Expectation

WILEY SERIES IN PROBABILITY AND STATISTICS

Established by *Walter A. Shewhart and Samuel S. Wilks*

Editors: *David J. Balding, Noel A. C. Cressie, Garrett M. Fitzmaurice, Geof H. Givens, Harvey Goldstein, Geert Molenberghs, David W. Scott, Adrian F. M. Smith, Ruey S. Tsay*

Editors Emeriti: *J. Stuart Hunter, Iain M. Johnstone, Joseph B. Kadane, Jozef L. Teugels*

The *Wiley Series in Probability and Statistics* is well established and authoritative. It covers many topics of current research interest in both pure and applied statistics and probability theory. Written by leading statisticians and institutions, the titles span both state-of-the-art developments in the field and classical methods.

Reflecting the wide range of current research in statistics, the series encompasses applied, methodological and theoretical statistics, ranging from applications and new techniques made possible by advances in computerized practice to rigorous treatment of theoretical approaches. This series provides essential and invaluable reading for all statisticians, whether in academia, industry, government, or research.

A complete list of titles in this series can be found at
http://www.wiley.com/go/wsps

Probability and Conditional Expectation

Fundamentals for the Empirical Sciences

Rolf Steyer

Institute of Psychology, University of Jena, Germany

Werner Nagel

Institute of Mathematics, University of Jena, Germany

WILEY

Registered office
John Wiley & Sons Ltd, The Atrium, Southern Gate, Chichester, West Sussex, PO19 8SQ, United Kingdom

For details of our global editorial offices, for customer services and for information about how to apply for permission to reuse the copyright material in this book please see our website at www.wiley.com.

Library of Congress Cataloging-in-Publication Data

Names: Steyer, Rolf, 1950– | Nagel, Werner, 1952–.
Title: Probability and conditional expectation : fundamentals for the empirical sciences / Rolf Steyer, Werner Nagel.
Description: Chichester, West Sussex : John Wiley & Sons, Inc., 2017. |
    Includes index.
Identifiers: LCCN 2016025874 | ISBN 9781119243526 (cloth) | ISBN 9781119243489
    (epub) | ISBN 9781119243502 (epdf)
Subjects: LCSH: Conditional expectations (Mathematics) | Random variables. | Independence (Mathematics) | Dependence (Statistics) | Measure theory. | Probability and statistics. | Correlation (Statistics) | Multivariate analysis. | Regression analysis. | Logistic regression analysis. | Measure algebras. | Probabilities.
Classification: LCC QA273 .S75325 2017 | DDC 519.2–dc23
LC record available at https://lccn.loc.gov/2016025874

A catalogue record for this book is available from the British Library.

Set in 10/12pt Times by Aptara Inc., New Delhi, India
Printed and bound in Malaysia by Vivar Printing Sdn Bhd

10 9 8 7 6 5 4 3 2 1

For my wonderful wife and children.
—RST

Contents

Preface

Why another book on probability?

This book has two titles. The subtitle, 'Fundamentals for the Empirical Sciences', reflects the intentions and the motivation of the first author for writing this book. He received his academic training in psychology but considers himself a methodologist. His scientific interest is in explicating fundamental concepts of empirical research (such as causal effects and latent variables) in terms of a language that is *precise* and at the same time *compatible with the statistical models* used in the analysis of empirical data. Applying statistical models aims at estimating and testing hypotheses about parameters such as expectations, variances, covariances, and so on (or of functions of these parameters, such as differences between expectations, ratios of variances, regression coefficients, etc.), all of which are terms of probability theory. Precision is necessary for securing logical consistency of theories, whereas compatibility of theories about real-world phenomena with statistical models is crucial for probing the empirical validity of theoretical propositions via statistical inference.

Much empirical research uses some kind of regression in order to investigate how the expectation of one random variable depends on the values of one or more other random variables. This is true for analysis of variance, regression analysis, applications of the general linear model and the generalized linear model, factor analysis, structural equation modeling, hierarchical linear modeling, and the analysis of qualitative data. Using these methods, we aim at learning about specific regressions. A regression is a synonym for what, in probability theory, is called a factorization of a conditional expectation, provided that the regressor is numerical. This explains the main title of this book, 'Probability and Conditional Expectation'.

What is it about?

Since the seminal book of Kolmogoroff (1933–1977), the fundamental concepts of probability theory are considered to be special concepts of measure theory. A probability measure is a special finite measure, random variables are special measurable mappings, and expectations of random variables are integrals of measurable mappings with respect to a probability measure. This motivates Part I of this book with three chapters on the measure-theoretical foundations of probability theory. Although at first sight this part seems to be far off from practical applications, the contrary is true. This part is indispensable for probability theory and for its applications in empirical sciences. This applies not only to the concepts of a measure and an integral but also, in particular, to the concept of a measurable mapping, although we concede that the full relevance of this concept will become apparent only in the chapters on conditional expectations. The relevance of measurable mappings is also the reason why chapter 2 is more detailed than the corresponding chapters in other books on measure theory.

Part II of the book is fairly conventional. The material covered – probability, random variable, expectation, variance, covariance, and some distributions – is found in many books on probability and statistics.

Part III is not only the longest; it is also the core of the book that distinguishes it from other books on probability or on probability and statistics. Only a few of these other books contain detailed chapters on conditional expectations. Exceptions are Billingsley (1995), Fristedt and Gray (1997), and Hoffmann-Jørgensen (1994). Our book does not cover any statistical model. However, we treat in much detail *what* we are estimating and which the hypotheses are that we test or evaluate in statistical modeling. *How* we are estimating is important, but *what* we are estimating is of most interest from the empirical scientist point of view, and this point is typically neglected in books on statistics and in books on probability theory such as Bauer (1996) or Klenke (2013). A simple example is the meaning of the coefficient β_1 in the equation $E(Y \mid X, Z) = \beta_0 + \beta_1 X + \beta_2 Z + \beta_3 ZX$. Oftentimes, this coefficient is misinterpreted as the 'main effect' of X. However, sometimes β_1 has no autonomous meaning at all, for example if $P(Z = 0) = 0$. In general, this coefficient is just a component of the function $g_1(Z) = \beta_1 + \beta_3 Z$ that can be used to compute the conditional effects of X on Y for various values z of Z (see chapter 15 for more details). The crucial point is that such concepts can be treated most clearly within probability theory, without referring to a statistical model, sample, estimation, or testing.

This also includes exemplifying the limitations of conditional expectations. Simple examples show that conditional expectations do not necessarily serve the purpose of the empirical researcher, which often is to evaluate the effects of an intervention on an outcome variable. But even in these situations, conditional expectations are indispensable for the definition of parameters and other terms of substantive interest (see, e.g., chapter 14).

There is much overlap of Parts II and III with Steyer (2003). However, that book is written in German, and the mathematics is considerably less rigorous. Aside from mathematical precision, the two books also differ in the definition of an important concept: In Steyer (2003), the term *regression* is used as a synonym of a conditional expectation, whereas in this book we use it as a synonym for the factorization of a conditional expectation $E(Y \mid X)$, provided that the codomain of X is $\mathbb{R}^n$.

In chapter 9, the first chapter of Part III, we gently introduce conditional expectation values and discrete conditional expectations. In chapter 10, we then present the general theory of conditional expectations that has been introduced by Kolmogoroff (1933–1977) and since that time has been treated in many books on probability theory – although much too briefly in order to be intelligible to researchers in empirical sciences. Our chapter on conditional expectations contains many more details and is supplemented by a number of other chapters on important special aspects and special cases.

Such a special aspect is the concept of a residual with respect to a conditional expectation (see chapter 11). Residuals have many interesting properties, and they are used in order to introduce the concepts of conditional variance and covariance, as well as the notion of a partial correlation. We then turn to specific parameterizations of a conditional expectation, including the concepts of a linear regression (chapter 12) and a linear logistic regression (chapter 13). Note that these concepts are introduced as probabilistic concepts. As mentioned, they are what we aim at estimating in applying the corresponding statistical models.

Chapters 14 to 16 provide the probabilistic foundations of the analysis of conditional and average effects of treatments, interventions, or expositions to potentially harmful or beneficial environments. To our knowledge, this material is not found in any other textbook. Note, however, that although these two chapters provide important concepts, they do not cover the theory of causal effects, which is another book project of the first author.

Part IV uses conditional expectations in order to introduce conditional independence (chapter 16) and conditional distributions (chapter 17). Although these two chapters are more extensive than comparable chapters or sections in other books, the material is found in other books on probability theory as well.

For whom is it?

This book has been written for two kinds of readers. The first are applied statisticians and empirical researchers who want to understand in a proper language (i.e., in terms of probability theory) what they estimate and test in their empirical studies. The second kind of readers are mathematicians who want to understand in terms of probability theory what applied statisticians and empirical researchers estimate and test in their research. Both kinds of readers are potential contributors to the methodology of empirical sciences.

Many exercises and their solutions provide extensive material for assignments in courses, but they also facilitate independent learning. At the same time, these exercises and their solutions help streamline the main text.

Note that we do not provide all proofs, in particular in the chapters on measure, integral, and distributions. In these cases, we refer to other textbooks instead. We decided to include only those proofs that may help to increase understanding of the background and to learn important mathematical procedures. Of course, we provide proofs of all propositions for which we did not find an appropriate reference.

Prerequisites

We assume that the reader is familiar with the elementary concepts of logic, sets, functions, sequences, and matrices, as presented for example in chapters 1 and 2 of Rosen (2012). We try to stick to his notation as closely as possible.

One of the exceptions is the symbol for the implication, for which we use $\Rightarrow$ instead of $\rightarrow$. Another exception is the symbol for the equivalence, for which we use $\Leftrightarrow$ instead of $\leftrightarrow$.

Box 0.1 summarizes the most important notation to start with. The concepts referred to by these symbols are defined, for example, in Rosen (2012) or in Ellis and Gulick (2006). For a rich collection of mathematical formulas, we recommend the handbooks of Harris and Stocker (1998) and Bronshtein *et al.* (2015).

Box 0.1   A first list of symbols.

$\neg$	not
$\wedge$	and
$\vee$	or
$\Rightarrow$	implies
$\Leftrightarrow$	equivalent to
$\exists$	there is (there exists)
$\forall$	for all
$a \in A$	a is an element of the set A
$\emptyset$	empty set
I	nonempty finite, countable, or uncountable index set
$(A_i, i \in I)$	family of sets $A_i, i \in I$
$A \cup B$	union of the sets A and B
$\bigcup_{i \in I} A_i$	union of the sets $A_i, i \in I$
$A \cap B$	intersection of the sets A and B
$\bigcap_{i \in I} A_i$	intersection of the sets $A_i, i \in I$
$A \setminus B$	set difference of the set A and the set B
$A^c := \Omega \setminus A$	complement of a set $A \subset \Omega$ with respect to a set Ω
$A \subset B$	A is a subset of the set B; $A \subset B$ includes $A = B$
$A \times B$	Cartesian product (product set) of A and B
$\bigtimes_{i=1}^{n} A_i$	family of sets $A_i, i = 1, \dots, n$
$f : A \to B$	mapping f assigning to each $a \in A$ (the domain) one and only one element $b \in B$ (the codomain)
$\sum_{i=1}^{n} a_i$	sum of the real numbers $a_1, \dots, a_n$
$\prod_{i=1}^{n} a_i$	product of the real numbers $a_1, \dots, a_n$
$\lim_{n \to \infty} a_n$	limit of a sequence $a_1, a_2, \dots$ of real numbers
$\sum_{i=1}^{\infty} a_i$	$\lim_{n \to \infty} \sum_{i=1}^{n} a_i$ where $a_1, a_2, \dots$ as well as $\sum_{i=1}^{1} a_i, \sum_{i=1}^{2} a_i, \dots$ are sequences of real numbers

Acknowledgements

This book could not have been written without the help of many. First of all, we thank Ivailo Partchev, who prepared the LaTeX framework and many of the figures, tables, and boxes. Some of the figures have been produced by Désirée Thielemann and Julie Toussaint, who also cared for references, read some of the chapters, and hinted at errors. For supporting us with respect to LaTeX, finding errors, or suggesting other improvements, we also thank Karoline Bading, Marcel Bauer, Sonja Hahn, Gregor Kappler, Christoph Kiefer, Andreas Neudecker, Axel Mayer, Erik Sengewald, Jan Plötner, Carolin Rebekka Scheifele, and Tom Landes. Thanks are also due to Ernesto San Martin for suggesting section 1.7 and proposition (iv) of Theorem 16.37. The proof of Lemma 12.38 is due to Peter Vogel. Finally, we would like to thank our students who kept us thinking on how to improve the text.

About the companion website

This book is accompanied by a companion website:

http://www.probability-and-conditional-expectation.de

This website includes:

- Errata
- Videos
- Slides
- Teaching tools
- Datasets.

Part I

MEASURE-THEORETICAL FOUNDATIONS OF PROBABILITY THEORY

1

Measure

In this chapter, we introduce the concept of a measure and other closely related notions. We start with some examples and then introduce the concept of a σ-*algebra*, which is crucial in measure theory and probability theory. At first glance this concept seems to be a pure technical construction, which is usually not dealt with in textbooks on 'Probability and Statistics' for empirical sciences. However, a σ-algebra turned out to be the natural domain for a measure, including probability measures. Moreover, in probability theory, a σ-algebra is not only the domain of probability measures. The σ-algebra generated by a random variable can be interpreted as the set of events that is represented by this random variable. This is treated in more detail in chapter 2 on measurable mappings, which provides the general theory of random variables because random variables are measurable mappings. The virtues of σ-algebras will become fully apparent in chapter 10 on conditional expectations and its subsequent chapters. The pair $(\Omega, \mathcal{A})$ consisting of a nonempty set Ω and a σ-algebra $\mathcal{A}$ on Ω is called a *measurable space*. Such a measurable space is crucial for the definition of a *measure*. Next, we treat some important examples of measures, including the *counting measure*, the *Dirac measure*, and the *Lebesgue measure*. Finally, we turn to *continuity* and *uniqueness* properties of a measure.

1.1 Introductory examples

Consider Figure 1.1 showing the set Ω of all points (x, y) inside the rectangle and the sets A and B of all points (x, y) inside the two ellipses, respectively. These three sets are subsets of the plane $\mathbb{R}^2 := \mathbb{R} \times \mathbb{R}$, where $\mathbb{R}$ denotes the *set of all real numbers*, and $\mathbb{R} \times \mathbb{R} := \{(a, b): a, b \in \mathbb{R}\}$ is the set of all ordered pairs (a, b) with $a, b \in \mathbb{R}$, called the *Cartesian product* or *product set* of $\mathbb{R}$ with itself. In Figure 1.1, the sets A and B have a nonempty intersection. Now let $area(A)$ and $area(B)$ denote their areas and $area(A \cap B)$ the area of their intersection. Inspecting this figure reveals:

$$area(A \cup B) = area(A) + area(B) - area(A \cap B).$$

Probability and Conditional Expectation: Fundamentals for the Empirical Sciences, First Edition. Rolf Steyer and Werner Nagel.
© 2017 John Wiley & Sons, Ltd. Published 2017 by John Wiley & Sons, Ltd.
Companion website: http://www.probability-and-conditional-expectation.de

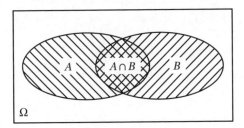

Figure 1.1   A Venn diagram of two sets and their intersection.

This example illustrates three important points:

(a) A measure such as *area* is a function on a *set system on* Ω, (i.e., on a *set of subsets* of a set Ω such as A, B, and $A \cap B$).

(b) If *area* is defined for the *subsets* $A, B \subset \Omega$, then it is also defined for their *intersection* $A \cap B$ and for their *union* $A \cup B$.

(c) Measures are *additive*. In other words, if A and B are *disjoint* subsets of Ω (i.e., if $A \cap B = \emptyset$), then $area\,(A \cup B) = area\,(A) + area\,(B)$.

Note that, in the example presented in Figure 1.1, the sets A and B are *not disjoint*, and this is why $area\,(A \cap B)$ has to be subtracted in the equation displayed above. Points (a) to (c) also apply to other measures such as *length* and *volume* as well as to *probability measures*. Therefore, we adopt a more general language and talk about subsets A, B of a set Ω (or *measurable sets* A, B) and their *measure* μ instead of lines and their lengths, rectangles and their areas, cubes and their volume, or events and their probabilities.

For example, if $\Omega = \{1, \dots, 6\}$ denotes the set of possible outcomes of tossing a fair dice, $A = \{1, 6\}$ and $B = \{2, 4, 6\}$ denote the events of tossing a 1 *or* a 6 and tossing an *even number*, respectively. Furthermore, $A \cap B = \{6\}$ and the probability of tossing a 1 or a 6 or an even number – the event $A \cup B$ – is

$$P(A \cup B) = P(A) + P(B) - P(A \cap B) = \frac{2}{6} + \frac{3}{6} - \frac{1}{6} = \frac{4}{6}.$$

In the first example, the measure *area* assigns a real number to a subset of $\mathbb{R}^2$. In the second example, the measure P assigns a real number to a subset of $\Omega = \{1, \dots, 6\}$. This suggests that a measure should be defined such that it assigns a real number *to all subsets* of a set (i.e., to all elements of the power set). Unfortunately, this may lead to contradictions (see, e.g., Georgii, 2008). In contrast, when defining a measure on a σ-algebra, such contradictions can be avoided.

1.2   σ-Algebra and measurable space

In Definition 1.1, we consider a set system $\mathscr{A}$ on Ω, a sequence $A_1, A_2, \dots$ of subsets of Ω, and their countable union. Remember, a *set system on a set* Ω is a set of subsets of Ω presuming that Ω is not empty. A *sequence of subsets of a set* Ω is a function from the set $\mathbb{N}_0 = \{0, 1, 2, \dots\}$

or $\mathbb{N} = \{1, 2, \ldots\}$ or a subset of these sets to $\mathcal{P}(\Omega)$, the *power set* of Ω. Furthermore, the *finite union* of the sets $A_1, \ldots, A_n$ and the *countable union* of the sets $A_1, A_2, \ldots$ are defined by

$$\bigcup_{i=1}^{n} A_i := \{a \in \Omega \colon \exists\, i \in \{1, \ldots, n\} \colon a \in A_i\} \tag{1.1}$$

and

$$\bigcup_{i=1}^{\infty} A_i := \{a \in \Omega \colon \exists\, i \in \mathbb{N} \colon a \in A_i\}, \tag{1.2}$$

respectively. Hence, by definition, $\bigcup_{i=1}^{n} A_i$ is the set of all elements that are an element of at least one of the sets A_i, $i = 1, \ldots, n$, and $\bigcup_{i=1}^{\infty} A_i$ is the set of all elements that are an element of at least one of the sets A_i, $i \in \mathbb{N}$. Finally, $A^c := \Omega \setminus A$ denotes the complement of A (with respect to Ω).

Definition 1.1 [σ-Algebra]
A set $\mathcal{A}$ of subsets of a nonempty set Ω is called a σ-algebra (or σ-field) on Ω, if the following three conditions hold:

(a) $\Omega \in \mathcal{A}$.

(b) If $A \in \mathcal{A}$, then $A^c \in \mathcal{A}$.

(c) If $A_1, A_2, \ldots \in \mathcal{A}$, then $\bigcup_{i=1}^{\infty} A_i \in \mathcal{A}$.

An element of a σ-algebra is called a measurable set.

Remark 1.2 [Closure with respect to set operations] Condition (c) postulates that σ-algebras are closed with respect to *countable* unions of sets $A_1, A_2, \ldots \in \mathcal{A}$. However, in conjunction with (a) and (b), this implies that a σ-algebra is also closed with respect to *finite* unions of sets $A_1, \ldots, A_n \in \mathcal{A}$, because every finite union of sets $A_1, \ldots, A_n \in \mathcal{A}$ can be represented as a countable union of the sets that are elements of $\mathcal{A}$, for example:

$$\bigcup_{i=1}^{n} A_i = A_1 \cup \ldots \cup A_n \cup \emptyset \cup \emptyset \cup \ldots . \tag{1.3}$$

Note that (a) and (b) imply $\emptyset \in \mathcal{A}$, because $\Omega^c = \emptyset$.

Furthermore, although condition (c) only requires explicitly that σ-algebras are closed with respect to countable unions, Definition 1.1 implies that a σ-algebra is closed also with respect to intersections such as $A_1 \cap A_2$ and set differences $A_1 \setminus A_2$. In other words, if A_1 and A_2 are elements of $\mathcal{A}$, then $A_1 \cup A_2$, $A_1 \cap A_2$, and $A_1 \setminus A_2$ are elements of $\mathcal{A}$ as well, provided that $\mathcal{A}$ is a σ-algebra. The same is true for countable intersections $A_1 \cap A_2 \cap \ldots$ of elements of $\mathcal{A}$. In more formal terms: If $\mathcal{A}$ is a σ-algebra, then,

$$A_1, A_2, \ldots \in \mathcal{A} \Rightarrow \bigcap_{i=1}^{\infty} A_i \in \mathcal{A} \tag{1.4}$$

(see Exercise 1.1), where $\bigcap_{i=1}^{\infty} A_i = A_1 \cap A_2 \cap \ldots$ is defined by

$$\bigcap_{i=1}^{\infty} A_i := \{a \in \Omega : \forall\, i \in \mathbb{N} : a \in A_i\}. \tag{1.5}$$

Because

$$\bigcap_{i=1}^{n} A_i = A_1 \cap \ldots \cap A_n \cap \Omega \cap \Omega \cap \ldots , \tag{1.6}$$

we can also conclude

$$A_1, \ldots, A_n \in \mathcal{A} \Rightarrow \bigcap_{i=1}^{n} A_i \in \mathcal{A}, \tag{1.7}$$

where $\bigcap_{i=1}^{n} A_i$, the *finite intersection* of the sets $A_1, \ldots, A_n$, is defined by

$$\bigcap_{i=1}^{n} A_i := \{a \in \Omega : \forall\, i \in \{1, \ldots, n\} : a \in A_i\}. \tag{1.8}$$

◁

Remark 1.3 [Countable and uncountable unions]   Defining a σ-*algebra,* we use the symbol σ in order to emphasize that unions of finitely or countably many sets are considered, *but not other unions of sets.* For example, the *closed interval* $[a, b] := \{x \in \mathbb{R} : a \leq x \leq b, a, b \in \mathbb{R}\}$ on the real axis is identical to the union of singletons $\{x\}$ that contain only one single element $x \in \mathbb{R}$, that is,

$$[a, b] = \bigcup_{a \leq x \leq b} \{x\}. \tag{1.9}$$

This union is neither finite nor countable. Hence, condition (c) of Definition 1.1 does *not* imply that this union is necessarily an element of a σ-algebra $\mathcal{A}$ on $\mathbb{R}$, even if all singletons $\{x\}, x \in \mathbb{R}$, are elements of $\mathcal{A}$.

◁

The following notion of a *measurable space* proves to be convenient in measure theory.

Definition 1.4 [Measurable space]
If Ω is a nonempty set and $\mathcal{A}$ a σ-algebra on Ω, then the pair $(\Omega, \mathcal{A})$ is called a measurable space.

Example 1.5 [The smallest σ-algebra]   The smallest σ-algebra on a nonempty set Ω is $\mathcal{A} = \{\Omega, \emptyset\}$. It contains only the elements Ω and the empty set $\emptyset$. As is easily seen, $\Omega \cup \emptyset = \Omega$, $\Omega^c = \emptyset$, and $\emptyset^c = \Omega$ are elements of $\mathcal{A}$. This shows that $\mathcal{A} = \{\Omega, \emptyset\}$ is closed with respect to union and complement.

◁

Example 1.6 [Power set]   The *power set* $\mathscr{P}(\Omega)$ of a nonempty set Ω (i.e., the *set of all subsets* of Ω) is a σ-algebra on Ω. It is the largest σ-algebra on a nonempty set Ω. All other σ-algebras on Ω are subsets of $\mathscr{P}(\Omega)$.   ◁

Example 1.7 [A small σ-algebra]   If A is a subset of a nonempty set Ω, then $\mathscr{A} = \{\Omega, \emptyset, A, A^c\}$ is always a σ-algebra on Ω (see Exercise 1.2). Again, it is easily seen that this set system is closed with respect to union and complement.   ◁

Remark 1.8 [Motivation for σ-algebras]   These examples show that there can be many different σ-algebras on a nonempty set Ω. Why not simply always use the largest one, the power set $\mathscr{P}(\Omega)$? In fact, this would be possible as long as Ω is finite or countable. There are at least three reasons for using σ-algebras. First, there are important sets Ω (e.g., $\Omega = \mathbb{R}$) such that measures of interest (e.g., *length* — which is the Lebesgue measure pertaining to $\Omega = \mathbb{R}$) cannot be defined on $\mathscr{P}(\Omega)$ (see e.g., Wise and Hall, 1993, counterexample 1.25). These measures can be defined, however, on other σ-algebras, such as the Borel-σ-algebra [see Eq. (1.18)]. (For an example in which the power set is 'too large', see Georgii, 2008.) Second, in some sense, σ-algebras contain those elements of a larger set system that are relevant for a particular question. In probability theory, together with Ω and a probability measure, each σ-algebra on Ω represents a random experiment that is in some sense contained in an (often larger) random experiment. For example, if we consider the random experiment of tossing a dice, then we may focus on whether or not the number of points is even. Together with Ω and the probability measure, the corresponding σ-algebra represents a 'new' random experiment contained in the random experiment of tossing a dice (see Exercise 1.3). Third, using different σ-algebras is indispensable for introducing conditional expectations, conditional independence, and conditional distributions (see chs. 9 to 17).   ◁

Example 1.9 [Joe and Ann]   Consider the following random experiment: First, we sample a unit u from the set $\Omega_U := \{Joe, Ann\}$. Second, each unit receives (*yes*) or does not receive a treatment (*no*). Third, it is observed whether (+) or not (−) a success criterion is reached (see Fig. 1.2). Defining $\Omega_X := \{yes, no\}$ and $\Omega_Y := \{+, -\}$, the Cartesian product

$$\Omega := \Omega_U \times \Omega_X \times \Omega_Y = \{(Joe, no, -), (Joe, no, +), \ldots, (Ann, yes, +)\}$$

is the set of possible outcomes ω of this random experiment. It has eight elements, namely the triples $(Joe, no, -)$, $(Joe, no, +)$, …, $(Ann, yes, +)$ (see all eight leaves of Fig. 1.2 for a complete list of these elements).

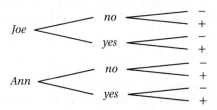

Figure 1.2   Example of a tree representation of a Cartesian product.

In this example, a first σ-algebra $\mathscr{A}$ we may consider is *the set of all subsets of* Ω, the power set $\mathscr{P}(\Omega)$. This set has $2^8 = 256$ elements, where 8 is the number of elements, that is, the cardinality of Ω (see Kheyfits, 2010, Th. 1.1.37). Among these elements is the set

$$A := \{(Joe, no, -), (Joe, no, +), (Joe, yes, -), (Joe, yes, +)\} = \{Joe\} \times \Omega_X \times \Omega_Y.$$

In the context of probability theory, it is also called the event that *Joe is drawn*. Other elements of $\mathscr{A}$ are the events

$$B := \{(Joe, yes, -), (Joe, yes, +), (Ann, yes, -), (Ann, yes, +)\} = \Omega_U \times \{yes\} \times \Omega_Y$$

that the *drawn person is treated*, and

$$C := \{(Joe, no, +), (Joe, yes, +), (Ann, no, +), (Ann, yes, +)\} = \Omega_U \times \Omega_X \times \{+\}$$

that *success* (+) occurs, irrespective of which person is drawn and whether or not the person is treated.

Aside from the power set of Ω, we could also consider the σ-algebras $\mathscr{A}_1 := \{\Omega, \emptyset, A, A^c\}$, $\mathscr{A}_2 := \{\Omega, \emptyset, B, B^c\}$, and $\mathscr{A}_3 := \{\Omega, \emptyset, C, C^c\}$, to name just three. (For another one, see Exercise 1.4.) In a sense, $\mathscr{A}_1$ represents the information regarding which person is drawn. In contrast, $\mathscr{A}_2$ contains the information regarding whether or not the drawn person is treated, and $\mathscr{A}_3$ whether or not the drawn person is successful. Of course, all these σ-algebras are subsets of $\mathscr{P}(\Omega)$, the power set of Ω.   ◁

Example 1.10 [Trace of a set system and trace σ-algebra]   Let Ω and Ω_0 be nonempty sets. If $\mathscr{E}$ is a set system on Ω and $\Omega_0 \subset \Omega$, then

$$\mathscr{E}|_{\Omega_0} := \{\Omega_0 \cap A : A \in \mathscr{E}\}$$

is a set system on Ω_0. It is called the *trace of $\mathscr{E}$ in* Ω_0. Furthermore, if $\mathscr{A}$ is a σ-algebra on Ω and $\Omega_0 \subset \Omega$, then the set system

$$\mathscr{A}|_{\Omega_0} := \{\Omega_0 \cap A : A \in \mathscr{A}\}$$

is a σ-algebra on Ω_0 (see Exercise 1.5). If $\Omega \neq \Omega_0$, then the trace $\mathscr{A}|_{\Omega_0}$ is a σ-algebra on Ω_0, but not on Ω, because $\Omega \notin \mathscr{A}|_{\Omega_0}$.   ◁

Example 1.11 [Joe and Ann – continued]   In Example 1.9, we defined the event A that Joe is drawn, the event B that the drawn person is treated, and the σ-algebra $\mathscr{A}_2 = \{\Omega, \emptyset, B, B^c\}$. The trace of $\mathscr{A}_2$ in A is

$$\mathscr{A}_2|_A = \{A, \emptyset, A \cap B, A \cap B^c\}.$$

Obviously, just like all elements of $\mathscr{A}_2$ are subsets of Ω, all elements of $\mathscr{A}_2|_A$ are subsets of A. From an application point of view, considering $\mathscr{A}_2|_A$ means to presume that Joe is drawn and consider the events that he is treated or not treated, respectively.   ◁

1.2.1 σ-Algebra generated by a set system

The concept of a σ-*algebra generated by a set system* is useful in order to define important σ-algebras. It is also useful for specifying certain measures (see section 1.6). Theorem 1.12 prepares Definition 1.13. Reading this theorem, remember that a σ-algebra on a set Ω is itself a set (of subsets of Ω), so that we can consider the intersection of σ-algebras.

Theorem 1.12 [Intersection of σ-algebras is a σ-algebra]
Let I be a nonempty (finite, countable, or uncountable) index set, and let all $\mathcal{A}_i$, $i \in I$, be σ-algebras on Ω. Then, $\bigcap_{i \in I} \mathcal{A}_i$ is also a σ-algebra on Ω.

(Proof p. 28)

This theorem allows us to define the σ-*algebra generated by a set system* on Ω.

Definition 1.13 [σ-Algebra generated by a set system]
Let $\mathcal{E}$ be a set system on a nonempty set Ω, and let $(\mathcal{A}_i, i \in I)$ be the family of all σ-algebras on Ω that contain $\mathcal{E}$ as a subset. Then, we define

$$\sigma(\mathcal{E}) := \bigcap_{i \in I} \mathcal{A}_i \tag{1.10}$$

and call it the σ-algebra generated by $\mathcal{E}$. The set $\mathcal{E}$ is also called a generating system of $\sigma(\mathcal{E})$.

Remark 1.14 [Smallest σ-algebra containing $\mathcal{E}$ as a subset] According to Theorem 1.12, every set system $\mathcal{E}$ on Ω generates a uniquely defined σ-algebra $\sigma(\mathcal{E})$ on Ω. Note that the σ-algebra $\sigma(\mathcal{E})$ is the *smallest* σ-algebra on Ω containing $\mathcal{E}$ as a subset, that is,

$$\mathcal{C} \text{ is a σ-algebra on } \Omega \text{ and } \mathcal{E} \subset \mathcal{C} \;\Rightarrow\; \sigma(\mathcal{E}) \subset \mathcal{C}. \tag{1.11}$$

Furthermore,

$$\sigma[\sigma(\mathcal{E})] = \sigma(\mathcal{E}). \tag{1.12}$$

◁

Lemma 1.15 immediately follows from (1.11). It can be used in proofs of the identity of two σ-algebras.

Lemma 1.15 [Smallest σ-algebra containing $\mathcal{E}$ as a subset]
Let $(\Omega, \mathcal{A})$ be a measurable space and $\mathcal{E}$ a set system on Ω with $\sigma(\mathcal{E}) = \mathcal{A}$. If $\mathcal{C}$ is a σ-algebra on Ω with $\mathcal{E} \subset \mathcal{C} \subset \mathcal{A}$, then $\mathcal{C} = \mathcal{A}$.

(Proof p. 29)

Remark 1.16 [σ-Algebra generated by unions of set systems] Let $\mathscr{D}$, $\mathscr{E}$, $\mathscr{F}$ be set systems on a nonempty set Ω. Then,

$$\sigma(\mathscr{D} \cup \mathscr{E} \cup \mathscr{F}) = \sigma[\mathscr{D} \cup \sigma(\mathscr{E} \cup \mathscr{F})] \tag{1.13}$$

(see Exercise 1.6).   ◁

Example 1.17 [Several set systems may generate the same σ-algebra] If A is a subset of a nonempty set Ω, then the set system $\{A\}$ generates the σ-algebra $\{\Omega, \emptyset, A, A^c\}$. Note that $\{\Omega, \emptyset, A, A^c\}$ is also generated by the set systems $\{A^c\}$ and $\{A, A^c\}$, for instance. Hence,

$$\sigma(\{A\}) = \sigma(\{A^c\}) = \sigma(\{A, A^c\}) = \sigma(\{\Omega, \emptyset, A, A^c\}) = \{\Omega, \emptyset, A, A^c\}.$$

In contrast, if $\emptyset \neq A \neq \Omega$, then the σ-algebra $\{\Omega, \emptyset, A, A^c\}$ is neither generated by the set system $\{\Omega\}$ nor by $\{\Omega, \emptyset\}$. Instead,

$$\sigma(\{\emptyset\}) = \sigma(\{\Omega\}) = \sigma(\{\Omega, \emptyset\}) = \{\Omega, \emptyset\},$$

that is, $\{\Omega\}$, $\{\emptyset\}$, and $\{\Omega, \emptyset\}$ generate the σ-algebra $\{\Omega, \emptyset\}$.   ◁

Example 1.18 [A generator of the power set] Let $\Omega \neq \emptyset$ be finite or countable, and let $\mathscr{E} := \{\{\omega\}: \omega \in \Omega\}$. Then, $\sigma(\mathscr{E}) = \mathcal{P}(\Omega)$ (see Exercise 1.7).   ◁

This example is generalized in Lemma 1.20.

Remark 1.19 [Partition] Reading Lemma 1.20, remember that a set system $\mathscr{E}$ on a nonempty set Ω is called a *partition* of Ω if

(a) $\forall B \in \mathscr{E}:\ B \neq \emptyset$.

(b) $\forall B, C \in \mathscr{E}: B \neq C \ \Rightarrow\ B \cap C = \emptyset$.

(c) $\bigcup_{B \in \mathscr{E}} B = \Omega$.

                                                                                     ◁

Lemma 1.20 [An element of a σ-algebra generated by a partition]
Let $\mathscr{E} := \{B_1, \dots, B_n\}$ or $\mathscr{E} := \{B_1, B_2, \dots\}$ be a finite or countable partition of Ω, respectively. Then, for all $C \in \sigma(\mathscr{E})$, there is an $I(C) \subset \mathbb{N}$ such that

$$C = \bigcup_{i \in I(C)} B_i = \bigcup_{B_i \subset C} B_i, \tag{1.14}$$

where, by convention, $\bigcup_{i \in \emptyset} B_i := \emptyset$.

                                                                       (Proof p. 29)

Remark 1.21 [Constructing a σ-algebra] If $\mathscr{E} = \{A_1, \dots, A_m\}$ is a finite set of subsets of Ω, then there is a finite partition $\mathscr{F} = \{B_1, \dots, B_n\}$ of Ω with $\sigma(\mathscr{E}) = \sigma(\mathscr{F})$. Furthermore, if

$\mathscr{E}$ is a finite set of subsets of Ω, then each element of $\sigma(\mathscr{E})$ is obtained by finitely many unions, intersections, or complements of elements of $\mathscr{E}$ (see Exercise 1.8).   ◁

Example 1.22 [Joe and Ann – continued]   In Example 1.11, we already considered the event A that Joe is drawn and noted that the trace of the σ-algebra $\mathscr{A}_2 = \{\Omega, \emptyset, B, B^c\}$ in A is $\mathscr{A}_2|_A = \{A, \emptyset, A \cap B, A \cap B^c\}$. In contrast, the σ-algebra on Ω generated by the trace $\mathscr{A}_2|_A$ is

$$\sigma(\mathscr{A}_2|_A) = \{\Omega, \emptyset, A, A^c, A \cap B, A \cap B^c, (A \cap B) \cup A^c, (A \cap B^c) \cup A^c\},$$

where $(A \cap B) \cup A^c = A^c \cup B$ and $(A \cap B^c) \cup A^c = A^c \cup B^c$.   ◁

Remark 1.23 [Monotonicity of generated σ-algebras]   Let $\mathscr{E}_1, \mathscr{E}_2$ be set systems on a nonempty set Ω with $\mathscr{E}_1 \subset \mathscr{E}_2$. Then, $\sigma(\mathscr{E}_1) \subset \sigma(\mathscr{E}_2)$ (see Exercise 1.9).   ◁

An important kind of σ-algebras are those for which there is a countable set system that generates them.

Definition 1.24 [Countably generated σ-algebra]
Let $(\Omega, \mathscr{A})$ be a measurable space. Then, $\mathscr{A}$ is called countably generated if there is a finite or countable set $\mathscr{E} \subset \mathscr{A}$ such that $\sigma(\mathscr{E}) = \mathscr{A}$.

Example 1.25 [Some countably generated σ-algebras]   Examples of countably generated σ-algebras are:

(a) All σ-algebras on a finite nonempty set Ω.

(b) $\mathscr{P}(\mathbb{N}_0^n), n \in \mathbb{N}$.

(For a proof, see Exercise 1.10. For another example, see Remark 1.28.)   ◁

Remark 1.26 [A caveat]   Note that there are countably generated σ-algebras for which not all of their elements can be constructed by countably many unions, intersections, or complements of elements of the generating system. An example in case are Borel σ-algebras on $\mathbb{R}$ or $\mathbb{R}^n$ (see Rem. 1.28 and Michel, 1978, sect. I.4).   ◁

Lemma 1.27 [σ-Algebra generated by the trace of a set system]
Let $A \subset \Omega$ be nonempty, $\mathscr{E} \subset \mathscr{P}(\Omega)$, and $\mathscr{E}|_A := \{C \cap A : C \in \mathscr{E}\}$. Then,

$$\sigma(\mathscr{E}|_A) = \sigma(\mathscr{E})|_A, \tag{1.15}$$

where $\sigma(\mathscr{E}|_A)$ denotes the σ-algebra generated on A, whereas $\sigma(\mathscr{E})$ is a σ-algebra on Ω. Furthermore, if $\mathscr{C}$ is a σ-algebra on Ω and $A \in \mathscr{E}$ such that

$$\forall C \in \mathscr{C} : \quad C \neq A \implies A \cap C = \emptyset, \tag{1.16}$$

(i.e., A does not intersect with any other element of $\mathscr{E}$), then

$$\sigma(\mathscr{C} \cup \mathscr{E})|_A = \mathscr{C}|_A. \tag{1.17}$$

(Proof p. 30)

Hence, according to Equation (1.15), the σ-algebra generated by the trace of a set system $\mathscr{E}$ is the trace of the σ-algebra generated by $\mathscr{E}$; and, according to Equation (1.17), the trace of the σ-algebra $\sigma(\mathscr{C} \cup \mathscr{E})$ in the set A is identical to the trace of the σ-algebra $\mathscr{C}$ in A, if (1.16) holds.

1.2.2   σ-Algebra of Borel sets on $\mathbb{R}^n$

For $a, b \in \mathbb{R}$ with $a < b$, let us consider a *half-open interval* $]a, b]$ in $\mathbb{R}$, which is defined by

$$]a, b] := \{x \in \mathbb{R}: a < x \le b\},$$

and the *set system*

$$\mathscr{I}_1 := \{]a, b]: a, b \in \mathbb{R} \text{ and } a < b\}$$

of all half-open intervals in $\mathbb{R}$. The σ-algebra generated by this set system is called the *Borel σ-algebra* on $\mathbb{R}$. It is denoted by $\mathscr{B}$ or $\mathscr{B}_1$. The elements of $\mathscr{B}$ are called the *Borel sets* of $\mathbb{R}$. In formal terms,

$$\mathscr{B} := \mathscr{B}_1 := \sigma(\mathscr{I}_1). \tag{1.18}$$

Note that there are several sets systems generating the Borel σ-algebra (see, e.g., Klenke, 2013, Th. 1.23). In particular,

$$\mathscr{B}_1 = \sigma(\{]-\infty, b]: b \in \mathbb{R}\}) \tag{1.19}$$

(see Georgii, 2008). Similarly, we define the *Borel σ-algebra on* $\mathbb{R}^2 = \mathbb{R} \times \mathbb{R}$ to be the σ-algebra generated by the set system $\mathscr{I}_2$ of all half-open *rectangles* in $\mathbb{R}^2$, whose sides are parallel to the axes (see Fig. 1.3). These rectangles are defined by

$$]a_1, b_1] \times]a_2, b_2] = \{(x_1, x_2) \in \mathbb{R}^2: a_1 < x_1 \le b_1, a_2 < x_2 \le b_2\}.$$

The σ-algebra $\sigma(\mathscr{I}_2)$ is denoted by $\mathscr{B}_2$, that is, $\mathscr{B}_2 := \sigma(\mathscr{I}_2)$, and its elements are called the *Borel sets of* $\mathbb{R}^2$.

This definition is easily generalized: The *Borel σ-algebra on* $\mathbb{R}^n$ is defined by $\mathscr{B}_n := \sigma(\mathscr{I}_n)$, $n \in \mathbb{N}$, where $\mathscr{I}_n$ is the system of all half-open *cuboids* in $\mathbb{R}^n$, whose sides are parallel to the axes. Such a cuboid is a set

$$]a_1, b_1] \times \ldots \times]a_n, b_n] = \{(x_1, \ldots, x_n) \in \mathbb{R}^n : a_1 < x_1 \le b_1, \ldots, a_n < x_n \le b_n\}, \tag{1.20}$$

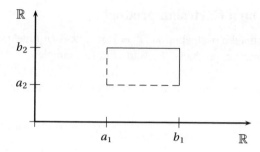

Figure 1.3 A half-open rectangle in the plane $\mathbb{R}^2$.

where $a_1, \ldots, a_n, b_1, \ldots, b_n \in \mathbb{R}$. Just like $\mathcal{B}_1$, the σ-algebra $\mathcal{B}_n$ has several generating systems, one of which is used in the equation

$$\mathcal{B}_n = \sigma(\{]-\infty, b_1] \times \ldots \times \,]-\infty, b_n]: b_1, \ldots, b_n \in \mathbb{R}\}) \tag{1.21}$$

(see Exercise 1.11).

Note that not every subset of $\mathbb{R}^n$ is a Borel set. In other words, $\mathcal{B}_n$ is not the power set of $\mathbb{R}^n$ (see Rem. 1.60). However, for each $x = (x_1, \ldots, x_n) \in \mathbb{R}^n$, the singleton $\{x\}$ *is* a Borel set of $\mathbb{R}^n$, that is,

$$\{x\} \in \mathcal{B}_n, \quad \forall x \in \mathbb{R}^n \tag{1.22}$$

(see Exercise 1.12).

Furthermore, if $\overline{\mathbb{R}} = \mathbb{R} \cup \{-\infty, +\infty\}$ denotes the *extended set of real numbers*, then

$$\overline{\mathcal{B}} := \sigma(\mathcal{B} \cup \{\{-\infty\}, \{+\infty\}\})$$

is a σ-algebra on $\overline{\mathbb{R}}$, and it is called the *Borel σ-algebra on* $\overline{\mathbb{R}}$. Similarly, $\overline{\mathcal{B}}_n$ is called the *Borel σ-algebra on* $\overline{\mathbb{R}}^n$. It is defined as the product σ-algebra of $\overline{\mathcal{B}}$ with itself (n times) (see Def. 1.31). Finally, we may sometimes consider $\overline{\mathcal{B}}_n|_{\Omega_0}$, the *trace of the Borel σ-algebra on* $\overline{\mathbb{R}}^n$ in $\Omega_0 \subset \overline{\mathbb{R}}^n$.

Remark 1.28 [The Borel σ-algebra is countably generated] Note that

$$\mathcal{B} = \sigma(\{]a, b]: a, b \in \mathbb{Q}, a < b\}),$$

where $\mathbb{Q}$ denotes the set of rational numbers. Because $\mathbb{Q}$ is countable, the set of intervals $\{]a, b]: a, b \in \mathbb{Q}, a < b\}$ is countable as well. Therefore, the Borel σ-algebra $\mathcal{B}$ is countably generated. This also holds for $\mathcal{B}_n, n \in \mathbb{N}$ (see Klenke, 2013, Th. 1.23). ◁

Remark 1.29 [Trace of the Borel σ-algebra in a countable subset of $\mathbb{R}$] Let $\mathcal{B}$ denote the Borel σ-algebra on $\mathbb{R}$. If $\Omega_0 \subset \mathbb{R}$ is finite or countable, then $\mathcal{B}|_{\Omega_0} = \mathcal{P}(\Omega_0)$, where $\mathcal{B}|_{\Omega_0}$ is the trace of the Borel σ-algebra on $\mathbb{R}$ in $\Omega_0 \subset \mathbb{R}$ (see Exercise 1.13). ◁

1.2.3 σ-Algebra on a Cartesian product

In section 1.2.2, we defined a σ-algebra on $\mathbb{R}^n = \mathbb{R} \times \ldots \times \mathbb{R}$ (*n*-times). Now we consider σ-algebras on general Cartesian products. We start with an example.

Example 1.30 [Joe and Ann – continued]   In Example 1.9, we already considered the Cartesian product

$$\Omega := \Omega_U \times \Omega_X \times \Omega_Y,$$

which consists of the eight triples (*Joe, no, −*), (*Joe, no, +*), ... , (*Ann, yes, +*) (see again Fig. 1.2). Now consider the σ-algebras $\mathcal{A}_1 := \mathcal{P}(\Omega_U)$, $\mathcal{A}_2 := \mathcal{P}(\Omega_X)$, and $\mathcal{A}_3 := \mathcal{P}(\Omega_Y)$, as well as the set

$$\mathcal{E} := \{A_1 \times A_2 \times A_3 : A_1 \in \mathcal{A}_1, A_2 \in \mathcal{A}_2, A_3 \in \mathcal{A}_3\},$$

which is a set system on Ω consisting of $4 \cdot 4 \cdot 4 = 64$ elements. For example, the set system $\mathcal{E}$ contains the elements

$$A := \{Joe\} \times \{no\} \times \{-\} = \{(Joe, no, -)\}$$

and

$$B := \{Ann\} \times \{yes\} \times \{+\} = \{(Ann, yes, +)\}.$$

However, $\mathcal{E}$ does not contain

$$A \cup B = \{(Joe, no, -), (Ann, yes, +)\}$$

as an element. The only product set $A_1 \times A_2 \times A_3$ with $A_1 \in \mathcal{A}_1, A_2 \in \mathcal{A}_2, A_3 \in \mathcal{A}_3$ that contains $A \cup B$ as a subset is $\Omega_U \times \Omega_X \times \Omega_Y = \Omega$. However, $A \cup B \neq \Omega$. Therefore, $\mathcal{E}$ is not a σ-algebra [cf. condition (c) of Rem. 1.2]. In this example, the σ-algebra generated by $\mathcal{E}$ is the power set of Ω, that is, $\sigma(\mathcal{E}) = \mathcal{P}(\Omega)$. It consists of $2^8 = 256$ elements. According to the following definition, $\sigma(\mathcal{E})$ is denoted by $\mathcal{A}_1 \otimes \mathcal{A}_2 \otimes \mathcal{A}_3$ and called the *product σ-algebra of* $\mathcal{A}_1, \mathcal{A}_2,$ *and* $\mathcal{A}_3$.   ◁

Definition 1.31 [Product σ-algebra]
Let $(\Omega_1, \mathcal{A}_1), \ldots, (\Omega_n, \mathcal{A}_n)$ *be measurable spaces and* $\Omega := \Omega_1 \times \ldots \times \Omega_n$. *Then*

$$\mathcal{A}_1 \otimes \ldots \otimes \mathcal{A}_n := \bigotimes_{i=1}^{n} \mathcal{A}_i := \sigma \left(\left\{ \underset{i=1}{\overset{n}{\times}} A_i : A_i \in \mathcal{A}_i, i = 1, \ldots, n \right\} \right) \qquad (1.23)$$

is called the product σ-algebra of the σ-algebras $\mathcal{A}_i, i = 1, \ldots, n.$

Note that the product σ-algebra $\mathcal{A}_1 \otimes \ldots \otimes \mathcal{A}_n$ is *not* the Cartesian product $\mathcal{A}_1 \times \ldots \times \mathcal{A}_n$. Instead, the product σ-algebra is generated by the set system of all Cartesian products of elements of the σ-algebras $\mathscr{A}_1, \ldots, \mathscr{A}_n$. In Lemma 2.42, we give an equivalent specification of a product σ-algebra, using projection mappings.

Lemma 1.32 provides a relationship between the generating systems of the σ-algebras $\mathscr{A}_i$, $i = 1, \ldots, n$, and the generating system of the product σ-algebra.

Lemma 1.32 [Generating system of a product σ-algebra]
For $i = 1, \ldots, n$, let $(\Omega_i, \mathscr{A}_i)$ be measurable spaces and $\mathscr{E}_i \subset \mathscr{A}_i$ with $\sigma(\mathscr{E}_i) = \mathscr{A}_i$. Then,

$$\bigotimes_{i=1}^{n} \mathscr{A}_i = \sigma\left(\left\{ \underset{i=1}{\overset{n}{\times}} A_i : A_i \in \mathscr{E}_i, i = 1, \ldots, n \right\}\right). \tag{1.24}$$

For a proof, see Klenke [2013, Th. 14.12 (i)].

This lemma implies

$$\mathscr{B}_n = \bigotimes_{i=1}^{n} \mathscr{B} = \mathscr{B} \otimes \ldots \otimes \mathscr{B} \ (n\text{-times})$$

for the Borel σ-algebra on $\mathbb{R}^n$. This lemma also implies the following corollary:

Corollary 1.33 [Countable generating system of a product σ-algebra]
Let $(\Omega_i, \mathscr{A}_i)$, $i = 1, \ldots, n$, be measurable spaces, where all $\mathscr{A}_i$ are countably generated. Then $\bigotimes_{i=1}^{n} \mathscr{A}_i$ is countably generated as well.

Example 1.34 [Countable sets and product σ-algebra]   Let $\Omega_1, \ldots, \Omega_n$ be finite or countable nonempty sets and $\mathscr{A}_1, \ldots, \mathscr{A}_n$ be their power sets. Then,

$$\bigotimes_{i=1}^{n} \mathscr{A}_i = \mathscr{P}\left(\underset{i=1}{\overset{n}{\times}} \Omega_i \right),$$

that is, $\bigotimes_{i=1}^{n} \mathscr{A}_i$ is the power set of $\Omega := \Omega_1 \times \ldots \times \Omega_n$ (see Exercise 1.14).   ◁

Remark 1.35 [Complement of a Cartesian product]   Let $(\Omega_1 \times \Omega_2, \mathscr{A}_1 \otimes \mathscr{A}_2)$ be a measurable space, $A \in \mathscr{A}_1$, and $B \in \mathscr{A}_2$. Then $(A \times B)^c \in \mathscr{A}_1 \otimes \mathscr{A}_2$, and this set can be written as:

$$(A \times B)^c = (A^c \times B) \cup (\Omega_1 \times B^c), \tag{1.25}$$

which is a union of disjoint sets (see Exercise 1.15).   ◁

1.2.4   ∩-Stable set systems that generate a σ-algebra

For many proofs, generating set systems are useful, which are ∩-stable.

Definition 1.36 [∩-Stability]

Let Ω denote a nonempty set. A set $\mathcal{E}$ of subsets of Ω is called $\cap$-stable (or $\cap$-closed) if $A \cap B \in \mathcal{E}$ for all $A, B \in \mathcal{E}$.

Example 1.37 [Set system with one single element]   A set system $\{A\}$ that has only a single element $A \subset \Omega \neq \emptyset$ is $\cap$-stable (cf. Example 1.17).   ◁

Example 1.38 [Partition and ∩-stability]   If $\mathcal{E}$ is a partition of the set Ω, then $\mathcal{D} := \mathcal{E} \cup \{\emptyset\}$ is $\cap$-stable.   ◁

Example 1.39 [A ∩-stable generating system of a product σ-algebra]   Consider the measurable spaces $(\Omega_i, \mathcal{A}_i)$, $i = 1, \ldots, n$. The set

$$\{A_1 \times \ldots \times A_n : A_i \in \mathcal{A}_i, \ i = 1, \ldots, n\},$$

is a $\cap$-stable generating system of $\bigotimes_{i=1}^n \mathcal{A}_i$ (see Exercise 1.16).   ◁

Another type of a set system is a Dynkin system. It can be used to show that a specific set system is a σ-algebra.

Definition 1.40 [Dynkin system]

A set $\mathcal{D}$ of subsets of a set Ω is called a Dynkin system on Ω, if the following three conditions hold:

(a) *$\Omega \in \mathcal{D}$.*

(b) *If $A \in \mathcal{D}$, then $A^c \in \mathcal{D}$.*

(c) *If $A_1, A_2, \ldots \in \mathcal{D}$ and they are pairwise disjoint, then $\bigcup_{i=1}^\infty A_i \in \mathcal{D}$.*

In the definition of a σ-algebra $\mathcal{A}$, we require $\bigcup_{i=1}^\infty A_i \in \mathcal{A}$ for all sequences $A_1, A_2, \ldots \in \mathcal{A}$, whereas for a Dynkin system the corresponding requirement is only made for all sequences $A_1, A_2, \ldots \in \mathcal{D}$ of pairwise disjoint sets. Analogously to Definition 1.13, for a set system $\mathcal{E}$ on Ω, $\delta(\mathcal{E})$ is defined as the Dynkin system generated by $\mathcal{E}$, that is, as the intersection of all Dynkin systems containing $\mathcal{E}$. According to Theorem 1.41, a Dynkin system is also a σ-algebra if and only if it is $\cap$-stable.

Theorem 1.41 [Dynkin system and σ-algebra]

Let Ω be a nonempty set.

(i) *A Dynkin system $\mathcal{D}$ on Ω is a σ-algebra if and only if it is $\cap$-stable.*

(ii) *If $\mathcal{E}$ is a $\cap$-stable set of subsets of Ω, then $\delta(\mathcal{E}) = \sigma(\mathcal{E})$.*

For a proof, see Bauer (2001, Ths. 2.3 and 2.4). According to proposition (i) of this theorem, we can prove that a set system is a σ-algebra by showing that it is a ∩-stable Dynkin system, and proposition (ii) can be applied to show that the Dynkin system generated by a ∩-stable set system is a σ-algebra.

1.3   Measure and measure space

A measure assigns to all elements of a σ-algebra an element of the closed interval

$$[0, \infty] := \{x \in \mathbb{R}: 0 \leq x\} \cup \{\infty\},$$

that is, a nonnegative real number or the element ∞.

Example 1.42 [A first example]   Let $\Omega = \mathbb{R}$, and assume that the closed interval $[3, 9] = \{x \in \mathbb{R}: 3 \leq x \leq 9\}$ as well as the union $[3, 9] \cup [10, 12]$ are elements of a σ-algebra on Ω. If the measure is *length*, then

$$length\,([3, 9]) = 9 - 3 = 6$$

and

$$length\,([3, 9] \cup [10, 12]) = length\,([3, 9]) + length\,([10, 12])$$
$$= (9 - 3) + (12 - 10) = 6 + 2 = 8,$$

because the two intervals are disjoint (i.e., their intersection is the empty set $\emptyset$). In this case, the lengths of the intervals $[3, 9]$ and $[10, 12]$ add up to the length of their union $[3, 9] \cup [10, 12]$. In Definition 1.43 (c), we require not only additivity but also σ-additivity.   ◁

Reading Definition 1.43, remember that, for a sequence $a_1, a_2, \ldots$ of nonnegative real numbers, $\sum_{i=1}^{\infty} a_i$ is defined by

$$\sum_{i=1}^{\infty} a_i := \lim_{n \to \infty} \sum_{i=1}^{n} a_i.$$

Definition 1.43 [Measure and measure space]
Let $(\Omega, \mathscr{A})$ be a measurable space. A function $\mu: \mathscr{A} \to \overline{\mathbb{R}}$ is called a measure and the triple $(\Omega, \mathscr{A}, \mu)$ is called a measure space, if

(a) $\mu(\emptyset) = 0.$

(b) $\mu(A) \geq 0, \ \forall A \in \mathscr{A}.$ *(nonnegativity)*

(c) *If $A_1, A_2, \ldots \in \mathscr{A}$ are pairwise disjoint, then $\mu\left(\bigcup_{i=1}^{\infty} A_i\right) = \sum_{i=1}^{\infty} \mu(A_i)$.*
     (σ-additivity)

1.3.1   σ-Additivity and related properties

Remark 1.44 [σ-Additivity implies finite additivity] Note that σ-additivity of a measure implies finite additivity, that is, it implies

$$\mu \left(\bigcup_{i=1}^{n} A_i \right) = \sum_{i=1}^{n} \mu(A_i), \quad \text{if } A_1, \dots, A_n \in \mathscr{A} \text{ are pairwise disjoint} \qquad (1.26)$$

[see Rule (ii) of Box 1.1 and its proof in Exercise 1.18]. ◁

Remark 1.45 [σ-Additivity] Using the term σ-*additivity* signals that unions of finitely or countably many sets are considered, but not other unions of sets. If, instead of σ-additivity, we would require additivity for *any kind of unions*, including uncountable unions, then the Lebesgue measure λ on $(\mathbb{R}, \mathscr{B})$ – the measure representing *length* – could not be constructed anymore. This is explained in more detail in Remark 1.71. ◁

Remark 1.46 [Representation of a union as a union of pairwise disjoint sets] Let $(\Omega, \mathscr{A})$ be a measurable space. If $A_1, A_2, \dots \in \mathscr{A}$ is a sequence of subsets of Ω, then there is a sequence $B_1, B_2, \dots \in \mathscr{A}$ of pairwise disjoint sets with

$$\bigcup_{i=1}^{\infty} A_i = \bigcup_{i=1}^{\infty} B_i. \qquad (1.27)$$

One way to construct $B_1, B_2, \dots$ is to define $B_1 := A_1$ and

$$B_i := A_i \setminus \left(\bigcup_{j=1}^{i-1} A_j \right), \quad \text{for } i > 1, \qquad (1.28)$$

(see Exercise 1.17). ◁

Remark 1.47 [Additivity of measures for partitions] Let $(\Omega, \mathscr{A}, \mu)$ be a measure space, $B \in \mathscr{A}$, and assume

(a) $A_1, \dots, A_n \in \mathscr{A}$ are pairwise disjoint,

(b) $B \subset \bigcup_{i=1}^{n} A_i$.

Then,

$$\mu(B) = \sum_{i=1}^{n} \mu(B \cap A_i). \qquad (1.29)$$

Analogously, if

(c) $A_1, A_2, \dots \in \mathscr{A}$ are pairwise disjoint,

(d) $B \subset \bigcup_{i=1}^{\infty} A_i,$

then

$$\mu(B) = \sum_{i=1}^{\infty} \mu(B \cap A_i) \tag{1.30}$$

(see Exercise 1.19). ◁

1.3.2   Other properties

Other important properties of a measure are displayed in Box 1.1. Some of these properties can intuitively be understood by inspecting the Venn diagram presented in Figure 1.1. These

---

Box 1.1   Rules of computation for measures.

Let $(\Omega, \mathscr{A}, \mu)$ be a measure space.
  If $A_1, A_2, \ldots \in \mathscr{A}$ are pairwise disjoint, then,

$$\mu\left(\bigcup_{i=1}^{\infty} A_i\right) = \sum_{i=1}^{\infty} \mu(A_i). \qquad \text{(\sigma-additivity)   (i)}$$

$$\mu\left(\bigcup_{i=1}^{n} A_i\right) = \sum_{i=1}^{n} \mu(A_i), \quad \forall\, n \in \mathbb{N}. \qquad \text{(finite additivity) (ii)}$$

If $A, B \in \mathscr{A}$, then,

$$\mu(A) = \mu(A \cap B) + \mu(A \setminus B). \tag{iii}$$

$$\mu(\Omega) = \mu(B) + \mu(B^c). \tag{iv}$$

$$\mu(A) \leq \mu(B), \quad \text{if } A \subset B. \qquad \text{(monotonicity) (v)}$$

$$\mu(A \setminus B) = \mu(A) - \mu(A \cap B), \quad \text{if } \mu(A \cap B) < \infty. \tag{vi}$$

$$\mu(A \cup B) = \mu(A) + \mu(B) - \mu(A \cap B), \quad \text{if } \mu(A \cap B) < \infty. \tag{vii}$$

$$\mu(A) = \mu(\Omega) < \infty \;\Rightarrow\; \mu(A \cap B) = \mu(B). \tag{viii}$$

$$\mu(A) = 0 \;\Rightarrow\; \mu(A \cup B) = \mu(B). \tag{ix}$$

Let $A \in \mathscr{A}$ and let $\Omega_0 \subset \Omega$ and be finite or countable with $\mu(\Omega \setminus \Omega_0) = 0$.
  If, for all $\omega \in \Omega_0$, $\{\omega\} \in \mathscr{A}$, then

$$\mu(A) = \sum_{\omega \in A \cap \Omega_0} \mu(\{\omega\}). \tag{x}$$

If $A_1, A_2, \ldots \in \mathscr{A}$, then

$$\mu\left(\bigcup_{i=1}^{\infty} A_i\right) \leq \sum_{i=1}^{\infty} \mu(A_i). \qquad \text{(\sigma-subadditivity) (xi)}$$

---

properties always hold with the conventions $+\infty + \infty = +\infty$ and $\alpha + \infty = +\infty$, for $\alpha \in \mathbb{R}$. However, note that the term $+\infty + (-\infty)$ or $+\infty - (+\infty)$ cannot meaningfully be defined. Therefore, properties (vi) and (vii) only hold if we assume $\mu(A \cap B) < \infty$. For proofs of all these properties, see Exercise 1.18.

Remark 1.48 [Finite additivity and σ-additivity applied to singletons] If Ω is finite or countable, then each $A \subset \Omega$ is finite or countable as well. Hence, for any measure μ on the measurable space $(\Omega, \mathscr{P}(\Omega))$,

$$\mu(A) = \mu \left(\bigcup_{\omega \in A} \{\omega\} \right) = \sum_{\omega \in A} \mu(\{\omega\}), \quad \forall\, A \subset \Omega. \tag{1.31}$$

This means that a measure on $(\Omega, \mathscr{P}(\Omega))$ is already uniquely defined if its values $\mu(\{\omega\})$ are uniquely defined for all $\omega \in \Omega$, provided that Ω is finite or countable. Rule (x) of Box 1.1 extends this result to a more general measure space $(\Omega, \mathscr{A}, \mu)$. This rule shows that a measure on $(\Omega, \mathscr{A})$ is already uniquely defined if its values $\mu(\{\omega\})$ are uniquely defined for all $\omega \in \Omega_0$, provided that Ω_0 is finite or countable with $\mu(\Omega \setminus \Omega_0) = 0$ and $\{\omega\} \in \mathscr{A}$ for all $\omega \in \Omega_0$.   ◁

1.4   Specific measures

Now we consider some examples of measures, all of which are used later on in this volume in order to introduce still other measures. For some of these examples, we use the *indicator* of a set A.

Definition 1.49 [Indicator]
Let Ω be a set and $A \subset \Omega$. Then, the function $1_A \colon \Omega \to \mathbb{R}$ defined by

$$1_A(\omega) = \begin{cases} 1, & \text{if } \omega \in A \\ 0, & \text{if } \omega \notin A, \end{cases} \tag{1.32}$$

is called the indicator *of A.*

Remark 1.50 [Sums and products of indicators] If $1_A, 1_B \colon \Omega \to \mathbb{R}$ are the indicators of two sets $A, B \subset \Omega$, then,

$$1_A \cdot 1_B = 1_{A \cap B} \tag{1.33}$$

and

$$1_A + 1_B - 1_{A \cap B} = 1_A + 1_B - 1_A \cdot 1_B = 1_{A \cup B}. \tag{1.34}$$

Equation (1.33) immediately implies

$$1_A + 1_B = 1_{A \cup B}, \quad \text{if } A \cap B = \emptyset. \tag{1.35}$$

More generally, if $A_1, \ldots, A_n$ is a finite sequence of pairwise disjoint subsets of Ω, then,

$$\sum_{i=1}^{n} 1_{A_i} = 1_{\bigcup_{i=1}^{n} A_i}, \tag{1.36}$$

that is, then the sum of the indicators of the sets $A_1, \ldots, A_n$ is the indicator of the union $\bigcup_{i=1}^{n} A_i$. Finally, if $A_1, A_2, \ldots$ is a sequence of pairwise disjoint subsets of Ω, then,

$$\sum_{i=1}^{\infty} 1_{A_i} = 1_{\bigcup_{i=1}^{\infty} A_i}. \tag{1.37}$$

◁

Remark 1.51 [Indicators of products sets] Let Ω_1, Ω_2 be nonempty sets, $A \subset \Omega_1$ and $B \subset \Omega_2$. Then,

$$1_A(\omega_1) \cdot 1_B(\omega_2) = 1_{A \times B}(\omega_1, \omega_2), \quad \forall (\omega_1, \omega_2) \in \Omega_1 \times \Omega_2. \tag{1.38}$$

This equation follows from the definitions of the product set and the indicator.   ◁

1.4.1   Dirac measure and counting measure

Example 1.52 [Dirac measure] Let $(\Omega, \mathscr{A})$ be a measurable space, let $\omega \in \Omega$, and consider the function $\delta_\omega \colon \mathscr{A} \to \{0, 1\}$ defined by

$$\delta_\omega(A) := 1_A(\omega), \quad \forall A \in \mathscr{A}. \tag{1.39}$$

Then δ_ω is a measure on $(\Omega, \mathscr{A})$ (see Exercise 1.20).   ◁

Definition 1.53 [Dirac measure]
The function δ_ω defined by Equation (1.39) is called the Dirac measure at (point) ω.

Example 1.54 [Counting measure] Let $(\Omega, \mathscr{A})$ be a measurable space, and define the function $\mu_\# \colon \mathscr{A} \to \overline{\mathbb{R}}$ by

$$\mu_\#(A) := \begin{cases} \displaystyle\sum_{\omega \in \Omega} 1_A(\omega), & \text{if } A \text{ is finite} \\[2mm] \infty, & \text{if } A \text{ is infinite,} \end{cases} \quad \forall A \in \mathscr{A}. \tag{1.40}$$

Then $\mu_\#$ is a measure on $(\Omega, \mathscr{A})$ (see Exercise 1.21).   ◁

Definition 1.55 [Counting measure]
The function $\mu_{\#}$ defined by Equation (1.40) is called the counting measure on $(\Omega, \mathcal{A})$.

Remark 1.56 [Cardinality of a set]   If A is finite, then $\mu_{\#}(A)$ is called the *cardinality* of A, that is, $\mu_{\#}(A)$ simply counts the number of elements ω of the set A. Furthermore, for finite or countable Ω and $A \subset \Omega$,

$$\mu_{\#}(A) = \sum_{\omega \in \Omega} 1_A(\omega) = \sum_{\omega \in \Omega} \delta_\omega(A). \tag{1.41}$$

$\triangleleft$

Example 1.57 [Sum of Dirac measures]   Let $(\Omega, \mathcal{A})$ be a measurable space. If $B \subset \Omega$ is finite or countable and δ_ω is the Dirac measure on $(\Omega, \mathcal{A})$ at point ω, then $\sum_{\omega \in B} \delta_\omega \colon \mathcal{A} \to [0, \infty]$ defined by

$$\left(\sum_{\omega \in B} \delta_\omega \right)(A) := \sum_{\omega \in B} \delta_\omega(A), \quad \forall A \in \mathcal{A}, \tag{1.42}$$

is a measure on $(\Omega, \mathcal{A})$ (see Exercise 1.22). Hence, if Ω itself is finite or countable, then $\sum_{\omega \in \Omega} \delta_\omega$ is a measure on $(\Omega, \mathcal{A})$, and it is identical to the counting measure defined in Example 1.54, because, for $A \in \mathcal{A}$,

$$\begin{aligned} \left(\sum_{\omega \in \Omega} \delta_\omega \right)(A) &= \sum_{\omega \in \Omega} \delta_\omega(A) && [(1.42)] \\ &= \sum_{\omega \in \Omega} 1_A(\omega) && [(1.39)] \\ &= \mu_{\#}(A). && [(1.41)] \end{aligned} \tag{1.43}$$

$\triangleleft$

1.4.2   Lebesgue measure

Consider the *half-open interval* $]a, b]$. Then,

$$\lambda_1(]a, b]) = b - a \tag{1.44}$$

is the *length* of the interval $]a, b]$. Next consider a *rectangle* $]a_1, b_1] \times]a_2, b_2]$ in $\mathbb{R}^2$ with $a_1 < b_1$ and $a_2 < b_2$. This set can be visualized by the set of all points inside the rectangle presented in Figure 1.3 (excluding the lower and left boundary). Obviously,

$$\lambda_2(]a_1, b_1] \times]a_2, b_2]) = (b_1 - a_1) \cdot (b_2 - a_2) \tag{1.45}$$

is the *area* of this rectangle.

According to Theorem 1.58, there is one and only one measure on $(\mathbb{R}, \mathcal{B})$ satisfying (1.44) for all such intervals. This measure is called the *Lebesgue measure on* $(\mathbb{R}, \mathcal{B})$ and is denoted by λ or λ_1. Similarly, there is one and only one measure on $(\mathbb{R}^2, \mathcal{B}_2)$ satisfying (1.45) for all such rectangles. It is called the *Lebesgue measure on* $(\mathbb{R}^2, \mathcal{B}_2)$ and is denoted by λ_2. Theorem 1.58 deals with the general case.

Theorem 1.58 [Existence and uniqueness of the Lebesgue measure]
For all $n \in \mathbb{N}$, there is a uniquely defined measure λ_n on $(\mathbb{R}^n, \mathscr{B}_n)$ satisfying

$$\lambda_n(]a_1, b_1] \times \ldots \times]a_n, b_n]) = \prod_{i=1}^{n} (b_i - a_i), \tag{1.46}$$

$$\forall\, a_i, b_i \in \mathbb{R} \text{ with } a_i < b_i, i = 1, \ldots, n.$$

For a proof, see Klenke (2013, Th. 1.55).

Definition 1.59 [Lebesgue measure]
The measure λ_n satisfying Equation (1.46) is called the Lebesgue measure on $(\mathbb{R}^n, \mathscr{B}_n)$.

Remark 1.60 [Sets of real numbers that are not Lebesgue measurable] Hence, the Lebesgue measure λ_n is defined on $(\mathbb{R}^n, \mathscr{B}_n)$. Note, however, that this measure space $(\mathbb{R}^n, \mathscr{B}_n, \lambda_n)$ can be completed by additionally including all subsets of sets $A \in \mathscr{B}_n$ with $\lambda_n(A) = 0$. In Wise and Hall (1993, counterexample 1.25), it is shown for $n = 1$ that there are subsets $B \subset \mathbb{R}$ that are not elements of the completed σ-algebra. Therefore, $B \notin \mathscr{B}$, and this implies $\mathscr{B} \neq \mathscr{P}(\mathbb{R})$. ◁

1.4.3 Other examples of a measure

Example 1.61 [Restriction of a measure to a sub-σ-algebra] Suppose $(\Omega, \mathscr{A}, \mu)$ is a measure space and $\mathscr{C} \subset \mathscr{A}$ a σ-algebra. Then the function $\nu \colon \mathscr{C} \to \overline{\mathbb{R}}$ defined by

$$\nu(A) := \mu(A), \quad \forall\, A \in \mathscr{C}, \tag{1.47}$$

is a measure on $(\Omega, \mathscr{C})$ (see Exercise 1.23). ◁

Example 1.62 [Weighted sum of measures] If $\mu_1, \mu_2, \ldots$ are measures on $(\Omega, \mathscr{A})$ and $0 \leq \alpha_1, \alpha_2, \ldots \in \mathbb{R}$, then $\sum_{i=1}^{\infty} \alpha_i \mu_i \colon \mathscr{A} \to [0, \infty]$ defined by

$$\left(\sum_{i=1}^{\infty} \alpha_i \mu_i \right)(A) := \sum_{i=1}^{\infty} \alpha_i \mu_i(A), \quad \forall\, A \in \mathscr{A}, \tag{1.48}$$

is again a measure on $(\Omega, \mathscr{A})$ (see Exercise 1.24). For $0 = \alpha_{n+1} = \alpha_{n+2} = \ldots$ this implies: If $\mu_1, \ldots, \mu_n$ are measures on $(\Omega, \mathscr{A})$ and $\alpha_1, \ldots, \alpha_n$ are nonnegative, then the function $\sum_{i=1}^{n} \alpha_i \mu_i$ defined by

$$\left(\sum_{i=1}^{n} \alpha_i \mu_i \right)(A) := \sum_{i=1}^{n} \alpha_i \mu_i(A), \quad \forall\, A \in \mathscr{A}, \tag{1.49}$$

is also a measure on $(\Omega, \mathscr{A})$. ◁

1.4.4   Finite and σ-finite measures

A measure μ on a measurable space $(\Omega, \mathscr{A})$ is called *finite* if $\mu(\Omega) < \infty$. Otherwise, it is called *infinite*. Within the class of infinite measures, there is a subclass with an important property, called σ-*finiteness*. Many fundamental propositions of measure and integration theory only hold for measures that are σ-finite.

Definition 1.63 [σ-Finite measure]
Let μ be a measure on a measurable space $(\Omega, \mathscr{A})$. Then μ is called σ-finite if there is a sequence $A_1, A_2, \ldots \in \mathscr{A}$ with $\bigcup_{i=1}^{\infty} A_i = \Omega$ and, for all $i = 1, 2, \ldots$, $\mu(A_i) < \infty$.

To emphasize, even if $\mu(\Omega) = \infty$, the measure μ can be σ-finite (see Examples 1.64 and 1.65). Note that any finite measure is also σ-finite.

Example 1.64 [σ-Finiteness of the Lebesgue measure]   The Lebesgue measure λ on $(\mathbb{R}, \mathscr{B})$ is σ-finite, because $\mathbb{R} = \bigcup_{i=1}^{\infty} [-i, i]$ and $\lambda([-i, i]) = 2 \cdot i < \infty$, for all $i \in \mathbb{N}$.   ◁

Example 1.65 [A σ-finite counting measure]   Consider the measurable space $(\mathbb{R}, \mathscr{B})$ and the measure $\mu \colon \mathscr{B} \to [0, \infty]$, where $\mu = \sum_{i=0}^{\infty} \delta_i$ and δ_i denotes the Dirac measure at i on $(\mathbb{R}, \mathscr{B})$ with $\delta_i(A) = 1_A(i)$, $A \in \mathscr{B}$, $i \in \mathbb{N}_0$ (see Example 1.57). Then μ is σ-finite because $\mathbb{R} = \bigcup_{n=1}^{\infty} [-n, n]$ and $\mu([-n, n]) = n + 1$, for all $n \in \mathbb{N}_0$. This measure simply counts the number of elements $i \in \mathbb{N}_0$ in a Borel set A. In other words, for all finite $A \in \mathscr{B}$, $\mu(A)$ is the cardinality of the set $A \cap \mathbb{N}_0$.   ◁

1.4.5   Product measure

In section 1.4.2, we considered the Lebesgue measure on $(\mathbb{R}^n, \mathscr{B}_n)$ that is specified for n-dimensional cuboids by Equation (1.46) using the *product* of one-dimensional Lebesgue measures on $(\mathbb{R}, \mathscr{B})$. Now we introduce the general concept of a product measure. Lemma 1.66 shows that σ-finiteness of measures is sufficient for the existence and uniqueness of such a measure. Hence, this lemma shows that presuming finite measures is sufficient but not necessary for the definition of the product measure.

Lemma 1.66 [Existence and uniqueness]
Let $(\Omega_i, \mathscr{A}_i, \mu_i)$ be measure spaces with σ-finite measures μ_i, $i = 1, \ldots, n$. Then there is a uniquely defined measure, denoted $\mu_1 \otimes \ldots \otimes \mu_n$, on the product space

$$\left(\underset{i=1}{\overset{n}{\times}} \Omega_i, \underset{i=1}{\overset{n}{\bigotimes}} \mathscr{A}_i \right),$$

satisfying

$$\forall\, (A_1, \ldots, A_n) \in \mathscr{A}_1 \times \ldots \times \mathscr{A}_n:$$
$$\mu_1 \otimes \ldots \otimes \mu_n(A_1 \times \ldots \times A_n) = \mu_1(A_1) \cdot \ldots \cdot \mu_n(A_n). \tag{1.50}$$

This measure is σ-finite as well.

For a proof, see Bauer (2001, Th. 23.9). Hence, $\mu := \mu_1 \otimes \dots \otimes \mu_n$ is a measure on the product space $\left(\times_{i=1}^n \Omega_i, \otimes_{i=1}^n \mathscr{A}_i \right)$ with

$$\mu(A_1 \times \dots \times A_n) := \mu_1(A_1) \cdot \dots \cdot \mu_n(A_n), \quad \forall \, (A_1, \dots, A_n) \in (\mathscr{A}_1 \times \dots \times \mathscr{A}_n). \tag{1.51}$$

> **Definition 1.67 [Product measure]**
> The measure $\mu_1 \otimes \dots \otimes \mu_n$ defined by Equation (1.50) is called the *product measure* of $\mu_1, \dots, \mu_n$.

1.5   Continuity of a measure

The term σ-additivity refers to *countable* unions of pairwise disjoint sets and it implies finite additivity, which involves *finite* unions of pairwise disjoint sets. Furthermore, σ-additivity implies the following continuity properties of a measure, which are essential for the definition of the integral (see ch. 3).

> **Theorem 1.68 [Continuity of a measure]**
> Let $(\Omega, \mathscr{A}, \mu)$ be a measure space, and let $A_1, A_2, \dots \in \mathscr{A}$.
>
>   (i) If $A_1 \subset A_2 \subset \dots$, then,
>
> $$\lim_{i \to \infty} \mu(A_i) = \mu \left(\bigcup_{i=1}^\infty A_i \right). \qquad \textit{(continuity from below)}$$
>
>   (ii) If $A_1 \supset A_2 \supset \dots$ and there is an $n \in \mathbb{N}$ with $\mu(A_n) < \infty$, then,
>
> $$\lim_{i \to \infty} \mu(A_i) = \mu \left(\bigcap_{i=1}^\infty A_i \right). \qquad \textit{(continuity from above)}$$

For a proof, see Klenke (2013, Theorem 1.36).

Remark 1.69 [Finite case] If $A_1, \dots, A_n \in \mathscr{A}$ is a finite sequence with $A_1 \subset \dots \subset A_n$, then $\bigcup_{i=1}^n A_i = A_n$ and

$$\mu \left(\bigcup_{i=1}^n A_i \right) = \mu(A_n). \tag{1.52}$$

This is a trivial case of Theorem 1.68 (i) (with $A_n = A_{n+1} = A_{n+2} = \dots$). ◁

Example 1.70 [Geometric examples] Figures 1.4 and 1.5 illustrate this theorem for the Lebesgue measure λ_2 on $(\mathbb{R}^2, \mathscr{B}_2)$, the *area* of a set O and the sets $A_i, i \in \mathbb{N}$. In this example, A_1 is the open rectangle in the open (i.e., the set without its boundary) egg-shaped set O displayed in Figure 1.4, A_2 the union of A_1 with two other rectangles in the middle

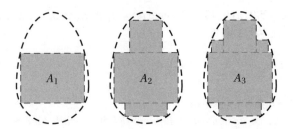

Figure 1.4   Approximation of an open egg-shaped set O from below.

figure, and A_3 the union of A_2 with two additional rectangles in the right figure. Adding more and more rectangles, it is plausible that $A_1 \subset A_2 \subset \ldots \subset O$ and that their union approximates O (i.e., $\bigcup_{i=1}^{\infty} A_i = O$). Under these premises, Theorem 1.68 (i) yields the conclusion $\lim_{i \to \infty} \lambda_2(A_i) = \lambda_2\left(\bigcup_{i=1}^{\infty} A_i\right) = \lambda_2(O)$. Figure 1.5 illustrates the same principle. However, now the area of the egg-shaped set O is approximated from above by subtracting the areas of appropriate rectangles.

As a second example, consider the Lebesgue measure λ on $(\mathbb{R}, \mathscr{B})$ and the intervals $A_i =]x - \frac{1}{i}, x]$, $i \in \mathbb{N}$. Obviously, $A_1 \supset A_2 \supset \ldots$ and $\lambda(A_i) = \frac{1}{i} < \infty$, for all $i \in \mathbb{N}$ (see also Exercise 1.12). Hence, for all $x \in \mathbb{R}$,

$$\lambda(\{x\}) = \lambda\left(\bigcap_{i=1}^{\infty}]x - \frac{1}{i}, x]\right) = \lim_{i \to \infty} \lambda\left(]x - \frac{1}{i}, x]\right) = \lim_{i \to \infty} \frac{1}{i} = 0. \tag{1.53}$$

This is an implication of continuity from above, and it implies

$$\forall\, a, b \in \mathbb{R}: a < b \;\Rightarrow\; \lambda(]a, b]) = \lambda([a, b]) = \lambda([a, b[) = \lambda(]a, b[)$$
$$= b - a. \tag{1.54}$$

◁

Remark 1.71 [A motivation for σ-additivity]   As already mentioned in Remark 1.45, σ-additivity refers to unions of finitely or countably many sets. Now consider $\bigcup_{1 \leq x \leq 2} \{x\} = [1, 2] \in \mathscr{B}$ [see Eq. (1.9)]. According to Equation (1.53), $\lambda(\{x\}) = 0$, for all $x \in [1, 2]$, and hence $\lambda(\{x \in [1, 2]: x \in \mathbb{Q}\}) = 0$, because the set of rational numbers is countable. In other words, the Lebesgue measure λ of the set of all rational numbers in the closed interval $[1, 2]$ is zero, and this is not a contradiction to

$$\lambda\left(\bigcup_{1 \leq x \leq 2} \{x\}\right) = \lambda([1, 2]) = 2 - 1 = 1,$$

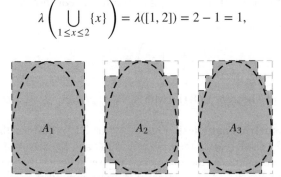

Figure 1.5   Approximation of an open egg-shaped set O from above.

because $\bigcup_{1 \leq x \leq 2} \{x\}$ is an uncountable union. This illustrates that additivity for uncountable unions can be meaningless.   ◁

1.6   Specifying a measure via a generating system

Given a measurable space $(\Omega, \mathscr{A})$, a measure is a function that is defined on $\mathscr{A}$. In many situations, such as when $\mathscr{A} = \sigma(\mathscr{E})$ can only be described by a generating set system $\mathscr{E}$ (e.g., the set system $\mathscr{I}_1$ generating the Borel σ-algebra on $\mathbb{R}$), it is important to answer the following questions:

(a) *Existence*: If there is a set function $\tilde{\mu} \colon \mathscr{E} \to \overline{\mathbb{R}}$, is there also a measure $\mu \colon \sigma(\mathscr{E}) \to \overline{\mathbb{R}}$ such that $\mu(A) = \tilde{\mu}(A), \forall A \in \mathscr{E}$?

(b) *Uniqueness*: Is a measure μ on $(\Omega, \sigma(\mathscr{E}))$ already uniquely defined by its values $\mu(A)$, $A \in \mathscr{E}$?

(Sufficient conditions for the existence of such a measure μ are formulated in Klenke, 2013, Theorem 1.53.)

The following uniqueness theorem for finite measures provides an answer to these questions, which suffices for our purposes. (A more general formulation for σ-finite measures with additional assumptions and a proof of Theorem 1.72 is found in Klenke, 2013, Lemma 1.42.)

Theorem 1.72 [Generating system and uniqueness of a measure]
Let $(\Omega, \mathscr{A})$ be a measurable space and let $\mathscr{E} \subset \mathscr{A}$, where $\mathscr{E}$ is ∩-stable and $\sigma(\mathscr{E}) = \mathscr{A}$. If μ_1 and μ_2 are finite measures on $(\Omega, \mathscr{A})$ (i.e., measures with $\mu_1(\Omega), \mu_2(\Omega) < \infty$), then,

$$\forall A \in \mathscr{E} \colon \mu_1(A) = \mu_2(A) \;\Rightarrow\; \forall A \in \mathscr{A} \colon \mu_1(A) = \mu_2(A).$$

Example 1.73 [Countable Ω]   Let Ω be a nonempty finite or countable set, and let $\mathscr{A} = \mathscr{P}(\Omega)$. Then the set system

$$\mathscr{E}_1 = \{\emptyset\} \cup \{\{\omega\} \colon \omega \in \Omega\}$$

is ∩-stable and $\sigma(\mathscr{E}_1) = \mathscr{A}$. As already noted in Remark 1.48, a finite measure μ on $(\Omega, \mathscr{A})$ is uniquely defined by its values $\mu(\{\omega\})$, $\omega \in \Omega$.   ◁

Example 1.74 [Measures on $(\mathbb{R}, \mathscr{B})$]   The set system

$$\mathscr{E}_2 = \{\,]a, b] \colon a < b, a, b \in \mathbb{R}\} \cup \{\emptyset\}$$

is ∩-stable and $\sigma(\mathscr{E}_2) = \mathscr{B}$ [see Eq. (1.18) and section 1.2.4]. Another ∩-stable set system $\mathscr{E}_3$ with $\sigma(\mathscr{E}_3) = \mathscr{B}$ is

$$\mathscr{E}_3 = \{\,]-\infty, b] \colon b \in \mathbb{R}\}$$

(cf. Klenke, 2013). This set system is crucial for the definition of a cumulative distribution function (see section 5.7.1).   ◁

1.7   σ-Algebra that is trivial with respect to a measure

All σ-algebras treated in section 1.2 have been defined without reference to a measure. Now we define the concept of a *trivial σ-algebra*, which is defined referring to a measure. We start with a lemma about the set of all subsets of a set Ω with $\mu(A) = 0$ or $\mu(A) = \mu(\Omega)$ (i.e., the set of all sets that are *trivial* with respect to the measure μ). Hence, the set of μ-trivial sets includes all *null sets* that is, all sets $A \in \mathscr{A}$ with $\mu(A) = 0$, and all sets $A \in \mathscr{A}$ with $\mu(A) = \mu(\Omega)$.

Lemma 1.75 [The set of all trivial sets is a σ-algebra]
Let $(\Omega, \mathscr{A}, \mu)$ be a measure space, and assume that μ is finite. Then,

$$\mathscr{T}_\mu := \{A \in \mathscr{A}: \mu(A) = 0 \text{ or } \mu(A) = \mu(\Omega)\} \tag{1.55}$$

is a σ-algebra.

(Proof p. 30)

This lemma allows for Definition 1.76:

Definition 1.76 [Trivial σ-algebra with respect to a measure]
Let $(\Omega, \mathscr{A}, \mu)$ be a measure space, assume that μ is finite, and let $\mathscr{T}_\mu$ be defined by (1.55). Then each σ-algebra $\mathscr{C} \subset \mathscr{T}_\mu$ is called a μ-trivial σ-algebra and its elements μ-trivial sets.

Obviously, $\{\Omega, \emptyset\}$ is a trivial σ-algebra with respect to all measures on $(\Omega, \mathscr{A})$. Hence, we can call it a *trivial σ-algebra* without reference to a specified measure.

1.8   Proofs

Proof of Theorem 1.12

(a)

$$\forall\, i \in I: \mathscr{A}_i \text{ is a σ-algebra on } \Omega \;\Rightarrow\; \forall\, i \in I: \Omega \in \mathscr{A}_i \quad \text{[Def. 1.1 (a)]}$$
$$\Rightarrow \Omega \in \bigcap_{i \in I} \mathscr{A}_i.$$

(b)

$$A \in \bigcap_{i \in I} \mathscr{A}_i \;\Rightarrow\; \forall\, i \in I: A \in \mathscr{A}_i$$
$$\Rightarrow \forall\, i \in I: A^c \in \mathscr{A}_i \quad \text{[Def. 1.1 (b)]}$$
$$\Rightarrow A^c \in \bigcap_{i \in I} \mathscr{A}_i.$$

(c)

$$A_1, A_2, \ldots \in \bigcap_{i \in I} \mathscr{A}_i \Rightarrow \forall i \in I: A_1, A_2, \ldots \in \mathscr{A}_i$$

$$\Rightarrow \forall i \in I: \bigcup_{j=1}^{\infty} A_j \in \mathscr{A}_i \qquad [\text{Def. 1.1 (c)}]$$

$$\Rightarrow \bigcup_{j=1}^{\infty} A_j \in \bigcap_{i \in I} \mathscr{A}_i.$$

Proof of Lemma 1.15

If $\mathscr{C}$ is a σ-algebra with $\mathscr{E} \subset \mathscr{C}$ and $\mathscr{A} = \sigma(\mathscr{E})$, then (1.11) and the assumption $\mathscr{C} \subset \mathscr{A}$ imply $\mathscr{A} = \sigma(\mathscr{E}) \subset \mathscr{C} \subset \mathscr{A}$. Hence, $\mathscr{C} = \mathscr{A}$.

Proof of Lemma 1.20

Define $\mathscr{D} := \left\{ C = \bigcup_{i \in I(C)} B_i : I(C) \subset \mathbb{N} \right\}$.

$\mathscr{E} \subset \mathscr{D}$: For $B_j \in \mathscr{E}$, choose $I(B_j) = \{j\}$. Then, $B_j = \bigcup_{i \in I(B_j)} B_i$.

$\mathscr{D} \subset \sigma(\mathscr{E})$: Because $\mathbb{N}$ is countable, any $I(C) \subset \mathbb{N}$ is finite or countable, and this implies that $C = \bigcup_{i \in C} B_i$ is an element of $\sigma(\mathscr{E})$ [see Def. 1.1 (c), (1.3)].

Checking the three conditions defining a σ-algebra (see Def. 1.1), we show that $\mathscr{D}$ is a σ-algebra.

(a)

$$\Omega = \begin{cases} \bigcup_{i=1}^{n} B_i, & \text{if } \mathscr{E} = \{B_1, \ldots, B_n\} \\ \bigcup_{i=1}^{\infty} B_i, & \text{if } \mathscr{E} = \{B_1, B_2, \ldots\}, \end{cases}$$

because $\mathscr{E}$ is assumed to be a partition. This shows that $\Omega \in \mathscr{D}$.

(b) The equation for Ω in (a) also implies $I(C^c) = I(C)^c$. Therefore, $C^c \in \mathscr{D}$ if $C \in \mathscr{D}$.

(c) If $C_1, C_2, \ldots \in \mathscr{D}$, then,

$$\bigcup_{j=1}^{\infty} C_j = \bigcup_{j=1}^{\infty} \bigcup_{i \in I(C_j)} B_i = \bigcup_{i \in \bigcup_{j=1}^{\infty} I(C_j)} B_i \in \mathscr{D},$$

because $\bigcup_{j=1}^{\infty} I(C_j) \subset \mathbb{N}$.

Finally, we prove the second equation in (1.14). If $j \in I(C)$ and $C = \bigcup_{i \in I(C)} B_i$, then $B_j \subset C$, which implies

$$\bigcup_{i \in I(C)} B_i \subset \bigcup_{B_i \subset C} B_i.$$

Vice versa, if $B_j \subset C$, then $j \in I(C)$, because for any $\omega \in B_j$, there is no $i \neq j$ such that $\omega \in B_i$ [see condition (b) of Rem. 1.19]. Hence,

$$\bigcup_{B_i \subset C} B_i \subset \bigcup_{i \in I(C)} B_i,$$

which proves the second equation in (1.14).

Proof of Lemma 1.27

In this proof, we use $\sigma_\Omega(\mathscr{E})$ to denote the σ-algebra on Ω generated by $\mathscr{E} \subset \mathscr{P}(\Omega)$. Similarly, $\sigma_A(\mathscr{D})$ denotes the σ-algebra on A generated by $\mathscr{D} \subset \mathscr{P}(A)$.

(1.15) $\sigma_\Omega(\mathscr{E})$ is a σ-algebra on Ω and $\mathscr{E} \subset \sigma_\Omega(\mathscr{E})$, by definition of $\sigma_\Omega(\mathscr{E})$. Hence, $\mathscr{E}|_A \subset \sigma_\Omega(\mathscr{E})|_A$, and $\sigma_\Omega(\mathscr{E})|_A$ is a σ-algebra on A (see Exercise 1.5). Therefore, the definition (1.10) yields

$$\sigma_A(\mathscr{E}|_A) \subset \sigma_\Omega(\mathscr{E})|_A.$$

Furthermore, $\mathscr{E} \subset \sigma_\Omega(\mathscr{E}|_A \cup \mathscr{E}|_{A^c})$, which implies

$$
\begin{aligned}
\sigma_\Omega(\mathscr{E}) &\subset \sigma_\Omega(\mathscr{E}|_A \cup \mathscr{E}|_{A^c}) && \text{[Rem. 1.23]}\\
&\subset \sigma_\Omega(\sigma_A(\mathscr{E}|_A) \cup \sigma_{A^c}(\mathscr{E}|_{A^c})) && \text{[Rem. 1.23]}\\
&= \{C \cup D: C \in \sigma_A(\mathscr{E}|_A),\ D \in \sigma_{A^c}(\mathscr{E}|_{A^c})\}. && \text{[This set system is a σ-algebra]}
\end{aligned}
$$

Therefore,

$$
\begin{aligned}
\sigma_\Omega(\mathscr{E})|_A &\subset \{C \cup D: C \in \sigma_A(\mathscr{E}|_A),\ D \in \sigma_{A^c}(\mathscr{E}|_{A^c})\}|_A\\
&= \{(C \cup D) \cap A: C \in \sigma_A(\mathscr{E}|_A),\ D \in \sigma_{A^c}(\mathscr{E}|_{A^c})\}\\
&= \{C \cap A: C \in \sigma_A(\mathscr{E}|_A)\} && [D \subset A^c]\\
&= \sigma_A(\mathscr{E}|_A). && [C \subset A]
\end{aligned}
$$

Hence, we have shown $\sigma_A(\mathscr{E}|_A) \subset \sigma_\Omega(\mathscr{E})|_A$ and $\sigma_\Omega(\mathscr{E})|_A \subset \sigma_A(\mathscr{E}|_A)$, which is equivalent to $\sigma_A(\mathscr{E}|_A) = \sigma_\Omega(\mathscr{E})|_A$.

(1.17)

$$
\begin{aligned}
\sigma_\Omega(\mathscr{C} \cup \mathscr{E})|_A &= \sigma_A(\mathscr{C} \cup \mathscr{E}|_A) && \text{[(1.15)]}\\
&= \sigma_A(\mathscr{C}|_A \cup \mathscr{E}|_A) && \text{[See def. of the trace in Example 1.10]}\\
&= \sigma_A(\mathscr{C}|_A \cup \{\emptyset, A\}) && \text{[(1.16)]}\\
&= \sigma_A(\mathscr{C}|_A) && [\{\emptyset, A\} \subset \mathscr{C}|_A]\\
&= \mathscr{C}|_A. && \text{[Exercise 1.5, (1.12)]}
\end{aligned}
$$

Proof of Lemma 1.75

(a) $\Omega \in \mathscr{T}_\mu$ by definition of $\mathscr{T}_\mu$.

(b) If $A \in \mathscr{T}_\mu$, then Rules (iv), (v) of Box 1.1 and finiteness of μ yield

$$
\mu(A^c) = \mu(\Omega) - \mu(A) = \begin{cases} \mu(\Omega), & \text{if } \mu(A) = 0\\ 0, & \text{if } \mu(A) = \mu(\Omega), \end{cases}
$$

which implies $A^c \in \mathscr{T}_\mu$.

(c) Let $A_1, A_2, \ldots \in \mathcal{A}$. We consider two cases. *First*, if $\mu(A_i) = 0$, for all A_i, $i \in \mathbb{N}$, then Rule (xi) of Box 1.1 yields $\left(\bigcup_{i=1}^{\infty} A_i\right) \leq \sum_{i=1}^{\infty} \mu(A_i) = 0$ (i.e., $\bigcup_{i=1}^{\infty} A_i \in \mathcal{T}_\mu$). *Second*, if there is a $j \in \mathbb{N}$ such that $\mu(A_j) = \mu(\Omega)$, then Rule (v) of Box 1.1 yields

$$\mu(\Omega) = \mu(A_j) \leq \mu\left(\bigcup_{i=1}^{\infty} A_i\right) \leq \mu(\Omega),$$

which implies $\mu\left(\bigcup_{i=1}^{\infty} A_i\right) = \mu(\Omega)$. Therefore, $\bigcup_{i=1}^{\infty} A_i \in \mathcal{T}_\mu$.

Exercises

1.1   Let $\mathcal{A}$ be a σ-algebra of subsets of a nonempty set Ω, and let $A_1, A_2, \ldots \in \mathcal{A}$. Show: (a) $A_1 \cap A_2 \cap \ldots \in \mathcal{A}$, (b) $A_1 \cap A_2 \in \mathcal{A}$, and (c) $A_1 \setminus A_2 \in \mathcal{A}$.

1.2   Show that the set system $\mathcal{A} = \{\Omega, \emptyset, A, A^c\}$ is stable (closed) with respect to union of elements of $\mathcal{A}$.

1.3   Consider the set $\Omega = \{\omega_1, \ldots, \omega_6\}$ representing the set of all possible outcomes of tossing a dice and the power set $\mathcal{P}(\Omega)$, which, in probability theory, represents the set of all possible events (including the 'impossible' event $\emptyset$) in this random experiment. Specify the σ-algebra on Ω that represents all possible events if we only distinguish between even and uneven number of points.

1.4   Consider the random experiment that has been described in Example 1.9. Aside from the power set of Ω, we already considered the σ-algebras $\mathcal{A}_1 = \{\Omega, \emptyset, A, A^c\}$, $\mathcal{A}_2 = \{\Omega, \emptyset, B, B^c\}$, and $\mathcal{A}_3 = \{\Omega, \emptyset, C, C^c\}$. Define another σ-algebra not yet mentioned.

1.5   Prove: If $\mathcal{A}$ is a σ-algebra on Ω and $\Omega_0 \subset \Omega$, then $\mathcal{A}|_{\Omega_0} = \{\Omega_0 \cap A : A \in \mathcal{A}\}$ is a σ-algebra on Ω_0.

1.6   Prove the proposition of Remark 1.16.

1.7   Show that $\sigma(\mathcal{E}) = \mathcal{P}(\Omega)$ if Ω is finite or countable and $\mathcal{E} := \{\{\omega\} : \omega \in \Omega\}$.

1.8   Prove the proposition of Remark 1.21.

1.9   Let $\mathcal{E}_1, \mathcal{E}_2$ be set systems on Ω with $\mathcal{E}_1 \subset \mathcal{E}_2$. Show that $\sigma(\mathcal{E}_1) \subset \sigma(\mathcal{E}_2)$.

1.10   Prove propositions (a) and (b) of Example 1.25.

1.11   Prove Equation (1.21).

1.12   Show that $\{x\} \in \mathcal{B}_n$ for all $x \in \mathbb{R}^n$, where $\mathcal{B}_n$ is the Borel σ-algebra on $\mathbb{R}^n$.

1.13   Let $\mathcal{B}$ be the Borel σ-algebra on $\mathbb{R}$, and let $\Omega_0 \subset \mathbb{R}$ be finite or countable. Show that $\mathcal{B}|_{\Omega_0} = \mathcal{P}(\Omega_0)$.

1.14   Prove the proposition of Example 1.34.

1.15   Prove the proposition of Remark 1.35.

1.16   Let $(\Omega_i, \mathcal{A}_i)$, $i = 1, \ldots, n$, be measurable spaces. Show that the set system $\mathcal{E} := \{A_1 \times \ldots \times A_n : A_i \in \mathcal{A}_i, i = 1, \ldots, n\}$ is $\cap$-stable.

1.17 Prove the proposition of Remark 1.46.

1.18 Prove the rules of Box 1.1.

1.19 Prove the propositions of Remark 1.47.

1.20 Show that $\delta_\omega\colon \mathcal{A} \to \{0, 1\}$ in Example 1.52 is a measure.

1.21 Prove that the function defined by Equation (1.40) is a measure on $(\Omega, \mathcal{A})$.

1.22 Show that $\sum_{\omega \in B} \delta_\omega$ in Example 1.57 is a measure.

1.23 Show that $\nu\colon \mathcal{C} \to \overline{\mathbb{R}}$ defined in Example 1.61 is a measure on $(\Omega, \mathcal{C})$.

1.24 Prove that the function $\sum_{i=1}^{\infty} \alpha_i \mu_i$ defined in Example 1.62 is a measure on $(\Omega, \mathcal{A})$.

Solutions

1.1 (a) If $A_1, A_2, \ldots \in \mathcal{A}$, then $A_1^c, A_2^c, \ldots \in \mathcal{A}$ [see Def. 1.1 (b)]. Hence,

$$\bigcap_{i=1}^{\infty} A_i = \left[\left(\bigcap_{i=1}^{\infty} A_i\right)^c\right]^c = \left[\bigcup_{i=1}^{\infty} A_i^c\right]^c \qquad \text{[de Morgan]}$$

$$\in \mathcal{A}. \qquad \text{[Def. 1.1 (c), (b)]}$$

(b) Let $A_1, A_2 \in \mathcal{A}$ and choose $A_3, A_4, \ldots$ such that $\Omega = A_i$, for all $i \geq 3$, $i \in \mathbb{N}$. Then, according to Definition 1.1 (a),

$$A_1 \cap A_2 = A_1 \cap A_2 \cap \Omega = \bigcap_{i=1}^{\infty} A_i \in \mathcal{A}.$$

(c) $A_1 \setminus A_2 = A_1 \cap A_2^c \in \mathcal{A}$ [see (b) and Def. 1.1 (b)].

1.2 The unions $\Omega \cup A = \Omega$, $\Omega \cup A^c = \Omega$, and $\Omega \cup \emptyset = \Omega$ are all elements of $\mathcal{A}$, and the same is true for $\emptyset \cup A = A$, $\emptyset \cup A^c = A^c$, and $A \cup A^c = \Omega$. Furthermore, $B \cup B = B$ for all $B \in \mathcal{A}$.

1.3 The σ-algebra on Ω that only distinguishes between an even and uneven number of points is $\mathcal{A}_1 := \{\{\omega_1, \omega_3, \omega_5\}, \{\omega_2, \omega_4, \omega_6\}, \Omega, \emptyset\}$. This is a sub-$\sigma$-algebra of $\mathcal{P}(\Omega)$. Therefore, $\mathcal{A}_1$ represents the set of all possible events of a random experiment that is, in a sense, contained in the original random experiment.

1.4 Consider the set system that contains as elements A, A^c, B, B^c, Ω, $\emptyset$, all unions and all intersections of these sets as well as the unions and intersections of the resulting sets such as $(A^c \cup B^c) \cap (A \cup B)$ and $(A^c \cup B^c) \cup (A \cup B)$. Altogether, these are 16 sets. This is $\sigma(\mathcal{A}_1 \cup \mathcal{A}_2)$, the σ-algebra generated by $\mathcal{A}_1 \cup \mathcal{A}_2 = \{A, A^c, B, B^c, \Omega, \emptyset\}$ (see Def. 1.13 and Rem. 1.21).

1.5 (a) $\Omega_0 \cap \Omega = \Omega_0$. This implies $\Omega_0 \in \mathcal{A}|_{\Omega_0}$.

(b)

$$A^* \in \mathcal{A}|_{\Omega_0} \;\Rightarrow\; \exists\, A \in \mathcal{A}\colon A^* = \Omega_0 \cap A.$$

With this set A and using B^c for the complement of a set B with respect to Ω,

$$
\begin{aligned}
\Omega_0 \setminus A^* &= \Omega_0 \setminus (\Omega_0 \cap A) \\
&= \Omega_0 \cap (\Omega_0 \cap A)^c \\
&= \Omega_0 \cap (\Omega_0^c \cup A^c) \\
&= (\Omega_0 \cap \Omega_0^c) \cup (\Omega_0 \cap A^c) \\
&= \Omega_0 \cap A^c \in \mathscr{A}|_{\Omega_0}.
\end{aligned}
$$

(c)

$$
A_1^*, A_2^*, \ldots \in \mathscr{A}|_{\Omega_0} \;\Rightarrow\; \exists\, A_1, A_2, \ldots \in \mathscr{A}\colon A_i^* = \Omega_0 \cap A_i,\, i \in \mathbb{N}.
$$

Hence,

$$
A_1^* \cup A_2^* \cup \ldots = (\Omega_0 \cap A_1) \cup (\Omega_0 \cap A_2) \cup \ldots = \Omega_0 \cap (A_1 \cup A_2 \cup \ldots) \in \mathscr{A}|_{\Omega_0}.
$$

1.6 If $\mathscr{G}$ is a σ-algebra on Ω, then

$$
\mathscr{E} \cup \mathscr{F} \subset \mathscr{G} \;\Leftrightarrow\; \sigma(\mathscr{E} \cup \mathscr{F}) \subset \mathscr{G}. \qquad [(1.11)] \qquad (1.56)
$$

Furthermore, for three sets A, B, C,

$$
A \cup B \subset C \;\Leftrightarrow\; A \subset C \wedge B \subset C. \qquad (1.57)
$$

Hence,

$$
\begin{aligned}
\mathscr{D} \cup \mathscr{E} \cup \mathscr{F} \subset \mathscr{G} &\Leftrightarrow (\mathscr{D} \subset \mathscr{G}) \wedge (\mathscr{E} \cup \mathscr{F} \subset \mathscr{G}) && [(1.57)] \\
&\Leftrightarrow (\mathscr{D} \subset \mathscr{G}) \wedge (\sigma(\mathscr{E} \cup \mathscr{F}) \subset \mathscr{G}) && [(1.56)] \\
&\Leftrightarrow \mathscr{D} \cup \sigma(\mathscr{E} \cup \mathscr{F}) \subset \mathscr{G}. && [(1.57)]
\end{aligned}
$$

Now Definition 1.13 yields the proposition.

1.7 If Ω is finite or countable, then each of its subsets A is finite or countable as well. Therefore,

$$
\forall A \subset \Omega\colon A = \bigcup_{\omega \in A} \{\omega\} \in \sigma(\mathscr{E}). \qquad [\text{Def. 1.1 (c), Rem. 1.2}]
$$

Because each element A of $\mathscr{P}(\Omega)$ is a union $\bigcup_{\omega \in A} \{\omega\}$ of singletons $\{\omega\}$, $\omega \in A$, this implies $\mathscr{P}(\Omega) \subset \sigma(\mathscr{E})$. Hence, $\mathscr{E} \subset \mathscr{P}(\Omega) \subset \sigma(\mathscr{E})$. Therefore, Lemma 1.15 implies $\sigma(\mathscr{E}) = \mathscr{P}(\Omega)$.

1.8 Suppose that $\mathscr{E} = \{A_1, \ldots, A_m\}$ and $A_j^1 := A_j$ and let A_j^c denote the complement of A_j. Then, for all $(k_1, \ldots, k_m) \in \{1, c\}^m$ define

$$
B_{(k_1, \ldots, k_m)} := \bigcap_{j=1}^{m} A_j^{k_j}.
$$

Then

$$\mathscr{F} := \{B_{(k_1,\ldots,k_m)}: (k_1, \ldots, k_m) \in \{1, c\}^m, B_{(k_1,\ldots,k_m)} \neq \emptyset\}$$

is a finite partition of Ω. Note that $\mathscr{F}$ contains all nonempty intersections of sets A_j or their complements, respectively, where $j = 1, \ldots, m$. Now Lemma 1.20 implies the proposition.

1.9 If $\mathscr{E}_1 \subset \mathscr{E}_2 \subset \mathcal{P}(\Omega)$, then for any σ-algebra $\mathscr{A}$ on Ω with $\mathscr{E}_2 \subset \mathscr{A}$ also $\mathscr{E}_1 \subset \mathscr{A}$. Remember, if $J \subset I$, then $\bigcap_{i \in I} B_i \subset \bigcap_{i \in J} B_i$, for any sets $B_i, i \in I$. Therefore, $\sigma(\mathscr{E}_1)$, which is the intersection of all σ-algebras containing $\mathscr{E}_1$, is a subset of the intersection of all σ-algebras containing $\mathscr{E}_2$, which is $\sigma(\mathscr{E}_2)$.

1.10 (a) If Ω is finite, then $\mathcal{P}(\Omega)$ is a finite set system. Therefore, each σ-algebra $\mathscr{A}$ on Ω is a finite set system. Because $\mathscr{A} = \sigma(\mathscr{A})$, this σ-algebra is countably generated.

(b) The set $\mathbb{N}_0$ is countable and therefore also $\mathbb{N}_0^n$ for $n \in \mathbb{N}$. Example 1.18 then implies that $\mathcal{P}(\mathbb{N}_0^n)$ is countably generated.

1.11 Let $\mathscr{H}_n = \{]-\infty, b_1] \times \ldots \times]-\infty, b_n]: b_1, \ldots, b_n \in \mathbb{R}\}$.
(i) For all $(b_1, \ldots, b_n) \in \mathbb{R}^n$ and all $m \in \mathbb{N}$ with $m < b_i, i = 1, \ldots, n$,

$$B_m :=]-m, b_1] \times \ldots \times]-m, b_n] \in \mathscr{I}_n.$$

According to Definition 1.1 (c) this implies

$$\bigcup_{\substack{m \in \mathbb{N} \\ m < b_i, i = 1, \ldots, n}} B_m =]-\infty, b_1] \times \ldots \times]-\infty, b_n] \in \sigma(\mathscr{I}_n).$$

Hence, $\mathscr{H}_n \subset \sigma(\mathscr{I}_n)$, which, according to (1.11) and (1.12), implies

$$\sigma(\mathscr{H}_n) \subset \sigma(\mathscr{I}_n) = \mathscr{B}_n.$$

(ii) For all $a_1, \ldots, a_n, b_1, \ldots, b_n \in \mathbb{R}$, with $a_i < b_i, i = 1, \ldots, n$,

$$]a_1, b_1] \times \ldots \times]a_n, b_n] =]-\infty, b_1] \times \ldots \times]-\infty, b_n] \setminus \left(\bigcup_{j=1}^n H_j \right),$$

where $H_j :=]-\infty, b_1] \times \ldots \times]-\infty, b_{j-1}] \times]-\infty, a_j] \times]-\infty, b_{j+1}] \times \ldots \times]-\infty, b_n]$. Hence, according to Remark 1.2, $]a_1, b_1] \times \ldots \times]a_n, b_n] \in \sigma(\mathscr{H}_n)$ and $\mathscr{I}_n \subset \sigma(\mathscr{H}_n)$, which, according to (1.11) and (1.12), implies

$$\mathscr{B}_n = \sigma(\mathscr{I}_n) \subset \sigma(\mathscr{H}_n).$$

1.12 If $x \in \mathbb{R}$, then $\{x\} = \bigcap_{i=1}^{\infty}]x - 1/i, x]$. According to Equation (1.18), the intervals $]x - 1/i, x]$ are elements of the generating set system of $\mathscr{B}$, the Borel σ-algebra on

$\mathbb{R}$. Therefore, their countable intersection is an element of $\mathcal{B}$. If $x = (x_1, \ldots, x_n) \in \mathbb{R}^n$, then

$$\{x\} = \bigcap_{i=1}^{\infty} \left(\underset{j=1}{\overset{n}{\times}}]x_j - \frac{1}{i}, x_j] \right).$$

According to Equation (1.20), the cuboids $\underset{j=1}{\overset{n}{\times}}]x_j - \frac{1}{i}, x_j]$ are elements of the set system $\mathscr{I}_n$ and $\sigma(\mathscr{I}_n) = \mathscr{B}_n$.

1.13 Because $\{x\} \in \mathscr{B}$ for all $x \in \mathbb{R}$ (see Exercise 1.12), we can conclude: $\{x\} \in \mathscr{B}|_{\Omega_0}$ for all $x \in \Omega_0$. Hence, if Ω_0 is finite or countable, Example 1.18 implies $\mathscr{B}|_{\Omega_0} = \mathcal{P}(\Omega_0)$.

1.14 Let $\Omega_1, \ldots, \Omega_n$ be finite or countable sets, and let $\mathscr{A}_1, \ldots, \mathscr{A}_n$ be their power sets. Then $\omega_1 \in \Omega_1, \ldots, \omega_n \in \Omega_n$ implies $\{\omega_1\} \in \mathscr{A}_1, \ldots, \{\omega_n\} \in \mathscr{A}_n$. Therefore,

$$\{(\omega_1, \ldots, \omega_n)\} = \{\omega_1\} \times \ldots \times \{\omega_n\} \in \left\{ \underset{i=1}{\overset{n}{\times}} A_i \colon A_i \in \mathscr{A}_i, i \in \{1, \ldots, n\} \right\}.$$

Hence,

$$\sigma(\{(\omega_1, \ldots, \omega_n)\} \colon \omega_1 \in \Omega_1, \ldots, \omega_n \in \Omega_n) \subset \bigotimes_{i=1}^{n} \mathscr{A}_i.$$

With Ω_i being finite or countable, $\Omega = \Omega_1 \times \ldots \times \Omega_n$ is finite or countable. Therefore,

$$\sigma(\{(\omega_1, \ldots, \omega_n)\} \colon \omega_1 \in \Omega_1, \ldots, \omega_n \in \Omega_n) = \mathcal{P}(\Omega)$$

(see Example 1.18). Because $\bigotimes_{i=1}^{n} \mathscr{A}_i \subset \mathcal{P}(\Omega)$, we can conclude

$$\bigotimes_{i=1}^{n} \mathscr{A}_i = \mathcal{P}(\Omega_1 \times \ldots \times \Omega_n) = \mathcal{P}\left(\underset{i=1}{\overset{n}{\times}} \Omega_i \right).$$

1.15

$$(A \times B)^c = \{(\omega_1, \omega_2) \in \Omega_1 \times \Omega_2 \colon \omega_1 \notin A \text{ or } \omega_2 \notin B\}$$
$$= \{(\omega_1, \omega_2) \in \Omega_1 \times \Omega_2 \colon (\omega_1 \notin A, \omega_2 \in B) \text{ or } \omega_2 \notin B\}$$
$$= (A^c \times B) \cup (\Omega_1 \times B^c)$$

and

$$(A^c \times B) \cap (\Omega_1 \times B^c)$$
$$= \{(\omega_1, \omega_2) \in \Omega_1 \times \Omega_2 \colon \omega_1 \notin A, \omega_2 \in B, \omega_2 \notin B\}$$
$$= \{(\omega_1, \omega_2) \in \Omega_1 \times \Omega_2 \colon \omega_1 \notin A, \omega_2 \in B \cap B^c = \emptyset\}$$
$$= \emptyset.$$

1.16 Remember that $(a \in A, b \in B)$ means $(a \in A \text{ and } b \in B)$ and that $(a \in A \text{ and } b \in B)$ and $(b \in B \text{ and } a \in A)$ are equivalent. Let $A_1, B_1 \in \mathscr{A}_1, \ldots, A_n, B_n \in \mathscr{A}_n$. Then $A_1 \cap B_1 \in$

$\mathscr{A}_1, \ldots, A_n \cap B_n \in \mathscr{A}_n$. Hence, $A_1 \times \ldots \times A_n \in \mathscr{E}$, $B_1 \times \ldots \times B_n \in \mathscr{E}$ and $(A_1 \cap B_1) \times \ldots \times (A_n \cap B_n) \in \mathscr{E}$. Furthermore,

$$(A_1 \times \ldots \times A_n) \cap (B_1 \times \ldots \times B_n)$$
$$= \{(\omega_1, \ldots, \omega_n): \omega_1 \in A_1, \ldots, \omega_n \in A_n, \omega_1 \in B_1, \ldots, \omega_n \in B_n\}$$
$$= \{(\omega_1, \ldots, \omega_n): \omega_1 \in (A_1 \cap B_1), \ldots, \omega_n \in (A_n \cap B_n)\}$$
$$= (A_1 \cap B_1) \times \ldots \times (A_n \cap B_n) \in \mathscr{E}.$$

1.17   Let B_i denote the sets defined in Remark 1.46.
   (i) $B_1 = A_1 \in \mathscr{A}$. For all $i \in \mathbb{N}, i > 1, B_i \in \mathscr{A}$:

$$B_i = A_i \setminus \left(\bigcup_{j=1}^{i-1} A_j \right) = A_i \cap \left(\bigcup_{j=1}^{i-1} A_j \right)^c \in \mathscr{A}. \qquad \text{[Def. 1.1 (b), Rem. 1.2]}$$

   (ii)   For any sequence $C_1, C_2, \ldots \subset \Omega$, define

$$\bigcup_{j=m}^{n} C_j := \varnothing, \quad \text{if } m > n, \quad \text{and} \quad \bigcap_{j=m}^{n} C_j := \Omega, \quad \text{if } m > n.$$

Then, using associativity and commutativity of the intersection, for $1 \le k < l$,

$$B_k \cap B_l = \left[A_k \setminus \left(\bigcup_{j=1}^{k-1} A_j \right) \right] \cap \left[A_l \setminus \left(\bigcup_{j=1}^{l-1} A_j \right) \right]$$
$$= A_k \cap \left(\bigcup_{j=1}^{k-1} A_j \right)^c \cap A_l \cap \left(\bigcup_{j=1}^{l-1} A_j \right)^c \qquad [A \setminus B = A \cap B^c]$$
$$= A_k \cap \left(\bigcap_{j=1}^{k-1} A_j^c \right) \cap A_l \cap \left(\bigcap_{j=1}^{l-1} A_j^c \right) \qquad \text{[de Morgan]}$$
$$= A_k \cap A_l \cap \left(\bigcap_{j=1}^{k-1} A_j^c \right) \cap \left(\bigcap_{j=1}^{k-1} A_j^c \right) \cap A_k^c \cap \left(\bigcap_{j=k+1}^{l-1} A_j^c \right)$$
$$= \varnothing. \qquad [A_k \cap A_k^c = \varnothing]$$

   (iii)   The sets B_i are defined such that $B_i \subset A_i$, for all $i \in I$. Therefore, $\bigcup_{i=1}^{\infty} B_i \subset \bigcup_{i=1}^{\infty} A_i$. Furthermore, for all $\omega \in \Omega$,

$$\omega \in \bigcup_{i=1}^{\infty} A_i \; \Rightarrow \; \exists\, i \in \mathbb{N}: \omega \in A_i \wedge (\forall j < i: \omega \notin A_j)$$
$$\Rightarrow \; \exists\, i \in \mathbb{N}: \omega \in A_1^c \cap \ldots \cap A_{i-1}^c \cap A_i = B_i$$
$$\Rightarrow \; \omega \in \bigcup_{i=1}^{\infty} B_i.$$

Hence, $\bigcup_{i=1}^{\infty} A_i \subset \bigcup_{i=1}^{\infty} B_i$, and this implies $\bigcup_{i=1}^{\infty} B_i = \bigcup_{i=1}^{\infty} A_i$.

1.18   (i)   This is condition (c) of Definition 1.43.

(ii)   If $A_1, \ldots, A_n \in \mathscr{A}$ are pairwise disjoint, then $A_1, A_2, \ldots$ with $\varnothing = A_{n+1} = A_{n+2} = \ldots$ is a sequence of pairwise disjoint measurable sets. Therefore, conditions (a) and (c) of Def. 1.43 imply

$$\mu\left(\bigcup_{i=1}^{n} A_i\right) = \mu\left(\bigcup_{i=1}^{\infty} A_i\right) = \sum_{i=1}^{\infty} \mu(A_i) = \sum_{i=1}^{n} \mu(A_i) + \sum_{i=n+1}^{\infty} \mu(\varnothing) = \sum_{i=1}^{n} \mu(A_i).$$

(iii)   For $A, B \subset \Omega$,

$$A = (A \cap B) \cup (A \cap B^c) = (A \cap B) \cup (A \setminus B)$$

and

$$(A \cap B) \cap (A \cap B^c) = A \cap B \cap B^c = \varnothing.$$

Hence, for sets $A, B \in \mathscr{A}$, Rule (ii) (finite additivity of μ) implies proposition (iii).

(iv)   This proposition is a special case of (iii) with $A = \Omega$.

(v)   Exchanging the roles of A and B in (iii), we obtain

$$\mu(B) = \mu(A \cap B) + \mu(B \setminus A).$$

If $A \subset B$, then $A \cap B = A$; and, because $\mu(B \setminus A) \geq 0$,

$$\mu(A) = \mu(A \cap B) \leq \mu(A \cap B) + \mu(B \setminus A) = \mu(B).$$

(vi)   This rule immediately follows from proposition (iv) for $\mu(A \cap B) < \infty$. [Note that $\mu(A) - \mu(A \cap B)$ is not defined if $\mu(A) = \mu(A \cap B) = \infty$.]

(vii)   For $A, B \subset \Omega$,

$$A \cup B = (A \setminus B) \cup (A \cap B) \cup (B \setminus A).$$

Because the right-hand side is a union of pairwise disjoint sets, finite additivity of μ yields

$$\mu(A \cup B) + \mu(A \cap B) = \mu(A \setminus B) + \mu(A \cap B) + \mu(B \setminus A) + \mu(A \cap B)$$
$$= \mu(A) + \mu(B). \qquad \text{[Box 1.1 (iii)]}$$

(viii) $\mu(\Omega) = \mu(A \cup A^c) = \mu(A) + \mu(A^c)$. Hence, if $\mu(\Omega) = \mu(A) < \infty$, then $\mu(A^c) = 0$. Therefore, for all $B \in \mathscr{A}$, (v) implies $\mu(A^c \cap B) = 0$. Furthermore, $B = (A \cap B) \cup (A^c \cap B)$ and $(A \cap B) \cap (A^c \cap B) = \varnothing$. Hence, $\mu(B) = \mu(A \cap B) + \mu(A^c \cap B) = \mu(A \cap B)$. Note that, in general, $\mu(A) = \mu(\Omega)$ does not imply $A = \Omega$.

(ix) $\mu(A) = 0$ implies

$$\begin{aligned}
\mu(B) &= \mu(A) + \mu(B) \\
&\geq \mu(A \cup B) \qquad &[(xi)] \\
&\geq \mu(B). \qquad &[(v)]
\end{aligned}$$

Note that, in general, $\mu(A) = 0$ does not imply $A = \emptyset$.

(x)   Let $B := \Omega \setminus \Omega_0$. Then $\mu(B) = 0$ as well as $\mu(A \cap B) = 0$ for all $A \in \mathcal{A}$ [see Box 1.1 (v)]. Furthermore, for $A \in \mathcal{A}$: $A = (A \cap \Omega_0) \cup (A \cap B)$, where $A \cap \Omega_0$ and $A \cap B$ are disjoint. Now, the sets $A \cap \Omega_0$, $A \in \mathcal{A}$, are the elements of the trace σ-algebra and $(\Omega_0, \mathcal{A}\,|_{\Omega_0}) = [\Omega_0, \mathcal{P}(\Omega_0)]$. Therefore, we can apply Equation (1.31). Hence, for all $A \in \mathcal{A}$,

$$\begin{aligned}
\mu(A) &= \mu(A \cap \Omega_0) + \mu(A \cap B) \qquad &[\text{Box 1.1 (ii)}] \\
&= \sum_{\omega \in A \cap \Omega_0} \mu(\{\omega\}) + \mu(A \cap B) \qquad &[(1.31)] \\
&= \sum_{\omega \in A \cap \Omega_0} \mu(\{\omega\}). \qquad &[\mu(A \cap B) = 0]
\end{aligned}$$

(xi)   Let $A_1, A_2, \ldots \in \mathcal{A}$ and define $B_1, B_2, \ldots \in \mathcal{A}$ by $B_1 = A_1$, and $B_i = A_i \setminus \bigcup_{j=1}^{i-1} B_j$ for $i > 1$ (see Rem. 1.46). Then $B_1, B_2, \ldots$ is a sequence of pairwise disjoint sets with $B_i \subset A_i$ for all $i \in \mathbb{N}$ and $\bigcup_{i=1}^{\infty} B_i = \bigcup_{i=1}^{\infty} A_i$. Hence,

$$\begin{aligned}
\mu\left(\bigcup_{i=1}^{\infty} A_i\right) &= \mu\left(\bigcup_{i=1}^{\infty} B_i\right) \\
&= \sum_{i=1}^{\infty} \mu(B_i) \qquad &[\text{Def. 1.43 (c)}] \\
&\leq \sum_{i=1}^{\infty} \mu(A_i). \qquad &[\text{Box 1.1 (v)}]
\end{aligned}$$

1.19   If the $A_1, \ldots, A_n \in \mathcal{A}$ are pairwise disjoint and $B \in \mathcal{A}$, then, for $i \neq j$, $i, j = 1, \ldots, n$,

$$(B \cap A_i) \cap (B \cap A_j) = B \cap (A_i \cap A_j) = B \cap \emptyset = \emptyset.$$

Hence, the sets $B \cap A_1, \ldots, B \cap A_n$ are pairwise disjoint. Furthermore, condition (b) of Remark 1.47 implies

$$\bigcup_{i=1}^{n} (B \cap A_i) = B \cap \bigcup_{i=1}^{n} A_i = B.$$

Therefore, additivity of μ yields

$$\mu(B) = \mu\left(\bigcup_{i=1}^{n} (B \cap A_i)\right) = \sum_{i=1}^{n} \mu(B \cap A_i),$$

which is Equation (1.29). The proof of Equation (1.30) is literally the same except for replacing $\bigcup_{i=1}^{n}$ by $\bigcup_{i=1}^{\infty}$, $\sum_{i=1}^{n}$ by $\sum_{i=1}^{\infty}$, and additivity of μ by σ-additivity.

1.20 Let $\omega \in \Omega$.

(a) According to Equation (1.32), $\delta_\omega(\emptyset) = 1_\emptyset(\omega) = 0$.

(b) According to Equation (1.32), $\delta_\omega(A) = 1_A(\omega) \in \{0, 1\}$, for all $A \in \mathscr{A}$, and this implies $\delta_\omega(A) \geq 0$, for all $A \in \mathscr{A}$.

(c) If $A_1, A_2, \ldots \in \mathscr{A}$ are pairwise disjoint, then

$$
\begin{aligned}
\delta_\omega\left(\bigcup_{i=1}^{\infty} A_i\right) &= 1_{\bigcup_{i=1}^{\infty} A_i}(\omega) & [(1.32)] \\
&= \sum_{i=1}^{\infty} 1_{A_i}(\omega) & [(1.37)] \\
&= \sum_{i=1}^{\infty} \delta_\omega(A_i). & [(1.32)]
\end{aligned}
$$

1.21 (a) According to Equation (1.40), $\mu_\#(\emptyset) = \sum_{\omega \in \Omega} 1_\emptyset(\omega) = 0$.

(b) According to Equation (1.40), $\mu_\#(A) = \sum_{\omega \in \Omega} 1_A(\omega)$, for all finite $A \in \mathscr{A}$, and $\mu_\#(A) = \infty$, if A is infinite. This implies $\mu_\#(A) \geq 0$, for all $A \in \mathscr{A}$.

(c) If $A_1, A_2, \ldots \in \mathscr{A}$ are pairwise disjoint and all A_i are finite, then

$$
\begin{aligned}
\mu_\#\left(\bigcup_{i=1}^{\infty} A_i\right) &= \sum_{\omega \in \Omega} 1_{\bigcup_{i=1}^{\infty} A_i}(\omega) & [(1.40)] \\
&= \sum_{\omega \in \Omega} \sum_{i=1}^{\infty} 1_{A_i}(\omega) & [(1.37)] \\
&= \sum_{i=1}^{\infty} \sum_{\omega \in \Omega} 1_{A_i}(\omega) \\
&= \sum_{i=1}^{\infty} \mu_\#(A_i). & [(1.40)]
\end{aligned}
$$

Note that the set $\bigcup_{i=1}^{\infty} A_i$ can be countably infinite, even if all A_i are finite. In this case, $\mu_\#\left(\bigcup_{i=1}^{\infty} A_i\right) = \infty = \sum_{i=1}^{\infty} \mu_\#(A_i)$. If at least one of the A_i is infinite, then $\bigcup_{j=1}^{\infty} A_j \supset A_i$ is an infinite set and $\mu_\#\left(\bigcup_{j=1}^{\infty} A_j\right) \geq \mu_\#(A_i)$ is infinite as well.

1.22 (a) Using Equations (1.42) and (1.39),

$$
\left(\sum_{\omega \in B} \delta_\omega\right)(\emptyset) = \sum_{\omega \in B} \delta_\omega(\emptyset) = \sum_{\omega \in B} 1_\emptyset(\omega) = \sum_{\omega \in B} 0 = 0.
$$

(b) Using Equations (1.42) and (1.39),

$$\forall A \in \mathscr{A} : \left(\sum_{\omega \in B} \delta_\omega \right)(A) = \sum_{\omega \in B} \delta_\omega(A) = \sum_{\omega \in B} 1_A(\omega) \geq 0.$$

(c) If $A_1, A_2, \ldots \in \mathscr{A}$ are pairwise disjoint, then

$$\left(\sum_{\omega \in B} \delta_\omega \right)\left(\bigcup_{i=1}^{\infty} A_i \right) = \sum_{\omega \in B} \delta_\omega \left(\bigcup_{i=1}^{\infty} A_i \right) \qquad [(1.42)]$$

$$= \sum_{\omega \in B} 1_{\bigcup_{i=1}^{\infty} A_i}(\omega) \qquad [(1.39)]$$

$$= \sum_{\omega \in B} \sum_{i=1}^{\infty} 1_{A_i}(\omega) \qquad [(1.37)]$$

$$= \sum_{\omega \in B} \sum_{i=1}^{\infty} \delta_\omega(A_i) \qquad [(1.39)]$$

$$= \sum_{i=1}^{\infty} \left(\left(\sum_{\omega \in B} \delta_\omega \right)(A_i) \right). \qquad [(1.42)]$$

1.23 (a) Equation (1.47) yields: $\nu(\emptyset) = \mu(\emptyset) = 0$.

(b) Equation (1.47) also yields: $\nu(A) = \mu(A) \geq 0$, for all $A \in \mathscr{C}$.

(c) If $A_1, A_2, \ldots \in \mathscr{C}$ are pairwise disjoint, then

$$\nu\left(\bigcup_{i=1}^{\infty} A_i \right) = \mu\left(\bigcup_{i=1}^{\infty} A_i \right) \qquad [\text{Def. 1.1 (c), (1.47)}]$$

$$= \sum_{i=1}^{\infty} \mu(A_i) \qquad [\text{Def. 1.43 (c)}]$$

$$= \sum_{i=1}^{\infty} \nu(A_i). \qquad [(1.47)]$$

1.24 (a) Using Equation (1.48) and Definition 1.43 (a) yields

$$\left(\sum_{i=1}^{\infty} \alpha_i \mu_i \right)(\emptyset) = \sum_{i=1}^{\infty} \alpha_i \mu_i(\emptyset) = \sum_{i=1}^{\infty} 0 = 0.$$

(b) Similarly, using Equation (1.48) yields, for all $A \in \mathscr{A}$,

$$\left(\sum_{i=1}^{\infty} \alpha_i \mu_i \right)(A) = \sum_{i=1}^{\infty} \alpha_i \mu_i(A) = \lim_{n \to \infty} \sum_{i=1}^{n} \alpha_i \mu_i(A) \geq 0,$$

because $\mu_i(A) \geq 0$, and we assume $\alpha_i \geq 0$.

(c) If $A_1, A_2, \ldots \in \mathscr{A}$ are pairwise disjoint, then

$$\left(\sum_{i=1}^{\infty} \alpha_i \mu_i \right) \left(\bigcup_{j=1}^{\infty} A_j \right) = \sum_{i=1}^{\infty} \alpha_i \mu_i \left(\bigcup_{j=1}^{\infty} A_j \right) \qquad [(1.48)]$$

$$= \sum_{i=1}^{\infty} \alpha_i \sum_{j=1}^{\infty} \mu_i(A_j) \qquad [\text{Def. 1.43 (c)}]$$

$$= \sum_{j=1}^{\infty} \sum_{i=1}^{\infty} \alpha_i \mu_i(A_j)$$

$$= \sum_{j=1}^{\infty} \left(\left(\sum_{i=1}^{\infty} \alpha_i \mu_i \right) (A_j) \right). \qquad [(1.48)]$$

Note that the last but one equation holds, because rearranging summands does not change the sum if the terms α_i and $\mu_i(A_j)$ are nonnegative.

2

Measurable mapping

In chapter 1, we treated the concepts of a σ-*algebra* and a σ-*algebra generated by a set system* on a set Ω. An element A of a σ-algebra $\mathscr{A}$ has been called a *measurable set*. We also introduced the concept of a *measure*, which assigns a nonnegative real number or ∞ to all elements of a σ-algebra. This chapter is devoted to the concept of a *measurable mapping*, related concepts such as the σ-*algebra generated by a mapping*, and the *image measure* of μ under f, the measure induced by a measurable mapping f on its codomain space. All these concepts play an important role in integration and probability theory. In probability theory, a measurable set is called an *event*, a measurable mapping f is called a *random variable*, and the image measure of the probability measure P under f is called the *distribution of f*.

2.1 Image and inverse image

Two key concepts of this chapter are the *image* of a set $A \subset \Omega$ and the *inverse image* of a set $A' \subset \Omega'$ under a mapping $f \colon \Omega \to \Omega'$. We start with the formal definitions and then illustrate these concepts in section 2.2.

> **Definition 2.1 [Image and inverse image]**
> Let Ω, Ω' denote two nonempty sets and $f \colon \Omega \to \Omega'$ a mapping. Then we call
>
> $$f(A) := \{f(\omega) \colon \omega \in A\}, \quad A \subset \Omega, \tag{2.1}$$
>
> *the image of A under f, and*
>
> $$f^{-1}(A') := \{\omega \in \Omega \colon f(\omega) \in A'\}, \quad A' \subset \Omega', \tag{2.2}$$
>
> *the inverse image of A' under f.*

Probability and Conditional Expectation: Fundamentals for the Empirical Sciences, First Edition. Rolf Steyer and Werner Nagel.
© 2017 John Wiley & Sons, Ltd. Published 2017 by John Wiley & Sons, Ltd.
Companion website: http://www.probability-and-conditional-expectation.de

Whereas the image $f(A)$ is a subset of Ω', the inverse image $f^{-1}(A')$ is the set of all elements of the *domain* Ω for which f takes on a value in the subset A' of its *codomain* Ω'. For convenience, we also use the notation

$$\{f \in A'\} := f^{-1}(A') \qquad \text{and} \qquad \{f = \omega'\} := f^{-1}(\{\omega'\}). \tag{2.3}$$

Remark 2.2 [Properties of inverse images] Let $f \colon \Omega \to \Omega'$ be a mapping, I be an index set, $A' \subset \Omega'$, and $(A'_i, i \in I)$ a family of subsets A'_i of Ω'. Then

$$f^{-1}[(A')^c] = [f^{-1}(A')]^c, \tag{2.4}$$

$$f^{-1}\left(\bigcap_{i \in I} A'_i\right) = \bigcap_{i \in I} f^{-1}(A'_i), \tag{2.5}$$

$$f^{-1}\left(\bigcup_{i \in I} A'_i\right) = \bigcup_{i \in I} f^{-1}(A'_i) \tag{2.6}$$

(see Exercise 2.1). Note that, in general, the corresponding properties do not necessarily hold for the image $f(A)$, $A \subset \Omega$.   ◁

2.2   Introductory examples

2.2.1   Example 1: Rectangles

Our first example deals with rectangles, their *images*, and their *inverse images* under a mapping f.

The measurable space

Let $[a, b]$, $a, b \in \mathbb{R}$, denote the closed-interval between a and b, inclusively, and consider the two rectangles

$$\Omega = [0, 10] \times [0, 6] \qquad \text{and} \qquad A = [2, 7] \times [2, 5]$$

depicted on the left-hand side of Figure 2.1. The elements of Ω and A are points $x = (x_1, x_2)$ in these rectangles with coordinates x_1 on the horizontal axis and x_2 on the vertical axis. Furthermore, let us consider a σ-algebra on Ω,

$$\mathscr{A} = \{\Omega, \varnothing, A, A^c\}.$$

The mapping and the image

Consider the set $\Omega' = \Omega$ and the function $f \colon \Omega \to \Omega'$ defined by

$$f(x) = \frac{3}{4} \cdot x = \left(\frac{3}{4} \cdot x_1, \frac{3}{4} \cdot x_2\right), \qquad \forall x \in \Omega. \tag{2.7}$$

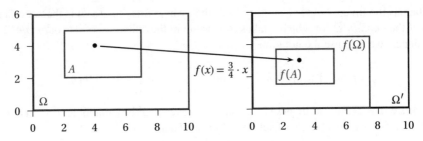

Figure 2.1 Rectangles and their images under a function.

Hence, f maps all points $x = (x_1, x_2) \in \Omega$ to the points $f(x_1, x_2) \in \Omega'$. This is illustrated by Figure 2.1 for the point $x = (4, 4)$, which is mapped to $f(x) = (3, 3)$. The right-hand side of Figure 2.1 also depicts the *image of A under f* (i.e., $f(A) = \{ f(x) : x \in A \}$), as well as the image $f(\Omega)$ of Ω under f.

The inverse images

We specify the σ-algebra

$$\mathscr{A}' = \{\Omega', \emptyset, B', (B')^c\}$$

on Ω', where

$$B' = \,]4.5, 7.5] \times [0, 4.5]$$

is the rectangle depicted on the right-hand side of Figure 2.2, and $(B')^c = \Omega' \setminus B'$ is its complement.

Now we consider the *inverse image* of B' under f [see Eq. (2.7)], that is,

$$f^{-1}(B') = \,]6, 10] \times [0, 6]$$

(see Fig. 2.2). It is the rectangle on the right side of Ω. For further examples, see Exercises 2.2 and 2.3.

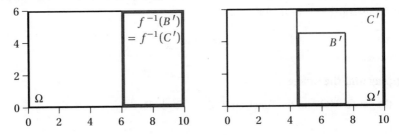

Figure 2.2 Rectangles and their inverse images under a function.

Also consider the inverse image of the rectangle

$$C' = \ ]4.5, 10] \times [0, 6]$$

(see Fig. 2.2). Its inverse image under f is identical to the inverse image $f^{-1}(B')$, that is,

$$f^{-1}(C') = f^{-1}(B'),$$

which follows from

$$
\begin{aligned}
f^{-1}(C') &= f^{-1}[B' \cup (C' \setminus B')] & [B' \subset C', \text{Fig. 2.2}] \\
&= f^{-1}(B') \cup f^{-1}(C' \setminus B') & [(2.6)] \\
&= f^{-1}(B') \cup \varnothing = f^{-1}(B').
\end{aligned}
$$

Note that $f^{-1}(C' \setminus B') = \varnothing$, because f has been defined on $\Omega = [0, 10] \times [0, 6]$. If we would define f on $\Omega = \mathbb{R}^2$, then the set $f^{-1}(C' \setminus B')$ would *not* be empty. (See also Exercise 2.4.)

2.2.2   Example 2: Flipping two coins

Now we consider the random experiment of *flipping two coins*.

The measurable space

In this random experiment, the set of possible outcomes is

$$\Omega = \{(h, h), (h, t), (t, h), (t, t)\}.$$

This set consists of four elements (pairs). For example, the first component of the pair (h, t) represents the outcome of flipping $h = heads$ with the first coin, and the second component represents the outcome of flipping $t = tails$ with the second coin. As a σ-algebra on Ω, we consider the power set $\mathscr{A} = \mathscr{P}(\Omega)$.

The mapping

Consider the function $X: \Omega \to \Omega' = \{0, 1, 2\}$ defined by

$$X[(t, t)] = 0, \qquad X[(t, h)] = 1, \qquad X[(h, t)] = 1, \qquad \text{and} \qquad X[(h, h)] = 2.$$

Looking at this assignment rule shows that this function may be called *number of flipping heads*. Again, we consider the *image of a set* $A \subset \Omega$ under X, that is, $X(A) = \{X(\omega): \omega \in A\}$, $A \subset \Omega$. For example, for $A = \{(h, h), (h, t)\}$, the image under X is $X(A) = \{1, 2\}$ (see Fig. 2.3).

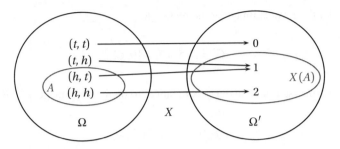

Figure 2.3   A set and its image under a function.

The inverse images

Suppose $\mathscr{A}' = \mathscr{P}(\Omega')$ is the power set of $\Omega' = \{0, 1, 2\}$. In this example, there are $2^3 = 8$ *inverse images* $X^{-1}(A') = \{\omega \in \Omega : X(\omega) \in A'\}$, $A' \in \mathscr{A}'$. Three of these eight inverse images are:

$$X^{-1}(\{0\}) = \{(t, t)\}, \quad X^{-1}(\{1\}) = \{(h, t), (t, h)\}, \quad X^{-1}(\{2\}) = \{(h, h)\}.$$

These are the events that X takes on the values 0, 1, and 2, respectively. (In order to identify the inverse images listed above, trace back the arrows from right to left in Figure 2.4.) Furthermore, consider the inverse images

$$X^{-1}(\{0, 1\}) = \{(t, t), (h, t), (t, h)\},$$
$$X^{-1}(\{0, 2\}) = \{(t, t), (h, h)\},$$
$$X^{-1}(\{1, 2\}) = \{(h, t), (t, h), (h, h)\}.$$

These are the events that X takes on a value in the sets $\{0, 1\}$, $\{0, 2\}$, and $\{1, 2\}$, respectively. One of these inverse images, namely $X^{-1}(B')$, with $B' := \{1, 2\}$, is represented in Figure 2.4. Finally,

$$X^{-1}(\Omega') = \Omega \quad \text{and} \quad X^{-1}(\emptyset) = \emptyset.$$

Hence, we listed all eight inverse images $X^{-1}(A')$, $A' \in \mathscr{A}'$. They are the eight measurable sets that can be represented by the mapping X and the σ-algebra $\mathscr{A}' = \mathscr{P}(\Omega')$. These sets are listed in Table 2.1, using the notation $\{X \in A'\} := X^{-1}(A'), A' \in \mathscr{A}'$, and $\{X = x\} := X^{-1}(\{x\})$, $\{x\} \in \mathscr{A}'$ [see Eq. (2.3)].

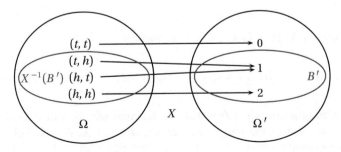

Figure 2.4   A set and its inverse image under a function.

Table 2.1   Example of measurable sets represented by a mapping X.

$\{X \in \Omega'\}$	$= X^{-1}(\Omega')$	$= \Omega$	0, 1, or 2 heads are flipped.
$\{X \in \emptyset\}$	$= X^{-1}(\emptyset)$	$= \emptyset$	Neither 0, 1, nor 2 *heads* are flipped.
$\{X = 0\}$	$= X^{-1}(\{0\})$	$= \{(t, t)\}$	No *heads* are flipped.
$\{X = 1\}$	$= X^{-1}(\{1\})$	$= \{(h, t), (t, h)\}$	*Heads* are flipped exactly once.
$\{X = 2\}$	$= X^{-1}(\{2\})$	$= \{(h, h)\}$	Two heads are flipped.
$\{X \in \{0, 1\}\}$	$= X^{-1}(\{0, 1\})$	$= \{(h, t), (t, h), (t, t)\}$	Not more than one *heads* are flipped.
$\{X \in \{0, 2\}\}$	$= X^{-1}(\{0, 2\})$	$= \{(h, h), (t, t)\}$	Either two heads or no *heads* at all are flipped.
$\{X \in \{1, 2\}\}$	$= X^{-1}(\{1, 2\})$	$= \{(h, h), (h, t), (t, h)\}$	At least one *heads* is flipped.

2.3   Measurable mapping

Now we define the concept of a *measurable mapping* and related concepts such as the *σ-algebra generated by a mapping* and *measurability of a mapping with respect to a mapping*.

Remark 2.3 [Mapping]   Remember, a mapping $f: \Omega \to \Omega'$ assigns to *all* $\omega \in \Omega$ a unique $f(\omega) \in \Omega'$. Hence, f is, by definition, a subset of the Cartesian product $\Omega \times \Omega'$, (i.e., $f = \{(\omega, f(\omega)): \omega \in \Omega\}$). This implies that, instead of $f: \Omega \to \Omega'$, we can also write $f: \Omega \to \Omega''$ for the *same mapping*, provided that $f(\Omega) \subset \Omega''$.   ◁

Remark 2.4 [Identical mappings]   If $f, g: \Omega \to \Omega'$ are two mappings, then,

$$f = g \iff \{(\omega, f(\omega)): \omega \in \Omega\} = \{(\omega, g(\omega)): \omega \in \Omega\}. \tag{2.8}$$

If $f = g$, we say that the two mappings are *identical*. Hence, even if $f: \Omega \to \Omega'$ and $g: \Omega \to \Omega''$ are mappings with $\Omega' \neq \Omega''$, it is still possible that f and g are identical. Note that (2.8) also implies: If, for $f: \Omega \to \Omega'$ and $g: \Omega \to \Omega''$, we write $f, g: \Omega \to \Omega'''$ with $\Omega''' := \Omega' \cup \Omega''$, then f and g remain unchanged.   ◁

2.3.1   Measurable mapping

Now the core concept of this chapter is defined as follows:

Definition 2.5 [Measurable mapping]
Let $(\Omega, \mathcal{A})$, $(\Omega', \mathcal{A}')$ be measurable spaces, and let $f: \Omega \to \Omega'$ be a mapping. Then f is called $(\mathcal{A}, \mathcal{A}')$-measurable if

$$f^{-1}(A') \in \mathcal{A}, \quad \forall A' \in \mathcal{A}'.$$

Remark 2.6 [Notation] We use the notation

$$f\colon (\Omega, \mathcal{A}) \to (\Omega', \mathcal{A}')$$

to express that the mapping $f\colon \Omega \to \Omega'$ is $(\mathcal{A}, \mathcal{A}')$-measurable. If there is no ambiguity about $\mathcal{A}'$, then we also say that f is $\mathcal{A}$-*measurable* or *measurable with respect to* $\mathcal{A}$. ◁

Examples

Example 2.7 [Rectangles – continued] In Example 2.2.1, we considered the mapping $f\colon \Omega \to \Omega' = \Omega$ defined by $f(x) = \frac{3}{4}x$. Furthermore, we considered the rectangle $B' = \,]4.5, 7.5] \times [0, 4.5]$ and the inverse image

$$f^{-1}(B') = \,]6, 10] \times [0, 6].$$

If $A = [2, 7] \times [2, 5]$, then the inverse image $f^{-1}(B')$ is not an element of the σ-algebra $\mathcal{A} = \{\Omega, \emptyset, A, A^c\}$. In this example, we also specified the σ-algebra $\mathcal{A}' = \{\Omega', \emptyset, B', (B')^c\}$. Hence, f is not $(\mathcal{A}, \mathcal{A}')$-measurable. However, if we specify a σ-algebra $\mathcal{C}$ such that $f^{-1}(B') \in \mathcal{C}$, then f is $(\mathcal{C}, \mathcal{A}')$-measurable. As we see later on in this chapter, $f^{-1}(B') \in \mathcal{C}$ is sufficient for f to be $(\mathcal{C}, \mathcal{A}')$-measurable (see Th. 2.20). ◁

Example 2.8 [Flipping two coins – continued] In Example 2.2.2, we considered the mapping $X = $ *number of flipping heads*, and in Table 2.1 we listed all inverse images $X^{-1}(A'), A' \in \mathcal{A}' = \mathcal{P}(\{0, 1, 2\})$. Of course, $\mathcal{A} = \mathcal{P}(\Omega)$ ensures that all inverse images $X^{-1}(A'), A' \in \mathcal{A}'$, are elements of $\mathcal{A}$.

However, instead of $\mathcal{A} = \mathcal{P}(\Omega)$, we might consider the σ-algebra

$$\mathcal{A}_0 = \{\Omega, \emptyset, \{(h, h), (h, t)\}, \{(t, h), (t, t)\}\}.$$

The element $\{(h, h), (h, t)\}$ represents the event that *heads* are flipped in the first flip, and $\{(t, h), (t, t)\}$ is the event that *tails* are flipped in the first flip. Hence, the σ-algebra $\mathcal{A}_0$ contains the events that refer to the outcome of the *first flip* only, whereas X represents the number of heads in *both coin flips*. If we choose $\mathcal{A}'$ to be the power set of $\Omega' = \{0, 1, 2\}$, then it is *not* true that all eight inverse images $X^{-1}(A'), A' \in \mathcal{A}'$, are elements of $\mathcal{A}_0$. The inverse image $X^{-1}(\{2\}) = \{(h, h)\}$, for example, is not an element of $\mathcal{A}_0$. Hence, if we consider the measurable spaces $(\Omega, \mathcal{A}_0)$ and $(\Omega', \mathcal{P}(\Omega'))$, then the mapping X is *not* $(\mathcal{A}_0, \mathcal{P}(\Omega'))$-measurable. Hence, in some sense, $\mathcal{A}_0$ is 'not well-adapted' to X. ◁

Example 2.9 [Two trivial cases] If (a) $\mathcal{A} = \mathcal{P}(\Omega)$ is the power set of Ω or if (b) $\mathcal{A}' = \{\Omega', \emptyset\}$, then every mapping $f\colon \Omega \to \Omega'$ is $(\mathcal{A}, \mathcal{A}')$-measurable. This is easily seen as follows: (a) If $\mathcal{A} = \mathcal{P}(\Omega)$ is the power set of Ω, then all inverse images $f^{-1}(A'), A' \subset \Omega'$, are elements in $\mathcal{A} = \mathcal{P}(\Omega)$, because it is the set of *all* subsets of Ω. (b) If $\mathcal{A}' = \{\Omega', \emptyset\}$, then every mapping $f\colon \Omega \to \Omega'$ is $(\mathcal{A}, \mathcal{A}')$-measurable, because $f^{-1}(\Omega') = \Omega$ and $f^{-1}(\emptyset) = \emptyset$. Again, the inverse images Ω and $\emptyset$ are both elements in every σ-algebra on Ω. Hence, in both cases, (a) and (b), every mapping $f\colon \Omega \to \Omega'$ is $(\mathcal{A}, \mathcal{A}')$-measurable. ◁

Example 2.10 [Constant mapping]   A constant mapping $f: \Omega \to \Omega'$ is defined by

$$f(\omega) = \omega', \quad \forall \, \omega \in \Omega,$$

where ω' is a fixed element of Ω'. Such a constant mapping is $(\mathcal{A}, \mathcal{A}')$-measurable for any σ-algebra $\mathcal{A}$ on Ω and any σ-algebra $\mathcal{A}'$ on Ω'. This is true, because for all subsets A' of Ω': If $\omega' \in A'$, then $f^{-1}(A') = \Omega$. If, in contrast, $\omega' \notin A'$, then $f^{-1}(A') = \emptyset$. However, Ω and $\emptyset$ are elements of *all* σ-algebras on Ω. ◁

Example 2.11 [Identity mapping]   The identity mapping $id: \Omega \to \Omega$ defined by

$$id(\omega) = \omega, \quad \forall \, \omega \in \Omega,$$

is $(\mathcal{A}, \mathcal{A}_0)$-measurable for any pair of σ-algebras on Ω with $\mathcal{A}_0 \subset \mathcal{A}$. This is easily seen as follows:

$$id^{-1}(A) = A, \quad \forall \, A \in \mathcal{A}_0.$$

Because we assume $\mathcal{A}_0 \subset \mathcal{A}$, we can conclude that id is $(\mathcal{A}, \mathcal{A}_0)$-measurable. ◁

Example 2.12 [Indicator of a measurable set]   Let $(\Omega, \mathcal{A})$, $(\Omega', \mathcal{A}')$ be two measurable spaces, where $\mathcal{A}'$ is any σ-algebra on $\Omega' \subset \mathbb{R}$ with $\{0\}, \{1\} \in \mathcal{A}'$. Then the indicator $1_A: \Omega \to \Omega'$ is $(\mathcal{A}, \mathcal{A}')$-measurable *if and only if* $A \in \mathcal{A}$. Note that the requirement $\{0\}, \{1\} \in \mathcal{A}'$ is satisfied not only by $(\Omega', \mathcal{A}') := (\{0, 1\}, \mathscr{P}(\{0, 1\}))$, but also by $(\Omega', \mathcal{A}') = (\mathbb{R}, \mathscr{B})$ and by $(\Omega', \mathcal{A}') = (\overline{\mathbb{R}}, \overline{\mathscr{B}})$, where $\mathscr{B}$ denotes the Borel σ-algebra on $\mathbb{R}$ and $\overline{\mathscr{B}}$ the Borel σ-algebra on $\overline{\mathbb{R}}$. ◁

Example 2.13 [Indicators of unions and intersections]   If $(\Omega, \mathcal{A})$ is a measurable space and $A, B \in \mathcal{A}$, then $1_{A \cap B}$ and $1_{A \cup B}$ are $(\mathcal{A}, \mathscr{B})$-measurable. This follows from the fact that $A \cap B \in \mathcal{A}$ and $A \cup B \in \mathcal{A}$. For the same reason, $A_1, A_2, \ldots \in \mathcal{A}$ implies that $1_{\bigcup_{i=1}^{\infty} A_i}$ is $(\mathcal{A}, \mathscr{B})$-measurable. ◁

Example 2.14 [Constant mapping]   Assume that $(\Omega, \mathcal{A})$ and $(\Omega', \mathcal{A}')$ are measurable spaces such that $\{\omega'\} \in \mathcal{A}'$, for all $\omega' \in \Omega'$. Furthermore, let $f: \Omega \to \Omega'$. If $\mathcal{A} = \{\Omega, \emptyset\}$, then f is $(\mathcal{A}, \mathcal{A}')$-measurable if and only if f is a constant mapping, that is, if and only if there is an $\omega' \in \Omega'$ such that $f(\omega) = \omega'$, for all $\omega \in \Omega$ (see Exercise 2.5). Note that for $(\Omega', \mathcal{A}') = (\overline{\mathbb{R}}, \overline{\mathscr{B}})$, $\{x\} \in \overline{\mathscr{B}}$, for all $x \in \overline{\mathbb{R}}$. ◁

Example 2.15 [Dichotomous function]   If $\mathcal{A} = \{\Omega, \emptyset, A, A^c\}$ with $A \subset \Omega$, then $f: \Omega \to \mathbb{R}$ is $(\mathcal{A}, \mathscr{B})$-measurable if and only if $f = \alpha_1 1_A + \alpha_2 1_{A^c}$ for $\alpha_1, \alpha_2 \in \mathbb{R}$ (see Exercise 2.6). ◁

Step function

Another important example of a measurable function is a step function, which is defined as follows:

Definition 2.16 [Step function]
Let $A_1, \ldots, A_n$, $n \in \mathbb{N}$, be a finite sequence of subsets of a set Ω. Then a finite linear combination

$$f = \sum_{i=1}^{n} \alpha_i 1_{A_i}, \quad \alpha_1, \ldots, \alpha_n \in \mathbb{R}, \tag{2.9}$$

is called a step function.

Remark 2.17 [Step function and a partition of Ω]   Assume that the sets $A_1, \ldots, A_n$ are pairwise disjoint, and define $A_{n+1} := \Omega \setminus (\bigcup_{i=1}^{n} A_i)$, then $\{A_1, \ldots, A_n, A_{n+1}\}$ is a finite partition of Ω. If f satisfies (2.9) and $\alpha_{n+1} := 0$, then, for all $A' \subset \mathbb{R}$,

$$f^{-1}(A') = \bigcup_{\substack{i=1, \ldots, n+1 \\ \alpha_i \in A'}} A_i \tag{2.10}$$

(see Exercise 2.7).   ◁

Remark 2.18 [Measurability of a step function]   If $(\Omega, \mathcal{A})$ is a measurable space and $A_1, \ldots, A_n \in \mathcal{A}$, then the step function $f \colon \Omega \to \mathbb{R}$ defined by Equation (2.9) is $(\mathcal{A}, \mathcal{B})$-measurable (see Exercise 2.8).   ◁

Lemma 2.19 [Measurability if $\mathcal{A}$ is countably generated]
Let $(\Omega, \mathcal{A})$ be a measurable space and let $\mathcal{A} = \sigma(\mathcal{E})$, where $\mathcal{E}$ is a finite (i.e., $\mathcal{E} = \{A_1, \ldots, A_n\}$) or countable (i.e., $\mathcal{E} = \{A_1, A_2, \ldots\}$) partition of Ω. Then $f \colon \Omega \to \overline{\mathbb{R}}$ is $(\mathcal{A}, \overline{\mathcal{B}})$-measurable if and only if there are $\alpha_1, \alpha_2, \ldots \in \overline{\mathbb{R}}$ such that $f = \sum_i \alpha_i 1_{A_i}$, where $i \in \{1, \ldots, n\}$ if $\mathcal{E}$ is finite, and $i \in \mathbb{N}$ if $\mathcal{E}$ is countable.

(Proof p. 72)

A necessary and sufficient condition of measurability

Let $(\Omega', \mathcal{A}')$ be a measurable space and $\mathcal{E}' \subset \mathcal{A}'$. Then we denote

$$f^{-1}(\mathcal{E}') := \{f^{-1}(A') \colon A' \in \mathcal{E}'\}. \tag{2.11}$$

This notation is used in the following theorem, which can be utilized for proving $(\mathcal{A}, \mathcal{A}')$-measurability of a mapping $f \colon \Omega \to \Omega'$.

Theorem 2.20 [Measurable mapping and generating systems]
Let $(\Omega, \mathscr{A})$, $(\Omega', \mathscr{A}')$ denote measurable spaces, let $\mathscr{E}' \subset \mathscr{A}'$, and $f \colon \Omega \to \Omega'$. Then

$$\sigma[f^{-1}(\mathscr{E}')] = f^{-1}[\sigma(\mathscr{E}')]. \tag{2.12}$$

Furthermore, if $\sigma(\mathscr{E}') = \mathscr{A}'$, then f is $(\mathscr{A}, \mathscr{A}')$-measurable if and only if $f^{-1}(A') \in \mathscr{A}$, for all $A' \in \mathscr{E}'$.

For a proof, see Klenke (2013, Theorem 1.81).

Now consider a finite or countable set Ω'. Then Theorem 2.20 and Example 1.18 immediately imply the following corollary:

Corollary 2.21 [Finite or countable generating systems]
Let $(\Omega, \mathscr{A})$, $(\Omega', \mathscr{P}(\Omega'))$ be measurable spaces, where Ω' is finite or countable, and let $\mathscr{E}' = \{\{\omega'\} \colon \omega' \in \Omega'\}$. Then a mapping $f \colon \Omega \to \Omega'$ is $(\mathscr{A}, \mathscr{P}(\Omega'))$-measurable if and only if $f^{-1}(\{\omega'\}) \in \mathscr{A}$, for all $\omega' \in \Omega'$.

Example 2.22 [Rectangles – continued] In Example 2.2.1, we considered the mapping $f \colon \Omega \to \Omega' = \Omega$ defined by $f(x) = \frac{3}{4}x$. Furthermore, we considered the rectangle $B' = {]4.5, 7.5]} \times [0, 4.5]$. The set system

$$\mathscr{E}' = \{B'\},$$

which contains B' as the only element, generates the σ-algebra

$$\mathscr{A}' = \{\Omega', \emptyset, B', (B')^c\}.$$

Hence according to Theorem 2.20, the mapping f is $(\mathscr{A}, \mathscr{A}')$-measurable provided that $f^{-1}(B') \in \mathscr{A}$.   ◁

Example 2.23 [Flipping two coins – continued] In Example 2.2.2, we defined the mapping $X = $ *number of flipping heads* with codomain $\Omega' = \{0, 1, 2\}$. Now consider the system

$$\mathscr{E}' = \{\{0\}, \{1\}\}$$

of subsets of Ω'. First, note that $\sigma(\mathscr{E}') = \mathscr{P}(\Omega')$. Therefore, Theorem 2.20 implies that X is $(\mathscr{A}, \mathscr{P}(\Omega'))$-measurable for each σ-algebra $\mathscr{A}$ on

$$\Omega = \{(h, h), (h, t), (t, h), (t, t)\}$$

for which

$$X^{-1}(\{0\}) = \{(t, t)\} \in \mathscr{A} \quad \text{and} \quad X^{-1}(\{1\}) = \{(h, t), (t, h)\} \in \mathscr{A}.$$

This holds not only for $\mathscr{A}_1 = \mathscr{P}(\Omega)$, but also for the σ-algebra

$$\mathscr{A}_2 = \{\Omega, \emptyset, \ \{(t, t)\}, \{(h, t), (t, h)\}, \{(h, h)\},$$
$$\{(h, h), (h, t), (t, h)\}, \ \{(h, h), (t, t)\}, \ \{(h, t), (t, h), (t, t)\}\}.$$

As mentioned before, $\mathscr{A}_2$ contains all events that can be represented by X (see Table 2.1). In contrast, this does not hold for the σ-algebra

$$\mathscr{A}_0 = \{\Omega, \emptyset, \{(h, h), (h, t)\}, \ \{(t, h), (t, t)\}\}$$

(see Example 2.8). Hence, X is measurable with respect to $\mathscr{A}_1$ and $\mathscr{A}_2$, but it is not measurable with respect to $\mathscr{A}_0$. In this application, this means that the events $\{(h, h), (h, t)\}$ and $\{(t, h), (t, t)\}$ cannot be formulated in terms of X. Furthermore, some of the events that *can* be formulated in terms of X are not elements of $\mathscr{A}_0$. For example, $X^{-1}(\{0\}) = \{(t, t)\}$ is *not* an element of $\mathscr{A}_0$. ◁

2.3.2 σ-Algebra generated by a mapping

Let us consider again Example 2.2.2 and the mapping $X = $ *number of flipping heads*. The set that consists of the eight inverse images $X^{-1}(A')$, $A' \in \mathscr{A}'$, is again a σ-algebra on Ω. In a sense, this σ-algebra carries the information associated with the mapping X; it contains all events that can be represented by X (see Table 2.1). In Theorem 2.24, we formulate the general proposition.

Theorem 2.24 [σ-Algebra generated by a mapping]
Let $f: \Omega \to \Omega'$ be a mapping, and let $(\Omega', \mathscr{A}')$ be a measurable space. Then

$$f^{-1}(\mathscr{A}') := \{f^{-1}(A'): A' \in \mathscr{A}'\} \tag{2.13}$$

is a σ-algebra on Ω.

For a proof, see Klenke (2013, Theorem 1.81).

Remark 2.25 [Smallest σ-algebra] Note that $f^{-1}(\mathscr{A}')$ is the smallest σ-algebra $\mathscr{C}$ on Ω such that f is $(\mathscr{C}, \mathscr{A}')$-measurable, that is,

$$\mathscr{C} \text{ is a σ-algebra on } \Omega \text{ and } f \text{ is } (\mathscr{C}, \mathscr{A}')\text{-measurable} \ \Rightarrow \ f^{-1}(\mathscr{A}') \subset \mathscr{C}.$$

◁

The set $f^{-1}(\mathscr{A}')$ contains all sets in $\mathscr{A}$ that can be represented by f and elements of $\mathscr{A}'$. Because $f^{-1}(\mathscr{A}')$ is important, it has its own name and an alternative notation, which is sometimes more convenient.

Definition 2.26 [σ-Algebra generated by a mapping]
The set $f^{-1}(\mathscr{A}')$ defined by Equation (2.13) is called the *σ-algebra generated by f and $\mathscr{A}'$. If there is no ambiguity about $\mathscr{A}'$, then we also say that $f^{-1}(\mathscr{A}')$ is generated by f and use the notation*

$$\sigma(f) := f^{-1}(\mathscr{A}'). \tag{2.14}$$

Remark 2.27 [Monotonicity]   Note that, for two set systems $\mathscr{C}' \subset \mathscr{A}'$,

$$f^{-1}(\mathscr{C}') \subset f^{-1}(\mathscr{A}'), \tag{2.15}$$

because $f^{-1}(\mathscr{C}') = \{f^{-1}(A'): A' \in \mathscr{C}'\} \subset \{f^{-1}(A'): A' \in \mathscr{A}'\} = f^{-1}(\mathscr{A}')$.   ◁

Corollary 2.28 immediately follows from Definition 2.26 and the definition of $(\mathscr{A}, \mathscr{A}')$-measurability (see Def. 2.5).

Corollary 2.28 [A condition equivalent to measurability]
Let $f: \Omega \to \Omega'$ be a mapping, and let $(\Omega', \mathscr{A}')$ be a measurable space. Then f is $(\mathscr{A}, \mathscr{A}')$-measurable if and only if $\sigma(f) \subset \mathscr{A}$.

In the following lemma and the subsequent remark, we treat a ∩-stable generating system (see Def. 1.36). For a σ-algebra $\mathscr{C}$ and a measurable mapping $f: (\Omega, \mathscr{A}) \to (\Omega', \mathscr{A}')$, we use the notation $\sigma(\mathscr{C}, f) := \sigma(\mathscr{C} \cup f^{-1}(\mathscr{A}'))$.

Lemma 2.29 [∩-Stable generating system]
Let $(\Omega, \mathscr{A})$ be a measurable space, and let $\mathscr{C} \subset \mathscr{A}$ be a σ-algebra. Furthermore, assume that Ω' is finite or countable and let $f: (\Omega, \mathscr{A}) \to (\Omega', \mathscr{P}(\Omega'))$ be a measurable mapping. Then the set

$$\mathscr{D} := \{C \cap f^{-1}(\{\omega'\}): \omega' \in \Omega' \text{ and } C \in \mathscr{C}\}$$

is a ∩-stable generating system of $\sigma(\mathscr{C}, f) := \sigma(\mathscr{C} \cup f^{-1}[\mathscr{P}(\Omega')])$.

(Proof p. 72)

Remark 2.30 [A special case]   Let us consider the special case in which $\mathscr{C} = \{\Omega, \emptyset\}$. In this case, Lemma 2.29 simplifies as follows: Let $(\Omega, \mathscr{A})$ be a measurable space and let $f: (\Omega, \mathscr{A}) \to (\Omega', \mathscr{P}(\Omega'))$ be a measurable mapping, where Ω' is finite or countable. Then the set $\{f^{-1}(\{\omega'\}): \omega' \in \Omega'\} \cup \{\emptyset\}$ is a ∩-stable generating system of $\sigma(f) := f^{-1}[\mathscr{P}(\Omega')]$.   ◁

Example 2.31 [σ-Algebra generated by an indicator]   Let $1_A: (\Omega, \mathscr{A}) \to (\mathbb{R}, \mathscr{B})$ be the indicator of $A \in \mathscr{A}$. Then $\sigma(1_A) = \{\Omega, \emptyset, A, A^c\}$. The same σ-algebra is generated by $1_A: (\Omega, \mathscr{A}) \to (\{0, 1\}, \mathscr{P}(\{0, 1\}))$ (see Remark 2.33 for the general proposition).   ◁

Example 2.32 [Flipping two coins – continued]   In Example 2.2.2, we considered flipping two coins and the measurable mapping $X = $ *number of flipping heads* with codomain $\Omega' = \{0, 1, 2\}$. In this example, all elements of $X^{-1}[\mathscr{P}(\Omega')]$ have been listed in Table 2.1 as the inverse images $X^{-1}(A')$ of the eight sets $A' \in \mathscr{P}(\Omega')$. Furthermore, $X^{-1}[\mathscr{P}(\Omega')] = \mathscr{A}_2$, where $\mathscr{A}_2$ is the σ-algebra defined in Example 2.23.

Instead of choosing $\Omega' = \{0, 1, 2\}$ as the codomain of X, we may also choose the set $\mathbb{R}$ of real numbers, (i.e., $X\colon \Omega \to \mathbb{R}$ is then considered to be a function into $\mathbb{R}$). In this case, we use the *Borel σ-algebra* $\mathscr{B}$ on $\mathbb{R}$. However, according to the following remark, the σ-algebra $X^{-1}(\mathscr{B})$ generated by X and $\mathscr{B}$ is the same as the σ-algebra $X^{-1}[\mathscr{P}(\Omega')]$ generated by X and the power set of $\Omega' = \{0, 1, 2\}$.   ◁

Remark 2.33 [σ-Algebra generated by a function into a countable set]   Let us consider a function $f\colon \Omega \to \Omega' \subset \mathbb{R}$, and let $\mathscr{B}$ denote the Borel σ-algebra on $\mathbb{R}$. If Ω' is finite or countable, then $f^{-1}[\mathscr{P}(\Omega')] = f^{-1}(\mathscr{B})$ (see Exercise 2.9).   ◁

Example 2.34 [Joe and Ann – continued]   Table 2.2 displays mappings on $(\Omega, \mathscr{A})$, all components of which have already been specified in Example 1.9. The first mapping displayed in Table 2.2 is the *person variable U* that assigns to each possible outcome $\omega \in \Omega$ the value *Joe* if $\omega \in \{Joe\} \times \Omega_X \times \Omega_Y$ and the value *Ann* if $\omega \in \{Ann\} \times \Omega_X \times \Omega_Y$. Hence, $U\colon \Omega \to \Omega_U$ is a mapping with domain $\Omega = \Omega_U \times \Omega_X \times \Omega_Y$ and codomain Ω_U. It projects the first component u of $\omega = (u, \omega_X, \omega_Y)$ onto the set Ω_U. Therefore, it is also called the first *projection mapping*.

The second mapping in this table is the *treatment variable X*. It assigns to each possible outcome $\omega \in \Omega$ the value 0 if $\omega \in \Omega_U \times \{no\} \times \Omega_Y$ and the value 1 if $\omega \in \Omega_U \times \{yes\} \times \Omega_Y$. Hence, $X\colon \Omega \to \Omega'$ is a function with domain Ω and codomain $\Omega' = \{0, 1\}$.

The third mapping is the *outcome variable Y*. It assigns to each $\omega \in \Omega$ the value 0 if $\omega \in \Omega_U \times \Omega_X \times \{-\}$ and the value 1 if $\omega \in \Omega_U \times \Omega_X \times \{+\}$. Therefore, $Y\colon \Omega \to \Omega'$ is a function with domain Ω and codomain $\Omega' = \{0, 1\}$. Hence, all three mappings U, X, and Y have the same domain Ω.

Table 2.2   Joe and Ann with randomized assignment and measurable mappings.

Elements of Ω				Measurable mappings		
Unit	Treatment	Success	$P(\{\omega\})$	Person variable U	Treatment variable X	Outcome variable Y
(*Joe, no, −*)			.09	*Joe*	0	0
(*Joe, no, +*)			.21	*Joe*	0	1
(*Joe, yes, −*)			.04	*Joe*	1	0
(*Joe, yes, +*)			.16	*Joe*	1	1
(*Ann, no, −*)			.24	*Ann*	0	0
(*Ann, no, +*)			.06	*Ann*	0	1
(*Ann, yes, −*)			.12	*Ann*	1	0
(*Ann, yes, +*)			.08	*Ann*	1	1

Considering $U: \Omega \to \Omega_U$ and the σ-algebra $\mathscr{A}_U := \{\Omega_U, \emptyset, \{Joe\}, \{Ann\}\}$, the σ-algebra $U^{-1}(\mathscr{A}_U)$ consists of the following four inverse images: the event

$$U^{-1}(\{Joe\}) = \{(Joe, no, -), (Joe, no, +), (Joe, yes, -), (Joe, yes, +)\}$$

that *Joe is drawn*, the event

$$U^{-1}(\{Ann\}) = \{(Ann, no, -), \ (Ann, no, +), \ (Ann, yes, -), \ (Ann, yes, +)\}$$

that *Ann is drawn*, the sure event $U^{-1}(\Omega_U) = \Omega$ that *Joe or Ann are drawn*, and the impossible event $U^{-1}(\emptyset) = \emptyset$ that *neither Joe nor Ann are drawn*. ◁

2.3.3   Final σ-algebra

Consider the mapping $f: \Omega \to \Omega'$. As noted in Remark 2.25, for a σ-algebra $\mathscr{A}'$ on Ω', $\sigma(f) = f^{-1}(\mathscr{A}')$ is the smallest σ-algebra on Ω for which f is measurable. In contrast, now we consider a σ-algebra $\mathscr{C}$ on Ω and look for the largest σ-algebra $\mathscr{C}'$ on Ω' such that f is $(\mathscr{C}, \mathscr{C}')$-measurable. This σ-algebra is specified in the following lemma. It is called the *final σ-algebra*.

Lemma 2.35 [Final σ-algebra]
Let $f: \Omega \to \Omega'$ be a mapping and $\mathscr{C}$ a σ-algebra on Ω.

  (i) Then

$$\mathscr{C}'_f := \{A' \subset \Omega': f^{-1}(A') \in \mathscr{C}\} \tag{2.16}$$

  is a σ-algebra on Ω'.

  (ii) Furthermore, if $f: (\Omega, \mathscr{C}) \to (\Omega', \mathscr{A}')$ is a measurable mapping, then $\mathscr{A}' \subset \mathscr{C}'_f$.

(Proof p. 73)

Note that (ii) is a formal way of saying that $\mathscr{C}'_f$ is the largest σ-algebra on Ω' such that f is $\mathscr{C}$-measurable.

Definition 2.36 [Final σ-algebra]
The σ-algebra $\mathscr{C}'_f$ defined by Equation (2.16) is called the final σ-algebra of $\mathscr{C}$ under f.

2.3.4   Multivariate mapping

Now consider the measurable space $(\times_{i=1}^n \Omega'_i, \otimes_{i=1}^n \mathscr{A}'_i)$, and note that the definitions of measurable mappings and of the σ-algebra generated by a mapping also apply to n-variate mappings $f: \Omega \to \Omega'_1 \times \ldots \times \Omega'_n$ and in particular to functions for which $\Omega'_1 \times \ldots \times \Omega'_n = \mathbb{R}^n$.

Lemma 2.37 [σ-Algebra generated by a multivariate mapping]
Let Ω be a nonempty set, let $(\Omega_i', \mathscr{A}_i')$, $i = 1, \ldots, n$, $n \in \mathbb{N}$, be measurable spaces, and $f = (f_1, \ldots, f_n)$ be a multivariate mapping with $f_i \colon \Omega \to \Omega_i', i = 1, \ldots, n$, that is, $f \colon \Omega \to \bigtimes_{i=1}^{n} \Omega_i'$. Then,

$$\sigma(f_1, \ldots, f_n) := \sigma(f) = f^{-1}\left(\bigotimes_{i=1}^{n} \mathscr{A}_i'\right) = \sigma\left(\bigcup_{i=1}^{n} \sigma(f_i)\right). \qquad (2.17)$$

(Proof p. 74)

According to the following theorem, a multivariate mapping is measurable if and only if all its components are measurable.

Theorem 2.38 [Measurability of multivariate mappings]
Under the assumptions of Lemma 2.37, the following two propositions are equivalent to each other:

(a) $f \colon (\Omega, \mathscr{A}) \to \left(\bigtimes_{i=1}^{n} \Omega_i', \bigotimes_{i=1}^{n} \mathscr{A}_i'\right)$ *is a measurable mapping.*

(b) $\forall\, i = 1, \ldots, n \colon f_i \colon (\Omega, \mathscr{A}) \to (\Omega_i', \mathscr{A}_i')$ *is a measurable mapping.*

(Proof p. 75)

Remark 2.39 [σ-Algebra generated by a family of mappings] Let I be a nonempty (finite, countable, or uncountable) index set and let $(f_i, i \in I)$ be a family of mappings $f_i \colon (\Omega, \mathscr{A}) \to (\Omega_i', \mathscr{A}_i')$. The σ-algebra generated by this family is defined as

$$\sigma(f_i, i \in I) := \sigma\left(\bigcup_{i \in I} \sigma(f_i)\right). \qquad (2.18)$$

Equation (2.17) implies

$$\sigma(f) = \sigma(f_i, i \in I), \quad \text{where } I = \{i = 1, \ldots, n\}. \qquad (2.19)$$

◁

Example 2.40 [Joe and Ann – continued] In Example 2.34, we already considered the function $X \colon \Omega \to \mathbb{R}$ indicating with its values 1 and 0 whether or not the drawn person is treated and the function $Y \colon \Omega \to \mathbb{R}$ indicating with its values 1 and 0 whether or not the drawn person is successful. If we specify the σ-algebra $\mathscr{A}$ on Ω such that X and Y are both $(\mathscr{A}, \mathscr{B})$-measurable, then the bivariate function $(X, Y) \colon \Omega \to \mathbb{R}^2$ is $(\mathscr{A}, \mathscr{B}_2)$-measurable. And vice versa, if we specify the σ-algebra $\mathscr{A}$ on Ω such that the bivariate function $(X, Y) \colon \Omega \to \mathbb{R}^2$ is $(\mathscr{A}, \mathscr{B}_2)$-measurable, then X and Y are both $(\mathscr{A}, \mathscr{B})$-measurable. In this example X, Y, and (X, Y) are measurable with respect to $\mathscr{A}$ whenever the two inverse images $X^{-1}(\{1\})$ and $Y^{-1}(\{1\})$ are elements of $\mathscr{A}$ (see Exercise 2.10). ◁

Remark 2.41 [Lower dimensional multivariate mappings]   Lemma 2.37 and Remark 1.23 imply

$$\sigma(f_i, i \in J) \subset f^{-1}\left(\bigotimes_{i=1}^{n} \mathscr{A}'_i\right), \quad \forall J \subset \{1, \dots, n\}.$$

Furthermore, Theorem 2.38 implies: If

$$f = (f_1, \dots, f_n): (\Omega, \mathscr{A}) \to \left(\bigtimes_{i=1}^{n} \Omega'_i, \bigotimes_{i=1}^{n} \mathscr{A}'_i\right)$$

is a measurable mapping and $J = \{i_1, \dots, i_k\} \subset \{1, \dots, n\}, k \le n$, then

$$f_J := (f_{i_1}, \dots, f_{i_k}): (\Omega, \mathscr{A}) \to \left(\bigtimes_{j=1}^{k} \Omega'_{i_j}, \bigotimes_{j=1}^{k} \mathscr{A}'_{i_j}\right)$$

is measurable as well.   ◁

2.3.5   Projection mapping

In Definition 1.31, we introduced the product σ-algebra $\bigotimes_{i=1}^{n} \mathscr{A}_i$ for a finite number of measurable spaces $(\Omega_i, \mathscr{A}_i)$. Now we give an equivalent characterization. Let $(\Omega_i, \mathscr{A}_i), i = 1, \dots, n$, be measurable spaces. Then, for $j = 1, \dots, n$, the jth *projection mapping* $\pi_j \colon \bigtimes_{i=1}^{n} \Omega_i \to \Omega_j$ is defined by

$$\pi_j(\omega_1, \dots, \omega_n) = \omega_j, \quad \forall (\omega_1, \dots, \omega_n) \in \bigtimes_{i=1}^{n} \Omega_i. \tag{2.20}$$

The inverse images are

$$\pi_j^{-1}(A_j) = \Omega_1 \times \dots \times \Omega_{j-1} \times A_j \times \Omega_{j+1} \times \dots \times \Omega_n, \quad \text{for } A_j \subset \Omega_j. \tag{2.21}$$

Lemma 2.42 [Product σ-algebra]
If $(\Omega_i, \mathscr{A}_i), i = 1, \dots, n$, are measurable spaces, then

$$\bigotimes_{i=1}^{n} \mathscr{A}_i = \sigma(\pi_1, \dots, \pi_n). \tag{2.22}$$

(Proof p. 75)

2.3.6   Measurability with respect to a mapping

In Definition 2.43, we consider two mappings and the concept of a mapping being measurable with respect to another mapping.

Definition 2.43 [Measurability with respect to a mapping]
Let $f \colon \Omega \to \Omega'$ and $h \colon \Omega \to \Omega''$ be mappings, and let $(\Omega', \mathscr{A}')$ and $(\Omega'', \mathscr{A}'')$ be measurable spaces. Then h is called measurable with respect to f (or f-measurable) if

$$h^{-1}(\mathscr{A}'') \subset f^{-1}(\mathscr{A}'). \tag{2.23}$$

If Ω' is finite or countable, then the following corollary provides a representation for all functions that are measurable with respect to f.

Corollary 2.44 [Measurability with respect to a discrete function]
Let $f \colon \Omega \to \Omega'$ be a mapping; let $(\Omega', \mathscr{P}(\Omega'))$ be a measurable space, where Ω' is finite or countable; and let $h \colon \Omega \to \overline{\mathbb{R}}$ be a function. Then h is measurable with respect to f if and only if for all $\omega' \in \Omega'$ there are $\alpha_{\omega'} \in \overline{\mathbb{R}}$ such that

$$h = \sum_{\omega' \in \Omega'} \alpha_{\omega'} \cdot 1_{f^{-1}(\{\omega'\})}. \tag{2.24}$$

(Proof p. 75)

Example 2.45 [Flipping two coins – continued]   Consider the mapping $X = $ *number of flipping heads* with codomain $\Omega' = \{0, 1, 2\}$, let $H := \{(h, t), (h, h), (t, h)\}$, and let $1_H \colon \Omega \to \Omega''$ denote the indicator of H, with $\Omega'' = \{0, 1\}$. Hence, 1_H indicates with its values 1 and 0 whether or not at least one *heads* is flipped. If we consider the σ-algebra $\mathscr{A}' = \mathscr{P}(\Omega')$ on Ω' and the σ-algebra $\mathscr{A}'' = \mathscr{P}(\Omega'')$ on Ω'', then

$$\begin{aligned}
X^{-1}(\mathscr{A}') = \ & \{\Omega, \varnothing, \{(h, h)\}, \{(h, t), (t, h)\}, \{(t, t)\}, \\
& \{(h, h), (h, t), (t, h)\}, \{(h, h), (t, t)\}, \{(h, t), (t, h), (t, t)\}\}
\end{aligned}$$

and

$$1_H^{-1}(\mathscr{A}'') = \{\Omega, \varnothing, \{(h, h), (h, t), (t, h)\}, \{(t, t)\}\}.$$

Obviously, $1_H^{-1}(\mathscr{A}'') \subset X^{-1}(\mathscr{A}')$. Therefore, 1_H is measurable with respect to X, but not vice versa. That is, X represents a more detailed information about the outcome of the random experiment than 1_H. Hence, if the value of X is known, then we can compute the value of 1_H, but not vice versa. In our example, Figure 2.5 shows: if $X(\omega) = 1$, then $1_H(\omega) = 1$. However, if $1_H(\omega) = 1$, then $X(\omega) = 1$ or $X(\omega) = 2$. (For a more general presentation of this property, see Lemma 2.52.)   ◁

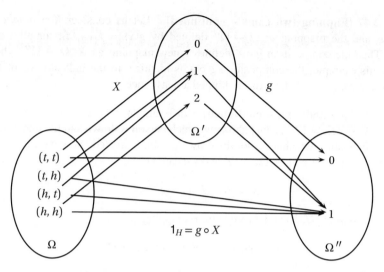

Figure 2.5   A composition of two mappings.

2.4   Theorems on measurable mappings

In this section, we consider compositions of mappings, which are defined as follows: Let Ω, Ω', and Ω'' be nonempty sets and let $f: \Omega \to \Omega'$ and $g: \Omega' \to \Omega''$ be mappings. Then the *composition of f and g* is the mapping $g \circ f: \Omega \to \Omega''$ defined by:

$$g \circ f(\omega) := g[f(\omega)], \quad \forall\, \omega \in \Omega, \tag{2.25}$$

(see Fig. 2.5), where $g \circ f(\omega)$ denotes the value of the mapping $g \circ f$ for the argument ω. Instead of $g \circ f$, we often use the notation $g(f)$ and say that $g(f)$ *is a function of f*. Using this notation, Equation (2.25) can be written as:

$$g(f)(\omega) = g[f(\omega)], \quad \forall\, \omega \in \Omega. \tag{2.26}$$

Lemma 2.46 [Compositions with a finite or countable number of values]
Let $f: \Omega \to \Omega'$ be a mapping, where Ω' is finite or countable, and let $g: \Omega' \to \overline{\mathbb{R}}$ be a function. Furthermore, for $\omega' \in \Omega'$, define $1_{f=\omega'} := 1_{f^{-1}(\{\omega'\})}$. Then,

$$g \circ f = g(f) = \sum_{\omega' \in \Omega'} g(\omega') \cdot 1_{f^{-1}(\{\omega'\})} = \sum_{\omega' \in \Omega'} g(\omega') \cdot 1_{f=\omega'}. \tag{2.27}$$

(Proof p. 76)

Hence, under the assumptions of Lemma 2.46, for all $\omega \in \Omega$,

$$g \circ f(\omega) = g[f(\omega)] = \sum_{\omega' \in \Omega'} g(\omega') \cdot 1_{f^{-1}(\{\omega'\})}(\omega) = \sum_{\omega' \in \Omega'} g(\omega') \cdot 1_{f=\omega'}(\omega). \tag{2.28}$$

Example 2.47 [Flipping two coins – continued]   Let us consider $X = $ *number of flipping heads* and the mapping $g \colon \Omega' \to \Omega''$ defined by $g(x) := 1_{\{1,2\}}(x)$, for all $x \in \Omega'$ (see Fig. 2.5). Then the composition $g \circ X$ defines a new mapping $g \circ X \colon \Omega \to \Omega''$, where $\Omega'' = \{0, 1\}$. In this example, the composition $g \circ X$ is identical to the indicator 1_H of the event $H = \{(h, h), (t, h), (h, t)\}$ that heads are flipped at least once.   ◁

Example 2.48 [Joe and Ann – continued]   In Example 2.34, we already considered the mapping $U \colon \Omega \to \Omega_U = \{Joe, Ann\}$ showing which person is drawn and the mapping $X \colon \Omega \to \Omega' = \{0, 1\}$ indicating whether or not the drawn person is treated. Now we can consider the bivariate mapping $(U, X) \colon \Omega \to \Omega_U \times \Omega'$ and we can write

$$X = g \circ (U, X) = g(U, X)$$

as the composition of (U, X) and a (projection) mapping g,

$$g[(u, x)] = x, \quad \forall \, (u, x) \in \Omega_U \times \Omega'.$$

◁

2.4.1   Measurability of a composition

Theorem 2.49 shows that measurability is preserved by the composition of mappings.

Theorem 2.49 [Measurability of a composition]
If $f \colon (\Omega, \mathscr{A}) \to (\Omega', \mathscr{A}')$ and $g \colon (\Omega', \mathscr{A}') \to (\Omega'', \mathscr{A}'')$ are measurable mappings, then the composition $g \circ f$ is $(\mathscr{A}, \mathscr{A}'')$-measurable.

(Proof p. 76)

Remark 2.50 [σ-Algebra generated by a composition]   Note that

$$(g \circ f)^{-1}(\mathscr{A}'') = f^{-1}[g^{-1}(\mathscr{A}'')] \tag{2.29}$$

(see the proof of Theorem 2.49).   ◁

Example 2.51 [Flipping two coins – continued]   Figure 2.5 illustrates Theorem 2.49. If (a) X is $(\mathscr{A}, \mathscr{A}')$-measurable and (b) g is $(\mathscr{A}', \mathscr{A}'')$-measurable, then $1_H = g \circ X$ is $(\mathscr{A}, \mathscr{A}'')$-measurable. Suppose $\mathscr{A}' = \mathscr{P}(\Omega')$ and $\mathscr{A}'' = \mathscr{P}(\Omega'')$, where $\Omega' = \{0, 1, 2\}$ and $\Omega'' = \{0, 1\}$. Then the premise '(a) and (b)' is satisfied if $\mathscr{A}$ is such that $X^{-1}(\mathscr{A}') \subset \mathscr{A}$. If the premise '(a) and (b)' is *not* satisfied, then we cannot conclude that 1_H is $(\mathscr{A}, \mathscr{A}'')$-measurable. Note that in this example 1_H can be $(\mathscr{A}, \mathscr{A}'')$-measurable even if (a) and (b) do not hold. A sufficient requirement is that $\{(t, t)\}$ and $\{(t, h), (h, t), (h, h)\}$, the inverse images of $\{0\}$ and $\{1\}$ under 1_H, respectively, are elements of $\mathscr{A}$ (see Cor. 2.21).   ◁

If a mapping h is measurable with respect to a mapping f, then each element in the σ-algebra generated by h is an element in the σ-algebra generated by f. If h is measurable with respect to f, then, in a sense, the information represented by h is already contained in f (cf. section 2.3.2). This is expressed in more formal terms in the following lemma, which is crucial, such as in the general definition of conditional expectation values $E(Y \mid X = x)$ (see ch. 10).

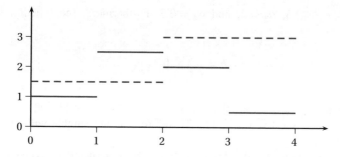

Figure 2.6   The two step functions f and h in Example 2.55.

Lemma 2.52 [Factorization lemma of measurable functions]
Let $f: \Omega \to \Omega'$ be a mapping, let $(\Omega', \mathscr{A}')$ be a measurable space, and let $h: \Omega \to \overline{\mathbb{R}}$ be a function. Then h is measurable with respect to f, that is, $h^{-1}(\overline{\mathscr{B}}) \subset f^{-1}(\mathscr{A}')$, if and only if there is a measurable function $g: (\Omega', \mathscr{A}') \to (\overline{\mathbb{R}}, \overline{\mathscr{B}})$ such that

$$h = g \circ f \tag{2.30}$$

is the composition of f and g. We call g a factorization of h with respect to f.

For a proof, see Klenke (2013, Corollary 1.97).

If, instead of $(\overline{\mathbb{R}}, \overline{\mathscr{B}})$ we consider a measurable space $(\Omega'', \mathscr{P}(\Omega''))$, where Ω'' is finite or countable, then the elements $\omega'' \in \Omega''$ can be renamed by real numbers such as 1, 2, and so on. Renaming is a one-to-one measurable function, because the σ-algebra on Ω'' is the power set of Ω'' (see Example 2.9). Hence, Lemma 2.52 implies the following corollary.

Corollary 2.53 [Factorization of a mapping into a finite or countable set]
Let $f: \Omega \to \Omega'$ be a mapping, $(\Omega', \mathscr{A}')$ a measurable space, and $h: \Omega \to \Omega''$ a mapping, where Ω'' is finite or countable. Then h is measurable with relation to f, that is, $h^{-1}[\mathscr{P}(\Omega'')] \subset f^{-1}(\mathscr{A}')$, if and only if there is a measurable mapping $g: (\Omega', \mathscr{A}') \to (\Omega'', \mathscr{P}(\Omega''))$ such that $h = g \circ f$.

Example 2.54 [Flipping two coins – continued]   If we specify $\Omega' = \{0, 1, 2\}$, the σ-algebra $\mathscr{A}' = \mathscr{P}(\Omega')$, the set $\Omega'' = \{0, 1\}$, the σ-algebra $\mathscr{A}'' = \mathscr{P}(\Omega'')$, and the function $h = 1_H$, then the example depicted in Figure 2.5 can be used to illustrate this corollary. The mapping g in this figure is such that $1_H = g \circ X$.   ◁

Example 2.55 [Two step functions]   Figure 2.6 presents an example in which $\Omega = [0, 4]$, $A_1 = [0, 1], A_2 =]1, 2], A_3 =]2, 3]$, and $A_4 =]3, 4]$. Note that the sets $A_1, \ldots, A_4$ are pairwise disjoint. The measurable function $f: (\Omega, \mathscr{A}) \to (\mathbb{R}, \mathscr{B})$ is defined by

$$f = \sum_{i=1}^{4} \alpha_i 1_{A_i},$$

where $\alpha_1 = 1$, $\alpha_2 = 2.5$, $\alpha_3 = 2$, and $\alpha_4 = 0.5$. Furthermore, the function $h: \Omega \to \mathbb{R}$ is defined by

$$h = \sum_{j \in \{1,3\}} \beta_j 1_{A_j \cup A_{j+1}}$$

with $\beta_1 = 1.5$ and $\beta_3 = 3$. Note that $\sigma(h) = \sigma(\{A_j \cup A_{j+1}: j \in \{1,3\}\})$, whereas $\sigma(f) = \sigma(\{A_i: i = 1, \dots, 4\})$ (see Exercise 2.11). Therefore, h is measurable with respect to f, that is, $\sigma(h) \subset \sigma(f)$.

According to Lemma 2.52, there is a function $g: \mathbb{R} \to \mathbb{R}$ such that $h = g \circ f$. In fact, if we define g by

$$g(x) = \sum_{j \in \{1,3\}} \beta_j 1_{\{\alpha_j, \alpha_{j+1}\}}(x), \quad \forall x \in \mathbb{R},$$

then $h = g \circ f$. The function g takes on the value 1.5 if $x = \alpha_1 = 1$ or $x = \alpha_2 = 2.5$ and the value 3 if $x = \alpha_3 = 2$ or $x = \alpha_4 = 0.5$. For all other $x \in \mathbb{R}$, the value of g is 0.   ◁

Example 2.56 [Square of a real-valued function] Suppose $f: (\Omega, \mathcal{A}) \to (\mathbb{R}, \mathcal{B})$ is a real-valued measurable function and $f^2(\omega) := f(\omega)^2$, for all $\omega \in \Omega$.

(i) If f is *nonnegative*, that is, if $f(\omega) \geq 0$, for all $\omega \in \Omega$, then f and f^2 are measurable with respect to each other, that is, $\sigma(f) = \sigma(f^2)$.

(ii) If there are $\omega_1, \omega_2 \in \Omega$ with $f(\omega_1) < 0 < f(\omega_2)$ and $f^2(\omega_1) = f^2(\omega_2)$, then $\sigma(f^2) \subset \sigma(f)$, but $\sigma(f) \neq \sigma(f^2)$.

(See Exercise 2.13.) In a sense, $\sigma(f) = \sigma(f^2)$ means that f and f^2 contain the same information, whereas $\sigma(f^2) \subset \sigma(f)$, $\sigma(f) \neq \sigma(f^2)$ means that f^2 contains less information than f. If, for example, $f^2(\omega) = 4$, then $f(\omega) = 2$ or $f(\omega) = -2$.   ◁

2.4.2 Theorems on measurable functions

In the first theorem, we consider sums and differences as well as products and ratios of measurable functions. The *sum* of two functions $f, h: \Omega \to \mathbb{R}^n$ is again a function $(f + h): \Omega \to \mathbb{R}^n$ defined by

$$(f + h)(\omega) := \begin{pmatrix} f_1 + h_1 \\ \vdots \\ f_n + h_n \end{pmatrix}(\omega) := \begin{pmatrix} f_1(\omega) + h_1(\omega) \\ \vdots \\ f_n(\omega) + h_n(\omega) \end{pmatrix}, \quad \forall \omega \in \Omega.$$

The first parentheses in the term $(f + h)(\omega)$ are used to make clear that $f + h$ is a symbol of a new function on Ω. Of course, the difference $f - h$ is defined in the same way as $f + h$ replacing $+$ by $-$.

Similarly, the *product* $f \cdot h$ of two functions $f, h: \Omega \to \mathbb{R}$ is again a function $(f \cdot h): \Omega \to \mathbb{R}$ defined by

$$(f \cdot h)(\omega) := f(\omega) \cdot h(\omega), \quad \forall \omega \in \Omega.$$

Correspondingly, $f/h: \Omega \to \mathbb{R}$ is defined by

$$(f/h)(\omega) := f(\omega)/h(\omega), \quad \forall\, \omega \in \Omega,$$

provided that $h(\omega) \neq 0$ for all $\omega \in \Omega$.

Theorem 2.57 [Sums and products of measurable functions]
If $f, h: (\Omega, \mathcal{A}) \to (\mathbb{R}^n, \mathcal{B}_n)$ are measurable functions, then $f + h$ and $f - h$ are $(\mathcal{A}, \mathcal{B}_n)$-measurable as well. Furthermore, if $f, h: (\Omega, \mathcal{A}) \to (\mathbb{R}, \mathcal{B})$ are measurable functions, then $f \cdot h$ and f/h (with $h(\omega) \neq 0$, for all $\omega \in \Omega$) are also $(\mathcal{A}, \mathcal{B})$-measurable.

For a proof, see Klenke (2013, Theorem 1.91).

Remark 2.58 [Squared function]   If $f: (\Omega, \mathcal{A}) \to (\mathbb{R}, \mathcal{B})$ is a measurable function, then $f^2 = f \cdot f$ is also $(\mathcal{A}, \mathcal{B})$-measurable. Obviously, this also applies to $f^n, n \in \mathbb{N}$. Hence, if f is $(\mathcal{A}, \mathcal{B})$-measurable, then f^n is also $(\mathcal{A}, \mathcal{B})$-measurable.   ◁

Example 2.59 [Scaling transformations and translations]   Remember that a constant real number can always be interpreted as a measurable function (see Example 2.10). Therefore, Theorem 2.57 implies that, for all $\alpha \in \mathbb{R}$, the functions $f + \alpha$, $f - \alpha$, and $\alpha \cdot f$ are $(\mathbb{R}, \mathcal{B})$-measurable if $f: (\Omega, \mathcal{A}) \to (\mathbb{R}, \mathcal{B})$ is a measurable function.   ◁

Example 2.60 [Number of flipping *heads*]   Consider flipping a coin n times, let $\Omega = \{h, t\}^n$, and let $1_{A_i}: \Omega \to \mathbb{R}$ denote the indicators of flipping *heads* at the ith flip of the coin. Then,

$$X = \sum_{i=1}^{n} 1_{A_i}$$

is the *number of flipping heads*. If $\mathcal{A} = \mathcal{P}(\Omega)$, then $(\Omega, \mathcal{A})$ is a measurable space and X is measurable for any σ-algebra on $\mathbb{R}$ (see Example 2.9). In the case $\mathcal{A} = \mathcal{P}(\Omega)$, it is not necessary to apply Theorem 2.57.   ◁

Example 2.61 [Linear combination of two functions]   Let $f, h: (\Omega, \mathcal{A}) \to (\mathbb{R}, \mathcal{B})$ be measurable functions and $\alpha, \beta \in \mathbb{R}$. Then, according to Theorem 2.57, the function $(\alpha \cdot f + \beta \cdot g): (\Omega, \mathcal{A}) \to (\mathbb{R}, \mathcal{B})$ defined by

$$(\alpha \cdot f + \beta \cdot h)(\omega) = \alpha \cdot f(\omega) + \beta \cdot h(\omega), \quad \forall\, \omega \in \Omega. \tag{2.31}$$

is $(\mathcal{A}, \mathcal{B})$-measurable.   ◁

Remark 2.62 [Positive and negative parts of a function]   In Theorem 2.66, we consider the positive and the negative parts of a function $f: \Omega \to \overline{\mathbb{R}}$. The *positive part* $f^+: \Omega \to \overline{\mathbb{R}}$ is defined by

$$f^+(\omega) := \max(f(\omega), 0), \quad \forall\, \omega \in \Omega,$$

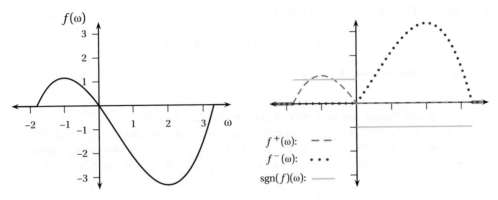

Figure 2.7 Positive and negative parts of a function, and its sign function.

and the *negative part* $f^-: \Omega \to \overline{\mathbb{R}}$ by

$$f^-(\omega) := -\min(f(\omega), 0), \quad \forall \, \omega \in \Omega.$$

Hence, the value $f^+(\omega)$ of the positive part of f is defined to be the *greater* one of the two numbers $f(\omega)$ and 0 if they differ and $f^+(\omega) = 0$ if $f(\omega) = 0$. In contrast, the value $f^-(\omega)$ of the negative part of f is defined to be the *smaller* one of the two numbers $f(\omega)$ and 0 *multiplied by* -1 if they differ and $f^-(\omega) = 0$ if $f(\omega) = 0$. Note that f^+ and f^- are both nonnegative functions and that

$$f = f^+ - f^-.$$

◁

Example 2.63 [Positive and negative parts of a function] The positive and negative parts of a function are illustrated by Figure 2.7 showing the graph of the function $f: \mathbb{R} \to \mathbb{R}$ defined by

$$f(x) = \begin{cases} \dfrac{x^3}{3} - \dfrac{x^2}{2} - 2x, & \text{if } -1.81 < x < 3.315 \\ 0, & \text{otherwise.} \end{cases}$$

The positive part f^+ takes on the value 0 if $x \leq 0$ (see the dashed line on the horizontal axis), whereas negative part f^- takes on the value 0 if $x \geq 0$ (see the dotted line on the horizontal axis).

◁

Remark 2.64 [Absolute value function] Furthermore, we consider the *absolute value function* $|f|: \Omega \to \overline{\mathbb{R}}$ defined by

$$|f|(\omega) := |f(\omega)| := \begin{cases} f(\omega), & \text{if } f(\omega) \geq 0 \\ -f(\omega), & \text{if } f(\omega) < 0. \end{cases}$$

Note that $|f| = f^+ + f^- = \max(f^+, f^-)$. Hence, in Figure 2.7, the absolute value function is represented by the dashed and dotted lines *above* (if $-1.81 < x < 3.315$) or *on* (if $x < -1.81$ or $x > 3.315$) the horizontal axis.   ◁

Remark 2.65 [Sign function]   In Theorem 2.66, we also refer to $\text{sgn}(f)\colon \Omega \to \mathbb{R}$, called the *sign function*, which is defined by

$$\text{sgn}(f)(\omega) = \begin{cases} 1, & \text{if } f(\omega) > 0 \\ 0, & \text{if } f(\omega) = 0 \\ -1, & \text{if } f(\omega) < 0. \end{cases}$$

In Figure 2.7, the graph of this function is represented by the four solid lines above, below and on the horizontal axis, and by the big point with coordinates $(0, 0)$.   ◁

Theorem 2.66 [Positive and negative parts of a function]
Let $(\Omega, \mathscr{A})$ be a measurable space. If $f\colon \Omega \to \overline{\mathbb{R}}$ is $(\mathscr{A}, \overline{\mathscr{B}})$-measurable, then the functions $f^+, f^-, |f|$, and $\text{sgn}(f)$ are $(\mathscr{A}, \overline{\mathscr{B}})$-measurable as well.

For a proof, see Klenke (2013, Corollary 1.89). The positive part f^+, the negative part f^-, and the absolute value function $|f|$ of a function f play important roles in integration theory (see ch. 3).

Another implication of Theorem 2.57 on the measurability of some sets that are often used is formulated in the following remark.

Remark 2.67 [Some important measurable sets]   Let $(\Omega, \mathscr{A})$ be a measurable space, and let $f, g\colon (\Omega, \mathscr{A}) \to (\overline{\mathbb{R}}, \overline{\mathscr{B}})$ be measurable functions. Then,

(a) $\{\omega \in \Omega\colon f(\omega \geq g(\omega)\} \in \mathscr{A}$.

(b) $\{\omega \in \Omega\colon f(\omega > g(\omega)\} \in \mathscr{A}$.

(c) $\{\omega \in \Omega\colon f(\omega = g(\omega)\} \in \mathscr{A}$.

(See Exercise 2.12.)   ◁

2.5   Equivalence of two mappings with respect to a measure

Now we study some properties of mappings $f\colon \Omega \to \Omega'$ involving a measure space $(\Omega, \mathscr{A}, \mu)$. In this case, we use the notation

$$f\colon (\Omega, \mathscr{A}, \mu) \to \Omega'$$

to express that $f\colon \Omega \to \Omega'$ is a mapping and that μ is a measure on the measurable space $(\Omega, \mathscr{A})$. If there is also a σ-algebra $\mathscr{A}'$ on Ω', then we use the notation

$$f\colon (\Omega, \mathscr{A}, \mu) \to (\Omega', \mathscr{A}')$$

to express that the mapping $f: \Omega \to \Omega'$ is $(\mathcal{A}, \mathcal{A}')$-measurable and that μ is a measure on the measurable space $(\Omega, \mathcal{A})$.

Remember, two mappings f and g are *identical*, that is, $f = g$, if and only if

$$\{\omega \in \Omega: f(\omega) \neq g(\omega)\} = \emptyset.$$

A less restrictive concept is their *equivalence with respect to a measure.*

Definition 2.68 [Equivalence of two mappings with respect to a measure]
Let $f, g: (\Omega, \mathcal{A}, \mu) \to \Omega'$ be mappings. Then f and g are called μ-equivalent, denoted by

$$f \underset{\mu}{=} g,$$

if there is an $A \in \mathcal{A}$ with $\mu(A) = 0$ and, for all $\omega \in \Omega \setminus A$, $f(\omega) = g(\omega)$.

Because $\mu(\emptyset) = 0$, $f = g$ implies $f \underset{\mu}{=} g$.

Remark 2.69 [A note on notation] If $f: (\Omega, \mathcal{A}, \mu) \to \Omega'_f$ and $g: (\Omega, \mathcal{A}, \mu) \to \Omega'_g$ are mappings, then we can choose $\Omega' = \Omega'_f \cup \Omega'_g$ and write $f, g: (\Omega, \mathcal{A}, \mu) \to \Omega'$ (see Rem. 2.4).   ◁

Remark 2.70 [An alternative notation] If $f \underset{\mu}{=} g$, we also say that $f = g$, μ-almost everywhere (μ-a.e.). Furthermore, we also write

$$f(\omega) = g(\omega), \quad \text{for } \mu\text{-almost all } \omega \in \Omega, \tag{2.32}$$

and use $f(\omega) \underset{\mu\text{-}a.a.}{=} g(\omega)$ as a shortcut.   ◁

Remark 2.71 [Singleton with a positive value of a measure] If $f \underset{\mu}{=} g$ or, equivalently, $f(\omega) \underset{\mu\text{-}a.a.}{=} g(\omega)$, and $\{\omega^*\} \in \mathcal{A}$ with $\mu(\{\omega^*\}) > 0$, then

$$f(\omega^*) = g(\omega^*).$$ ◁

Remark 2.72 [μ-Equivalence, restricted functions, and compositions] Let $f, g: (\Omega, \mathcal{A}, \mu) \to \Omega'$ be mappings.

(i) If $\Omega' = \mathbb{R}$, then

$$f \underset{\mu}{=} g \Rightarrow 1_A \cdot f \underset{\mu}{=} 1_A \cdot g, \quad \forall A \in \mathcal{A}. \tag{2.33}$$

(ii) If $h: \Omega' \to \Omega''$ is also a mapping, then

$$f \underset{\mu}{=} g \Rightarrow h \circ f \underset{\mu}{=} h \circ g \tag{2.34}$$

(see Exercise 2.14).   ◁

Remark 2.73 [Equivalence relation]   If $\mathcal{M}$ is a set of mappings $(\Omega, \mathcal{A}, \mu) \to \Omega'$, then $\underset{\mu}{=}$ is an *equivalence relation* on $\mathcal{M}$ (see Exercise 2.15). In other words, if $f, g, h \in \mathcal{M}$, then

(i) $f \underset{\mu}{=} f$ (reflexivity).

(ii) $g \underset{\mu}{=} f$ if and only if $f \underset{\mu}{=} g$ (symmetry).

(iii) If $f \underset{\mu}{=} g$ and $g \underset{\mu}{=} h$, then $f \underset{\mu}{=} h$ (transitivity).

◁

Definition 2.74 [Equivalence class with respect to a measure]
Let $\mathcal{M}$ be a set of mappings $(\Omega, \mathcal{A}, \mu) \to \Omega'$ and let $f \in \mathcal{M}$. Then,

$$C(f) := \{g \in \mathcal{M} : g \underset{\mu}{=} f\}$$

is called the μ-equivalence class of f in $\mathcal{M}$ and f a representative of the class $C(f)$.

Remark 2.75 [A partition of the set $\mathcal{M}$]   If $\mathcal{M}$ is a set of mappings $(\Omega, \mathcal{A}, \mu) \to \Omega'$, then the set $\{C(f) : f \in \mathcal{M}\}$ is a *partition* of $\mathcal{M}$, that is,

(a) $\forall f \in \mathcal{M} : C(f) \neq \emptyset$.

(b) $\forall f, g \in \mathcal{M} : C(f) = C(g)$ or $C(f) \cap C(g) = \emptyset$.

(c) $\displaystyle\bigcup_{f \in \mathcal{M}} C(f) = \mathcal{M}$.

(See Exercise 2.16.)

◁

Remark 2.76 [Other properties of μ-equivalence]

(i) Let $f, g : (\Omega, \mathcal{A}, \mu) \to \Omega'$. If $\mu(\Omega) > 0$, then

$$\forall \alpha \in \Omega', \forall \beta \in \Omega' : f \underset{\mu}{=} \alpha \ \wedge \ g \underset{\mu}{=} \beta \ \wedge \ f \underset{\mu}{=} g \ \Rightarrow \ \alpha = \beta. \tag{2.35}$$

(ii) If $f, g, f^*, g^* : (\Omega, \mathcal{A}, \mu) \to \mathbb{R}$, then

$$\begin{aligned}
f \underset{\mu}{=} f^* \wedge g \underset{\mu}{=} g^* \Rightarrow f + g &\underset{\mu}{=} f^* + g^*, \\
f - g &\underset{\mu}{=} f^* - g^*, \\
f \cdot g &\underset{\mu}{=} f^* \cdot g^*.
\end{aligned} \tag{2.36}$$

Furthermore, suppose $\mu(\{\omega \in \Omega : g(\omega) = 0\}) = 0$, and define $\dfrac{f}{g} : \Omega \to \mathbb{R}$ by

$$
\frac{f}{g}(\omega) := \begin{cases} \dfrac{f(\omega)}{g(\omega)}, & \text{if } g(\omega) \neq 0, \\ 0, & \text{otherwise}, \end{cases} \quad \forall\, \omega \in \Omega,
$$

and let $\dfrac{f^*}{g^*}$ be defined analogously. Then

$$
f \underset{\mu}{=} f^* \wedge g \underset{\mu}{=} g^* \;\Rightarrow\; \frac{f}{g} \underset{\mu}{=} \frac{f^*}{g^*}. \tag{2.37}
$$

(iii) If $f_i, f_i^* : (\Omega, \mathscr{A}, \mu) \to \mathbb{R}$ and $\alpha_i \in \mathbb{R}$, $i = 1, \ldots, n$, then

$$
(\forall\, i = 1, \ldots, n : f_i \underset{\mu}{=} f_i^*) \;\Rightarrow\; \sum_{i=1}^{n} \alpha_i f_i \underset{\mu}{=} \sum_{i=1}^{n} \alpha_i f_i^*. \tag{2.38}
$$

(iv) If $f_1, f_2, \ldots, f_1^*, f_2^*, \ldots : (\Omega, \mathscr{A}, \mu) \to \mathbb{R}$ and $\alpha_1, \alpha_2, \ldots \in \mathbb{R}$, then

$$
(\forall\, i = 1, 2, \ldots : f_i \underset{\mu}{=} f_i^*) \;\Rightarrow\; \sum_{i=1}^{\infty} \alpha_i f_i \underset{\mu}{=} \sum_{i=1}^{\infty} \alpha_i f_i^*, \tag{2.39}
$$

provided that the limits denoted by the infinite sums (see Box 0.1) exist.

For proofs, see Exercise 2.17.   ◁

Remark 2.77 [Order relations for real-valued functions] For two mappings $f, g : \Omega \to \mathbb{R}$, we write $f < g$, if and only if

$$
\{\omega \in \Omega : f(\omega) \geq g(\omega)\} = \varnothing.
$$

The notation $f > g$, $f \leq g$, and $f \geq g$ is used correspondingly.   ◁

Remark 2.78 [Order relations with respect to a measure μ] For functions $f, g, h :$ $(\Omega, \mathscr{A}, \mu) \to \mathbb{R}$, we also use the notation

$$
f \underset{\mu}{<} g,
$$

if there is an $A \in \mathscr{A}$ with $f(\omega) < g(\omega)$ for all $\omega \in \Omega \setminus A$ and $\mu(A) = 0$. The notation $f \underset{\mu}{>} g$, $f \underset{\mu}{\leq} g$, and $f \underset{\mu}{\geq} g$ is used correspondingly. Furthermore,

$$
f \underset{\mu}{<} g \quad \text{and} \quad g \underset{\mu}{=} h \;\Rightarrow\; f \underset{\mu}{<} h. \tag{2.40}
$$

The analog propositions hold for $\underset{\mu}{>}$, $\underset{\mu}{\leq}$, and $\underset{\mu}{\geq}$ (see Exercise 2.18).   ◁

2.6   Image measure

In the definition of a measurable mapping $f: (\Omega, \mathscr{A}) \to (\Omega', \mathscr{A}')$ we required $f^{-1}(A') \in \mathscr{A}$, for all $A' \in \mathscr{A}'$. Because a measure μ assigns a value to *all* elements $A \in \mathscr{A}$, the measure μ also assigns a value to each $f^{-1}(A') := \{\omega \in \Omega: f(\omega) \in A'\}$. This is the reason for choosing the term *measurable mapping*: If μ is a measure on $\mathscr{A}$ and f is $(\mathscr{A}, \mathscr{A}')$-measurable, then there is a value $\mu[f^{-1}(A')]$ for all inverse images $f^{-1}(A'), A' \in \mathscr{A}'$.

According to the following theorem, a measurable mapping $f: (\Omega, \mathscr{A}, \mu) \to (\Omega', \mathscr{A}')$ induces a measure on the codomain space $(\Omega', \mathscr{A}')$.

Theorem 2.79 [Image measure]

Let $f: (\Omega, \mathscr{A}, \mu) \to (\Omega', \mathscr{A}')$ be a measurable mapping. Then the function $\mu_f: \mathscr{A}' \to \overline{\mathbb{R}}$ defined by

$$\mu_f(A') := \mu[f^{-1}(A')], \quad \forall A' \in \mathscr{A}', \tag{2.41}$$

is a measure on the measurable space $(\Omega', \mathscr{A}')$.

(Proof p. 76)

Definition 2.80 [Image measure]

If $f: (\Omega, \mathscr{A}, \mu) \to (\Omega', \mathscr{A}')$ is a measurable mapping, then $\mu_f: \mathscr{A}' \to \overline{\mathbb{R}}$ defined by Equation (2.41) is called the image measure of μ under f.

Example 2.81 [Rectangles – continued]   Now we consider a measure $\mu: \mathscr{A} \to \mathbb{R}$, which is specified by

$$\mu(A) = (7 - 2) \cdot (5 - 2) = 15$$

and

$$\mu(\Omega) = (10 - 0) \cdot (6 - 0) = 60.$$

This specification determines the areas of all four sets in $\mathscr{A}$, because $\mu(A^c) = \mu(\Omega) - \mu(A) = 60 - 15 = 45$ and $\mu(\emptyset) = 0$. Hence, the measure space $(\Omega, \mathscr{A}, \mu)$ is completely determined. Note that μ is the restriction of the Lebesgue measure λ_2 to the σ-algebra $\mathscr{A}$ (i.e., $\mu(A) = \lambda_2(A)$, for all $A \in \mathscr{A}$).

In Example 2.2.1, we considered the mapping $f: \Omega \to \Omega' = \Omega$ defined by $f(x) = \frac{3}{4}x$. Furthermore, we considered the rectangle $B' = \,]4.5, 7.5] \times [0, 4.5]$ and the σ-algebra $\mathscr{A}' = \{\Omega', \emptyset, B', (B')^c\}$. If we specify $\mathscr{A}$ such that $f^{-1}(B') \in \mathscr{A}$, then f is $(\mathscr{A}, \mathscr{A}')$-measurable. In this case, all inverse images $f^{-1}(A')$ of sets $A' \in \mathscr{A}'$ are elements of the σ-algebra $\mathscr{A}$. Therefore, the areas $\lambda_2[f^{-1}(A')]$ of these inverse images are defined by the measure λ_2 on

$\mathscr{A}$ that assigns the area to *all* elements of $\mathscr{A}$. If we specify $\mathscr{A} = \{\Omega, \emptyset, f^{-1}(B'), f^{-1}[(B')^c]\}$, then

$$\lambda_{2_f}(B') = \lambda_2[f^{-1}(B')] = \lambda_2(]6, 10] \times [0, 6]) = (10 - 6) \cdot (6 - 0) = 24,$$

$$\lambda_{2_f}((B')^c) = \lambda_2(f^{-1}[(B')^c]) = \lambda_2([0, 6] \times [0, 6]) = (6 - 0) \cdot (6 - 0) = 36,$$

$\lambda_{2_f}(\Omega') = 60$, and $\lambda_{2_f}(\emptyset) = 0$. Then the function $\lambda_{2_f} : \mathscr{A}' \to \mathbb{R}$ defined by

$$\lambda_{2_f}(B') = \lambda_2[f^{-1}(B')], \quad \forall\, B' \in \mathscr{A}',$$

is again a measure, the *image measure* of λ_2 under f. Therefore, $(\Omega', \mathscr{A}', \lambda_{2_f})$ is a measure space.

Note that the image measure λ_{2_f} on the σ-algebra $\mathscr{A}'$ differs from the area measure on $\mathscr{A}'$. In fact, the area of B' is $(7.5 - 4.5) \cdot 4.5 = 13.5$, and the area of $(B')^c$ is $60 - 13.5 = 46.5$. $\triangleleft$

Remark 2.82 [Cumulation of the values $\mu(\{\omega\})$]   If $\{\omega\} \in \mathscr{A}$, for all $\omega \in \Omega$, then

$$\mu_f(\{\omega'\}) = \sum_{\omega:\, f(\omega)=\omega'} \mu(\{\omega\}), \text{ if } \{\omega'\} \in \mathscr{A}' \tag{2.42}$$

provided that the sum is over a finite or countable number of summands. The measure μ assigns to the singletons and other elements $A \in \mathscr{A}$ a nonnegative number $\mu(A)$, and f maps each element $\omega \in \Omega$ to an element ω' in Ω'. Thereby, it translates the values $\mu(A)$ of the measure μ to their images $f(A)$. In particular, this applies to the singletons $\{\omega\}$. This is illustrated in the following example. $\triangleleft$

Example 2.83 [Flipping two coins – continued]   In this example,

$$P(\{\omega\}) = \frac{1}{4}, \quad \forall\, \omega \in \Omega, \tag{2.43}$$

uniquely defines a measure $P: \mathscr{P}(\Omega) \to \mathbb{R}$ and the measure space $(\Omega, \mathscr{P}(\Omega), P)$. The reason is that the singletons $\{\omega\}$ are pairwise disjoint and Rule (x) of Box 1.1 implies

$$P(A) = P\left(\bigcup_{\omega \in A} \{\omega\} \right) = \sum_{\omega \in A} P(\{\omega\}), \quad \forall\, A \in \mathscr{A}.$$

For instance, the set $A = $ *flipping one and only one head* is the union $A = \{(h, t)\} \cup \{(t, h)\} = \{(h, t), (t, h)\}$. Hence,

$$P(A) = \sum_{\omega \in A} P(\{\omega\}) = P(\{(h, t)\}) + P(\{(t, h)\}) = \frac{1}{4} + \frac{1}{4} = \frac{1}{2}.$$

Now consider Figure 2.4 and realize that each arrow translates the value $\mu(\{\omega\}) = \frac{1}{4}$ from left to right. According to Equation (2.42), this yields

$$P_X(\{0\}) = P[X^{-1}(\{0\})] = P[\{(t, t)\}] = \frac{1}{4},$$

$$P_X(\{1\}) = P[X^{-1}(\{1\})] = P[\{(t, h), (h, t)\}] = \frac{2}{4},$$

and

$$P_X(\{2\}) = P[X^{-1}(\{2\})] = P[\{(h, h)\}] = \frac{1}{4}.$$

◁

Example 2.84 [Image measure under a step function]   If $f: (\Omega, \mathscr{A}, \mu) \to (\mathbb{R}, \mathscr{B})$ is measurable such that $f = \sum_{i=1}^{n} \alpha_i 1_{A_i}$ with pairwise different $\alpha_1, \dots, \alpha_n \in \mathbb{R}$, $\alpha_i \neq 0$, and pairwise disjoint $A_i \in \mathscr{A}$, $i = 1, \dots, n$, and if we define $A_{n+1} := \Omega \setminus (\bigcup_{i=1}^{n} A_i)$ and $\alpha_{n+1} := 0$, then the image measure is

$$\mu_f = \sum_{i=1}^{n+1} \mu(A_i) \cdot \delta_{\alpha_i} \tag{2.44}$$

(see Exercise 2.19). Equation (2.44) generalizes Equation (2.42): For all $\omega \in A_i$

$$f(\omega) = \alpha_i \cdot 1_{A_i}(\omega) = \alpha_i.$$

Hence, f translates the value $\mu(A_i)$ to $\alpha_i \in \mathbb{R}$ and μ_f assigns the value $\mu(A_i)$ to the singleton $\{\alpha_i\}$, $i = 1, \dots, n+1$.

◁

Our next theorem deals with the image measures of μ-equivalent measurable mappings. As a random variable is a particular measurable mapping and the distribution of a random variable a particular image measure (see section 5.1), this theorem has important implications on all concepts that in some sense describe properties of distributions of random variables such as expectations, variances, covariances, and so on.

Theorem 2.85 [μ-Equivalence implies equality of image measures]
If $f, g: (\Omega, \mathscr{A}, \mu) \to (\Omega', \mathscr{A}')$ are measurable mappings, then

$$f \underset{\mu}{=} g \implies \mu_f = \mu_g. \tag{2.45}$$

(Proof p. 77)

In Theorem 2.86, we present a necessary and sufficient condition for μ-equivalence of two compositions $g \circ f$ and $g^* \circ f$.

Theorem 2.86 [μ-Equivalence of compositions]
If $f\colon (\Omega, \mathscr{A}, \mu) \to (\Omega', \mathscr{A}')$ and $g, g^*\colon (\Omega', \mathscr{A}') \to (\overline{\mathbb{R}}, \overline{\mathscr{B}})$ are measurable mappings, then

$$g \underset{\mu_f}{=} g^* \Leftrightarrow g \circ f \underset{\mu}{=} g^* \circ f. \tag{2.46}$$

(Proof p. 77)

2.7   Proofs

Proof of Lemma 2.19

(a) If $f = \sum_i \alpha_i 1_{A_i}$, then for all $B \in \overline{\mathscr{B}}$,

$$f^{-1}(B) = \bigcup_{i:\, \alpha_i \in B} A_i \in \mathscr{A},$$

because $\mathscr{A}$ is closed with respect to finite and countable unions.

(b) Assume that there are no $\alpha_1, \alpha_2, \ldots \in \overline{\mathbb{R}}$ such that $f = \sum_i \alpha_i 1_{A_i}$. Then there are an i and elements $\omega_1, \omega_2 \in A_i$ with $f(\omega_1) \neq f(\omega_2)$. Applying Equation (2.5) yields

$$f^{-1}(\{f(\omega_1)\}) \cap f^{-1}(\{f(\omega_2)\}) = \varnothing.$$

Furthermore, because $\omega_j \in f^{-1}(\{f(\omega_j)\}), j = 1, 2$,

$$f^{-1}(\{f(\omega_1)\}) \cap A_i \neq \varnothing, \quad \text{and} \quad f^{-1}(\{f(\omega_2)\}) \cap A_i \neq \varnothing.$$

Therefore, we conclude: $f^{-1}(\{f(\omega_1)\}) \notin \mathscr{A}$ and $f^{-1}(\{f(\omega_2)\}) \notin \mathscr{A}$. Because $\{f(\omega_1)\}$, $\{f(\omega_2)\} \in \overline{\mathscr{B}}$, it follows that f is not $(\mathscr{A}, \overline{\mathscr{B}})$-measurable.

Proof of Lemma 2.29

(i) $\cap$-stability of $\mathscr{D}$. If $C_1, C_2 \in \mathscr{C}$ and $\omega_1', \omega_2' \in \Omega'$, then

$$
\begin{aligned}
&[C_1 \cap f^{-1}(\{\omega_1'\})] \cap [C_2 \cap f^{-1}(\{\omega_2'\})] \\
&= (C_1 \cap C_2) \cap [f^{-1}(\{\omega_1'\}) \cap f^{-1}(\{\omega_2'\})] \qquad [\cap \text{ is associative and commutative}] \\
&= \begin{cases} (C_1 \cap C_2) \cap f^{-1}(\{\omega_1'\}), & \text{if } \omega_1' = \omega_2' \\ \varnothing, & \text{if } \omega_1' \neq \omega_2' \end{cases} \qquad\qquad [(2.5)]
\end{aligned}
$$

is an element of $\mathscr{D}$, because $C_1 \cap C_2 \in \mathscr{C}$ and $\varnothing \in \mathscr{C}$, which follows from the definition of a σ-algebra.

(ii)   Denote $\mathscr{A}' = \mathscr{P}(\Omega')$, and define $\sigma(\mathscr{C}, f) := \sigma[\mathscr{C} \cup f^{-1}(\mathscr{A}')]$.

   (a) $\sigma(\mathscr{D}) \subset \sigma[\mathscr{C} \cup f^{-1}(\mathscr{A}')]$. Obviously, $\{f^{-1}(\{\omega'\}) : \omega' \in \Omega'\} \subset f^{-1}(\mathscr{A}')$. Therefore,

$$\begin{aligned}
\mathscr{D} &= \{C \cap f^{-1}(\{\omega'\}) : \omega' \in \Omega', C \in \mathscr{C}\} \\
&\subset \{C \cap f^{-1}(A') : A' \in \mathscr{A}', C \in \mathscr{C}\} \\
&\subset \sigma[\mathscr{C} \cup f^{-1}(\mathscr{A}')]. \qquad\qquad \text{[Rem. 1.2]}
\end{aligned}$$

Hence, according to Remark 1.23, $\sigma(\mathscr{D}) \subset \sigma[\mathscr{C} \cup f^{-1}(\mathscr{A}')]$.

   (b) $\sigma[\mathscr{C} \cup f^{-1}(\mathscr{A}')] \subset \sigma(\mathscr{D})$. Because $\Omega \in \mathscr{C}$ and $\Omega \in f^{-1}(\mathscr{A}')$, all $C \in \mathscr{C}$ and all $f^{-1}(A')$ are elements of $\sigma(\mathscr{D})$ (see Def. 1.1, Ω' is finite or countable). Therefore, $\mathscr{C} \cup f^{-1}(\mathscr{A}') \subset \sigma(\mathscr{D})$. Proposition (1.11) then implies $\sigma[\mathscr{C} \cup f^{-1}(\mathscr{A}')] \subset \sigma(\mathscr{D})$.

Proof of Lemma 2.35

(i)   We have to show that $\mathscr{C}'_f$ satisfies conditions (a) to (c) of Definition 1.1.

   (a)

$$\Omega = f^{-1}(\Omega') \in \mathscr{C} \;\Rightarrow\; \Omega' \in \mathscr{C}'_f. \qquad \text{[(2.16)]}$$

   (b)

$$\begin{aligned}
A' \in \mathscr{C}'_f &\Rightarrow f^{-1}(A') \in \mathscr{C} && \text{[(2.16)]} \\
&\Rightarrow f^{-1}(A')^c = f^{-1}[(A')^c] \in \mathscr{C} && \text{[Def. 1.1(b), (2.4)]} \\
&\Rightarrow (A')^c \in \mathscr{C}'_f. && \text{[(2.16)]}
\end{aligned}$$

   (c)

$$\begin{aligned}
A'_1, A'_2, \ldots \in \mathscr{C}'_f &\Rightarrow f^{-1}(A'_1), f^{-1}(A'_2), \ldots \in \mathscr{C} && \text{[(2.16)]} \\
&\Rightarrow \bigcup_{i=1}^{\infty} f^{-1}(A'_i) = f^{-1}\left(\bigcup_{i=1}^{\infty} A'_i\right) \in \mathscr{C} && \text{[Def. 1.1(c), (2.6)]} \\
&\Rightarrow \bigcup_{i=1}^{\infty} A'_i \in \mathscr{C}'_f. && \text{[(2.16)]}
\end{aligned}$$

(ii)   For all $A' \in \mathscr{A}'$,

$$\begin{aligned}
A' \in \mathscr{A}' &\Rightarrow f^{-1}(A') \in \mathscr{C} && \text{[$(\mathscr{C}, \mathscr{A}')$-measurability of f]} \\
&\Rightarrow A' \in \mathscr{C}'_f. && \text{[(2.16)]}
\end{aligned}$$

Hence, $(\mathscr{C}, \mathscr{A}')$-measurability of f implies $\mathscr{A}' \subset \mathscr{C}'_f$.

Proof of Lemma 2.37

First, note that, for $A_i' \in \mathscr{A}_i'$, $i = 1, \dots, n$,

$$
\begin{aligned}
f^{-1}(A_1' \times \dots \times A_n') &= \{\omega \in \Omega : f(\omega) \in A_1' \times \dots \times A_n'\} \\
&= \{\omega \in \Omega : (f_1(\omega), \dots, f_n(\omega)) \in A_1' \times \dots \times A_n'\} \\
&= \{\omega \in \Omega : f_1(\omega) \in A_1', \dots, f_n(\omega) \in A_n'\} \\
&= \bigcap_{i=1}^{n} \{\omega \in \Omega : f_i(\omega) \in A_i'\} \\
&= \bigcap_{i=1}^{n} f_i^{-1}(A_i').
\end{aligned}
\tag{2.47}
$$

Hence,

$$
\begin{aligned}
\sigma(f) &= \left\{ f^{-1}(A') : A' \in \bigotimes_{i=1}^{n} \mathscr{A}_i' \right\} && \text{[Def. 2.26]} \\
&= \sigma(\{ f^{-1}(A_1' \times \dots \times A_n') : A_i' \in \mathscr{A}_i', i = 1, \dots, n \}) && \text{[Th. 2.20, Defs. 1.13, 1.31]} \\
&= \sigma(\{ f_1^{-1}(A_1') \cap \dots \cap f_n^{-1}(A_n') : A_i' \in \mathscr{A}_i', i = 1, \dots, n \}) && \text{[(2.47)]} \\
&\supset \sigma \left(\bigcup_{i=1}^{n} \left\{ f_i^{-1}(A_i') \cap \bigcap_{j=1, j \neq i}^{n} f_j^{-1}(\Omega_j') : A_i' \in \mathscr{A}_i', i = 1, \dots, n \right\} \right) && \text{[Rem. 1.23]} \\
&= \sigma \left(\bigcup_{i=1}^{n} \{ f_i^{-1}(A_i') : A_i' \in \mathscr{A}_i', i = 1, \dots, n \} \right) && \text{[}f_j^{-1}(\Omega_j') = \Omega\text{]} \\
&= \sigma \left(\bigcup_{i=1}^{n} \sigma(f_i) \right). && \text{[Def. 2.26]}
\end{aligned}
$$

Furthermore,

$$
\begin{aligned}
&\{ f_1^{-1}(A_1') \cap \dots \cap f_n^{-1}(A_n') : A_i' \in \mathscr{A}_i', i = 1, \dots, n \} \\
&\subset \sigma \left(\bigcup_{i=1}^{n} \{ f_i^{-1}(A_i') : A_i' \in \mathscr{A}_i', i = 1, \dots, n \} \right). && \text{[Rem. 1.2, finite intersections]}
\end{aligned}
$$

Therefore,

$$
\begin{aligned}
\sigma \left(\bigcup_{i=1}^{n} \sigma(f_i) \right) &= \sigma \left(\bigcup_{i=1}^{n} \{ f_i^{-1}(A_i') : A_i' \in \mathscr{A}_i', i = 1, \dots, n \} \right) && \text{[Def. 2.26]} \\
&\supset \sigma(\{ f_1^{-1}(A_1') \cap \dots \cap f_n^{-1}(A_n') : A_i' \in \mathscr{A}_i', i = 1, \dots, n \}) && \text{[Rem. 1.23]} \\
&= \sigma(f).
\end{aligned}
$$

Hence, $\sigma(f) = \sigma\left(\bigcup_{i=1}^{n} \sigma(f_i) \right)$.

Proof of Theorem 2.38

(b) $\Rightarrow$ (a) For all $i = 1, \dots, n$: Let $A'_i \in \mathscr{A}'_i$. If f_i is measurable, then $f_i^{-1}(A'_i) \in \mathscr{A}$. Hence,

$$f^{-1}(A'_1 \times \dots \times A'_n) = \{\omega \in \Omega : f(\omega) \in A'_1 \times \dots \times A'_n\}$$

$$= \bigcap_{i=1}^{n} f_i^{-1}(A'_i) \in \mathscr{A}.$$

Because $\{A'_1 \times \dots \times A'_n : A'_i \in \mathscr{A}'_i, i = 1, \dots, n\}$ is a generating system of $\bigotimes_{i=1}^{n} \mathscr{A}'_i$, Theorem 2.20 implies that f is measurable.

(a) $\Rightarrow$ (b) If f is measurable, then for all $i = 1, \dots, n$,

$$f_i^{-1}(\mathscr{A}'_i) = \{f_i^{-1}(A'_i) : A'_i \in \mathscr{A}'_i\}$$

$$= \left\{ f_i^{-1}(A'_i) \cap \bigcap_{j=1, j \neq i}^{n} f_j^{-1}(\Omega'_j) : A'_i \in \mathscr{A}'_i \right\}$$

$$= \{f^{-1}(\Omega'_1 \times \dots \times \Omega'_{i-1} \times A'_i \times \Omega'_{i+1} \times \dots \times \Omega'_n) : A'_i \in \mathscr{A}'_i\} \quad [(2.47)]$$

$$\subset f^{-1}\left(\bigotimes_{i=1}^{n} \mathscr{A}'_i \right) \subset \mathscr{A}.$$

Proof of Lemma 2.42

Consider the projection mappings $\pi_1, \dots, \pi_n$ defined by Equation (2.20) and the mapping

$$\pi = (\pi_1, \dots, \pi_n): \left(\bigtimes_{i=1}^{n} \Omega_i, \bigotimes_{i=1}^{n} \mathscr{A}_i \right) \to \left(\bigtimes_{i=1}^{n} \Omega_i, \bigotimes_{i=1}^{n} \mathscr{A}_i \right).$$

Analogously to the proof of Lemma 2.37 and using Definition 1.31,

$$\bigotimes_{i=1}^{n} \mathscr{A}_i = \sigma(\{A_1 \times \dots \times A_n : A_i \in \mathscr{A}_i, i = 1, \dots, n\})$$

$$= \sigma(\{\pi^{-1}(A_1 \times \dots \times A_n) : A_i \in \mathscr{A}_i, i = 1, \dots, n\}) \qquad [(2.20), (2.21)]$$

$$= \sigma(\pi) \qquad\qquad [\text{Th. } 2.20, (2.12)]$$

$$= \sigma(\pi_i, i = 1, \dots, n). \qquad\qquad [\text{Lem. } 2.37, (2.18)]$$

Proof of Corollary 2.44

If Ω' is finite or countable and we consider the measurable space $(\Omega', \mathscr{P}(\Omega'))$, then $\sigma(f) = \sigma[\{f^{-1}(\{\omega'\}): \omega' \in \Omega'\}]$ [see Lemma 2.29 with $\mathscr{C} = \{\Omega, \emptyset\}$]. Because $\{f^{-1}(\{\omega'\}): \omega' \in \Omega'\}$ is a finite or countable partition of Ω, this corollary is an immediate implication of Lemma 2.19.

Proof of Lemma 2.46

For all $\omega \in \Omega$ and all $\omega' \in \Omega'$,

$$g[f(\omega)] \cdot 1_{f=\omega'}(\omega) = g(\omega') \cdot 1_{f=\omega'}(\omega) = \begin{cases} 0, & \text{if } f(\omega) \neq \omega' \\ g(\omega'), & \text{if } f(\omega) = \omega'. \end{cases} \qquad (2.48)$$

This equation is equivalent to

$$g(f) \cdot 1_{f=\omega'} = g(\omega') \cdot 1_{f=\omega'}. \qquad (2.49)$$

Because the set $\{f^{-1}(\{\omega'\}) \colon \omega' \in \Omega'\}$ is a finite or countable partition of Ω we can conclude: $1_{\Omega} = \sum_{\omega' \in \Omega'} 1_{f=\omega'}$. Therefore,

$$g(f) = g(f) \cdot 1_{\Omega} = \sum_{\omega' \in \Omega'} g(f) \cdot 1_{f=\omega'} = \sum_{\omega' \in \Omega'} g(\omega') \cdot 1_{f=\omega'},$$

and this implies Equation (2.27).

Proof of Theorem 2.49

If $f \colon (\Omega, \mathscr{A}) \to (\Omega', \mathscr{A}')$, $g \colon (\Omega', \mathscr{A}') \to (\Omega'', \mathscr{A}'')$ are measurable mappings, then, according to Corollary 2.28, $f^{-1}(\mathscr{A}') \subset \mathscr{A}$ and $g^{-1}(\mathscr{A}'') \subset \mathscr{A}'$. Hence, for all $A'' \in \mathscr{A}''$,

$$\begin{aligned} (g \circ f)^{-1}(A'') &= \{\omega \in \Omega \colon g[f(\omega)] \in A''\} \\ &= \{\omega \in \Omega \colon f(\omega) \in g^{-1}(A'')\} \\ &= f^{-1}[g^{-1}(A'')]. \end{aligned}$$

Furthermore,

$$\begin{aligned} f^{-1}[g^{-1}(\mathscr{A}'')] &\subset f^{-1}(\mathscr{A}') && [(2.15)] \\ &\subset \mathscr{A}. && [\text{Def. 2.5}] \end{aligned}$$

Proof of Theorem 2.79

We show that μ_f has the properties (a) to (c) required in Definition 1.43. For each property of μ_f we use the corresponding property of μ.

(a) $\mu_f(\emptyset) = \mu[f^{-1}(\emptyset)] = \mu(\emptyset) = 0$.

(b) For all $A' \in \mathscr{A}' \colon \mu_f(A') = \mu[f^{-1}(A')] \geq 0$.

(c) If $A'_1, A'_2, \ldots \in \mathscr{A}'$ are pairwise disjoint, then, according to Equation (2.5), for $i \neq j$,

$$f^{-1}(A'_i) \cap f^{-1}(A'_j) = f^{-1}(A'_i \cap A'_j) = f^{-1}(\emptyset) = \emptyset,$$

that is, the inverse images $f^{-1}(A'_1), f^{-1}(A'_2), \ldots$ are pairwise disjoint as well. Therefore,

$$\mu_f\left(\bigcup_{i=1}^{\infty} A'_i\right) = \mu\left(f^{-1}\left(\bigcup_{i=1}^{\infty} A'_i\right)\right) \qquad [(2.41)]$$

$$= \mu\left(\bigcup_{i=1}^{\infty} f^{-1}(A'_i)\right) \qquad [(2.6)]$$

$$= \sum_{i=1}^{\infty} \mu(f^{-1}(A'_i)) \qquad [\text{Def. 1.43(c)}]$$

$$= \sum_{i=1}^{\infty} \mu_f(A'_i). \qquad [(2.41)]$$

Proof of Theorem 2.85

If $f \underset{\mu}{=} g$, then there is a set $A \in \mathscr{A}$ satisfying

$$f(\omega) = g(\omega), \quad \forall\, \omega \in \Omega \setminus A \text{ and } \mu(A) = 0.$$

Monotonicity of μ implies $\mu(\{\omega \in A: f(\omega) \in A'\}) = 0 = \mu(\{\omega \in A: g(\omega) \in A'\})$ for all $A' \in \mathscr{A}'$. Hence, using additivity of μ,

$$\begin{aligned}
\mu_f(A') &= \mu[f^{-1}(A')] \\
&= \mu(\{\omega \in \Omega \setminus A: f(\omega) \in A'\}) + \mu(\{\omega \in A: f(\omega) \in A'\}) \\
&= \mu(\{\omega \in \Omega \setminus A: g(\omega) \in A'\}) + \mu(\{\omega \in A: g(\omega) \in A'\}) \\
&= \mu[g^{-1}(A')] = \mu_g(A').
\end{aligned}$$

Proof of Theorem 2.86

For measurable functions $g, g^*: (\Omega', \mathscr{A}') \to (\overline{\mathbb{R}}, \overline{\mathscr{B}})$ define $A' := \{\omega' \in \Omega': g(\omega') \neq g^*(\omega')\}$. Note that $A' \in \mathscr{A}'$ [see Rem. 2.67 (c)]. Then

$$\begin{aligned}
f^{-1}(A') &= \{\omega \in \Omega: f(\omega) \in A'\} \\
&= \{\omega \in \Omega: g[f(\omega)] \neq g^*[f(\omega)]\} \\
&= \{\omega \in \Omega: (g \circ f)(\omega) \neq (g^* \circ f)(\omega)\}.
\end{aligned}$$

Hence, $g \underset{\mu_f}{=} g^* \Leftrightarrow \mu_f(A') = 0 \Leftrightarrow \mu[f^{-1}(A')] = 0 \Leftrightarrow g \circ f \underset{\mu}{=} g^* \circ f$.

Exercises

2.1 Prove Equations (2.4) to (2.6).

2.2 Consider Example 2.2.1 and compute the inverse images of the sets $\{(4.5, 0)\}$, $\{(7.5, 0)\}$, $\{(7.5, 4.5)\}$, and $\{(4.5, 4.5)\}$ under the function

$$f(x_1, x_2) = \frac{3}{4} \cdot (x_1, x_2) = \left(\frac{3}{4} \cdot x_1, \frac{3}{4} \cdot x_2\right).$$

2.3 Consider Example 2.2.1 and specify the inverse images of the rectangles $[8, 10] \times [0, 2]$ and $[3, 7.5] \times [0, 3]$ under the function $f: \Omega \to \Omega'$ defined by $f(x_1, x_2) = \frac{3}{4} \cdot (x_1, x_2)$.

2.4 Consider Example 2.2.1 and use Equation (2.4) to determine the inverse image $f^{-1}[(C')^c]$.

2.5 Prove the proposition of Example 2.14.

2.6 Prove the proposition of Example 2.15.

2.7 Prove the proposition of Example 2.17.

2.8 Prove the proposition of Remark 2.18.

2.9 Prove the proposition formulated in Remark 2.33.

2.10 Consider Example 2.40 and show that X, Y, and (X, Y) are measurable with respect to $\mathscr{A}$ whenever the two inverse images $X^{-1}(\{1\})$ and $Y^{-1}(\{1\})$ are elements of $\mathscr{A}$.

2.11 In Example 2.55, we considered $\Omega = [0, 4]$, $A_1 = [0, 1]$, $A_2 =]1, 2]$, $A_3 =]2, 3]$, and $A_4 =]3, 4]$. There, we also defined the functions f and h. Show that $\sigma(h) = \sigma(\{A_j \cup A_{j+1}: j \in J\})$ and $\sigma(f) = \sigma(\{A_i: i = 1, \dots, 4\})$. Furthermore, show $\sigma(h) \subset \sigma(f)$.

2.12 Prove the propositions of Remark 2.12.

2.13 Prove the proposition of Example 2.56.

2.14 Prove the propositions of Remark 2.72.

2.15 Consider Remark 2.73 and show: If $\mathscr{M}$ is a set of mappings $(\Omega, \mathscr{A}, \mu) \to \Omega'$, then $\underset{\mu}{=}$ is an equivalence relation on $\mathscr{M}$.

2.16 Show that $\{C(f): f \in \mathscr{M}\}$ is a partition of $\mathscr{M}$ (see Remark 2.75).

2.17 Prove the propositions of Remark 2.76.

2.18 Prove proposition (2.40).

2.19 Prove the proposition of Example 2.84.

Solutions

2.1 Equation (2.4):

$$f^{-1}[(A')^c] = \{\omega \in \Omega: f(\omega) \in (A')^c\} = \{\omega \in \Omega: f(\omega) \notin A'\} = [f^{-1}(A')]^c.$$

Equation (2.5):

$$f^{-1}\left(\bigcap_{i\in I} A_i'\right) = \left\{\omega \in \Omega: f(\omega) \in \bigcap_{i\in I} A_i'\right\} = \{\omega \in \Omega: f(\omega) \in A_i', \forall i \in I\}$$

$$= \bigcap_{i\in I} \{\omega \in \Omega: f(\omega) \in A_i'\} = \bigcap_{i\in I} f^{-1}(A_i').$$

Equation (2.6):

$$f^{-1}\left(\bigcup_{i\in I} A_i'\right) = \left\{\omega \in \Omega: f(\omega) \in \bigcup_{i\in I} A_i'\right\} = \{\omega \in \Omega: \exists\, i \in I: f(\omega) \in A_i'\}$$

$$= \bigcup_{i\in I} \{\omega \in \Omega: f(\omega) \in A_i'\} = \bigcup_{i\in I} f^{-1}(A_i').$$

2.2   The inverse images are the sets $f^{-1}[\{(4.5, 0)\}] = \{(6, 0)\}, f^{-1}[\{(7.5, 0)\}] = \{(10, 0)\}$, $f^{-1}[\{(7.5, 4.5)\}] = \{(10, 6)\}$, and $f^{-1}[\{(4.5, 4.5)\}] = \{(6, 6)\}$.

2.3 $f^{-1}([8, 10] \times [0, 2]) = \emptyset$ and $f^{-1}([3, 7.5] \times [0, 3]) = [4, 10] \times [0, 4]$.

2.4   According to Equation (2.4), the inverse image of $(C')^c$ under f is

$$f^{-1}[(C')^c] = [f^{-1}(C')]^c$$
$$= \Omega \setminus (]6, 10] \times [0, 6])$$
$$= ([0, 10] \times [0, 6]) \setminus (]6, 10] \times [0, 6])$$
$$= [0, 6] \times [0, 6].$$

2.5   If $f: \Omega \to \Omega'$ is constant, then, according to Example 2.10, it is $(\mathscr{A}, \mathscr{A}')$-measurable for $\mathscr{A} = \{\Omega, \emptyset\}$. Now, assume that f is not constant, that is, $\exists\, \omega_1, \omega_2 \in \Omega: f(\omega_1) \neq f(\omega_2)$. According to our assumptions,

$$\{f(\omega_1)\}, \{f(\omega_2)\} \in \mathscr{A}'.$$

Furthermore, $\omega_i \in f^{-1}[f\{(\omega_i)\}]$, for $i = 1, 2$, that is, the inverse images are nonempty sets. Now, $f(\omega_1) \neq f(\omega_2)$ implies

$$\{f(\omega_1)\} \cap \{f(\omega_2)\} = \emptyset,$$

and, using Equation (2.5),

$$f^{-1}(\{f(\omega_1)\}) \cap f^{-1}(\{f(\omega_2)\}) = f^{-1}(\{f(\omega_1)\} \cap \{f(\omega_2)\}) = f^{-1}(\emptyset) = \emptyset.$$

Hence, the inverse images are nonempty disjoint sets, and therefore none of them is in $\mathscr{A} = \{\Omega, \emptyset\}$. This implies that f is not $(\mathscr{A}, \mathscr{A}')$-measurable if it is not constant.

2.6   We consider $\{A, A^c, \Omega, \emptyset\}$. If $f = \alpha_1 1_A + \alpha_2 1_{A^c}$, then for all $A' \in \mathcal{B}$,

$$f^{-1}(A') = \begin{cases} \emptyset, & \text{if } \alpha_1 \notin A', \alpha_2 \notin A' \\ A, & \text{if } \alpha_1 \in A', \alpha_2 \notin A' \\ A^c, & \text{if } \alpha_1 \notin A', \alpha_2 \in A' \\ \Omega, & \text{if } \alpha_1 \in A', \alpha_2 \in A'. \end{cases}$$

Hence, f is $(\mathcal{A}, \mathcal{B})$-measurable. (Note that this also holds if $A = \emptyset$ or $A = \Omega$, and also if $\alpha_1 = \alpha_2$.)

Now assume that f is an $(\mathcal{A}, \mathcal{B})$-measurable function.

(a) If f takes on only one single value, say α, then

$$f = \alpha 1_\Omega = \alpha_1 1_A + \alpha_2 1_{A^c}, \quad \text{with } \alpha_1 = \alpha_2 = \alpha.$$

(b) If f takes on exactly two different values $\beta_1 \neq \beta_2$, then $f^{-1}(\{\beta_1, \beta_2\}) = f^{-1}(\{\beta_1\}) \cup f^{-1}(\{\beta_2\}) = \Omega$, and according to Equation (2.5), $f^{-1}(\{\beta_1\}) \cap f^{-1}(\{\beta_2\}) = \emptyset$, and $f^{-1}(\{\beta_i\}) \neq \emptyset$, for $i = 1, 2$. Hence, f is $(\mathcal{A}, \mathcal{B})$-measurable if and only if

$$f^{-1}(\{\beta_1\}) = A \quad \text{or} \quad f^{-1}(\{\beta_1\}) = A^c \quad \text{and} \quad A, A^c \neq \emptyset.$$

This implies

$$f = \beta_1 1_A + \beta_2 1_{A^c} \quad \text{or} \quad f = \beta_2 1_A + \beta_1 1_{A^c},$$

respectively.

(c) If f takes on *three or more* pairwise different values, then, using the same kind of argument as in (a), we can conclude that there are at least three pairwise disjoint and nonempty inverse images under f, say $A_1, A_2, A_3 \subset \Omega$. Hence, in this case f is not $(\mathcal{A}, \mathcal{B})$-measurable.

2.7   If $A_1, \ldots, A_n \in \mathcal{A}$ are pairwise disjoint, we define $A_{n+1} := \Omega \setminus (\bigcup_{i=1}^n A_i)$, and $\alpha_{n+1} := 0$, then

$$f = \sum_{i=1}^n \alpha_i 1_{A_i} = \sum_{i=1}^{n+1} \alpha_i 1_{A_i}.$$

Because $A_1, \ldots, A_{n+1}$ are pairwise disjoint and $\bigcup_{i=1}^{n+1} A_i = \Omega$, there is, for all $\omega \in \Omega$, exactly one $i \in \{1, \ldots, n+1\}$ such that $\omega \in A_i$, and therefore $f(\omega) = \alpha_i$. Hence, the codomain of f is $\{\alpha_1, \ldots, \alpha_{n+1}\}$. Vice versa, for all α_i, $i = 1, \ldots, n+1$, we obtain the inverse image

$$f^{-1}(\{\alpha_i\}) = \{\omega \in \Omega : f(\omega) = \alpha_i\} = \bigcup_{j : \alpha_i = \alpha_j} A_j \tag{2.50}$$

(If the $\alpha_1, \dots, \alpha_{n+1}$ are pairwise different, then $f^{-1}(\{\alpha_i\}) = A_i$.) Now, for all $A' \subset \mathbb{R}$,

$$f^{-1}(A') = \{\omega \in \Omega : f(\omega) \in A'\} \qquad \text{[Def. 2.1]}$$

$$= \left\{\omega \in \Omega : f(\omega) \in \bigcup_{i:\, \alpha_i \in A'} \{\alpha_i\}\right\} \qquad \text{[codomain of } f \text{ is } \{\alpha_1, \dots, \alpha_{n+1}\}]$$

$$= f^{-1}\left(\bigcup_{i:\, \alpha_i \in A'} \{\alpha_i\}\right) \qquad \text{[Def. 2.1]}$$

$$= \bigcup_{i:\, \alpha_i \in A'} f^{-1}(\{\alpha_i\}) \qquad \text{[(2.6)]}$$

$$= \bigcup_{i:\, \alpha_i \in A'} A_i. \qquad \text{[(2.50)]}$$

2.8 If $A_1, \dots, A_n \in \mathscr{A}$ are pairwise disjoint, we define $A_{n+1} := \Omega \setminus (\bigcup_{i=1}^n A_i)$, and $\alpha_{n+1} := 0$, then

$$\forall A' \in \mathscr{B}: f^{-1}(A') = \bigcup_{\substack{i=1, \dots, n+1 \\ \alpha_i \in A'}} A_i \in \mathscr{A},$$

[see Eq. (2.10)].

If $A_1, \dots, A_n \in \mathscr{A}$ are *not* pairwise disjoint, define the 2^n sets

$$B_j := A_1^{c_1(j)} \cap \dots \cap A_n^{c_n(j)} \in \mathscr{A}, \quad j = 1, \dots, 2^n,$$

with $(c_1(j), \dots, c_n(j)) \in \{0, 1\}^n$ and

$$A_i^0 := A_i, \qquad A_i^1 := A_i^c.$$

Note that some of the sets B_j can be empty. Then

$$f = \sum_{i=1}^n \alpha_i 1_{A_i} = \sum_{j=1}^{2^n} \beta_j 1_{B_j},$$

with $\beta_j = \sum_{i:\, c_i(j)=0} \alpha_i$. Because $B_1, \dots, B_{2^n}$ are pairwise disjoint and $\bigcup_{j=1}^{2^n} B_j = \Omega$, the function f is $(\mathscr{A}, \mathscr{B})$-measurable (see the first part of this solution).

2.9 If all values of f are elements of Ω', then

$$f^{-1}(B) = f^{-1}(\Omega' \cap B), \quad \forall B \in \mathscr{B}.$$

Therefore, $f^{-1}(\mathscr{B}) = f^{-1}(\mathscr{B}|_{\Omega'})$, where $\mathscr{B}|_{\Omega'}$ denotes the trace of $\mathscr{B}$ in Ω' (see Example 1.10). Note that $\mathscr{B}|_{\Omega'} = \mathscr{P}(\Omega')$ (see Exercise 1.13). Hence, $f^{-1}(\mathscr{B}) = f^{-1}(\mathscr{B}|_{\Omega'}) = f^{-1}[\mathscr{P}(\Omega')]$.

2.10 First of all, note that X, Y are measurable with respect to $\mathscr{A}$ if and only if and (X, Y) is measurable with respect to $\mathscr{A}$ (see Th. 2.38). Hence, it suffices to show that X is measurable with respect to $\mathscr{A}$ if $X^{-1}(\{1\}) \in \mathscr{A}$. Because $X \colon \Omega \to \mathbb{R}$ is an indicator, $X^{-1}(\{0\}) = X^{-1}(\{1\}^c) = (X^{-1}(\{1\}))^c$ [see Eq. (2.5)]. Hence, if $X^{-1}(\{1\}) \in \mathscr{A}$, then $(X^{-1}(\{1\}))^c = X^{-1}(\{0\}) \in \mathscr{A}$. Furthermore, for all $B \in \mathscr{B}$,

$$
X^{-1}(B) = \begin{cases}
\varnothing, & \text{if } 0 \notin B,\, 1 \notin B \\
X^{-1}(\{1\}), & \text{if } 0 \notin B,\, 1 \in B \\
\Omega \setminus X^{-1}(\{1\}), & \text{if } 0 \in B,\, 1 \notin B \\
\Omega, & \text{if } \{0, 1\} \subset B.
\end{cases}
$$

(The proof for Y is analog.)

2.11 Because the codomain of f is $\mathbb{R}$,

$$
\begin{aligned}
\sigma(f) &= \sigma\left(f^{-1}(\mathscr{P}(\{1, 2.5, 2, 0.5\}))\right) & \text{[Rem. 2.33]} \\
&= \sigma\left(\{f^{-1}(\{1\}), f^{-1}(\{2.5\}), f^{-1}(\{2\}), f^{-1}(\{0.5\})\}\right) & \text{[(2.12)]} \\
&= \sigma(\{A_1, A_2, A_3, A_4\}).
\end{aligned}
$$

Analogously,

$$
\begin{aligned}
\sigma(h) &= \sigma\left(h^{-1}(\mathscr{P}(\{1.5, 3\}))\right) & \text{[Rem. 2.33]} \\
&= \sigma\left(\{h^{-1}(\{1.5\}), h^{-1}(\{3\})\}\right) & \text{[(2.12)]} \\
&= \sigma(\{A_1 \cup A_2,\, A_3 \cup A_4\}).
\end{aligned}
$$

Because

$$
\{A_1 \cup A_2,\, A_3 \cup A_4\} \subset \sigma(\{A_1,\, A_2, A_3,\, A_4\}), \qquad \text{[Rem. 1.2]}
$$

monotonicity of the generated σ-algebras (see Remark 1.23) implies $\sigma(h) \subset \sigma(f)$.

2.12 Denote $A_\infty := \{\omega \in \Omega \colon f(\omega) = \infty\}$, $A_{-\infty} := \{\omega \in \Omega \colon f(\omega) = -\infty\}$, $B_\infty := \{\omega \in \Omega \colon g(\omega) = \infty\}$, and $B_{-\infty} := \{\omega \in \Omega \colon g(\omega) = -\infty\}$. Because $\{\infty\}, \{-\infty\} \in \overline{\mathscr{B}}$, all four sets defined above are elements of $\mathscr{A}$. Furthermore,

$$
A := \{\omega \in \Omega \colon -\infty < f(\omega) < \infty\} = f^{-1}(\mathbb{R}) \in \mathscr{A}
$$

and

$$
B := \{\omega \in \Omega \colon -\infty < g(\omega) < \infty\} = g^{-1}(\mathbb{R}) \in \mathscr{A}.
$$

(a) Now

$$\{\omega \in \Omega : f(\omega \geq g(\omega)\}$$
$$= A_{\infty} \cup \{\omega \in A \cap B : f(\omega) \geq g(\omega)\} \cup B_{-\infty}$$
$$= A_{\infty} \cup \{\omega \in A \cap B : f(\omega) - g(\omega) \geq 0\} \cup B_{-\infty}$$
$$= A_{\infty} \cup \left(1_{A\cap B}^{-1}(\{1\})\right) \cap \left[1_{A\cap B} \cdot (f - g)\right]^{-1}([0, \infty[) \cup B_{-\infty} \qquad \text{[Def. 2.1]}$$
$$\in \mathcal{A}. \qquad\qquad\qquad\qquad\qquad\qquad\qquad\qquad\qquad\qquad\quad \text{[Th. 2.57]}$$

(b) Analogously,

$$\{\omega \in \Omega : f(\omega > g(\omega)\}$$
$$= (A_{\infty} \cap (B \cup B_{-\infty})) \cup \{\omega \in A \cap B : f(\omega > g(\omega)\}$$
$$= (A_{\infty} \cap (B \cup B_{-\infty})) \cup \left[1_{A\cap B} \cdot (f - g)\right]^{-1}(]0, \infty[) \qquad \text{[Def. 2.1]}$$
$$\in \mathcal{A}. \qquad\qquad\qquad\qquad\qquad\qquad\qquad\qquad\qquad\qquad\quad \text{[Th. 2.57]}$$

(c) Finally,

$$\{\omega \in \Omega : f(\omega = g(\omega)\} = \{\omega \in \Omega : f(\omega \geq g(\omega)\} \setminus \{\omega \in \Omega : f(\omega > g(\omega)\}$$
$$\in \mathcal{A}. \qquad\qquad\qquad\qquad\qquad\qquad\qquad\qquad \text{[Rem. 1.2]}$$

2.13 For any real-valued measurable function $f : (\Omega, \mathcal{A}) \to (\mathbb{R}, \mathcal{B})$, Lemma 2.52 yields $\sigma(f^2) \subset \sigma(f)$, because $f^2 = g(f)$ for the measurable function $g : (\mathbb{R}, \mathcal{B}) \to (\mathbb{R}, \mathcal{B})$ with $g(x) = x^2$, for all $x \in \mathbb{R}$. [Note that g is a continuous function that is $\mathcal{B}$-measurable (see Klenke, 2013, Th. 1.88).]

(i)   We assume that f is nonnegative and measurable. Then $f^2(\omega) = x$ if and only if $f(\omega) = \sqrt{x}$, for all $x \geq 0$. Hence, for all $A \in \mathcal{A}$,

$$A \in \sigma(f) \Rightarrow \exists B_1 \in \mathcal{B} : A = f^{-1}(B_1)$$
$$\Rightarrow \exists B_2 \in \mathcal{B} : A = (f^2)^{-1}(B_2) \qquad [\text{choose } B_2 := g^{-1}(B_1)]$$
$$\Rightarrow A \in \sigma(f^2).$$

This implies $\sigma(f) \subset \sigma(f^2)$.

(ii) Assume that there are $\omega_1, \omega_2 \in \Omega$ with $f(\omega_1) < 0 < f(\omega_2)$ and $f^2(\omega_1) = f^2(\omega_2)$. Then $A := f^{-1}(]-\infty, 0[)$ implies $A \in f^{-1}(\mathcal{B})$, and $\omega_1 \in A$ and $\omega_2 \notin A$. Furthermore, for all $B \in \mathcal{B}$:

$$\{\omega_1, \omega_2\} \cap (f^2)^{-1}(B), \quad \text{if } f^2(\omega_1) \in B$$

and

$$\{\omega_1, \omega_2\} \cap (f^2)^{-1}(B) = \emptyset, \quad \text{if } f^2(\omega_1) \notin B.$$

Hence, $A \notin (f^2)^{-1}(\mathcal{B})$.

2.14 (i)   If $f \underset{\mu}{=} g$, then there is a set $B \in \mathcal{A}$ with $\mu(B) = 0$ and $f(\omega) = g(\omega)$ for all $\omega \in \Omega \setminus B$. Hence, $1_A(\omega) \cdot f(\omega) = 1_A(\omega) \cdot g(\omega)$ for all $\omega \in \Omega \setminus B$. According to Definition 2.68, $1_A f \underset{\mu}{=} 1_A g$.

(ii)   Note that $A_h := \{\omega \in \Omega: h[f(\omega)] \neq h[g(\omega)]\} \subset \{\omega \in \Omega: f(\omega) \neq g(\omega)\} =: A$. Therefore, $\mu(A) = 0$ implies $\mu(A_h) = 0$ [see Box 1.1 (v)].

2.15   Reflexivity. $\mu(\{\omega \in \Omega: f(\omega) \neq f(\omega)\}) = \mu(\emptyset) = 0$. Hence, $f \underset{\mu}{=} f$.

Symmetry. Assume that $f, g \in \mathcal{M}$ and $f \underset{\mu}{=} g$. Then

$$\mu(\{\omega \in \Omega: g(\omega) \neq f(\omega)\}) = \mu(\{\omega \in \Omega: f(\omega) \neq g(\omega)\}) = 0.$$

Hence, $g \underset{\mu}{=} f$.

Transitivity. Assume that $f, g, h \in \mathcal{M}, f \underset{\mu}{=} g$, and $g \underset{\mu}{=} h$. Then transitivity of $=$ and subadditivity of μ yield

$$\mu(\{\omega \in \Omega: f(\omega) \neq h(\omega)\}) \leq \mu(\{\omega \in \Omega: f(\omega) \neq g(\omega)\} \cup \{\omega \in \Omega: g(\omega) \neq h(\omega)\})$$
$$\leq \mu(\{\omega \in \Omega: f(\omega) \neq g(\omega)\}) + \mu(\{\omega \in \Omega: g(\omega) \neq h(\omega)\})$$
$$= 0 + 0 \qquad [f \underset{\mu}{=} g, g \underset{\mu}{=} h]$$
$$= 0.$$

Therefore, $f \underset{\mu}{=} h$.

2.16   (a) $\forall f \in \mathcal{M}: f \underset{\mu}{=} f$ (reflexivity). This implies: $\forall f \in \mathcal{M}: f \in C(f)$ and therefore $\forall f \in \mathcal{M}: C(f) \neq \emptyset$.

(b)   Let $f, g \in \mathcal{M}$. We consider two cases, $f \underset{\mu}{=} g$ and $\neg(f \underset{\mu}{=} g)$.

$f \underset{\mu}{=} g$. Transitivity implies: $\forall h \in \mathcal{M}: f \underset{\mu}{=} h$ if and only if $g \underset{\mu}{=} h$. Hence, $\forall h \in \mathcal{M}: h \in C(f)$ if and only if $h \in C(g)$. This means that $C(f) = C(g)$.

$\neg(f \underset{\mu}{=} g)$. We show $C(f) \cap C(g) = \emptyset$ by contraposition. Assume: $\exists h \in \mathcal{M}: h \in C(f) \cap C(g)$.

Then $f \underset{\mu}{=} h, h \underset{\mu}{=} g$, and transitivity implies: $f \underset{\mu}{=} g$, which is a contradiction to $\neg(f \underset{\mu}{=} g)$.

(c)   In part (a) we showed: $\forall f \in \mathcal{M}: f \in C(f)$. Therefore, $\forall f \in \mathcal{M}: f \in \bigcup_{f^* \in \mathcal{M}} C(f^*)$.

2.17   (i)   Suppose that $\mu \neq 0$, $\alpha, \beta \in \Omega'$ and that $f \underset{\mu}{=} \alpha, g \underset{\mu}{=} \beta, f \underset{\mu}{=} g$. If $\alpha \neq \beta$, then subadditivity and monotonicity of μ yield

$$0 < \mu(\Omega)$$
$$= \mu(\{\omega \in \Omega: f(\omega) = \alpha \wedge g(\omega) = \beta\})$$
$$\quad + \mu(\{\omega \in \Omega: f(\omega) \neq \alpha \vee g(\omega) \neq \beta\}) \quad [\text{Box 1.1 (iv)}]$$
$$= \mu(\{\omega \in \Omega: f(\omega) = \alpha \wedge g(\omega) = \beta\}) \qquad [f \underset{\mu}{=} \alpha, g \underset{\mu}{=} \beta]$$
$$\leq \mu(\{\omega \in \Omega: f(\omega) \neq g(\omega)\}) \qquad\qquad [\alpha \neq \beta]$$
$$= 0, \qquad\qquad\qquad\qquad\qquad\qquad [f \underset{\mu}{=} g]$$

which proves (i) by contraposition.

(ii)   If $f \underset{\mu}{=} f^*$ and $g \underset{\mu}{=} g^*$ and $A_0 := \{\omega \in \Omega : f(\omega) \neq f^*(\omega) \vee g(\omega) \neq g^*(\omega)\}$, then

$$
\begin{aligned}
&\mu(A_0) \\
&= \mu(\{\omega \in \Omega : f(\omega) \neq f^*(\omega)\} \cup \{\omega \in \Omega : g(\omega) \neq g^*(\omega)\}) \\
&\leq \mu(\{\omega \in \Omega : f(\omega) \neq f^*(\omega)\}) + \mu(\{\omega \in \Omega : g(\omega) \neq g^*(\omega)\}) \quad [\text{Box 1.1 (vii), (v)}] \\
&= 0. \qquad\qquad\qquad\qquad\qquad\qquad\qquad\qquad\qquad\qquad\qquad [f \underset{\mu}{=} f^*, g \underset{\mu}{=} g^*\,]
\end{aligned}
$$

Note that $\{\omega \in \Omega : f(\omega) + g(\omega) \neq f^*(\omega) + g^*(\omega)\} \subset A_0$, which also holds for the corresponding sets for the difference, product, and ratio. This implies Equations (2.36) and (2.37).

(iii), (iv)   Define $I := \{1, \ldots, n\}$ for (iii) and $I := \mathbb{N}$ for (iv), respectively. Furthermore, define

$$
A_0 := \bigcup_{i \in I} \{\omega \in \Omega : f_i(\omega) \neq f_i^*(\omega)\} = \{\omega \in \Omega : \exists\, i \in I : f_i(\omega) \neq f_i^*(\omega)\}.
$$

Then

$$
\mu(A_0) \leq \sum_{i \in I} \{\omega \in \Omega : f_i(\omega) \neq f_i^*(\omega)\} \qquad [\text{Box 1.1 (xi)}]
$$

$$
= 0, \quad \text{if } f_i \underset{\mu}{=} f_i^*, \forall\, i \in I.
$$

Hence,

$$
\mu\left(\left\{\omega \in \Omega : \sum_{i \in I} \alpha_i f_i(\omega) \neq \sum_{i \in I} \alpha_i f_i^*(\omega)\right\}\right) \leq \mu(A_0) \qquad [\text{Box 1.1 (v)}]
$$

$$
= 0, \quad \text{if } f_i \underset{\mu}{=} f_i^*, \forall\, i \in I.
$$

2.18

$$
\forall\, \omega \in \Omega : (f(\omega) < g(\omega) \wedge g(\omega) = h(\omega)) \Rightarrow f(\omega) < h(\omega),
$$

which, by contraposition, is equivalent to

$$
\forall\, \omega \in \Omega : f(\omega) \geq h(\omega) \Rightarrow (f(\omega) \geq g(\omega) \vee g(\omega) \neq h(\omega)).
$$

Therefore,

$$
\{\omega \in \Omega : f(\omega) \geq h(\omega)\} \subset \{\omega \in \Omega : f(\omega) \geq g(\omega)\} \cup \{\omega \in \Omega : g(\omega) \neq h(\omega)\}.
$$

Now $(f \underset{\mu}{\leq} g \wedge g \underset{\mu}{=} h)$ implies

$$\mu(\{\omega \in \Omega: f(\omega) \geq h(\omega)\})$$
$$\leq \mu(\{\omega \in \Omega: f(\omega) \geq g(\omega)\}) + \mu(\{\omega \in \Omega: g(\omega) \neq h(\omega)\}) \qquad \text{[Box 1.1 (vii)]}$$
$$= 0 + 0 = 0. \qquad\qquad [f \underset{\mu}{\leq} g, g \underset{\mu}{=} h]$$

Because a measure is nonnegative, this implies $\mu(\{\omega \in \Omega: f(\omega) \geq h(\omega)\}) = 0$, which is equivalent to $f \underset{\mu}{\leq} h$.

2.19   For all $A' \in \mathscr{A}'$,

$$\mu_f(A') = \mu[f^{-1}(A')] \qquad\qquad \text{[(2.41)]}$$

$$= \mu \left(\bigcup_{\substack{i=1,\,\ldots,\,n+1 \\ \alpha_i \in A'}} A_i \right) \qquad\qquad \text{[Def. 2.16]}$$

$$= \sum_{\substack{i=1,\,\ldots,\,n+1 \\ \alpha_i \in A'}} \mu(A_i) \qquad\qquad \text{[Def. 1.43 (c)]}$$

$$= \sum_{i=1}^{n+1} \mu(A_i) \cdot \delta_{\alpha_i}(A'). \qquad\qquad \text{[(1.39)]}$$

3

Integral

In the preceding chapters, we introduced the most important concepts of measure theory related to the concepts of a measure and a measurable mapping. In this chapter, we introduce the *integral* of measurable functions. This concept is fundamental also for probability theory, because the expectation of a numerical random variable with respect to a probability measure is the *integral* of a measurable function with respect to a probability measure. In chapters 6 and 7, we shall see that this also applies to variances, covariances, and correlations. We start defining the integral of a measurable function with respect to a measure μ. Then we study the most important rules of computation and other properties of integrals, introduce the concept of a *measure with density*, and treat the relationship between the Riemann integral and the integral with respect to the Lebesgue measure. The next section is on *absolute continuity* and the *Radon-Nikodym theorem*. Both issues are crucial for conditional expectations (see ch. 10). A section on the integral with respect to a product measure concludes this chapter.

3.1 Definition

At first we define the integral for *nonnegative step functions*, then we extend the integral to *nonnegative measurable functions*, and finally we introduce the integral for *measurable functions* that may take on negative or nonnegative values.

3.1.1 Integral of a nonnegative step function

In this subsection, we introduce the integral of a *nonnegative step function*, also called *a nonnegative simple function* or *elementary function*.

Probability and Conditional Expectation: Fundamentals for the Empirical Sciences, First Edition. Rolf Steyer and Werner Nagel.
© 2017 John Wiley & Sons, Ltd. Published 2017 by John Wiley & Sons, Ltd.
Companion website: http://www.probability-and-conditional-expectation.de

Nonnegative step function

> **Definition 3.1 [Nonnegative step function and normal representation]**
> Let $(\Omega, \mathscr{A})$ be a measurable space. Then $f\colon \Omega \to \mathbb{R}$ is called a *nonnegative step function*, if there is a finite sequence $A_1, \ldots, A_n \in \mathscr{A}$ and a finite sequence $\alpha_1, \ldots, \alpha_n \in \mathbb{R}$, $\alpha_i \geq 0$, $i = 1, \ldots, n$, such that
>
> $$f = \sum_{i=1}^{n} \alpha_i \, 1_{A_i}. \qquad (3.1)$$
>
> If $A_1, \ldots, A_n \in \mathscr{A}$ are pairwise disjoint, then $f = \sum_{i=1}^{n} \alpha_i \, 1_{A_i}$ is called a *normal representation of* f.

Remark 3.2 [Step functions take on finitely many values]   A nonnegative step function $f = \sum_{i=1}^{n} \alpha_i \, 1_{A_i}$ is a measurable function $f\colon (\Omega, \mathscr{A}) \to (\mathbb{R}, \mathscr{B})$ taking on only a finite number of *nonnegative* values. These values are not necessarily $\alpha_1, \ldots, \alpha_n$. However, note:

(i) If $\mathscr{E} = \{A_1, \ldots, A_n\}$ is a partition of Ω, then $\alpha_1, \ldots, \alpha_n$ are the values of f.

(ii) If $A_1, \ldots, A_n$ are pairwise disjoint but $\mathscr{E}$ is not a partition of Ω, that is, $A_{n+1} := \Omega \setminus \bigcup_{i=1}^{n} A_i \neq \emptyset$, then

$$f = \sum_{i=1}^{n} \alpha_i \, 1_{A_i} + 0 \cdot 1_{A_{n+1}}.$$

This implies: $f(\omega) = 0$, for all $\omega \in A_{n+1}$.

(iii) If $A_1, \ldots, A_n$ are pairwise disjoint and additionally $\alpha_1, \ldots, \alpha_n$ are pairwise different and not 0, then $A_i = f^{-1}(\{\alpha_i\})$, $i = 1, \ldots, n$.

(iv) If $A_1, \ldots, A_n$ are pairwise disjoint, then, for all $\alpha_i \neq 0$,

$$f^{-1}(\{\alpha_i\}) = \bigcup_{j\colon \alpha_j = \alpha_i} A_j.$$

Hence in this case, the inverse image of the set $\{\alpha_i\}$ under f is the union of all sets A_j, $j \in \{1, \ldots, n\}$, for which $\alpha_j = \alpha_i$.   ◁

Remark 3.3 [Different representations of a nonnegative step function]   Note that nonnegative step functions can have different representations and also different normal representations (see Example 3.7).   ◁

Example 3.4 [Indicator function]   Let $(\Omega, \mathscr{A})$ be a measurable space and $A \in \mathscr{A}$. The indicator function 1_A, which has already been introduced in Example 2.12, can also be written as $1 \cdot 1_A + 0 \cdot 1_{A^c}$. Hence, because $A \in \mathscr{A}$ and 1 is a real number, 1_A is a nonnegative step function. Note that $1 \cdot 1_A$ is also a normal representation of a nonnegative step function.   ◁

Example 3.5 [Two nonnegative step functions]   In Example 2.55, we already presented two nonnegative step functions f and h and an illustrating figure (see Fig. 2.6). The representations of both functions are normal.   ◁

Example 3.6 [Tossing a dice]   Consider the set $\Omega = \{\omega_1, \ldots, \omega_6\}$ of possible outcomes of tossing a dice. Furthermore, let $\mathscr{A} = \mathscr{P}(\Omega)$ be the power set of Ω, and define $X\colon \Omega \to \mathbb{R}$ by

$$X(\omega_i) = i, \quad \forall\, \omega_i \in \Omega,$$

where i, and therefore $X(\omega_i)$ is the *number of dots*. Considering the elements $\{\omega_1\}, \ldots, \{\omega_6\}$ of $\mathscr{A}$, and

$$X = \sum_{i=1}^{6} i \cdot 1_{\{\omega_i\}}$$

shows that X has a normal representation of a nonnegative step function. (For a related example, see Exercise 3.1.)   ◁

Example 3.7 [Several representations of nonnegative step functions]   Consider the measurable space $(\mathbb{R}, \mathscr{B})$ and the nonnegative function $f\colon \mathbb{R} \to \mathbb{R}$ defined by

$$f(x) = \begin{cases} 2, & \text{if } x \in [0, 1[\\ 5, & \text{if } x \in [1, 2] \\ 4, & \text{if } x \in \,]2, 3] \\ 1, & \text{if } x \in \,]3, 4] \\ 0, & \text{otherwise.} \end{cases}$$

This function can also be represented by

$$\begin{aligned} f &= 2 \cdot 1_{[0,1[} + 5 \cdot 1_{[1,2]} + 4 \cdot 1_{]2,3]} + 1 \cdot 1_{]3,4]} \\ &= 2 \cdot 1_{[0,.5]} + 2 \cdot 1_{].5,1]} + 5 \cdot 1_{[1,2]} + 4 \cdot 1_{]2,3]} + 1 \cdot 1_{]3,4]} \\ &= 2 \cdot 1_{[0,2]} + 3 \cdot 1_{[1,3]} + 1 \cdot 1_{[2,4]} \\ &= 1 \cdot 1_{[0,4]} + 1 \cdot 1_{[0,3]} + 2 \cdot 1_{[1,3]} + 1 \cdot 1_{[1,2]}. \end{aligned} \tag{3.2}$$

The first two representations are normal, and the latter two are nonnormal representations of f.   ◁

Remark 3.8 [Existence of a normal representation]   For every nonnegative step function, there exists a normal representation (see Exercise 3.2).

If $f = \sum_{i=1}^{n} \alpha_i\, 1_{A_i}$ is a normal representation of a nonnegative step function, then there may be another sequence $C_1, \ldots, C_m$ of pairwise disjoint elements of $\mathscr{A}$ and another sequence $\gamma_1, \ldots, \gamma_m$ of nonnegative real numbers such that

$$f = \sum_{i=1}^{n} \alpha_i\, 1_{A_i} = \sum_{i=1}^{m} \gamma_i\, 1_{C_i}.$$

Both sum terms are normal representations. The first two representations of f in Equation (3.2) provide an example. ◁

Integral of a nonnegative step function

The following uniqueness property holds for two normal representations of a nonnegative step function:

Lemma 3.9 [A uniqueness property]
Let $(\Omega, \mathcal{A}, \mu)$ be a measure space. If $f: \Omega \to \mathbb{R}$ is a nonnegative step function and $f = \sum_{i=1}^{n} \alpha_i \, 1_{A_i} = \sum_{i=1}^{m} \gamma_i \, 1_{C_i}$ are two normal representations, then

$$\sum_{i=1}^{n} \alpha_i \, \mu(A_i) = \sum_{i=1}^{m} \gamma_i \, \mu(C_i). \tag{3.3}$$

For a proof, see Klenke (2013, Lemma 4.1). Note, by convention, $0 \cdot \infty = 0$.

According to this lemma, the number $\sum_{i=1}^{n} \alpha_i \, \mu(A_i)$ assigned to a nonnegative step function f does not depend on the specific normal representation of f (for an illustration, see Exercise 3.3). This property allows us to define the *integral* of a nonnegative step function with respect to a measure μ as follows:

Definition 3.10 [Integral of a nonnegative step function]
Let $(\Omega, \mathcal{A}, \mu)$ be a measure space and let $f = \sum_{i=1}^{n} \alpha_i \, 1_{A_i}$ be a normal representation of a nonnegative step function $f: \Omega \to \mathbb{R}$. Then the number

$$\int f \, d\mu = \sum_{i=1}^{n} \alpha_i \, \mu(A_i) \tag{3.4}$$

is called the integral of f (over Ω) with respect to μ.

Remark 3.11 [Integral of a constant]   Let $(\Omega, \mathcal{A}, \mu)$ be a measure space. If $f = \alpha, \alpha \in \mathbb{R}$, then Equation (3.4) immediately implies

$$\int \alpha \, d\mu = \alpha \cdot \mu(\Omega). \tag{3.5}$$
◁

Remark 3.12 [Integral over a subset of Ω]   Let $A \in \mathcal{A}$. If $f = \sum_{i=1}^{n} \alpha_i 1_{A_i}$ is a normal representation of a nonnegative step function, then the product $1_A \cdot f$ is a nonnegative step function as well and can be written as:

$$1_A \cdot f = \sum_{i=1}^{n} \alpha_i \, 1_{A \cap A_i}, \tag{3.6}$$

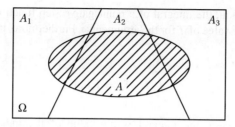

Figure 3.1 A partition and a subset of Ω.

which is a normal representation of $1_A \cdot f$ (see Fig. 3.1 and Exercise 3.4). Hence, we may also consider the integral $\int 1_A \cdot f \, d\mu$ and define the *integral of f over a subset A of Ω* by

$$\int_A f \, d\mu = \int 1_A \cdot f \, d\mu. \tag{3.7}$$

◁

The following corollary is a special case of Equation (3.7) if $f = \alpha$, $\alpha \in \mathbb{R}$. (For a proof, see Exercise 3.5.)

Corollary 3.13 [Constants]
Let $(\Omega, \mathcal{A}, \mu)$ be a measure space and $\alpha \in \mathbb{R}$. If $A \in \mathcal{A}$, then

$$\int_A \alpha \, d\mu = \alpha \, \mu(A). \tag{3.8}$$

Examples

Example 3.14 [Indicator function] Consider a measure space $(\Omega, \mathcal{A}, \mu)$ and the indicator 1_A of $A \in \mathcal{A}$. Then $1_A = 1 \cdot 1_A$ is a normal representation of 1_A. Therefore,

$$\int_A d\mu = \int 1_A \, d\mu = 1 \cdot \mu(A) = \mu(A). \tag{3.9}$$

◁

Example 3.15 [Nonnegative step function and Dirac measure] Let $(\Omega, \mathcal{A})$ be a measurable space, and for $\omega \in \Omega$ let δ_ω denote the *Dirac measure* at ω (see Example 1.52). Furthermore, consider a normal representation $f = \sum_{i=1}^{m} \alpha_i \, 1_{A_i}$ of a nonnegative step function. Its integral with respect to the Dirac measure is

$$\int f \, d\delta_\omega = \sum_{i=1}^{m} \alpha_i \, \delta_\omega(A_i) = \sum_{i=1}^{m} \alpha_i \, 1_{A_i}(\omega) = f(\omega). \tag{3.10}$$

According to this equation, the integral of a nonnegative step function f with respect to the Dirac measure at ω is the value of f for the argument ω. Furthermore, if $f = 1_A$ is the indicator of $A \in \mathcal{A}$, then

$$\int 1_A \, d\delta_\omega = 1_A(\omega). \tag{3.11}$$

Hence, in this special case, the integral is the value of the indicator 1_A for the argument ω. ◁

Example 3.16 [Nonnegative step function and counting measure] Suppose $\Omega = \{1, \dots, n\}$, $n \in \mathbb{N}$. For the measurable space $(\Omega, \mathcal{P}(\Omega))$, the counting measure $\mu_\#$ on the power set $\mathcal{P}(\Omega)$ is defined by

$$\mu_\#(A) = \sum_{\omega=1}^{n} 1_A(\omega), \quad \forall \, A \subset \Omega, \tag{3.12}$$

(see Example 1.54). Hence, $\mu_\#(A)$ is simply the number of elements (i.e., the cardinality of A). Now consider a nonnegative step function with normal representation $f = \sum_{i=1}^{m} \alpha_i \, 1_{A_i}$. According to Equations (3.4) and (3.12), its integral with respect to the counting measure is

$$\begin{aligned}
\int f \, d\mu_\# &= \sum_{i=1}^{m} \alpha_i \, \mu_\#(A_i) = \sum_{i=1}^{m} \alpha_i \sum_{\omega=1}^{n} 1_{A_i}(\omega) = \sum_{\omega=1}^{n} \sum_{i=1}^{m} \alpha_i 1_{A_i}(\omega) \\
&= \sum_{\omega=1}^{n} f(\omega).
\end{aligned} \tag{3.13}$$

Hence, the integral of a nonnegative step function f with respect to the counting measure $\mu_\#$ is the sum over all values of f (see also Exercise 3.6). Using Equations (1.41) and (3.10), this integral can also be written as:

$$\int f \, d\mu_\# = \int f \, d \left(\sum_{\omega=1}^{n} \delta_\omega \right) = \sum_{\omega=1}^{n} \int f \, d\delta_\omega. \tag{3.14}$$

◁

3.1.2 Integral of a nonnegative measurable function

In this section, we extend the concept of an integral to nonnegative measurable functions. Before introducing the definition, we consider a theorem according to which every nonnegative measurable function can be represented as a limit of an increasing sequence of nonnegative step functions. We begin with an example.

Example 3.17 [Increasing sequence of nonnegative step functions] Consider the measurable space $(\mathbb{R}, \mathcal{B})$ and the function $f \colon \mathbb{R} \to \mathbb{R}$ defined by

$$f(x) = \begin{cases} 1 - x^2, & \forall \, x \in [0, 1] \\ 0, & \text{otherwise.} \end{cases} \tag{3.15}$$

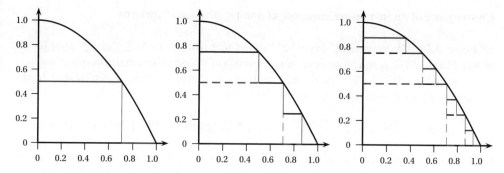

Figure 3.2   Increasing nonnegative step functions.

Now we construct three functions $f_i: \mathbb{R} \to \mathbb{R}$, $i = 1, 2, 3$, with $f_1 \leq f_2 \leq f_3 \leq f$ that approximate f (see Fig. 3.2). Let us start with f_1 defined by

$$f_1(x) = \begin{cases} \alpha_1 = .50, & \text{if } x \in A_1 := [0, (1 - .50)^{1/2}] \\ \alpha_2 = 0, & \text{if } x \in A_1^c, \end{cases}$$

where $[0, (1 - .50)^{1/2}]$ denotes the closed interval between 0 and $(1 - .50)^{1/2} \approx .707$. Because A_1 is an element of $\mathscr{B}$ and .50 is a nonnegative real number, $f_1 = \alpha_1 \, 1_{A_1}$ is a nonnegative step function. Next consider f_2 defined by

$$f_2(x) = \begin{cases} \beta_1 = .75, & \text{if } x \in B_1 := [0, (1 - .75)^{1/2}] \\ \beta_2 = .50, & \text{if } x \in B_2 :=]\,(1 - .75)^{1/2}, (1 - .50)^{1/2}] \\ \beta_3 = .25, & \text{if } x \in B_3 :=]\,(1 - .50)^{1/2}, (1 - .25)^{1/2}] \\ \beta_4 = 0, & \text{if } x \in (B_1 \cup B_2 \cup B_3)^c. \end{cases}$$

Because B_1, B_2, B_3 are elements of $\mathscr{B}$ and .75, .50, .25 are nonnegative real numbers, $f_2 = \sum_{i=1}^{3} \beta_i \, 1_{B_i}$ is a nonnegative step function. Finally, let f_3 be defined by

$$f_3(x) = \begin{cases} \gamma_1 = .875, & \text{if } x \in C_1 := [0, (1 - .875)^{1/2}] \\ \gamma_2 = .750, & \text{if } x \in C_2 :=]\,(1 - .875)^{1/2}, (1 - .750)^{1/2}] \\ \gamma_3 = .625, & \text{if } x \in C_3 :=]\,(1 - .750)^{1/2}, (1 - .625)^{1/2}] \\ \gamma_4 = .500, & \text{if } x \in C_4 :=]\,(1 - .625)^{1/2}, (1 - .500)^{1/2}] \\ \gamma_5 = .375, & \text{if } x \in C_5 :=]\,(1 - .500)^{1/2}, (1 - .375)^{1/2}] \\ \gamma_6 = .250, & \text{if } x \in C_6 :=]\,(1 - .375)^{1/2}, (1 - .250)^{1/2}] \\ \gamma_7 = .125, & \text{if } x \in C_7 :=]\,(1 - .250)^{1/2}, (1 - .125)^{1/2}] \\ \gamma_8 = 0, & \text{if } x \in (C_1 \cup \ldots \cup C_7)^c. \end{cases}$$

Again, $C_1, \ldots, C_7$ is a sequence of elements of $\mathscr{B}$ and .875, .750, .625, .500, .375, .250, .125 is a sequence of nonnegative real numbers. Therefore, $f_3 = \sum_{i=1}^{7} \gamma_i \, 1_{C_i}$ is a nonnegative step function. The integrals of the functions f_1 and f_2 are computed in Exercise 3.7.   ◁

Convergence of an increasing sequence of nonnegative step functions

Example 3.18 [Convergence] Figure 3.2 shows that $f_1(\omega) \leq f_2(\omega) \leq f_3(\omega) \leq f(\omega)$ for all $\omega \in \Omega$. Hence, f_1, f_2, f_3 is a *finite increasing sequence* of nonnegative step functions. The interval $[0, 1]$ on the vertical axis is partitioned, and these partitions are refined step by step. In our example, we started with the partition $\{[0, .50[, [.50, 1]\}$. Then we partitioned

$$[0, .50[\text{ to } \{[0, .25[, [.25, .50[\} \quad \text{and} \quad [.50, 1] \text{ to } \{[.50, .75[, [.75, 1]\}, \quad \text{etc.}$$

Following this idea, we can define functions $f_4, f_5, \ldots$ such that $f_1, f_2, \ldots$ is an infinite sequence of nonnegative step functions with $f_1(\omega) \leq f_2(\omega) \leq \ldots \leq f(\omega)$, for all $\omega \in \Omega$, and $\lim_{n \to \infty} f_n(\omega) = f(\omega)$, for all $\omega \in \Omega$. According to Theorem 3.19, this holds for *all* nonnegative measurable functions $f: (\Omega, \mathcal{A}) \to (\overline{\mathbb{R}}, \overline{\mathcal{B}})$. ◁

Theorem 3.19 [Approximation of nonnegative functions]
Let $(\Omega, \mathcal{A})$ be a measurable space and $f: (\Omega, \mathcal{A}) \to (\overline{\mathbb{R}}, \overline{\mathcal{B}})$ a nonnegative measurable function. Then,

(i) *There is a sequence $f_1, f_2, \ldots$ of nonnegative step functions such that*

$$f_1(\omega) \leq f_2(\omega) \leq \ldots, \quad \forall \, \omega \in \Omega \tag{3.16}$$

and

$$\lim_{n \to \infty} f_n(\omega) = f(\omega), \quad \forall \, \omega \in \Omega. \tag{3.17}$$

(ii) *There is a sequence of sets $A_1, A_2, \ldots \in \mathcal{A}$ and a sequence of nonnegative real numbers $\alpha_1, \alpha_2, \ldots$ such that*

$$f = \sum_{i=1}^{\infty} \alpha_i \, 1_{A_i}. \tag{3.18}$$

For a proof, see Klenke (2013, Theorem 1.96).

Remark 3.20 [Infinite sums] Equation (3.18) can be visualized by Figure 3.2. The function f_3 on the right-hand side of this figure is already close to f. Partitioning the intervals *on the vertical axis* again and again leads to better approximations of f. Note that the horizontal axis does not have to be a subset of $\mathbb{R}$; instead, it can be any nonempty set Ω.

Remember, the right-hand side of Equation (3.18) is just a symbol for the corresponding limit, that is,

$$\sum_{i=1}^{\infty} \alpha_i \, 1_{A_i} = \lim_{n \to \infty} \sum_{i=1}^{n} \alpha_i \, 1_{A_i}. \tag{3.19}$$

Note that, for $\alpha_i \geq 0$, this limit always exists. ◁

Before turning to the definition of the integral of a nonnegative measurable function let us use the properties (3.16) and (3.17) to define the concepts *increasing sequence of nonnegative step functions* and *pointwise convergence*.

> **Definition 3.21 [Increasing sequence of nonnegative step functions]**
> *Let $(\Omega, \mathcal{A})$ be a measurable space and $f\colon (\Omega, \mathcal{A}) \to (\overline{\mathbb{R}}, \overline{\mathcal{B}})$ a nonnegative measurable function. A sequence $f_1, f_2, \ldots$ of nonnegative step functions satisfying (3.16) is called increasing. If it also satisfies (3.17), then we say that $f_1, f_2, \ldots$ converges pointwise to f and denote it by $f_n \uparrow f$.*

Uniqueness of the limits of an integral

In Theorem 3.19 we have seen that every nonnegative measurable function f can be represented by the limit $\lim_{n \to \infty} f_n$ of an increasing sequence $f_1, f_2, \ldots$ of nonnegative step functions, that is,

$$f = \lim_{n \to \infty} f_n. \tag{3.20}$$

The definition of the integral of nonnegative step functions implies that the *integrals* of the functions f_n are increasing as well, that is,

$$f_n \leq f_{n+1} \quad \Rightarrow \quad \int f_n \, d\mu \leq \int f_{n+1} \, d\mu, \quad \forall n \in \mathbb{N}$$

[see Bauer, 2001, proposition (10.7)]. Hence, the sequence of the integrals either converges to a (finite) real number or diverges to $+\infty$.

In Figure 3.2, we presented the first three nonnegative step functions $f_1, f_2,$ and f_3 of such an increasing sequence $f_1, f_2, \ldots$ that approximates the function $f\colon \mathbb{R} \to \mathbb{R}$ defined by Equation (3.15). Figure 3.3 visualizes the convergence of the integrals $\int f_n \, d\lambda$ with respect to the Lebesgue measure λ on $(\mathbb{R}, \mathcal{B})$ (see the shaded areas in Fig. 3.3).

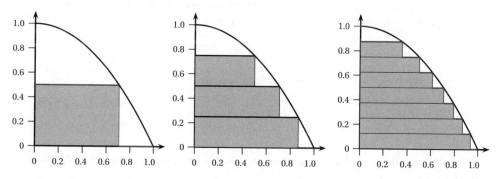

Figure 3.3   Integral of nonnegative step functions with respect to the Lebesgue measure.

It should be noted, however, that there is not only one single increasing sequence of non-negative step functions whose limit is f. This is illustrated in the following example.

Example 3.22 [Uniqueness] For an example, construct a sequence $g_1, g_2, \ldots$ analogously to the sequence $f_1, f_2, \ldots$ in Example 3.17, using other partitions of the interval $[0, 1]$ on the vertical axis, such as

$$\{[0, .40[, [.40, 1]\} \quad \text{and} \quad \{[0, .20[, [.20, .40[, [.40, .80[, [.80, 1]\}, \quad \text{and so on.}$$

Then $g_1, g_2, \ldots$ is a second increasing sequence that also approximates f. Figure 3.3 suggests that the specific choice of an increasing sequence of nonnegative step functions is irrelevant for the limit of their integrals. And in fact, according to the following theorem, this applies not only to our example and to the Lebesgue measure λ on $\mathcal{B}$, but to *any* nonnegative measurable function and *any* measure μ.   ◁

Theorem 3.23 [Uniqueness of the limits of integrals]
If $f_1, f_2, \ldots$ and $g_1, g_2, \ldots$ are two increasing sequences of nonnegative step functions $f_n, g_n: (\Omega, \mathcal{A}, \mu) \to (\overline{\mathbb{R}}, \overline{\mathcal{B}})$, then $\lim_{n \to \infty} f_n = \lim_{n \to \infty} g_n$ implies

$$\lim_{n \to \infty} \int f_n \, d\mu = \lim_{n \to \infty} \int g_n \, d\mu. \tag{3.21}$$

For a proof, see Bauer (2001, Corollary 11.2).

According to this theorem, if we consider two increasing sequences of nonnegative step functions with identical limits, then we know that the limits of their integrals are identical.

Definition of the integral of a nonnegative measurable function

Based on the result of Theorem 3.23, we define the integral of any nonnegative measurable function.

Definition 3.24 [Integral of a nonnegative measurable function]
Assume that $f: (\Omega, \mathcal{A}, \mu) \to (\overline{\mathbb{R}}, \overline{\mathcal{B}})$ is a nonnegative measurable function and let $f = \lim_{n \to \infty} f_n$ be a representation of f as the limit of an increasing sequence $f_1, f_2, \ldots$ of nonnegative step functions. Then

$$\int f \, d\mu := \lim_{n \to \infty} \int f_n \, d\mu \tag{3.22}$$

is called the integral of f (over Ω) with respect to μ.

Note that the integral of a nonnegative measurable function is either a nonnegative real number or $+\infty$.

Example 3.25 [Integral with respect to a Dirac measure]   Suppose the assumptions of Definition 3.24 hold. Then, for $\omega \in \Omega$,

$$
\int f \, d\delta_\omega = \lim_{n \to \infty} \int f_n \, d\delta_\omega \qquad [(3.22)]
$$
$$
= \lim_{n \to \infty} f_n(\omega) \qquad [(3.10)] \qquad\qquad (3.23)
$$
$$
= f(\omega). \qquad [(3.20)]
$$

Hence, the integral of a nonnegative measurable function f with respect to the Dirac measure at ω is the value of f for ω.   ◁

We conclude this section by the following lemma on monotonicity of the integrals of nonnegative measurable functions.

Lemma 3.26 [Monotonicity]
If $f, g: (\Omega, \mathscr{A}) \to (\overline{\mathbb{R}}, \overline{\mathscr{B}})$ are nonnegative and measurable, then

$$
f \leq g \quad \Rightarrow \quad \int f \, d\mu \leq \int g \, d\mu. \qquad (3.24)
$$

For a proof, see Bauer [2001, Eq. (11.8)].

Remark 3.27 [Bounds of the Integral of a bounded function]   Let $f: (\Omega, \mathscr{A}) \to (\overline{\mathbb{R}}, \overline{\mathscr{B}})$ be nonnegative and measurable and $\alpha \in \mathbb{R}$. Then, for $g = \alpha$, Equations (3.24) and (3.5) imply

$$
f \leq \alpha \quad \Rightarrow \quad \int f \, d\mu \leq \alpha \cdot \mu(\Omega), \qquad (3.25)
$$

and

$$
f \geq \alpha \quad \Rightarrow \quad \int f \, d\mu \geq \alpha \cdot \mu(\Omega). \qquad (3.26)
$$

◁

3.1.3   Integral of a measurable function

Now we define the integral of a measurable function $f: (\Omega, \mathscr{A}) \to (\overline{\mathbb{R}}, \overline{\mathscr{B}})$ using the positive part f^+ and the negative part f^- of f that have been introduced in section 2.4.2. According to Theorem 2.66, f^+ and f^- are both nonnegative measurable functions. Reading the following definition, remember the conventions: $\infty + \infty = \infty$, $-\infty - \infty = -\infty$, $x + \infty = \infty$, for all $x \in \mathbb{R}$, $x - \infty = -\infty$, for all $x \in \mathbb{R}$. Also note that $\infty - \infty$ is *not defined*, which has to be observed whenever integrals are not necessarily finite.

Definition 3.28 [Integral of a measurable function]
Let $f: (\Omega, \mathcal{A}, \mu) \to (\overline{\mathbb{R}}, \overline{\mathcal{B}})$ be a measurable function. If $\int f^+ \, d\mu$ or $\int f^- \, d\mu$ are finite, then f is called quasi-integrable with respect to μ, or simply quasi-μ-integrable, and

$$\int f \, d\mu = \int f^+ \, d\mu - \int f^- \, d\mu \tag{3.27}$$

is called the integral of f (over Ω) with respect to μ. If $\int f^+ \, d\mu$ and $\int f^- \, d\mu$ are both finite, then f is called integrable with respect to μ, or simply μ-integrable.

Remark 3.29 [Integrability and quasi-integrability]   Of course, every integrable measurable function is quasi-integrable and each nonnegative function is also quasi-integrable. Furthermore, assuming that a function $f: (\Omega, \mathcal{A}, \mu) \to (\overline{\mathbb{R}}, \overline{\mathcal{B}})$ is integrable or quasi-integrable includes the assumption that f is measurable. Finally, if f is μ-integrable, then

$$-\infty < \int f \, d\mu < +\infty,$$

that is, the integral is *finite*, taking a value in $\mathbb{R}$. If f is quasi-μ-integrable, then the integral may also be *infinite* (i.e., it may also take on the values $+\infty$ or $-\infty$).   ◁

Remark 3.30 [A standard method for proofs]   The integral of a quasi-integrable function has been defined in three steps, for *nonnegative measurable step functions*, for *nonnegative measurable functions*, and for *quasi-integrable functions*. Oftentimes, these steps are also followed in proofs of propositions involving integrals. That is, in a first step, it is shown that the proposition holds for nonnegative measurable step functions. In a second step, using Equation (3.22), it is proven for nonnegative measurable functions, and finally, Equation (3.27) is applied to complete the proof for all quasi-integrable functions. An example is the proof of Theorem 3.36. Oftentimes, we only detail the first step, in particular if the remaining two steps are straightforward.   ◁

Example 3.31 [Integral with respect to the Lebesgue measure λ]   Figure 3.4 displays the integral of a function $f: (\mathbb{R}, \mathcal{B}, \lambda) \to (\overline{\mathbb{R}}, \overline{\mathcal{B}})$ with respect to the Lebesgue measure λ. Because f^+ and f^- are both nonnegative (see Rem. 2.62), the integrals $\int f^+ \, d\mu$ and $\int f^- \, d\mu$ are positive and identical to the areas marked $+$ and $-$ in Figure 3.4. According to Equation (3.27), the integral of f is the difference between the area $\int f^+ \, d\mu$ and the area $\int f^- \, d\mu$.   ◁

Remark 3.32 [An alternative notation]   An alternative notation for the integral of f is

$$\int f \, d\mu = \int f(\omega) \, \mu(d\omega) = \int_\Omega f(\omega) \, \mu(d\omega), \tag{3.28}$$

which explicitly uses the values $f(\omega)$ of f. This notation conveys the idea that the values $f(\omega)$ of f are weighted by the measure of $d\omega$. If $\Omega = \mathbb{R}$, then $d\omega$ symbolizes the length of an infinitesimal interval between two elements in $\mathbb{R}$. If Ω is finite or countable, then $\mu(d\omega)$ symbolizes the value of μ for the singleton $\{\omega\}$, and the integral can be written as a sum (see Example 3.16).   ◁

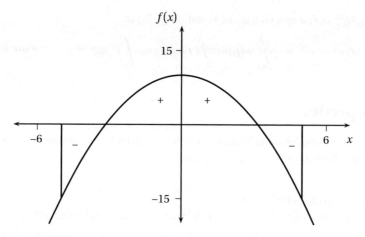

Figure 3.4   Lebesgue integral of a function from −5 to 5.

Lemma 3.33 [Integrability carries over to restrictions of functions]

   (i) *If* $f\colon (\Omega, \mathcal{A}, \mu) \to (\overline{\mathbb{R}}, \overline{\mathcal{B}})$ *is quasi-μ-integrable and $A \in \mathcal{A}$, then $1_A f$ is quasi-μ-integrable.*

   (ii) *If* $f\colon (\Omega, \mathcal{A}, \mu) \to (\overline{\mathbb{R}}, \overline{\mathcal{B}})$ *is μ-integrable and $A \in \mathcal{A}$, then $1_A f$ is μ-integrable.*

                                                   (Proof p. 115)

Remark 3.34 [Integral of $1_A f$]   Lemma 3.33 (ii) means: If $f\colon (\Omega, \mathcal{A}, \mu) \to (\overline{\mathbb{R}}, \overline{\mathcal{B}})$ is quasi-μ-integrable and $A \in \mathcal{A}$, then

$$\int f \, d\mu \text{ is finite} \quad \Rightarrow \quad \int 1_A f \, d\mu \text{ is finite.} \tag{3.29}$$

                                            ◁

If $f\colon (\Omega, \mathcal{A}, \mu) \to (\overline{\mathbb{R}}, \overline{\mathcal{B}})$ is quasi-μ-integrable and $A \in \mathcal{A}$, then Lemma 3.33 implies that the integral $\int 1_A f \, d\mu$ is well-defined. Hence, we can now introduce the integral of f over a subset A of Ω as follows:

Definition 3.35 [Integral over a subset A of Ω]
If $f\colon (\Omega, \mathcal{A}, \mu) \to (\overline{\mathbb{R}}, \overline{\mathcal{B}})$ *is quasi-μ-integrable and $A \in \mathcal{A}$, then*

$$\int_A f \, d\mu := \int 1_A f \, d\mu \tag{3.30}$$

is called the integral of f over A with respect to μ.

Because $1_\Omega f = f$, a special case of Equation (3.30) is

$$\int_\Omega f \, d\mu = \int 1_\Omega f \, d\mu = \int f \, d\mu. \tag{3.31}$$

3.2 Properties

In this section, we consider some important properties and rules of computation for the integral of a measurable function $f: (\Omega, \mathcal{A}, \mu) \to (\overline{\mathbb{R}}, \overline{\mathcal{B}})$.

Theorem 3.36 [Linearity]
Consider the functions $f: (\Omega, \mathcal{A}, \mu) \to (\overline{\mathbb{R}}, \overline{\mathcal{B}})$ and $g: (\Omega, \mathcal{A}, \mu) \to (\mathbb{R}, \mathcal{B})$.

(i) If f is quasi-μ-integrable and $\alpha \in \mathbb{R}$, then αf is quasi-μ-integrable and

$$\int \alpha f \, d\mu = \alpha \int f \, d\mu. \tag{3.32}$$

(ii) If f is quasi-μ-integrable and g is μ-integrable, then $f + g$ is quasi-μ-integrable, and

$$\int (f + g) \, d\mu = \int f \, d\mu + \int g \, d\mu. \tag{3.33}$$

(Proof p. 115)

Combining propositions (i) and (ii) of Theorem 3.36 immediately yields Corollary 3.37.

Corollary 3.37 [Linearity]
Consider the functions $f: (\Omega, \mathcal{A}, \mu) \to (\overline{\mathbb{R}}, \overline{\mathcal{B}})$, $g: (\Omega, \mathcal{A}, \mu) \to (\mathbb{R}, \mathcal{B})$ and let $\alpha, \beta \in \mathbb{R}$. If f is quasi-μ-integrable and g is μ-integrable, then $\alpha f + \beta g$ is quasi-μ-integrable, and

$$\int (\alpha f + \beta g) \, d\mu = \alpha \int f \, d\mu + \beta \int g \, d\mu. \tag{3.34}$$

Linearity can also be used to prove the following corollary on the equivalence of integrability of a measurable function f and finiteness of the integral of the absolute value function $|f|$.

Corollary 3.38 [Integrability and absolute value function]
The function $f: (\Omega, \mathcal{A}, \mu) \to (\overline{\mathbb{R}}, \overline{\mathcal{B}})$ is μ-integrable if and only if

$$\int |f| \, d\mu < \infty.$$

(Proof p. 119)

Example 3.39 [Integral over the union of two sets] If $f : (\Omega, \mathscr{A}, \mu) \to (\overline{\mathbb{R}}, \overline{\mathscr{B}})$ is μ-integrable and $A, B \in \mathscr{A}$, then $1_{A \cup B} f$ is μ-integrable and

$$\int_{A \cup B} f \, d\mu = \int 1_{A \cup B} f \, d\mu = \int_A f \, d\mu + \int_B f \, d\mu - \int_{A \cap B} f \, d\mu. \qquad (3.35)$$

If $A \cap B = \varnothing$ and f is quasi-μ-integrable, then

$$\int_{A \cup B} f \, d\mu = \int 1_{A \cup B} f \, d\mu = \int_A f \, d\mu + \int_B f \, d\mu. \qquad (3.36)$$

(see Exercise 3.8). ◁

Lemma 3.40 [Measures that are identical on a sub-σ-algebra]
Assume that $f : (\Omega, \mathscr{A}, \mu) \to (\overline{\mathbb{R}}, \overline{\mathscr{B}})$ is nonnegative or μ-integrable. Furthermore, let $\mathscr{C} \subset \mathscr{A}$ be a σ-algebra, let f be $\mathscr{C}$-measurable, and suppose that $\nu(A) = \mu(A)$, for all $A \in \mathscr{C}$. Then, $(\Omega, \mathscr{C}, \nu)$ is a measure space and

$$\int f \, d\nu = \int f \, d\mu. \qquad (3.37)$$

(Proof p. 119)

Hence, the integral $\int f \, d\mu$ only depends on the values of μ on the σ-algebra $\sigma(f)$, the σ-algebra generated by f.

Lemma 3.41 [Integrable functions are μ-almost everywhere real-valued]
Let $f : (\Omega, \mathscr{A}, \mu) \to (\overline{\mathbb{R}}, \overline{\mathscr{B}})$ be measurable. If f is μ-integrable, then f is real-valued μ-almost everywhere. If f is quasi-μ-integrable, then

$$\mu(\{\omega \in \Omega : f(\omega) = \infty\}) > 0 \quad \Rightarrow \quad \int f \, d\mu = \infty, \qquad (3.38)$$

$$\mu(\{\omega \in \Omega : f(\omega) = -\infty\}) > 0 \quad \Rightarrow \quad \int f \, d\mu = -\infty. \qquad (3.39)$$

(Proof p. 120)

Remark 3.42 [Integrable functions are assumed to be real-valued] Contraposition of (3.38) and (3.39) yields: If $\int f \, d\mu$ is finite, then $f(\omega) \in \mathbb{R}$ (i.e., $-\infty < f(\omega) < \infty$), for μ-almost all $\omega \in \Omega$ (see Def. 2.68 and Remark 2.70). In this case, there is a *real-valued* measurable function $f^* : (\Omega, \mathscr{A}, \mu) \to (\mathbb{R}, \mathscr{B})$ with $f^* \underset{\mu}{=} f$. (For example, define $f^* := 1_A \cdot f + 1_{A^c} \cdot 0 = 1_A \cdot f$ for $A := \{\omega \in \Omega : f(\omega) \in \mathbb{R}\}$.) Therefore, without substantial loss of generality, for simplicity, we often assume that a function is real-valued if it has a finite integral. ◁

3.2.1   Integral of μ-equivalent functions

The concept of equivalence of two measurable functions with respect to a measure has already been introduced in section 2.5. Now we treat the relationship of this concept to the integrals of two numerical functions, that is functions with codomain $\overline{\mathbb{R}}$.

Theorem 3.43 [A condition equivalent to $f \underset{\mu}{=} 0$]

If $f: (\Omega, \mathcal{A}, \mu) \to (\overline{\mathbb{R}}, \overline{\mathcal{B}})$ is a nonnegative measurable function, then

$$\int f \, d\mu = 0 \quad \Leftrightarrow \quad f \underset{\mu}{=} 0. \tag{3.40}$$

For a proof, see Bauer (2001, Theorem 13.2).

Lemma 3.44 [Integral of a positive function]

If $f: (\Omega, \mathcal{A}, \mu) \to (\overline{\mathbb{R}}, \overline{\mathcal{B}})$ is quasi-μ-integrable, and there is an $A \in \mathcal{A}$ such that $\mu(A) > 0$ and $f(\omega) > 0$, for all $\omega \in A$, then

$$\int 1_A \cdot f \, d\mu > 0. \tag{3.41}$$

(Proof p. 120)

If $(\Omega, \mathcal{A}, \mu)$ is a measure space, then a set $A \in \mathcal{A}$ with $\mu(A) = 0$ is called a *null set* with respect to μ. In Lemma 3.45, we consider the integral over such a null set (see Exercise 3.9).

Lemma 3.45 [Integral over a null set]

Let $f: (\Omega, \mathcal{A}, \mu) \to (\overline{\mathbb{R}}, \overline{\mathcal{B}})$ be measurable. If $A \in \mathcal{A}$, with $\mu(A) = 0$, then $1_A \cdot f$ is μ-integrable and

$$\int_A f \, d\mu = \int 1_A \cdot f \, d\mu = 0. \tag{3.42}$$

(Proof p. 120)

Remark 3.46 [Integration over null sets can be neglected]   The conjunction of Equations (3.36) and (3.42) implies: If f is quasi-μ-integrable and $A \in \mathcal{A}$ with $\mu(A) = 0$, then

$$\int f \, d\mu = \int_\Omega f \, d\mu = \int_{\Omega \setminus A} f \, d\mu + \int_A f \, d\mu = \int_{\Omega \setminus A} f \, d\mu. \tag{3.43}$$

◁

Lemma 3.47 [Integrals of μ-equivalent functions]
Let $f, g: (\Omega, \mathscr{A}, \mu) \to (\overline{\mathbb{R}}, \overline{\mathscr{B}})$ be quasi-μ-integrable. Then

$$f \underset{\mu}{=} g \quad \Rightarrow \quad \int f \, d\mu = \int g \, d\mu. \tag{3.44}$$

(Proof p. 121)

Theorem 3.48, presents a condition that is necessary and sufficient for μ-equivalence of f and g.

Theorem 3.48 [Identity of integrals of μ-equivalent functions]
If $f, g: (\Omega, \mathscr{A}, \mu) \to (\overline{\mathbb{R}}, \overline{\mathscr{B}})$ are μ-integrable, then

$$f \underset{\mu}{=} g \quad \Leftrightarrow \quad \int_A f \, d\mu = \int_A g \, d\mu, \quad \forall A \in \mathscr{A}. \tag{3.45}$$

(Proof p. 121)

In section 3.4, we shall see that, if f and g are μ-integrable and nonnegative, then it is sufficient to consider the integrals over all sets A in a $\cap$-stable generating system of $\mathscr{A}$ in order to show μ-equivalence of f and g (see Th. 3.68).

Remark 3.49 [A counter-example]   Note that Equation (3.45) does not hold if f, g are non-negative but not μ-integrable measurable functions. This is exemplified as follows: Consider $f, g: (\mathbb{R}, \mathscr{A}, \mu) \to (\overline{\mathbb{R}}, \overline{\mathscr{B}})$, where $\mathscr{A} = \{\mathbb{R}, \varnothing\}$, $\mu(\mathbb{R}) = \infty$, $f = 1$, and $g = 2$. Then $\int_{\mathbb{R}} f \, d\mu = \int_{\mathbb{R}} g \, d\mu = \infty$ and $\int_\varnothing f \, d\mu = \int_\varnothing g \, d\mu = 0$. Hence, $\int_A f \, d\mu = \int_A g \, d\mu$, for all $A \in \mathscr{A}$, but f and g are *not equivalent* with respect to μ.   ◁

Remark 3.50 [Some special cases]   Theorem 3.48 implies: If $f: (\Omega, \mathscr{A}, \mu) \to (\overline{\mathbb{R}}, \overline{\mathscr{B}})$ is a measurable function, then $f \underset{\mu}{=} \alpha$, $\alpha \in \mathbb{R}$, is equivalent to

$$\int_A f \, d\mu = \int_A \alpha \, d\mu = \int 1_A \, \alpha \, d\mu = \alpha \cdot \int 1_A \, d\mu = \alpha \cdot \mu(A), \quad \forall A \in \mathscr{A}. \tag{3.46}$$

Furthermore, $f \underset{\mu}{=} 0$ is equivalent to

$$\int_A f \, d\mu = 0, \quad \forall A \in \mathscr{A}, \tag{3.47}$$

using the convention $0 \cdot \infty = 0$, if necessary.
An immediate implication of Equation (3.46) for $A = \Omega$ is

$$f \underset{\mu}{=} \alpha, \quad \alpha \in \mathbb{R} \quad \Rightarrow \quad \int f \, d\mu = \alpha \, \mu(\Omega). \tag{3.48}$$

For $\alpha = 0$, this yields

$$f \underset{\mu}{=} 0 \quad \Rightarrow \quad \int f \, d\mu = 0. \tag{3.49}$$

$\triangleleft$

Remark 3.51 [Almost everywhere]   The notion of μ-equivalence of f and g is an example of a property that holds for all $\omega \in \Omega \setminus A$ with $\mu(A) = 0$. We also say that such a property holds *μ-almost everywhere* (μ-a.e.) (see Rem. 2.70). Another example is the property

$$f(\omega) \leq g(\omega), \quad \forall \, \omega \in \Omega \setminus A \text{ and } \mu(A) = 0,$$

which is denoted by $f \underset{\mu}{\leq} g$. The proposition of Lemma 3.47 analogously holds for the relations $\underset{\mu}{\leq}$ and $\underset{\mu}{<}$. $\triangleleft$

The following theorem generalizes Lemma 3.26.

Theorem 3.52 [Monotonicity]
Let $f, g: (\Omega, \mathcal{A}, \mu) \to (\overline{\mathbb{R}}, \overline{\mathcal{B}})$ be measurable functions.

(i) If f and g are quasi-μ-integrable, then

$$f \underset{\mu}{\leq} g \quad \Rightarrow \quad \int f \, d\mu \leq \int g \, d\mu. \qquad \text{(monotonicity)} \tag{3.50}$$

(ii) If $\mu(\Omega) > 0$ and f, g are μ-integrable, then

$$f \underset{\mu}{<} g \quad \Rightarrow \quad \int f \, d\mu < \int g \, d\mu. \qquad \text{(strict monotonicity)} \tag{3.51}$$

(Proof p. 122)

3.2.2   Integral with respect to a weighted sum of measures

In Example 1.62, we already noted that a weighted sum of measures with nonnegative weights is again a measure. As a special case, if μ is a measure on $(\Omega, \mathcal{A})$ and α is a nonnegative number, then $\alpha \cdot \mu$ is a measure on $(\Omega, \mathcal{A})$ as well. Furthermore, if f is μ-integrable, then

$$\int f \, d(\alpha \, \mu) = \int \alpha f \, d\mu = \alpha \int f \, d\mu \tag{3.52}$$

(see Exercise 3.10). This is generalized in the following theorem.

Theorem 3.53 [Integral with respect to a weighted sum of measures]
If $f\colon (\Omega, \mathcal{A}) \to (\overline{\mathbb{R}}, \overline{\mathcal{B}})$ is measurable and nonnegative, $\mu_1, \mu_2, \ldots$ are measures on $(\Omega, \mathcal{A})$, and $\alpha_1, \alpha_2, \ldots \in \mathbb{R}$ are nonnegative, then

$$\int f \, d\left(\sum_{i=1}^{\infty} \alpha_i \, \mu_i\right) = \sum_{i=1}^{\infty} \alpha_i \int f \, d\mu_i. \tag{3.53}$$

For a proof, see Equation (3.52) and Bauer (2001, Example 3).

If we consider a *finite* weighted sum of measures, the assumption that f is nonnegative can be replaced by integrability of f. In Theorem 3.54, we consider a weighted sum of two measures. In Remark 3.55, we extend the result to a finite weighted sum of measures.

Theorem 3.54 [Integral with respect to a weighted sum of two measures]
Let μ_1, μ_2 be measures on $(\Omega, \mathcal{A})$. If $f\colon (\Omega, \mathcal{A}) \to (\overline{\mathbb{R}}, \overline{\mathcal{B}})$ is integrable with respect to μ_1 and μ_2, and $0 \leq \alpha_1, \alpha_2 \in \mathbb{R}$, then f is integrable with respect to $\alpha_1 \, \mu_1 + \alpha_2 \, \mu_2$, and

$$\int f \, d(\alpha_1 \, \mu_1 + \alpha_2 \, \mu_2) = \alpha_1 \int f \, d\mu_1 + \alpha_2 \int f \, d\mu_2. \tag{3.54}$$

For a proof, see Equation (3.52) and Bauer (2001, Example 5).

Remark 3.55 [Integral with respect to a finite weighted sum of measures]   By induction, Theorem 3.54 yields, for nonnegative $\alpha_1, \ldots, \alpha_n \in \mathbb{R}$,

$$\int f \, d\left(\sum_{i=1}^{n} \alpha_i \, \mu_i\right) = \sum_{i=1}^{n} \alpha_i \int f \, d\mu_i, \tag{3.55}$$

provided that f is integrable with respect to all measures $\mu_1, \ldots, \mu_n$.   ◁

Example 3.56 [Integral with respect to the weighted sum of Dirac measures]   Let $(\Omega, \mathcal{A})$ be a measurable space and, for $i \in \mathbb{N}$, let $\omega_i \in \Omega$, $\alpha_i \in \mathbb{R}$, $\alpha_i \geq 0$, and δ_{ω_i} denote the Dirac measure at ω_i. Then

$$\mu = \sum_{i=1}^{\infty} \alpha_i \, \delta_{\omega_i} \tag{3.56}$$

defined by $\mu(A) = \sum_{i=1}^{\infty} \alpha_i \delta_{\omega_i}(A)$, for all $A \in \mathcal{A}$, is a measure on $(\Omega, \mathcal{A})$ (see Example 1.62). For any nonnegative measurable function $f: \Omega \to \overline{\mathbb{R}}$, we obtain

$$\int f \, d\mu = \int f \, d\left(\sum_{i=1}^{\infty} \alpha_i \delta_{\omega_i}\right) \qquad [(3.56)]$$

$$= \sum_{i=1}^{\infty} \alpha_i \int f \, d\delta_{\omega_i} \qquad [(3.53)] \qquad (3.57)$$

$$= \sum_{i=1}^{\infty} \alpha_i f(\omega_i). \qquad [(3.23)]$$

For $\mu(A) = \sum_{i=1}^{n} \alpha_i \delta_{\omega_i}(A)$, for all $A \in \mathcal{A}$, Equation (3.57) with $\alpha_i = 0$ for $i > n$ yields

$$\int f \, d\mu = \sum_{i=1}^{n} \alpha_i f(\omega_i). \qquad (3.58)$$

Hence, the integral of a nonnegative measurable function f with respect to a finite or countable weighted sum of Dirac measures with nonnegative weights is a weighted sum of values of f. ◁

3.2.3   Integral with respect to an image measure

The next theorem is relevant whenever we consider the integral of a composition $g \circ f$ of a mapping f with a numerical function g [see Eq. (2.25)] or the integral with respect to the image measure μ_f of μ under f [see Def. 2.80].

Theorem 3.57 [Transformation theorem]

Let $f: (\Omega, \mathcal{A}, \mu) \to (\Omega', \mathcal{A}')$ and $g: (\Omega', \mathcal{A}') \to (\overline{\mathbb{R}}, \overline{\mathcal{B}})$ be measurable.

(i) If g is nonnegative or integrable with respect to μ_f, then

$$\int g \, d\mu_f = \int g \circ f \, d\mu. \qquad (3.59)$$

(ii) g is integrable with respect to μ_f if and only if $g \circ f$ is μ-integrable.

For a proof, see Bauer (2001, Corollary 19.2.1).

If $f: (\Omega, \mathcal{A}, \mu) \to (\overline{\mathbb{R}}, \overline{\mathcal{B}})$ is a numerical measurable function and we replace g by the identity function $id: (\overline{\mathbb{R}}, \overline{\mathcal{B}}) \to (\overline{\mathbb{R}}, \overline{\mathcal{B}})$, then Theorem 3.57 implies Corollary 3.58.

Corollary 3.58 [An implication of the transformation theorem]

If $f: (\Omega, \mathcal{A}, \mu) \to (\overline{\mathbb{R}}, \overline{\mathcal{B}})$ is nonnegative or μ-integrable, then

$$\int id \, d\mu_f = \int_{\overline{\mathbb{R}}} id \, d\mu_f = \int f \, d\mu = \int_{\Omega} f \, d\mu. \qquad (3.60)$$

Using the alternative notation of an integral introduced in Remark 3.32, Equation (3.59) can also be written as:

$$\int_{\overline{\mathbb{R}}} g(x) \, \mu_f(dx) = \int_{\Omega} g[f(\omega)] \, \mu(d\omega). \tag{3.61}$$

Correspondingly, Equation (3.60) can also be written as:

$$\int_{\overline{\mathbb{R}}} x \, \mu_f(dx) = \int_{\Omega} f(\omega) \, \mu(d\omega). \tag{3.62}$$

In Definition 3.10, we considered the case in which $f = \sum_{i=1}^{n} \alpha_i \, 1_{A_i}$ is a nonnegative step function and defined its integral by $\int f \, d\mu = \sum_{i=1}^{n} \alpha_i \, \mu(A_i)$, presuming that $A_1, \ldots, A_n$ are pairwise disjoint. Now we consider a measurable function f with a finite number of values, which can be 0, positive, or negative.

Corollary 3.59 [Integral of a function with a finite number of values]
If $(\Omega, \mathscr{A}, \mu)$ is a measure space and $f = \sum_{i=1}^{n} \alpha_i \, 1_{A_i}$ with pairwise different $\alpha_1, \ldots, \alpha_n \in \mathbb{R}$, $\alpha_i \neq 0$, and pairwise disjoint $A_1, \ldots, A_n \in \mathscr{A}$, then f is μ-integrable if and only if $\mu(A_i) < \infty$ for all $i = 1, \ldots, n$. If f is μ-integrable, then

$$\int f \, d\mu = \sum_{i=1}^{n} \alpha_i \, \mu(A_i) = \sum_{i=1}^{n} \alpha_i \, \mu_f(\{\alpha_i\}). \tag{3.63}$$

(Proof p. 123)

3.2.4   Convergence theorems

The next two theorems deal with convergence of integrals. In the first one, we assume that $f_1, f_2, \ldots$ is an increasing sequence of measurable functions that converge to f.

Theorem 3.60 [Monotone convergence; B. Levi]
Let the functions $f_n : (\Omega, \mathscr{A}, \mu) \to (\overline{\mathbb{R}}, \overline{\mathscr{B}})$ be measurable, for all $n \in \mathbb{N}$.

(i) *If the sequence $f_1, f_2, \ldots$ is increasing with $\lim_{n \to \infty} f_n = f$ and the functions f_n are nonnegative for all $n \in \mathbb{N}$ or μ-integrable for all $n \in \mathbb{N}$, then*

$$\int f \, d\mu = \lim_{i \to \infty} \int f_n \, d\mu. \tag{3.64}$$

(ii) *If the functions f_i are nonnegative for all $i \in \mathbb{N}$, then*

$$\int \left(\sum_{i=1}^{\infty} f_i \right) d\mu = \sum_{i=1}^{\infty} \int f_i \, d\mu. \tag{3.65}$$

The integrals on both sides are finite or $+\infty$.

For a proof of (i), assuming nonnegativity, see Bauer (2001, Theorem 11.4). For a proof of (i), assuming integrability, see Klenke (2013, Theorem 4.20). For a proof of (ii), see Bauer (2001, Corollary 11.5).

Note that, by definition of an 'infinite sum' (see Box 0.1), Equation (3.65) is equivalent to

$$\int \left(\lim_{n \to \infty} \sum_{i=1}^{n} f_i \right) d\mu = \lim_{n \to \infty} \sum_{i=1}^{n} \int f_i \, d\mu. \tag{3.66}$$

In Theorem 3.61, we replace the assumption that $f_1, f_2, \ldots$ is increasing by the assumption that there is a μ-integrable function g dominating the absolute value functions of all f_n.

Theorem 3.61 [Dominated convergence; Lebesgue convergence theorem]
If $g, f_n : (\Omega, \mathcal{A}, \mu) \to (\overline{\mathbb{R}}, \overline{\mathcal{B}})$, $n \in \mathbb{N}$, are μ-integrable and there is a measurable function $f : (\Omega, \mathcal{A}, \mu) \to (\overline{\mathbb{R}}, \overline{\mathcal{B}})$ with $\lim_{n \to \infty} f_n = f$, and $|f_n| \leq g$ for all $n \in \mathbb{N}$, then

$$\int f \, d\mu = \lim_{n \to \infty} \int f_n \, d\mu, \tag{3.67}$$

and this integral is finite.

For a proof, see Bauer (2001, Theorem 15.6).

3.3   Lebesgue and Riemann integral

The Lebesgue measures λ_n on $(\mathbb{R}^n, \mathcal{B}_n)$, $n = 1, 2, 3$, represent *length*, *area*, and *volume*, respectively. As the examples illustrated by Figure 3.4 show, the integral of the $(\mathcal{B}, \mathcal{B})$-measurable function $f : \mathbb{R} \to \mathbb{R}$ with respect to the Lebesgue measure $\lambda = \lambda_1$ (i.e., the Lebesgue integral) yields the difference between the *areas* marked by + and the areas marked by −, respectively.

It is useful to know conditions under which the Lebesgue integral and the Riemann integral are identical, because a lot of tools are available for Riemann integration (see, e.g., Ellis & Gulick, 2006). The following theorem is proved in Klenke (2013, Theorem 4.23), who also provides a brief definition of the Riemann integral and Riemann integrability.

Theorem 3.62 [Lebesgue integral and Riemann integral]
Let λ denote the Lebesgue measure on $(\mathbb{R}, \mathcal{B})$ and let $[a, b]$, $a, b \in \mathbb{R}$, $a < b$, be a closed interval. If $f : [a, b] \to \mathbb{R}$ is Riemann integrable on $[a, b]$, then f is λ-integrable, and

$$\int_a^b f(x) \, dx = \int_{[a,b]} f \, d\lambda = \int_{[a,b]} f(x) \, \lambda(dx) = \int 1_{[a,b]} \cdot f \, d\lambda, \tag{3.68}$$

where $\displaystyle\int_a^b f(x) \, dx$ denotes the Riemann integral from a to b.

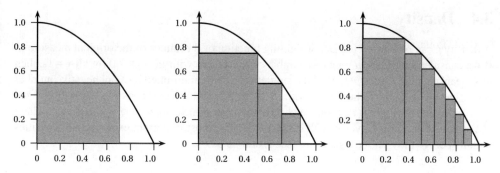

Figure 3.5   Illustrating the construction of the Riemann integral.

Note that

$$\int_a^b f(x)\, dx = F(x)\Big|_a^b := F(b) - F(a),$$
(3.69)

where F is an antiderivative of f.

Remark 3.63 [Lebesgue vs. Riemann integral]   If we want to define the integral of a measurable function $f: (\Omega, \mathscr{A}) \to (\mathbb{R}, \mathscr{B})$, where the set Ω is not necessarily a subset of $\mathbb{R}$, then this means that the traditional Riemann integral cannot be used. The *Riemann integral* is constructed by partitioning the *domain* of f, the set $\mathbb{R}$ of real numbers, into small intervals and adding the area of the rectangles on these intervals in order to approximate the area under the function $f: \mathbb{R} \to \mathbb{R}$ (see Fig. 3.5). If $\Omega \not\subset \mathbb{R}$, then this idea does not work any more. Instead, the *Lebesgue integral* is constructed by partitioning the *codomain* of f, which is the set $\mathbb{R}$ of real numbers, into small intervals (see Fig. 3.3). This is also possible if $\Omega \not\subset \mathbb{R}$, and in this aspect, the Lebesgue integral is more general than the Riemann integral.   ◁

Note, however, that even if the domain of f is a subset of the set of real numbers, there are functions for which the Lebesgue integral exists and the Riemann integral does not exist (see, e.g., Klenke, 2013, Example 4.24). Also note that there are functions that are Riemann integrable on a half-open or unbounded interval but not Lebesgue integrable (see, e.g., Klenke, 2013, Remark 4.25).

Example 3.64 [Using the Riemann integral]   As a simple application of Theorem 3.62, consider the function f defined by $f(x) = 10 - x^2$ on a closed interval $[a, b]$. Because f is a continuous function, it is Riemann integrable. Hence, we can apply Equation (3.68). For $a = -5$ and $b = 5$, this equation yields

$$\int_{[-5,\, 5]} f\, d\lambda = \int_{-5}^5 f(x)\, dx = 16.\bar{6}$$

(see Exercise 3.11). This integral is the difference between the areas marked by $+$ and the areas marked by $-$ in Figure 3.4.   ◁

3.4 Density

A density f can be interpreted as a weighting function of the values of the original measure μ. If we consider a measure μ on a measurable space $(\Omega, \mathscr{A})$ with $\mu(\{x\}) > 0$ for all $x \in \Omega$, then this means that the values $\mu(\{x\})$ of the singletons $\{x\}$ are multiplied by a nonnegative number $f(x)$. If we consider the Lebesgue measure λ on $\mathbb{R}$, then, intuitively speaking, any infinitesimal interval dx gets a weight $f(x)$. Using such a density, a new measure ν on $\mathscr{A}$ is introduced, where $\nu(A)$ is the integral of f over A with respect to μ. The most important examples are densities with respect to the Lebesgue measure (see Example 3.69).

Theorem 3.65 [Measure with density]

Let $f\colon (\Omega, \mathscr{A}, \mu) \to (\overline{\mathbb{R}}, \overline{\mathscr{B}})$ be a nonnegative measurable function. Then the function $\nu\colon \mathscr{A} \to \overline{\mathbb{R}}$ defined by

$$\nu(A) = \int_A f \, d\mu, \quad \forall \, A \in \mathscr{A}, \tag{3.70}$$

is a measure, called the measure with density f with respect to μ. It is denoted by $f \odot \mu$, that is, $f \odot \mu := \nu$.

For a proof, see Bauer (2001, Theorem 17.1).

The notation $f \odot \mu$ has been adopted from Elstrodt (2007). Using this notation, Equation (3.70) can also be written as:

$$f \odot \mu(A) = \int_A f \, d\mu, \quad \forall \, A \in \mathscr{A}. \tag{3.71}$$

Definition 3.66 [Density]

Let ν be a measure on $(\Omega, \mathscr{A})$. If $f\colon (\Omega, \mathscr{A}, \mu) \to (\overline{\mathbb{R}}, \overline{\mathscr{B}})$ is a nonnegative measurable function satisfying Equation (3.70), then it is called a density of ν with respect to μ.

The following theorem generalizes Equation (3.52).

Theorem 3.67 [Integral with respect to a measure with density]

Let $(\Omega, \mathscr{A}, \mu)$ be a measure space and $f\colon (\Omega, \mathscr{A}) \to (\overline{\mathbb{R}}, \overline{\mathscr{B}})$ a nonnegative measurable function. Furthermore, let $f \odot \mu\colon \mathscr{A} \to \overline{\mathbb{R}}$ be the measure with density f with respect to μ and let $g\colon (\Omega, \mathscr{A}) \to (\mathbb{R}, \mathscr{B})$ be measurable.

   (i) If g is nonnegative, then

$$\int g \, d f \odot \mu = \int g \cdot f \, d\mu. \tag{3.72}$$

   (ii) g is integrable with respect to $f \odot \mu$ if and only if $g \cdot f$ is μ-integrable.

   (iii) If g is integrable with respect to $f \odot \mu$, then Equation (3.72) holds.

For a proof, see Bauer (2001, Theorem 17.3).

In Theorem 3.68, we summarize some necessary and sufficient conditions for μ-equivalence of measurable functions on a measure space.

Theorem 3.68 [Necessary and sufficient conditions of μ-equivalence]
Let $f, g: (\Omega, \mathcal{A}, \mu) \to (\overline{\mathbb{R}}, \overline{\mathcal{B}})$ be measurable functions, let $\mathcal{E} \subset \mathcal{A}$, and consider:

(a) $f \underset{\mu}{=} g$.

(b) $\displaystyle\int_A f \, d\mu = \int_A g \, d\mu, \quad \forall A \in \mathcal{A}$.

(c) $f \odot \mu = g \odot \mu$.

(d) $\displaystyle\int_A f \, d\mu = \int_A g \, d\mu, \quad \forall A \in \mathcal{E}$.

Then,

(i) $(a) \Rightarrow (b)$, *if f and g are quasi-μ-integrable.*

(ii) $(a) \Leftrightarrow (b)$, *if f, g are μ-integrable.*

(iii) $(a) \Leftrightarrow (c)$, *if f, g are μ-integrable and nonnegative.*

(iv) $(a) \Leftrightarrow (b) \Leftrightarrow (c) \Leftrightarrow (d)$, *if f, g are μ-integrable, nonnegative, and $\mathcal{E} \subset \mathcal{A}$ is $\cap$-stable with $\sigma(\mathcal{E}) = \mathcal{A}$.*

(Proof p. 123)

Example 3.69 [A density of the normal distribution]   As a special case of Equation (3.70), consider

$$\nu(A) = \int_A f \, d\lambda, \quad \forall A \in \mathcal{A}, \tag{3.73}$$

with

$$f(x) = \frac{1}{\sqrt{2\pi}} \cdot \exp\left(\frac{-x^2}{2}\right), \quad \forall x \in \mathbb{R}. \tag{3.74}$$

In this case, the measure $\nu = f \odot \lambda$ is a probability measure, and it is called the *standard normal distribution*. For an interval $[a, b]$, Theorem 3.67 yields

$$\nu([a, b]) = \int 1_{[a,b]} \, d\nu = \int 1_{[a,b]} \, df \odot \lambda = \int 1_{[a,b]} f \, d\lambda = \int_a^b f(x) \, dx, \tag{3.75}$$

because f is Riemann-integrable (see Th. 3.62). According to this equation, the value $\nu([a, b])$ of the interval $[a, b]$ can be represented as the area between the density and the x-axis above

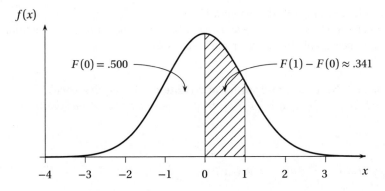

Figure 3.6   Integral of a density for two intervals.

$[a, b]$. Figure 3.6 illustrates this fact for the interval $[0, 1]$. In this figure,

$$F(\alpha) = \int_{-\infty}^{\alpha} f(x)\, dx, \quad \forall\, \alpha \in \mathbb{R}, \tag{3.76}$$

denotes the corresponding distribution function (see Def. 5.81), which is a special antiderivative of f [see Eq. (3.69)].   ◁

3.5   Absolute continuity and the Radon-Nikodym theorem

Let μ and ν be measures on a measurable space $(\Omega, \mathcal{A})$. A necessary and sufficient condition
for the existence of a density of ν with respect to μ is formulated in the Radon-Nikodym
theorem (see Th. 3.72), which is used not only for densities but also for introducing conditional
expectations (see ch. 10). The following definition prepares this theorem.

Definition 3.70 [Absolute continuity]
Let μ and ν be measures on a measurable space $(\Omega, \mathcal{A})$.

(i) The measure ν is called absolutely continuous *with respect to μ, denoted*
    $\nu \underset{\mathcal{A}}{\ll} \mu$, *if*

$$\forall\, A \in \mathcal{A}: \quad \mu(A) = 0 \quad \Rightarrow \quad \nu(A) = 0. \tag{3.77}$$

(ii) The measures μ and ν are called null-set equivalent, *denoted $\nu \underset{\mathcal{A}}{\approx} \mu$, if $\nu \underset{\mathcal{A}}{\ll} \mu$*
    and $\mu \underset{\mathcal{A}}{\ll} \nu$, that is, if

$$\forall\, A \in \mathcal{A}: \quad \mu(A) = 0 \quad \Leftrightarrow \quad \nu(A) = 0. \tag{3.78}$$

If there is ambiguity about the measurable space, we use the terms absolutely continuous *on $(\Omega, \mathcal{A})$ and* null-set equivalent *on $(\Omega, \mathcal{A})$.*

Remark 3.71 [An implication]   If there is a density f of ν with respect to μ, then $\nu \underset{\mathscr{A}}{\ll} \mu$. This is a straightforward implication of Lemma 3.45 and (3.72) (see Exercise 3.12).   ◁

Vice versa, if $\nu \underset{\mathscr{A}}{\ll} \mu$, then, according to the following theorem, there is a density f of ν with respect to μ, provided that μ and ν are σ-finite (see Definition 1.63).

Theorem 3.72 [Radon-Nikodym]
Let μ and ν be σ-finite measures on a measurable space $(\Omega, \mathscr{A})$.

(i) Then ν has a measurable density with respect to μ if and only if $\nu \underset{\mathscr{A}}{\ll} \mu$. This density is denoted by $\frac{d\nu}{d\mu}$ and called a Radon-Nikodym derivative.

(ii) If $\nu \underset{\mathscr{A}}{\ll} \mu$, then $\frac{d\nu}{d\mu}$ is real-valued μ-almost everywhere.

For a proof, see Klenke (2013, Corollary 7.34) or Bauer (2001, Theorem 17.10 and Theorem 17.11).

Remark 3.73 [μ-Equivalence of densities]   Note that, for σ-finite measures, all densities of ν with respect to μ are pairwise μ-equivalent (for a proof, see Bauer, 2001, Theorem 17.11). If ν is a finite measure, which is equivalent to μ-integrability of the density $\frac{d\nu}{d\mu}$, then Theorem 3.68 (iii) implies μ-equivalence of all densities of ν with respect to μ. The premise that ν is finite holds in particular if ν is a probability measure.   ◁

Remark 3.74 [An implication of the Radon-Nikodym theorem]   Theorem 3.72 implies for σ-finite measures ν and μ: If ν and μ are null-set equivalent, then $\frac{d\nu}{d\mu}$ and $\frac{d\mu}{d\nu}$ both exist.   ◁

The Radon-Nikodym theorem is used to prove the existence of the conditional expectation [see the proof of Theorem 10.9 (Bauer, 1996, Theorem 15.1)]. Corollary 3.75 immediately follows from Theorem 3.67 and 3.72.

Corollary 3.75 [An implication of the Radon-Nikodym theorem]
Let μ and ν be σ-finite measures on a measurable space $(\Omega, \mathscr{A})$, and suppose $\nu \underset{\mathscr{A}}{\ll} \mu$. Furthermore, let $g\colon (\Omega, \mathscr{A}) \to (\overline{\mathbb{R}}, \overline{\mathscr{B}})$ be a measurable function.

(i) If g is nonnegative or ν-integrable, then

$$\int g \, d\nu = \int g \cdot \frac{d\nu}{d\mu} \, d\mu. \tag{3.79}$$

(ii) g is ν-integrable if and only if $g \cdot \frac{d\nu}{d\mu}$ is μ-integrable.

3.6   Integral with respect to a product measure

Theorem 3.76 shows that integration with respect to a product measure can be decomposed into a twofold iterated integration where the order of integration is arbitrary.

Theorem 3.76 [Fubini]
Assume that $(\Omega_i, \mathcal{A}_i, \mu_i)$, $i = 1, 2$, are σ-finite measure spaces and let $f : \Omega_1 \times \Omega_2 \to \overline{\mathbb{R}}$ be $(\mathcal{A}_1 \otimes \mathcal{A}_2, \overline{\mathcal{B}})$-measurable. Furthermore, let $f_i \colon \Omega_i \to \overline{\mathbb{R}}$, $i = 1, 2$, be defined by

$$f_1(\omega_1) := \int f(\omega_1, \omega_2)\, \mu_2(d\omega_2) \quad \text{and} \quad f_2(\omega_2) := \int f(\omega_1, \omega_2)\, \mu_1(d\omega_1).$$

If f is nonnegative or integrable with respect to the product measure $\mu_1 \otimes \mu_2$, then the functions f_i are $(\mathcal{A}_i, \overline{\mathcal{B}})$-measurable, $i = 1, 2$. Furthermore,

$$\int_{\Omega_1 \times \Omega_2} f\, d(\mu_1 \otimes \mu_2) = \int_{\Omega_1 \times \Omega_2} f(\omega_1, \omega_2)\, \mu_1 \otimes \mu_2\, [d(\omega_1, \omega_2)]$$

$$= \int_{\Omega_1} \left(\int_{\Omega_2} f(\omega_1, \omega_2)\, \mu_2(d\omega_2) \right) \mu_1(d\omega_1) \qquad (3.80)$$

$$= \int_{\Omega_2} \left(\int_{\Omega_1} f(\omega_1, \omega_2)\, \mu_1(d\omega_1) \right) \mu_2(d\omega_2).$$

For a proof, see Klenke (2013, Th. 14.16). If $f = 1_C$ for $C \in \mathcal{A}_1 \otimes \mathcal{A}_2$, then this theorem and (3.9) immediately yield the following corollary:

Corollary 3.77 [Indicators]
Let $(\Omega_i, \mathcal{A}_i, \mu_i)$, $i = 1, 2$, be σ-finite measure spaces, let $C \in \mathcal{A}_1 \otimes \mathcal{A}_2$, and define

$$\forall\, \omega_1 \in \Omega_1 \colon \quad C_{\omega_1} := \{\omega_2 \in \Omega_2 \colon (\omega_1, \omega_2) \in C\}$$

and

$$\forall\, \omega_2 \in \Omega_2 \colon \quad C_{\omega_2} := \{\omega_1 \in \Omega_1 \colon (\omega_1, \omega_2) \in C\}.$$

Then

$$\mu_1 \otimes \mu_2(C) = \int \mu_2(C_{\omega_1})\, \mu_1(d\omega_1) = \int \mu_1(C_{\omega_2})\, \mu_2(d\omega_2). \qquad (3.81)$$

(Proof p. 124)

Remark 3.78 [A special case] Choosing $C = A_1 \times A_2$, Equation (3.81) yields

$$\mu_1 \otimes \mu_2(A_1 \times A_2) = \int \mu_2(A_2) \cdot 1_{A_1}(\omega_1)\, \mu_1(d\omega_1) = \mu_2(A_2) \cdot \int 1_{A_1}(\omega_1)\, \mu_1(d\omega_1)$$
$$= \mu_1(A_1) \cdot \mu_2(A_2),$$

which is consistent with Equation (1.50).   ◁

3.7   Proofs

Proof of Lemma 3.33

(i) If f is measurable and $A \in \mathcal{A}$, then 1_A is measurable as well (see Th. 2.57 and Example 2.12). Suppose that f is quasi-μ-integrable, that is, suppose that $\int f^+ d\mu$ or $\int f^- d\mu$ are finite. Because

$$(1_A f)^+ = 1_A f^+ \quad \text{and} \quad (1_A f)^- = 1_A f^-$$

as well as

$$0 \le 1_A f^+ \le f^+ \quad \text{and} \quad 0 \le 1_A f^- \le f^-,$$

monotonicity of the integral of nonnegative measurable functions (Lemma 3.26) yields

$$0 \le \int (1_A f)^+\, d\mu = \int 1_A f^+ d\mu \le \int f^+ d\mu$$

and

$$0 \le \int (1_A f)^-\, d\mu = \int 1_A f^- d\mu \le \int f^- d\mu,$$

which implies that $\int (1_A f)^+\, d\mu$ or $\int (1_A f)^-\, d\mu$ is finite. Hence, $1_A f$ is quasi-μ-integrable.

(ii) If f is μ-integrable, then $\int f^+ d\mu < \infty$ and $\int f^- d\mu < \infty$. Just like in the proof of (i), this implies $\int (1_A f)^+\, d\mu < \infty$ and $\int (1_A f)^-\, d\mu < \infty$. Hence, $1_A f$ is μ-integrable.

Proof of Theorem 3.36

(i) Step 1: Let $\alpha \ge 0$, let f be a nonnegative step function, and $f = \sum_{i=1}^{n} \alpha_i 1_{A_i}$ a normal representation (see Rem. 3.8). Then, according to (3.4),

$$\int \alpha f\, d\mu = \int \alpha \sum_{i=1}^{n} \alpha_i 1_{A_i}\, d\mu \qquad [\text{Rem. 3.8}]$$
$$= \int \sum_{i=1}^{n} (\alpha\, \alpha_i)\, 1_{A_i}\, d\mu$$

$$= \sum_{i=1}^{n} (\alpha \, \alpha_i) \, \mu(A_i) \qquad [(3.4)]$$

$$= \alpha \sum_{i=1}^{n} \alpha_i \, \mu(A_i)$$

$$= \alpha \int f \, d\mu. \qquad [(3.4)]$$

Step 2: Let $\alpha \geq 0$, f be a nonnegative measurable function, and $f_1 \leq f_2 \leq \ldots$ an increasing sequence of nonnegative step functions with $\lim_{n \to \infty} f_n = f$ (see Th. 3.19). Then, according to Equation (3.22),

$$\int \alpha f \, d\mu = \int \alpha \lim_{n \to \infty} f_n \, d\mu$$

$$= \int \lim_{n \to \infty} (\alpha f_n) \, d\mu$$

$$= \lim_{n \to \infty} \int \alpha f_n \, d\mu \qquad [(3.22)]$$

$$= \lim_{n \to \infty} \alpha \int f_n \, d\mu \qquad [\text{Step 1}]$$

$$= \alpha \lim_{n \to \infty} \int f_n \, d\mu$$

$$= \alpha \int f \, d\mu. \qquad [(3.22)]$$

Step 3: Assume that $\alpha \geq 0$ and that f is quasi-μ-integrable. Because $\alpha f = \alpha \, (f^+ - f^-) = \alpha f^+ - \alpha f^-$,

$$\int \alpha f \, d\mu = \int \alpha f^+ \, d\mu - \int \alpha f^- \, d\mu \qquad [(3.27)]$$

$$= \alpha \int f^+ \, d\mu - \alpha \int f^- \, d\mu \qquad [\text{Step 2}]$$

$$= \alpha \left(\int f^+ \, d\mu - \int f^- \, d\mu \right)$$

$$= \alpha \int f \, d\mu. \qquad [(3.27)]$$

This proves Equation (3.32) for $\alpha \geq 0$. For $\alpha < 0$, note that

$$(\alpha f)^+ = -\alpha f^- \quad \text{and} \quad (\alpha f)^- = -\alpha f^+. \qquad (3.82)$$

Therefore,

$$\int \alpha f\, d\mu = \int (\alpha f)^+ \, d\mu - \int (\alpha f)^- \, d\mu \qquad [(3.27)]$$

$$= \int (-\alpha) f^- \, d\mu - \int (-\alpha) f^+ \, d\mu \qquad [(3.82)]$$

$$= -\alpha \int f^- \, d\mu - (-\alpha) \int f^+ \, d\mu \qquad [-\alpha > 0, \text{ first part of Step 3}]$$

$$= \alpha \left(\int f^+ \, d\mu - \int f^- \, d\mu \right)$$

$$= \alpha \int f\, d\mu. \qquad [(3.27)]$$

This shows that $\int \alpha f\, d\mu = \alpha \int f\, d\mu$ holds for all $\alpha \in \mathbb{R}$, all quasi-μ-integrable f, and therefore also all integrable f. This also implies that αf is quasi-μ-integrable or μ-integrable if f is quasi-μ-integrable or μ-integrable, respectively.

(ii) Step 1: Let f and g be nonnegative step functions and let $f = \sum_{i=1}^{n} \alpha_i 1_{A_i}$, $g = \sum_{j=1}^{m} \beta_j 1_{B_j}$ be normal representations (see Rem. 3.8) with $\bigcup_{i=1}^{n} A_i = \Omega$ and $\bigcup_{j=1}^{m} B_j = \Omega$. (Note that these latter requirements can always be fulfilled using $A_n := \Omega \setminus \bigcup_{i=1}^{n-1} A_i$ and $\alpha_n := 0$, if $f = \sum_{i=1}^{n-1} \alpha_i 1_{A_i}$ is already a normal representation.) Then $f + g = \sum_{i=1}^{n} \alpha_i 1_{A_i} + \sum_{j=1}^{m} \beta_j 1_{B_j}$ is again a nonnegative step function (see Def. 3.10) and

$$f + g = \sum_{i=1}^{n} \sum_{j=1}^{m} \gamma_{ij} 1_{C_{ij}}$$

is a normal representation, where $C_{ij} := A_i \cap B_j$ and $\gamma_{ij} := \alpha_i + \beta_j$. Note that some of these sets C_{ij} may be empty. Now,

$$\int f + g \, d\mu = \sum_{i=1}^{n} \sum_{j=1}^{m} \gamma_{ij}\, \mu(C_{ij}) \qquad [(3.4)]$$

$$= \sum_{i=1}^{n} \sum_{j=1}^{m} (\alpha_i + \beta_j)\, \mu(A_i \cap B_j)$$

$$= \sum_{i=1}^{n} \sum_{j=1}^{m} \alpha_i\, \mu(A_i \cap B_j) + \sum_{i=1}^{n} \sum_{j=1}^{m} \beta_j\, \mu(A_i \cap B_j)$$

$$= \sum_{i=1}^{n} \alpha_i \sum_{j=1}^{m} \mu(A_i \cap B_j) + \sum_{j=1}^{m} \beta_j \sum_{i=1}^{n} \mu(A_i \cap B_j)$$

$$= \sum_{i=1}^{n} \alpha_i\, \mu(A_i) + \sum_{j=1}^{m} \beta_j\, \mu(B_j) \qquad [\text{Rem. 1.47}]$$

$$= \int f\, d\mu + \int g\, d\mu. \qquad [(3.4)]$$

Step 2: Let f, g be nonnegative measurable functions and $f_1 \leq f_2 \leq \dots$, $g_1 \leq g_2 \leq \dots$ increasing sequences of nonnegative step functions with $\lim_{n \to \infty} f_n = f$ and $\lim_{n \to \infty} g_n = g$, respectively (see Th. 3.19). Then $f_1 + g_1 \leq f_2 + g_2 \leq \dots$ is an increasing sequence of nonnegative step functions with $\lim_{n \to \infty} (f_n + g_n) = \lim_{n \to \infty} f_n + \lim_{n \to \infty} g_n = f + g$. Then

$$
\begin{aligned}
\int f + g \, d\mu &= \int \lim_{n \to \infty} (f_n + g_n) \, d\mu \\
&= \lim_{n \to \infty} \int (f_n + g_n) \, d\mu && [(3.22)] \\
&= \lim_{n \to \infty} \left(\int f_n \, d\mu + \int g_n \, d\mu \right) && [\text{Step 1}] \\
&= \lim_{n \to \infty} \int f_n \, d\mu + \lim_{n \to \infty} \int g_n \, d\mu \\
&= \int f \, d\mu + \int g \, d\mu. && [(3.22)]
\end{aligned}
$$

Step 3: Assume that f is quasi-μ-integrable and g is μ-integrable. Then

$$
\begin{aligned}
f + g &= f^+ - f^- + g^+ - g^-, \\
f + g &= (f + g)^+ - (f + g)^-.
\end{aligned}
$$

This implies

$$
(f + g)^+ - (f + g)^- = f^+ - f^- + g^+ - g^-,
$$

which is equivalent to

$$
(f + g)^+ + f^- + g^- = (f + g)^- + f^+ + g^+.
$$

Applying the result of Step 2 yields

$$
\int (f + g)^+ \, d\mu + \int f^- \, d\mu + \int g^- \, d\mu = \int (f + g)^- \, d\mu + \int f^+ \, d\mu + \int g^+ \, d\mu.
$$

$$(3.83)$$

If g is μ-integrable, then $\int g^+ \, d\mu$ and $\int g^- \, d\mu$ are finite, and if f is quasi-μ-integrable, then at most one of $\int f^+ d\mu$ and $\int f^- d\mu$ is infinite, and the other one is finite. Furthermore, $(f + g)^+ \leq f^+ + g^+$ and $(f + g)^- \leq f^- + g^-$. Hence, Lemma 3.26 implies

$$
\int (f + g)^+ \, d\mu \leq \int f^+ + g^+ \, d\mu = \int f^+ \, d\mu + \int g^+ \, d\mu \qquad (3.84)
$$

and

$$\int (f+g)^- \, d\mu \le \int f^- + g^- \, d\mu = \int f^- \, d\mu + \int g^- \, d\mu.$$

Therefore, at most one of the integrals $\int (f+g)^+ \, d\mu$ and $\int (f+g)^- \, d\mu$ is infinite, and this implies that $f + g$ is quasi-μ-integrable. If $\int (f+g)^+ \, d\mu = \infty$, then

$$\int (f+g) \, d\mu = \int (f+g)^+ \, d\mu - \int (f+g)^- \, d\mu = \infty,$$

and, according to (3.84), $\int f^+ \, d\mu = \infty$. This implies

$$\int f \, d\mu + \int g \, d\mu = \int f^+ d\mu - \int f^- d\mu + \int g \, d\mu = \infty.$$

Analogously, if $\int (f+g)^- \, d\mu = \infty$, then

$$\int (f+g) \, d\mu = -\infty = \int f \, d\mu + \int g \, d\mu.$$

If both, $\int (f+g)^+ \, d\mu$ and $\int (f+g)^- \, d\mu$ are finite, then (3.83) is equivalent to

$$\int (f+g)^+ \, d\mu - \int (f+g)^- \, d\mu = \int f^+ \, d\mu - \int f^- \, d\mu + \int g^+ \, d\mu - \int g^- \, d\mu,$$

which in turn is equivalent to

$$\int (f+g) \, d\mu = \int f \, d\mu + \int g \, d\mu.$$

Proof of Corollary 3.38

Because $|f| = f^+ + f^-$, this proposition immediately follows from the definition of integrability (see Def. 3.28) and linearity of the integral [see Eq. (3.34)].

Proof of Lemma 3.40

If f is $\mathscr{C}$-measurable, then f^+ and f^- are $\mathscr{C}$-measurable as well (see Th. 2.66). Furthermore, f^+ and f^- can be represented as limits of increasing sequences of nonnegative step functions on $(\Omega, \mathscr{C})$ [see Th. 3.19 (i)]. Hence, according to Equations (3.22) and (3.4), the values of the integrals $\int f^+ \, d\mu$ and $\int f^- \, d\mu$ only depend on the values $\mu(A)$, $A \in \mathscr{C}$. Therefore, if μ and ν are identical on $\mathscr{C}$, then $\int f \, d\mu = \int f \, d\nu$, for all $\mathscr{C}$-measurable functions that are nonnegative or μ-integrable.

Proof of Lemma 3.41

Define

$$A_+ := \{\omega \in \Omega : f(\omega) = \infty\} \in \mathscr{A} \quad \text{and} \quad A_- := \{\omega \in \Omega : f(\omega) = -\infty\} \in \mathscr{A}.$$

If $\mu(A_+) > 0$, then define the increasing sequence $g_n \colon \Omega \to [0, \infty)$, $n \in \mathbb{N}$, by $g_n = n \cdot 1_{A_+}$. Because $1_{A_+} \cdot f^+ = \lim_{n \to \infty} g_n$,

$$\int f^+ \, d\mu = \int 1_{A_+} \cdot f^+ \, d\mu + \int 1_{\Omega \backslash A_+} \cdot f^+ \, d\mu \qquad [(3.36)]$$

$$= \lim_{n \to \infty} \int g_n \, d\mu + \int 1_{\Omega \backslash A_+} \cdot f^+ \, d\mu \qquad [\text{Def. } 3.24]$$

$$= \lim_{n \to \infty} \int n \cdot 1_{A_+} \, d\mu + \int 1_{\Omega \backslash A_+} \cdot f^+ \, d\mu$$

$$= \lim_{n \to \infty} n \cdot \mu(A_+) + \int 1_{\Omega \backslash A_+} \cdot f^+ \, d\mu \qquad [(3.4)]$$

$$= \infty.$$

Analogously we can prove that $\int f^- \, d\mu = \infty$, if $\mu(A_-) > 0$ replacing f^+ by f^- and A_+ by A_-. Therefore, if f is quasi-μ-integrable and $\int f^+ \, d\mu = \infty$, then $\int f^- \, d\mu$ is finite (see Def. 3.28) and $\int f \, d\mu = \int f^+ \, d\mu - \int f^- \, d\mu = \infty$. This proves (3.38). Analogously, if f is quasi-μ-integrable and $\int f^- \, d\mu = \infty$, then $\int f^+ \, d\mu$ is finite (see Def. 3.28) and $\int f \, d\mu = \int f^+ \, d\mu - \int f^- \, d\mu = -\infty$, which proves (3.39). Finally, if $\mu(A_+ \cup A_-) > 0$ and hence $\mu(A_+) > 0$ or $\mu(A_-) > 0$, then, according to (3.38) and (3.39), $\int f \, d\mu$ is not defined or not finite. Thus, by contraposition, if f is μ-integrable, then $\mu(A_+ \cup A_-) = 0$, that is, f is real-valued, μ-almost everywhere.

Proof of Lemma 3.44

If $f(\omega) > 0$, for all $\omega \in A$, then $1_A \cdot f \colon (\Omega, \mathscr{A}, \mu) \to (\overline{\mathbb{R}}, \overline{\mathscr{B}})$ is a nonnegative measurable function (see Th. 2.57). Hence, $\int 1_A \cdot f \, d\mu \geq 0$ (see Defs. 3.24 and 3.10). Because $\mu(\{\omega \in \Omega : (1_A \cdot f(\omega) > 0\}) = \mu(A)$, the assumption $\mu(A) > 0$ implies that $1_A \cdot f \underset{\mu}{=} 0$ does *not* hold. Therefore, according to Equation (3.40), $\int 1_A \cdot f \, d\mu \neq 0$, and we can conclude: $\int 1_A \cdot f \, d\mu > 0$.

Proof of Lemma 3.45

If $f \colon (\Omega, \mathscr{A}, \mu) \to (\overline{\mathbb{R}}, \overline{\mathscr{B}})$ is measurable and $A \in \mathscr{A}$ with $\mu(A) = 0$, then $1_A \cdot f$ is measurable (see Th. 2.57) and $1_A \cdot f \underset{\mu}{=} 0$ (see Exercise 3.9). This implies $(1_A \cdot f)^+ \underset{\mu}{=} 0$ and $(1_A \cdot f)^- \underset{\mu}{=} 0$. Now Equation (3.40) yields $\int (1_A \cdot f)^+ \, d\mu = 0$ and $\int (1_A \cdot f)^- \, d\mu = 0$. Hence, $\int 1_A \cdot f \, d\mu$ exists (see Def. 3.28) and Equation (3.27) implies

$$\int 1_A \cdot f \, d\mu = \int (1_A \cdot f)^+ d\mu - \int (1_A \cdot f)^- d\mu = 0.$$

Proof of Lemma 3.47

Define $A := \{\omega \in \Omega: f(\omega) \neq g(\omega)\}$ and note that $A \in \mathscr{A}$ [see Rem. 2.67 (c)]. Therefore, $f \underset{\mu}{=} g$ implies $\mu(A) = 0$. Hence,

$$\int f \, d\mu = \int_{\Omega \setminus A} f \, d\mu \qquad [(3.43)]$$

$$= \int_{\Omega \setminus A} g \, d\mu \qquad [\text{Def. of } A]$$

$$= \int g \, d\mu. \qquad [(3.43)]$$

Proof of Theorem 3.48

(a) $\quad f, g$ are μ-integrable and $f \underset{\mu}{=} g$

$\Rightarrow \forall A \in \mathscr{A}: \quad 1_A \cdot f \underset{\mu}{=} 1_A \cdot g$ and $1_A \cdot f, 1_A \cdot g$ are μ-integrable $\qquad [(2.33), (3.29)]$

$\Rightarrow \forall A \in \mathscr{A}: \quad \int_A f \, d\mu = \int_A g \, d\mu. \qquad [(3.44)]$

(b) If f, g are μ-integrable, then f, g are real-valued μ-a.e. (see Lemma 3.41). Hence, for

$$B := \{\omega \in \Omega: f(\omega) \in \{-\infty, \infty\}\} \cup \{\omega \in \Omega: g(\omega) \in \{-\infty, \infty\}\},$$

$B \in \mathscr{A}$ and $\mu(B) = 0$. Now define

$$A_> := \{\omega \in \Omega: f(\omega) > g(\omega)\} \quad \text{and} \quad A_< := \{\omega \in \Omega: f(\omega) < g(\omega)\}.$$

According to Remark 2.67 (b) and (a), $A_> \in \mathscr{A}$ and $A_< \in \mathscr{A}$. Then

$\forall A \in \mathscr{A}: \displaystyle\int_A f \, d\mu = \int_A g \, d\mu$

$\Rightarrow \forall A \in \mathscr{A}: \displaystyle\int_{A \cap B^c} f \, d\mu = \int_{A \cap B^c} g \, d\mu \qquad [(3.43)]$

$\Rightarrow \forall A \in \mathscr{A}: \displaystyle\int_{A \cap B^c} (f - g) \, d\mu = 0 \qquad [f, g \ \mu\text{-integrable}, (3.34)]$

$\Rightarrow \displaystyle\int_{A_> \cap B^c} (f - g) \, d\mu = 0 \quad \text{and} \quad \int_{A_< \cap B^c} (f - g) \, d\mu = 0 \qquad [A_>, A_< \in \mathscr{A}]$

$\Rightarrow \mu(A_> \cap B^c) = 0$ and $\mu(A_< \cap B^c) = 0 \qquad [(3.41)]$

$\Rightarrow \quad \mu(A_> \cup A_<)$

$\quad = \mu((A_> \cap B^c) \cup (A_> \cap B) \cup (A_< \cap B^c) \cup (A_< \cap B)) = 0 \qquad [\text{Box 1.1 (ii)}]$

$\Rightarrow f \underset{\mu}{=} g. \qquad [\text{Def. 2.68}]$

Proof of Theorem 3.52

(i) Define $A := \{\omega \in \Omega: f(\omega) > g(\omega)\}$. According to Remark 2.67 (b), $A \in \mathcal{A}$. Therefore, $f \underset{\mu}{\leq} g$ implies $\mu(A) = 0$. Furthermore, define

$$A_{-\infty} := \{\omega \in \Omega \setminus A: f(\omega) = -\infty\} \quad \text{and} \quad A_{\infty} := \{\omega \in \Omega \setminus A: g(\omega) = \infty\}.$$

If $\mu(A_{-\infty}) > 0$, then $\int f \, d\mu = -\infty$ [see (3.39)]. Therefore, $\int f \, d\mu \leq \int g \, d\mu$. If $\mu(A_{\infty}) > 0$, then $\int g \, d\mu = \infty$ [see (3.38)], and therefore $\int f \, d\mu \leq \int g \, d\mu$.

Now define $B := (\Omega \setminus A) \setminus (A_{\infty} \cup A_{-\infty})$, which implies $\mu(\Omega \setminus B) = 0$, and

$$\forall \, \omega \in B: f(\omega), g(\omega) \text{ are finite}, f(\omega) \leq g(\omega).$$

If $\int (1_B f)^- \, d\mu = \infty$, then $\int_B f \, d\mu = -\infty$. Hence, $\int f \, d\mu \leq \int g \, d\mu$. If $\int (1_B f)^+ \, d\mu = \infty$, then $1_B f \leq 1_B g$ implies $(1_B f)^+ \leq (1_B g)^+$ and $\int (1_B g)^+ \, d\mu = \infty$ [see Eq. (3.24)]. Therefore, $\int f \, d\mu = \int g \, d\mu = \infty$, which implies $\int f \, d\mu \leq \int g \, d\mu$.

Now, if all $(1_B f)^+, (1_B f)^-, (1_B g)^+, (1_B g)^-$ are μ-integrable, then

$$\int g \, d\mu = \int_B g \, d\mu \qquad\qquad [(3.43)]$$

$$= \int_B (f + g - f) \, d\mu \qquad\qquad [1_B f, 1_B g \text{ are real-valued}]$$

$$= \int_B f \, d\mu + \int_B (g - f) \, d\mu \qquad\qquad [(3.34)]$$

$$\geq \int_B f \, d\mu \qquad\qquad [1_B(g - f) \geq 0, (3.24)]$$

$$= \int f \, d\mu. \qquad\qquad [(3.43)]$$

(ii) Define

$$B := \{\omega \in \Omega: f(\omega) \in \{-\infty, \infty\}\} \cup \{\omega \in \Omega: g(\omega) \in \{-\infty, \infty\}\}.$$

If f, g are μ-integrable, then Lemma 3.41 implies $\mu(B) = 0$. Furthermore, define $A := \{\omega \in \Omega \setminus B: f(\omega) \geq g(\omega)\}$. Then $f \underset{\mu}{<} g$ implies $\mu(A) = 0$. Hence, $\mu(A \cup B) = 0$ [see Box 1.1 (xi)]. Now,

$$\int g \, d\mu = \int_{\Omega \setminus (A \cup B)} g \, d\mu \qquad\qquad [(3.43)]$$

$$= \int_{\Omega \setminus (A \cup B)} (f + g - f) \, d\mu \qquad [1_{\Omega \setminus (A \cup B)} f, 1_{\Omega \setminus (A \cup B)} g \text{ are real-valued}]$$

$$= \int_{\Omega \setminus (A \cup B)} f \, d\mu + \int_{\Omega \setminus (A \cup B)} (g - f) \, d\mu \qquad\qquad [(3.34)]$$

$$> \int_{\Omega \setminus (A \cup B)} f \, d\mu. \qquad [1_{\Omega \setminus (A \cup B)}(g - f) > 0, \mu(\Omega \setminus (A \cup B)) > 0, \text{Lem. 3.44}]$$

Proof of Corollary 3.59

Assume that $f: (\Omega, \mathscr{A}) \to (\overline{\mathbb{R}}, \overline{\mathscr{B}})$ is measurable with a finite number of *positive* values $\alpha_1, \ldots, \alpha_m > 0$ and a finite number of *negative* values $\alpha_{m+1}, \ldots, \alpha_n < 0$. By convention, if $m = n$, then $\sum_{i=m+1}^{n} \alpha_i 1_{A_i} = 0$, and if $m = 0$, then $\sum_{i=1}^{m} \alpha_i 1_{A_i} = 0$. Then

$$f = \sum_{i=1}^{n} \alpha_i 1_{A_i} = \sum_{i=1}^{m} \alpha_i 1_{A_i} + \sum_{i=m+1}^{n} \alpha_i 1_{A_i}$$

and $f^+ = \sum_{i=1}^{m} \alpha_i 1_{A_i}$ as well as $f^- = -\sum_{i=m+1}^{n} \alpha_i 1_{A_i} = \sum_{i=m+1}^{n} -\alpha_i 1_{A_i}$. Therefore,

$$\int f^+ \, d\mu = \sum_{i=1}^{m} \alpha_i \, \mu(A_i) \quad \text{and} \quad \int f^- \, d\mu = \sum_{i=m+1}^{n} -\alpha_i \mu(A_i),$$

and $\int f^+ \, d\mu$ as well as $\int f^- \, d\mu$ are finite if and only if $\mu(A_i) < \infty$, for all $i = 1, \ldots, n$. Now, $\mu(A_i) < \infty$, for all $i = 1, \ldots, n$, implies

$$\int f \, d\mu = \int f^+ \, d\mu - \int f^- \, d\mu \qquad \text{[Def. 3.28]}$$

$$= \sum_{i=1}^{m} \alpha_i \, \mu(A_i) + \sum_{i=m+1}^{n} \alpha_i \, \mu(A_i)$$

$$= \sum_{i=1}^{n} \alpha_i \, \mu(A_i)$$

$$= \sum_{i=1}^{n} \alpha_i \, \mu_f(\{\alpha_i\}). \qquad \text{[Def. 2.80, (2.10)]}$$

Note that, in the last equation, we used the assumption that the $\alpha_1, \ldots, \alpha_n$ are pairwise different.

Proof of Theorem 3.68

(i) This is the proposition of Lemma 3.47.

(ii) This proposition is Theorem 3.48.

(iii) If f, g are μ-integrable and nonnegative, then it suffices to show: (b) $\Leftrightarrow$ (c) [see (ii)]. Now

$$\int_A f \, d\mu = \int_A g \, d\mu, \qquad \forall A \in \mathscr{A}$$

$$\Leftrightarrow \int 1_A f \, d\mu = \int 1_A g \, d\mu, \qquad \forall A \in \mathscr{A} \qquad \text{[Def. 3.35]}$$

$$\Leftrightarrow \int 1_A \, df \odot \mu = \int 1_A \, dg \odot \mu, \quad \forall A \in \mathscr{A} \qquad \text{[Th. 3.67 (i)]}$$

$$\Leftrightarrow f \odot \mu(A) = g \odot \mu(A), \qquad \forall A \in \mathscr{A} \qquad \text{[(3.9)]}$$

$$\Leftrightarrow f \odot \mu = g \odot \mu.$$

(iv) If f, g are μ-integrable and nonnegative, then the equivalence of (a), (b), and (c) follows from (ii) and (iii). Hence, it suffices to show: (c) $\Leftrightarrow$ (d). Because μ-integrability of f and g implies that $f \odot \mu$ and $g \odot \mu$ are finite measures, applying Theorem 1.72 completes the proof.

Proof of Corollary 3.77

Note that

$$\forall\, (\omega_1, \omega_2) \in \Omega_1 \times \Omega_2\colon 1_C(\omega_1, \omega_2) = 1_{C_{\omega_1}}(\omega_2) = 1_{C_{\omega_2}}(\omega_1). \tag{3.85}$$

Now,

$$
\begin{aligned}
\mu_1 \otimes \mu_2(C) &= \int 1_C \, d(\mu_1 \otimes \mu_2) && [(3.9)]\\
&= \int \int 1_C(\omega_1, \omega_2)\, \mu_2(d\omega_2)\, \mu_1(d\omega_1) && [(3.80)]\\
&= \int \int 1_{C_{\omega_1}}(\omega_2)\, \mu_2(d\omega_2)\, \mu_1(d\omega_1) && [(3.85)]\\
&= \int \mu_2(C_{\omega_1})\, \mu_1(d\omega_1). && [(3.9)]
\end{aligned}
$$

The proof of the second equation is analog.

Exercises

3.1 Construct a representation of the identity function on $\Omega = \{1, \dots, n\}$ as a weighted sum of indicators of elements of $\mathscr{A}$, where $n \in \mathbb{N}$ and $(\Omega, \mathscr{A})$ with $\mathscr{A} = \mathscr{P}(\Omega)$.

3.2 Prove that, for every nonnegative step function, there exists a normal representation (see Rem. 3.8).

3.3 Consider the measure space $(\mathbb{R}, \mathscr{B}, \lambda)$, where λ is the Lebesgue measure. Show that the number $\sum_{i=1}^{n} \alpha_i\, \lambda(A_i)$ assigned to the nonnegative step function f defined in Example 3.7 is identical for the four specified representations of f, two of which are nonnormal representations.

3.4 Let $(\Omega, \mathscr{A})$ be a measurable space and let $A \in \mathscr{A}$. Show that if $f = \sum_{i=1}^{n} \alpha_i 1_{A_i}$ is a normal representation of a nonnegative step function, then the product $1_A f$ of the indicator 1_A and f is also a normal representation of a nonnegative step function, and $1_A f = \sum_{i=1}^{n} \alpha_i 1_{A \cap A_i}$.

3.5 Prove Equation (3.8).

3.6 Compute the integral of the identity mapping $id\colon \Omega \to \Omega$ with respect to the counting measure $\mu_\#$ on $\mathscr{P}(\Omega)$, where $\Omega = \{1, \dots, n\}$. Then look at it for $n = 5$.

3.7 Compute the integrals $\int f_1 \, d\lambda$ and $\int f_2 \, d\lambda$ of the functions f_1 and f_2 defined in Example 3.17.

3.8   Prove the propositions of Example 3.39.

3.9   Let $f: (\Omega, \mathscr{A}, \mu) \to (\overline{\mathbb{R}}, \overline{\mathscr{B}})$ be measurable and $A \in \mathscr{A}$. Show that $1_A f \underset{\mu}{=} 0$ if $\mu(A) = 0$.

3.10   Prove Equation (3.52).

3.11   Compute the integral of the function $f(x) = 10 - x^2$ considered in Example 3.64.

3.12   Prove the proposition of Remark 3.71.

Solutions

3.1   The identity function $id: \Omega \to \Omega$ on $\Omega = \{1, \dots, n\}$ is defined by

$$id(i) = i, \quad \forall i \in \Omega.$$

According to Example 2.9, it is $(\mathscr{A}, \mathscr{A}_0)$-measurable for all σ-algebras $\mathscr{A}_0 \subset \mathscr{A}$. Now consider the set $\{1, \dots, n\}$ of values of id and the partition $\{\{1\}, \dots, \{n\}\}$ of Ω. Then

$$id = \sum_{i=1}^{n} i \cdot 1_{\{i\}} = 1 \cdot 1_{\{1\}} + \cdots + n \cdot 1_{\{n\}}.$$

3.2   Let $f: (\Omega, \mathscr{A}) \to (\mathbb{R}, \mathscr{B})$ be a nonnegative step function, with $f = \sum_{i=1}^{n} \alpha_i 1_{A_i}$. Define, for all nonempty $J \subset \{1, \dots, n\}$,

$$B_J := \left(\bigcap_{i \in J} A_i \right) \cap \left(\bigcap_{i \notin J} A_i^c \right).$$

These are $2^n - 1$ sets, where several of them may be empty, and all are pairwise disjoint. Then

$$f = \sum_{J:\, B_J \neq \varnothing} \left(\sum_{i \in J} \alpha_i \right) \cdot 1_{B_J}$$

is a normal representation of f.

3.3   We compute the sum for all four representations of f. The first one is:

$$\sum_{i=1}^{4} \alpha_i\, \lambda(A_i) = 2 \cdot (1 - 0) + 5 \cdot (2 - 1) + 4 \cdot (3 - 2) + 1 \cdot (4 - 3)$$

$$= 2 \cdot 1 + 5 \cdot 1 + 4 \cdot 1 + 1 \cdot 1 = 12.$$

The sum for the second representation of f is:

$$\sum_{i=1}^{5} \gamma_i\, \lambda(C_i) = 2 \cdot (.5 - 0) + 2 \cdot (1 - .5) + 5 \cdot (2 - 1) + 4 \cdot (3 - 2) + 1 \cdot (4 - 3)$$

$$= 2 \cdot .5 + 2 \cdot .5 + 5 \cdot 1 + 4 \cdot 1 + 1 \cdot 1 = 12.$$

The sum for the third representation of f, which is nonnormal, is:

$$\sum_{i=1}^{3} \beta_i \, \lambda(B_i) = 2 \cdot (2 - 0) + 3 \cdot (3 - 1) + 1 \cdot (4 - 2)$$

$$= 2 \cdot 2 + 3 \cdot 2 + 1 \cdot 2 = 12.$$

The sum for the fourth representation of f, which is also nonnormal, is:

$$\sum_{i=1}^{4} \delta_i \, \lambda(D_i) = 1 \cdot (4 - 0) + 1 \cdot (3 - 0) + 2 \cdot (3 - 1) + 1 \cdot (2 - 1)$$

$$= 1 \cdot 4 + 1 \cdot 3 + 2 \cdot 2 + 1 \cdot 1 = 12.$$

Obviously, all four sums are identical.

3.4 Let $f = \sum_{i=1}^{n} \alpha_i \mathbf{1}_{A_i}$, where $A_1, \ldots, A_n \in \mathcal{A}$. This implies $A \cap A_1, \ldots, A \cap A_n \in \mathcal{A}$, for $A \in \mathcal{A}$. Therefore, and because of $\mathbf{1}_A \cdot f = \sum_{i=1}^{n} \alpha_i \mathbf{1}_{A \cap A_i}$, the function $\mathbf{1}_A \cdot f$ is a nonnegative step function. If $f = \sum_{i=1}^{n} \alpha_i \mathbf{1}_{A_i}$ is a normal representation, then $A_i \cap A_j = \emptyset$ for $i \neq j$, which implies $(A \cap A_i) \cap (A \cap A_j) = A \cap (A_i \cap A_j) = A \cap \emptyset = \emptyset$, for $i \neq j$. Therefore, $\mathbf{1}_A \cdot f = \sum_{i=1}^{n} \alpha_i \mathbf{1}_{A \cap A_i}$ is a normal representation as well.

3.5 $\int_A \alpha \, d\mu = \int \alpha \, \mathbf{1}_A \, d\mu = \alpha \, \mu(A)$ [see Eq. (3.4)].

3.6 Consider the elements $\{1\}, \ldots, \{n\}$ of $\mathcal{A} = \mathcal{P}(\Omega)$ and $id = \sum_{\omega=1}^{n} \omega \cdot \mathbf{1}_{\{\omega\}}$. According to Definition 3.10,

$$\int id \, d\mu_\# = \sum_{\omega=1}^{n} \omega \cdot \mu_\#(\{\omega\})$$

$$= 1 \cdot \mu_\#(\{1\}) + 2 \cdot \mu_\#(\{2\}) + \cdots + n \cdot \mu_\#(\{n\}) \qquad (3.86)$$

$$= \sum_{i=1}^{n} i = \frac{n(n+1)}{2}$$

is the integral of id over Ω with respect to the measure $\mu_\#$. Hence, in this example, the integral $\int id \, d\mu_\#$ is the sum over all elements in Ω. For $n = 5$, this formula yields $\int id \, d\mu_\# = 15$.

3.7 In Example 3.17, we considered the measure space $(\mathbb{R}, \mathcal{B}, \lambda)$, where λ is the Lebesgue (or length) measure on $\mathcal{B}$. Remember, the Lebesgue measure satisfies

$$\lambda(]a, b]) = \lambda([a, b]) = b - a,$$

for $a < b$ [see Eq. (1.54)]. We also considered $f_1 = \alpha_1 \mathbf{1}_{A_1}$ with $A_1 = [0, (1 - .50)^{1/2}]$ and $\alpha_1 = .50$. Hence, $f_1 = .50 \cdot \mathbf{1}_{A_1}$. Therefore,

$$\int f_1 \, d\lambda = \alpha_1 \cdot \lambda(A_1) = .50 \cdot \lambda(A_1) = .50 \cdot (1 - .50)^{1/2} = .50 \cdot .50^{1/2} \approx .3536.$$

This is the area shaded in the left part of Figure 3.3.

Similarly, in Example 3.17, we also considered $f_2 = \sum_{i=1}^{3} \beta_i 1_{B_i}$ with the three intervals

$$B_1 = [0, (1-.75)^{1/2}], \quad B_2 =](1-.75)^{1/2}, (1-.50)^{1/2}],$$
$$B_3 =](1-.50)^{1/2}, (1-.25)^{1/2}].$$

Again note that B_1, B_2, B_3 is a sequence of elements of $\mathscr{A}$. Furthermore, $f_2 = \sum_{i=1}^{3} \beta_i 1_{B_i}$ with $\beta_1 = .75$, $\beta_2 = .50$, and $\beta_3 = .25$. Hence, the integral of $f_2 = \sum_{i=1}^{3} \beta_i 1_{B_i}$ with respect to λ is

$$\int f_2 \, d\lambda = \sum_{i=1}^{3} \beta_i \cdot \lambda(B_i) = .75 \cdot \lambda(B_1) + .50 \cdot \lambda(B_2) + .25 \cdot \lambda(B_3)$$

$$= .75 \cdot [(1-.75)^{1/2}] + .50 \cdot [(1-.50)^{1/2} - (1-.75)^{1/2}]$$
$$+ .25 \cdot [(1-.25)^{1/2} - (1-.50)^{1/2}]$$

$$\approx .75 \cdot .50 + .50 \cdot .2071 + .25 \cdot .1589 \approx 0.3750 + .1036 + .0397 = .5183.$$

This is the area shaded in the middle part of Figure 3.3. The integral of $f_3 = \sum_{i=1}^{7} \gamma_i 1_{C_i}$ can be computed correspondingly. It is the area shaded in the right part of Figure 3.3.

3.8 If $f: (\Omega, \mathscr{A}, \mu) \to (\overline{\mathbb{R}}, \overline{\mathscr{B}})$ is μ-integrable and $A, B \subset \mathscr{A}$, then $1_{A \cup B} f$ is μ-integrable (see Lemma 3.33) and

$$\int_{A \cup B} f \, d\mu = \int 1_{A \cup B} f \, d\mu \qquad [(3.30)]$$

$$= \int (1_A + 1_B - 1_{A \cap B}) f \, d\mu \qquad [(1.34)]$$

$$= \int (1_A f + 1_B f - 1_{A \cap B} f) \, d\mu$$

$$= \int 1_A f \, d\mu + \int 1_B f \, d\mu - \int 1_{A \cap B} f \, d\mu \qquad [(3.34)]$$

$$= \int_A f \, d\mu + \int_B f \, d\mu - \int_{A \cap B} f \, d\mu. \qquad [(3.30)]$$

If $A \cap B = \emptyset$ and $f: (\Omega, \mathscr{A}, \mu) \to (\overline{\mathbb{R}}, \overline{\mathscr{B}})$ is quasi-μ-integrable, then $1_{A \cup B} \cdot f = 1_A \cdot f + 1_B \cdot f$, and the functions $1_{A \cup B} f$, $1_A f$, $1_B f$ are quasi-μ-integrable (see Lem. 3.33). If $\int f^+ \, d\mu$ is finite, then $\int 1_{A \cup B} \cdot f^+ \, d\mu$, $\int 1_A \cdot f^+ \, d\mu$, and $\int 1_B \cdot f^+ \, d\mu$ are finite as well (see Lem. 3.33). If $\int f^+ \, d\mu$ is infinite, then quasi-μ-integrability of f implies that $\int f^- \, d\mu$ is finite and, according to Lemma 3.33, also the integrals $\int 1_{A \cup B} \cdot f^- \, d\mu$, $\int 1_A \cdot f^- \, d\mu$, and $\int 1_B \cdot f^- \, d\mu$. Hence, in both cases,

$$\int 1_{A \cup B} \cdot f \, d\mu = \int 1_{A \cup B} \cdot f^+ \, d\mu - \int 1_{A \cup B} \cdot f^- \, d\mu \quad [(3.27)]$$

$$= \int 1_A \cdot f^+ \, d\mu + \int 1_B \cdot f^+ \, d\mu$$

$$- \left(\int 1_A \cdot f^- \, d\mu + \int 1_B \cdot f^- \, d\mu \right) \quad [(3.34)]$$

$$= \left(\int 1_A \cdot f^+ \, d\mu - \int 1_A \cdot f^- \, d\mu \right)$$

$$+ \left(\int 1_B \cdot f^+ \, d\mu - \int 1_B \cdot f^- \, d\mu \right)$$

$$= \int 1_A \cdot f \, d\mu + \int 1_B \cdot f \, d\mu. \qquad\qquad [(3.27)]$$

3.9
$$1_A(\omega) \cdot f(\omega) = \begin{cases} 0, & \text{if } \omega \notin A \\ f(\omega), & \text{if } \omega \in A. \end{cases}$$

Therefore, $\{\omega \in \Omega: 1_A(\omega) \cdot f(\omega) \neq 0\} \subset A$. Hence, $\mu(\{\omega \in \Omega: 1_A(\omega) \cdot f(\omega) \neq 0\}) \leq \mu(A) = 0$.

3.10 Let $\alpha \geq 0$ and remember that the measure $\alpha \mu$ on $(\Omega, \mathcal{A})$ is defined by $(\alpha \mu)(A)) = \alpha \mu(A)$, for all $A \in \mathcal{A}$. The proof is conducted in three steps: (a) for a nonnegative step function, (b) for a nonnegative numerical measurable function, and (c) for a μ-integrable numerical function (see Rem. 3.30).
  (a) If f is a nonnegative step function and $f = \sum_{i=1}^{n} \alpha_i 1_{A_i}$ a normal representation, then

$$\int f \, d(\alpha \mu) = \sum_{i=1}^{n} \alpha_i \, (\alpha \mu)(A_i) = \alpha \sum_{i=1}^{n} \alpha_i \, \mu(A_i) = \alpha \int f \, d\mu.$$

  (b) If f is a nonnegative numerical measurable function and f_i, $i \in \mathbb{N}$, is an increasing sequence of nonnegative step functions with $\lim_{i \to \infty} f_i = f$, then

$$\int f \, d(\alpha \mu) = \lim_{i \to \infty} \int f_i \, d(\alpha \mu)$$

$$= \lim_{i \to \infty} \alpha \int f_i \, d\mu \qquad\qquad [(a)]$$

$$= \alpha \lim_{i \to \infty} \int f_i \, d\mu = \alpha \int f \, d\mu.$$

  (c) If f is a μ-integrable numerical function, then

$$\int f \, d(\alpha \mu) = \int f^+ \, d(\alpha \mu) - \int f^- \, d(\alpha \mu)$$

$$= \alpha \int f^+ \, d\mu - \alpha \int f^- \, d\mu \qquad\qquad [(b)]$$

$$= \alpha \int f \, d\mu.$$

3.11 Because the derivative $g'(x)$ of a function $g(x) = \alpha + \beta x + \gamma x^n$, $\alpha, \beta, \gamma \in \mathbb{R}$, $n \in \mathbb{N}$, is $g'(x) = \beta + \gamma n x^{n-1}$, the indefinite integral of $f(x)$ is

$$\int f(x) \, dx = F(x) = 10 x - \frac{x^3}{3} + c, \qquad c \in \mathbb{R},$$

and therefore,

$$\int_a^b f(x)\, dx = F(x)\Big|_a^b := F(b) - F(a) = \left(10\,x - \frac{x^3}{3}\right)\Big|_a^b.$$

For $a = -5$ and $b = 5$, this equation yields

$$\int_{-5}^5 f(x)\, dx = \left(10\,x - \frac{x^3}{3}\right)\Big|_{-5}^5 = \left(50 - \frac{125}{3}\right) - \left(-50 + \frac{125}{3}\right) = 100 - \frac{250}{3} = 16.\bar{6}.$$

Hence,

$$\int_{[-5,\,5]} f\, d\lambda = \int_{-5}^5 f(x)\, dx = 16.\bar{6}.$$

3.12 Let $A \in \mathscr{A}$ with $\mu(A) = 0$ and let f be a density f of ν with respect to μ (i.e., $\nu = f \odot \mu$). Then

$$\nu(A) = \int 1_A\, d\nu = \int 1_A\, df \odot \mu \qquad [(3.8),\ \nu = f \odot \mu]$$

$$= \int 1_A \cdot f\, d\mu = 0. \qquad [(3.72),\ (3.42)]$$

Hence, $\nu \underset{\mathscr{A}}{\ll} \mu$.

Part II

PROBABILITY, RANDOM VARIABLE, AND ITS DISTRIBUTION

4

Probability measure

In chapter 1, we introduced the concept of a *measure*, and we treated various examples of measures and some of their properties. In this chapter, we turn to a special class of examples called *probability measures*. We start with the definition of a probability measure, then turn to conditional probabilities and the most important theorems related to conditional probability: the *multiplication rule*, the *theorem of total probability*, and *Bayes' theorem*. Furthermore, we introduce the concept of a *conditional-probability measure*. Next, we define *independence of events* and *independence of sets of events* with respect to a probability measure. A section on *conditional independence given an event* concludes this chapter.

4.1 Probability measure and probability space

Now we introduce the concept of a *probability measure* as defined by Kolmogorov (?/1977) (for the English version of this book, see Kolmogorov, 1956). As we shall see, a probability measure is a special finite measure that is standardized.

4.1.1 Definition

Definition 4.1 [Probability measure]
Let $(\Omega, \mathcal{A})$ be a measurable space. Then the function $P\colon \mathcal{A} \to [0, 1]$ is called a proba-bility measure on $(\Omega, \mathcal{A})$, if the following conditions hold:

(a) $P(\Omega) = 1$ (standardization).

(b) $P(A) \geq 0, \ \forall A \in \mathcal{A}$ (nonnegativity).

(c) $A_1, A_2, \ldots \in \mathcal{A}$ are pairwise disjoint $\Rightarrow P\left(\bigcup\limits_{i=1}^{\infty} A_i\right) = \sum\limits_{i=1}^{\infty} P(A_i)$ (σ-additivity).

Probability and Conditional Expectation: Fundamentals for the Empirical Sciences, First Edition. Rolf Steyer and Werner Nagel.
© 2017 John Wiley & Sons, Ltd. Published 2017 by John Wiley & Sons, Ltd.
Companion website: http://www.probability-and-conditional-expectation.de

Remark 4.2 [Probability and probability space]   Let P be a probability measure on $(\Omega, \mathscr{A})$. Then the triple $(\Omega, \mathscr{A}, P)$ is called a *probability space* and the value $P(A)$ of P is called the *probability* of A.   ◁

Remark 4.3 [Elementary event and event]   Let $(\Omega, \mathscr{A}, P)$ be a probability space. Then $A \in \mathscr{A}$ is called an *event*, and a singleton $\{\omega\}$, $\omega \in \Omega$, is called an *elementary event*, if $\{\omega\} \in \mathscr{A}$. Note the distinction between an *outcome* $\omega \in \Omega$ and an *elementary event* $\{\omega\} \in \mathscr{A}$ (see Exercise 4.1). Also note that the term *event* is only used in the context of a probability space $(\Omega, \mathscr{A}, P)$. Otherwise, $A \in \mathscr{A}$ is called a *measurable set*.   ◁

4.1.2   Formal and substantive meaning of probabilistic terms

We distinguish between the mathematical or formal meaning of probabilistic terms and the meaning of these terms if used in an application of probability theory to a concrete real-world phenomenon. Often, such a real-world phenomenon is a random experiment such as flipping a coin. In such a case, the set Ω represents the set of possible outcomes of the random experiment considered; and, in this sense, Ω has a real-world interpretation. In other words, in this case Ω is not only an abstract set anymore. This *real-world meaning* or *substantive meaning* is additional to their mathematical or *formal* meaning, namely being elements of the (abstract) set Ω.

The terms *probability measure, probability of an event*, and so on hint at an important area of application of probability theory: real-world phenomena called random experiments. However, formally speaking, a probability measure is simply a label for a measure on a measurable space $(\Omega, \mathscr{A})$ satisfying $P(\Omega) = 1$. If Ω is not just an abstract set but represents a concrete random experiment, then the probability of an event $A \in \mathscr{A}$ corresponds to the common language meaning of the term *probability*.

Remark 4.4 [No time order between events]   The intuitive concept of an event often implies that events are ordered with respect to time. That is, in common language an 'event' is prior, simultaneous, or posterior to another 'event'. In contrast, events as defined in probability theory are not necessarily ordered with respect to time. However, time order between events and sets of events can be introduced with respect to a filtration (see Def. 4.17, Fig. 4.1, and Example 4.19).   ◁

Remark 4.5 [A priori perspective]   If we apply probability theory to real-world phenomena, then we consider random experiments from the *a priori perspective*. Hence, the possible outcomes of a random experiment and events are considered *before* they happen. Even if an event already happened, we do *as if* it did not happen when we talk about its probability (see also Rem. 4.13).   ◁

4.1.3   Properties of a probability measure

Comparing conditions (a) to (c) of the definition of a probability measure to the conditions defining a measure (see Def. 1.43) shows that (b) and (c) are identical; only condition (a) differs. However, $P(\Omega) = 1$ implies $P(\emptyset) = 0$, because σ-additivity of P yields

$$P(\Omega) = P(\Omega \cup \emptyset \cup \emptyset \cup \ldots) = P(\Omega) + \sum_{i=1}^{\infty} P(\emptyset).$$

Hence, $\sum_{i=1}^{\infty} P(\emptyset) = P(\Omega) - P(\Omega) = 0$, and this yields $P(\emptyset) = 0$. This proves Corollary 4.6:

Corollary 4.6 [A probability measure is a measure]
A measure P on $(\Omega, \mathscr{A})$ is a probability measure on $\mathscr{A}$ if and only if $P(\Omega) = 1$.

A direct implication of this corollary is that all rules of computation for a measure (see Box 1.1) also hold for a probability measure. For convenience, these rules are explicitly formulated for probability measures in Box 4.1 using the additional property $P(\Omega) = 1$.

Remark 4.7 [Distribution vs. probability measure] A probability measure on $(\Omega, \mathscr{A})$ is also called a *distribution on $(\Omega, \mathscr{A})$*. Although this term is preferably used in the context of a random variable (see Def. 5.3), the term *distribution* is well defined without referring to a random variable.   ◁

4.1.4   Examples

Example 4.8 [Continuous uniform distribution] Let $\mathscr{B}_2$ denote the Borel σ-algebra on $\mathbb{R}^2$, and consider a probability space $(\Omega, \mathscr{A}, P)$, where $\Omega \in \mathscr{B}_2$, $\mathscr{A} = \mathscr{B}_2|_{\Omega} := \{\Omega \cap A : A \in \mathscr{B}_2\}$ is the trace of $\mathscr{B}_2$ in Ω (see Example 1.10). Furthermore, let λ_2 denote the Lebesgue measure on $(\mathbb{R}^2, \mathscr{B}_2)$, assume $0 < \lambda_2(\Omega) < \infty$, and define

$$P(A) = \frac{\lambda_2(A)}{\lambda_2(\Omega)}, \quad \forall A \in \mathscr{A}. \tag{4.1}$$

Then P is the *continuous uniform distribution* over Ω. The relative size of the set $A \in \mathscr{A}$ represents the probability $P(A)$, and Figure 1.1 can be used to illustrate some of its properties, for example, Rules (iii) to (ix) of Box 4.1. This example is generalized in section 8.2.1.   ◁

Example 4.9 [Joe and Ann with randomized assignment – continued] In Example 1.9, we specified the set

$$\Omega = \{(Joe, no, -), (Joe, no, +), \ldots, (Ann, yes, +)\},$$

which is also presented in the first column of Table 4.1. In this table, we also specify the probability measure P on $\mathscr{A} = \mathscr{P}(\Omega)$ by the probabilities of the eight elementary events $\{\omega\} \in \Omega$. Except for the empty set, which has probability $P(\emptyset) = 0$, all $2^8 = 256$ elements of $\mathscr{A}$ are either one of the eight elementary events $\{(Joe, no, -)\}$, $\{(Joe, no, +)\}$, $\ldots$, $\{(Ann, yes, +)\}$ or a union of some of these elementary events. Note that elementary events are always pairwise *disjoint* (i.e., $\{\omega_i\} \cap \{\omega_j\} = \emptyset$, if $\omega_i \neq \omega_j$). Therefore, the probabilities of their unions can

Box 4.1   Rules of computation for probabilities.

Let $(\Omega, \mathcal{A}, P)$ be a probability space.
If $A_1, A_2, \ldots \in \mathcal{A}$ are pairwise disjoint, then

$$P\left(\bigcup_{i=1}^{\infty} A_i\right) = \sum_{i=1}^{\infty} P(A_i) \qquad \text{(σ-additivity)} \qquad \text{(i)}$$

$$P\left(\bigcup_{i=1}^{n} A_i\right) = \sum_{i=1}^{n} P(A_i), \quad \forall\, n \in \mathbb{N}. \qquad \text{(finite additivity)} \qquad \text{(ii)}$$

If $A, B \in \mathcal{A}$, then,

$$P(A) = P(A \cap B) + P(A \setminus B) \qquad \text{(iii)}$$
$$P(A^c) = 1 - P(A) \qquad \text{(iv)}$$
$$P(A) \leq P(B), \quad \text{if } A \subset B \qquad \text{(monotonicity)} \qquad \text{(v)}$$
$$P(A \setminus B) = P(A) - P(A \cap B) \qquad \text{(vi)}$$
$$P(A \cup B) = P(A) + P(B) - P(A \cap B) \qquad \text{(vii)}$$
$$P(A) = 1 \Rightarrow P(A \cap B) = P(B) \qquad \text{(viii)}$$
$$P(A) = 0 \Rightarrow P(A \cup B) = P(B). \qquad \text{(ix)}$$

Let $A \in \mathcal{A}$ and let $\Omega_0 \subset \Omega$ be finite or countable with $P(\Omega_0) = 1$.
If, for all $\omega \in \Omega_0$, $\{\omega\} \in \mathcal{A}$, then

$$P(A) = \sum_{\omega \in A \cap \Omega_0} P(\{\omega\}). \qquad \text{(x)}$$

If $A_1, A_2, \ldots \in \mathcal{A}$, then

$$P\left(\bigcup_{i=1}^{\infty} A_i\right) \leq \sum_{i=1}^{\infty} P(A_i). \qquad \text{(σ-subadditivity)} \qquad \text{(xi)}$$

easily be computed using finite additivity of the probability measure [see Rule (ii) of Box 4.1].
In order to illustrate this point, consider the event that *Joe is drawn*,

$$A = \{(Joe, no, -), (Joe, no, +), (Joe, yes, -), \ (Joe, yes, +)\},$$

and the event that *the drawn person is successful*,

$$C = \{(Joe, no, +), (Joe, yes, +)\}, (Ann, no, +), (Ann, yes, +)\}.$$

The event A has the probability

$$P(A) = P[\{(Joe, no, -)\}] + P[\{(Joe, no, +)\}] + P[\{(Joe, yes, -)\}] + P[\{(Joe, yes, +)\}]$$
$$= .09 + .21 + .04 + .16 = .5.$$

Table 4.1 Joe and Ann with randomized assignment: probability measures.

Unit, Treatment, Success	$P(\{\omega\})$	$P^B(\{\omega\})$	$P^A(\{\omega\})$	$P^{A^c}(\{\omega\})$
(*Joe, no, −*)	.09	0	.18	0
(*Joe, no, +*)	.21	0	.42	0
(*Joe, yes, −*)	.04	.1	.08	0
(*Joe, yes, +*)	.16	.4	.32	0
(*Ann, no, −*)	.24	0	0	.48
(*Ann, no, +*)	.06	0	0	.12
(*Ann, yes, −*)	.12	.3	0	.24
(*Ann, yes, +*)	.08	.2	0	.16

Note: P, P^B, P^A, and P^{A^c} are probability measures on $(\Omega, \mathscr{A})$.

Similarly, the event C has the probability

$$P(C) = P[\{(Joe, no, +)\}] + P[\{(Joe, yes, +)\}] + P[\{(Ann, no, +)\}] + P[\{(Ann, yes, +)\}]$$
$$= .21 + .16 + .06 + .08 = .51,$$

and the event *Joe is drawn and is successful*, $A \cap C = \{(Joe, no, +), (Joe, yes, +)\}$, has the probability

$$P(A \cap C) = P[\{(Joe, no, +)\}] + P[\{(Joe, yes, +)\}] = .21 + .16 = .37.$$

The probability measures specified in the last three columns of Table 4.1 are treated in Examples 4.34 and 4.35. ◁

Remark 4.10 [Other examples] In section 8.1, probability measures on the measurable space $(\mathbb{N}_0, \mathscr{P}(\mathbb{N}_0))$ are considered, such as the *binomial distribution* (see Def. 8.7), the *Poisson distribution* (see Def. 8.14), and the *geometric distribution* (see Def. 8.20). In all these examples, a probability measure on $(\mathbb{N}_0, \mathscr{P}(\mathbb{N}_0))$ is uniquely defined, if the probabilities of the elementary events $\{x\}$ are determined for all $x \in \mathbb{N}_0$ [see Box 4.1 (x) for $\Omega_0 = \mathbb{N}_0$].

The example of the Poisson distribution shows that, even for the countably infinite set $\mathbb{N}_0$, there are probability measures P_λ on $(\mathbb{N}_0, \mathscr{P}(\mathbb{N}_0))$ with $P_\lambda(\{x\}) > 0$, for all $x \in \mathbb{N}_0$, and

$$P_\lambda(\mathbb{N}_0) = \sum_{x \in \mathbb{N}_0} P_\lambda(\{x\}) = \sum_{x=0}^{\infty} P_\lambda(\{x\}) = 1.$$

For another example, see Exercise 4.2. ◁

Example 4.11 [Finite mixture of probability measures]   In Example 1.62 we already noted that the weighted sum of measures on a measurable space $(\Omega, \mathcal{A})$ is again a measure on $(\Omega, \mathcal{A})$. With an additional assumption, this also applies to probability measures. More precisely, if $P_1, \ldots, P_n$ are probability measures on $(\Omega, \mathcal{A})$, $\alpha_i \geq 0$, $i = 1, \ldots, n$, and we additionally assume $\sum_{i=1}^{n} \alpha_i = 1$, then $\sum_{i=1}^{n} \alpha_i P_i$ is again a probability measure on $(\Omega, \mathcal{A})$. It is called a *finite mixture* of $P_1, \ldots, P_n$. Such a finite mixture of probability measures is illustrated by Example 4.35 using conditional-probability measures.   ◁

4.2   Conditional probability

Conditional probabilities can be used to describe *dependencies* between two events $A, B \in \mathcal{A}$ with respect to a probability measure P on $\mathcal{A}$. In section 4.2.7, we also use this concept in order to introduce the concept of a *conditional-probability measure*.

4.2.1   Definition

Definition 4.12 [Conditional probability]
Let $(\Omega, \mathcal{A}, P)$ be a probability space, let $A, B \in \mathcal{A}$, and let $P(B) > 0$. Then,

$$P(A \mid B) := \frac{P(A \cap B)}{P(B)} \tag{4.2}$$

is called the conditional probability of A given B with respect to P.

Remark 4.13 [A priori perspective]   The conditional probability $P(A \mid B)$ is the probability of the event A if it is known that the event B occurred. In order to compute $P(A \mid B)$ according to Equation (4.2), we need the (unconditional) probability $P(B)$. The fact that B occurred is reflected by $P(B \mid B) = 1$.   ◁

Remark 4.14 [Continuous uniform distribution – continued]   In Example 4.8, we defined the continuous uniform distribution on $(\Omega, \mathcal{A})$ by Equation (4.1). Using the area of the ellipses presented in Figure 1.1, the conditional probability $P(A \mid B)$ corresponds to the area of the intersection $A \cap B$ divided by the area of B.   ◁

Example 4.15 [Flipping a coin two times]   Consider the random experiment of flipping a coin two times, the measurable space $(\Omega, \mathcal{A})$ of which is the same as in subsection 2.2.2; the probability measure has been specified by Equation (2.43). The conditional probability $P(B \mid A)$ that we flip *heads* in the second flip (B) given that we flip *heads* in the first flip (A) is $1/2$, which is equal to the *unconditional* probability $P(B)$ of flipping *heads* in the second flip. In such a case, the two events A and B are *independent* (see section 4.3). Note that the conditional probability $P(A \mid B)$ that we flip *heads* in the first flip (A) given that we flip *heads* in the second flip (B) is also equal to the unconditional probability $P(A)$ of flipping *heads* in the first flip. This example shows that we may condition on events that occur later in time and that a conditional probability does not necessarily describe a causal dependence. Note, however, that conditional probabilities *can* be used to describe causal dependencies, provided that additional assumptions hold (see Examples 4.16 and 4.36).

As another example, consider the event *flipping at least one heads* (A) and the event *no heads are flipped in the first flip* (B). In this case,

$$P(A \mid B) = \frac{1}{2} \neq P(A) = \frac{3}{4},$$

and the two events are not independent (see section 4.3).   ◁

Example 4.16 [Joe and Ann with randomized assignment – continued]   Consider again Table 4.1, define $\Omega_U = \{Joe, Ann\}$ and $\Omega_X = \{yes, no\}$, and let

$$C = \Omega_U \times \Omega_X \times \{+\} = \{(Joe, no, +), (Joe, yes, +), (Ann, no, +), (Ann, yes, +)\}$$

be the event that the *drawn person is successful*. Furthermore, let

$$B = \Omega_U \times \{yes\} \times \Omega_Y = \{(Joe, yes, -), (Joe, yes, +), (Ann, yes, -), (Ann, yes, +)\}$$

denote the event that the *drawn person is treated*. Then, Equation (4.2) yields:

$$P(C \mid B) = \frac{P(C \cap B)}{P(B)} = \frac{P(\Omega_U \times \{yes\} \times \{+\})}{P(\Omega_U \times \{yes\} \times \Omega_Y)} = \frac{.16 + .08}{.04 + .16 + .12 + .08} = .6.$$

Conditioning on the event B^c that the *drawn person is not treated* yields

$$P(C \mid B^c) = \frac{P(C \cap B^c)}{P(B^c)} = \frac{P(\Omega_U \times \{no\} \times \{+\})}{P(\Omega_U \times \{no\} \times \Omega_Y)} = \frac{.21 + .06}{.09 + .21 + .24 + .06} = .45.$$

In this example, the difference $P(C \mid B) - P(C \mid B^c) = .6 - .45$ *can* be used to evaluate the average effect of the treatment. This is substantiated in more detail in Example 4.36.   ◁

4.2.2   Filtration and time order between events and sets of events

As mentioned in Remark 4.4, the definition of an event in probability theory does not presume that there is a time order between events and sets of events. However, in many applications of probability theory, such a time order is important. In Example 4.16, for instance, it is crucial that the event C is *posterior* to the event B. Such a time order is formalized in the theory of stochastic processes (see, e.g., Bauer, 1996; Klenke, 2013) and in the theory of causal effects (see, e.g., Steyer *et al.*, 2014; Mayer *et al.*, 2014).

Definition 4.17 [Filtration]
Let $(\Omega, \mathcal{A})$ be a measurable space and $T \subset \mathbb{R}$. A family $(\mathcal{F}_t, t \in T)$ of sub-σ-algebras $\mathcal{F}_t$ of $\mathcal{A}$ is called a filtration in $\mathcal{A}$, if $\mathcal{F}_s \subset \mathcal{F}_t$ for all $s, t \in T$ with $s \leq t$.

Referring to such a filtration, time order between events can be introduced as follows.

Remark 4.18 [Event A is prior, simultaneous, and posterior to event B]   Let $(\Omega, \mathscr{A}, P)$ be a probability space, $(\mathscr{F}_t, t \in T)$ a filtration in $\mathscr{A}$, and $A, B \in \mathscr{A}$.

(i) The event A is called *prior to B* (and B is called *posterior to A*) in $(\mathscr{F}_t, t \in T)$, if there is an $s \in T$ such that $A \in \mathscr{F}_s$, $B \notin \mathscr{F}_s$, and a $t \in T$, $t > s$, such that $B \in \mathscr{F}_t$.

(ii) Assume that T is finite. Then the event A is called *simultaneous to B* in $(\mathscr{F}_t, t \in T)$, if there is a $t \in T$ such that $A, B \in \mathscr{F}_t$ and no $s \in T$, $s < t$, such that $A \in \mathscr{F}_s$ or $B \in \mathscr{F}_s$.

Note that the concept of simultaneity of events can also be extended to cases in which T is not finite. For simplicity, we confine ourselves to the finite case.   ◁

Example 4.19 [Joe and Ann with randomized assignment – continued]   In the random experiment described by Table 4.1, the event A that Joe is drawn (see Example 4.9) is prior to the event

$$B = \{(Joe, yes, +), (Joe, yes, -), (Ann, yes, +), (Ann, yes, -)\}$$

that the drawn person is treated, which itself is prior to the event C that the drawn person is successful. This time order in the real-world can be represented formally by the following filtration:

$$\mathscr{F}_1 := \sigma(\{A\}), \qquad \mathscr{F}_2 := \sigma(\{A, B\}), \qquad \mathscr{F}_3 := \sigma(\{A, B, C\}) = \mathscr{P}(\Omega),$$

using the concept of a σ-algebra generated by a set system (see Def. 1.13). With respect to the filtration $(\mathscr{F}_t, t \in T)$, $T = \{1, 2, 3\}$, the event A is prior to B, because $A \in \mathscr{F}_1$, $B \notin \mathscr{F}_1$, but $B \in \mathscr{F}_2$ (see Fig. 4.1, Rem. 4.18, and Exercise 4.3).   ◁

Remark 4.20 [Formal and substantive meaning of time order]   As noted in section 4.2.1, we distinguish between the mathematical or formal meaning of a probabilistic term on one side and the meaning of these terms if used in an application of probability theory to a concrete real-world phenomenon on the other side. This also applies to the terms *prior, simultaneous*, and *posterior* with respect to a filtration. In applications in which the elements of the set T represent time points, these terms not only have a formal meaning that is specified by their

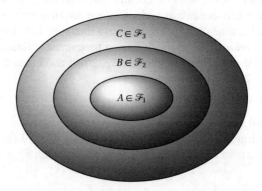

Figure 4.1   A filtration with $T = \{1, 2, 3\}$.

mathematical definition, but also a substantive meaning: In the *real world*, an event A that is prior to B with respect to $(\mathscr{F}_t, t \in T)$ occurs or does not occur before the event B. In other applications, the term *prior* may not express the actual real-world meaning. Of course, this applies to the terms *simultaneous* and *posterior* as well.   ◁

4.2.3   Multiplication rule

Now we treat some theorems involving conditional probabilities. The first one shows how the probability $P(A_1 \cap \ldots \cap A_n)$ can be factorized into a product of an unconditional probability and conditional probabilities.

Remark 4.21 [Multiplication rule for two and for three events]   For two events A_1 and A_2, the multiplication rule is

$$P(A_1 \cap A_2) = P(A_1) \cdot P(A_2 \mid A_1), \tag{4.3}$$

provided that $P(A_1) > 0$. This equation directly follows from the definition of the conditional probability $P(A_2 \mid A_1)$ [see Eq. (4.2)]. For three events A_1, A_2, and A_3, the multiplication rule is

$$P(A_1 \cap A_2 \cap A_3) = P(A_1) \cdot P(A_2 \mid A_1) \cdot P(A_3 \mid A_1 \cap A_2), \tag{4.4}$$

provided that $P(A_1 \cap A_2) > 0$. This equation follows from the definition of the conditional probability

$$P(A_3 \mid A_1 \cap A_2) = \frac{P(A_1 \cap A_2 \cap A_3)}{P(A_1 \cap A_2)}, \tag{4.5}$$

inserting Equation (4.3) for $P(A_1 \cap A_2)$, and solving the resulting equation for $P(A_1 \cap A_2 \cap A_3)$.   ◁

For n events $A_1, \ldots, A_n$, the multiplication rule is formulated in Theorem 4.22.

Theorem 4.22 [Multiplication rule]
Let $(\Omega, \mathscr{A}, P)$ be a probability space and $A_1, \ldots, A_n \in \mathscr{A}$, where $2 \leq n \in \mathbb{N}$. If $P(\bigcap_{i=1}^{n-1} A_i) > 0$, then,

$$P\left(\bigcap_{i=1}^{n} A_i\right) = P(A_1) \cdot \prod_{j=2}^{n} P\left(A_j \,\middle|\, \bigcap_{i=1}^{j-1} A_i\right). \tag{4.6}$$

(Proof p. 154)

4.2.4   Examples

Example 4.23 [Joe and Ann with randomized assignment – continued]   Consider again
the example presented in Table 4.1, and let

$$A = \{(Joe, no, -), (Joe, no, +), (Joe, yes, -), (Joe, yes, +)\}$$

denote the event that Joe is drawn,

$$B = \{(Joe, yes, -), (Joe, yes, +), (Ann, yes, -), (Ann, yes, +)\}$$

denote the event that the drawn person is treated, and

$$C = \{(Joe, no, +), (Joe, yes, +), (Ann, no, +), (Ann, yes, +)\}$$

denote the event that there is success, irrespective of the drawn person and treatment received.
Then,

$$A \cap B \cap C = \{(Joe, yes, +)\}$$

is the event that Joe is drawn, receives the treatment, and is successful. According to Equation
(4.4), the probability of this event can be computed by

$$P(A \cap B \cap C) = P(A) \cdot P(B \mid A) \cdot P(C \mid A \cap B)$$

$$= (.09 + .21 + .04 + .16) \cdot \frac{.04 + .16}{.09 + .21 + .04 + .16} \cdot \frac{.16}{.04 + .16}$$

$$= .5 \cdot .4 \cdot .8 = .16$$

(see Exercise 4.4). Of course, Equation (4.4) can also be applied to the other seven sets $A \cap B \cap$
C^c to $A^c \cap B^c \cap C^c$ in Figure 4.2. In this example, $P(A \cap B \cap C) = P(\{(Joe, yes, +)\}) = .16$ is
the probability of an elementary event (see Table 4.1).   ◁

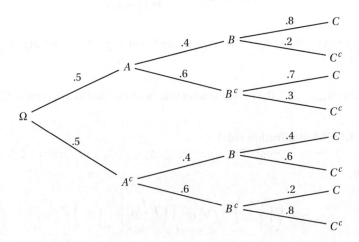

Figure 4.2   Probability tree illustrating the multiplication rule.

Example 4.24 [Drawing three balls] Consider drawing three balls without replacement from an urn containing two white balls and four black balls. Furthermore, let us consider the three events A_i to *draw a black ball at time i*, where $i = 1, 2, 3$. According to Theorem 4.22, the probability of drawing three black balls is

$$P(A_1 \cap A_2 \cap A_3) = P(A_1) \cdot P(A_2 \mid A_1) \cdot P(A_3 \mid A_1 \cap A_2),$$

where $P(A_1) = 4/6$, $P(A_2 \mid A_1) = 3/5$, and $P(A_3 \mid A_1 \cap A_2) = 2/4$. Hence,

$$P(A_1 \cap A_2 \cap A_3) = \frac{4}{6} \cdot \frac{3}{5} \cdot \frac{2}{4} = \frac{24}{120} = \frac{1}{5}.$$

◁

4.2.5   Theorem of total probability

In Theorem 4.25, called the *theorem of total probability*, we show how the probability of an event $B \subset A_1 \cup \ldots \cup A_n$ can additively be decomposed into the products $P(B \mid A_i) \cdot P(A_i)$ of conditional and unconditional probabilities. In this theorem, we assume that the events $A_1, \ldots, A_n$ are *pairwise disjoint* (i.e., we assume $A_i \cap A_j = \emptyset$, for all $i, j = 1, \ldots, n$, with $i \neq j$).

Theorem 4.25 [Theorem of total probability]
Let $(\Omega, \mathcal{A}, P)$ be a probability space and $B \in \mathcal{A}$.

(i) *If*

   (a) $A_1, \ldots, A_n \in \mathcal{A}$ *are pairwise disjoint, and*
   (b) $B \subset \bigcup_{i=1}^{n} A_i$,

   then

$$P(B) = \sum_{i=1}^{n} P(B \cap A_i). \tag{4.7}$$

(ii) *If (a) and (b) of (i) hold as well as*

   (c) $P(A_i) > 0$, $\quad \forall i = 1, \ldots, n$,

   then

$$P(B) = \sum_{i=1}^{n} P(B \mid A_i) \cdot P(A_i). \tag{4.8}$$

(iii) *If*

   (a) $A_1, A_2, \ldots \in \mathcal{A}$ *are pairwise disjoint, and*
   (b) $B \subset \bigcup_{i=1}^{\infty} A_i$,

then

$$P(B) = \sum_{i=1}^{\infty} P(B \cap A_i).$$ (4.9)

(iv) If (a) and (b) of (iii) hold as well as
   (c) $P(A_i) > 0, \quad \forall\, i = 1, 2, \ldots,$
then

$$P(B) = \sum_{i=1}^{\infty} P(B \mid A_i) \cdot P(A_i).$$ (4.10)

(Proof p. 154)

Equation (4.7) can be illustrated by Figure 4.3. If we assume that $(\Omega, \mathcal{A}, P)$ is the probability space specified in Example 4.8, then the figure visualizes that $P(B) = P(B \cap A_1) + P(B \cap A_2) + P(B \cap A_3)$. The crucial points are:

(a) If the events $A_1, \ldots, A_n$ are pairwise disjoint, then $B \cap A_1, \ldots, B \cap A_n$ are pairwise disjoint as well.

(b) The probability measure P is additive.

4.2.6   Bayes' theorem

Our next theorem, called *Bayes' theorem*, reveals how the conditional probabilities $P(B \mid A_i)$ are related to the conditional probabilities $P(A_i \mid B)$. Using the definitions of the conditional probabilities $P(A_i \mid B)$ and $P(B \mid A_i)$ yields

$$P(A_i \mid B) = \frac{P(B \mid A_i) \cdot P(A_i)}{P(B)}.$$ (4.11)

Inserting Equation (4.8) for $P(B)$ then proves Theorem 4.26.

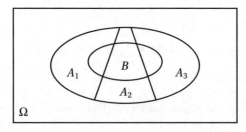

Figure 4.3   Venn diagram illustrating a partition of a set.

Theorem 4.26 [Bayes' theorem]
Let $(\Omega, \mathscr{A}, P)$ be a probability space, $B \in \mathscr{A}$, and $P(B) > 0$. Under the assumptions (a) to (c) of Theorem 4.25 (i) and (ii),

$$P(A_i \mid B) = \frac{P(B \mid A_i) \cdot P(A_i)}{\sum_{j=1}^{n} P(B \mid A_j) \cdot P(A_j)}, \quad \forall i = 1, \dots, n. \tag{4.12}$$

Analogously, under the assumptions (a) to (c) of Theorem 4.25 (iii) and (iv),

$$P(A_i \mid B) = \frac{P(B \mid A_i) \cdot P(A_i)}{\sum_{j=1}^{\infty} P(B \mid A_j) \cdot P(A_j)}, \quad \forall i \in \mathbb{N}. \tag{4.13}$$

Example 4.27 [Joe and Ann with randomized assignment – continued] Let

$$A = \{(Joe, no, -), (Joe, no, +), (Joe, yes, -), (Joe, yes, +)\}$$

denote the event that Joe is drawn,

$$A^c = \{(Ann, no, -), (Ann, no, +), (Ann, yes, -), \ (Ann, yes, +)\}$$

denote the event that Ann is drawn, and

$$B = \{(Joe, yes, -), (Joe, yes, +), (Ann, yes, -), (Ann, yes, +)\}$$

denote the event that the *drawn person is treated*. Then,

$$P(A \mid B) = \frac{P(B \mid A) \cdot P(A)}{P(B \mid A) \cdot P(A) + P(B \mid A^c) \cdot P(A^c)}$$

$$= \frac{.4 \cdot .5}{.4 \cdot .5 + .4 \cdot .5} = .5$$

is the conditional probability that *Joe is drawn* given that the *drawn person is treated* (see Table 4.1). The corresponding probability that *Ann is drawn* given that the *drawn person is treated* is identical in this example, that is, $P(A^c \mid B) = .5$. Hence, given treatment, each person has the same probability to be drawn. This is the *sampling perspective* of a randomized experiment supplementing the *assignment perspective*, according to which the treatment probability is the same for each person, that is, $P(B \mid A) = P(B \mid A^c) = .4$ (see again Table 4.1). ◁

4.2.7 Conditional-probability measure

Just like probabilities, conditional probabilities of events $A \in \mathscr{A}$ given B are values of a probability measure.

Theorem 4.28 [Conditional-probability measure]
Let $(\Omega, \mathcal{A}, P)$ be a probability space. If $B \in \mathcal{A}$ and $P(B) > 0$, then the function $P^B \colon \mathcal{A} \to [0, 1]$ defined by

$$P^B(A) = P(A \mid B), \quad \forall\, A \in \mathcal{A}, \tag{4.14}$$

is a probability measure on $(\Omega, \mathcal{A})$.

(Proof p. 155)

According to this theorem, for each $B \in \mathcal{A}$ with $P(B) > 0$, the triple $(\Omega, \mathcal{A}, P^B)$ is a probability space.

Definition 4.29 [Conditional-probability measure]
Let $(\Omega, \mathcal{A}, P)$ be a probability space, let $B \in \mathcal{A}$, and let $P(B) > 0$. Then the function P^B defined by (4.14) is called the B-conditional-probability measure on $(\Omega, \mathcal{A})$.

In the Lemma 4.30, we consider the relationship between conditional probabilities with respect to the measures P^B and P.

Lemma 4.30 [Conditional probabilities with respect to P^B]
Let $(\Omega, \mathcal{A}, P)$ be a probability space. If $A, B, C \in \mathcal{A}$ and $P(B \cap C) > 0$, then,

$$P^B(A \mid C) = P(A \mid B \cap C). \tag{4.15}$$

(Proof p. 155)

Remark 4.31 [Total conditional probability] Suppose $A, B, C \in \mathcal{A}$, $P(B \cap C) > 0$, and $P(B \cap C^c) > 0$. This implies $P(B) > 0$ and $P^B(C) = P(C \mid B) = P(C \cap B)/P(B) > 0$. Applying Equation (4.8) to the measure P^B then yields

$$P^B(A) = P^B(A \mid C) \cdot P^B(C) + P^B(A \mid C^c) \cdot P^B(C^c), \tag{4.16}$$

and Equations (4.14) and (4.15) imply

$$P(A \mid B) = P(A \mid B \cap C) \cdot P(C \mid B) + P(A \mid B \cap C^c) \cdot P(C^c \mid B). \tag{4.17}$$

$\triangleleft$

According to Lemma 4.32, P^B is *absolutely continuous with respect to P,* that is.

$$\forall\, A \in \mathcal{A} \colon P(A) = 0 \;\Rightarrow\; P^B(A) = 0. \tag{4.18}$$

This is denoted by $P^B \underset{\mathcal{A}}{\ll} P$ [see Def. 3.70 (i)]. In contrast, $P \underset{\mathcal{A}}{\ll} P^B$ does *not* always hold.

Lemma 4.32 [Absolute continuity of the conditional-probability measure]
Let $(\Omega, \mathscr{A}, P)$ be a probability space, $B \in \mathscr{A}$, and $P(B) > 0$. Then,

(i) $P^B \underset{\mathscr{A}}{\ll} P.$

(ii) $P^B = \left(\dfrac{1}{P(B)} \cdot 1_B \right) \odot P.$

(Proof p. 156)

Remark 4.33 [P^B is a measure with density]   Proposition (ii) of Lemma 4.32 implies that P^B is a measure with density $1_B/P(B)$ with respect to P. The following equations show how $P^B(A)$ can be written as an integral in various ways:

$$
\forall\, A \in \mathscr{A}: P^B(A) = \int_A dP^B \qquad\qquad [(3.8)]
$$

$$
= \int 1_A \, d\left(\frac{1}{P(B)} \cdot 1_B \odot P \right) \qquad [(3.30),\ \text{Lem. 4.32 (ii)}]
$$

$$
= \int 1_A \cdot 1_B \frac{1}{P(B)}\, dP \qquad\qquad [(3.72)]
$$

$$
= \frac{1}{P(B)} \cdot \int 1_{A \cap B}\, dP. \qquad\qquad [(1.33),\ (3.32)]
$$

(4.19)

Note that, according to Theorem 3.72 (i), the density $1_B/P(B)$ can be written as a Radon-Nikodym derivate of P^B with respect to P, that is,

$$
\frac{1}{P(B)} \cdot 1_B = \frac{dP^B}{dP}. \qquad\qquad (4.20)
$$

◁

Example 4.34 [Joe and Ann with randomized assignment – continued]   Consider the example presented in Table 4.1. We specify the B-conditional-probability measure $P^B \colon \mathscr{A} \to [0, 1]$ for the event that the *drawn person is treated*, that is, for

$$
B = \{(Joe, yes, -), (Joe, yes, +), (Ann, yes, -), (Ann, yes, +)\}.
$$

For the first two elementary events, $P^B(\{(Joe, no, +)\}) = P^B(\{(Joe, no, -)\}) = 0$, because the intersections $\{(Joe, no, -)\} \cap B$ and $\{(Joe, no, +)\} \cap B$ are empty. For the next two elementary events, the B-conditional probabilities are

$$
P^B(\{(Joe, yes, -)\}) = \frac{P(\{(Joe, yes, -)\} \cap B)}{P(B)} = \frac{.04}{.4} = .1
$$

and

$$P^B(\{(Joe, yes, +)\}) = \frac{P(\{(Joe, yes, +)\} \cap B)}{P(B)} = \frac{.16}{.4} = .4.$$

For the next two elementary events, $P^B(\{(Ann, no, -)\}) = P^B(\{(Ann, no, +)\}) = 0$, because the intersections $\{(Ann, no, -)\} \cap B$ and $\{(Ann, no, +)\} \cap B$ are again empty. Finally, for the last two elementary events, the B-conditional probabilities are

$$P^B(\{(Ann, yes, -)\}) = \frac{P(\{(Ann, yes, -)\} \cap B)}{P(B)} = \frac{.12}{.4} = .3$$

and

$$P^B(\{(Ann, yes, +)\}) = \frac{P(\{(Ann, yes, +)\} \cap B)}{P(B)} = \frac{.08}{.4} = .2.$$

These probabilities are summarized in the third column of Table 4.1. Except for $\emptyset$, all other events are unions of these elementary events. Because the elementary events are *disjoint*, the probabilities of their unions can easily be computed using finite additivity of the probability measure [see Rule (ii) of Box 4.1 and Exercise 4.5].   ◁

Example 4.35 [Joe and Ann with randomized assignment – continued]   Two other conditional-probability measures on $(\Omega, \mathscr{A})$ are P^A and P^{A^c}, where A is the event

$$A = \{(Joe, no, -), (Joe, no, +), (Joe, yes, -), (Joe, yes, +)\}$$

that Joe is sampled and A^c the event

$$A^c = \{(Ann, no, -), (Ann, no, +), (Ann, yes, -), (Ann, yes, +)\}$$

that Ann is sampled. The values of these conditional-probability measures are presented in the last two columns of Table 4.1. These measures can also be used to illustrate a *mixture of two probability measures*. As is easily seen,

$$P = .5 \cdot P^A + .5 \cdot P^{A^c},$$

that is, the measure P is a mixture of the two conditional-probability measures P^A and P^{A^c} (see Examples 4.11 and 1.62).   ◁

Example 4.36 [Joe and Ann with randomized assignment – continued]   In Example 4.16, we computed the two conditional probabilities $P(C \mid B) = .6$ and $P(C \mid B^c) = .45$ of success given treatment and no treatment, respectively. These are conditional probabilities with respect to the measure P. Let us now consider the *individual treatment effects* of Joe and of Ann. These individual effects can be computed using the P^A- and P^{A^c}-conditional-probability measures, respectively. For Joe, the individual treatment effect is

$$P^A(C \mid B) - P^A(C \mid B^c) = \frac{P^A(\Omega_U \times \{yes\} \times \{+\})}{P^A(\Omega_U \times \{yes\} \times \Omega_Y)} - \frac{P^A(\Omega_U \times \{no\} \times \{+\})}{P^A(\Omega_U \times \{no\} \times \Omega_Y)}$$

$$= \frac{.32 + 0}{.08 + .32 + 0 + 0} - \frac{.32 + 0}{.18 + .42 + 0 + 0}$$

$$= .8 - .7 = .1,$$

and for Ann it is

$$P^{A^c}(C \mid B) - P^{A^c}(C \mid B^c) = \frac{P^{A^c}(\Omega_U \times \{yes\} \times \{+\})}{P^{A^c}(\Omega_U \times \{yes\} \times \Omega_Y)} - \frac{P^{A^c}(\Omega_U \times \{no\} \times \{+\})}{P^{A^c}(\Omega_U \times \{no\} \times \Omega_Y)}$$

$$= \frac{.16 + 0}{.24 + .16 + 0 + 0} - \frac{.12 + 0}{.48 + .12 + 0 + 0}$$

$$= .4 - .2 = .2.$$

Hence, the treatment effect $P(C \mid B) - P(C \mid B^c) = .15$ (see Example 4.16) is just the weighted average $.5 \cdot .1 + .5 \cdot .2 = .15$ of the two individual treatment effects, where the weights are .5 for Joe and for Ann (see Example 4.35). Note that this property does not always hold [see Table 11.2 and Example 11.28]. ◁

4.3 Independence

4.3.1 Independence of events

Independence of two events A and B means that the conditional and unconditional probabilities are the same (i.e., $P(A \mid B) = P(A)$ and $P(B \mid A) = P(B)$). This definition presupposes that $P(A), P(B) > 0$, because otherwise the two conditional probabilities are not defined. The following definition does not rest on this requirement and extends the concept of independence to more than two events.

Definition 4.37 [Independence of events]
Let $(\Omega, \mathcal{A}, P)$ be a probability space.

(i) *Two events $A, B \in \mathcal{A}$ are called P-independent, denoted by $A \underset{P}{\perp\!\!\!\perp} B$, if*

$$P(A \cap B) = P(A) \cdot P(B). \tag{4.21}$$

(ii) Let I be a nonempty set and let $A_i \in \mathcal{A}$, $i \in I$. Then $(A_i, i \in I)$ is called a family of P-independent events, denoted by $\underset{P}{\perp\!\!\!\perp} (A_i, i \in I)$, if

$$P\left(\bigcap_{i \in I_0} A_i\right) = \prod_{i \in I_0} P(A_i), \quad \forall \text{ finite } I_0 \subset I. \tag{4.22}$$

Remark 4.38 [Pairwise and triple-wise independence] For n events $A_1, \ldots, A_n$, P-independence will also be denoted by

$$\underset{P}{\perp\!\!\!\perp} A_1, \ldots, A_n.$$

For three events, for instance, it means that

$$P(A_i \cap A_j) = P(A_i) \cdot P(A_j), \quad i \neq j, \quad i, j = 1, 2, 3, \tag{4.23}$$

(pairwise P-independence) and

$$P(A_1 \cap A_2 \cap A_3) = P(A_1) \cdot P(A_2) \cdot P(A_3) \tag{4.24}$$

(triple-wise P-independence) hold.

Note that pairwise P-independence of more than two events does not imply P-independence of these events. Furthermore, triple-wise P-independence, for instance, does not imply pairwise P-independence. For more propositions on independence of events, see Box 4.2.   ◁

Remark 4.39 [Independence of any event A with Ω and $\emptyset$] For any probability space $(\Omega, \mathcal{A}, P)$,

$$\forall A \in \mathcal{A}: A \underset{P}{\perp\!\!\!\perp} \Omega \quad \text{and} \quad A \underset{P}{\perp\!\!\!\perp} \emptyset. \tag{4.25}$$

(see Exercise 4.7).   ◁

4.3.2   Independence of set systems

Now we extend the concept of P-independence to *set systems* (i.e., to sets of events), and illustrate independence by an example.

Definition 4.40 [Family of independent set systems]
Let $(\Omega, \mathcal{A}, P)$ be a probability space and $\mathcal{E}_i \subset \mathcal{A}$, $i \in I \neq \emptyset$. Then, $(\mathcal{E}_i, i \in I)$ is called a family of P-independent set systems, denoted by $\underset{P}{\perp\!\!\!\perp} (\mathcal{E}_i, i \in I)$, if $\underset{P}{\perp\!\!\!\perp} (A_i, i \in I)$ holds for all families $(A_i, i \in I)$ with $A_i \in \mathcal{E}_i$, $i \in I$. If $I = \{1, 2\}$, we also use the notation $\mathcal{E}_1 \underset{P}{\perp\!\!\!\perp} \mathcal{E}_2$ instead of $\underset{P}{\perp\!\!\!\perp} (\mathcal{E}_i, i \in I)$.

Remark 4.41 [Independence of an event and a set system] Let $(\Omega, \mathscr{A}, P)$ be a probability space. An event $A \in \mathscr{A}$ and a set system $\mathscr{E} \subset \mathscr{A}$ are called *P-independent*, denoted by $A \underset{P}{\perp\!\!\!\perp} \mathscr{E}$, if $\{A\} \underset{P}{\perp\!\!\!\perp} \mathscr{E}$.   ◁

Remark 4.42 [Independence of σ-algebras] Note that σ-algebras are special set systems referred to in Definition 4.40. Hence, a family $(\mathscr{A}_i, i \in I)$ of sub-σ-algebras of $\mathscr{A}$ can be *P*-independent as well. This fact will be used when introducing the concept of *P-independence of random variables* (see section 5.4).   ◁

Example 4.43 [Joe and Ann with randomized assignment – continued] Suppose $A = \{Joe\} \times \Omega_X \times \Omega_Y$ denotes the event that Joe is sampled and $B = \Omega_U \times \{yes\} \times \Omega_Y$ the event that the person sampled is treated. Then A and B are independent, because

$$P(A \cap B) = P(\{Joe\} \times \{yes\} \times \Omega_Y) = .04 + .16 = .2$$

and

$$P(A) \cdot P(B) = P(\{Joe\} \times \Omega_X \times \Omega_Y) \cdot P(\Omega_U \times \{yes\} \times \Omega_Y)$$
$$= (.09 + .21 + .04 + .16) \cdot (.04 + .16 + .12 + .08)$$
$$= .5 \cdot .4 = .2.$$

Hence, $P(A \cap B) = P(A) \cdot P(B)$. This implies that the σ-algebras $\{A, A^c, \Omega, \emptyset\}$ and $\{B, B^c, \Omega, \emptyset\}$ are independent as well [see Box 4.2 (iii)]. In fact, this is a special case of the following theorem, because the set systems $\mathscr{E}_1 := \{A\}$ and $\mathscr{E}_2 := \{B\}$ are ∩-stable (see Def. 1.36) and $\sigma(\mathscr{E}_1) = \{A, A^c, \Omega, \emptyset\}$ and $\sigma(\mathscr{E}_2) = \{B, B^c, \Omega, \emptyset\}$ are the σ-algebras generated by $\mathscr{E}_1$ and $\mathscr{E}_2$, respectively (see Def. 1.13 and Example 1.17).   ◁

According to Theorem 4.44, it is sufficient to check *P*-independence of a family of ∩-stable generating systems in order to check *P*-independence of a family of σ-algebras. In this theorem, $(\sigma(\mathscr{E}_i), i \in I)$ denotes the family of σ-algebras generated by the set systems $\mathscr{E}_i, i \in I$.

Theorem 4.44 [∩-Stable set systems and independence]
If $(\Omega, \mathscr{A}, P)$ is a probability space and $\mathscr{E}_i \subset \mathscr{A}, i \in I$, are ∩-stable, then,

$$\underset{P}{\perp\!\!\!\perp} (\mathscr{E}_i, i \in I) \quad \Rightarrow \quad \underset{P}{\perp\!\!\!\perp} (\sigma(\mathscr{E}_i), i \in I). \tag{4.26}$$

For a proof, see Georgii (2008, Theorem 3.19).

4.4 Conditional independence given an event

Now we extend the concept of independence of events and of sets of events by introducing *conditional independence of events* and *of sets of events given an event*.

4.4.1   Conditional independence of events given an event

Definition 4.45 [Conditional independence of two events]
*Let $(\Omega, \mathcal{A}, P)$ be a probability space, $A, B, C \in \mathcal{A}$, and $P(B) > 0$. Then the events A and C
are called B-conditionally P-independent, denoted by $A \underset{P}{\perp\!\!\!\perp} C \mid B$, if*

$$P(A \cap C \mid B) = P(A \mid B) \cdot P(C \mid B). \tag{4.27}$$

Remark 4.46 [A condition equivalent to conditional independence] Suppose $P(B \cap C)$
> 0. Then Equation (4.27) is equivalent to

$$P(A \mid B \cap C) = P(A \mid B) \tag{4.28}$$

[see Box 4.2 (xii)]. Exchanging A and C immediately yields: If $P(A \cap B) > 0$, then Equation
(4.27) is equivalent to

$$P(C \mid A \cap B) = P(C \mid B). \tag{4.29}$$

$\triangleleft$

Remark 4.47 [Independence of events with respect to P^B] Using the conditional-probability measure P^B defined by (4.14), we can rewrite Equation (4.27) as:

$$P^B(A \cap C) = P^B(A) \cdot P^B(C). \tag{4.30}$$

This equation shows that B-conditional P-independence of A and C is equivalent to P^B-
independence of A and C, which will also be denoted by $A \underset{P^B}{\perp\!\!\!\perp} C$. $\triangleleft$

Remark 4.48 [Independence and conditional independence] Assume that $B \in \mathcal{A}$ with
$P(B) > 0$. Then P-independence of A and C neither implies nor is implied by B-conditional
P-independence of A and C (see Exercise 4.8). However, P-independence of A, B, and C *does*
imply B-conditional P-independence of A and C [see Box 4.2 (x)]. For more propositions on
conditional independence of events, see Box 4.2, which is proved in Exercise 4.9. $\triangleleft$

4.4.2   Conditional independence of set systems given an event

Now we extend the concept of conditional P-independence to *set systems*. In Remark 4.47,
we already noted that B-conditional P-independence of two events A and C is equivalent to
P^B-independence of A and C. Correspondingly, B-conditional P-independence of a family
$(\mathcal{E}_i, i \in I)$ of events is defined as P^B-independence of $(\mathcal{E}_i, i \in I)$.

Definition 4.49 [Family of conditionally independent set systems]
Let $(\Omega, \mathcal{A}, P)$ be a probability space, $B \in \mathcal{A}$ with $P(B) > 0$, and $\mathcal{E}_i \subset \mathcal{A}$, $i \in I$. Then
$(\mathcal{E}_i, i \in I)$ is called a family of B-conditionally P-independent set systems, denoted
by $\underset{P}{\perp\!\!\!\perp} (\mathcal{E}_i, i \in I) \mid B$, if $\underset{P^B}{\perp\!\!\!\perp} (\mathcal{E}_i, i \in I)$.

Box 4.2   Independence and conditional independence of events.

Let $(\Omega, \mathcal{A}, P)$ be a probability space and $A, B, C \in \mathcal{A}$. Then,

$$A \underset{P}{\perp\!\!\!\perp} B \quad \Leftrightarrow \quad P(A \cap B) = P(A) \cdot P(B) \tag{i}$$

$$A \underset{P}{\perp\!\!\!\perp} B \quad \Leftrightarrow \quad A^c \underset{P}{\perp\!\!\!\perp} B \tag{ii}$$

$$A \underset{P}{\perp\!\!\!\perp} B \quad \Leftrightarrow \quad \sigma(\{A\}) \underset{P}{\perp\!\!\!\perp} \sigma(\{B\}). \tag{iii}$$

$$\underset{P}{\perp\!\!\!\perp} A, B, C \quad \Leftrightarrow \quad P(A \cap B) = P(A) \cdot P(B), \tag{iv}$$
$$P(A \cap C) = P(A) \cdot P(C),$$
$$P(B \cap C) = P(B) \cdot P(C),$$
$$P(A \cap B \cap C) = P(A) \cdot P(B) \cdot P(C).$$

$$\underset{P}{\perp\!\!\!\perp} A, B, C \quad \Rightarrow \quad A \underset{P}{\perp\!\!\!\perp} B, A \underset{P}{\perp\!\!\!\perp} C, B \underset{P}{\perp\!\!\!\perp} C. \tag{v}$$

If $P(B) > 0$, then,

$$A \underset{P}{\perp\!\!\!\perp} B \quad \Leftrightarrow \quad P(A \mid B) = P(A) \tag{vi}$$

$$A \underset{P}{\perp\!\!\!\perp} C \mid B \quad \Leftrightarrow \quad P(A \cap C \mid B) = P(A \mid B) \cdot P(C \mid B) \tag{vii}$$

$$A \underset{P}{\perp\!\!\!\perp} C \mid B \quad \Leftrightarrow \quad A \underset{P^B}{\perp\!\!\!\perp} C \tag{viii}$$

$$A \underset{P}{\perp\!\!\!\perp} C \mid B \quad \Leftrightarrow \quad A \underset{P}{\perp\!\!\!\perp} C^c \mid B \tag{ix}$$

$$\underset{P}{\perp\!\!\!\perp} A, B, C \quad \Rightarrow \quad A \underset{P}{\perp\!\!\!\perp} C \mid B. \tag{x}$$

If $P(B), P(B^c) > 0$, then,

$$A \underset{P}{\perp\!\!\!\perp} B \quad \Leftrightarrow \quad P(A \mid B) = P(A \mid B^c). \tag{xi}$$

If $P(B \cap C) > 0$, then,

$$A \underset{P}{\perp\!\!\!\perp} C \mid B \quad \Leftrightarrow \quad P(A \mid B \cap C) = P(A \mid B). \tag{xii}$$

If $P(B \cap C^c) > 0$, then,

$$A \underset{P}{\perp\!\!\!\perp} C \mid B \quad \Leftrightarrow \quad P(A \mid B \cap C^c) = P(A \mid B). \tag{xiii}$$

If $P(B \cap C), P(B \cap C^c) > 0$, then,

$$A \underset{P}{\perp\!\!\!\perp} C \mid B \quad \Leftrightarrow \quad P(A \mid B \cap C) = P(A \mid B \cap C^c) \tag{xiv}$$

$$B \underset{P}{\perp\!\!\!\perp} C \quad \Rightarrow \quad P(A \mid B) = P(A \mid B \cap C) \cdot P(C) + P(A \mid B \cap C^c) \cdot P(C^c) \tag{xv}$$

$$A \underset{P}{\perp\!\!\!\perp} C \mid B \quad \Rightarrow \quad P(A \mid B) = P(A \mid B \cap C) \cdot P(C) + P(A \mid B \cap C^c) \cdot P(C^c). \tag{xvi}$$

Remark 4.50 [Conditional independence of σ-algebras]   Again, σ-algebras can be such set systems referred to in Definition 4.49. Hence, a family $(\mathcal{A}_i, i \in I)$ of sub-σ-algebras of $\mathcal{A}$ can be B-conditionally P-independent as well.   ◁

Remark 4.51 [Independence of set systems with respect to P^B]   According to Theorem 4.44, under the assumptions of Definition 4.49, ∩-stability of the set systems $\mathcal{E}_i, i \in I$, implies

$$\underset{P^B}{\perp\!\!\!\perp} (\mathcal{E}_i, i \in I) \quad \Rightarrow \quad \underset{P^B}{\perp\!\!\!\perp} (\sigma(\mathcal{E}_i), i \in I). \tag{4.31}$$

◁

Together with Definition 4.49, this remark immediately implies Corollary 4.52.

Corollary 4.52 [∩-Stable set systems and conditional independence]
If $(\Omega, \mathcal{A}, P)$ is a probability space, $B \in \mathcal{A}$ with $P(B) > 0$, and $(\mathcal{E}_i, i \in I)$ is a family of ∩-stable set systems $\mathcal{E}_i \subset \mathcal{A}$, then,

$$\underset{P}{\perp\!\!\!\perp} (\mathcal{E}_i, i \in I) \mid B \quad \Rightarrow \quad \underset{P}{\perp\!\!\!\perp} (\sigma(\mathcal{E}_i), i \in I) \mid B. \tag{4.32}$$

4.5   Proofs

Proof of Theorem 4.22

In Remark 4.21, we have already shown that Equation (4.6) holds for $n = 2$ [see Box 4.1 (v)]. Hence, for an induction over n, it suffices to show that (4.6) holds for $A_1, \ldots, A_n$ if it holds for $A_1, \ldots, A_{n-1}$. Note that $P(\bigcap_{i=1}^{n-1} A_i) > 0$ implies $P(\bigcap_{i=1}^{j-1} A_i) > 0$ for $2 \le j \le n$. Hence,

$$
\begin{aligned}
P\left(\bigcap_{i=1}^n A_i \right) &= P\left(\bigcap_{i=1}^{n-1} A_i \cap A_n \right) \\
&= P\left(\bigcap_{i=1}^{n-1} A_i \right) \cdot P\left(A_n \,\middle|\, \bigcap_{i=1}^{n-1} A_i \right) &&\text{[(4.6)]} \\
&= P(A_1) \cdot \left[\prod_{j=2}^{n-1} P\left(A_j \,\middle|\, \bigcap_{i=1}^{j-1} A_i \right) \right] \cdot P\left(A_n \,\middle|\, \bigcap_{i=1}^{n-1} A_i \right) &&\text{[(4.6), ass. of induction]} \\
&= P(A_1) \cdot \prod_{j=2}^n P\left(A_j \,\middle|\, \bigcap_{i=1}^{j-1} A_i \right).
\end{aligned}
$$

Proof of Theorem 4.25

(i) This equation immediately follows from (1.29).

(ii) If $P(A_i) > 0$, then $P(B \cap A_i) = P(B \mid A_i) \cdot P(A_i)$ [see Eq. (4.2)]. Hence, (4.7) immediately implies

$$P(B) = \sum_{i=1}^n P(B \mid A_i) \cdot P(A_i).$$

(iii) This proposition immediately follows from (1.30).

(iv) This proposition immediately follows from (iii) inserting $P(B \cap A_i) = P(B \mid A_i) \cdot P(A_i)$ [see Eq. (4.2)].

Proof of Theorem 4.28

We show that the conditions (a) to (c) of Definition 4.1 hold for P^B.

(a)

$$P^B(\Omega) = \frac{P(B \cap \Omega)}{P(B)} \qquad [(4.2)]$$

$$= \frac{P(B)}{P(B)} \qquad [B \subset \Omega]$$

$$= 1.$$

(b) We assume $P(B) > 0$. Therefore, $P(A \cap B) \geq 0$, for all $A \in \mathcal{A}$, implies that $P^B(A) = P(A \cap B)/P(B) \geq 0$, for all $A \in \mathcal{A}$.

(c) If $A_1, A_2, \ldots$ are pairwise disjoint, then $A_1 \cap B, A_2 \cap B, \ldots$ are pairwise disjoint. Therefore,

$$P^B\left(\bigcup_{i=1}^{\infty} A_i\right) = \frac{P\left[\left(\bigcup_{i=1}^{\infty} A_i\right) \cap B\right]}{P(B)} \qquad [(4.2)]$$

$$= \frac{P\left[\bigcup_{i=1}^{\infty} (A_i \cap B)\right]}{P(B)}$$

$$= \frac{\sum_{i=1}^{\infty} P(A_i \cap B)}{P(B)} \qquad [\text{Def. 4.1 (c)}]$$

$$= \sum_{i=1}^{\infty} P^B(A_i). \qquad [(4.2)]$$

Proof of Lemma 4.30

$$P(A \mid B \cap C) = \frac{P(A \cap B \cap C)}{P(B \cap C)} \qquad [(4.2)]$$

$$= \frac{P(A \cap C \mid B) \cdot P(B)}{P(C \mid B) \cdot P(B)} \qquad [(4.2)]$$

$$= \frac{P^B(A \cap C)}{P^B(C)} \qquad [(4.14)]$$

$$= P^B(A \mid C). \qquad [(4.2)]$$

Proof of Lemma 4.32

(i) For all $A \in \mathcal{A}$,

$$P(A) = 0 \quad \Rightarrow \quad P(A \cap B) = 0 \qquad \text{[Box 4.1 (v)]}$$

$$\Rightarrow \quad \frac{P(A \cap B)}{P(B)} = 0 \qquad [P(B) > 0]$$

$$\Rightarrow \quad P(A \mid B) = 0 \qquad \text{[(4.2)]}$$

$$\Rightarrow \quad P^B(A) = 0. \qquad \text{[(4.14)]}$$

Hence, $P^B \underset{\mathcal{A}}{\ll} P$ (see Def. 3.70).

(ii) For all $A \in \mathcal{A}$,

$$P^B(A) = \frac{P(A \cap B)}{P(B)} \qquad \text{[(4.14), (4.2)]}$$

$$= \frac{1}{P(B)} \int 1_{A \cap B} \, dP \qquad \text{[(3.9)]}$$

$$= \frac{1}{P(B)} \int 1_A \cdot 1_B \, dP \qquad \text{[(1.33)]}$$

$$= \int_A \frac{1}{P(B)} \cdot 1_B \, dP. \qquad \text{[(3.30), (3.32)]}$$

According to Theorem 3.65, this means $P^B = \left(\dfrac{1}{P(B)} \cdot 1_B \right) \odot P$.

Exercises

4.1 Consider flipping a coin n times and the event $A_1 = $ *flipping heads at the first flip*. Specify the set Ω of possible outcomes of this random experiment and the set A_1 as a subset of Ω. How many elements has Ω? How many elements has the event A_1?

4.2 Draw the interval $[0, 1]$, cut it in two halves, cut the right-hand piece in two halves, cut the remaining rightmost part in two halves, and so on. In this way, you can visualize the sequence $1/2, 1/4, 1/8, \ldots$ by lengths of intervals. This sequence can also be written as: $1/2^i, i \in \mathbb{N}$. Note that all terms $1/2^i$ of this sequence are positive (i.e., $1/2^i > 0$ for all $i \in \mathbb{N}$). Determine

$$\sum_{i=1}^{\infty} \frac{1}{2^i} = \lim_{n \to \infty} \sum_{i=1}^{n} \frac{1}{2^i}.$$

4.3 Consider Example 4.19, and list all elements of the σ-algebras $\mathcal{F}_1$ and $\mathcal{F}_2$ referring explicitly to the elements of Ω.

4.4 Compute the probabilities $P(A)$ and $P(C \mid A \cap B)$ of the events defined in Example 4.23.

4.5   Compute the *B*-conditional probability of the event $\{(Ann, yes, -), (Ann, yes, +)\}$, that is, the event that *Ann is sampled and treated*. Use the results already obtained in Example 4.34.

4.6   In Example 4.16, we computed the conditional probabilities $P(C \mid B) = .6$ and $P(C \mid B^c) = .45$, where *C* is the event that the *drawn person is successful* and *B* is the event that the *drawn person is treated*. What is the conclusion regarding the effect of the treatment if we compare these two conditional probabilities to each other?

4.7   Prove the proposition of Remark 4.39.

4.8   Show by examples that, for $B \in \mathscr{A}$ with $P(B) > 0$, *P*-independence of *A* and *C* neither implies nor is implied by *B*-conditional *P*-independence of *A* and *C*.

4.9   Prove the propositions of Box 4.2.

Solutions

4.1   The set of possible outcomes is $\Omega = \{h, t\}^n = \{h, t\} \times \ldots \times \{h, t\}$ (*n*-times). The event *flipping heads at time 1* is $A_1 = \{h\} \times \{h, t\}^{n-1}$. The set Ω has 2^n elements, and A_1 has $2^n/2 = 2^{n-1}$ elements.

4.2   The picture of this interval is

and this illustrates that $\displaystyle\sum_{i=1}^{\infty} \frac{1}{2^i} = \lim_{n \to \infty} \sum_{i=1}^{n} \frac{1}{2^i} = 1$.

4.3

$$\mathscr{F}_1 = \sigma(\{A\}) = \{A, A^c, \Omega, \emptyset\}$$
$$= \{\{(Joe, no, -), (Joe, no, +), (Joe, yes, -), (Joe, yes, +)\},$$
$$\{(Ann, no, -), (Ann, no, +), (Ann, yes, -), (Ann, yes, +)\}, \Omega, \emptyset\}.$$

$$\mathscr{F}_2 = \sigma(\{A, B\})$$
$$= \{A, A^c, B, B^c, (A \cap B) \cup (A^c \cap B^c), (A \cap B^c) \cup (A^c \cap B),$$
$$A \cap B, A^c \cap B, A \cap B^c, A^c \cap B^c,$$
$$A \cup B, A^c \cup B, A \cup B^c, A^c \cup B^c, \Omega, \emptyset\}$$

where

$$A = \{(Joe, yes, -), (Joe, yes, +), (Joe, no, -), (Joe, no, +)\},$$
$$A^c = \{(Ann, yes, -), (Ann, yes, +), (Ann, no, -), (Ann, no, +)\},$$
$$B = \{(Joe, yes, -), (Joe, yes, +), (Ann, yes, -), (Ann, yes, +)\},$$
$$B^c = \{(Joe, no, -), (Joe, no, +), (Ann, no, -), (Ann, no, +)\},$$
$$(A \cap B) \cup (A^c \cap B^c) = \{(Joe, yes, -), (Joe, yes, +), (Ann, no, -), (Ann, no, +)\},$$
$$(A \cap B^c) \cup (A^c \cap B) = \{(Joe, no, -), (Joe, no, +), (Ann, yes, -), (Ann, yes, +)\},$$

$$A \cap B = \{(Joe, yes, -), (Joe, yes, +)\},$$
$$A^c \cap B = \{(Ann, yes, -), (Ann, yes, +)\},$$
$$A \cap B^c = \{(Joe, no, -), (Joe, no, +)\},$$
$$A^c \cap B^c = \{(Ann, no, -), (Ann, no, +)\},$$

$A \cup B = \{(Joe, yes, -), (Joe, yes, +), (Joe, no, -), (Joe, no, +), (Ann, yes, -), (Ann, yes, +)\}$,
$A^c \cup B = \{(Ann, yes, -), (Ann, yes, +), (Ann, no, -), (Ann, no, +), (Joe, yes, -),$
$\qquad (Joe, yes, +)\}$,
$A \cup B^c = \{(Joe, yes, -), (Joe, yes, +), (Joe, no, -), (Joe, no, +), (Ann, no, -), (Ann, no, +)\}$,
$A^c \cup B^c = \{(Ann, yes, -), (Ann, yes, +), (Ann, no, -), (Ann, no, +), (Joe, no, -),$
$\qquad (Joe, no, +)\}$.

4.4 Because the four events $\{(Joe, no, -)\}, \dots, \{(Joe, yes, +)\}$ are pairwise disjoint, we can simply add their probabilities. Hence, $P(A) = .09 + .21 + .04 + .16 = .5$ (see the second column of Table 4.1). In order to compute $P(C \mid A \cap B)$, note that $A \cap B = \{(Joe, yes, -), (Joe, yes, +)\}$ is the event that *Joe is drawn and treated*. Again, because the two elementary events involved are disjoint, $P(A \cap B) = .04 + .16 = 0.2$. Furthermore, $A \cap B \cap C = \{(Joe, yes, +)\}$ is the event that *Joe is drawn, treated, and successful*. Its probability is $P(A \cap B \cap C) = P(\{(Joe, yes, +)\}) = .16$. Hence,

$$P(C \mid A \cap B) = \frac{P(A \cap B \cap C)}{P(A \cap B)} = \frac{.16}{.04 + .16} = .8.$$

4.5 In Example 4.34, we already computed the two B-conditional probabilities $P^B(\{(Ann, yes, -)\}) = .3$ and $P^B(\{(Ann, yes, +)\}) = .2$. Because these elementary events are *disjoint*, the probabilities of their union can be computed easily using the additivity property of the probability measure P^B. Hence, $P^B(\{(Ann, yes, -), (Ann, yes, +)\}) = .3 + .2 = .5$.

4.6 Although this question and the concepts needed for an answer are beyond the scope of this book, the difference $P(C \mid B) - P(C \mid B^c) = .6 - .45 = .15$ is the *average total treatment effect* (see Steyer *et al.*, 2014). It is the average of the two *individual total treatment effects* of Joe and of Ann. For Joe, this individual treatment effect is $.8 - .7 = .1$ (probability of *success* given *Joe* and *treatment* minus probability of *success* given *Joe* and no *treatment*), whereas it is $.4 - .2 = .2$ for Ann.

4.7 Let $(\Omega, \mathcal{A}, P)$ be a probability space. Then, for all $A \in \mathcal{A}$: $P(\Omega \cap A) = P(A) = 1 \cdot P(A) = P(\Omega) \cdot P(A)$ and $P(\emptyset \cap A) = P(\emptyset) = 0 = 0 \cdot P(A) = P(\emptyset) \cdot P(A)$.

4.8 Consider the example in subsection 2.2.2, and let $A = \{(h, t), (h, h)\}$ denote the event to flip heads with the first coin, $B = \{(t, t), (h, h)\}$ the event to flip tails or to flip heads with both coins, and $C = \{(t, h), (h, h)\}$ the event to flip heads with the second coin. All three events have the same probability $P(A) = P(B) = P(C) = .5$. Now,

$$P(A \cap C) = P(\{(h, h)\}) = .25 = .5 \cdot .5 = P(A) \cdot P(C)$$

and

$$P(B \cap C) = P(\{(h, h)\}) = .25 = .5 \cdot .5 = P(B) \cdot P(C).$$

Hence, A and C as well as B and C are P-independent, which implies $P(A \mid B) = .5$ and $P(C \mid B) = .5$. However,

$$P(A \cap C \mid B) = \frac{P(A \cap C \cap B)}{P(B)} = \frac{.25}{.5} = .5$$
$$\neq .25 = .5 \cdot .5 = P(A \mid B) \cdot P(C \mid B),$$

which shows that A and C are not B-conditionally P-independent.

Now we present an example in which A and C are B-conditionally P-independent but not (unconditionally) P-independent. Consider flipping three coins. This random experiment is represented by the probability space $(\Omega, \mathcal{A}, P)$, where $\Omega = \{h, t\}^3$, $\mathcal{A} = \mathcal{P}(\Omega)$, and $P: \mathcal{A} \to [0, 1]$, satisfying $P(\{\omega\}) = .125$ for all $\omega \in \Omega$. Furthermore, let $A = \{(t, t, t), (t, t, h)\}$ denote the event to flip tails with the first two coins, $B = \{(t, t, t), (t, t, h), (t, h, t), (t, h, h)\}$ the event to flip tails with the first coin, and $C = \{(t, t, h), (t, h, h)\}$ the event to flip tails with the first coin and heads with the third coin. The two events A and C have the same probability $P(A) = P(C) = .25$ and $P(B) = .5$. Because

$$P(A \cap C) = P(\{(t, t, h)\}) = .125 \neq .25 \cdot .25 = P(A) \cdot P(C),$$

A and C are *not P-independent*. Furthermore, $P(A \cap B) = P(\{(t, t, t), (t, t, h)\}) = .25$, $P(C \cap B) = P(\{(t, t, h), (t, h, h)\}) = .25$, and

$$P(A \cap C \mid B) = \frac{P(A \cap B \cap C)}{P(B)} = \frac{.125}{.5} = .25$$
$$= .5 \cdot .5 = \frac{P(A \cap B)}{P(B)} \cdot \frac{P(B \cap C)}{P(B)} = P(A \mid B) \cdot P(C \mid B).$$

This shows that A and C are B-conditionally P-independent.

4.9 (i)   This is the definition of $A \underset{P}{\perp\!\!\!\perp} B$.

(ii)
$$\begin{aligned}
P(A^c \cap B) &= P(B \setminus A) \\
&= P(B) - P(A \cap B) &&\text{[Box 4.1 (vi)]} \\
&= P(B) - P(A) \cdot P(B) &&[A \underset{P}{\perp\!\!\!\perp} B] \\
&= [1 - P(A)] \cdot P(B) \\
&= P(A^c) \cdot P(B) &&\text{[Box 4.1 (iv)]},
\end{aligned}$$

which is $A^c \underset{P}{\perp\!\!\!\perp} B$.

(iii)   We have to show that $A \underset{P}{\perp\!\!\!\perp} B$ implies

$$P(A_i \cap B_j) = P(A_i) \cdot P(B_j), \quad \forall\, A_i \in \{A, A^c, \Omega, \emptyset\} \text{ and } \forall\, B_j \in \{B, B^c, \Omega, \emptyset\}.$$

Whenever A_i or B_j is Ω or $\emptyset$, this equation holds [see (4.25)]. Furthermore, $P(A \cap B) = P(A) \cdot P(B)$ is equivalent to our premise $A \underset{P}{\perp\!\!\!\perp} B$, and $P(A^c \cap B) =$

$P(A^c) \cdot P(B)$ is proposition (ii). The corresponding argument holds for $P(A \cap B^c) = P(A) \cdot P(B^c)$ and $P(A^c \cap B^c) = P(A^c) \cdot P(B^c)$, exchanging the roles of A and B.

(iv)   This is the definition of $\underset{P}{\perp\!\!\!\perp} A, B, C$.

(v)   This proposition immediately follows from (iv) and (i).

(vi)   We assume $P(B) > 0$. Then,

$$
\begin{aligned}
A \underset{P}{\perp\!\!\!\perp} B \quad &\Leftrightarrow \quad P(A \cap B) = P(A) \cdot P(B) && \text{[(i)]} \\
&\Leftrightarrow \quad \frac{P(A \cap B)}{P(B)} = P(A) \\
&\Leftrightarrow \quad P(A \mid B) = P(A). && \text{[(4.2)]}
\end{aligned}
$$

(vii)   This is the definition of $A \underset{P}{\perp\!\!\!\perp} C \mid B$.

(viii)   We assume $P(B) > 0$. Therefore,

$$
\begin{aligned}
A \underset{P}{\perp\!\!\!\perp} C \mid B \quad &\Leftrightarrow \quad P(A \cap C \mid B) = P(A \mid B) \cdot P(C \mid B) && \text{[(vii)]} \\
&\Leftrightarrow \quad P^B(A \cap C) = P^B(A) \cdot P^B(C) && \text{[(4.14)]} \\
&\Leftrightarrow \quad A \underset{P^B}{\perp\!\!\!\perp} C. && \text{[(i)]}
\end{aligned}
$$

(ix)   We assume $P(B) > 0$. Therefore,

$$
\begin{aligned}
A \underset{P}{\perp\!\!\!\perp} C \mid B \quad &\Leftrightarrow \quad A \underset{P^B}{\perp\!\!\!\perp} C && \text{[(viii)]} \\
&\Leftrightarrow \quad A \underset{P^B}{\perp\!\!\!\perp} C^c. && \text{[(ii)]}
\end{aligned}
$$

(x)   We assume $P(B) > 0$. Then,

$$
\begin{aligned}
P(A \cap C \mid B) &= \frac{P(A \cap B \cap C)}{P(B)} && \text{[(4.2)]} \\
&= \frac{P(A) \cdot P(B) \cdot P(C)}{P(B)} && \text{[}\underset{P}{\perp\!\!\!\perp} A, B, C, \text{(iv)]} \\
&= P(A \mid B) \cdot P(C \mid B). && \text{[(v), (vi)]}
\end{aligned}
$$

(xi)   We assume $P(B), P(B^c) > 0$. Then,

$$
\begin{aligned}
P(A \mid B) = P(A \mid B^c) \quad &\Leftrightarrow \quad \frac{P(A \cap B)}{P(B)} = \frac{P(A \cap B^c)}{1 - P(B)} && \text{[(4.2), Box 4.1 (iv)]} \\
&\Leftrightarrow \quad P(A \cap B) \cdot [1 - P(B)] = P(A \cap B^c) \cdot P(B) \\
&\Leftrightarrow \quad P(A \cap B) = [P(A \cap B) + P(A \cap B^c)] \cdot P(B) \\
&\Leftrightarrow \quad P(A \cap B) = P(A) \cdot P(B) && \text{[(4.7)]} \\
&\Leftrightarrow \quad A \underset{P}{\perp\!\!\!\perp} B. && \text{[(i)]}
\end{aligned}
$$

(xii)   We assume $P(B \cap C) > 0$. This implies $P(B) > 0$ and

$$A \underset{P}{\perp\!\!\!\perp} C \mid B \quad \Leftrightarrow \quad P(A \cap C \mid B) = P(A \mid B) \cdot P(C \mid B) \qquad [(vii)]$$

$$\Leftrightarrow \quad \frac{P(A \cap B \cap C)}{P(B)} = \frac{P(A \cap B)}{P(B)} \cdot \frac{P(B \cap C)}{P(B)} \qquad [(4.2)]$$

$$\Leftrightarrow \quad \frac{P(A \cap B \cap C)}{P(B \cap C)} = \frac{P(A \cap B)}{P(B)}$$

$$\Leftrightarrow \quad P(A \mid B \cap C) = P(A \mid B). \qquad [(4.2)]$$

(xiii) We assume $P(B \cap C^c) > 0$. This implies $P(B) > 0$ and

$$A \underset{P}{\perp\!\!\!\perp} C \mid B \quad \Leftrightarrow \quad A \underset{P^B}{\perp\!\!\!\perp} C \qquad\qquad\qquad [(viii)]$$

$$\Leftrightarrow \quad A \underset{P^B}{\perp\!\!\!\perp} C^c \qquad\qquad\qquad [(ii)]$$

$$\Leftrightarrow \quad P^B(A \mid C^c) = P^B(A) \qquad\qquad [(vi)]$$

$$\Leftrightarrow \quad P(A \mid B \cap C^c) = P(A \mid B) \qquad [(4.15),\ (4.14)]$$

(xiv)   We assume $P(B \cap C),\ P(B \cap C^c) > 0$.

$$P(A \mid B \cap C) = P(A \mid B \cap C^c) \quad \Leftrightarrow \quad P^B(A \mid C) = P^B(A \mid C^c) \qquad [(4.15)]$$
$$\Leftrightarrow \quad A \underset{P^B}{\perp\!\!\!\perp} C \qquad\qquad\qquad [(xi)]$$
$$\Leftrightarrow \quad A \underset{P}{\perp\!\!\!\perp} C \mid B. \qquad\qquad [(viii)]$$

(xv)   We assume $P(B \cap C),\ P(B \cap C^c) > 0$.

$$B \underset{P}{\perp\!\!\!\perp} C \quad \Rightarrow \quad P(C \mid B) = P(C),\ \ P(C^c \mid B) = P(C^c) \qquad\qquad [(vi),\ (ii)]$$
$$\Rightarrow \quad P(A \mid B) = P(A \mid B \cap C) \cdot P(C) + P(A \mid B \cap C^c) \cdot P(C^c). \quad [(4.17)]$$

(xvi)   We assume $P(B \cap C),\ P(B \cap C^c) > 0$, and $A \underset{P}{\perp\!\!\!\perp} C \mid B$.

$$P(A \mid B) = P(A \mid B \cap C) \qquad\qquad\qquad [A \underset{P}{\perp\!\!\!\perp} C \mid B,\ (xii)]$$
$$= P(A \mid B \cap C) \cdot [P(C) + P(C^c)] \qquad\qquad [\text{Box 4.1 (iv)}]$$
$$= P(A \mid B \cap C) \cdot P(C) + P(A \mid B \cap C) \cdot P(C^c)$$
$$= P(A \mid B \cap C) \cdot P(C) + P(A \mid B \cap C^c) \cdot P(C^c). \qquad\qquad [(xiv)]$$

5

Random variable, distribution, density, and distribution function

In chapter 4, we translated the concepts *measure* and *measure space* to probability theory, introducing the notions *probability measure* and *probability space*. In this chapter, we define a *random variable* as a measurable mapping and its *distribution* as the image measure of a measurable mapping with respect to a probability measure (see ch. 2). The distribution of a random variable contains the comprehensive information about its properties. It informs us about the probability of each event that can be represented by this random variable. Expectation, variance, and other moments of a random variable are determined by its distribution (see ch. 6). For a multivariate random variable, the (joint) distribution also contains the information about the dependencies between its components. It also determines the conditional expectations (see ch. 10). In this chapter, we apply the concept of independence of families of events in order to introduce *independence of random variables* and *families of random variables*. Finally, the last sections of this chapter are devoted to the concept of a *probability function* and, for a real-valued random variable, the notions of a *distribution function* and a *probability density*, which are very useful for describing a distribution, for calculations (see, e.g., ch. 6), and for providing instructive illustrations of the underlying distributions (see ch. 8).

5.1 Random variable and its distribution

In section 2.6, we introduced the notation

$$f: (\Omega, \mathscr{A}, \mu) \to (\Omega', \mathscr{A}'),$$

which expresses that $f: \Omega \to \Omega'$ is an $(\mathscr{A}, \mathscr{A}')$-measurable mapping and that μ is a measure on the measurable space $(\Omega, \mathscr{A})$. If μ is a probability measure, then a measurable mapping is also called a *random variable*, and its image measure μ_f is also called its *distribution*. This change of terms goes along with a change in notation. Instead of f, g, and h, we preferably use letters such as X, Y, and Z.

Probability and Conditional Expectation: Fundamentals for the Empirical Sciences, First Edition. Rolf Steyer and Werner Nagel.
© 2017 John Wiley & Sons, Ltd. Published 2017 by John Wiley & Sons, Ltd.
Companion website: http://www.probability-and-conditional-expectation.de

Definition 5.1 [Random variable]

If $(\Omega, \mathcal{A}, P)$ is a probability space and $X: (\Omega, \mathcal{A}) \to (\Omega'_X, \mathcal{A}'_X)$ a measurable mapping, that is, if $X: \Omega \to \Omega'_X$ satisfies

$$X^{-1}(A') \in \mathcal{A}, \quad \forall A' \in \mathcal{A}'_X, \tag{5.1}$$

then X is called a random variable on $(\Omega, \mathcal{A}, P)$ with values in $(\Omega'_X, \mathcal{A}'_X)$. If $(\Omega'_X, \mathcal{A}'_X) = (\mathbb{R}, \mathcal{B})$, then X is called real-valued, and if $(\Omega'_X, \mathcal{A}'_X) = (\overline{\mathbb{R}}, \overline{\mathcal{B}})$, then X is called numerical.

Remark 5.2 [Measurability of inverse images] Equation (5.1) implies that all inverse images

$$X^{-1}(A') := \{\omega \in \Omega: X(\omega) \in A'\}, \quad A' \in \mathcal{A}'_X,$$

are elements of the σ-algebra $\mathcal{A}$ on Ω. Because the measure $P: \mathcal{A} \to [0, 1]$ assigns a probability to *all* elements of $\mathcal{A}$, the probabilities $P[X^{-1}(A')]$ of these inverse images are determined by P (see Exercises 5.1 and 5.2).   ◁

Definition 5.3 [Distribution of a random variable]

Let $X: (\Omega, \mathcal{A}, P) \to (\Omega'_X, \mathcal{A}'_X)$ be a random variable. Then the function $P_X: \mathcal{A}'_X \to [0, 1]$ defined by

$$P_X(A') = P[X^{-1}(A')], \quad \forall A' \in \mathcal{A}'_X, \tag{5.2}$$

is called the distribution of X (with respect to P).

Remark 5.4 [Notation $P(X \in A')$ and $P(X=x)$]   If $A' \in \mathcal{A}'_X$, we use the notation

$$P(X \in A') := P[X^{-1}(A')] \tag{5.3}$$

for the probability of the event $\{X \in A'\} = X^{-1}(A')$, that is, the event that X takes on a value in the subset A' of Ω'_X. If $\{x\} \in \mathcal{A}'_X$,

$$P(X=x) := P[X^{-1}(\{x\})] \tag{5.4}$$

for the probability of the event $\{X=x\} := X^{-1}(\{x\}) = \{\omega \in \Omega: X(\omega) = x\}$. If we write $P(X=x)$, then we always assume $\{x\} \in \mathcal{A}'_X$, even if not mentioned explicitly.   ◁

Remark 5.5 [A new probability space]   Definition 5.1 implies that *every* random variable X on a probability space $(\Omega, \mathcal{A}, P)$ has a distribution P_X. Furthermore, $P_X: \mathcal{A}'_X \to [0, 1]$ is also a measure, the *image measure of P under X* (see Th. 2.79 and Def. 2.80). Because

$P_X(\Omega'_X) = P(\Omega) = 1$, we can conclude that P_X is a probability measure, and $(\Omega'_X, \mathscr{A}'_X, P_X)$ is also a probability space. Therefore, we use the notation

$$X: (\Omega, \mathscr{A}, P) \to (\Omega'_X, \mathscr{A}'_X, P_X)$$

expressing:

(a) $X: \Omega \to \Omega'_X$ is a random variable on the probability space $(\Omega, \mathscr{A}, P)$.

(b) $\mathscr{A}'_X$ is a σ-algebra on Ω'_X.

(c) P_X is the distribution of X.

◁

Definition 5.6 [Identically distributed random variables]
Let $X: (\Omega^{(1)}, \mathscr{A}^{(1)}, P^{(1)}) \to (\Omega', \mathscr{A}')$ *and* $Y: (\Omega^{(2)}, \mathscr{A}^{(2)}, P^{(2)}) \to (\Omega', \mathscr{A}')$ *be random variables. If* $P_X = P_Y$*, then we say that* X *and* Y *are* **identically distributed.**

Note that, Definition 5.6 does not preclude, $(\Omega^{(1)}, \mathscr{A}^{(1)}, P^{(1)}) = (\Omega^{(2)}, \mathscr{A}^{(2)}, P^{(2)})$. Now we consider the distribution of a composition $g(X)$ of a random variable $X: (\Omega, \mathscr{A}, P) \to (\Omega'_X, \mathscr{A}'_X)$ and a measurable function $g: (\Omega'_X, \mathscr{A}'_X) \to (\Omega', \mathscr{A}')$. According to Remark 5.5, the mapping g is a random variable on the probability space $(\Omega'_X, \mathscr{A}'_X, P_X)$. Furthermore, according to Lemma 5.7, $g(X)$ is a random variable on $(\Omega, \mathscr{A}, P)$ and the distribution of $g(X)$ is the image measure of P_X under g. The notation of this image measure is $(P_X)_g$.

Lemma 5.7 [Distribution of a composition]
Let $X: (\Omega, \mathscr{A}, P) \to (\Omega'_X, \mathscr{A}'_X)$ *be a random variable and* $g: (\Omega'_X, \mathscr{A}'_X) \to (\Omega', \mathscr{A}')$ *a measurable function. Then the composition* $g(X)$ *is a random variable on* $(\Omega, \mathscr{A}, P)$ *with values in* $(\Omega', \mathscr{A}')$ *and*

$$(P_X)_g = P_{g(X)} . \tag{5.5}$$

(Proof p. 196)

Example 5.8 [Indicator (variable) of an event] If $(\Omega, \mathscr{A}, P)$ is a probability space and $A \in \mathscr{A}$, then the mapping $1_A: (\Omega, \mathscr{A}, P) \to (\{0, 1\}, \mathscr{P}(\{0, 1\}))$ is a random variable. It is called the *indicator (variable) of* A. The distribution of 1_A is

$$P_{1_A}(\{0\}) = P(A^c), \quad P_{1_A}(\{1\}) = P(A), \quad P_{1_A}(\{0, 1\}) = P(\Omega) = 1, \quad P_{1_A}(\emptyset) = P(\emptyset) = 0.$$

If we consider the same event A and the measurable space $(\mathbb{R}, \mathscr{B})$, then we can also write $1_A: (\Omega, \mathscr{A}, P) \to (\mathbb{R}, \mathscr{B})$ in order to express that 1_A is also $(\mathscr{A}, \mathscr{B})$-measurable. Note, however, that now the distribution of 1_A is a probability measure on $(\mathbb{R}, \mathscr{B})$, and for all $B \in \mathscr{B}$,

$$P_{1_A}(B) = P[1_A^{-1}(B)] = P[\{\omega \in \Omega : 1_A(\omega) \in B\}] \qquad [(5.2), (2.2)]$$

$$= \begin{cases} P(\emptyset) = 0, & \text{if } 0 \notin B, 1 \notin B \\ P(A), & \text{if } 0 \notin B, 1 \in B \\ P(A^c), & \text{if } 0 \in B, 1 \notin B \\ P(\Omega) = 1, & \text{if } \{0, 1\} \subset B. \end{cases} \qquad \triangleleft$$

Example 5.9 [Indicator of an inverse image] If $(\Omega, \mathscr{A}, P)$ is a probability space, $X: (\Omega, \mathscr{A}, P) \rightarrow (\Omega_X', \mathscr{A}_X')$ a random variable, and $A' \in \mathscr{A}_X'$, then $1_{X^{-1}(A')}: (\Omega, \mathscr{A}, P) \rightarrow (\{0, 1\}, \mathscr{P}(\{0, 1\}))$ is a random variable on $(\Omega, \mathscr{A}, P)$ and

$$1_{X \in A'} := 1_{X^{-1}(A')} = 1_{A'}(X) = 1_{A'} \circ X \qquad (5.6)$$

(see Exercise 5.3). The distribution of $1_{X \in A'}$ is

$$P_{1_{X \in A'}}(\{0\}) = P(X \notin A'), \quad P_{1_{X \in A'}}(\{1\}) = P(X \in A'),$$
$$P_{1_{X \in A'}}(\{0, 1\}) = P(\Omega) = 1, \quad P_{1_{X \in A'}}(\emptyset) = P(\emptyset) = 0. \qquad \triangleleft$$

Example 5.10 [Dichotomous random variable] Let $X: (\Omega, \mathscr{A}, P) \rightarrow (\Omega_X', \mathscr{A}_X')$ be a random variable on $(\Omega, \mathscr{A}, P)$. Then X is called *dichotomous with values x_1 and x_2* if $\{x_1\}, \{x_2\} \in \mathscr{A}_X'$, $P(X \in \{x_1, x_2\}) = 1$ and $0 < P(X = x_1) < 1$. If X is dichotomous with values 0 and 1, then $X \underset{P}{=} 1_{X=1}$. $\qquad \triangleleft$

Example 5.11 [Flipping two coins – continued] In the example of section 2.2.2 and in Example 2.83, we considered flipping two coins and defined $X: (\Omega, \mathscr{A}, P) \rightarrow (\Omega_X', \mathscr{P}(\Omega_X'))$, a random variable representing with its values the *number of flipping heads*. Its possible values are 0, 1, or 2. Hence, we can choose $\Omega_X' := \{0, 1, 2\}$, and

$$P(X = 0) = P_X(\{0\}) = P[X^{-1}(\{0\})] = P[\{(t, t)\}] = \frac{1}{4},$$

$$P(X = 1) = P_X(\{1\}) = P[X^{-1}(\{1\})] = P[\{(h, t), (t, h)\}] = \frac{1}{2},$$

$$P(X = 2) = P_X(\{2\}) = P[X^{-1}(\{2\})] = P[\{(h, h)\}] = \frac{1}{4}$$

are the probabilities assigned to the singletons $\{0\}$, $\{1\}$, and $\{2\}$, whereas

$$P(X \in \{0, 1\}) = P_X(\{0, 1\}) = P[X^{-1}(\{0, 1\})] = P[\{(t, t), (h, t), (t, h)\}] = \frac{3}{4},$$

$$P(X \in \{0, 2\}) = P_X(\{0, 2\}) = P[X^{-1}(\{0, 2\})] = P[\{(t, t), (h, h)\}] = \frac{2}{4},$$

$$P(X \in \{1, 2\}) = P_X(\{1, 2\}) = P[X^{-1}(\{1, 2\})] = P[\{(h, t), (t, h)(h, h)\}] = \frac{3}{4}$$

are the probabilities assigned to the sets $\{0, 1\}$, $\{0, 2\}$, and $\{1, 2\}$, which consist of two elements of Ω_X'. Finally, $P_X(\Omega_X') = P[X^{-1}(\Omega_X')] = P(\Omega) = 1$ and $P_X(\emptyset) = P[X^{-1}(\emptyset)] = P(\emptyset) = 0$. $\qquad \triangleleft$

Example 5.12 [Tom, Jim, and Kate]   Now we consider an example that is similar to the experiment with Joe and Ann. However, the set of persons is now $\Omega_U := \{Tom, Jim, Kate\}$, and we consider three treatments, the elements of the set $\Omega_X := \{Con, BTh, PTh\}$, where *Con* could be *no treatment*. The random experiment consists of: drawing a unit u from the set Ω_U, assigning it to one of the three treatment conditions *Con*, *BTh*, or *PTh*, and observing whether ($+$) or not ($-$) a success criterion is reached. Hence, the set of possible outcomes of this random experiment is

$$\Omega := \Omega_U \times \Omega_X \times \Omega_Y = \{(Tom, Con, -), (Tom, Con, +), \ldots, (Kate, PTh, +)\}.$$

It consists of the $3 \cdot 3 \cdot 2 = 18$ triples (u, ω_X, ω_Y) listed in the first column of Table 5.1. As the set of possible events $\mathcal{A}$, we consider the power set $\mathcal{P}(\Omega)$. This set has $2^{18} = 262{,}144$ elements, where 18 is the number of elements of Ω. The probabilities of the 18 elementary events $\{\omega\}$, $\omega \in \Omega$, are displayed in the second column of the table. With these specifications, the probabilities $P(A)$ of all 2^{18} elements $A \in \mathcal{A}$ are determined [see Rule (x) of Box 4.1]. Hence, the probability space $(\Omega, \mathcal{A}, P)$ is completely specified.

Table 5.1   Tom, Jim, and Kate.

Elements of Ω				Random variables		
Unit	Treatment	Success	Probabilities of elementary events $P(\{\omega\})$	Person variable U	Treatment variable X	Outcome variable Y
(*Tom, Con, −*)			10/99	*Tom*	0	0
(*Tom, Con, +*)			10/99	*Tom*	0	1
(*Tom, BTh, −*)			2/99	*Tom*	1	0
(*Tom, BTh, +*)			6/99	*Tom*	1	1
(*Tom, PTh, −*)			1/99	*Tom*	2	0
(*Tom, PTh, +*)			4/99	*Tom*	2	1
(*Jim, Con, −*)			5/99	*Jim*	0	0
(*Jim, Con, +*)			15/99	*Jim*	0	1
(*Jim, BTh, −*)			3/99	*Jim*	1	0
(*Jim, BTh, +*)			5/99	*Jim*	1	1
(*Jim, PTh, −*)			2/99	*Jim*	2	0
(*Jim, PTh, +*)			3/99	*Jim*	2	1
(*Kate, Con, −*)			12/99	*Kate*	0	0
(*Kate, Con, +*)			8/99	*Kate*	0	1
(*Kate, BTh, −*)			5/99	*Kate*	1	0
(*Kate, BTh, +*)			3/99	*Kate*	1	1
(*Kate, PTh, −*)			4/99	*Kate*	2	0
(*Kate, PTh, +*)			1/99	*Kate*	2	1

Table 5.1 also displays the values of the three random variables U: $(\Omega, \mathcal{A}, P) \rightarrow$ $[\Omega_U, \mathcal{P}(\Omega_U)]$, X: $(\Omega, \mathcal{A}, P) \rightarrow [\Omega'_X, \mathcal{P}(\Omega'_X)]$, and Y: $(\Omega, \mathcal{A}, P) \rightarrow [\Omega'_Y, \mathcal{P}(\Omega'_Y)]$, where $\Omega'_X :=$ $\{0, 1, 2\}$ and $\Omega'_Y := \{0, 1\}$. For the singletons $\{x\}$, $x \in \Omega'_X$, the values $P_X(\{x\}) = P[X^{-1}(\{x\})]$ of the distribution of X are

$$P_X(\{0\}) = 60/99, \quad P_X(\{1\}) = 24/99, \quad P_X(\{2\}) = 15/99,$$

for the sets that consist of two elements of Ω'_X, they are

$$P_X(\{0, 1\}) = 84/99, \quad P_X(\{0, 2\}) = 75/99, \quad P_X(\{1, 2\}) = 39/99,$$

and for Ω'_X and $\emptyset$, they are $P_X(\Omega'_X) = 1$ and $P_X(\emptyset) = 0$.

For the singletons $\{u\}$, $u \in \Omega_U$, the values $P_U(\{u\}) = P[U^{-1}(\{u\})]$ of the distribution of U are

$$P_U(\{Tom\}) = P_U(\{Jim\}) = P_U(\{Kate\}) = 1/3,$$

for the sets that consist of two elements of Ω_U, they are

$$P_U(\{Tom, Jim\}) = P_U(\{Tom, Kate\}) = P_U(\{Jim, Kate\}) = 2/3,$$

and for Ω_U and $\emptyset$, they are $P_U(\Omega_U) = 1$ and $P_U(\emptyset) = 0$.   ◁

Time order between random variables

In Example 5.12 and also in the examples with Joe and Ann, there is a time order between the random variables involved. Obviously, the person variable U represents events that are *prior* to the events represented by the treatment variable X and to the events represented by the outcome variable Y. In Definition 5.13, we extend the definitions introduced in section 4.2.2 to random variables.

Definition 5.13 [X is prior, simultaneous, posterior to Y]
Let X: $(\Omega, \mathcal{A}, P) \rightarrow (\Omega'_X, \mathcal{A}'_X)$, Y: $(\Omega, \mathcal{A}, P) \rightarrow (\Omega'_Y, \mathcal{A}'_Y)$ be random variables and $(\mathcal{F}_t, t \in T)$ a filtration in $\mathcal{A}$. Then,

(i) *X is called prior to Y (and Y posterior to X) in $(\mathcal{F}_t, t \in T)$, if there is an $s \in T$ such that $\sigma(X) \subset \mathcal{F}_s$, $\sigma(Y) \not\subset \mathcal{F}_s$, and a $t \in T$, $t > s$, such that $\sigma(Y) \in \mathcal{F}_t$.*

(ii) *Assume that T is finite. Then X is called simultaneous to Y in $(\mathcal{F}_t, t \in T)$, if there is a $t \in T$ such that $\sigma(X), \sigma(Y) \subset \mathcal{F}_t$ and no $s \in T$, $s < t$, such that $\sigma(X) \subset \mathcal{F}_s$ or $\sigma(Y) \subset \mathcal{F}_s$.*

Note that the concept of simultaneity of random variables can also be extended to cases in which T is not finite. For simplicity, we confine ourselves to the finite case.

Remark 5.14 [Filtration generated by a family of random variables]   Definition 5.13 does not presume that all pairs of random variables can be ordered and compared to each other with respect to a filtration $(\mathscr{F}_t, t \in T)$. However, given a family $(X_i, i \in I)$ of random variables where $I \subset \mathbb{R}$, we can define a filtration $(\mathscr{F}_i, i \in I)$ by $\mathscr{F}_i := \sigma(X_j, j \leq i)$ for all $i \in I$. This filtration is called the *filtration generated by* $(X_i, i \in I)$ (see Klenke, 2013). In this filtration, X_i is prior to X_j if and only if $i < j$, where $i, j \in I$.   ◁

Example 5.15 [Flipping a coin twice]   If we consider the random experiment of flipping the same coin twice, the probability space $(\Omega, \mathscr{A}, P)$ is identical to the one specified in section 2.2.2 and Example 2.83, where we considered the random experiment of flipping two coins. Hence, the set of possible outcomes is

$$\Omega = \{(h, h), (h, t), (t, h), (t, t)\}.$$

The possible outcome (t, h) represents obtaining tails in the first flip and heads in the second flip. Now, for $i = 1, 2$, define the random variables $X_i \colon \Omega \to \{0, 1\}$ by

$$X_i[(a_1, a_2)] := \begin{cases} 1, & \text{if } a_i = h \\ 0, & \text{if } a_i = t, \end{cases} \quad \forall\, (a_1, a_2) \in \Omega. \tag{5.7}$$

Hence, the value 1 of X_i indicates that the outcome of the ith flip is heads. If we define the filtration $(\mathscr{F}_1, \mathscr{F}_2)$ by $\mathscr{F}_1 := \sigma(X_1)$ and $\mathscr{F}_2 := \sigma(X_1, X_2)$, then X_1 is prior to X_2 in the filtration $(\mathscr{F}_1, \mathscr{F}_2)$. Hence, this filtration serves to introduce time order between the first and the second flips of the coin. It can be shown, for example, that X_1 is also prior to $X_1 \cdot X_2$ in $(\mathscr{F}_1, \mathscr{F}_2)$ (see Exercise 5.4).   ◁

Example 5.16 [Joe and Ann with randomized assignment – continued]   In Example 5.37, the random variable U is prior to X in the filtration $(\mathscr{F}_t, t \in \{1, 2, 3\})$ specified in Example 4.19, because $\sigma(U) \subset \mathscr{F}_1$, $\sigma(X) \not\subset \mathscr{F}_1$, and $\sigma(X) \subset \mathscr{F}_2$. Analogously, it can be shown that, in this example, X is prior to Y.   ◁

5.2   Equivalence of two random variables with respect to a probability measure

5.2.1   Identical and P-equivalent random variables

Let $X, Y \colon (\Omega, \mathscr{A}, P) \to (\Omega', \mathscr{A}')$ be two random variables. Then X and Y are called *identical* if

$$\forall\, \omega \in \Omega \colon X(\omega) = Y(\omega). \tag{5.8}$$

Remark 5.17 [P-equivalent random variables]   Let $X, Y \colon (\Omega, \mathscr{A}, P) \to (\Omega', \mathscr{A}')$ be two random variables. Then X and Y are *almost surely identical* with respect to P or *P-equivalent*, denoted by $X \underset{P}{=} Y$, if

$$\exists\, A \in \mathscr{A} \colon (\forall\, \omega \in \Omega \setminus A \colon X(\omega) = Y(\omega) \quad \text{and} \quad P(A) = 0) \tag{5.9}$$

(see Def. 2.68). Another notation for $X \underset{P}{=} Y$ is $X(\omega) \underset{P\text{-}a.a.}{=} Y(\omega)$, which is a shortcut for

$$X(\omega) = Y(\omega), \quad \text{for } P\text{-a.a. } \omega \in \Omega, \tag{5.10}$$

meaning that the values of X and Y are identical for P-*almost all* $\omega \in \Omega$ (see Rem. 2.70).   ◁

Remark 5.18 [Singleton with a positive probability] If $X \underset{P}{=} Y$ or, equivalently, if $X(\omega) \underset{P\text{-}a.a.}{=} Y(\omega)$, and $\{\omega^*\} \in \mathscr{A}$, with $P(\{\omega^*\}) > 0$, then $X(\omega^*) = Y(\omega^*)$ [see Rem. 2.71].   ◁

Example 5.19 [Indicator of a null set] Let $(\Omega, \mathscr{A}, P)$ be a probability space and $A \in \mathscr{A}$. If $P(A) = 0$, then,

$$1_A \underset{P}{=} 0 \quad \text{and} \quad 1_{A^c} = 1 - 1_A \underset{P}{=} 1 \tag{5.11}$$

(see Example 5.8).   ◁

Remark 5.20 [Q-equivalence] Note that the definition of equivalence of two random variables X and Y with respect to a probability measure only presumes that X and Y are measurable with respect to a σ-algebra on Ω and that the measure considered is a probability measure on this σ-algebra. Hence, we can consider the equivalence of X and Y with respect to different probability measures, say P and Q, and study their relationship.   ◁

In Lemma 5.21, we consider the relationship between P-equivalence and Q-equivalence, presuming $Q \underset{\mathscr{C}}{\ll} P$ (absolute continuity), that is, presuming

$$\forall\, C \in \mathscr{C} \colon P(C) = 0 \;\Rightarrow\; Q(C) = 0$$

(see Def. 3.70).

Lemma 5.21 [P-equivalence and Q-equivalence]
Let $X, Y \colon (\Omega, \mathscr{A}, P) \to (\Omega', \mathscr{A}')$ be random variables. If Q is a probability measure on $(\Omega, \mathscr{A})$ such that $Q \underset{\mathscr{A}}{\ll} P$, then $X \underset{P}{=} Y$ implies $X \underset{Q}{=} Y$.

(Proof p. 196)

According to Lemma 4.32 (i), $P^B \underset{\mathscr{A}}{\ll} P$, provided that $B \in \mathscr{A}$ is an event for which $P(B) > 0$. Hence, Lemma 5.21 immediately implies Corollary 5.22.

Corollary 5.22 [P-equivalence implies P^B-equivalence]
Let $X, Y \colon (\Omega, \mathscr{A}, P) \to (\Omega', \mathscr{A}')$ be two random variables, and let $B \in \mathscr{A}$ with $P(B) > 0$. Then, $X \underset{P}{=} Y$ implies $X \underset{P^B}{=} Y$.

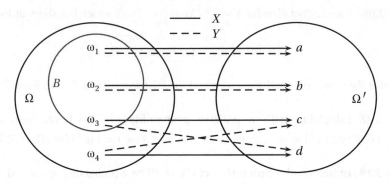

Figure 5.1   Two random variables that are P^B-equivalent if $P(\{\omega_3\}) = 0$.

Example 5.23 [P^B-equivalence does not imply P-equivalence] Consider the set $\Omega = \{\omega_1, \omega_2, \omega_3, \omega_4\}$ with the σ-algebra $\mathscr{A} = \mathscr{P}(\Omega)$, and the set $\Omega' = \{a, b, c, d\}$ with the σ-algebra $\mathscr{A}' = \mathscr{P}(\Omega')$. Furthermore, let $P\colon \mathscr{A} \to [0, 1]$ satisfy $P(\{\omega_1\}) = .25$, $P(\{\omega_2\}) = .25$, $P(\{\omega_3\}) = 0$, and $P(\{\omega_4\}) = .50$. Finally, define $X, Y\colon (\Omega, \mathscr{A}, P) \to (\Omega', \mathscr{A}')$ by

$$
X(\omega) = \begin{cases} a, & \text{if } \omega = \omega_1 \\ b, & \text{if } \omega = \omega_2 \\ c, & \text{if } \omega = \omega_3 \\ d, & \text{if } \omega = \omega_4 \end{cases}
\qquad
Y(\omega) = \begin{cases} a, & \text{if } \omega = \omega_1 \\ b, & \text{if } \omega = \omega_2 \\ d, & \text{if } \omega = \omega_3 \\ c, & \text{if } \omega = \omega_4 \end{cases}
$$

(see Fig. 5.1). If $B = \{\omega_1, \omega_2, \omega_3\}$, then $X \underset{P^B}{=} Y$, but neither $X = Y$ nor $X \underset{P}{=} Y$ (see Exercise 5.5). Therefore, equivalence with respect to P^B does not imply equivalence with respect to P. ◁

Theorem 2.85 on the equivalence of image measures immediately implies Corollary 5.24 on the identity of the distributions of two P-equivalent random variables:

> **Corollary 5.24 [P-equivalence implies identical distributions]**
> Let $X, Y\colon (\Omega, \mathscr{A}, P) \to (\Omega', \mathscr{A}')$ be random variables with distributions P_X and P_Y, respectively. If $X \underset{P}{=} Y$, then $P_X = P_Y$.

In other words, if X and Y are P-equivalent, then they are identically distributed. Note, however, that identical distributions of X and Y do not imply that X and Y are P-equivalent.

In chapter 6, we shall see that Corollary 5.24 also implies that the expectations, variances, and other moments of X and Y are identical if $X, Y\colon (\Omega, \mathscr{A}, P) \to (\overline{\mathbb{R}}, \overline{\mathscr{B}})$ are P-equivalent numerical random variables, provided that the expectations, variances, and other moments of X and Y exist.

Corollary 5.25 is an immediate implication of Theorem 2.86.

Corollary 5.25 [P-equivalence of compositions]

Let $X: (\Omega, \mathscr{A}, P) \to (\Omega'_X, \mathscr{A}'_X)$ be a random variable with distribution P_X, and let $g, g^*: (\Omega'_X, \mathscr{A}'_X) \to (\overline{\mathbb{R}}, \overline{\mathscr{B}})$ be measurable functions. Then,

(i) $g(X) \underset{P}{=} g^*(X) \Leftrightarrow g \underset{P_X}{=} g^*$.

(ii) $g(X) \underset{P}{<} g^*(X) \Leftrightarrow g \underset{P_X}{<} g^*$.

(iii) $g(X) \underset{P}{\leq} g^*(X) \Leftrightarrow g \underset{P_X}{\leq} g^*$.

(Proof p. 196)

Remark 5.26 [Alternative notation]   Note that:

$$g \underset{P_X}{=} g^* \Leftrightarrow g(x) = g^*(x), \quad \text{for } P_X\text{-a.a. } x \in \Omega'_X, \tag{5.12}$$

$$g \underset{P_X}{<} g^* \Leftrightarrow g(x) < g^*(x), \quad \text{for } P_X\text{-a.a. } x \in \Omega'_X, \tag{5.13}$$

$$g \underset{P_X}{\leq} g^* \Leftrightarrow g(x) \leq g^*(x), \quad \text{for } P_X\text{-a.a. } x \in \Omega'_X. \tag{5.14}$$

◁

5.2.2 P-equivalence, P^B-equivalence, and absolute continuity

Now we consider the relationship between equivalence of two random variables $X, Y: (\Omega, \mathscr{A}, P) \to (\Omega', \mathscr{A}')$ with respect to P and P^B, and absolute continuity of P_X with respect to P_X^B, the distribution of X with respect to the conditional-probability measure P^B. Remember, for $B \in \mathscr{A}$ and $P(B) > 0$, we defined the B-conditional probability measure P^B (see Def. 4.29). Referring to such a measure, $X \underset{P^B}{=} Y$ means

$$\exists A \in \mathscr{A}: (\forall \omega \in \Omega \setminus A: X(\omega) = Y(\omega) \quad \text{and} \quad P^B(A) = 0) \tag{5.15}$$

[see (5.9)]. If B denotes the event $\{X = x\} = \{\omega \in \Omega: X(\omega) = x\}$, then we define $P^{X=x} := P^B$ and call it the $(X=x)$-conditional probability measure on $(\Omega, \mathscr{A})$.

Lemma 5.27 [An implication of absolute continuity]

Let $X, Y: (\Omega, \mathscr{A}, P) \to (\Omega', \mathscr{A}')$ be random variables that are measurable with respect to the σ-algebra $\mathscr{C} \subset \mathscr{A}$, and assume $\{X \neq Y\} \in \mathscr{C}$. Furthermore, let $B \in \mathscr{A}$ with $P(B) > 0$. If $X \underset{P^B}{=} Y$ and $P \underset{\mathscr{C}}{\ll} P^B$, then $X \underset{P}{=} Y$.

(Proof p. 196)

Example 5.28 [No treatment for Joe]   Consider Table 5.2. In this example, we define the set $\Omega_U = \{Joe, Jim, Ann\}$ and

$$\mathscr{A}_U = \mathscr{P}(\Omega_U) = \{\{Joe\}, \{Jim\}, \{Ann\}, \{Joe, Jim\}, \{Joe, Ann\}, \{Jim, Ann\}, \Omega_U, \emptyset\}.$$

Table 5.2 No treatment for Joe.

Unit	Treatment	Success	$P(\{\omega\})$	Person variable U	Person variable U^*	Treatment variable X	Outcome variable Y	$P^{X=0}(\{\omega\})$ (rounded)	$P^{X=1}(\{\omega\})$
(Joe, no, −)			.152	Joe	Joe	0	0	.245	0
(Joe, no, +)			.348	Joe	Joe	0	1	.561	0
(Joe, yes, −)			0	Joe	Jim	1	0	0	0
(Joe, yes, +)			0	Joe	Jim	1	1	0	0
(Ann, no, −)			.096	Ann	Ann	0	0	.155	0
(Ann, no, +)			.024	Ann	Ann	0	1	.039	0
(Ann, yes, −)			.228	Ann	Ann	1	0	0	.60
(Ann, yes, +)			.152	Ann	Ann	1	1	0	.40

Using these sets, not only $U: (\Omega, \mathscr{A}, P) \to (\Omega_U, \mathscr{A}_U)$ is a random variable, but also $U^*: (\Omega, \mathscr{A}, P) \to (\Omega_U, \mathscr{A}_U)$ defined in Table 5.2. Now the distribution of U is specified by $P_U(\{Joe\}) = .5$, $P_U(\{Jim\}) = 0$, and $P_U(\{Ann\}) = .5$. The probabilities of the other five elements of $\mathscr{A}_U$ are obtained using Rule (x) of Box 4.1. Furthermore, $P_{U^*} = P_U$.

Considering the measure $P^{X=0}$, we find $U \underset{P^{X=0}}{=} U^*$, because

$$P^{X=0}(\{U \neq U^*\}) = P^{X=0}(\{(Joe, yes, +), (Joe, yes, -)\}) = 0.$$

Furthermore, there are only two sets $A \in \sigma(U)$ with $P^{X=0}(A) = 0$. These are the sets $U^{-1}(\{Jim\}) = \{(Joe, yes, +), (Joe, yes, -)\}$ and $\emptyset$, and for these sets we find $P(U^{-1}(\{Jim\})) = P(\emptyset) = 0$. Hence, $P \underset{\sigma(U)}{\ll} P^{X=0}$, and according to Lemma 5.27 this implies $U \underset{P}{=} U^*$. In fact, we find

$$P(\{U \neq U^*\}) = P(\{(Joe, yes, +), (Joe, yes, -)\}) = 0.$$

◁

Lemma 5.29 [Absolute continuity]

Let $X: (\Omega, \mathscr{A}, P) \to (\Omega_X', \mathscr{A}_X')$ be a random variable and $B \in \mathscr{A}$ with $P(B) > 0$. Then,

$$P \underset{\sigma(X)}{\ll} P^B \Leftrightarrow P_X \underset{\mathscr{A}_X'}{\ll} P_X^B. \tag{5.16}$$

(Proof p. 196)

Example 5.30 [No treatment for Joe – continued]   In Example 5.28, we already found

$$P \underset{\sigma(U)}{\ll} P^{X=0}.$$

There are only two sets $A' \in \mathscr{A}_U$ with $P_U^{X=0}(A') = 0$, the sets $\{Jim\}$ and $\emptyset$, and for these sets we find $P_U(\{Jim\}) = P_U(\emptyset) = 0$. Hence, in this example,

$$P_U \underset{\mathscr{A}_U}{\ll} P_U^{X=0}$$

holds as well.   ◁

Lemmas 5.29 and 4.32 immediately imply Corollary 5.31.

> **Corollary 5.31 [Null-set equivalence]**
> *Let $X: (\Omega, \mathscr{A}, P) \to (\Omega_X', \mathscr{A}_X')$ be a random variable and $B \in \mathscr{A}$ with $P(B) > 0$. Then P and P^B are null-set equivalent on $(\Omega, \sigma(X))$ if and only if P_X and P_X^B are null-set equivalent on $(\Omega_X', \mathscr{A}_X')$.*

According to Lemma 4.32, absolute continuity of P_X^B with respect to P_X always holds. In other words, $P_X^B \underset{\mathscr{A}_X'}{\ll} P_X$, which is equivalent to

$$\forall A' \in \mathscr{A}_X': \ P_X(A') = 0 \ \Rightarrow \ P_X^B(A') = 0, \tag{5.17}$$

always holds. In contrast, $P_X \underset{\mathscr{A}_X'}{\ll} P_X^B$ is *not necessarily true*.

Example 5.32 [No treatment for Joe – continued]   Table 5.2 displays an example illustrating absolute continuity of P_U with respect to P_U^B for a discrete random variable U. Consider the event $B = \{X=1\} = \{\omega \in \Omega: X(\omega) = 1\}$. Using this notation, P_U is not absolutely continuous with respect to $P_U^{X=1}$, (i.e., $P_U \underset{\mathscr{A}_U}{\ll} P_U^{X=1}$ does *not hold*). In contrast, $P_U \underset{\mathscr{A}_U}{\ll} P_U^{X=0}$ does hold.

In this example, the eight elements of Ω are listed in the first column of the table. Furthermore, we choose $\mathscr{A} = \mathscr{P}(\Omega)$, and the probability measure on $(\Omega, \mathscr{A})$ is specified by the probabilities of the singletons $\{\omega\}$ specified in the second column of the table [see Box 4.1 (x)]. The random variables $U: (\Omega, \mathscr{A}, P) \to (\Omega_U, \mathscr{A}_U)$, with $\Omega_U = \{Joe, Ann\}$, $\mathscr{A}_U = \mathscr{P}(\Omega_U)$, and $X, Y: (\Omega, \mathscr{A}, P) \to (\Omega', \mathscr{P}(\Omega'))$ with $\Omega' = \{0, 1\}$, are specified in columns 3, 5, and 6 of Table 5.2. (The random variable U^* has been used in Example 5.28.) Note that the distribution of U is:

$$P_U(\{Joe\}) = P(\{(Joe, no, -), (Joe, no, +), (Joe, yes, +), (Joe, yes, -)\}) = .5,$$
$$P_U(\{Ann\}) = P(\{(Ann, no, -), (Ann, no, +), (Ann, yes, +), (Ann, yes, -)\}) = .5,$$
$$P_U(\Omega_U) = 1, \quad P_U(\emptyset) = 0.$$

Now, we compute the $(X=1)$-conditional probabilities of the elementary events:

$$P^{X=1}(\{\omega_1\}) = P^{X=1}(\{(Joe, no, -)\}) = \frac{P(\{(Joe, no, -)\} \cap \{X=1\})}{P(X=1)}$$

$$= 0/(.228 + .152) = 0,$$

and the same result is obtained for ω_2 to ω_6. In contrast,

$$P^{X=1}(\{\omega_7\}) = P^{X=1}(\{(Ann, yes, -)\}) = \frac{P(\{(Ann, yes, -)\} \cap \{X=1\})}{P(X=1)}$$

$$= .228/(.228 + .152) = .6,$$

and

$$P^{X=1}(\{\omega_8\}) = P^{X=1}(\{(Ann, yes, +)\}) = \frac{P(\{(Ann, yes, +)\} \cap \{X=1\})}{P(X=1)}$$

$$= .152/(.228 + .152) = .4.$$

These results are displayed in the last column of Table 5.2, and the last but one column shows the probabilities $P^{X=0}(\{\omega\})$ of the singletons with respect to $P^{X=0}$.

Now consider the set $\{Joe\} \in \mathcal{A}_U$. Inspecting the last and the second columns of Table 5.2 shows that

$$P_U^{X=1}(\{Joe\}) = 0 \quad \text{and} \quad P_U(\{Joe\}) = .5.$$

According to Definition 3.70 (i), this implies that $P_U \underset{\mathcal{A}_U}{\ll} P_U^{X=1}$ does *not* hold. In contrast, none of the four elements $A' \in \mathcal{A}_U$ satisfies

$$P_U^{X=0}(A') = 0 \quad \text{and} \quad P_U(A') \neq 0.$$

Therefore, in this example, $P_U \underset{\mathcal{A}_U}{\ll} P_U^{X=0}$ does hold.   ◁

5.3   Multivariate random variable

Univariate random variables take on their values in sets such as $\Omega' \subset \overline{\mathbb{R}}$, $\Omega' = \{male, female\}$, or $\Omega' = \{low, medium, high\}$, whereas bivariate random variables take on their values in sets such as $\Omega' \subset \overline{\mathbb{R}}^2$ or

$$\Omega' = \{male, female\} \times \{low, medium, high\}.$$

The values of bivariate random variables are pairs such as $(5, 8)$ or $(male, low)$. The values of n-variate random variables are n-tuples. If X takes on values such as $male$ or $(male, low)$, then we call X *qualitative*. If X takes on values in $\mathbb{R}^n$, $n \in \mathbb{N}$, we call it *n-variate real-valued*. If X takes on values in $\overline{\mathbb{R}}^n$, $n \in \mathbb{N}$, we call it *n-variate numerical*.

Remark 5.33 [Joint and marginal distributions]   Definition 5.1 also applies to an n-variate random variable X, that is, to a random variable

$$X = (X_1, \ldots, X_n) \colon (\Omega, \mathcal{A}, P) \to \left(\underset{i=1}{\overset{n}{\times}} \Omega_i', \underset{i=1}{\overset{n}{\bigotimes}} \mathcal{A}_i' \right) \tag{5.18}$$

that consists of n random variables $X_i \colon (\Omega, \mathcal{A}, P) \to (\Omega_i', \mathcal{A}_i')$. Hence,

$$X(\omega) = [X_1(\omega), \ldots, X_n(\omega)], \quad \forall \, \omega \in \Omega . \tag{5.19}$$

The distribution $P_X = P_{X_1, \ldots, X_n}$ of X is also called the *joint distribution* of the random variables $X_i, i = 1, \ldots, n$.
  Because $\pi_i(X_1, \ldots, X_n) = X_i$,

$$P_{X_i} = P_{\pi_i(X_1, \ldots, X_n)}, \quad i = 1, \ldots, n, \tag{5.20}$$

[see Eq. (2.20) defining the projection π_i]. In this context, P_{X_i} is called the (one-dimensional) *marginal distribution of X_i*. Equation (5.20) shows that the joint distribution uniquely determines all marginal distributions, but not vice versa! More specifically, for $i = 1, \ldots, n$,

$$P_{X_i}(A_i') = P_{X_1, \ldots, X_n}(\Omega_1' \times \ldots \times \Omega_{i-1}' \times A_i' \times \Omega_{i+1}' \times \ldots \times \Omega_n'), \quad \forall \, A_i' \in \mathcal{A}_i' . \tag{5.21}$$

Analogously, we may also describe the marginal distribution of $(X_{i_1}, \ldots, X_{i_m})$, where $\{i_1, \ldots, i_m\} \subset \{1, \ldots, n\}$.   ◁

Remark 5.34 [Joint distribution vs. other quantities]   The joint distribution of a multivariate random variable contains the essential information about the random variables $X_1, \ldots, X_n$. All other quantities such as expectations $E(X_i)$, variances $Var(X_i)$, covariances $Cov(X_i, X_j)$, or conditional expectations such as $E(X_1 \mid X_2, \ldots, X_n)$, which are introduced in succeeding chapters, are determined by the joint distribution, and usually they contain less information. Nevertheless, these other quantities often reveal certain properties of a multivariate random variable more clearly than the joint distribution.   ◁

Example 5.35 [Flipping two coins – continued]   In the example of section 2.2.2 and in Example 2.83, we considered the random experiment of flipping two coins and defined the random variable X representing with its values the number of flipping heads. Additional to X, we may also define the random variables $X_1, X_2 \colon (\Omega, \mathcal{A}, P) \to (\{0, 1\}, \mathcal{P}(\{0, 1\}))$ by

$$X_1(\omega) = \begin{cases} 1, & \text{if } \omega \in \{(h, t), (h, h)\} \\ 0, & \text{if } \omega \in \{(t, h), (t, t)\} \end{cases} \tag{5.22}$$

and

$$X_2(\omega) = \begin{cases} 1, & \text{if } \omega \in \{(t, h), (h, h)\} \\ 0, & \text{if } \omega \in \{(h, t), (t, t)\} . \end{cases} \tag{5.23}$$

They indicate with their value 1 if *heads* are flipped at the first and second flip, respectively. Obviously, $X = X_1 + X_2$. Furthermore,

$$(X_1, X_2): (\Omega, \mathcal{A}, P) \rightarrow (\{0, 1\} \times \{0, 1\}, \mathcal{P}(\{0, 1\}) \otimes \mathcal{P}(\{0, 1\}))$$

is a two-dimensional random variable with values $(0, 0), (0, 1), (1, 0)$, and $(1, 1)$. The joint distribution P_{X_1, X_2} is uniquely defined by

$$P_{X_1, X_2}(\{(i, j)\}) = \frac{1}{4}, \quad \forall\, i, j = 0, 1.$$

The marginal distribution of X_1 is

$$P_{X_1}(\{i\}) = P_{X_1, X_2}(\{(i, 0)\}) + P_{X_1, X_2}(\{(i, 1)\}) = \frac{1}{4} + \frac{1}{4} = \frac{1}{2}, \quad i = 0, 1,$$

$P_{X_1}(\{0, 1\}) = 1$, and $P_{X_1}(\emptyset) = 0$. Obviously, P_{X_1} is completely determined by the joint distribution P_{X_1, X_2}, and the same applies to the marginal distribution P_{X_2}.   ◁

Example 5.36 [Tom, Jim, and Kate – continued]   The second column of Table 5.1 also displays the probabilities $P_{U,X,Y}(\{(u, x, y)\}) = P(\{\omega\})$ of the three-dimensional random variable (U, X, Y) that maps the elements $\omega \in \Omega$ onto the set

$$\Omega' = \{Tom, Jim, Kate\} \times \{0, 1, 2\} \times \{0, 1\}$$

on which we consider the σ-algebra

$$\mathcal{A}' = \mathcal{P}(\Omega') = \mathcal{P}(\{Tom, Jim, Kate\}) \otimes \mathcal{P}(\{0, 1, 2\}) \otimes \mathcal{P}(\{0, 1\}).$$

The probabilities $P_{U,X,Y}(\{(u, x, y)\})$, $(u, x, y) \in \Omega'$, uniquely determine the joint distribution $P_{U,X,Y}$ as well as the one-dimensional marginal distributions P_U, P_X, and P_Y, and the two-dimensional marginal distributions $P_{U,X}$, $P_{U,Y}$, and $P_{X,Y}$.   ◁

For another example of a joint distribution, which refers to Example 2.34, see Exercise 5.6.

5.4   Independence of random variables

The concepts of independence of events and of set systems (i.e., of sets of events), which have been introduced in Definition 4.40, can be used to define *stochastic independence of random variables*. Remember that

$$\sigma(X) := X^{-1}(\mathcal{A}') := \{X^{-1}(A'): A' \in \mathcal{A}'\}$$

is a σ-algebra on Ω, called the *σ-algebra generated by X* (see Def. 2.26). Hence, we can define the random variables $X_1: (\Omega, \mathcal{A}, P) \rightarrow (\Omega_1', \mathcal{A}_1')$ and $X_2: (\Omega, \mathcal{A}, P) \rightarrow (\Omega_2', \mathcal{A}_2')$ to be

P-independent if $X_1^{-1}(\mathscr{A}_1')$ and $X_2^{-1}(\mathscr{A}_2')$ are *P*-independent. In other words, X_1 and X_2 are defined to be *P*-independent, if and only if

$$P(A \cap B) = P(A) \cdot P(B), \quad \forall (A, B) \in X_1^{-1}(\mathscr{A}_1') \times X_2^{-1}(\mathscr{A}_2'). \tag{5.24}$$

Using the notation introduced in Remark 5.4 and

$$P(X_1 \in A', X_2 \in B') := P(\{X_1 \in A'\} \cap \{X_2 \in B'\}), \tag{5.25}$$

this equation is equivalent to

$$P(X_1 \in A', X_2 \in B') = P(X_1 \in A') \cdot P(X_2 \in B'), \quad \forall (A', B') \in \mathscr{A}_1' \times \mathscr{A}_2'. \tag{5.26}$$

Independence of the random variables X_1 and X_2 with respect to P is denoted by $X_1 \underset{P}{\perp\!\!\!\perp} X_2$.

Example 5.37 [Joe and Ann with randomized assignment – continued]   In Table 2.2, we presented the random experiment of drawing a person from a set of persons, $\Omega_U = \{Joe, Ann\}$, performing a randomized assignment of the drawn person to one of two treatment conditions represented by the elements of the set $\Omega_X = \{yes, no\}$, and observing success or failure, represented by the elements of the set $\Omega_Y = \{-, +\}$. Hence, the set of possible outcomes of this random experiment is

$$\Omega = \Omega_U \times \Omega_X \times \Omega_Y,$$

which consists of the eight triples listed in the first column of Table 2.2. In that table, we considered the random variables

$$U: (\Omega, \mathscr{P}(\Omega)) \to (\Omega_U, \mathscr{P}(\Omega_U)) \quad \text{and} \quad X, Y: (\Omega, \mathscr{P}(\Omega)) \to (\Omega', \mathscr{A}'),$$

where $\Omega' = \{0, 1\}$ and $\mathscr{A}' = \{\{0\}, \{1\}, \Omega', \emptyset\}$. In order to check if Equation (5.26) actually holds, we choose the two sets $\{Joe\} \in \mathscr{P}(\Omega_U)$ and $\{0\} \in \mathscr{A}'$ and compare the probability

$$P(X=0, U=Joe) = P(\{(Joe, no, -), (Joe, no, +)\}) = .3$$

(see the first two rows of Table 2.2) to the product of the two probabilities

$$P(X=0) = P(\{(Joe, no, -), (Joe, no, +), (Ann, no, -), (Ann, no, +)\}) = .6$$

and

$$P(U=Joe) = P(\{(Joe, no, -), (Joe, no, +), (Joe, yes, -), (Joe, yes, +)\}) = .5.$$

Obviously, Equation (5.26) holds for the pair $(\{0\}, \{Joe\}) \in \mathscr{A}' \times \mathscr{P}(\Omega_U)$. Repeating the corresponding comparisons for all elements (pairs) of $\mathscr{A}' \times \mathscr{P}(\Omega_U)$ shows that Equation (5.26) actually holds in this example (see also Exercise 5.7).   ◁

Remark 5.38 [A methodological note on randomized assignment]   In random experiments such as the one presented in Example 5.37, with a randomized assignment of the drawn person to one of several treatment conditions, we create independence of X and the person variable U. According to Equation (5.26), this implies that we create independence of X and all U-measurable mappings $f(U)$, because $\sigma[f(U)] \subset \sigma(U)$ (see Cor. 2.53). More generally, randomized assignment of an observational unit (e.g., a person) creates independence of X and all pretreatment variables.   ◁

Using Definition 4.40, Definition 5.39 extends the concept of independence of two random variables to a family of random variables. This includes a finite sequence of random variables X_i, $i \in I = \{1, \ldots, n\}$, an infinite sequence of random variables X_i, $i \in I = \{1, 2, \ldots\}$, and a family $(X_i, i \in I)$ of random variables in which the index set I may be *any* set, including, for example, $I \subset \mathbb{R}$.

Definition 5.39 [Family of independent random variables]
A family $(X_i, i \in I)$ of random variables $X_i \colon (\Omega, \mathcal{A}, P) \to (\Omega_i', \mathcal{A}_i')$ is called P-independent, denoted by $\underset{P}{\perp\!\!\!\perp} (X_i, i \in I)$, if $(X_i^{-1}(\mathcal{A}_i'), i \in I)$ is a family of P-independent σ-algebras.

Remark 5.40 [Independence of three random variables]   Hence, three random variables X_1, X_2, and X_3 are independent with respect to P, denoted by $\underset{P}{\perp\!\!\!\perp} X_1, X_2, X_3$, if and only if

$$P(X_1 \in A', X_2 \in B', X_3 \in C') = P(X_1 \in A') \cdot P(X_2 \in B') \cdot P(X_3 \in C'),$$
$$\forall\, (A', B', C') \in \mathcal{A}_1' \times \mathcal{A}_2' \times \mathcal{A}_3' \tag{5.27}$$

(see Rem. 4.38). Note that pairwise independence of X_1, X_2, and X_3 follows from choosing $A' = \Omega_1'$, $B' = \Omega_2'$, or $C' = \Omega_3'$, respectively.   ◁

Remark 5.41 [Independence of n random variables]   Correspondingly, the random variables $X_1, \ldots, X_n$ are *independent*, denoted by $\underset{P}{\perp\!\!\!\perp} X_1, \ldots, X_n$, if and only if

$$P(X_1 \in A_1', \ldots, X_n \in A_n') = P(X_1 \in A_1') \cdot \ldots \cdot P(X_n \in A_n'),$$
$$\forall\, (A_1', \ldots, A_n') \in \mathcal{A}_1' \times \ldots \times \mathcal{A}_n'. \tag{5.28}$$
                                                                                    ◁

Remark 5.42 [Random sample]   Oftentimes, we assume that $X_1, \ldots, X_n$ is a sequence of independent and identically distributed (abbreviated i.i.d.) random variables (see, e.g., chs. 6 and 8). In statistics, a sequence $X_1, \ldots, X_n$ of i.i.d. random variables is called a *random sample*. An important example of i.i.d. random variables is treated in the section on Bernoulli trials (see section 8.1.2).   ◁

Remark 5.43 [Independence with respect to a probability measure]   If there is no ambiguity, we also use the term *independence* of events, sets of events, random variables, and sets of random variables. Note that, if Q is another probability measure on $(\Omega, \mathcal{A})$, then events, sets of events, and random variables can be P-independent, although they are not Q-independent. ◁

Remark 5.44 [A random variable and a set system] Independence of a set system and a random variable is defined in the same way. A set system $\mathscr{C} \subset \mathscr{A}$ and a random variable $X: (\Omega, \mathscr{A}, P) \to (\Omega'_X, \mathscr{A}'_X)$ are called *independent*, denoted by $\mathscr{C} \underset{P}{\perp\!\!\!\perp} X$, if $\mathscr{C}$ and $\sigma(X)$ are independent. Of course, $\mathscr{C}$ can also be a sub-σ-algebra of $\mathscr{A}$. ◁

Remark 5.45 [A random variable and a family of random variables] Independence of a random variable X and a family $(Y_i, i \in I)$ of random variables, denoted by $X \underset{P}{\perp\!\!\!\perp} (Y_i, i \in I)$, is defined by $X \underset{P}{\perp\!\!\!\perp} \sigma(Y_i, i \in I)$ [see Eq. (2.18)]. Note that $X \underset{P}{\perp\!\!\!\perp} (Y_i, i \in I)$ implies $X \underset{P}{\perp\!\!\!\perp} \sigma(Y_i)$, for all $i \in I$. ◁

Remark 5.46 [Equivalent propositions] Let $(\Omega, \mathscr{A}, P)$ be a probability space, $B \in \mathscr{A}$, and $\mathscr{C} \subset \mathscr{A}$. Then the following propositions are equivalent to each other:

$$1_B \underset{P}{\perp\!\!\!\perp} \mathscr{C}, \quad \sigma(\{B\}) \underset{P}{\perp\!\!\!\perp} \mathscr{C}, \quad \{B\} \underset{P}{\perp\!\!\!\perp} \mathscr{C}, \quad B \underset{P}{\perp\!\!\!\perp} \mathscr{C}$$

(see Rem. 4.41 and Exercise 5.8). ◁

In Corollary 5.24, we noted that P-equivalent random variables have identical distributions. According to Lemma 5.47, this also has implications for independence of random variables.

Lemma 5.47 [P-equivalence and independence]
Let $X_i: (\Omega, \mathscr{A}, P) \to (\Omega', \mathscr{A}')$, $i = 1, 2$, and $Y: (\Omega, \mathscr{A}, P) \to (\Omega'_Y, \mathscr{A}'_Y)$ be random variables. Then,

$$\left(X_1 \underset{P}{=} X_2 \wedge X_1 \underset{P}{\perp\!\!\!\perp} Y \right) \Rightarrow X_2 \underset{P}{\perp\!\!\!\perp} Y. \tag{5.29}$$

(Proof p. 197)

Now we consider the probability measure P and the B-conditional-probability measure P^B on $(\Omega, \mathscr{A})$ (see Def. 4.29). In Lemma 4.32, we have shown that P^B is absolutely continuous (see Def. 3.70) with respect to P on $(\Omega, \mathscr{A})$. In Lemma 5.48, we show that P is absolutely continuous with respect to P^B on $(\Omega, \mathscr{C})$, $\mathscr{C} \subset \mathscr{A}$, provided that B and $\mathscr{C}$ are independent.

Lemma 5.48 [Independence and absolute continuity]
Let $(\Omega, \mathscr{A}, P)$ be a probability space, let $\mathscr{C} \subset \mathscr{A}$ be a σ-algebra, and $B \in \mathscr{A}$ with $P(B) > 0$. Then $1_B \underset{P}{\perp\!\!\!\perp} \mathscr{C}$ implies $P \underset{\mathscr{C}}{\ll} P^B$.

(Proof p. 197)

In Lemma 5.49, $P_{X_1} \otimes \ldots \otimes P_{X_n}$ denotes the product measure of the marginal distributions (see Def. 1.67 and Rem. 5.33).

Lemma 5.49 [Independence and product measure]

Let $X = (X_1, \ldots, X_n)$ be an n-variate random variable as specified in (5.18). Then,

$$\underset{P}{\amalg} \, X_1, \ldots, X_n \Leftrightarrow P_{X_1, \ldots, X_n} = P_{X_1} \otimes \ldots \otimes P_{X_n}. \tag{5.30}$$

(Proof p. 197)

Example 5.50 [Tom, Jim, and Kate – continued]   In example 5.12, we considered the random variables X and U, which have been constructed such that they are independent. All $8 \cdot 8 = 64$ pairs (A, B) of elements $A \in X^{-1}(\mathscr{A}_X')$ and $B \in U^{-1}(\mathscr{A}_U)$ satisfy $P(A \cap B) = P(A) \cdot P(B)$. Let us consider, for example, $A_1 = X^{-1}(\{0\})$ and $B_1 = U^{-1}(\{Tom\})$, $B_2 = U^{-1}(\{Jim\})$, and $B_3 = U^{-1}(\{Kate\})$. Then,

$$P(A_1 \cap B_j) = \frac{20}{99}, \quad j = 1, 2, 3$$

and

$$P(A_1) \cdot P(B_j) = \frac{60}{99} \cdot \frac{1}{3} = \frac{20}{99}, \quad j = 1, 2, 3.$$

Similarly, considering the events $A_2 = X^{-1}(\{1\})$ and B_j,

$$P(A_2 \cap B_j) = \frac{8}{99}, \quad j = 1, 2, 3$$

and

$$P(A_2) \cdot P(B_j) = \frac{24}{99} \cdot \frac{1}{3} = \frac{8}{99}, \quad j = 1, 2, 3.$$

Finally, considering the events $A_3 = X^{-1}(\{2\})$ and B_j yields

$$P(A_3 \cap B_j) = \frac{5}{99}, \quad j = 1, 2, 3$$

and

$$P(A_3) \cdot P(B_j) = \frac{15}{99} \cdot \frac{1}{3} = \frac{5}{99}, \quad j = 1, 2, 3.$$

Because $\varnothing$ and all sets $A \in \mathscr{A}$ are independent, this implies that independence holds for all pairs $(A, B) \in \{A_1, A_2, A_3, \varnothing\} \times \{B_1, B_2, B_3, \varnothing\}$. Furthermore, because

(a) $\mathscr{E}_1 = \{A_1, A_2, A_3, \varnothing\}$ and $\mathscr{E}_2 = \{B_1, B_2, B_3, \varnothing\}$ are ∩-stable set systems with $\mathscr{E}_1$, $\mathscr{E}_2 \subset \mathscr{A}$, and

(b) $\sigma(\mathscr{E}_1) = X^{-1}(\mathscr{A}_X')$ and $\sigma(\mathscr{E}_2) = U^{-1}(\mathscr{A}_U)$, we can conclude that $P(A \cap B) = P(A) \cdot P(B)$ holds for *all* elements $A \in X^{-1}(\mathscr{A}_X')$ and $B \in U^{-1}(\mathscr{A}_U)$ (see Th. 4.44). Therefore, according to Equation (5.24), X and U are independent.   ◁

> **Lemma 5.51 [Independence of a constant and a set of events]**
> Let $X: (\Omega, \mathscr{A}, P) \to (\Omega_X', \mathscr{A}_X')$ be a random variable and $\mathscr{C} \subset \mathscr{A}$. If $X \underset{P}{=} \alpha$ and $\{\alpha\} \in \mathscr{A}_X'$,
> then X and $\mathscr{C}$ are independent.
>
> *(Proof p. 198)*

Now we consider mappings of independent random variables. If two random variables $X_i: (\Omega, \mathscr{A}, P) \to (\Omega_i', \mathscr{A}_i'), i = 1, 2,$ are independent and $f_i: (\Omega_i', \mathscr{A}_i') \to (\Omega_i'', \mathscr{A}_i''), i = 1, 2,$ are measurable mappings, then the two random variables $f_1(X_1)$ and $f_2(X_2)$ are independent as well. More generally, if $f_i: (\Omega_i', \mathscr{A}_i') \to (\Omega_i'', \mathscr{A}_i''), i = 1, \ldots, n,$ is a sequence of measurable mappings, then,

$$\underset{P}{\perp\!\!\!\perp} \, f_1(X_1), \ldots, f_n(X_n),$$

that is, then $f_1(X_1), \ldots, f_n(X_n)$ is a sequence of independent random variables on $(\Omega, \mathscr{A}, P)$, provided that $X_1, \ldots, X_n$ are independent. In Theorem 5.5.2, we generalize this proposition.

> **Theorem 5.52 [Mappings of families of independent random variables]**
> Let $X_i: (\Omega, \mathscr{A}, P) \to (\Omega_i', \mathscr{A}_i'), \; i = 1, \ldots, n,$ be random variables and, for $m \in \mathbb{N}$, let
> $I_1 = \{1, \ldots, i_1\}, I_2 = \{i_1+1, \ldots, i_2\}, \ldots, I_m = \{i_{m-1}+1, \ldots, n\}$. Furthermore, let
>
> $$f_j: \left(\underset{i \in I_j}{\times} \Omega_i', \underset{i \in I_j}{\bigotimes} \mathscr{A}_i' \right) \to (\Omega_j'', \mathscr{A}_j''), \quad j = 1, \ldots, m,$$
>
> be measurable mappings. If $X_1, \ldots, X_n$ are independent, then
>
> $$f_1(X_1, \ldots, X_{i_1}), f_2(X_{i_1+1}, \ldots, X_{i_2}), \ldots, f_m(X_{i_{m-1}+1}, \ldots, X_n)$$
>
> are independent.

For a generalization and a proof, see Bauer (1996, Theorem 9.6).

Example 5.53 [Sums of independent random variables]   Let $X_1, \ldots, X_{2n}, n \in \mathbb{N}$, be independent real-valued random variables. Then the n random variables

$$X_1 + X_2, X_3 + X_4, \ldots, X_{2n-1} + X_{2n}$$

are independent as well (see Th. 2.57).   ◁

Example 5.54 [Tom, Jim, and Kate – continued] In Example 5.50, we showed that the random variables X and U are independent. Now we consider the mappings $f: \{0, 1, 2\} \rightarrow \{0, 1\}$ and $g: \Omega_U \rightarrow \{male, female\}$ defined by

$$f(x) = \begin{cases} 0, & \text{if } x = 0 \\ 1, & \text{if } x = 1 \text{ or } x = 2 \end{cases}$$

and

$$g(u) = \begin{cases} male, & \text{if } u = Tom \text{ or } u = Jim \\ female, & \text{if } u = Kate, \end{cases}$$

respectively. According to Theorem 5.52, the mappings $f(X)$ (control vs. any of the two treatments) and $g(U)$ (*sex*) are independent as well (see Exercise 5.9). ◁

Remark 5.55 [Conditional independence of random variables given an event] In chapter 4, we also considered *conditional* independence of events and families of events *given an event B*. If, for random variables $X_1, \ldots, X_n$ (or, more generally, families of random variables), we consider the set systems $\sigma(X_1), \ldots, \sigma(X_n)$, then we can use Definition 4.49 in order to define conditional independence of $X_1, \ldots, X_n$ given an event B, presuming $P(B) > 0$. According to Definitions 4.49 and 5.39, conditional independence given B is equivalent to independence with respect to the probability measure P^B. In chapter 16, we generalize this concept and study it in more detail. ◁

5.5   Probability function of a discrete random variable

The distribution of a discrete random variable can be described by its *probability function* that is now introduced. Remember, if $X: (\Omega, \mathscr{A}, P) \rightarrow (\Omega'_X, \mathscr{A}'_X)$ is a random variable, then the distribution P_X of X is a probability measure on $(\Omega'_X, \mathscr{A}'_X)$.

Definition 5.56 [Discrete random variable and its probability function]
Let $X: (\Omega, \mathscr{A}, P) \rightarrow (\Omega'_X, \mathscr{A}'_X)$ be a random variable and assume that $\Omega'_0 \subset \Omega'_X$ is finite or countable with $P_X(\Omega'_0) = 1$ and $\{x\} \in \mathscr{A}'_X$ for all $x \in \Omega'_0$. Then X and its distribution P_X are called discrete, and the function $p_X: \Omega'_X \rightarrow [0, 1]$ defined by

$$p_X(x) = \begin{cases} P_X(\{x\}), & \text{if } x \in \Omega'_0, \\ 0, & \text{if } x \in \Omega'_X \setminus \Omega'_0, \end{cases} \tag{5.31}$$

is called the probability function of X.

Remark 5.57 [Notation] Note that $P(X=x) = p_X(x)$, using the notation introduced in Remark 5.4. ◁

Remark 5.58 [Probability function vs. distribution] The *distribution* P_X is defined for *every* random variable, whereas the *probability function* p_X only applies to *discrete* random variables. While P_X assigns probabilities to *subsets* of the codomain Ω'_X of X, the probability function p_X assigns a probability to each *element* x in Ω'_X. Note that p_X is a real-valued random variable on the probability space $(\Omega'_X, \mathscr{A}'_X, P_X)$.   ◁

Remark 5.59 [The probability function uniquely determines the distribution] Note that σ-additivity of the probability measure P_X implies that P_X is uniquely determined by the probability function p_X [see Rule (x) in Box 4.1]. Vice versa, according to Definition 5.56, P_X defines p_X. Hence, if $X, Y \colon (\Omega, \mathscr{A}, P) \to (\Omega', \mathscr{A}')$ are discrete random variables, then,

$$p_X = p_Y \Leftrightarrow P_X = P_Y. \tag{5.32}$$

◁

Remark 5.60 [Probability function of a discrete distribution] Note that (5.32) allows us to use the term *probability function of a discrete distribution* instead of *probability function of a discrete random variable*.   ◁

Lemma 5.61 [Characterizations of a discrete random variable]
Let $X \colon (\Omega, \mathscr{A}, P) \to (\Omega'_X, \mathscr{A}'_X)$ be a random variable.

(i) Then X is discrete if and only if there is a finite or countable $\Omega'_0 \subset \Omega'_X$ such that $\{x\} \in \mathscr{A}'_X$ for all $x \in \Omega'_0$, $P_X(\Omega'_0) = 1$, and

$$1_{X=x_0} \underset{P}{=} 1 - \sum_{x \in \Omega'_0 \setminus \{x_0\}} 1_{X=x}, \quad \forall\, x_0 \in \Omega'_0. \tag{5.33}$$

(ii) Now assume that $X \colon (\Omega, \mathscr{A}, P) \to (\mathbb{R}, \mathscr{B})$ is real-valued. Then X is discrete if and only if there is a finite or countable $\Omega' \subset \mathbb{R}$ such that

$$X \underset{P}{=} \sum_{x \in \Omega'} x \cdot 1_{X=x}. \tag{5.34}$$

(Proof p. 198)

Remark 5.62 [A caveat] Note that Equation (5.33) is equivalent to $P_X(\Omega'_0) = 1$. In proposition (ii), we can choose Ω' such that $0 \notin \Omega'$ even if $P(X=0) > 0$. In this case, the set Ω'_0 referred to in (i) can be chosen such that $\Omega'_0 := \Omega' \cup \{0\}$.   ◁

Corollary 5.63 [Discrete real-valued random variable]
Assume that $X \colon (\Omega, \mathscr{A}, P) \to (\mathbb{R}, \mathscr{B})$ is a real-valued random variable. Then X is discrete if and only if the following two conditions hold:

(a) $\Omega'_> := \{x \in \mathbb{R} \colon P(X=x) > 0\}$ is finite or countable.

(b) $X \underset{P}{=} \sum_{x \in \Omega'_>} x \cdot 1_{X=x}.$

(Proof p. 199)

Condition (b) may equivalently be written as:

$$X \underset{P}{=} \sum_{\substack{x \in \mathbb{R} \\ P(X=x)>0}} x \cdot 1_{X=x} . \tag{5.35}$$

Example 5.64 [Flipping two coins – continued] Consider again Example 5.35, and let $X: (\Omega, \mathcal{A}, P) \to (\Omega'_X, \mathcal{A}'_X)$ denote the *number of flipping heads*. If we assume that $P(\{\omega\}) = \frac{1}{4}$ for all $\omega \in \Omega$, then,

$$p_X(0) = P_X(\{0\}) = \frac{1}{4}, \quad p_X(1) = P_X(\{1\}) = \frac{1}{2}, \quad p_X(2) = P_X(\{2\}) = \frac{1}{4}$$

are the values of the probability function p_X of X. They are the probabilities of the events that X takes on the values 0, 1, and 2, respectively. For simplicity, we also denote these probabilities by $P(X=0)$, $P(X=1)$, and $P(X=2)$. In this example, we may choose different measurable spaces $(\Omega'_X, \mathcal{A}'_X)$. If we choose $(\Omega'_X, \mathcal{A}'_X) = (\{0, 1, 2\}, \mathcal{P}(\{0, 1, 2\}))$, then $\Omega'_0 = \Omega'_X$ (see Def. 5.56). If we choose $(\Omega'_X, \mathcal{A}'_X) = (\mathbb{R}, \mathcal{B})$, then $\Omega'_0 = \{0, 1, 2\}$ and $\mathcal{A}'_X|_{\Omega'_0} = \mathcal{B}|_{\Omega'_0} = \mathcal{P}(\Omega'_0)$ (see Remark 1.29). According to Equation (5.34),

$$X = 0 \cdot 1_{X=0} + 1 \cdot 1_{X=1} + 2 \cdot 1_{X=2} .$$

This example is a special case of a random variable with a binomial distribution. The general case is treated in Definition 8.7. Other examples of a discrete random variable and their probability function are random variables that have a *Poisson distribution* or a *geometric distribution*. In both cases, the random variable considered takes on an infinite and countable number of values, each of which has a probability greater than 0. These examples are treated in chapter 8 (see Defs. 8.14 and 8.20).   ◁

Example 5.65 [Flipping two coins – continued] In Example 5.35, we introduced the random variables X_1 and X_2, which indicate if we flip *heads* in the first and second trial, respectively. The probability function of the bivariate random variable $X = (X_1, X_2)$ is

$$p_{X_1, X_2}(x_1, x_2) = \frac{1}{4}, \quad \forall (x_1, x_2) \in \{0, 1\}^2 .$$

◁

Lemma 5.66 [Probability function of a marginal distribution]
Consider a multivariate random variable $X = (X_1, \ldots, X_n)$ as specified in (5.18), and assume that there is a finite or countable set $\Omega'_0 \subset \times_{i=1}^n \Omega'_i$ with $P(\Omega'_0) = 1$ and $\{x\} \in \bigotimes_{i=1}^n \mathcal{A}'_i$ for all $x \in \Omega'_0$. Furthermore, for all $x_i \in \Omega'_i$, define

$$\Omega'_{0, x_i} := \left\{ (x_1, \ldots, x_{i-1}, x_{i+1}, \ldots, x_n) \colon (x_1, \ldots, x_n) \in \Omega'_0 \right\} .$$

Then, for all $x_i \in \Omega'_i$,

$$p_{X_i}(x_i) = \sum_{(x_1, \ldots, x_{i-1}, x_{i+1}, \ldots, x_n) \in \Omega'_{0, x_i}} p_X(x_1, \ldots, x_n), \tag{5.36}$$

where p_{X_i} denotes the probability function of X_i, $i = 1, \ldots, n$, which is also called the marginal probability function of X_i.

(Proof p. 200)

Now we turn to a condition that is equivalent to independence of discrete random variables.

Remark 5.67 [Support sets of discrete random variables]   Under the assumptions of Lemma 5.66, we define the support sets of the distributions P_{X_i}, that is, the sets

$$\Omega'_{0,i} := \{x_i \in \Omega'_i \colon p_{X_i}(x_i) > 0\}, \quad i = 1, \ldots, n. \tag{5.37}$$

Obviously, $\Omega'_{0,i}$ is finite or countable for all $i = 1, \ldots, n$. Hence, $\Omega'_{sn} = \times_{i=1}^n \Omega'_{0,i}$ is finite or countable as well. Furthermore, $P(X \in \Omega'_{sn}) = 1$, because, for $(x_1, \ldots, x_n) \in \Omega'_0 \setminus \Omega'_{sn}$, there is at least one i such that $X_i \notin \Omega'_{0,i}$ and therefore $p_{X_i} = P(X_i = x_i) = 0$, which implies $P(X_1 = x_1, \ldots, X_i = x_i, \ldots, X_n = x_n) = 0$.   ◁

Lemma 5.68 [A condition equivalent to independence]
Let X be a multivariate random variable as specified in (5.18), and assume that there is a finite or countable set $\Omega'_0 \subset \times_{i=1}^n \Omega'_i$ with $P_X(\Omega'_0) = 1$ and $\{x\} \in \mathcal{A}'_X$, for all $x \in \Omega'_0$. Furthermore, let $p_X, p_{X_1}, \ldots, p_{X_n}$ denote the probability functions of $X, X_1, \ldots, X_n$, respectively, and let $\Omega'_{0,1}, \ldots, \Omega'_{0,n}$ be the sets defined in (5.37). Then, $X_1, \ldots, X_n$ are independent if and only if

$$p_X(x_1, \ldots, x_n) = p_{X_1}(x_1) \cdot \ldots \cdot p_{X_n}(x_n), \quad \forall \, (x_1, \ldots, x_n) \in \overset{n}{\underset{i=1}{\times}} \Omega'_{0,i}. \tag{5.38}$$

(Proof p. 200)

Note that, in Lemmas 5.66 and 5.68, the set Ω'_0 is not necessarily a Cartesian product. We only require that it is *a subset* of a Cartesian product.

In section 5.6, we shall see that a probability function is a special probability density (see Th. 5.77).

5.6   Probability density with respect to a measure

Some probability measures can also be described by a density with respect to the Lebesgue measure on $(\mathbb{R}, \mathcal{B})$ or the counting measure on $(\Omega, \mathcal{P}(\Omega))$, where Ω is a finite or countable set (see Th. 3.65). Such a density is useful for explicit numerical calculations and comparing distributions to each other. We start by translating some concepts and results of chapter 3 to probability measures.

5.6.1   General concepts and properties

According to Theorem 3.65 and Definition 3.66, a nonnegative measurable function $f: (\Omega, \mathcal{A}, \mu) \to (\overline{\mathbb{R}}, \overline{\mathcal{B}})$ is called a *density of ν with respect to μ*, if

$$\nu(A) = \int_A f \, d\mu, \quad \forall\, A \in \mathcal{A}. \tag{5.39}$$

The function $\nu: \mathcal{A} \to \overline{\mathbb{R}}$ defined by (5.39) is a measure, which is also denoted by $f \odot \mu$. Hence, $f \odot \mu(A) = \int_A f \, d\mu, \forall\, A \in \mathcal{A}$. Theorem 3.65 implies Corollary 5.69.

Corollary 5.69 [Probability measure with density]
Let $(\Omega, \mathcal{A}, \mu)$ be a measure space. If $f: (\Omega, \mathcal{A}, \mu) \to (\overline{\mathbb{R}}, \overline{\mathcal{B}})$ is a nonnegative μ-integrable function with $\int f \, d\mu = 1$, then $P = f \odot \mu$ is a probability measure on $(\Omega, \mathcal{A})$.

Corollary 5.69 justifies Definition 5.70.

Definition 5.70 [Probability density]
Let $(\Omega, \mathcal{A}, P)$ be a probability space and μ a measure on $(\Omega, \mathcal{A})$. If $P = f \odot \mu$, then f is called a probability density of P with respect to μ.

Remark 5.71 [Probability density of a random variable]   Consider the random variable $X: (\Omega, \mathcal{A}, P) \to (\Omega'_X, \mathcal{A}'_X)$, and let μ be a measure on $(\Omega'_X, \mathcal{A}'_X)$. If $P_X = f_X \odot \mu$, then f_X is also referred to as a *probability density of X with respect to μ*.   ◁

Applying Equation (5.39) yields the following corollary.

Corollary 5.72 [Characterizing the probability measure by a density]
Let $f: (\Omega, \mathcal{A}, \mu) \to (\mathbb{R}, \mathcal{B})$ be μ-integrable and nonnegative, and P a probability measure on $(\Omega, \mathcal{A})$. Then, f is a (probability) density of P with respect to μ if and only if it satisfies

$$P(A) = \int_A f \, d\mu, \quad \forall\, A \in \mathcal{A}. \tag{5.40}$$

Theorem 3.68 (a) and (c) imply Corollary 5.73:

Corollary 5.73 [Probability densities are μ-equivalent]
Let μ and P be measures on the measurable space $(\Omega, \mathcal{A})$, where P is a probability measure. If $f, f^*: (\Omega, \mathcal{A}, \mu) \to (\overline{\mathbb{R}}, \overline{\mathcal{B}})$ are probability densities of P with respect to μ, then $f \underset{\mu}{=} f^*$.

Now we translate the Radon-Nikodym theorem (see Th. 3.72), which yields Corollary 5.74.

Corollary 5.74 [An implication of the Radon-Nikodym theorem]
Let $(\Omega, \mathcal{A}, \mu)$ be a measure space. If μ is σ-finite, P is a probability measure on $(\Omega, \mathcal{A})$, and $P \underset{\mathcal{A}}{\ll} \mu$, then there is a probability density f of P with respect to μ (also called a Radon-Nikodym derivative), that is,

$$f = \frac{dP}{d\mu}. \tag{5.41}$$

Example 5.75 [Conditional-probability measure] In Lemma 4.32 (ii), we showed that $1_B/P(B)$ is a density of P^B with respect to P, and according to Remark 5.71, it is a probability density of P^B.   ◁

5.6.2   Density of a discrete random variable

As a special case, we consider a discrete random variable (see section 5.5).

Remark 5.76 [A sum of Dirac measures] Let the assumptions of Definition 5.56 hold, that is, let $X: (\Omega, \mathcal{A}, P) \to (\Omega'_X, \mathcal{A}'_X)$ be a random variable and suppose $\Omega'_0 \subset \Omega'_X$ is finite or countable with $P_X(\Omega'_0) = 1$ and $\{x\} \in \mathcal{A}'_X$ for all $x \in \Omega'_0$. Furthermore, define

$$\mu := \sum_{x \in \Omega'_0} \delta_x, \tag{5.42}$$

the sum of Dirac measures at x, $x \in \Omega'_0$. According to Example 1.57, μ is a measure on $(\Omega'_X, \mathcal{A}'_X)$, which is σ-finite.   ◁

Theorem 5.77 asserts that the probability function p_X is a density of P_X with respect to μ.

Theorem 5.77 [The probability function is a density]
Let the assumptions of Definition 5.56 be satisfied, and let μ be defined by Equation (5.42). Then,

(i) $P_X \underset{\mathcal{A}'_X}{\ll} \mu$.

(ii) *The probability function p_X is a density of P_X with respect to μ, that is,*

$$p_X = \frac{dP_X}{d\mu} \tag{5.43}$$

and, for all $A' \in \mathcal{A}'_X$,

$$P_X(A') = \int_{A'} p_X \, d\mu \tag{5.44}$$

$$= \sum_{x \in \Omega'_0} 1_{A'}(x) \cdot p_X(x) \tag{5.45}$$

$$= \sum_{x \in A'} p_X(x). \tag{5.46}$$

(Proof p. 201)

Hence, each probability $P_X(A')$, $A' \in \mathcal{A}'_X$, can be computed from the probability function p_X.

5.6.3   Density of a bivariate random variable

Now we consider bivariate random variables. However, extending the following notation and propositions to general multivariate random variables is straightforward.

Lemma 5.78 [Absolute continuity of marginal distributions]
Let $(X, Y): (\Omega, \mathcal{A}, P) \to (\Omega'_X \times \Omega'_Y, \mathcal{A}'_X \otimes \mathcal{A}'_Y)$ be a random variable, and suppose that $(\Omega'_X, \mathcal{A}'_X, \mu)$ and $(\Omega'_Y, \mathcal{A}'_Y, \nu)$ are σ-finite measure spaces. Then,

$$P_{X,Y} \underset{\mathcal{A}'_X \otimes \mathcal{A}'_Y}{\ll} \mu \otimes \nu \;\Rightarrow\; P_X \underset{\mathcal{A}'_X}{\ll} \mu \quad and \quad P_Y \underset{\mathcal{A}'_Y}{\ll} \nu.$$

(Proof p. 201)

Let

$$f_{X,Y} := \frac{dP_{X,Y}}{d\mu \otimes \nu}, \quad f_X := \frac{dP_X}{d\mu}, \quad f_Y := \frac{dP_Y}{d\nu}$$

denote Radon-Nikodym derivatives (see Th. 3.72). In Lemma 5.79, we use the notation '$\underset{\mu\text{-}a.a.}{=}$' introduced in Remark 2.70.

Lemma 5.79 [Marginal densities]
Let $(X, Y): (\Omega, \mathcal{A}, P) \to (\Omega'_X \times \Omega'_Y, \mathcal{A}'_X \otimes \mathcal{A}'_Y)$ be a random variable, $(\Omega'_X, \mathcal{A}'_X, \mu)$, $(\Omega'_Y, \mathcal{A}'_Y, \nu)$ be measure spaces, μ, ν be σ-finite, and assume $P_{X,Y} \underset{\mathcal{A}'_X \otimes \mathcal{A}'_Y}{\ll} \mu \otimes \nu$. Then,

$$f_X(x) \underset{\mu\text{-}a.a.}{=} \int f_{X,Y}(x, y) \, \nu(dy), \qquad (5.47)$$

and

$$f_Y(y) \underset{\nu\text{-}a.a.}{=} \int f_{X,Y}(x, y) \, \mu(dx). \qquad (5.48)$$

The functions f_X and f_Y are also called marginal densities of X and Y, respectively.
(Proof p. 202)

Remark 5.80 [Marginal and joint density]   Suppose that X and Y are real-valued. Then, for the Lebesgue measure $\mu = \nu = \lambda$ and a Riemann integrable density $f_{X,Y}$, Equations (5.47) and (5.48) yield

$$f_X(x) \underset{\lambda\text{-}a.a.}{=} \int_{-\infty}^{\infty} f_{X,Y}(x, y) \, dy \qquad (5.49)$$

and

$$f_Y(y) \underset{\lambda\text{-}a.a.}{=} \int_{-\infty}^{\infty} f_{X,Y}(x, y) \, dx, \qquad (5.50)$$

respectively (see Th. 3.62).   ◁

Note that, for a discrete random variable $X = (X_1, \ldots, X_n)$, the probability function p_X is a density of P_X with respect to the measure μ specified in Equation (5.42), and the marginal probability functions p_{X_i}, $i = 1, \ldots, n$, are marginal densities. In section 5.7.4, we consider multivariate densities with respect to the Lebesgue measure $\mu = \nu = \lambda$.

5.7 Uni- or multivariate real-valued random variable

The remaining sections of this chapter show how to describe distributions of *real-valued* uni- and multivariate random variables.

5.7.1 Distribution function of a univariate real-valued random variable

If we consider a univariate real-valued random variable X, then the *distribution function* F_X assigns to each $x \in \mathbb{R}$ the probability $P(X \leq x)$ of the event $\{X \leq x\} = \{\omega \in \Omega: X(\omega) \leq x\}$ that

X takes on a value *smaller than or equal* to x. As we shall see, the distribution function uniquely determines the distribution P_X.

Definition 5.81 [Distribution function]

Let $X: (\Omega, \mathscr{A}, P) \to (\mathbb{R}, \mathscr{B}, P_X)$ denote a real-valued random variable. Then the (cumulative) distribution function $F_X: \mathbb{R} \to [0, 1]$ of X is defined by:

$$F_X(x) := P_X(]-\infty, x]) = P(X \leq x), \quad \forall\, x \in \mathbb{R}. \tag{5.51}$$

Remark 5.82 [Probabilities of intervals]   This definition implies that we can compute the probability $P(a < X \leq b)$ of X taking a value in the interval $]a, b]$ by

$$P(a < X \leq b) = F_X(b) - F_X(a), \quad \text{if } a < b, \tag{5.52}$$

because

$$P(a < X \leq b) = P_X(]-\infty, b] \;\backslash\;]-\infty, a]) = P_X(]-\infty, b]) - P_X(]-\infty, a])$$

[see Rule (vi), Box 4.1]. ◁

Remark 5.83 [The distribution function determines the distribution]   Every random variable X has a distribution P_X. Therefore, the distribution function F_X exists for all real-valued random variables. The distribution function uniquely determines the distribution P_X of a real-valued random variable, because a finite measure on $(\Omega, \mathscr{A})$ is already uniquely specified by its values on a $\cap$-stable generating system of $\mathscr{A}$ (see Th. 1.72), and the set system $\{]-\infty, x]: x \in \mathbb{R}\}$ is a $\cap$-stable generating system of $\mathscr{B}$, the Borel σ-algebra on $\mathbb{R}$ [see Eq. (1.19)]. Hence, F_X uniquely determines P_X, which implies Theorem 5.84. ◁

Theorem 5.84 [Uniqueness]

Let P_X, P_Y denote the distributions and F_X, F_Y the distribution functions of two real-valued random variables $X, Y: (\Omega, \mathscr{A}, P) \to (\mathbb{R}, \mathscr{B})$. Then,

$$F_X = F_Y \;\Leftrightarrow\; P_X = P_Y. \tag{5.53}$$

This theorem facilitates describing distributions and calculations involving distributions considerably, because the distribution function is defined on the set $\mathbb{R}$ of real numbers, whereas P_X is defined on a much more complex domain, the Borel σ-algebra $\mathscr{B}$.

Example 5.85 [Flipping two coins – continued]   In Example 5.11, we considered flipping two coins and specified the distribution P_X of $X: (\Omega, \mathscr{A}, P) \to \big(\Omega'_X, \mathscr{P}(\Omega'_X)\big)$, representing with its values x the number of flipping heads. The distribution P_X assigns a probability to all $2^3 = 8$ subsets of $\Omega'_X := \{0, 1, 2\}$. Because $\{0, 1, 2\} \subset \mathbb{R}$, the random variable X is also a random variable $X: (\Omega, \mathcal{A}, P) \to (\mathbb{R}, \mathcal{B})$ with values in $\mathbb{R}$. In this case, F_X is a step function defined by

$$F_X(x) = P_X(]-\infty, x]) = P(X \leq x) = \begin{cases} 0, & \text{if } x < 0 \\ 1/4, & \text{if } 0 \leq x < 1 \\ 3/4, & \text{if } 1 \leq x < 2 \\ 1, & \text{if } x \geq 2. \end{cases}$$

According to Theorem 5.84, the distribution P_X is uniquely defined by these four values. In other words, if we know these four values, then we know the probabilities $P_X(A')$ for all elements A' of the Borel σ-algebra $\mathcal{B}$ (see Exercise 5.10).   ◁

Now we turn to the *quantile function*, which, in some cases, is the inverse of the distribution function. Sometimes this function is also called the *pseudo-inverse* of F_X. It assigns to each $p \in [0, 1]$ the smallest real number x for which $P(X \leq x) = F_X(x) \geq p$.

Definition 5.86 [Quantile function]
Let $X: (\Omega, \mathcal{A}, P) \to (\mathbb{R}, \mathcal{B})$ be a real-valued random variable with distribution function F_X. Then the quantile function $Q_X: [0, 1] \to \overline{\mathbb{R}}$ of X is defined by:

$$Q_X(p) = \inf\{x \in \mathbb{R}: F_X(x) \geq p\}, \qquad \forall p \in \,]0, 1[,$$

$$Q_X(0) = \begin{cases} \inf\{x \in \mathbb{R}: F_X(x) > 0\}, & \text{if } \exists\, x \in \mathbb{R} \text{ with } F_X(x) = 0 \\ -\infty, & \text{if } F_X(x) > 0, \forall\, x \in \mathbb{R}, \end{cases}$$ (5.54)

$$Q_X(1) = \begin{cases} \inf\{x \in \mathbb{R}: F_X(x) = 1\}, & \text{if } \exists\, x \in \mathbb{R} \text{ with } F_X(x) = 1 \\ \infty, & \text{if } F_X(x) < 1, \forall\, x \in \mathbb{R}. \end{cases}$$

A value $Q_X(p)$, $p \in [0, 1]$, is called the quantile of p with respect to F_X.

Remark 5.87 [Inverse function of F_X] If F_X is continuous and strictly monotone (i.e., if $x_1 < x_2$ implies $F_X(x_1) < F_X(x_2)$), then,

$$Q_X(p) = F_X^{-1}(p), \qquad \forall p \in \,]0, 1[.$$ (5.55)

where F_X^{-1} denotes the inverse function of F_X.   ◁

Example 5.88 [Flipping two coins – continued] In Example 5.85, we specified the distribution function of $X = $ *number of flipping heads* for the random experiment of flipping two coins. The corresponding quantile function takes on the following three values:

$$Q_X(p) = \begin{cases} 0, & \text{if } 0 \leq p \leq 1/4 \\ 1, & \text{if } 1/4 < p \leq 3/4 \\ 2, & \text{if } 3/4 < p \leq 1. \end{cases}$$

◁

5.7.2 Distribution function of a multivariate real-valued random variable

Now we extend the concept of a distribution function to the multivariate case. In Definition 5.89, we use the notation introduced in Equation (5.25).

Definition 5.89 [Joint distribution function]

Let $(X_1, \ldots, X_n)$: $(\Omega, \mathcal{A}, P) \to (\mathbb{R}^n, \mathcal{B}_n)$ be a random variable. Its distribution function $F_{X_1, \ldots, X_n}$: $\mathbb{R}^n \to [0, 1]$ is defined by

$$F_{X_1, \ldots, X_n}(x_1, \ldots, x_n) := P(X_1 \leq x_1, \ldots, X_n \leq x_n), \quad \forall (x_1, \ldots, x_n) \in \mathbb{R}^n. \quad (5.56)$$

It is also called the joint distribution function of $X_1, \ldots, X_n$.

Example 5.90 [Flipping two coins – continued] In example 5.35, we considered flipping two coins and defined the random variables X_1 and X_2 indicating whether or not we flip *heads* at first and second flip, respectively. In this example, the bivariate distribution function F_{X_1, X_2} takes on the values

$$F_{X_1, X_2}(x_1, x_2) = \begin{cases} 0, & \text{if } x_1 < 0 \text{ or } x_2 < 0 \\ 1/4, & \text{if } 0 \leq x_1 < 1, 0 \leq x_2 < 1 \\ 2/4, & \text{if } x_1 \geq 1, 0 \leq x_2 < 1 \\ 2/4, & \text{if } 0 \leq x_1 < 1, x_2 \geq 1 \\ 1, & \text{if } x_1 \geq 1, x_2 \geq 1. \end{cases}$$

◁

Just like in Theorem 5.84, we can prove uniqueness, using a ∩-stable generating system for $\mathcal{B}_n$, now referring to Equation (1.21).

Theorem 5.91 [Uniqueness]

Let P_X, P_Y denote the distributions and F_X, F_Y the distribution functions of two n-variate real-valued random variables X, Y: $(\Omega, \mathcal{A}, P) \to (\mathbb{R}^n, \mathcal{B}_n)$. Then,

$$F_X = F_Y \Leftrightarrow P_X = P_Y. \quad (5.57)$$

As an implication of Equation (5.21) with $A'_i =]-\infty, x_i]$ and $\Omega'_j = \mathbb{R}$, $i \neq j$, we obtain the Corollary 5.92. In the special case of a bivariate real-valued random variable (X_1, X_2), this corollary asserts the value of the *marginal distribution function of X_1* for the argument x_1 (i.e., $\lim_{x_2 \to \infty} F_{X_1, X_2}(x_1, x_2) = F_{X_1}(x_1)$). In this corollary,

$$\lim_{\substack{x_j \to \infty \\ j \neq i}} F_{X_1, \ldots, X_n}(x_1, \ldots, x_n), \quad (5.58)$$

denotes the limit of the distribution function of $(X_1, \ldots, X_n)$ for $x_j \to \infty$, for all $j = 1, \ldots, i-1$, $i+1, \ldots, n$. This limit is the value of the *marginal distribution function of X_i* for the argument x_i, and Corollary 5.92 asserts that this limit is identical to the value $F_{X_i}(x_i)$ of the distribution function of X_i for the argument x_i.

Corollary 5.92 [Joint and marginal distribution function]

Let $(X_1, \ldots, X_n)\colon (\Omega, \mathcal{A}, P) \to (\mathbb{R}^n, \mathcal{B}_n)$ be a random variable. Then,

$$F_{X_i}(x_i) = \lim_{\substack{x_j \to \infty \\ j \neq i}} F_{X_1, \ldots, X_n}(x_1, \ldots, x_n), \quad \forall\, x_i \in \mathbb{R}. \tag{5.59}$$

(Proof p. 202)

Corollary 5.93 shows how independence of the random variables $X_1, \ldots, X_n$ can be formulated in terms of their distribution functions.

Corollary 5.93 [Independence and joint distribution function]

Let $(X_1, \ldots, X_n)\colon (\Omega, \mathcal{A}, P) \to (\mathbb{R}^n, \mathcal{B}_n)$ be a random variable. Then the following two propositions are equivalent to each other:

(a) $X_1, \ldots, X_n$ are independent.

(b) $F_{X_1, \ldots, X_n}(x_1, \ldots, x_n) = F_{X_1}(x_1) \cdot \ldots \cdot F_{X_n}(x_n), \quad \forall\, (x_1, \ldots, x_n) \in \mathbb{R}^n.$

(Proof p. 203)

5.7.3   Density of a continuous univariate real-valued random variable

As a special case, we consider a random variable $X\colon (\Omega, \mathcal{A}, P) \to (\mathbb{R}, \mathcal{B})$ for which there is a nonnegative measurable function $f_X\colon (\mathbb{R}, \mathcal{B}, P_X) \to (\mathbb{R}, \mathcal{B})$ such that

$$P_X(B) = \int_B f_X \, d\lambda, \quad \forall\, B \in \mathcal{B}, \tag{5.60}$$

where λ denotes the Lebesgue measure on $(\mathbb{R}, \mathcal{B})$. According to Theorem 3.68 (ii), this equation is equivalent to $P_X = f_X \odot \lambda$.

Definition 5.94 is a special case of Definition 5.70.

Definition 5.94 [Continuous random variable and its density]

Let $X\colon (\Omega, \mathcal{A}, P) \to (\mathbb{R}, \mathcal{B})$ be a real-valued random variable with distribution P_X. We call X c o n t i n u o u s if there is a nonnegative function $f_X\colon \mathbb{R} \to \mathbb{R}$ that is integrable with respect to the Lebesgue measure λ and

$$P_X = f_X \odot \lambda. \tag{5.61}$$

A function f_X satisfying (5.61) is called a (p r o b a b i l i t y) d e n s i t y of X.

Note that Equation (5.61) is equivalent to

$$F_X(x) = \int_{]-\infty,\,x]} f_X \, d\lambda, \quad \forall\, x \in \mathbb{R}, \tag{5.62}$$

because

$$F_X(x) = P_X(]-\infty,\,x]) = \int 1_{]-\infty,\,x]} \, dP_X = \int_{]-\infty,\,x]} f_X \, d\lambda, \quad \forall\, x \in \mathbb{R}, \tag{5.63}$$

(see Th. 3.67). Furthermore, Theorem 3.62 immediately implies Corollary 5.95:

Corollary 5.95 [Riemann integral of the density]
If f_X is a density of the random variable $X: (\Omega, \mathcal{A}, P) \to (\mathbb{R}, \mathcal{B})$ and f_X is Riemann integrable, then,

$$F_X(x) = \int_{-\infty}^{x} f_X(t) \, dt, \quad \forall\, x \in \mathbb{R}. \tag{5.64}$$

Remark 5.96 [Interpretation of densities]   Note that the term $f_X(t)$ in Equation (5.64) is not a probability; instead, it is a value of the density for $t \in \mathbb{R}$. However, the probability $P(a < X \leq b)$ that X takes on a value in the interval $]a, b]$ can be computed using Equation (5.52) and the density f_X, provided that it exists and is Riemann integrable:

$$P(a < X \leq b) = F_X(b) - F_X(a) = \int_a^b f_X(x) \, dx, \quad \text{if } a < b. \tag{5.65}$$

This probability can be represented as the area between the density and the x-axis above the interval $[a, b]$ (see Fig. 3.6).   ◁

Remark 5.97 [Continuity of X implies $P(X=x) = 0$]   Consider a continuous random variable $X: (\Omega, \mathcal{A}, P) \to (\mathbb{R}, \mathcal{B})$. Definition 5.94 and Remark 3.71 imply $P_X \underset{\mathcal{B}}{\ll} \lambda$. Because $\lambda(\{x\}) = 0$ [see Eq. (1.53)], we can conclude $P(X=x) = 0$, for all $x \in \mathbb{R}$. Hence, additivity of P yields, for all $a, b \in \mathbb{R}, a < b$,

$$P(a < X \leq b) = P(a \leq X \leq b) = P(a \leq X < b) = P(a < X < b), \tag{5.66}$$

provided that X is continuous.   ◁

Example 5.98 [Continuous random variables and their densities]   In section 8.2, we present some examples of continuous random variables and their densities, such as the densities of normal distributions, central χ^2-distributions, central t-distributions, and central F-distributions.   ◁

5.7.4   Density of a continuous multivariate real-valued random variable

Remark 5.99 [Multivariate case]   Let $X: (\Omega, \mathscr{A}, P) \to (\mathbb{R}^n, \mathscr{B}_n)$ be a multivariate random variable with distribution P_X. If $f_X: \mathbb{R}^n \to \mathbb{R}$ is nonnegative and integrable with respect to the Lebesgue measure λ_n on $(\mathbb{R}^n, \mathscr{B}_n)$, and $P_X = f_X \odot \lambda_n$, then X is continuous with probability density f_X, and

$$F_X(x_1, \dots, x_n) = \int_B f_X \, d\lambda_n, \quad \forall\, (x_1, \dots, x_n) \in \mathbb{R}^n, \tag{5.67}$$

where $B := \times_{i=1}^n \,]{-}\infty, x_i]$. If f_X is Riemann integrable, then,

$$F_X(x_1, \dots, x_n) = \int_{-\infty}^{x_n} \dots \int_{-\infty}^{x_1} f_X(t_1, \dots, t_n) \, dt_1 \dots dt_n, \quad \forall\, (x_1, \dots, x_n) \in \mathbb{R}^n. \tag{5.68}$$

More generally, for any $B \in \mathscr{B}_n$,

$$P_X(B) = P(X \in B) = \int 1_B \cdot f_X \, d\lambda_n. \tag{5.69}$$

The probability $P_X(B)$ can be represented as the $(n+1)$-dimensional *volume* between the joint density and the $(x_1, \dots, x_n)$-hyperplane above B (see Fig. 8.8).   ◁

In analogy to Corollary 5.93, independence of continuous real-valued random variables can also be formulated in terms of probability densities, using the marginal densities $f_{X_1}, \dots, f_{X_n}$ (see Lemma 5.79).

Corollary 5.100 [Independence and probability densities]
Let $(X_1, \dots, X_n): (\Omega, \mathscr{A}, P) \to (\mathbb{R}^n, \mathscr{B}_n)$, $n > 1$, be a random variable, and suppose that all random variables X_i, $i = 1, \dots, n$, have a density f_{X_i} with respect to the Lebesgue measure λ on $(\mathbb{R}, \mathscr{B})$. Then, $X_1, \dots, X_n$ are independent if and only if

$$f_{X_1, \dots, X_n}(x_1, \dots, x_n) := f_{X_1}(x_1) \cdot \dots \cdot f_{X_n}(x_n), \quad \forall\, (x_1, \dots, x_n) \in \mathbb{R}^n, \tag{5.70}$$

is a joint density of $(X_1, \dots, X_n)$ with respect to λ_n.

(Proof p. 203)

Remark 5.101 [Independence, densities, and Riemann integrals]   If all densities f_{X_i}, $i = 1, \dots, n$, are Riemann integrable, then $f_{X_1, \dots, X_n}$ in (5.70) is Riemann integrable as well (see, e.g., Ellis & Gulick, 2006).   ◁

Remark 5.102 [Other random variables]   In this section, we considered univariate and multivariate real-valued random variables X. All these random variables have a distribution P_X and also a (cumulative) distribution function F_X. If X is discrete, then its distribution can be described by the probability function, and the distribution function F_X is a step function (see Example 5.90 and Fig. 8.1). If X is continuous, then its distribution can be specified by a density with respect to the Lebesgue measure. The distribution functions of continuous random variables do not have jumps (see Rem. 5.97 and Fig. 8.7 for an example). Note, however, that there are random variables that are neither discrete nor continuous. Their distribution functions have jumps but are not step functions.   ◁

5.8   Proofs

Proof of Lemma 5.7

According to Theorem 2.49, the composition $g(X)$ of X and g is $(\mathcal{A}, \mathcal{A}')$-measurable. Therefore, it is a random variable on $(\Omega, \mathcal{A}, P)$. For all $A \in \mathcal{A}$,

$$
\begin{aligned}
P_{g(X)}(A) &= P([g(X)]^{-1}(A)) && \text{[(5.2)]} \\
&= P(\{\omega \in \Omega : g[X(\omega)] \in A\}) && \text{[Def. 2.1]} \\
&= P(\{\omega \in \Omega : X(\omega) \in g^{-1}(A)\}) && \text{[Def. 2.1]} \\
&= P(X^{-1}[g^{-1}(A)]) && \text{[Def. 2.1]} \\
&= P_X[g^{-1}(A)] && \text{[(5.2)]} \\
&= (P_X)_g(A). && \text{[(5.2)]}
\end{aligned}
$$

Proof of Lemma 5.21

$$
\begin{aligned}
X \underset{P}{=} Y &\Rightarrow \exists A \in \mathcal{A} : (\forall \omega \in \Omega \setminus A : X(\omega) = Y(\omega) \text{ and } P(A) = 0) && \text{[(5.9)]} \\
&\Rightarrow \exists A \in \mathcal{A} : (\forall \omega \in \Omega \setminus A : X(\omega) = Y(\omega) \text{ and } Q(A) = 0) && [Q \underset{\mathcal{A}}{\ll} P] \\
&\Rightarrow X \underset{Q}{=} Y. && \text{[(5.9)]}
\end{aligned}
$$

Proof of Corollary 5.25

Proposition (i) is a special case of Theorem 2.86. Analogously to the proof of Theorem 2.86, we can prove Propositions (ii) and (iii) using Remark 2.67.

Proof of Lemma 5.27

Let $A := \{\omega \in \Omega : X(\omega) \neq Y(\omega)\} \in \mathcal{C}$. Hence, if $X \underset{P^B}{=} Y$, then the conjunction of $P^B(A) = 0$ and $P \underset{\mathcal{C}}{\ll} P^B$ implies $P(A) = 0$.

Proof of Lemma 5.29

$P \underset{\sigma(X)}{\ll} P^B \Rightarrow P_X \underset{\mathcal{A}'_X}{\ll} P^B_X$ If $P \underset{\sigma(X)}{\ll} P^B$, then, for all $C' \in \mathcal{A}'_X$,

$$
\begin{aligned}
P^B_X(C') = 0 &\Rightarrow P^B[X^{-1}(C')] = 0 && \text{[(5.2)]} \\
&\Rightarrow P[X^{-1}(C')] = 0 && [P \underset{\sigma(X)}{\ll} P^B] \\
&\Rightarrow P_X(C') = 0. && \text{[(5.2)]}
\end{aligned}
$$

$P_X \underset{\mathscr{A}'_X}{\ll} P_X^B \Rightarrow P \underset{\sigma(X)}{\ll} P^B$ If $P_X \underset{\mathscr{A}'_X}{\ll} P_X^B$ and $C \in \sigma(X)$, then there is a $C' \in \mathscr{A}'_X$ such that

$$C = X^{-1}(C').$$

Hence,

$$
\begin{aligned}
P^B(C) = 0 \;\Rightarrow\; & P^B[X^{-1}(C')] = 0 \\
\Rightarrow\; & P_X^B(C') = 0 && [(5.2)] \\
\Rightarrow\; & P_X(C') = 0 && [P_X \underset{\mathscr{A}'_X}{\ll} P_X^B] \\
\Rightarrow\; & P[X^{-1}(C')] = 0 && [(5.2)] \\
\Rightarrow\; & P(C) = 0.
\end{aligned}
$$

Proof of Lemma 5.47

If $X_1 \underset{P}{=} X_2$, then,

$$
\begin{aligned}
& X_1 \underset{P}{\perp\!\!\!\perp} Y \\
\Rightarrow\; & \forall\, (A', B') \in \mathscr{A}' \times \mathscr{A}'_Y: \\
& P[X_1^{-1}(A') \cap Y^{-1}(B')] = P[X_1^{-1}(A')] \cdot P[Y^{-1}(B')] && [(5.24)] \\
\Rightarrow\; & \forall\, (A', B') \in \mathscr{A}' \times \mathscr{A}'_Y: \\
& P[X_2^{-1}(A') \cap Y^{-1}(B')] = P[X_2^{-1}(A')] \cdot P[Y^{-1}(B')] && [X_1 \underset{P}{=} X_2, \text{Def. 5.3, Cor. 5.24}] \\
\Rightarrow\; & X_2 \underset{P}{\perp\!\!\!\perp} Y. && [(5.24)]
\end{aligned}
$$

Proof of Lemma 5.48

According to Remark 5.46, $1_B \underset{P}{\perp\!\!\!\perp} \mathscr{C} \Leftrightarrow B \underset{P}{\perp\!\!\!\perp} \mathscr{C}$. Hence, for all $C \in \mathscr{C}$,

$$
\begin{aligned}
P^B(C) = 0 \;\Rightarrow\; & P(B \cap C) = 0 \\
\Rightarrow\; & P(B) \cdot P(C) = 0 && [B \underset{P}{\perp\!\!\!\perp} C] \\
\Rightarrow\; & P(C) = 0. && [P(B) > 0]
\end{aligned}
$$

Proof of Lemma 5.49

$\Rightarrow$ If $X_1, \ldots, X_n$ are independent, then, for all $A'_i \in \mathscr{A}'_i$, $i = 1, \ldots, n$,

$$
\begin{aligned}
P_{X_1, \ldots, X_n}(A'_1 \times \ldots \times A'_n) &= P[X_1^{-1}(A'_1) \cap \ldots \cap X_n^{-1}(A'_n)] \\
&= P[X_1^{-1}(A'_1)] \cdot \;\ldots\; \cdot P[X_n^{-1}(A'_n)] \\
&= P_{X_1}(A'_1) \cdot \;\ldots\; \cdot P_{X_n}(A'_n) \\
&= (P_{X_1} \otimes \ldots \otimes P_{X_n})(A'_1 \times \ldots \times A'_n).
\end{aligned}
$$

Hence, according to the definition of the product measure (see Def. 1.67),

$$P_{X_1, \dots, X_n} = P_{X_1} \otimes \dots \otimes P_{X_n}.$$

$\Leftarrow$ If the right-hand side of (5.30) holds, then, for all $A'_i \in \mathscr{A}'_i$, $i = 1, \dots, n$,

$$P[X_1^{-1}(A'_1) \cap \dots \cap X_n^{-1}(A'_n)] = P_{X_1, \dots, X_n}(A'_1 \times \dots \times A'_n)$$
$$= P_{X_1}(A'_1) \cdot \dots \cdot P_{X_n}(A'_n)$$
$$= P[X_1^{-1}(A'_1)] \cdot \dots \cdot P[X_n^{-1}(A'_n)].$$

According to Equation (5.28), this implies independence of $X_1, \dots, X_n$.

Proof of Lemma 5.51

Assume that $X: (\Omega, \mathscr{A}, P) \rightarrow (\Omega'_X, \mathscr{A}'_X)$ is a random variable and that there is an $\alpha \in \Omega'_X$ and $\{\alpha\} \in \mathscr{A}'_X$ such that $X \underset{P}{=} \alpha$. If $A = \{\omega \in \Omega: X(\omega) \neq \alpha\}$, then $A = \Omega \setminus X^{-1}(\{\alpha\}) \in \mathscr{A}$, $P(A) = 0$ and $P(A^c) = 1$. This implies, for all $A' \in \mathscr{A}'_X$,

$$P[X^{-1}(A')] = P[X^{-1}(A') \cap A^c] \qquad \text{[Box 4.1 (viii)]}$$
$$= P[\{\omega \in \Omega: X(\omega) \in A', X(\omega) = \alpha\}]$$
$$= \begin{cases} P(\emptyset), & \text{if } \alpha \notin A' \\ P(A^c), & \text{if } \alpha \in A' \end{cases}$$
$$= \begin{cases} 0, & \text{if } \alpha \notin A' \\ 1, & \text{if } \alpha \in A'. \end{cases} \qquad (5.71)$$

This implies, for all $A' \in \mathscr{A}'_X$ and all $C \in \mathscr{C}$,

$$P[X^{-1}(A') \cap C] = \begin{cases} 0, & \text{if } \alpha \notin A' \\ P(C), & \text{if } \alpha \in A' \end{cases} \qquad \text{[Box 4.1 (v), (viii)]}$$
$$= P[X^{-1}(A')] \cdot P(C). \qquad \text{[(5.71)]}$$

Proof of Lemma 5.61

(i) $\Rightarrow$ If X is discrete, then there is a finite or countable $\Omega'_0 \subset \Omega'_X$ with $\{x\} \in \mathscr{A}'_X$ for all $x \in \Omega'_0$ and $P_X(\Omega'_0) = 1$. This implies

$$P\left(1 \neq \sum_{x \in \Omega'_0} 1_{X=x}\right) = P(X \notin \Omega'_0) = 0,$$

and therefore

$$1 \underset{P}{=} \sum_{x \in \Omega'_0} 1_{X=x}.$$

$\Leftarrow$ If Equation (5.33) holds and $\Omega_0' \subset \Omega_X'$ is finite or countable with $\{x\} \in \mathscr{A}_X'$ for all $x \in \Omega_0'$, then $1 \underset{P}{=} \sum_{x \in \Omega_0'} 1_{X=x}$ and therefore

$$1 = P\left(1 = \sum_{x \in \Omega_0'} 1_{X=x}\right) = P(X \in \Omega_0') = P_X(\Omega_0').$$

(ii) If X is real-valued, then X is discrete if and only if there is a finite or countable $\Omega' \subset \mathbb{R}$ such that

$$
\begin{aligned}
X &= X \cdot 1 \\
&\underset{P}{=} X \cdot \sum_{x \in \Omega'} 1_{X=x} && \text{[(i)]} \\
&= \sum_{x \in \Omega'} X \cdot 1_{X=x} \\
&= \sum_{x \in \Omega'} x \cdot 1_{X=x} \, . && [1_{X=x}(\omega) = 0 \text{ if } X(\omega) \neq x, (5.6)]
\end{aligned}
$$

Proof of Corollary 5.63

$\Rightarrow$ If X is discrete and $\Omega_0' \subset \Omega_X'$ (see Def. 5.56), that is, if Ω_0' is finite or countable and $P_X(\Omega_0') = 1$, then $\Omega_>' \subset \Omega_0'$, because $P_X(\Omega_X' \setminus \Omega_0') = 0$. Hence, $\Omega_>'$ is finite or countable as well, which proves (a). Furthermore, for finite or countable sets $\Omega_0', \Omega_>'$:

$$P_X(\Omega_0' \setminus \Omega_>') = \sum_{x \in \Omega_0' \setminus \Omega_>'} P\,(X=x) = 0. \qquad \text{[Def. of } \Omega_>']$$

Hence,

$$P_X(\Omega_X' \setminus \Omega_>') = P_X(\Omega_X' \setminus \Omega_0') + P_X(\Omega_0' \setminus \Omega_>') = 0$$

and, according to (5.11),

$$1_{X \in (\Omega_X' \setminus \Omega_>')} \underset{P}{=} 0. \qquad\qquad (5.72)$$

Therefore,

$$
\begin{aligned}
X &= 1_{X \in \Omega_>'} \cdot X + 1_{X \in (\Omega_X' \setminus \Omega_>')} \cdot X && [1 = 1_{X \in \Omega_>'} + 1_{X \in (\Omega_X' \setminus \Omega_>')}] \\
&\underset{P}{=} X \cdot \sum_{x \in \Omega_>'} 1_{X=x} && [(5.72), \ \Omega_>' \text{ is finite or countable}] \\
&= \sum_{x \in \Omega_>'} X \cdot 1_{X=x} \\
&= \sum_{x \in \Omega_>'} x \cdot 1_{X=x} \, . && [X \cdot 1_{X=x} = x \cdot 1_{X=x}]
\end{aligned}
$$

$\Leftarrow$ This is an immediate implication of Lemma 5.61 (ii).

Proof of Lemma 5.66

If $x_i \in \Omega_i'$ and $\Omega_{0, x_i}' = \emptyset$, then $p_{X_i}(x_i) = P(X_i = x_i) = 0$. If $\Omega_{0, x_i}' \neq \emptyset$, then it is finite or countable, because Ω_0' is finite or countable. Then, using $P_X(\Omega_0') = 1$,

$$p_{X_i}(x_i)$$

$$= P(X_i = x_i)$$

$$= P(X_1 \in \Omega_1', \ldots, X_{i-1} \in \Omega_{i-1}', X_i = x_i, X_{i+1} \in \Omega_{i+1}', \ldots, X_n \in \Omega_n') \qquad \text{[Box 4.1 (viii)]}$$

$$= P\left((X_1, \ldots, X_n) \in (\Omega_1' \times \ldots \times \Omega_{i-1}' \times \{x_i\} \times \Omega_{i+1}' \times \ldots \times \Omega_n')\right)$$

$$= P\left((X_1, \ldots, X_n) \in \left((\Omega_1' \times \ldots \times \Omega_{i-1}' \times \{x_i\} \times \Omega_{i+1}' \times \ldots \times \Omega_n') \cap \Omega_0'\right)\right) \qquad \text{[Box 4.1 (viii)]}$$

$$= \sum_{(x_1, \ldots, x_{i-1}, x_{i+1}, \ldots, x_n) \in \Omega_{0, x_i}'} P(X_1 = x_1, \ldots, X_{i-1} = x_{i-1}, X_i = x_i, X_{i+1} = x_{i+1}, \ldots, X_n = x_n) \qquad \text{[Box 4.1 (x)]}$$

$$= \sum_{(x_1, \ldots, x_{i-1}, x_{i+1}, \ldots, x_n) \in \Omega_{0, x_i}'} p_X(x_1, \ldots, x_{i-1}, x_i, x_{i+1}, \ldots, x_n). \qquad \text{[(5.31)]}$$

Proof of Lemma 5.68

If $X_1, \ldots, X_n$ are independent and $X = (X_1, \ldots, X_n)$, then, for all $(x_1, \ldots, x_n) \in \times_{i=1}^n \Omega_{0,i}'$,

$$p_X(x_1, \ldots, x_n) = P((X_1, \ldots, X_n) = (x_1, \ldots, x_n)) \qquad \text{[(5.31)]}$$

$$= P(X_1 = x_1, \ldots, X_n = x_n)$$

$$= P(X_1 = x_1) \cdot \ldots \cdot P(X_n = x_n) \qquad \text{[(5.28)]}$$

$$= p_{X_1}(x_1) \cdot \ldots \cdot p_{X_n}(x_n). \qquad \text{[(5.31)]}$$

For all $A_1' \in \mathscr{A}_1', \ldots, A_n' \in \mathscr{A}_n'$,

$$(A_1' \times \ldots \times A_n') \cap (\Omega_{0,1}' \times \ldots \times \Omega_{0,n}') = (A_1' \cap \Omega_{0,1}') \times \ldots \times (A_n' \cap \Omega_{0,n}'). \qquad (5.73)$$

Now assume that Equation (5.38) holds and define $\Omega_{sn}' := \times_{i=1}^n \Omega_{0,i}'$. Then,

$$P(X_1 \in A_1', \ldots, X_n \in A_n')$$

$$= P[(X_1, \ldots, X_n) \in (A_1' \times \ldots \times A_n')]$$

$$= P[(X_1, \ldots, X_n) \in (A_1' \times \ldots \times A_n') \cap \Omega_{sn}'] \qquad [P_X(\Omega_{sn}') = 1, \text{ Box 4.1, (viii)}]$$

$$= \sum_{(x_1, \ldots, x_n) \in (A_1' \times \ldots \times A_n') \cap \Omega_{sn}'} P[(X_1, \ldots, X_n) = (x_1, \ldots, x_n)] \qquad \text{[Box 4.1 (x)]}$$

$$= \sum_{(x_1, \ldots, x_n) \in (A_1' \times \ldots \times A_n') \cap \Omega_{sn}'} p_X(x_1, \ldots, x_n) \qquad \text{[(5.36)]}$$

$$= \sum_{(x_1, \ldots, x_n) \in (A_1' \times \ldots \times A_n') \cap \Omega_{sn}'} p_{X_1}(x_1) \cdot \ldots \cdot p_{X_n}(x_n) \qquad \text{[(5.38)]}$$

$$= \left(\sum_{x_1 \in A_1' \cap \Omega_{0,1}'} p_{X_1}(x_1) \right) \cdot \ldots \cdot \left(\sum_{x_n \in A_n' \cap \Omega_{0,n}'} p_{X_n}(x_n) \right) \qquad [(5.73)]$$

$$= P(X_1 \in A_1') \cdot \ldots \cdot P(X_n \in A_n'), \qquad [P_{X_i}(\Omega_{0,i}') = 1, \text{Box 4.1 (viii), (x)}]$$

which implies independence of $X_1, \ldots, X_n$ [see (5.28)].

Proof of Theorem 5.77

For $A' \in \mathscr{A}_X'$,

$$P_X(A') = \sum_{x \in A' \cap \Omega_0'} P_X(\{x\}) \qquad [P_X(\Omega_0') = 1, \text{Box 4.1 (viii), (x)}]$$

$$= \sum_{x \in A' \cap \Omega_0'} p_X(x) \qquad [(5.31)]$$

$$= \sum_{x \in \Omega_0'} p_X(x) \cdot 1_{A'}(x)$$

$$= \sum_{x \in \Omega_0'} \int p_X \cdot 1_{A'} \, d\delta_x \qquad [(3.23)]$$

$$= \int 1_{A'} \cdot p_X \, d\left(\sum_{x \in \Omega_0'} \delta_x \right) \qquad [(3.53)]$$

$$= \int_{A'} p_X \, d\left(\sum_{x \in \Omega_0'} \delta_x \right) \qquad [(3.30)]$$

$$= \int_{A'} p_X \, d\mu. \qquad [(5.42)]$$

Hence, according to Definition 5.70, the probability function p_X is a density of P_X with respect to the measure μ on the measurable space $(\Omega_X', \mathscr{A}_X')$ and $p_X = \dfrac{dP}{d\mu}$ (see Th. 3.72).

Proof of Lemma 5.78

Let $A' \in \mathscr{A}_X'$ with $\mu(A') = 0$. Then,

$$\mu \otimes \nu(A' \times \Omega_Y') = \mu(A') \cdot \nu(\Omega_Y') = 0 \cdot \nu(\Omega_Y') = 0,$$

and this holds even if $\nu(\Omega_Y') = \infty$. Together with $P_{X,Y} \underset{\mathscr{A}_X' \otimes \mathscr{A}_Y'}{\ll} \mu \otimes \nu$, this implies

$$P_X(A') = P(X \in A', Y \in \Omega_Y') = P_{X,Y}(A' \times \Omega_Y') = 0.$$

Therefore, $P_X \underset{\mathscr{A}_X'}{\ll} \mu$. The proof for P_Y is analogous.

Proof of Lemma 5.79

For all $A' \in \mathcal{A}'_X$:

$$\int 1_{A'}(x) \left(\int f_{X,Y}(x, y)\, \nu(dy) \right) \mu(dx)$$

$$= \int 1_{A'}(x) f_{X,Y}(x, y)\, \mu \otimes \nu[d(x, y)] \qquad \text{[Th. 3.76]}$$

$$= \int 1_{A'}(x)\, P_{X,Y}[d(x, y)] \qquad \text{[Th. 3.67]}$$

$$= P_{X,Y}(A' \times \Omega'_Y) \qquad \text{[(3.30), (3.8)]}$$

$$= P_X(A'). \qquad \text{[(5.21)]}$$

Theorem 3.65 then implies the lemma. The proof for f_Y is analogous.

Proof of Corollary 5.92

Note that

$$\mathbb{R}^{n-1} = \bigcup_{m=1}^{\infty} \,]-\infty, m]^{n-1}. \qquad (5.74)$$

For all $x_i \in \mathbb{R}$,

$$F_{X_i}(x_i) = P(X_i \leq x_i)$$

$$= P_{X_1, \ldots, X_n}(\mathbb{R} \times \ldots \times \mathbb{R} \times]-\infty, x_i] \times \mathbb{R} \times \ldots \times \mathbb{R}) \qquad \text{[(5.21)]}$$

$$= P(X_1 \in \mathbb{R}, \ldots, X_{i-1} \in \mathbb{R}, X_i \in]-\infty, x_i], X_{i+1} \in \mathbb{R}, \ldots, X_n \in \mathbb{R}) \qquad \text{[(5.2)]}$$

$$= P((X_1, \ldots, X_{i-1}, X_{i+1}, \ldots, X_n) \in \mathbb{R}^{n-1}, X_i \in]-\infty, x_i])$$

$$= P(X_j \in \bigcup_{m_j=1}^{\infty} \,]-\infty, m_j], j \in \{1, \ldots, i-1, i+1, \ldots, n\}, X_i \in]-\infty, x_i]) \qquad \text{[(5.74)]}$$

$$= \lim_{m_j \to \infty, j \neq i} P(X_j \in]-\infty, m_j], j \in \{1, \ldots, i-1, i+1, \ldots, n\}, X_i \in]-\infty, x_i])$$

$$\text{[Th. 1.68 (i)]}$$

$$= \lim_{m_j \to \infty, j \neq i} F_{X_1, \ldots, X_n}(m_1, \ldots, m_{i-1}, x_i, m_{i+1}, \ldots, m_n) \qquad \text{[(5.56)]}$$

$$= \lim_{x_j \to \infty, j \neq i} F_{X_1, \ldots, X_n}(x_1, \ldots, x_n).$$

The limits exist, because $F_{X_1, \ldots, X_n}$ is monotone in all coordinates.

Proof of Corollary 5.93

(a) $\Rightarrow$ (b) For all $(x_1, \ldots, x_n) \in \mathbb{R}^n$,

$$
\begin{aligned}
F_{X_1,\ldots,X_n}(x_1, \ldots, x_n) &= P_{X_1,\ldots,X_n}(]-\infty, x_1] \times \ldots \times]-\infty, x_n]) && [(5.56)] \\
&= P_{X_1} \otimes \ldots \otimes P_{X_n}(]-\infty, x_1] \times \ldots \times]-\infty, x_n]) && \text{[Lemma 5.49 and (a)]} \\
&= P_{X_1}(]-\infty, x_1]) \cdot \ldots \cdot P_{X_n}(]-\infty, x_n]) && [(1.50)] \\
&= F_{X_1}(x_1) \cdot \ldots \cdot F_{X_n}(x_n). && \text{[Def. 5.81]}
\end{aligned}
$$

(b) $\Rightarrow$ (a) For all $(x_1, \ldots, x_n) \in \mathbb{R}^n$,

$$
\begin{aligned}
&P_{X_1,\ldots,X_n}(]-\infty, x_1] \times \ldots \times]-\infty, x_n]) \\
&= F_{X_1,\ldots,X_n}(x_1, \ldots, x_n) && [(5.56)] \\
&= F_{X_1}(x_1) \cdot \ldots \cdot F_{X_n}(x_n) && [(b)] \\
&= P_{X_1}(]-\infty, x_1]) \cdot \ldots \cdot P_{X_n}(]-\infty, x_n]) && \text{[Def. 5.81]} \\
&= P_{X_1} \otimes \ldots \otimes P_{X_n}(]-\infty, x_1] \times \ldots \times]-\infty, x_n]). && [(1.50)]
\end{aligned}
$$

Because $\{]-\infty, x_1] \times \ldots \times]-\infty, x_n] : (x_1, \ldots, x_n) \in \mathbb{R}^n \}$ is a $\cap$-stable generating system of $\mathcal{B}_n$ [see (1.21)], Theorem 1.72 yields $P_{X_1,\ldots,X_n} = P_{X_1} \otimes \ldots \otimes P_{X_n}$. Applying Lemma 5.49 then completes the proof.

Proof of Corollary 5.100

(i) We prove: $(X_1, \ldots, X_n)$ are independent $\Rightarrow f_{X_1,\ldots,X_n}$ in (5.70) is a joint density.

Because $f_{X_1,\ldots,X_n}$ defined in (5.70) is nonnegative and integrable with respect to λ_n, we can conclude that $f_{X_1,\ldots,X_n} \odot \lambda_n$ defines a finite measure on $(\mathbb{R}^n, \mathcal{B}_n)$ (see Th. 3.65). Furthermore, if $X_1, \ldots, X_n$ are independent and f_{X_i} is a density of X_i for all $i = 1, \ldots, n$, then, for all $(x_1, \ldots, x_n) \in \mathbb{R}^n$ and $B = \times_{i=1}^n]-\infty, x_i]$,

$$
\begin{aligned}
&\int_B f_{X_1,\ldots,X_n}(t_1, \ldots, t_n) \, \lambda_n[d(t_1, \ldots, t_n)] \\
&= \int_B f_{X_1}(t_1) \cdot \ldots \cdot f_{X_n}(t_n) \, \lambda_n[d(t_1, \ldots, t_n)] && [(5.70)] \\
&= \prod_{i=1}^n \int_{]-\infty, x_i]} f_{X_i}(t_i) \, \lambda(dt_i) && \text{[Th. 3.76]} \\
&= \prod_{i=1}^n F_{X_i}(x_i) && [(5.62)] \\
&= F_{X_1,\ldots,X_n}(x_1, \ldots, x_n) && \text{[Cor. 5.93]} \\
&= P_{X_1,\ldots,X_n}(B).
\end{aligned}
$$

This shows that $f_{X_1,\ldots,X_n} \odot \lambda_n(B) = P_{X_1,\ldots,X_n}(B)$, and this implies that $f_{X_1,\ldots,X_n}$ defined in (5.70) is a density of $(X_1, \ldots, X_n)$ with respect to λ_n (see Def. 5.70).

(ii) We prove: $f_{X_1,\ldots,X_n}$ in (5.70) is a density of $(X_1, \ldots, X_n) \Rightarrow X_1, \ldots, X_n$ are independent. If Equation (5.70) holds and $B = \times_{i=1}^{n} \,]{-}\infty, x_i]$, then for all $(x_1, \ldots, x_n) \in \mathbb{R}^n$

$$
F_{X_1,\ldots,X_n}(x_1, \ldots, x_n)
$$

$$
= \int_B f_{X_1,\ldots,X_n}(t_1, \ldots, t_n)\, \lambda_n[d(t_1, \ldots, t_n)] \qquad [(5.67)]
$$

$$
= \int_{]-\infty, x_n]} \cdots \int_{]-\infty, x_1]} f_{X_1}(t_1) \cdot \ldots \cdot f_{X_n}(t_n)\, \lambda(dt_1) \ldots \lambda(dt_n) \qquad [\text{Th. 3.76}]
$$

$$
= \prod_{i=1}^{n} \int_{]-\infty, x_i]} f_{X_i}(t_i)\, \lambda(dt_i)
$$

$$
= \prod_{i=1}^{n} F_{X_i}(x_i). \qquad [(5.62)]
$$

Now, Corollary 5.93 implies that $X_1, \ldots, X_n$ are independent.

Exercises

5.1   Consider the random variable X defined in Example 2.34. Which are the elements ω in the inverse image $X^{-1}(\{1\})$, and which are the probabilities of the events $\{\omega\}$?

5.2   Consider again the random variable X defined in Example 2.34. What are the values of the distribution of X and the distribution of U?

5.3   Consider Example 5.9. Show that $1_{X \in A'}$ is a random variable on $(\Omega, \mathcal{A}, P)$ and that $1_{X \in A'} = 1_{A'}(X) = 1_{A'} \circ X$.

5.4   Consider Example 5.15, and show that X_1 is prior to $X_1 \cdot X_2$ in $(\mathcal{F}_1, \mathcal{F}_2)$.

5.5   Show that $X \underset{PB}{=} Y$ for $B = \{\omega_1, \omega_2, \omega_3\}$ and that $X \underset{P}{=} Y$ does *not* hold in Example 5.23.

5.6   Consider the random variable $(U, X)\colon \Omega \to \Omega_U \times \{0; 1\}$ defined in Example 2.34. Which are the elements ω in the inverse image $(U, X)^{-1}(\{(Joe, 1)\})$, and which are the probabilities of the events $\{\omega\}$?

5.7   Show that the random variables U and X presented in Table 2.2 are independent.

5.8   Let $(\Omega, \mathcal{A}, P)$ be a probability space, $B \in \mathcal{B}$, and $\mathcal{C} \subset \mathcal{A}$ a σ-algebra. Prove that the following propositions are equivalent to each other: $1_B \underset{P}{\perp\!\!\!\perp} \mathcal{C}$, $\sigma(\{B\}) \underset{P}{\perp\!\!\!\perp} \mathcal{C}$, $\{B\} \underset{P}{\perp\!\!\!\perp} \mathcal{C}$, and $B \underset{P}{\perp\!\!\!\perp} \mathcal{C}$.

5.9   In Example 5.50 we showed that X and U are independent, and in Example 5.54 we defined the mappings $f\colon \{0, 1, 2\} \to \{0, 1\}$ and $g\colon \Omega_U \to \{male, female\}$. Use Equation (5.24) to show that the mappings $f(X)$ (control vs. any of the two treatments) and $g(U)$ (*sex*) are independent as well.

5.10   In Example 5.85, we specified the values of the distribution function F_X for the random variable *number of flipping heads*. Use these values to compute all eight values of the distribution P_X.

Solutions

5.1 The inverse image of the set $\{1\}$ under X is

$$X^{-1}(\{1\}) = \{(Joe, yes, -), (Joe, yes, +), (Ann, yes, -) (Ann, yes, +)\}.$$

The events $\{\omega\}$, $\omega \in X^{-1}(\{1\})$ have the probabilities $P[\{(Joe, yes, -)\}] = .04$, $P[\{(Joe, yes, +)\}] = .16$, $P[\{(Ann, yes, -)\}] = .12$, and $P[\{(Ann, yes, +)\}] = .08$.

5.2 The random variable X (the treatment variable) has the following distribution:

$$P_X(\{1\}) = P[X^{-1}(\{1\})] = .40, \quad P_X(\{0\}) = P[X^{-1}(\{0\})] = .60,$$
$$P_X(\Omega') = P[X^{-1}(\Omega')] = 1, \quad P_X(\emptyset) = P[X^{-1}(\emptyset)] = 0,$$

whereas the distribution of the random variable U (the observational-unit variable) is:

$$P_U(\{Joe\}) = P[U^{-1}(\{Joe\})] = .50, \quad P_U(\{Ann\}) = P[U^{-1}(\{Ann\})] = .50,$$
$$P_U(\Omega_U) = P[U^{-1}(\Omega_U)] = 1, \quad P_U(\emptyset) = P[U^{-1}(\emptyset)] = 0.$$

5.3 Let $A' \in \mathscr{A}_X'$, and consider the indicator function $1_{X \in A'}$: $(\Omega, \mathscr{A}, P) \to (\mathbb{R}, \mathscr{B})$. Measurability: For all $B \in \mathscr{B}$,

$$
\begin{aligned}
(1_{X \in A'})^{-1}(B) &= (1_{X^{-1}(A')})^{-1}(B) && [(5.6)] \\
&= \{\omega \in \Omega : 1_{X^{-1}(A')}(\omega) \in B\} && [(2.2)] \\
&= \begin{cases} \emptyset, & \text{if } 0 \notin B, 1 \notin B \\ X^{-1}(A'), & \text{if } 0 \notin B, 1 \in B \\ \Omega \setminus X^{-1}(A'), & \text{if } 0 \in B, 1 \notin B \\ \Omega, & \text{if } \{0, 1\} \subset B, \end{cases}
\end{aligned}
$$

and all these sets are elements of $\mathscr{A}$, because X is assumed to be a random variable.
Furthermore, for all $\omega \in \Omega$,

$$
\begin{aligned}
1_{X \in A'}(\omega) &= 1_{X^{-1}(A')}(\omega) && [(5.6)] \\
&= \begin{cases} 1, & \text{if } \omega \in X^{-1}(A') \\ 0, & \text{if } \omega \notin X^{-1}(A') \end{cases} \\
&= \begin{cases} 1, & \text{if } X(\omega) \in A' \\ 0, & \text{if } X(\omega) \notin A' \end{cases} && [(2.2)] \\
&= 1_{A'}[X(\omega)] \\
&= 1_{A'} \circ X(\omega). && [(2.25)]
\end{aligned}
$$

5.4 The filtration $(\mathscr{F}_1, \mathscr{F}_2)$ is defined by $\mathscr{F}_1 := \sigma(X_1)$ and $\mathscr{F}_2 := \sigma(X_1, X_2)$. Hence,

$$\sigma(X_1) \subset \mathscr{F}_1 = \{\Omega, \emptyset, \{(h, h), (h, t)\}, \{(t, h), (t, t)\}\}.$$

In contrast,

$$\sigma(X_1 \cdot X_2) = \{\Omega, \emptyset, \{(h,h)\}, \{(h,t),(t,h),(t,t)\}\} \not\subset \mathscr{F}_1.$$

However, $\sigma(X_1 \cdot X_2) \subset \mathscr{F}_2$, because $\mathscr{F}_2 = \mathscr{P}(\Omega)$. Hence, according to Definition 5.13 (i), X_1 is prior to $X_1 \cdot X_2$ in the filtration $(\mathscr{F}_1, \mathscr{F}_2)$.

5.5 In order to prove $X \underset{P^B}{=} Y$ for $B = \{\omega_1, \omega_2, \omega_3\}$, it is sufficient to show $P^B(A) = 0$, where $A := \{\omega \in \Omega: X(\omega) \neq Y(\omega)\} \in \mathscr{A} = \mathscr{P}(\Omega)$. In this example, $A := \{\omega_3, \omega_4\}$. Now $P(\{\omega_3\}) = 0$, which implies

$$P^B(\{\omega_3\}) = \frac{P(B \cap \{\omega_3\})}{P(B)} = \frac{P(\{\omega_3\})}{P(B)} = \frac{0}{P(B)} = 0.$$

Furthermore,

$$P^B(\{\omega_4\}) = \frac{P(B \cap \{\omega_4\})}{P(B)} = \frac{P(\emptyset)}{P(B)} = \frac{0}{P(B)} = 0.$$

Additivity of the measure P^B then implies $P^B(A) = 0$.

In order to prove that $X \underset{P}{=} Y$ *does not hold*, it suffices to show $P(A) \neq 0$. Now $P(\{\omega_3\}) = 0$ and $P(\{\omega_4\}) = .50$. Therefore, $P(A) = P(\{\omega_4\}) = .50 > 0$.

5.6 The elements of the inverse image of $(U, X)^{-1}(\{Joe, 1\})$ are $(Joe, yes, -)$ and $(Joe, yes, +)$. The probabilities of the corresponding elementary events are $P(\{(Joe, yes, -)\}) = .04$ and $P(\{(Joe, yes, +)\}) = .16$.

5.7 In Example 4.43, we already showed that the events $A = \{Joe\} \times \Omega_X \times \Omega_Y$ (that Joe is sampled) and $B = \Omega_U \times \{yes\} \times \Omega_Y$ (that the person sampled is treated) are independent. According to Box 4.2 (iii), this implies that the σ-algebras $\{A, A^c, \Omega, \emptyset\}$ and $\{B, B^c, \Omega, \emptyset\}$ are independent as well. If $X: (\Omega, \mathscr{A}, P) \rightarrow (\Omega', \mathscr{A}')$ and $U: (\Omega, \mathscr{A}, P) \rightarrow (\Omega_U, \mathscr{A}_U)$ are the random variables defined in Table 2.2, then $X^{-1}(\mathscr{A}') = \{A, A^c, \Omega, \emptyset\}$ and $U^{-1}(\mathscr{A}_U) = \{B, B^c, \Omega, \emptyset\}$ are the σ-algebras generated by U and X, respectively. Hence, in this example, X and U are independent.

5.8 First of all,

$$\{B\} \underset{P}{\perp\!\!\!\perp} \mathscr{C} \Leftrightarrow B \underset{P}{\perp\!\!\!\perp} \mathscr{C}. \qquad \text{[Rem. 4.41]}$$

Furthermore, $\sigma(1_B) = \{\Omega, \emptyset, B, B^c\}$ (see Example 2.31) and $\sigma(\{B\}) = \{\Omega, \emptyset, B, B^c\}$. Hence, $\sigma(1_B) = \sigma(\{B\})$, and this implies

$$1_B \underset{P}{\perp\!\!\!\perp} \mathscr{C} \Leftrightarrow \sigma(\{B\}) \underset{P}{\perp\!\!\!\perp} \mathscr{C}. \qquad \text{[Rem. 5.44]}$$

Finally,

$$\sigma(\{B\}) \underset{P}{\perp\!\!\!\perp} \mathscr{C} \Leftrightarrow \{B\} \underset{P}{\perp\!\!\!\perp} \mathscr{C}. \qquad \text{[Box 4.2 (ii)]}$$

5.9   Consider the two events $A := f(X)^{-1}(\{0\})$ and $B := g(U)^{-1}(\{male\})$.

$$P(A \cap B) = \frac{40}{99}$$

and

$$P(A) \cdot P(B) = \frac{60}{99} \cdot \frac{2}{3} = \frac{40}{99}.$$

Because $f(X)^{-1}(\mathscr{P}(\{0, 1\})) = \{A, A^c, \Omega, \emptyset\}$ and $g(U)^{-1}(\mathscr{P}(\{male, female)\}) = \{B, B^c, \Omega, \emptyset\}$, this proves that $f(X)$ and $g(U)$ are independent (see Exercise 5.8).

5.10   We consider all elements $B \in \mathscr{B}$ of the Borel σ-algebra on $\mathbb{R}$ and assign the following values:

$$P_X(B) = \begin{cases} F_X(-1) = 0, & \text{if } 0, 1, 2 \notin B \\ F_X(0) = \frac{1}{4}, & \text{if } 0 \in B, 1, 2 \notin B \\ F_X(1) = \frac{3}{4}, & \text{if } 0, 1 \in B, 2 \notin B \\ F_X(2) = 1, & \text{if } 0, 1, 2 \in B \\ F_X(1) - F_X(0) = \frac{2}{4}, & \text{if } 1 \in B, 0, 2 \notin B \\ F_X(2) - F_X(0) = \frac{3}{4}, & \text{if } 1, 2 \in B, 0 \notin B \\ F_X(2) - F_X(1) = \frac{1}{4}, & \text{if } 2 \in B, 0, 1 \notin B \\ F_X(2) - F_X(1) + F_X(0) = \frac{1}{2}, & \text{if } 0, 2 \in B, 1 \notin B. \end{cases}$$

6

Expectation, variance, and other moments

In chapter 4 we introduced a probability measure as a special finite measure, and in chapter 5 we defined a random variable as a measurable mapping on a probability space. In this chapter, we will translate integration theory (see ch. 3) to probability theory introducing *expectations* of numerical random variables and other important concepts that are special expectations: *central* and *noncentral moments*, and *variances*. Even *covariances* and *correlations* are special expectations (see ch. 7). All these quantities describe important properties of random variables, although, in general, they do not determine the complete distribution.

6.1   Expectation

6.1.1   Definition

Reading the Definition 6.1, remember that a random variable Y is called *quasi-integrable* with respect to P if $\int Y^+ dP$ or $\int Y^- dP$ are finite, where Y^+ and Y^- denote the positive and negative parts of Y, respectively (see Rem. 2.62 and Def. 3.28).

Definition 6.1 [Expectation]
Let $Y: (\Omega, \mathcal{A}, P) \to (\overline{\mathbb{R}}, \overline{\mathcal{B}})$ be a numerical random variable that is quasi-integrable with respect to P. Then we define

$$E(Y) := \int Y \, dP, \qquad (6.1)$$

call it the expectation of Y (with respect to P), and say that it exists. Instead of expectation with respect to P, we also use the term P-expectation.

Probability and Conditional Expectation: Fundamentals for the Empirical Sciences, First Edition. Rolf Steyer and Werner Nagel.
© 2017 John Wiley & Sons, Ltd. Published 2017 by John Wiley & Sons, Ltd.
Companion website: http://www.probability-and-conditional-expectation.de

Remark 6.2 [Existence of the expectation] Note that $E(Y)$ can be infinite. Furthermore, if $E(Y)$ *exists*, we also say that Y is a random variable *with expectation $E(Y)$*. If Y is not quasi-integrable with respect to P and therefore also not P-integrable, then we say that the expectation of Y with respect to P does *not exist*.   ◁

Remark 6.3 [Notation and synonymous terms] A synonym for expectation is *expectation value*. The reference to the measure P is usually omitted if the context is unambiguous. If we consider the expectation with respect to another probability measure on $(\Omega, \mathscr{A})$, such as, the conditional-probability measure P^B (see Def. 4.29), then we adapt the notation correspondingly:

$$E^B(Y) := \int Y \, dP^B. \tag{6.2}$$

Expectation with respect to P^B is used synonymously with P^B-expectation.   ◁

Remark 6.4 [Random variables with finite expectations] Without substantial loss of generality, we often assume that a random variable Y is *real-valued* if its expectation is finite. According to Remark 3.42, if the random variable $Y: (\Omega, \mathscr{A}, P) \to (\overline{\mathbb{R}}, \overline{\mathscr{B}})$ has a finite expectation, then there is random variable $Y^*: (\Omega, \mathscr{A}, P) \to (\mathbb{R}, \mathscr{B})$ such that $Y \underset{P}{=} Y^*$.   ◁

6.1.2   Expectation of a discrete random variable

In this section, we use the notation $P(Y = y_i) := P[Y^{-1}(\{y_i\})]$ introduced in Remark 5.4.

Remark 6.5 [Random variable with a finite number of real values] Suppose $y_1, \dots, y_n \in \mathbb{R}$, $n \in \mathbb{N}$, denote all (negative, 0, or positive) values of a real-valued random variable $Y: (\Omega, \mathscr{A}, P) \to (\mathbb{R}, \mathscr{B})$, and let P_Y denote the distribution, and p_Y the probability function of Y. Then the expectation $E(Y)$ exists, and

$$E(Y) = \sum_{i=1}^{n} y_i \cdot P(Y = y_i) = \sum_{i=1}^{n} y_i \cdot P_Y(\{y_i\}) = \sum_{i=1}^{n} y_i \cdot p_Y(y_i) \tag{6.3}$$

[see Cor. 3.59, (3.45), Eq. (5.34), and Def. 5.56]. Hence, if Y has only a finite number n of values, then its expectation is simply the sum of its values, each one weighted by its probability $P(Y = y_i) = P_Y(\{y_i\}) = p_Y(y_i)$.   ◁

Example 6.6 [Expectation of an indicator] If $(\Omega, \mathscr{A}, P)$ is a probability space and 1_A is the indicator of $A \in \mathscr{A}$, then

$$E(1_A) = 0 \cdot P(1_A = 0) + 1 \cdot P(1_A = 1) = P(1_A = 1) = P(A) \tag{6.4}$$

is the expectation of 1_A (see also Example 3.14). Considering the event $\{Y = y\}$ and using the notation $1_{Y=y} := 1_{\{Y=y\}}$, this yields

$$E(1_{Y=y}) = P(Y = y). \tag{6.5}$$

◁

Example 6.7 [Joe and Ann with randomized assignment – continued] In Example 1.9, we defined the set

$$B = \{(Joe, yes, -), (Joe, yes, +), (Ann, yes, -),\ (Ann, yes, +)\} = \Omega_U \times \{yes\} \times \Omega_Y,$$

the event that the drawn person is treated, and the set

$$C = \{(Joe, no, +),\ (Joe, yes, +),\ (Ann, no, +),\ (Ann, yes, +)\} = \Omega_U \times \Omega_X \times \{+\},$$

the event that success (+) occurs, irrespective of which person is drawn and whether or not the person is treated. In Table 2.2, we assigned probabilities to each elementary event $\{\omega\}$, $\omega \in \Omega$ and defined $X = 1_B$, the treatment variable, as well as $Y = 1_C$, the outcome variable. Applying Equation (6.4) to the indicator 1_B yields:

$$E(X) = E(1_B) = P(B)$$
$$= P[\{(Joe, yes, -)\}] + P[\{(Joe, yes, +)\}] + P[\{(Ann, yes, -)\}] + P[\{(Ann, yes, +)\}]$$
$$= .04 + .16 + .12 + .08 = .4.$$

Similarly, for the indicator 1_C, we obtain

$$E(Y) = E(1_C) = P(C)$$
$$= P[\{(Joe, no, +)\}] + P[\{(Joe, yes, +)\}] + P[\{(Ann, no, +)\}] + P[\{(Ann, yes, +)\}]$$
$$= .21 + .16 + .06 + .08 = .51.$$
◁

Example 6.8 [Tossing a dice – continued] In Example 3.6, we considered the random variable $X = $ *number of dots*. In this example we specified the probability space $(\Omega, \mathcal{A}, P)$ by $\Omega = \{\omega_1, \ldots, \omega_6\}$, $\mathcal{A} = \mathcal{P}(\Omega)$, and $P(\{\omega_1\}) = \ldots = P(\{\omega_6\}) = 1/6$. Because $P(X=x_i) = P_X(\{i\}) = P(\{\omega_i\})$, $i = 1, \ldots, 6$, Equation (6.3) yields

$$E(X) = \sum_{i=1}^{6} i \cdot P(X = i) = 1 \cdot \frac{1}{6} + 2 \cdot \frac{1}{6} + \ldots + 6 \cdot \frac{1}{6} = 3.5.$$
◁

Remark 6.9 [Random variable with a countable number of real values] Let $y_1, y_2, \ldots \in \mathbb{R}$ denote the values of a real-valued random variable $Y: (\Omega, \mathcal{A}, P) \to (\mathbb{R}, \mathcal{B})$ and suppose that the expectation of Y exists. Then

$$E(Y) = \sum_{i=1}^{\infty} y_i \cdot P(Y=y_i) = \lim_{n \to \infty} \sum_{i=1}^{n} y_i \cdot P(Y=y_i) \qquad (6.6)$$

[see (3.45), Eq. (5.34), and Th. 3.60]. Examples in which the expectation of a random variable is the 'infinite sum' of its values weighted by their probabilities are random variables with a Poisson distribution and with a geometric distribution (see Theorems 8.16 and 8.22).   ◁

Example 6.10 [A discrete random variable without expectation] Suppose that $\Omega = \mathbb{N}_0 = \{0, 1, 2, \ldots\}$, and consider the random variable $Y: (\Omega, \mathcal{P}(\Omega), P) \to (\mathbb{R}, \mathcal{B})$ defined by

$$Y(i) = y_i = (-1)^i\, i!\, e, \quad \forall\, i \in \mathbb{N}_0, \qquad (6.7)$$

with

$$P(Y=y_i) = \frac{1}{e} \cdot \frac{1}{i!}, \quad \forall i \in \mathbb{N}_0. \tag{6.8}$$

Note that $e = \sum_{i=0}^{\infty} \frac{1}{i!}$. Dividing both sides by e yields $\sum_{i=0}^{\infty} \frac{1}{i!\,e} = 1$. Therefore, Equation (6.8) specifies a probability distribution. Now consider

$$\sum_{i=0}^{n} y_i \cdot P(Y=y_i) = \sum_{i=0}^{n} (-1)^i = \begin{cases} 1, & \text{if } n \text{ is even} \\ 0, & \text{if } n \text{ is odd.} \end{cases} \tag{6.9}$$

Obviously, in this example, the limit

$$\lim_{n \to \infty} \sum_{i=0}^{n} y_i \cdot P(Y=y_i)$$

occurring in Equation (6.6) does not exist. Therefore, according to Definitions 6.1 and 3.28, $E(Y)$ is not defined.   ◁

6.1.3   Computing the expectation using a density

According to the Theorem 6.11, the expectation of a continuous real-valued random variable can also be computed using its density f_Y with respect to the Lebesgue measure (see Def. 5.94) and the *Riemann integral*.

Theorem 6.11 [Expectation of a continuous random variable]
Let $Y: (\Omega, \mathcal{A}, P) \to (\mathbb{R}, \mathcal{B})$ *be a continuous random variable with expectation* $E(Y)$ *and a density* f_Y *that is Riemann integrable. Then,*

$$E(Y) = \int_{-\infty}^{\infty} y \cdot f_Y(y)\, dy. \tag{6.10}$$

(Proof p. 221)

Examples of continuous random variables and their expectations are treated in chapter 8, section 8.2.

Example 6.12 [A continuous random variable without expectation]   Consider the continuous random variable $Y: (\Omega, \mathcal{A}, P) \to (\mathbb{R}, \mathcal{B})$ with density

$$f_Y(y) = \frac{1}{\pi} \cdot \frac{1}{1+y^2}, \quad \forall y \in \mathbb{R}, \tag{6.11}$$

and distribution function

$$F_Y(y) = \frac{1}{2} + \frac{1}{\pi} \cdot \arctan y, \quad \forall y \in \mathbb{R}. \tag{6.12}$$

A distribution P_Y with density (6.11) is called a *standard Cauchy distribution* or *central t-distribution* with one degree of freedom.

The integral of the positive part of $y \cdot f_Y(y)$ is

$$\int_{-\infty}^{\infty} [y \cdot f_Y(y)]^+ \, dy = \int_0^{\infty} y \cdot \frac{1}{\pi} \cdot \frac{1}{1+y^2} \, dy = \frac{1}{2\pi} \ln(1+y^2) \Big|_0^{\infty} = \infty,$$

and the integral of the negative part is

$$\int_{-\infty}^{\infty} [y \cdot f_Y(y)]^- \, dy = \int_{-\infty}^0 -y \cdot \frac{1}{\pi} \cdot \frac{1}{1+y^2} \, dy = -\frac{1}{2\pi} \ln(1+y^2) \Big|_{-\infty}^0 = \infty$$

[for the notation cf. Eq. (3.69)]. Hence, $y \cdot f_Y(y)$ is not quasi-P-integrable on $\mathbb{R}$ (see Def. 3.28), and $E(Y)$ does not exist. ◁

6.1.4   Transformation theorem

Let $Y: (\Omega, \mathscr{A}, P) \to (\Omega'_Y, \mathscr{A}'_Y)$ and $g: (\Omega'_Y, \mathscr{A}'_Y, P_Y) \to (\mathbb{R}, \mathscr{B})$ be random variables. Then we denote the expectation of g with respect to the distribution P_Y by $E_Y(g)$. Theorem 6.13 immediately follows from Theorem 3.57, translating the measure theory terms to probability theory. Theorem 6.13 is relevant whenever we consider the expectation of a composition $g \circ Y = g(Y)$ of a random variable Y and a function g [see Eq. (2.25)] or the expectation of g with respect to the distribution P_Y [see Eq. (5.2)].

> **Theorem 6.13 [Transformation theorem]**
> Let $Y: (\Omega, \mathscr{A}, P) \to (\Omega'_Y, \mathscr{A}'_Y)$ be a random variable and $g: (\Omega'_Y, \mathscr{A}'_Y) \to (\mathbb{R}, \mathscr{B})$ be measurable.
>
> (i) If g is nonnegative or has a finite expectation $E_Y(g)$, then
>
> $$E_Y(g) = \int g \, dP_Y = \int g(y) \, P_Y(dy) = \int g(Y) \, dP = E[g(Y)]. \qquad (6.13)$$
>
> (ii) $E_Y(g)$ is finite if and only if $E[g(Y)]$ is finite.

The virtue of Equation (6.13) is that we do not have to know the distribution of $g(Y)$. Instead, the distribution of Y suffices.

Remark 6.14 [A special case]   If we consider the special case in which g is the identity function $id: \mathbb{R} \to \mathbb{R}$, defined by $id(y) = y$, for all $y \in \mathbb{R}$, then $id(Y) = Y$ and Equations (6.13) yield

$$E(Y) = \int Y \, dP = \int y \, P_Y(dy). \qquad (6.14)$$

◁

Remark 6.15 [Finite number of values] If Y takes on only a finite number of different values $y_1, \dots, y_n \in \mathbb{R}$, then Equation (6.13) simplifies to

$$E[g(Y)] = E_Y(g) = \int g \, dP_Y = \sum_{i=1}^{n} g(y_i) \cdot P(Y=y_i), \qquad (6.15)$$

where $P(Y=y_i) = P_Y(\{y_i\}) = p_Y(y_i)$, $i = 1, \dots, n$, and p_Y denotes the probability function of Y (see Def. 5.56). ◁

Remark 6.16 [Countable number of values] If Y takes on a countable number of different values $y_1, y_2, \dots \in \mathbb{R}$ and $\sum_{i=1}^{\infty} g^+(y_i) \cdot P(Y=y_i) < \infty$ or $\sum_{i=1}^{\infty} g^-(y_i) \cdot P(Y=y_i) < \infty$, then

$$E[g(Y)] = E_Y(g) = \int g \, dP_Y = \sum_{i=1}^{\infty} g(y_i) \cdot P(Y=y_i). \qquad (6.16)$$

Note that (6.16) also applies if g is nonnegative, because in this case $g^- = 0$ holds for the negative part of g, which implies $\sum_{i=1}^{\infty} g^-(y_i) = 0 < \infty$. ◁

Equation (6.13) immediately implies the Corollary 6.17 according to which the expectations of two random variables X and Y are identical if they have identical distributions, provided that the expectations exist (see also Remark 6.27).

Corollary 6.17 [Identical distributions imply identical expectations]
Let $X, Y\colon (\Omega, \mathcal{A}, P) \to (\Omega', \mathcal{A}')$ be random variables and $g\colon (\Omega', \mathcal{A}') \to (\overline{\mathbb{R}}, \overline{\mathcal{B}})$ a measurable function that is nonnegative or with expectation $E_X(g)$. If $P_X = P_Y$, then $E[g(X)] = E[g(Y)]$.

This property allows us to use the term *expectation of a distribution* instead of expectation of a random variable.

In the Lemma 6.18, we consider a bivariate random variable (X, Y) and a numerical function $g(X)$, that is, a function that only depends on X. According to Lemma 6.18, the expectation of g with respect to the joint distribution $P_{X,Y}$ is identical to the expectation of g with respect to the marginal distribution P_X.

Lemma 6.18 [Expectation with respect to joint and marginal distributions]
Let $(X, Y)\colon (\Omega, \mathcal{A}, P) \to (\Omega'_X \times \Omega'_Y, \mathcal{A}'_X \otimes \mathcal{A}'_Y)$ be a bivariate random variable with joint distribution $P_{X,Y}$, and let $g\colon (\Omega'_X, \mathcal{A}'_X) \to (\overline{\mathbb{R}}, \overline{\mathcal{B}})$ be a measurable function that is nonnegative or with expectation $E_X(g)$. Then,

$$E_{X,Y}(g) = E_X(g), \qquad (6.17)$$

which is equivalent to

$$\int g(x)\, P_{X,Y}[d(x,y)] = \int g(x)\, P_X(dx).$$ (6.18)

(Proof p. 221)

Example 6.19 [Flipping two coins – continued] Consider again the random variable $X =$ *number of flipping heads* and the indicator $1_H\colon \Omega \to \mathbb{R}$ of the event that *at least one heads is flipped*. In Example 2.47 we showed that $1_H = g \circ X$, where $g\colon \mathbb{R} \to \mathbb{R}$ is defined by

$$g(x) = \begin{cases} 1 & \text{if } x \in \{1, 2\} \\ 0, & \text{otherwise} \end{cases} \quad \text{for all } x \in \mathbb{R}.$$

According to (6.15),

$$E(1_H) = E[g(X)] = E_X(g) = 0 \cdot P(\{X{=}0\}) + 1 \cdot P(X \in \{1,2\}) = P(X \in \{1,2\}) = \frac{3}{4}.$$

◁

Example 6.20 [Expectation of Y^2] Let $Y\colon (\Omega, \mathcal{A}, P) \to (\mathbb{R}, \mathcal{B})$ be a real-valued random variable, let $g\colon (\mathbb{R}, \mathcal{B}) \to (\mathbb{R}, \mathcal{B})$ be measurable, and let $g(Y) := Y^2$. Then, according to Equation (6.13),

$$E(Y^2) = E[g(Y)] = E_Y(g) = \int g\, dP_Y = \int y^2\, P_Y(dy).$$

[Note that Y^2 is nonnegative and $E(Y^2)$ can be infinite.] This equation shows that the expectation of Y^2 solely depends on the distribution P_Y of Y, which illustrates Corollary 6.17. Using the integral $\int g\, dP_Y$ is often the most convenient way of computing the expectation $E(Y^2)$. If Y takes on only a finite number of values $y_1, \ldots, y_n \in \mathbb{R}$, then this equation simplifies to

$$E(Y^2) = \sum_{i=1}^{n} y_i^2 \cdot P_Y(\{y_i\}) = \sum_{i=1}^{n} y_i^2 \cdot P(Y{=}y_i).$$ (6.19)

These equations only involve the probabilities $P(Y{=}y_i) = P_Y(\{y_i\})$, not the probabilities $P(Y^2{=}y^2)$.

◁

Example 6.21 [Multiplication with indicators] Let $Y\colon (\Omega, \mathcal{A}, P) \to (\overline{\mathbb{R}}, \overline{\mathcal{B}})$ be a numerical random variable with expectation $E(Y)$. If $A \in \mathcal{A}$ and $P(A) = 0$, then $(1_A\, Y) \underset{P}{=} 0$, and Rule (i) of Box 6.1 implies

$$E(1_A\, Y) = 0.$$ (6.20)

If $C = A \cup B$, $A \cap B = \emptyset$, and $A, B \in \mathcal{A}$, then $1_C\, Y = 1_A\, Y + 1_B\, Y$ and Rule (vi) of Box 6.1 implies

$$E(1_C\, Y) = E(1_A\, Y) + E(1_B\, Y).$$ (6.21)

◁

Box 6.1   Rules of computation for expectations.

Let $Y: (\Omega, \mathcal{A}, P) \to (\overline{\mathbb{R}}, \overline{\mathcal{B}})$ be a random variable with expectation $E(Y)$ and let $\alpha \in \mathbb{R}$. Then,

$$Y \underset{P}{=} \alpha \;\Rightarrow\; E(Y) = \alpha. \tag{i}$$

$$E(\alpha + Y) = \alpha + E(Y). \tag{ii}$$

$$E(\alpha \cdot Y) = \alpha \cdot E(Y). \tag{iii}$$

Let $A, B \in \mathcal{A}$. Then,

$$E(1_A \cdot 1_B) = P(A \cap B). \tag{iv}$$

$$E(1_A \cdot Y) = 0, \quad \text{if } P(A) = 0. \tag{v}$$

If Y_1, Y_2 are nonnegative or real-valued with finite expectations $E(Y_1)$ and $E(Y_2)$, then,

$$E(Y_1 + Y_2) = E(Y_1) + E(Y_2). \tag{vi}$$

For $i = 1, \ldots, n$, let $Y_i: (\Omega, \mathcal{A}, P) \to (\mathbb{R}, \mathcal{B})$ be random variables with finite expectations $E(Y_i)$ and $\alpha_i \in \mathbb{R}$. Then,

$$E\left(\sum_{i=1}^{n} \alpha_i \cdot Y_i \right) = \sum_{i=1}^{n} \alpha_i \cdot E(Y_i). \tag{vii}$$

Let $X, Y: (\Omega, \mathcal{A}, P) \to (\overline{\mathbb{R}}, \overline{\mathcal{B}})$ be random variables that are nonnegative or with finite expectations $E(X)$ and $E(Y)$. Then,

$$X \underset{P}{=} Y \;\Rightarrow\; E(X) = E(Y). \tag{viii}$$

$$X \underset{P}{=} Y \;\Leftrightarrow\; \forall A \in \mathcal{A}: E(1_A X) = E(1_A Y). \tag{ix}$$

$$X \underset{P}{\perp\!\!\!\perp} Y \;\Rightarrow\; E(X \cdot Y) = E(X) \cdot E(Y). \tag{x}$$

The Corollary 6.22 shows how to compute the expectation of the composition $g(Y)$ using the density of Y. As mentioned, the virtue of Equation (6.22) is that we do not have to know the density of $g(Y)$; the density of Y suffices.

Corollary 6.22 [Transformation theorem, continuous random variable]
Let $Y: (\Omega, \mathcal{A}, P) \to (\mathbb{R}, \mathcal{B})$ be a continuous random variable with a Riemann integrable density f_Y. If $g: (\mathbb{R}, \mathcal{B}) \to (\overline{\mathbb{R}}, \overline{\mathcal{B}})$ is a measurable function that is nonnegative or numerical with finite expectation $E_Y(g) = \int g \, dP_Y$, then,

$$E[g(Y)] = E_Y(g) = \int_{-\infty}^{\infty} g(y) \cdot f_Y(y) \, dy. \tag{6.22}$$

(Proof p. 221)

6.1.5   Rules of computation

Some rules of computation for expectations are gathered in Box 6.1 (for proofs, see Exercise 6.1).

Example 6.23 [Expectation of a sample mean]   Let $Y_1, \ldots, Y_n$ be a random sample. This means that the random variables $Y_1, \ldots, Y_n$ are i.i.d. (see Rem. 5.42). Furthermore, if the $Y_1, \ldots, Y_n$ are real-valued, define

$$\overline{Y} = \frac{1}{n} \cdot \sum_{i=1}^{n} Y_i, \tag{6.23}$$

the *arithmetic mean*, which in statistics is also called the *sample mean*. If Y_1 is nonnegative or with finite expectation and

$$\mu_Y = E(Y_1) \tag{6.24}$$

denotes the identical expectations of the variables $Y_1, \ldots, Y_n$, then

$$E(\overline{Y}) = \mu_Y \tag{6.25}$$

(see Exercise 6.4).   ◁

Now we turn to a generalization of Rule (x) of Box 6.1.

Theorem 6.24 [Expectation of the product of random variables]
Let $Y_i: (\Omega, \mathscr{A}, P) \to (\mathbb{R}, \mathscr{B})$, $i = 1, \ldots, n$, be real-valued random variables that are non-negative or with finite expectations, and assume that the $Y_1, \ldots, Y_n$ are independent. Then,

$$E\left(\prod_{i=1}^{n} Y_i \right) = \prod_{i=1}^{n} E(Y_i). \tag{6.26}$$

If the expectations $E(Y_i)$, $i = 1, \ldots, n$, are finite, then $E\left(\prod_{i=1}^{n} Y_i \right)$ is finite, too.

For a proof, see, for example, Bauer (1996, Theorem 8.1). Later we will weaken the independence assumption [see Rem. 7.10 and Box 7.1 (1)]. However, if the variables Y_i are *not independent*, then Equation (6.26) does *not necessarily hold* (see Remark 7.10).

6.2   Moments, variance, and standard deviation

The expectation $E(Y)$ of a numerical random variable Y is also called the *first moment of Y*, provided that this expectation exists, whereas the expectation $E(Y^2)$ is called the *second moment of Y* (see Example 6.20). For second and higher moments, we distinguish between *moments* and *central moments*.

Remark 6.25 [Higher moments of Y] Analogously to Example 6.20, we may consider $g(Y) = |Y^n|$ or $g(Y) = Y^n$ with $n \geq 1$. Note that

(i) $E(|Y^n|) = E(|Y|^n)$ always exists, because $|Y^n|$ is nonnegative (see Def. 6.1).

(ii) If, for $n \in \mathbb{N}$, the expectation $E(|Y^n|)$ is finite, then $E(Y^n)$ exists and is finite (see Cor. 3.38).

(iii) If, for $n \in \mathbb{N}$, the expectation $E(|Y^n|)$ is finite, then $E(Y^m)$ exists and is finite as well for all m with $1 \leq m \leq n$ (see Exercise 6.2).

(iv) If n is even, then the random variable Y^n is nonnegative and $E(Y^n)$ exists. In contrast, if n is odd, then the expectation $E(Y^n)$ does not necessarily exist.   ◁

Definition 6.26 [Moments]
Let $Y: (\Omega, \mathcal{A}, P) \to (\mathbb{R}, \overline{\mathcal{B}})$ be a numerical random variable and let $n \in \mathbb{N}$.

   (i) The expectation $E(|Y|^n)$ is called the nth absolute moment of Y.

   (ii) If Y is nonnegative or such that $E(Y^n)$ is finite, then $E(Y^n)$ is called the nth moment of Y.

   (iii) If $E(Y^n)$ is finite or if Y is nonnegative and the expectation $E(Y)$ is finite, then we call $E([Y - E(Y)]^n)$ the nth central moment of Y.

Remark 6.27 [Moments under P-equivalence] If the expectations of Y^n and therefore also of $[Y - E(Y)]^n$ are finite, then they can be represented as expectations of functions $g(Y)$ of Y, where $g: (\mathbb{R}, \overline{\mathcal{B}}) \to (\mathbb{R}, \overline{\mathcal{B}})$ is a measurable function with finite expectation $E_Y(g)$ [see Proposition (ii) of Th. 6.13]. Therefore, according to Corollary 6.17, all moments (central or noncentral) of a numerical random variable Y solely depend on its distribution P_Y. Hence, if two random variables Y_1 and Y_2 have the same distribution $P_{Y_1} = P_{Y_2}$, then they have the same moments. For instance, if $Y_1 \underset{P}{=} Y_2$ and the expectations $E(Y_1^2)$ and $E(Y_2^2)$ are finite, then $E(Y_1) = E(Y_2)$ and $E(Y_1^2) = E(Y_2^2)$. This allows us to use the term *(central) moments of a distribution* instead of (central) moments of a random variable.   ◁

Variance and standard deviation are the most important parameters describing the *variability* of a random variable. They are defined as follows:

Definition 6.28 [Variance and standard deviation]
Let $Y: (\Omega, \mathcal{A}, P) \to (\mathbb{R}, \overline{\mathcal{B}})$ be a numerical random variable and assume that $E(Y^2) < \infty$. Then the variance of Y is defined by

$$Var(Y) := E([Y - E(Y)]^2), \tag{6.27}$$

and the standard deviation of Y by the positive square root of the variance, that is,

$$SD(Y) := \sqrt{Var(Y)}. \tag{6.28}$$

Box 6.2   Rules of computation for variances.

Let $X, Y: (\Omega, \mathscr{A}, P) \to (\overline{\mathbb{R}}, \overline{\mathscr{B}})$ be random variables with finite second moments and let $\alpha \in \mathbb{R}$. Then,

$$Var(Y) = E(Y^2) - E(Y)^2. \tag{i}$$

$$Var(\alpha + Y) = Var(Y). \tag{ii}$$

$$Var(\alpha \cdot Y) = \alpha^2 \cdot Var(Y). \tag{iii}$$

$$\exists\, \alpha \in \mathbb{R} : Y \underset{P}{=} \alpha \Leftrightarrow Var(Y) = 0. \tag{iv}$$

$$X \underset{P}{=} Y \Rightarrow Var(X) = Var(Y). \tag{v}$$

For $i = 1, \ldots, n$, let the random variables $Y_i: (\Omega, \mathscr{A}, P) \to (\mathbb{R}, \mathscr{B})$ be independent with finite second moments and $\alpha_i \in \mathbb{R}$. Then,

$$Var\left(\sum_{i=1}^{n} \alpha_i \cdot Y_i\right) = \sum_{i=1}^{n} \alpha_i^2 \cdot Var(Y_i). \tag{vi}$$

According to this definition, $Var(Y)$ is the expectation of the squared *mean centered* random variable $Y - E(Y)$. Hence, the variance of Y is the second central moment of Y. Note that variances and standard deviations are nonnegative. The variance of Y is also denoted by σ_Y^2 and the standard deviation by σ_Y. Box 6.2 summarizes some important properties of variances (see Exercise 6.3).

Example 6.29 [Location versus variability] Consider two random variables X, Y: $(\Omega, \mathscr{A}, P) \to (\mathbb{R}, \mathscr{B})$ with $P(X=-1) = P(X=1) = .5$ and $P(Y=-10) = P(Y=10) = .5$. Then $E(X) = E(Y) = 0$ but $Var(X) = 1 \neq Var(Y) = 100$. In contrast, if $P(X=-1) = P(X=1) = .5$ and $P(Y=9) = P(Y=11) = .5$, then $E(X) = 0 \neq E(Y) = 10$ but $Var(X) = Var(Y) = 1$. This illustrates that the expectation describes the 'location' of a random variable [see Rule (ii) of Box 6.1], while the variance is invariant with respect to translations [see Rule (ii) of Box 6.2]. In contrast, the variance describes the 'variability' of a random variable, whereas, in general, the expectation does not.   ◁

Example 6.30 [Variance of an indicator] Let $(\Omega, \mathscr{A}, P)$ be a probability space and let 1_A denote the indicator of $A \in \mathscr{A}$. Then,

$$
\begin{aligned}
Var(1_A) &= E\left(1_A^2\right) - [E(1_A)]^2 &&\text{[Box 6.2, (i)]}\\
&= E(1_A) - [E(1_A)]^2 &&[1_A^2 = 1_A]\\
&= E(1_A) \cdot [1 - E(1_A)]\\
&= P(A) \cdot [1 - P(A)]. &&[(6.4)]
\end{aligned}
\tag{6.29}
$$

According to Equation (6.4), the expectation of the indicator 1_A is $P(A)$, and Equation (6.29) shows that its variance is $P(A) \cdot [1 - P(A)]$. Obviously, the variance of an indicator variable does not contain any information additional to the expectation $E(1_A) = P(A)$. In fact, in this case, $E(1_A)$ contains the full information about the distribution of 1_A. This is not surprising because the distribution of 1_A is completely determined by the single parameter $P(A)$. Unlike

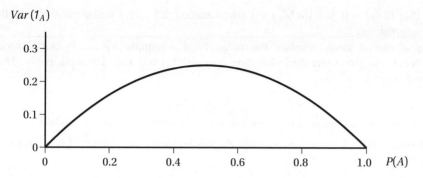

Figure 6.1   Variance of an indicator of an event as a function of its probability.

the expectation, the variance of 1_A *does not* comprise the full information about the distribution of 1_A. For example, $Var(1_A) = .1 \cdot .9 = .09$ if $P(A) = .1$ or $P(A) = .9$. The variance of 1_A has its maximum for $P(A) = 1/2$ and goes to 0 if $P(A)$ approaches 0 or 1 (see Fig. 6.1).   ◁

Example 6.31 [Joe and Ann with randomized assignment – continued]   In the example presented in Table 2.2, X is an indicator variable. Hence, its variance is most easily computed as follows:

$$Var(X) = P(X=1) \cdot [1 - P(X=1)] = .4 \cdot .6 = .24.$$

Similarly, the variance of Y is obtained by

$$Var(Y) = P(Y=1) \cdot [1 - P(Y=1)] = .51 \cdot (1 - .51) = 0.2499.$$

◁

Example 6.32 [Variance and standard error of the sample mean]   Let $Y_1, \ldots, Y_n$ be a sample (see Example 6.23), and $\overline{Y}$ the sample mean [see Eq. (6.23)]. If $E(Y_1{}^2) < \infty$, and

$$\sigma^2 := Var(Y_1) \tag{6.30}$$

denotes the identical variances of the $Y_1, \ldots, Y_n$, then,

$$\sigma_{\overline{Y}}^2 := Var(\overline{Y}) = \frac{\sigma^2}{n} \tag{6.31}$$

(see Exercise 6.4). Hence,

$$\sigma_{\overline{Y}} := \sqrt{\sigma_{\overline{Y}}^2} = SD(\overline{Y}) = \frac{\sigma}{\sqrt{n}}. \tag{6.32}$$

In statistics, $SD(\overline{Y})$ is also denoted $SE(\overline{Y})$ and called the *standard error of the sample mean*.   ◁

Remark 6.33 [Z-transformation]   Every real-valued random variable $Y \colon (\Omega, \mathscr{A}, P) \to (\mathbb{R}, \mathscr{B})$ with finite and positive variance $Var(Y)$ can be transformed by

$$Z_Y := \frac{Y - E(Y)}{SD(Y)}. \tag{6.33}$$

Note that $E(Z_Y) = 0$ and $Var(Z_Y) = 1$ (see Exercise 6.5). This transformation is called the *Z-transformation*.

As a special case, consider the mean $\overline{Y}$ of a sample $Y_1, \dots, Y_n$ with expectation $\mu := E(Y_1)$ and finite standard deviation $\sigma := SD(Y_1) > 0$ (see Example 6.32). Then the Z-transformation of $\overline{Y}$ is

$$Z_{\overline{Y}} = \sqrt{n} \cdot \frac{\overline{Y} - \mu}{\sigma} \tag{6.34}$$

(see Exercise 6.6). The random variable $Z_{\overline{Y}}$ will be used in the Central Limit Theorem (see Th. 8.34). ◁

Remark 6.34 [The expectation minimizes the mean squared error] If $E(Y_2) < \infty$, then the function $MSE(a) := E[(Y - a)^2]$, $a \in \mathbb{R}$, is minimized at $a = E(Y)$. Hence, the expectation $E(Y)$ minimizes the mean squared error (see Exercise 6.7). ◁

Definition 6.35 [Coefficient of variation]
Let $Y: (\Omega, \mathcal{A}, P) \to (\overline{\mathbb{R}}, \overline{\mathcal{B}})$ be a nonnegative numerical random variable. If $E(Y^2)$ is finite and $E(Y) \neq 0$, then,

$$CV(Y) := \frac{SD(Y)}{|E(Y)|} \tag{6.35}$$

is called the *coefficient of variation* of Y.

Remark 6.36 [Properties of the coefficient of variation] This coefficient is a nonnegative dimensionless quantity that expresses the variability of Y in units of its expectation. Note that sometimes the coefficient of variation is defined as $SD(Y)/E(Y)$. If $\alpha \neq 0$, then $CV(\alpha Y) = CV(Y)$ [see Box 6.2 (iii), (6.28), and Box 6.1 (iii)], that is, $CV(Y)$ is invariant under multiplication with (nonzero) constant. In contrast, if $\beta \neq -E(Y)$, then $CV(Y + \beta) = SD(Y)/|E(Y) + \beta|$ [see Box 6.2 (ii), (6.28), and Box 6.1 (ii)], that is, $CV(Y)$ is not invariant under translations. ◁

The following parameter quantifies, in a sense, the deviation of a distribution from symmetry around its expectation.

Definition 6.37 [Skewness]
Let $Y: (\Omega, \mathcal{A}, P) \to (\overline{\mathbb{R}}, \overline{\mathcal{B}})$ be a numerical random variable. If $E(Y^3)$ is finite and $Var(Y) > 0$, then,

$$E\left(\frac{[Y - E(Y)]^3}{SD(Y)^3}\right) \tag{6.36}$$

is called the *skewness* of Y.

6.3   Proofs

Proof of Theorem 6.11

$$E(Y) = \int Y \, dP \qquad\qquad\qquad [(6.1)]$$

$$= \int y \, P_Y(dy) \qquad\qquad\qquad [(3.62)]$$

$$= \int y \cdot f_Y(y) \, \lambda(dy) \qquad\qquad [(3.72), (5.61)]$$

$$= \int_{-\infty}^{\infty} y \cdot f_Y(y) \, dy. \qquad [\text{Riemann integrability of } f_Y, (3.68)]$$

Proof of Lemma 6.18

For all $A' \in \mathscr{A}_X'$ and $g = 1_{A'}$,

$$\int 1_{A'}(x) \, P_X(dx) = P_X(A') \qquad\qquad [(3.9)]$$

$$= P_{X,Y}(A' \times \Omega_Y') \qquad\qquad [(5.21)]$$

$$= \int 1_{A'}(x) \cdot 1_{\Omega_Y'}(y) \, P_{X,Y}[d(x, y)] \qquad [(3.9), (1.38)]$$

$$= \int 1_{A'}(x) \, P_{X,Y}[d(x, y)]. \qquad\qquad [1_{\Omega_Y'}(y) = 1]$$

Now the proposition follows, applying the standard methods of proofs described in Remark 3.30.

Proof of Corollary 6.22

$$E[g(Y)] = E_Y(g) = \int g(y) \, P_Y(dy) \qquad\qquad [(6.13)]$$

$$= \int g(y) \cdot f_Y(y) \, \lambda(dy) \qquad\qquad [(3.27), (5.61)]$$

$$= \int_{-\infty}^{\infty} y \cdot f_Y(y) \, dy. \qquad [\text{Riemann integrability of } y \cdot f_Y(y), (3.68)]$$

Exercises

6.1   Prove the rules of computation of Box 6.1.

6.2   Show: If, for $n \in \mathbb{N}$, the expectation $E(Y^n)$ exists and is finite, then $E(Y^m)$ exists and is finite as well for all $1 \leq m \leq n$.

6.3 Prove the rules of Box 6.2.

6.4 Prove Equations (6.25) and (6.31).

6.5 Let $Y: (\Omega, \mathscr{A}, P) \to (\overline{\mathbb{R}}, \overline{\mathscr{B}})$ be a numerical random variable with finite second moment $E(Y^2)$ and $Var(Y) > 0$. Show that the expectation of $Z := [Y - E(Y)]/SD(Y)$ is 0 and its variance is 1.

6.6 Let $Y_i: (\Omega, \mathscr{A}, P) \to (\mathbb{R}, \mathscr{B})$, $i = 1, 2, \ldots, n$, be a sample with expectations $E(Y_i) = \mu$ and finite and positive variances $Var(Y_i) = \sigma^2$, $i = 1, \ldots, n$. Furthermore, let $\overline{Y} := S/n$ be the arithmetic mean, where $S := \sum_{i=1}^{n} Y_i$, and

$$\overline{Z} := \frac{(\overline{Y} - \mu) \cdot \sqrt{n}}{\sigma}.$$

Show that $E(\overline{Z}) = 0$ and $Var(\overline{Z}) = 1$.

6.7 Assume $E(Y^2) < \infty$ and show that the function $MSE(a) := E[(Y - a)^2]$, $a \in \mathbb{R}$, is minimized at $a = E(Y)$.

Solutions

6.1 Because the expectation of a numerical random variable is defined as an integral, we simply can refer to the corresponding propositions of chapter 3.

(i)   This is Equation (3.8) with $A = \Omega$ and $\mu(A) = P(\Omega) = 1$.

(ii)   This is a special case of Equation (3.34).

(iii)   This is a special case of Equation (3.32).

(iv)   Note that $1_A \cdot 1_B = 1_{A \cap B}$. Hence, this rule follows from Equation (3.8), with $\alpha = 1$.

(v)   This is a special case of Lemma 3.45.

(vi)   If Y_1, Y_2 are nonnegative, then this equation is a special case of Equation (3.65). If Y_1 or Y_2 has a finite expectation, then this equation is a special case of Equation (3.33).

(vii)   This rule follows from Equation (3.34) and complete induction.

(viii)   This is a special case of Lemma 3.47.

(ix)   This is a special case of Theorem 3.48.

(x)   This is a special case of Theorem 6.24.

6.2 Assume that, for $n \in \mathbb{N}$, the expectation $E(Y^n)$ exists and is finite. Furthermore, let $A := \{\omega \in \Omega: |Y(\omega)| > 1\}$. Now, for all m with $1 \le m \le n$,

$$|Y(\omega)^m| \le |Y(\omega)^n|, \quad \forall \, \omega \in A \quad \text{and} \quad |Y(\omega)^m| \le 1, \quad \forall \, \omega \in A^c.$$

Therefore, applying Corollary 3.38, for $1 \leq m \leq n$:

$$E(|Y^m|) = \int |Y^m| \, dP \qquad\qquad [(6.1)]$$

$$= \int_A |Y^m| \, dP + \int_{A^c} |Y^m| \, dP \qquad\qquad [(3.36)]$$

$$\leq \int_A |Y^n| \, dP + \int_{A^c} 1 \, dP \qquad\qquad [\text{Lemma 3.26}]$$

$$\leq \int_A |Y^n| \, dP + 1 \qquad\qquad [(3.30), \, 1_{A^c} \leq 1, \text{Lemma 3.26}]$$

$$< \infty.$$

6.3   Rules (i), (ii), and (iii) are special cases of rules (i), (ii), and (iii) of Box 7.1, which are proved in Exercise 7.3.

(iv) $\Rightarrow$ This proposition follows from (vii) of Box 7.1.

$\Leftarrow$

$$\begin{aligned}
& Var(Y) = 0 \\
\Rightarrow\ & [Y - E(Y)]^2 \underset{P}{=} 0 && [(6.27), \text{Th. 3.43}] \\
\Rightarrow\ & Y - E(Y) \underset{P}{=} 0 && [[Y - E(Y)]^2 = 0 \Leftrightarrow Y - E(Y) = 0] \\
\Rightarrow\ & Y \underset{P}{=} E(Y) \in \mathbb{R}. && [E(Y)^2 < \infty \Rightarrow E(Y) \in \mathbb{R}, \text{Rem. 6.25 (iii)}]
\end{aligned}$$

(v) $X \underset{P}{=} Y \Rightarrow X^2 \underset{P}{=} Y^2$ [see (2.34)]. Hence, $X \underset{P}{=} Y$ implies $E(X) = E(Y)$ and $E(X^2) = E(Y^2)$ (see Cor. 6.17). Therefore, rule (i) of Box 6.2 yields $Var(X) = Var(Y)$.

(vi) This proposition follows from rules (ix) and (vi) of Box 7.1.

6.4   Equation (6.25) can be derived as follows:

$$E(\bar{Y}) = E\left(\frac{1}{n} \cdot \sum_{i=1}^{n} Y_i \right) \qquad\qquad [(6.23)]$$

$$= \frac{1}{n} \cdot \sum_{i=1}^{n} E(Y_i) \qquad\qquad [\text{Box 6.1 (vii)}]$$

$$= \frac{1}{n} \cdot n \cdot \mu = \mu. \qquad [Y_1, \dots, Y_n \text{ are identically distributed, (6.24)}]$$

Equation (6.31) can be derived as follows:

$$Var(\bar{Y}) = Var\left(\frac{1}{n} \cdot \sum_{i=1}^{n} Y_i \right) \qquad\qquad [(6.23)]$$

$$= \frac{1}{n^2} \cdot \sum_{i=1}^{n} Var(Y_i) \qquad\qquad [\underset{P}{\perp\!\!\!\perp} Y_1, \dots, Y_n, \text{Box 6.2 (vi)}]$$

$$= \frac{1}{n^2} \cdot n \cdot \sigma^2 = \frac{\sigma^2}{n}. \qquad [Y_1, \dots, Y_n \text{ are identically distributed, (6.30)}]$$

6.5   Let $\mu := E(Y)$ and $\sigma := SD(Y)$. Then,

$$E(Z) = E\left(\frac{Y-\mu}{\sigma}\right) = E\left(\frac{1}{\sigma} \cdot (Y-\mu)\right) = \frac{1}{\sigma} \cdot E(Y-\mu) \qquad \text{[Box 6.1 (iii)]}$$

$$= \frac{1}{\sigma} \cdot [E(Y) - E(\mu)] = \frac{1}{\sigma} \cdot (\mu - \mu) = 0. \qquad \text{[Box 6.1 (vi), (i)]}$$

$$Var(Z) = Var\left(\frac{Y-\mu}{\sigma}\right) = Var\left(\frac{1}{\sigma} \cdot (Y-\mu)\right)$$

$$= \frac{1}{\sigma^2} \cdot Var(Y) = \frac{1}{\sigma^2} \cdot \sigma^2 = 1. \qquad \text{[Box 6.2 (iii), (ii)]}$$

6.6   Using Equations (6.25) and (6.32), $E(Z_Y) = 0$, and $Var(Z_Y) = 1$ (see Exercise 6.5) yields

$$E(\bar{Z}) = E\left(\frac{\bar{X}-\mu}{\sigma/\sqrt{n}}\right) = 0 \quad \text{and} \quad Var(\bar{Z}) = Var\left(\frac{\bar{X}-\mu}{\sigma/\sqrt{n}}\right) = 1.$$

6.7   For all $a \in \mathbb{R}$, using Box 6.1 (iii) and (ii)

$$E[(Y-a)^2] = E([Y - E(Y) + E(Y) - a]^2)$$
$$= E([Y - E(Y)]^2) + [E(Y) - a]^2 + 2 \cdot E([Y - E(Y)] \cdot [E(Y) - a])$$
$$= E([Y - E(Y)]^2) + [E(Y) - a]^2$$
$$\geq E([Y - E(Y)]^2),$$

and '=' holds if and only if $E(Y) = a$.

7

Linear quasi-regression, covariance, and correlation

Expectation and variance are parameters that describe important properties of a univariate numerical random variable and its distribution. Now we consider *two* numerical random variables, say X and Y, and their joint distribution. In other words, we consider the distribution of the bivariate real-valued random variable $(X, Y) \colon (\Omega, \mathscr{A}, P) \to (\mathbb{R}^2, \mathscr{B}_2)$. We also introduce a new random variable that can be used to describe a specific kind of dependence of Y on X. It is the kind of dependence of Y on X that is represented by the best fitting linear function $\alpha_0 + \alpha_1 X$, 'best fitting' in terms of the minimal mean squared error. This function is the composition of X and the *linear quasi-regression* or the *linear least-squares regression*. *Covariance* and *correlation* are important parameters quantifying the strength of the kind of dependence that can be described by a linear quasi-regression.

7.1 Linear quasi-regression

Remark 7.1 [Implications of finite second moments] Reading the following definition, note that $E(X^2)$, $E(Y^2) < \infty$ implies that $E(X)$, $E(Y)$, and $E(X \cdot Y)$ are finite (see Klenke, 2013, Remark 5.2). Hence, according to Remark 3.42, there is no substantial loss of generality if we additionally assume that X and Y are real-valued. ◁

Definition 7.2 [Linear quasi-regression]
Let $X, Y \colon (\Omega, \mathscr{A}, P) \to (\mathbb{R}, \mathscr{B})$ be two real-valued random variables, and assume $E(X^2)$, $E(Y^2) < \infty$, and $\mathit{Var}(X) > 0$. Then the function $f \colon \mathbb{R} \to \mathbb{R}$ defined by

$$f(x) = \alpha_0 + \alpha_1 x, \quad \forall\, x \in \mathbb{R}, \tag{7.1}$$

Probability and Conditional Expectation: Fundamentals for the Empirical Sciences, First Edition. Rolf Steyer and Werner Nagel.
© 2017 John Wiley & Sons, Ltd. Published 2017 by John Wiley & Sons, Ltd.
Companion website: http://www.probability-and-conditional-expectation.de

where the pair (α_0, α_1) *minimizes the function MSE*: $\mathbb{R}^2 \rightarrow \mathbb{R}$ *with*

$$MSE(a_0, a_1) = E([Y - (a_0 + a_1 X)]^2), \quad \forall\, a_0, a_1 \in \mathbb{R}, \tag{7.2}$$

is called the linear quasi-regression *of Y on X. The composition of X and f is denoted by* $Q_{lin}(Y \mid X)$, *that is,*

$$Q_{lin}(Y \mid X) = f(X) = \alpha_0 + \alpha_1 X. \tag{7.3}$$

Remark 7.3 [Distinguishing between f and $f(X)$] To emphasize, the function $f: \mathbb{R} \rightarrow \mathbb{R}$ is called a linear quasi-regression. In this context, the random variable X is called the *regressor* and Y the *regressand*. Note that f is a function assigning a real number *to all real numbers*. This applies even if X only takes on two different real values. In contrast, the number of different values of the composition $Q_{lin}(Y \mid X) = f(X): \Omega \rightarrow \mathbb{R}$ is smaller than or equal to the number of values of X, provided that X takes on a finite number of values only. ◁

Remark 7.4 [Coefficient of determination] Under the assumptions of Definition 7.2 and $Var(Y) > 0$, we define

$$Q_{Y|X}^2 := \frac{Var[Q_{lin}(Y \mid X)]}{Var(Y)} \tag{7.4}$$

and call it the *coefficient of determination* of the linear quasi-regression $Q_{lin}(Y \mid X)$. In Remark 7.29, this definition is extended to the case in which X is an n-dimensional real-valued random variable. ◁

Example 7.5 [Discrete regressor with three different values] Let X and Y be real-valued random variables on $(\Omega, \mathcal{A}, P)$ with values 1, 2, 3 and 1, 2, respectively. Furthermore, assume that their distribution is specified by

$$P(X{=}1, Y{=}1) = .25, \quad P(X{=}2, Y{=}2) = .5, \quad P(X{=}3, Y{=}1) = .25.$$

Then the linear quasi-regression $f: \mathbb{R} \rightarrow \mathbb{R}$ is specified by

$$f(x) = \alpha_0 + \alpha_1 \cdot x = 1.5 + 0 \cdot x = 1.5, \quad \forall\, x \in \mathbb{R},$$

and the composition of X and f is

$$Q_{lin}(Y \mid X) = \alpha_0 + \alpha_1 \cdot X = 1.5 + 0 \cdot X = 1.5$$

(see Exercise 7.1). The black points in Figure 7.1 represent the three pairs of values of X and Y. All values of the linear quasi-regression are on the horizontal line, which, in this example, is parallel to the x-axis because its slope is 0. The circles on this line represent the values $f(x)$ for $x = 1, 2, 3$, that is, for those values of X with a nonzero probability $P_X(\{x\}) > 0$. In contrast, in this example, $P_X(\{x\}) = 0$, for all $x \in \mathbb{R} \setminus \{1, 2, 3\}$. Nevertheless, as mentioned, a linear quasi-regression f is a function assigning a real number *to all real numbers*. ◁

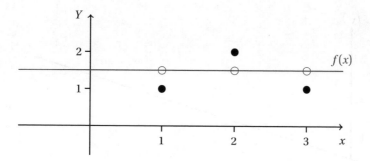

Figure 7.1   Linear quasi-regression.

Remark 7.6 [Linear quasi-regression versus regression] As the term 'linear quasi-regression' suggests, there is also a 'genuine' regression of Y on X (see Def. 10.25), and the two terms are not necessarily identical. As will be explained in more detail in Remark 10.27, the 'genuine' regression is a function $g: \mathbb{R} \to \mathbb{R}$ such that the composition $g(X)$ is X-measurable and minimizes the mean squared error $E([Y - g(X)]^2)$. In contrast to the conditional expectation $E(Y \mid X)$ [see Eq. (10.1)], the composition $Q_{lin}(Y \mid X) = f(X)$ minimizes the function *MSE* specified in Equation (7.2). Hence, f has to be a *linear* function even in those cases in which there are no $a_0, a_1 \in \mathbb{R}$, such that $E(Y \mid X) \underset{P}{=} a_0 + a_1 X$ (see Example 7.5).   ◁

Remark 7.7 [Intercept and slope]   Note that the composition $Q_{lin}(Y \mid X) = f(X) = \alpha_0 + \alpha_1 X$ is a random variable on $(\Omega, \mathcal{A}, P)$ (see Fig. 7.2). The coefficient α_0 is called the *intercept*, and α_1 the *slope* of (the linear quasi-regression) f (see Fig. 7.3). Obviously,

$$f(0) = \alpha_0 + \alpha_1 \cdot 0 = \alpha_0. \tag{7.5}$$

Furthermore, if $x_1, x_2 \in \mathbb{R}$ and $x_2 > x_1$, then,

$$\alpha_1 = \frac{1}{x_2 - x_1} \cdot [f(x_2) - f(x_1)] \tag{7.6}$$

(see Exercise 7.2). Equation (7.6) yields

$$\alpha_1 = f(x_2) - f(x_1), \quad \text{if } x_2 - x_1 = 1. \tag{7.7}$$

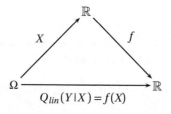

Figure 7.2   The regressor X, the linear quasi-regression f, and their composition $Q_{lin}(Y \mid X) = f(X)$.

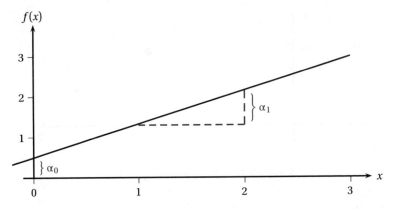

Figure 7.3 Intercept and slope of a linear function $f: \mathbb{R} \to \mathbb{R}$.

These equations justify calling α_0 the *intercept* and α_1 the *slope* of the linear quasi-regression f (see Fig. 7.3). Note that these equations also apply if $P(X=0) = P(X=x_1) = P(X=x_2) = 0$. They even apply if $0, x_1, x_2 \notin X(\Omega)$, because, by definition, f is a function on $\mathbb{R}$.

Figure 7.3 illustrates the intercept and the slope of a linear function such as the linear quasi-regression f. In this figure, $\alpha_0 = .5$ and $\alpha_1 = .85$. If X is discrete, then $Q_{lin}(Y \mid X) = f(X)$ is discrete as well. More precisely, the number of different values of $Q_{lin}(Y \mid X)$ is always smaller than or equal to the number of different values of X. In contrast, the linear quasi-regression $f: \mathbb{R} \to \mathbb{R}$ takes on uncountably many values unless its slope is 0. In this case, its sole value is α_0. ◁

7.2 Covariance

While the variance quantifies the variability of a numerical random variable, the covariance quantifies the degree of covariation of two numerical random variables, that is, the degree to which the two variables vary together in the following sense: If one variable takes on a large value (i.e., large positive deviation from its expectation), then the other one tends to take on a large value as well. Furthermore, if one variable takes on a small value (i.e., large negative deviation from its expectation), then the other one tends to take on a small value, too. In this case, the covariance will be positive. However, the covariance may also be a negative real number. In this case, the two random variables covary in the following sense: If one variable takes on a large value, then the other one tends to take on a small value. Furthermore, if one variable takes on a small value, then the other one tends to take on a large value.

Definition 7.8 [Covariance]
Let $X, Y: (\Omega, \mathcal{A}, P) \to (\overline{\mathbb{R}}, \overline{\mathcal{B}})$ be two numerical random variables with $E(X^2)$, $E(Y^2) < \infty$. Then the covariance of X and Y is defined by

$$Cov(X, Y) := E([X - E(X)] \cdot [Y - E(Y)]). \tag{7.8}$$

Comparing Equations (7.8) and (6.27) to each other shows that the variance is the covariance of a numerical random variable with itself.

Remark 7.9 [Correlated numerical random variables] According to this definition, the *covariance* of X and Y is the expectation of the product of the centered variables $X - E(X)$ and $Y - E(Y)$. Hence, a covariance can be negative, zero, or positive. If the covariance is different from zero, then we say that X and Y are *correlated*; otherwise, we say that they are *uncorrelated*.   ◁

Remark 7.10 [Rules of computation] The most important rules of computation for covariances are summarized in Box 7.1. Proofs are provided in Exercise 7.3. Rule (i) immediately implies

$$E(X \cdot Y) = E(X) \cdot E(Y) + Cov(X, Y). \tag{7.9}$$

Hence, X and Y are uncorrelated if and only if $E(X \cdot Y) = E(X) \cdot E(Y)$ [see Rule (vi)]. Furthermore, this equation and Theorem 6.24 imply that X and Y are uncorrelated if X and Y are independent.

Symmetry of the covariance [see Box 7.1 (v)] yields an alternative way to write Rule (viii) of Box 7.1:

$$Var\left(\sum_{i=1}^{n} \alpha_i Y_i\right) = \sum_{i=1}^{n} \alpha_i^2 \, Var(Y_i) + 2 \cdot \sum_{i=1}^{n-1} \sum_{j=i+1}^{n} \alpha_i \, \alpha_j \, Cov(Y_i, Y_j). \tag{7.10}$$

This equation simplifies to

$$Var\left(\sum_{i=1}^{n} \alpha_i Y_i\right) = \sum_{i=1}^{n} \alpha_i^2 \, Var(Y_i), \tag{7.11}$$

if $Y_1, \ldots, Y_n$ is a sequence of pairwise uncorrelated random variables. Note that independence of $Y_1, \ldots, Y_n$ implies $Cov(Y_i, Y_j) = 0$, for $i \neq j$ [see Rule (vi)].

For $n = 2$, Rule (viii) simplifies to

$$Var(\alpha_1 Y_1 + \alpha_2 Y_2) = \alpha_1^2 \, Var(Y_1) + \alpha_2^2 \, Var(Y_2) + 2\,\alpha_1\,\alpha_2 \, Cov(Y_1, Y_2). \tag{7.12}$$

Similarly, for $n = m = 2$, Rule (ix) simplifies to

$$\begin{aligned} Cov(\alpha_1 X_1 + \alpha_2 X_2, \ \beta_1 Y_1 + \beta_2 Y_2) \\ = \alpha_1 \beta_1 Cov(X_1, Y_1) + \alpha_1 \beta_2 Cov(X_1, Y_2) + \alpha_2 \beta_1 Cov(X_2, Y_1) + \alpha_2 \beta_2 Cov(X_2, Y_2). \end{aligned} \tag{7.13}$$
◁

Remark 7.11 [Covariance of indicators] For $A, B \in \mathcal{A}$, Rule (i) of Box 7.1 and Equations (1.33) and (6.4) yield

$$Cov(1_A, 1_B) = E(1_A \cdot 1_B) - E(1_A) \cdot E(1_B) \tag{7.14}$$
$$= P(A \cap B) - P(A) \cdot P(B). \tag{7.15}$$
◁

The following theorem helps to clarify the relationship between the covariance and the variances of two numerical random variables X and Y.

Box 7.1   Rules of computation for covariances.

Let X, Y be numerical random variables on the probability space $(\Omega, \mathscr{A}, P)$ with $E(X^2)$, $E(Y^2) < \infty$. Furthermore, let $\alpha, \beta \in \mathbb{R}$. Then,

$$Cov(X, Y) = E(X \cdot Y) - E(X) \cdot E(Y). \tag{i}$$

$$Cov(\alpha + X, \beta + Y) = Cov(X, Y). \tag{ii}$$

$$Cov(\alpha X, \beta Y) = \alpha \beta Cov(X, Y). \tag{iii}$$

$$Cov(X, X) = Var(X). \tag{iv}$$

$$Cov(X, Y) = Cov(Y, X). \tag{v}$$

$$X \underset{P}{\perp\!\!\!\perp} Y \Rightarrow Cov(X, Y) = 0. \tag{vi}$$

$$\exists\, \alpha \in \mathbb{R}\colon X \underset{P}{=} \alpha \Rightarrow Cov(X, Y) = 0. \tag{vii}$$

If Y_i are real-valued random variables on the probability space $(\Omega, \mathscr{A}, P)$ with $E(Y_i^2) < \infty$ and $\alpha_i \in \mathbb{R}$, $i = 1, \ldots, n$, then,

$$Var\left(\sum_{i=1}^{n} \alpha_i Y_i\right) = \sum_{i=1}^{n} \alpha_i^2 \, Var(Y_i) + \sum_{i=1}^{n} \sum_{j=1, i \neq j}^{n} \alpha_i \alpha_j \, Cov(Y_i, Y_j). \tag{viii}$$

If X_i, Y_j are real-valued random variables on the probability space $(\Omega, \mathscr{A}, P)$ with $E(X_i^2)$, $E(Y_j^2) < \infty$, and $\alpha_i, \beta_j \in \mathbb{R}$, for all $i = 1, \ldots, n$ and $j = 1, \ldots, m$, then,

$$Cov\left(\sum_{i=1}^{n} \alpha_i X_i, \sum_{j=1}^{m} \beta_j Y_j\right) = \sum_{i=1}^{n} \sum_{j=1}^{m} \alpha_i \beta_j \, Cov(X_i, Y_j). \tag{ix}$$

If $X_1 \underset{P}{=} X_2$ and $E(Y^2)$, $E(X_1^2)$, $E(X_2^2) < \infty$, then,

$$Cov(Y, X_1) = Cov(Y, X_2). \tag{x}$$

Theorem 7.12 [Cauchy-Schwarz inequality]

If $X, Y\colon (\Omega, \mathscr{A}, P) \to (\overline{\mathbb{R}}, \overline{\mathscr{B}})$ are random variables with $E(X^2)$, $E(Y^2) < \infty$, then,

$$Cov(X, Y)^2 \leq Var(X) \cdot Var(Y). \tag{7.16}$$

Furthermore, if $Cov(X, Y) \neq 0$, then,

$$Cov(X, Y)^2 = Var(X) \cdot Var(Y) \Leftrightarrow \exists\, a, b \in \mathbb{R}\colon Y \underset{P}{=} a + bX. \tag{7.17}$$

(Proof p. 240)

Remark 7.13 [Squared weighted sum of random variables]   If $X, Y : (\Omega, \mathcal{A}, P) \to (\mathbb{R}, \mathcal{B})$ are random variables with $E(X^2)$, $E(Y^2) < \infty$ and $\alpha, \beta \in \mathbb{R}$, then $E[(\alpha X + \beta Y)^2] < \infty$ (see Exercise 7.4).   ◁

In Theorem 7.14, we revisit the linear quasi-regression, studying three equivalent propositions. The first of these propositions deals with the residual variable $\epsilon := Y - f(X)$, where f is the linear quasi-regression of Y on X. Note that this residual is not necessarily identical to the residual with respect to a conditional expectation that will be treated in chapters 9 to 11.

Theorem 7.14 [Three characterizations of the linear quasi-regression]
Let $X, Y : (\Omega, \mathcal{A}, P) \to (\mathbb{R}, \mathcal{B})$ be two real-valued random variables with $E(X^2)$, $E(Y^2) < \infty$, and $Var(X) > 0$. Furthermore, let $\alpha_0, \alpha_1 \in \mathbb{R}$, $f(X) = \alpha_0 + \alpha_1 X$ be the composition of X and $f : \mathbb{R} \to \mathbb{R}$, and define $\epsilon := Y - f(X)$. Then the following three propositions are equivalent to each other:

(i) $E(\epsilon) = Cov(X, \epsilon) = 0.$

(ii) $\alpha_0 = E(Y) - \alpha_1 E(X)$ and $\alpha_1 = \dfrac{Cov(X, Y)}{Var(X)}.$

(iii) $f(X) = Q_{lin}(Y \mid X)$, *that is, α_0, α_1 minimize the function $MSE(a_0, a_1)$ defined by Equation (7.2).*

(Proof p. 241)

Remark 7.15 [Uniqueness]   Suppose that the assumptions of Theorem 7.14 hold and $f(X) = Q_{lin}(Y \mid X)$. Then proposition (ii) of this theorem implies that the coefficients α_0 and α_1 are uniquely defined. Because $Q_{lin}(Y \mid X) = \alpha_0 + \alpha_1 X$, the linear quasi-regression $f : \mathbb{R} \to \mathbb{R}$ is uniquely defined as well.   ◁

Remark 7.16 [Relationship between slope and covariance]   According to proposition (ii), a *zero covariance* between X and Y implies that the slopes of the linear quasi-regressions of Y on X and of X on Y are zero, provided that $Var(Y) > 0$. A *negative covariance* implies that the slopes of the linear quasi-regressions of Y on X and of X on Y are negative, and a *positive covariance* implies that the slopes of the linear quasi-regressions of Y on X and of X on Y are positive.   ◁

Example 7.17 [Discrete regressor with three different values – continued]   In Example 7.5, we specified the distribution of (X, Y). Now we use the equations in Theorem 7.14 (ii) in order to compute the coefficients α_0 and α_1. For this purpose, we have to compute the expectations of X and Y, the variance of X, and the covariance of X and Y. Hence, with $n = 3$, $m = 2$, $x_1 = 1, x_2 = 2, x_3 = 3$, and $y_1 = 1, y_2 = 2$,

$$E(X) = \sum_{i=1}^{n} x_i \cdot P(X = x_i) = 1 \cdot \frac{1}{4} + 2 \cdot \frac{1}{2} + 3 \cdot \frac{1}{4} = 2,$$

$$E(Y) = \sum_{i=1}^{m} y_i \cdot P(Y = y_i) = 1 \cdot \frac{1}{2} + 2 \cdot \frac{1}{2} = \frac{3}{2},$$

$$Var(X) = E(X^2) - E(X)^2 = \sum_{i=1}^{n} x_i^2 \cdot P(X=x_i) - E(X)^2 \qquad \text{[Box 6.2 (i), (6.19)]}$$

$$= 1^2 \cdot \frac{1}{4} + 2^2 \cdot \frac{1}{2} + 3^2 \cdot \frac{1}{4} - 2^2 = \frac{1}{2},$$

$$Cov(X, Y) = E(X \cdot Y) - E(X) \cdot E(Y) \qquad \text{[Box 7.1 (i)]}$$

$$= \sum_{i=1}^{n} \sum_{j=1}^{m} x_i \cdot y_j \cdot P\big((X, Y)=(x_i, y_j)\big) - E(X) \cdot E(Y) \qquad \text{[(6.3)]}$$

$$= 1 \cdot 1 \cdot \frac{1}{4} + 2 \cdot 2 \cdot \frac{1}{2} + 3 \cdot 1 \cdot \frac{1}{4} - 2 \cdot \frac{3}{2}$$

$$= \frac{1}{4} + \frac{8}{4} + \frac{3}{4} - \frac{12}{4} = 0.$$

Using the equations in Theorem 7.14 (ii) yields $\alpha_1 = Cov(X, Y)/Var(X) = \frac{0}{1/2} = 0$ and $\alpha_0 = E(Y) - \alpha_1 E(X) = \frac{3}{2} - 0 \cdot 2 = 1.5$, the same result as obtained in Exercise 7.1, in which we minimize the function $MSE(a_0, a_1)$. ◁

7.3 Correlation

As mentioned, the covariance between two numerical random variables quantifies the strength of the dependence that can be described by a linear quasi-regression. However, the covariance is not invariant under multiplication with constants [scale transformations; see Box 7.1 (iii)] of the random variables involved. In contrast, the correlation, which quantifies the strength of the same kind of dependence, *is invariant* under scale transformations (see Rem. 7.22).

Definition 7.18 [Correlation]
Let $X, Y: (\Omega, \mathcal{A}, P) \to (\overline{\mathbb{R}}, \overline{\mathcal{B}})$ be two numerical random variables with $E(X^2)$, $E(Y^2) < \infty$. Then the *correlation of* X and Y is defined by

$$Corr(X, Y) := \begin{cases} \dfrac{Cov(X, Y)}{SD(X) \cdot SD(Y)}, & if \ SD(X), \ SD(Y) > 0 \\[2mm] 0, & otherwise. \end{cases} \qquad (7.18)$$

Remark 7.19 [Correlation of a random variable with itself] Assume $Var(X) > 0$. Because $Cov(X, X) = Var(X) = SD(X) \cdot SD(X)$, Equation (7.18) implies that $Corr(X, X) = 1$. Similarly, because $Cov(X, -X) = -Var(X) = -SD(X) \cdot SD(X)$, Equation (7.18) implies that $Corr(X, -X) = -1$. ◁

Remark 7.20 [Range of the correlation] An implication of (7.16) is

$$-1 \leq Corr(X, Y) \leq 1, \qquad (7.19)$$

provided that $Corr(X, Y)$ exists, that is, provided that the assumptions hold under which the correlation is defined (see Exercise 7.5).   ◁

Remark 7.21 [Correlation and Z-transformed variables] If the standard deviations of X and Y are positive, then the correlation is also the expectation of the product of the Z-transformed variables [see Eq. (6.33)], that is,

$$Corr(X, Y) = E\left(\frac{X - E(X)}{SD(X)} \cdot \frac{Y - E(Y)}{SD(Y)}\right) \tag{7.20}$$

(see Exercise 7.6).   ◁

Remark 7.22 [An invariance property of the correlation] The correlation of linear transformations of X and Y is

$$Corr(a_0 + a_1 X, b_0 + b_1 Y) = \begin{cases} Corr(X, Y), & \text{if } a_1 \cdot b_1 > 0 \\ -Corr(X, Y), & \text{if } a_1 \cdot b_1 < 0 \\ 0, & \text{if } a_1 \cdot b_1 = 0, \end{cases} \tag{7.21}$$

where $a_0, a_1, b_0, b_1 \in \mathbb{R}$ (see Exercise 7.7). This equation implies that the correlation is invariant (up to change of signs) under linear transformations, which include *translations* ($a_1 = 1$ and $b_1 = 1$) and *scale transformations* ($a_0 = b_0 = 0$ and $a_1, b_1 \neq 0$).   ◁

Theorem 7.12 implies the following corollary about the cases $Corr(X, Y) = 1$ or $Corr(X, Y) = -1$, that is, the 'perfect' correlation.

Corollary 7.23 [Perfect correlation of two random variables]
Let the assumptions of Definition 7.8 be satisfied, and suppose that $Var(X), Var(Y) > 0$. Then $|Corr(X, Y)| = 1$ if and only if there are $a_0, a_1 \in \mathbb{R}$, $a_1 \neq 0$, such that $Y \underset{P}{=} a_0 + a_1 X$. In this case,

$$Corr(X, Y) = \begin{cases} 1, & \text{if } a_1 > 0 \\ -1, & \text{if } a_1 < 0. \end{cases} \tag{7.22}$$

Remark 7.24 [Covariance and standard deviations] Let $Y \underset{P}{=} a_0 + a_1 X$, $a_0, a_1 \in \mathbb{R}$. If $a_1 \geq 0$, then $Cov(X, Y) = SD(X) \cdot SD(Y)$. If $a_1 < 0$, then $Cov(X, Y) = -SD(X) \cdot SD(Y)$ (see Exercise 7.8).   ◁

Remark 7.25 [Slope of a linear quasi-regression and correlation] If α_1 is the slope of the linear quasi-regression of Y on X (see Def. 7.2), then,

$$\alpha_1 = \frac{Cov(X, Y)}{Var(X)} = Corr(X, Y) \cdot \frac{SD(Y)}{SD(X)} \tag{7.23}$$

[see proposition (ii) of Th. 7.14]. This equation shows that the slope α_1 of the linear quasi-regression has the same sign as the covariance and the correlation. The size of the absolute value of α_1 depends on the ratio $SD(Y)/SD(X)$ of the standard deviations. The smaller the standard deviation of X compared to the standard deviation of Y, the larger the absolute value of α_1. Furthermore, given a fixed variance $Var(X)$, this equation also shows that α_1 is proportional to $Cov(X, Y)$ and $Corr(X, Y)$. In this sense, all three parameters α_1, $Cov(X, Y)$, and $Corr(X, Y)$ quantify the strength of the dependence of Y on X described by a linear quasi-regression. Note, however, that α_1 and $Cov(X, Y)$ are not invariant under scale transformations of X and Y. This can be seen in the following equation for the slope α_1^* of the linear quasi-regression of bY on aX, $a, b \in \mathbb{R}$, $a, b \neq 0$:

$$\alpha_1^* = \frac{Cov(aX, bY)}{Var(aX)} = Corr(aX, bY) \cdot \frac{SD(bY)}{SD(aX)} = \frac{b}{a} \cdot \alpha_1. \tag{7.24}$$

Hence, the slope of the linear quasi-regression of bY on aX is identical to the slope of the linear quasi-regression of Y on X multiplied by $\frac{b}{a}$. In contrast, the slope of the linear quasi-regression is invariant under translations $c + X$, $d + Y$, $c, d \in \mathbb{R}$ (see Exercise 7.9).   ◁

Remark 7.26 [Correlation and coefficient of determination]   Under the assumptions of Definition 7.18 and $Var(X)$, $Var(Y) > 0$,

$$Q_{Y|X}^2 = \frac{Var[Q_{lin}(Y \mid X)]}{Var(Y)} = \frac{Var[\alpha_0 + \alpha_1 Var(X)]}{Var(Y)} = \frac{\alpha_1^2 Var(X)}{Var(Y)} \tag{7.25}$$

$$= Corr(X, Y)^2.$$

◁

Example 7.27 [Joe and Ann with randomized assignment – continued]   Consider the example presented in Table 2.2. In this example, the covariance of X and Y is most easily computed using

$$Cov(X, Y) = E(X \cdot Y) - E(X) \cdot E(Y) \qquad \text{[Box 7.1 (i)]}$$

$$= \sum_{(x,y)} (x \cdot y) \cdot P(X=x, Y=y) - P(X=1) \cdot P(Y=1) \qquad [(6.15)]$$

$$= P(X=1, Y=1) - P(X=1) \cdot P(Y=1)$$

$$= (.16 + .08) - .4 \cdot .51 = .036,$$

where $P(X=1) = E(X) = .4$ and $P(Y=1) = E(Y) = .51$ have been computed in Example 6.7. Note that $\sum_{(x,y)}$ is the sum over all pairs (x, y) of values of X and Y. In this example, there are four such pairs, only one of which, namely $(1, 1)$, yields a product $x \cdot y \neq 0$. Using the results of Example 6.31 on the variances of X and Y yields the correlation

$$Corr(X, Y) = \frac{Cov(X, Y)}{SD(X) \cdot SD(Y)} = \frac{.036}{\sqrt{.24} \cdot \sqrt{.2499}} \approx .147$$

Hence, treatment and outcome variables have a positive correlation. This is in accordance with comparing the conditional probability of success given treatment, $P(C \mid B) = .6$, to the conditional probability of success given no treatment, $P(C \mid B^c) = .45$ (see Example 4.16).

In this example,

$$Q_{lin}(Y \mid X) = \alpha_0 + \alpha_1 \cdot X$$

$$= \left(E(Y) - \frac{Cov(X, Y)}{Var(X)} \cdot E(X) \right) + \frac{Cov(X, Y)}{Var(X)} \cdot X \qquad \text{[Th. 7.14 (ii)]}$$

$$= \left(.51 - \frac{.036}{.24} \cdot .4 \right) + \frac{.036}{.24} \cdot X$$

$$= .45 + .15 \cdot X,$$

and the linear quasi-regression $f\colon \mathbb{R} \to \mathbb{R}$ of Y on X is specified by

$$f(x) = .45 + .15 \cdot x, \quad \forall \ x \in \mathbb{R}.$$

$\triangleleft$

7.4   Expectation vector and covariance matrix

7.4.1   Random vector and random matrix

Let $X = (X_1, \ldots, X_n)$ be an n-variate numerical random variable on a probability space $(\Omega, \mathcal{A}, P)$. In order to utilize matrix algebra, we consider the *column random vector*

$$x := \begin{bmatrix} X_1 \\ \vdots \\ X_n \end{bmatrix},$$

that is, the column vector of the random variables $X_1, \ldots, X_n$. Correspondingly, we consider the *row random vector* $x' := [X_1, \ldots, X_n]$, the *transpose* of x.

In this section, we also consider a *random matrix*, that is, a matrix

$$X := \begin{bmatrix} X_{11} & X_{12} & \cdots & X_{1m} \\ X_{21} & X_{22} & \cdots & X_{2m} \\ \vdots & \vdots & \ddots & \vdots \\ X_{n1} & X_{n2} & \cdots & X_{nm} \end{bmatrix} \qquad (7.26)$$

of type $n \times m$ of numerical random variables X_{ij} on a probability space $(\Omega, \mathcal{A}, P)$, $i = 1, \ldots, n$, $j = 1, \ldots, m$. Such a random matrix is an $n \times m$-array of an $(n \cdot m)$-variate random variable (see section 5.3).

7.4.2   Expectation of a random vector and a random matrix

The *expectation of a (row or column) random vector* is defined as the (row or column) vector of the expectations of its components, that is,

$$E(x') := [E(X_1), \ldots, E(X_n)] \qquad (7.27)$$

and $E(x) := [E(X_1), \ldots, E(X_n)]'$, provided that the expectations exist. Hence,

$$E(x') = (E(x))'. \qquad (7.28)$$

Box 7.2   Rules of computation for expectations of random matrices.

Let $X = (X_1, \ldots, X_n)$ be an n-variate and $Y = (Y_1, \ldots, Y_m)$ be an m-variate real-valued random variable on a probability space $(\Omega, \mathcal{A}, P)$ such that the expectations of X_i and Y_j are finite, for all $i = 1, \ldots, n$, $j = 1, \ldots, m$. Furthermore, let $x = [X_1, \ldots, X_n]'$ and $y = [Y_1, \ldots, Y_m]'$ denote column vectors, let $\mathbf{a} = [a_1, \ldots, a_n]'$ denote a column vector of real numbers, and let $\mathbf{A}$ and $\mathbf{B}$ be matrices of types $k \times n$ and $k \times m$, respectively, each of their components being a real number. Furthermore, let $\mathbf{C}$ and $\mathbf{D}$ be matrices of real numbers of types $l \times n$ and $r \times m$, respectively. Then,

$$x \underset{P}{=} \mathbf{a} \;\Rightarrow\; E(x) = \mathbf{a}. \tag{i}$$

$$E(\mathbf{a} + x) = \mathbf{a} + E(x). \tag{ii}$$

$$E(\mathbf{a}'x) = \mathbf{a}'E(x) = E(x)'\,\mathbf{a} = E(x'\mathbf{a}). \tag{iii}$$

$$E(\mathbf{A}\,x) = \mathbf{A}E(x). \tag{iv}$$

$$E(\mathbf{A}\,x + \mathbf{B}\,y) = \mathbf{A}E(x) + \mathbf{B}E(y). \tag{v}$$

Let X be an $(n \times k)$-matrix and Y an $(m \times k)$-matrix of real-valued random variables on $(\Omega, \mathcal{A}, P)$, all with finite second moments. Then,

$$X \underset{P}{=} \mathbf{A}' \;\Rightarrow\; E(X) = \mathbf{A}'. \tag{vi}$$

$$E(\mathbf{A}' + X) = \mathbf{A}' + E(X). \tag{vii}$$

$$E(\mathbf{C}X) = \mathbf{C}\,E(X). \tag{viii}$$

$$E(\mathbf{C}\,XY'\,\mathbf{D}') = \mathbf{C}\,E(XY')\,\mathbf{D}'. \tag{ix}$$

Analogously to Equation (7.27), the *expectation of an $n \times m$-random matrix* is defined as the $n \times m$-matrix of the expectations of its components, that is,

$$E\begin{bmatrix} X_{11} & X_{12} & \ldots & X_{1m} \\ X_{21} & X_{22} & \ldots & X_{2m} \\ \vdots & \vdots & \ddots & \vdots \\ X_{n1} & X_{n2} & \ldots & X_{nm} \end{bmatrix} := \begin{bmatrix} E(X_{11}) & E(X_{12}) & \ldots & E(X_{1m}) \\ E(X_{21}) & E(X_{22}) & \ldots & E(X_{2m}) \\ \vdots & \vdots & \ddots & \vdots \\ E(X_{n1}) & E(X_{n2}) & \ldots & E(X_{nm}) \end{bmatrix}, \tag{7.29}$$

provided that the expectations exist. Obviously, if X' denotes the transpose of the matrix X, then,

$$E(X') = (E(X))'. \tag{7.30}$$

In Box 7.2, we present some rules of computation for the expectations of random vectors and random matrices (for proofs, see Exercise 7.10). In this box, we use

$$\mathbf{a}'x := \sum_{i=1}^{n} a_i \cdot X_i, \tag{7.31}$$

the *inner product* of an n-vector $\mathbf{a} = [a_1, \ldots, a_n]'$ of real numbers and the random vector $\mathbf{x}$. Correspondingly,

$$\mathbf{A}\,x := \begin{bmatrix} \mathbf{a}_1' \, x \\ \vdots \\ \mathbf{a}_k' \, x \end{bmatrix}, \tag{7.32}$$

where $\mathbf{a}_l'$, $l = 1, \ldots, k$, denotes the lth row of the $(k \times n)$-matrix $\mathbf{A}$ of real numbers.

7.4.3   Covariance matrix of two multivariate random variables

Now we consider two multivariate numerical random variables $X = (X_1, \ldots, X_n)$ and $Y = (Y_1, \ldots, Y_m)$ on a probability space $(\Omega, \mathscr{A}, P)$. In particular, we assume that the second moments of all these random variables are finite and focus on their covariance matrix, again utilizing the representation of X and Y as row or column vectors that has been introduced at the beginning of section 7.4.1.

Note that $[x - E(x)]\,[y - E(y)]'$ is an $n \times m$-matrix of the random variables

$$[X_i - E(X_i)] \cdot [Y_j - E(Y_j)], \quad i = 1, \ldots, n, \; j = 1, \ldots, m.$$

Therefore, using (7.29), the *covariance matrix* Σ_{xy} is defined by

$$\Sigma_{xy} := E([x - E(x)]\,[y - E(y)]'). \tag{7.33}$$

In other words, the covariance matrix of x and y is the matrix of covariances, that is,

$$\Sigma_{xy} = \begin{bmatrix} \sigma_{X_1 Y_1} & \sigma_{X_1 Y_2} & \cdots & \sigma_{X_1 Y_m} \\ \sigma_{X_2 Y_1} & \sigma_{X_2 Y_2} & \cdots & \sigma_{X_2 Y_m} \\ \vdots & \vdots & \ddots & \vdots \\ \sigma_{X_n Y_1} & \sigma_{X_n Y_2} & \cdots & \sigma_{X_n Y_m} \end{bmatrix}, \tag{7.34}$$

where $\sigma_{X_i Y_j} := Cov(X_i, Y_j) = E([X_i - E(X_i)] \cdot [Y_j - E(Y_j)])$, $i = 1, \ldots, n$, $j = 1, \ldots, m$. If we assume that the second moments of the random variables X_i and Y_j are finite, then all covariances $Cov(X_i, Y_j)$ are finite as well, and we say that Σ_{xy} *exists*.

If we consider a univariate random variable Y, then $y = [Y]$ is also a vector consisting of a single component, the random variable Y. In this special case, Σ_{xy} is a matrix of type $n \times 1$, the column vector

$$\Sigma_{xy} = \begin{bmatrix} \sigma_{X_1 Y} \\ \vdots \\ \sigma_{X_n Y} \end{bmatrix}.$$

Another special case is $x = y$. The covariance matrix Σ_{xx} of x and x is called the *variance–covariance matrix of x* (and of X). Hence,

$$\Sigma_{xx} := E([x - E(x)]\,[x - E(x)]') \tag{7.35}$$

Box 7.3   Rules of computation for covariance matrices.

Let $X = (X_1, \ldots, X_n)$ be an n-variate and $Y = (Y_1, \ldots, Y_m)$ be an m-variate real-valued random variable on a probability space $(\Omega, \mathcal{A}, P)$ such that the second moments of X_i and Y_j are finite, for all $i = 1, \ldots, n, j = 1, \ldots, m$. Furthermore, let $x = [X_1, \ldots, X_n]'$ and $y = [Y_1, \ldots, Y_m]'$ denote column vectors, let $a = [a_1, \ldots, a_n]'$ and $b = [b_1, \ldots, b_m]'$ denote column vectors of real numbers, and let $\mathbf{A}$ and $\mathbf{B}$ be matrices of types $k \times n$ and $k \times m$, respectively, each of their components being a real number. Finally, let $\mathbf{O}$ denote the $(n \times m)$-null matrix. Then,

$$\Sigma_{xy} = E(x\,y') - E(x)\,E(y'). \tag{i}$$

$$\Sigma_{a+x, b+y} = \Sigma_{xy}. \tag{ii}$$

$$\Sigma_{\mathbf{A}x, \mathbf{B}y} = \mathbf{A}\,\Sigma_{xy}\,\mathbf{B}'. \tag{iii}$$

$$\Sigma_{xy} = \Sigma'_{yx}. \tag{iv}$$

$$X \underset{P}{\perp\!\!\!\perp} Y \Rightarrow \Sigma_{xy} = \mathbf{O}. \tag{v}$$

$$x \underset{P}{=} a \Rightarrow \Sigma_{xy} = \mathbf{O}. \tag{vi}$$

Additionally, let $w = [W_1, \ldots, W_r]'$ and $z = [Z_1, \ldots, Z_s]'$ be real-valued random column vectors on $(\Omega, \mathcal{A}, P)$ such that all their components have finite second moments. Furthermore, let $\mathbf{C}$ and $\mathbf{D}$ be matrices of real numbers of type $l \times r$ and $l \times s$. Then,

$$\Sigma_{\mathbf{A}x+\mathbf{B}y,\ \mathbf{C}w+\mathbf{D}z} = \mathbf{A}\,\Sigma_{xw}\,\mathbf{C}' + \mathbf{A}\,\Sigma_{xz}\,\mathbf{D}' + \mathbf{B}\,\Sigma_{yw}\,\mathbf{C}' + \mathbf{B}\,\Sigma_{yz}\,\mathbf{D}'. \tag{vii}$$

$$n = s \text{ and } x \underset{P}{=} z \Rightarrow \Sigma_{xy} = \Sigma_{zy}. \tag{viii}$$

and

$$\Sigma_{xx} = \begin{bmatrix} \sigma_{X_1}^2 & \sigma_{X_1 X_2} & \cdots & \sigma_{X_1 X_n} \\ \sigma_{X_2 X_1} & \sigma_{X_2}^2 & \cdots & \sigma_{X_2 X_n} \\ \vdots & \vdots & \ddots & \vdots \\ \sigma_{X_n X_1} & \sigma_{X_n X_2} & \cdots & \sigma_{X_n}^2 \end{bmatrix}. \tag{7.36}$$

The diagonal components of the matrix Σ_{xx} are the variances of the variables $X_1, \ldots, X_n$, because $\sigma_{X_i X_i} := Cov(X_i, X_i) = Var(X_i) = \sigma_{X_i}^2, i = 1, \ldots, n$.

In Box 7.3, we present some rules of computation for covariance matrices. They are proved in Exercise 7.11.

7.5   Multiple linear quasi-regression

In the following definition, we generalize the concept of a linear quasi-regression considering a multivariate regressor $X = (X_1, \ldots, X_n)$. We use the notation $x = [X_1, \ldots, X_n]'$ to denote the column vector of X, $\beta' = [\beta_1, \ldots, \beta_n]$ for the row vector of the real numbers $\beta_1, \ldots, \beta_n$, and $b' = [b_1, \ldots, b_n]$ for the row vector of the real numbers $b_1, \ldots, b_n$.

Definition 7.28 [Multiple linear quasi-regression]
Let $X_i: (\Omega, \mathscr{A}, P) \to (\mathbb{R}, \mathscr{B})$, $i = 1, \dots, n$, and $Y: (\Omega, \mathscr{A}, P) \to (\mathbb{R}, \mathscr{B})$ be real-valued random variables, define $X := (X_1, \dots, X_n)$, and assume $E(X_i^2)$, $E(Y^2) < \infty$, $i = 1, \dots, n$, that the inverse Σ_{xx}^{-1} exists. Finally, define the function $f: \mathbb{R}^n \to \mathbb{R}$ by

$$f(x) = \beta_0 + \sum_{i=1}^{n} \beta_i x_i, \quad \forall x = (x_1, \dots, x_n) \in \mathbb{R}^n, \tag{7.37}$$

where β_0, $\boldsymbol{\beta} = [\beta_1, \dots, \beta_n]'$ minimize the function $MSE: \mathbb{R}^{n+1} \to \mathbb{R}$ with

$$MSE(b_0, \mathbf{b}) = E([Y - (b_0 + \mathbf{b}'x)]^2), \quad \forall (b_0, \mathbf{b}) \in \mathbb{R}^{n+1}. \tag{7.38}$$

Then f is called the linear quasi-regression of Y on $X_1, \dots, X_n$. The composition of X and f is denoted by $Q_{lin}(Y \mid X)$ or $Q_{lin}(Y \mid X_1, \dots, X_n)$, that is,

$$Q_{lin}(Y \mid X) := f(X) = \beta_0 + \boldsymbol{\beta}'x = \beta_0 + \sum_{i=1}^{n} \beta_i X_i. \tag{7.39}$$

Remark 7.29 [Coefficient of determination]   Let the assumptions of Definition 7.28 hold, and assume $Var(Y) > 0$. Then,

$$Q_{Y|X}^2 := \frac{Var[Q_{lin}(Y \mid X)]}{Var(Y)} \tag{7.40}$$

is called the *coefficient of determination of the linear quasi-regression of Y on X.*   ◁

In the following theorem, we generalize Theorem 7.14 considering a multivariate real-valued regressor $X = (X_1, \dots, X_n)$. In this theorem, $\boldsymbol{\Sigma}_{x\epsilon}$ denotes the covariance vector of x and ϵ, which is defined by

$$\epsilon := Y - Q_{lin}(Y \mid X_1, \dots, X_n) \tag{7.41}$$

and called the *residual of Y with respect to its linear quasi-regression on X.*

Theorem 7.30 [Charactizations of the multiple linear quasi-regression]
Let $X_1, \dots, X_n, Y: (\Omega, \mathscr{A}, P) \to (\mathbb{R}, \mathscr{B})$ be real-valued random variables, $x = [X_1, \dots, X_n]'$, and assume $E(X_i^2)$, $E(Y^2) < \infty$ for all $i = 1, \dots, n$, and that the inverse Σ_{xx}^{-1} exists. Furthermore, let $f: \mathbb{R}^n \to \mathbb{R}$, with

$$f(X_1, \dots, X_n) = \beta_0 + \boldsymbol{\beta}'x, \quad \beta_0 \in \mathbb{R}, \boldsymbol{\beta} \in \mathbb{R}^n, \tag{7.42}$$

be the composition of $(X_1, \ldots, X_n)$ and f, and define $\epsilon := Y - f(X_1, \ldots, X_n)$. Then the following three propositions are equivalent to each other:

(i) $E(\epsilon) = 0$ and $\Sigma_{x\epsilon} = 0$.

(ii) $\beta_0 = E(Y) - \beta' E(x)$ and $\beta = \Sigma_{xx}^{-1} \Sigma_{xy}$.

(iii) $f(X_1, \ldots, X_n) = Q_{lin}(Y \mid X_1, \ldots, X_n)$.

(Proof p. 243)

Remark 7.31 [Uniqueness] Suppose that the assumptions of Theorem 7.30 hold and $f(X_1, \ldots, X_n) = Q_{lin}(Y \mid X)$. Then proposition (ii) of this theorem implies that the coefficients $\beta_0, \beta_1, \ldots, \beta_n$ are uniquely defined. Because $f(X_1, \ldots, X_n) = \beta_0 + \beta' x = Q_{lin}(Y \mid X)$, the linear quasi-regression f and $Q_{lin}(Y \mid X)$ are uniquely defined as well.   ◁

Corollary 7.32 [No correlation between linear quasi-regression and its residual]
Under the assumptions of Definition 7.28,

$$Cov\,[Q_{lin}(Y \mid X),\ \epsilon] = 0 \tag{7.43}$$

and

$$Var[Q_{lin}(Y \mid X)] = Cov\,[Y, Q_{lin}(Y \mid X)]. \tag{7.44}$$

(Proof p. 244)

Remark 7.33 [Correlation between linear quasi-regression and its regressand] Under the assumptions of Definition 7.28 and $Var(Y) > 0$,

$$Q_{Y|X}^2 = Corr\,[Y, Q_{lin}(Y \mid X)]^2 \tag{7.45}$$

(see Exercise 7.12).   ◁

7.6 Proofs

Proof of Theorem 7.12

Suppose $Var(Y) = 0$. Then rules (iv) of Box 6.2 and (vii) of Box 7.1 imply that $Cov(X, Y) = 0$. This shows that the Inequality (7.16) holds if $Var(Y) = 0$. Now suppose $Var(Y) > 0$. Then,

$$0 \leq Var\left(X - \frac{Cov(X, Y)}{Var(Y)} \cdot Y\right) \cdot Var(Y)$$

$$= \left(Var(X) - 2 \cdot \frac{Cov(X, Y)}{Var(Y)} Cov(X, Y) + \frac{Cov(X, Y)^2}{Var(Y)^2} Var(Y)\right) \cdot Var(Y) \qquad [(7.12)]$$

$$= Var(X) \cdot Var(Y) - Cov(X, Y)^2,$$

which is equivalent to (7.16). According to the first part of the proof,

$$Cov(X, Y)^2 = Var(X) \cdot Var(Y) \Leftrightarrow Var(Y) = 0 \text{ or } Var\left(X - \frac{Cov(X, Y)}{Var(Y)} \cdot Y\right) = 0.$$

Rule (iv) of Box 6.2 implies that this is equivalent to

$$\exists\, a \in \mathbb{R} \colon Y \underset{P}{=} a \quad \text{or} \quad \exists\, c \in \mathbb{R} \colon X - \frac{Cov(X, Y)}{Var(Y)} \cdot Y \underset{P}{=} c.$$

If $Cov(X, Y) \neq 0$, this is equivalent to

$$\exists\, a \in \mathbb{R} \colon Y \underset{P}{=} a + 0 \cdot X \quad \text{or} \quad \exists\, c \in \mathbb{R} \colon Y \underset{P}{=} \left(-\frac{c \cdot Var(Y)}{Cov(X, Y)}\right) + \left(\frac{Var(Y)}{Cov(X, Y)}\right) \cdot X.$$

Obviously, in both cases, there is a linear function of X that is P-equivalent to Y. Furthermore, if there are $a, b \in \mathbb{R}$ with $Y \underset{P}{=} a + bX$, which implies $Var(Y) = b^2 Var(X)$ [see Box 6.2 (ii), (iii)], then,

$$
\begin{aligned}
Cov(X, Y)^2 &= Cov(X, a + bX)^2 & \text{[Box 7.1 (x)]} \\
&= b^2 Cov(X, X)^2 & \text{[Box 7.1 (ii), (iii)]} \\
&= b^2 Var(X)^2 & \text{[Box 7.1 (iv)]} \\
&= Var(X) \cdot Var(Y). & \text{[Box 6.2 (ii), (iii)]}
\end{aligned}
$$

Proof of Theorem 7.14

The proof is organized as follows: (iii) $\Rightarrow$ (ii) $\Rightarrow$ (i) $\Rightarrow$ (iii), which will prove that (i), (ii), and (iii) are equivalent.

(iii) $\Rightarrow$ (ii) The first partial derivatives of

$$
\begin{aligned}
&MSE\,(a_0, a_1) \\
&= E([Y - (a_0 + a_1 X)]^2) \\
&= E(Y^2) + E[(a_0 + a_1 X)^2] - 2E[Y \cdot (a_0 + a_1 X)] & \text{[Box 6.1 (vi)]} \\
&= E(Y^2) + a_0^2 + a_1^2 E(X^2) + 2a_0 a_1 E(X) - 2a_0 E(Y) - 2a_1 E(X \cdot Y) & \text{[Box 6.1 (iii), (vi)]}
\end{aligned}
$$

with respect to a_0 and a_1 are

$$\frac{\partial MSE(a_0, a_1)}{\partial a_0} = 2a_0 + 2a_1 E(X) - 2E(Y)$$

and

$$\frac{\partial MSE(a_0, a_1)}{\partial a_1} = 2a_1 E(X^2) + 2a_0 E(X) - 2E(X \cdot Y).$$

If $f(X) = \alpha_0 + \alpha_1 X = Q_{lin}(Y \mid X)$, then

$$2\alpha_0 + 2\alpha_1 E(X) - 2E(Y) = 0$$

and

$$2\alpha_1 E(X^2) + 2\alpha_0 E(X) - 2E(X \cdot Y) = 0.$$

Solving the first equation for α_0 yields

$$\alpha_0 = E(Y) - \alpha_1 E(X).$$

Inserting this result into the second equation yields

$$\alpha_1 E(X^2) + E(Y) \cdot E(X) - \alpha_1 E(X)^2 - E(X \cdot Y) = 0.$$

Using $Cov(X, Y) = E(X \cdot Y) - E(X) \cdot E(Y)$ [see Box 7.1 (i)] and $Var(X) = E(X^2) - E(X)^2$ [see Box 6.2 (i)], we receive

$$\alpha_1 = \frac{Cov(X, Y)}{Var(X)}.$$

(ii) $\Rightarrow$ (i)

$$
\begin{aligned}
E(\epsilon) &= E[Y - f(X)] \\
&= E[Y - (\alpha_0 + \alpha_1 X)] & \text{[def. of } f(X)] \\
&= E(Y - [E(Y) - \alpha_1 E(X) + \alpha_1 X]) & \text{[(ii)]} \\
&= E(Y) - E(Y) + \alpha_1 E(X) - \alpha_1 \cdot E(X) & \text{[Box 6.1 (vii)]} \\
&= 0. \\
Cov(X, \epsilon) &= Cov(X, [Y - (\alpha_0 + \alpha_1 X)]) & \text{[def. of } \epsilon] \\
&= Cov(X, Y) - \alpha_1 Var(X) & \text{[Box 7.1 (ii), (iii)]} \\
&= Cov(X, Y) - \frac{Cov(X, Y)}{Var(X)} \cdot Var(X) & \text{[(ii)]} \\
&= 0.
\end{aligned}
$$

(i) $\Rightarrow$ (iii) Let $f(X) = \alpha_0 + \alpha_1 X$, $\alpha_0, \alpha_1 \in \mathbb{R}$, be a linear function of X with $E(\epsilon) = 0$ and $Cov(X, \epsilon) = 0$, where $\epsilon = Y - f(X)$. Then, for any linear function $h(X) = a_0 + a_1 X$, a_0, $a_1 \in \mathbb{R}$,

$$
\begin{aligned}
E((Y - f(X))[f(X) - h(X)]) &= E(\epsilon \cdot [f(X) - h(X)]) & \text{[def. of } \epsilon] \\
&= E(\epsilon \cdot [(\alpha_0 + \alpha_1 X) - (a_0 + a_1 X)]) & \text{[defs. of } f(X), h(X)] \\
&= E(\epsilon \cdot [(\alpha_0 - a_0) + (\alpha_1 - a_1)X]) \\
&= (\alpha_0 - a_0) \cdot E(\epsilon) + (\alpha_1 - a_1)E(\epsilon \cdot X) & \text{[Box 6.1 (vii)]} \\
&= 0. & \text{[(i), Box 7.1 (i)]}
\end{aligned}
$$

Using this result and considering

$$
\begin{aligned}
E([Y - h(X)]^2) &= E[([Y - f(X)] + [f(X) - h(X)])^2] \\
&= E([Y - f(X)]^2) + E([f(X) - h(X)]^2) + 2 \cdot E([Y - f(X)][f(X) - h(X)]) \\
&= E([Y - f(X)]^2) + E([f(X) - h(X)]^2) \\
&\geq E([Y - f(X)]^2).
\end{aligned}
$$

In this inequality, '=' holds if and only if $f(X) \underset{P}{=} h(X)$. Because $f(X) = \alpha_0 + \alpha_1 X$ and $h(X) = a_0 + a_1 X$, the property $f(X) \underset{P}{=} h(X)$ is equivalent to $f(X) = h(X)$.

Proof of Theorem 7.30

The proof is organized as follows: (iii) $\Rightarrow$ (ii) $\Rightarrow$ (i) $\Rightarrow$ (iii), which will prove that (i), (ii), and (iii) are equivalent.

(iii) $\Rightarrow$ (ii) The first partial derivative of

$$
\begin{aligned}
&MSE\,(b_0, b_1, \dots, b_n)\\
&= E([Y - (b_0 + \mathbf{b}'\mathbf{x})]^2) && [(7.38)]\\
&= E(Y^2) + E((b_0 + \mathbf{b}'\mathbf{x})^2) - 2E(Y \cdot (b_0 + \mathbf{b}'\mathbf{x})) && [\text{Box 6.1 (vi), (iii)}]\\
&= E(Y^2) + b_0^2 + E((\mathbf{b}'\mathbf{x})^2) + 2b_0 \cdot E(\mathbf{b}'\mathbf{x}) - 2b_0 E(Y) - 2E(\mathbf{b}'\mathbf{x} \cdot Y) && [\text{Box 6.1 (vi), (i), (iii)}]\\
&= E(Y^2) + b_0^2 + E((\mathbf{b}'\mathbf{x})^2) + 2b_0 \cdot \mathbf{b}'E(\mathbf{x}) - 2b_0 E(Y) - 2\mathbf{b}'E(\mathbf{x} \cdot Y) && [\text{Box 7.2 (iii)}]
\end{aligned}
$$

with respect to b_0 is

$$
\frac{\partial MSE\,(b_0, b_1, \dots, b_n)}{\partial b_0} = 2b_0 + 2\mathbf{b}'E(\mathbf{x}) - 2E(Y).
$$

If $f(X_1, \dots, X_n) = \beta_0 + \boldsymbol{\beta}'\mathbf{x} = Q_{lin}(Y \mid X_1, \dots, X_n)$, then,

$$
2\beta_0 + 2\boldsymbol{\beta}'E(\mathbf{x}) - 2E(Y) = 0.
$$

Dividing both sides by 2 and solving for β_0 yields

$$
\beta_0 = E(Y) - \boldsymbol{\beta}'E(\mathbf{x}).
$$

Gathering the first partial derivatives of $MSE\,(b_0, b_1, \dots, b_n)$ with respect to $b_1, \dots, b_n$ in a vector yields

$$
\left[\frac{\partial MSE\,(b_0, b_1, \dots, b_n)}{\partial b_1}, \dots, \frac{\partial MSE\,(b_0, b_1, \dots, b_n)}{\partial b_n} \right]' = 2E(\mathbf{x}\mathbf{x}')\,\mathbf{b} + 2b_0 E(\mathbf{x}) - 2E(\mathbf{x} \cdot Y).
$$

If $f(X_1, \dots, X_n) = \beta_0 + \boldsymbol{\beta}'\mathbf{x} = Q_{lin}(Y \mid X_1, \dots, X_n)$, then $2E(\mathbf{x}\mathbf{x}')\,\boldsymbol{\beta} + 2\beta_0 E(\mathbf{x}) - 2E(\mathbf{x} \cdot Y) = \mathbf{0}$, and dividing both sides by 2 yields

$$
E(\mathbf{x}\mathbf{x}')\,\boldsymbol{\beta} + \beta_0 E(\mathbf{x}) - E(\mathbf{x} \cdot Y) = \mathbf{0}.
$$

Inserting our result $\beta_0 = E(Y) - \boldsymbol{\beta}'E(\mathbf{x})$, using $\Sigma_{xy} = E(\mathbf{x} \cdot Y) - E(\mathbf{x}) \cdot E(Y)$ and $\Sigma_{xx} = E(\mathbf{x}\mathbf{x}') - E(\mathbf{x})\,E(\mathbf{x}')$ [see Box 7 (i)], yields

$$
\begin{aligned}
&E(\mathbf{x}\mathbf{x}')\,\boldsymbol{\beta} + (E(Y) - \boldsymbol{\beta}'E(\mathbf{x}))E(\mathbf{x}) - E(\mathbf{x} \cdot Y)\\
&= E(\mathbf{x}\mathbf{x}')\,\boldsymbol{\beta} + E(Y) \cdot E(\mathbf{x}) - E(\mathbf{x})E(\mathbf{x}')\,\boldsymbol{\beta} - E(\mathbf{x} \cdot Y)\\
&= (E(\mathbf{x}\mathbf{x}') - E(\mathbf{x})E(\mathbf{x}'))\,\boldsymbol{\beta} - (E(\mathbf{x} \cdot Y) - E(\mathbf{x}) \cdot E(Y))\\
&= \Sigma_{xx}\,\boldsymbol{\beta} - \Sigma_{xy} = \mathbf{0},
\end{aligned}
$$

which is a necessary condition for a minimum. (*Note*: In this proof, we do not have to check a sufficient condition.) Adding Σ_{xy} on both sides yields $\Sigma_{xx}\,\beta = \Sigma_{xy}$, and when multiplying both sides from the left by Σ_{xx}^{-1}, we receive

$$\beta = \Sigma_{xx}^{-1}\,\Sigma_{xy}.$$

(ii) $\Rightarrow$ (i)

$$
\begin{aligned}
E(\epsilon) &= E(Y - f(X)) && \text{[def. of ϵ]}\\
&= E(Y - (\beta_0 + \beta'x)) && \text{[def. of $f(X)$]}\\
&= E(Y - (E(Y) - \beta'E(x) + \beta'x)) && \text{[(ii)]}\\
&= E(Y) - E(Y) + \beta'E(x) - \beta'E(x) && \text{[Box 6.1 (vi), (i), Box 7.2 (iii)]}\\
&= 0.
\end{aligned}
$$

$$
\begin{aligned}
\Sigma_{x\epsilon} &= \Sigma_{x,Y-(\beta_0+\beta'x)} && \text{[def. of ϵ]}\\
&= \Sigma_{xy} - \Sigma_{xx}\,\beta && \text{[Box 7.3 (ii), (vii)]}\\
&= \Sigma_{xy} - \Sigma_{xx}\,\Sigma_{xx}^{-1}\,\Sigma_{xy} && \text{[(ii)]}\\
&= 0.
\end{aligned}
$$

(i) $\Rightarrow$ (iii) Let $f(X) = \beta_0 + \beta'x$, $\beta_0 \in \mathbb{R}$, $\beta \in \mathbb{R}^n$ such that $E(\epsilon) = 0$ and $\Sigma_{x\epsilon} = 0$, where $\epsilon = Y - f(X)$. Then, for any linear function $h(X) = b_0 + b'x$, $b_0 \in \mathbb{R}$, $b \in \mathbb{R}^n$,

$$
\begin{aligned}
E((Y &- f(X))[f(X) - h(X)])\\
&= E(\epsilon \cdot [f(X) - h(X)]) && \text{[def. of ϵ]}\\
&= E(\epsilon \cdot [(\beta_0 + \beta'x) - (b_0 + b'x)]) && \text{[defs. of $f(X), h(X)$]}\\
&= E(\epsilon \cdot [(\beta_0 - b_0) + (\beta - b)'x])\\
&= (\beta_0 - b_0) \cdot E(\epsilon) + (\beta - b)'E(\epsilon \cdot x) && \text{[Box 6.1 (vi), Box 7.2 (iii)]}\\
&= (\beta - b)'\,\Sigma_{x\epsilon} && \text{[(i), Box 7.3 (i)]}\\
&= 0. && \text{[(i)]}
\end{aligned}
$$

Using this result and considering

$$
\begin{aligned}
E([Y - h(X)]^2) &= E([[Y - f(X)] + [f(X) - h(X)]]^2)\\
&= E([Y - f(X)]^2) + E([f(X) - h(X)]^2) + 2 \cdot E([Y - f(X)][f(X) - h(X)])\\
&= E([Y - f(X)]^2) + E([f(X) - h(X)]^2)\\
&\ge E([Y - f(X)]^2).
\end{aligned}
$$

Hence, $f(X) = Q_{lin}(Y \mid X)$.

Proof of Corollary 7.32

$$
\begin{aligned}
Cov\,[Q_{lin}(Y \mid X), \epsilon] &= Cov(\beta_0 + \beta'x, \epsilon) && \text{[(7.39)]}\\
&= \beta'Cov\,(x, \epsilon) && \text{[Box 7.3 (ii), (iii)]}\\
&= 0. && \text{[$Cov(x, \epsilon) = \Sigma_{x\epsilon} = 0$, Th. 7.30]}
\end{aligned}
$$

Furthermore,

$$Var[Q_{lin}(Y \mid X)] = Cov[Q_{lin}(Y \mid X), Q_{lin}(Y \mid X)] \qquad \text{[Box 7.1 (iv)]}$$

$$= Cov\,[Q_{lin}(Y \mid X), \epsilon] + Cov[Q_{lin}(Y \mid X), Q_{lin}(Y \mid X)] \qquad \text{[(7.43)]}$$

$$= Cov\,[Q_{lin}(Y \mid X) + \epsilon, Q_{lin}(Y \mid X)] \qquad \text{[Box 7.1 (ix)]}$$

$$= Cov[Y, Q_{lin}(Y \mid X)]. \qquad \text{[(7.41)]}$$

Exercises

7.1 Use Definition 7.2 in order to determine the coefficients α_0 and α_1 of the linear quasi-regression in Example 7.5.

7.2 Consider the linear quasi-regression f with $f(x) = \alpha_0 + \alpha_1 x, x \in \mathbb{R}$. Prove: If $x_1, x_2 \in \mathbb{R}$ with $x_1 \neq x_2$, then $\alpha_0 = f(0)$ and $\alpha_1 = \dfrac{1}{x_2 - x_1}[f(x_2) - f(x_1)]$.

7.3 Prove the propositions of Box 7.1.

7.4 Prove the proposition of Remark 7.13.

7.5 Prove the proposition of Remark 7.20.

7.6 Prove the proposition of Remark 7.21.

7.7 Show that $Corr\,(a_0 + a_1 X, b_0 + b_1 Y) = Corr\,(X, Y)$, where $a_0, a_1, b_0, b_1 \in \mathbb{R}$.

7.8 Show

$$Cov\,(X, Y) = \begin{cases} SD(X) \cdot SD(Y), & \text{if } a_1 > 0 \\ -SD(X) \cdot SD(Y), & \text{if } a_1 < 0, \end{cases} \qquad (7.46)$$

provided that there are $a_0, a_1 \in \mathbb{R}$ with $Y \underset{P}{=} a_0 + a_1 X, a_1 \neq 0$.

7.9 Prove Equation (7.24) and that the slope is invariant under translations $c + X, d + Y,$ $c, d \in \mathbb{R}$.

7.10 Prove the rules of computation of Box 7.2.

7.11 Prove the rules of computation of Box 7.3.

7.12 Under the assumptions of Definition 7.28 and $Var(Y) > 0$, prove Equation (7.45).

Solutions

7.1

$$MSE\,(a_0, a_1) = E([Y - (a_0 + a_1 x)]^2) \qquad \text{[(7.2)]}$$

$$= \frac{1}{4} \cdot (1 - a_0 - a_1)^2 + \frac{1}{2} \cdot (2 - a_0 - 2a_1)^2 + \frac{1}{4} \cdot (1 - a_0 - 3a_1)^2 \quad \text{[(6.3)]}$$

$$= \frac{1}{4} \cdot \left(1 + a_0^2 + a_1^2 - 2a_0 - 2a_1 + 2a_0 a_1\right)$$

$$+ \frac{1}{2} \cdot \left(4 + a_0^2 + 4a_1^2 - 4a_0 - 8a_1 + 4a_0 a_1 \right)$$

$$+ \frac{1}{4} \cdot \left(1 + a_0^2 + 9a_1^2 - 2a_0 - 6a_1 + 6a_0 a_1 \right)$$

$$= \frac{5}{2} + a_0^2 + \frac{9}{2} \cdot a_1^2 - 3a_0 - 6a_1 + 4a_0 a_1.$$

The partial derivatives are

$$\frac{\partial MSE(a_0, a_1)}{\partial a_0} = 2a_0 - 3 + 4a_1 \qquad \text{and} \qquad \frac{\partial MSE(a_0, a_1)}{\partial a_1} = 9a_1 - 6 + 4a_0.$$

Fixing the partial derivatives to 0 and denoting the solutions by α_0 and α_1 yield

$$2\alpha_0 - 3 + 4\alpha_1 = 0 \qquad \text{and} \qquad 9\alpha_1 - 6 + 4\alpha_0 = 0.$$

The first equation implies $\alpha_0 = -2\alpha_1 + \frac{3}{2}$. Inserting this result into the second equation yields $9\alpha_1 - 6 - 8\alpha_1 + 6 = 0$, which implies $\alpha_1 = 0$ and $\alpha_0 = \frac{3}{2}$. The values $\alpha_1 = 0$ and $\alpha_0 = \frac{3}{2}$ satisfy a necessary condition for a minimum; now we check if they also satisfy a sufficient condition. The second partial derivatives are

$$\frac{\partial^2 MSE(a_0, a_1)}{\partial a_0^2} = 2 > 0, \qquad \frac{\partial^2 MSE(a_0, a_1)}{\partial a_1^2} = 9 > 0, \qquad \text{and} \qquad \frac{\partial^2 MSE(a_0, a_1)}{\partial a_0\, \partial a_1} = 4,$$

that is, in this case they are constant for all $a_0, a_1 \in \mathbb{R}$. Because $2 \cdot 9 - 4^2 > 0$, we can conclude that $MSE(a_0, a_1)$ has its minimum for $a_0 = \alpha_0 = \frac{3}{2}$ and $a_1 = \alpha_1 = 0$ (see Ellis & Gulick, 2006, Th. 13.21).

7.2 The equation $f(x) = \alpha_0 + \alpha_1 x, \ x \in \mathbb{R}$, yields $f(0) = \alpha_0$,

$$f(x_1) = \alpha_0 + \alpha_1 x_1, \qquad \text{and} \qquad f(x_2) = \alpha_0 + \alpha_1 x_2.$$

Hence,

$$f(x_2) - f(x_1) = \alpha_0 + \alpha_1 x_2 - (\alpha_0 + \alpha_1 x_1) = \alpha_1 (x_2 - x_1).$$

Multiplying both sides by $\frac{1}{x_2 - x_1}$ yields $\alpha_1 = \frac{1}{x_2 - x_1}[f(x_2) - f(x_1)]$, provided that $x_1 \neq x_2$.

7.3 (i)

$$\begin{aligned}
Cov(X, Y) &= E([X - E(X)] \cdot [Y - E(Y)]) && [(7.8)] \\
&= E[X \cdot Y - X \cdot E(Y) - E(X) \cdot Y + E(X) \cdot E(Y)] \\
&= E(X \cdot Y) - E[X \cdot E(Y)] - E[E(X) \cdot Y] + E(X) \cdot E(Y) && [\text{Box 6.1 (vi), (i)}] \\
&= E(X \cdot Y) - 2E(X) \cdot E(Y) + E(X) \cdot E(Y) && [\text{Box 6.1 (iii)}] \\
&= E(X \cdot Y) - E(X) \cdot E(Y).
\end{aligned}$$

(ii)

$$Cov\,(\alpha + X,\ \beta + Y) = E([\alpha + X - E(\alpha + X)]\ \cdot\ [\beta + Y - E(\beta + Y)]) \qquad [(7.8)]$$
$$= E([\alpha + X - \alpha - E(X)]\ \cdot\ [\beta + Y - \beta - E(Y)]) \qquad [\text{Box } 6.1\ (ii)]$$
$$= E([X - E(X)]\ \cdot\ [Y - E(Y)])$$
$$= Cov\,(X, Y)\,. \qquad [(7.8)]$$

(iii)

$$Cov\,(\alpha X,\ \beta Y) = E(\alpha X \cdot \beta Y)\ -\ E(\alpha X) \cdot E(\beta Y) \qquad [\text{Box } 7.1\ (i)]$$
$$= \alpha \beta E(X \cdot Y)\ -\ \alpha \beta E(X) \cdot E(Y) \qquad [\text{Box } 6.1\ (iii)]$$
$$= \alpha \beta [E(X \cdot Y)\ -\ E(X) \cdot E(Y)]$$
$$= \alpha \beta\, Cov\,(X, Y)\,. \qquad [\text{Box } 7.1\ (i)]$$

(iv)   This rule immediately follows from Equations (6.27) and (7.8).

(v)   This rule immediately follows from Equation (7.8).

(vi)   Independence of X and Y implies $E(X \cdot Y) = E(X) \cdot E(Y)$ (see Th. 6.24) and $Cov(X, Y) = 0$ [see Rule (i) of Box 7.1].

(vii)   According to Lemma 5.51, X and Y are independent if $X \underset{P}{=} \alpha$. This implies that $Cov(X, Y) = 0$ [see Rule (vi) of Box 7.1].

(viii)

$$Var\left(\sum_{i=1}^{n} \alpha_i\, Y_i\right)$$

$$= E\left(\left[\sum_{i=1}^{n} \alpha_i\, Y_i - E\left(\sum_{i=1}^{n} \alpha_i\, Y_i\right)\right]^2\right) \qquad [(6.27)]$$

$$= E\left[\left(\sum_{i=1}^{n} \alpha_i\, Y_i - \sum_{i=1}^{n} \alpha_i\, E(Y_i)\right)^2\right] \qquad [\text{Box } 6.1\ (vii)]$$

$$= E\left[\left(\sum_{i=1}^{n} \alpha_i\, [Y_i - E(Y_i)]\right)^2\right]$$

$$= E\left[\sum_{i=1}^{n} \alpha_i^2\, (Y_i - E(Y_i))^2 + \sum_{i=1}^{n} \sum_{j=1, j \neq i}^{n} \alpha_i\, \alpha_j\, (Y_i - E(Y_i))(Y_j - E(Y_j))\right]$$

$$= \sum_{i=1}^{n} \alpha_i^2\, E(Y_i - E(Y_i))^2 + \sum_{i=1}^{n} \sum_{j=1, j \neq i}^{n} \alpha_i\, \alpha_j\, E[(Y_i - E(Y_i))(Y_j - E(Y_j))] \qquad [\text{Box } 6.1\ (vii)]$$

$$= \sum_{i=1}^{n} \alpha_i^2\, Var(Y_i) + \sum_{i=1}^{n} \sum_{j=1, j \neq i}^{n} \alpha_i\, \alpha_j\, Cov\,(Y_i, Y_j)\,. \qquad [(6.27),\ (7.8)]$$

(ix)

$$Cov\left(\sum_{i=1}^{n} \alpha_i X_i, \sum_{j=1}^{m} \beta_j Y_j\right)$$

$$= E\left[\left(\sum_{i=1}^{n} \alpha_i X_i - E\left(\sum_{i=1}^{n} \alpha_i X_i\right)\right)\left(\sum_{j=1}^{m} \beta_j Y_j - E\left(\sum_{j=1}^{m} \beta_j Y_j\right)\right)\right] \qquad [(7.8)]$$

$$= E\left[\left(\sum_{i=1}^{n} \alpha_i [X_i - E(X_i)]\right)\left(\sum_{j=1}^{m} \beta_j [Y_j - E(Y_j)]\right)\right] \qquad [\text{Box 6.1 (vii)}]$$

$$= E\left[\sum_{i=1}^{n}\sum_{j=1}^{m} \alpha_i \beta_j [X_i - E(X_i)] [Y_j - E(Y_j)]\right]$$

$$= \sum_{i=1}^{n}\sum_{j=1}^{m} \alpha_i \beta_j\, Cov(X_i, Y_j). \qquad [\text{Box 6.1 (vii), (7.8)}]$$

(x)   If $X_1 \underset{P}{=} X_2$, then $[Y - E(Y)] \cdot [X_1 - E(X_1)] \underset{P}{=} [Y - E(Y)] \cdot [X_2 - E(X_2)]$. According to Corollary 5.24, these two product variables have the same distribution, and according to Corollary 6.17, the same expectation. However, the expectations of these product variables are the covariances. Hence, $Cov(Y, X_1) = Cov(Y, X_2)$.

7.4 If $X, Y\colon (\Omega, \mathcal{A}, P) \to (\mathbb{R}, \mathcal{B})$ are random variables, then

$$(\alpha X + \beta Y)^2 = \alpha^2 X^2 + \beta^2 Y^2 + 2\alpha\beta X \cdot Y$$

is also a random variable on $(\Omega, \mathcal{A}, P)$ (see Def. 5.1, Example 2.61, and Th. 2.57), and this implies

$$\begin{aligned} E[(\alpha X + \beta Y)^2] &= E(\alpha^2 X^2 + \beta^2 Y^2 + 2\alpha\beta X \cdot Y)\\ &= \alpha^2 E(X^2) + \beta^2 E(Y^2) + 2\alpha\beta E(X \cdot Y). \qquad [\text{Box 6.1 (vii)}]\end{aligned}$$

The terms $E(X^2)$ and $E(Y^2)$ are finite by assumption and, according to Remark 7.1, $E(X \cdot Y)$ is finite as well. This implies $E[(\alpha X + \beta Y)^2] < \infty$.

7.5 Inequality (7.16) implies

$$-\sqrt{Var(X) \cdot Var(Y)} \le Cov(X, Y) \le \sqrt{Var(X) \cdot Var(Y)}.$$

If $\sqrt{Var(X) \cdot Var(Y)} = SD(X) \cdot SD(Y) > 0$, then these inequalities yield $-1 \le Corr(X, Y) \le 1$. If $\sqrt{Var(X) \cdot Var(Y)} = SD(X) \cdot SD(Y) = 0$, then, by Definition 7.18, $Corr(X, Y) = 0$.

7.6 If $SD(X), SD(Y) > 0$, then

$$
\begin{aligned}
Corr\,(X, Y) &= \frac{Cov\,(X, Y)}{SD(X) \cdot SD(Y)} && [(7.18)] \\[2mm]
&= \frac{E([X - E(X)] \cdot [Y - E(Y)])}{SD(X) \cdot SD(Y)} && [(7.8)] \\[2mm]
&= E\left(\frac{[X - E(X)] \cdot [Y - E(Y)]}{SD(X) \cdot SD(Y)}\right) && [\text{Box 6.1 (iii)}] \\[2mm]
&= E\left(\frac{X - E(X)}{SD(X)} \cdot \frac{Y - E(Y)}{SD(Y)}\right).
\end{aligned}
$$

7.7 According to Rules (ii) and (iii) of Box 7.1, $Cov(a_0 + a_1 X, b_0 + b_1 Y) = a_1 b_1 Cov(X, Y)$. Similarly, according to Rules (ii) and (iii) of Box 6.2, $Var(a_0 + a_1 X) = a_1^2\, Var(X)$, which implies

$$
SD(a_0 + a_1 X) =
\begin{cases}
a_1 SD(X), & \text{if } a_1 > 0 \\[2mm]
-a_1 SD(X), & \text{if } a_1 < 0 \\[2mm]
0, & \text{if } a_1 = 0.
\end{cases}
$$

Hence, for $a_1, b_1 > 0$,

$$
\begin{aligned}
Corr\,(a_0 + a_1 X, b_0 + b_1 Y) &= \frac{Cov\,(a_0 + a_1 X,\ b_0 + b_1 Y)}{SD(a_0 + a_1 X) \cdot SD(b_0 + b_1 Y)} \\[2mm]
&= \frac{a_1\, b_1\, Cov\,(X, Y)}{a_1\, SD(X) \cdot b_1\, SD(Y)} \\[2mm]
&= \frac{Cov\,(X, Y)}{SD(X) \cdot SD(Y)} \\[2mm]
&= Corr\,(X, Y).
\end{aligned}
$$

The proofs for the other cases of $a_1, b_1 \neq 0$ are analogous. Note that we defined $Corr\,(X, Y) = 0$ if $SD(X) = 0$ or $SD(Y) = 0$.

7.8 According to Box 7.1 (x), $Cov(X, Y) = Cov\,(X, \alpha_0 + \alpha_1 X)$ if $Y \underset{P}{=} \alpha_0 + \alpha_1 X$. Box 6.2 (v), (ii), and (iii) imply $Var(Y) = \alpha_1^2 Var(X)$. Therefore,

$$
Corr\,(X, Y) = \frac{Cov\,(X, \alpha_0 + \alpha_1 X)}{SD(X) \cdot \sqrt{\alpha_1^2 Var(X)}} = \frac{\alpha_1\, Var(X)}{SD(X) \cdot |\alpha_1| \cdot SD(X)} = \frac{\alpha_1}{|\alpha_1|}.
$$

Hence, $Corr\,(X, Y) = 1$ if $\alpha_1 > 0$ and $Corr\,(X, Y) = -1$ if $\alpha_1 < 0$. Therefore,

$$
Corr\,(X, Y) := \frac{Cov\,(X, Y)}{SD(X) \cdot SD(Y)}
$$

yields (7.46).

7.9

$$\alpha_1^* = \frac{Cov\,(aX, bY)}{Var\,(aX)} \qquad \text{[Th. 7.14 (ii)]}$$

$$= \frac{ab \cdot Cov\,(X, Y)}{a^2 \cdot Var\,(X)} \qquad \text{[Box 7.1 (iii), Box 6.2 (iii)]}$$

$$= \frac{b}{a} \cdot \alpha_1. \qquad \text{[Th. 7.14 (ii)]}$$

According to Theorem 7.14 (ii), the slope of the linear quasi-regression of $c + Y$ on $d + X$ is

$$\frac{Cov\,(c + X, d + Y)}{Var\,(c + X)} = \frac{Cov\,(X, Y)}{Var\,(X)} \qquad \text{[Box 7.1 (ii), Box 6.2 (ii)]}$$

$$= \alpha_1. \qquad \text{[Th. 7.14 (ii)]}$$

7.10 (i) Equation (7.27) and Rule (i) of Box 6.1 imply

$$E(x) = [E(X_1), \ldots , E(X_n)]' = [a_1, \ldots , a_n]' = \mathbf{a}.$$

(ii) Equation (7.27) and Rule (ii) of Box 6.1 imply

$$E(\mathbf{a} + x) = \begin{bmatrix} E(a_1 + X_1) \\ \vdots \\ E(a_n + X_n) \end{bmatrix} = \begin{bmatrix} a_1 + E(X_1) \\ \vdots \\ a_n + E(X_n) \end{bmatrix} = \mathbf{a} + E(x).$$

(iii) Equation (7.31) and Rule (vii) of Box 6.1 imply

$$E(\mathbf{a}'x) = E\left(\sum_{i=1}^{n} a_i X_i \right) = \sum_{i=1}^{n} a_i \cdot E(X_i) = \mathbf{a}'\,E(x).$$

The other equations summarized in (iii) follow from the fact that $\mathbf{a}'x$ is a one-dimensional random variable (see Example 2.61) and $\mathbf{a}'x = x'\mathbf{a}$.

(iv) Let $\mathbf{a}_l'$ and $\mathbf{b}_l'$, $l = 1, \ldots , k$, denote the row vectors of $\mathbf{A}$ and $\mathbf{B}$, respectively. Applying Equation (7.27), Rule (vi) of Box 6.1, and Rule (iii) to the terms $\mathbf{a}_l'\,x$ and $\mathbf{b}_l'\,y$, $l = 1, 2 \ldots , k$, respectively, yields

$$E(\mathbf{A}x + \mathbf{B}y) = E\begin{bmatrix} \mathbf{a}_1'x + \mathbf{b}_1'y \\ \vdots \\ \mathbf{a}_k'x + \mathbf{b}_k'y \end{bmatrix} = \begin{bmatrix} E(\mathbf{a}_1'x + \mathbf{b}_1'y) \\ \vdots \\ E(\mathbf{a}_k'x + \mathbf{b}_k'y) \end{bmatrix} = \begin{bmatrix} E(\mathbf{a}_1'x) + E(\mathbf{b}_1'y) \\ \vdots \\ E(\mathbf{a}_k'x) + E(\mathbf{b}_k'y) \end{bmatrix}$$

$$= \begin{bmatrix} \mathbf{a}_1'\,E(x) + \mathbf{b}_1'\,E(y) \\ \vdots \\ \mathbf{a}_k'\,E(x) + \mathbf{b}_k'\,E(y) \end{bmatrix} = \mathbf{A}E(x) + \mathbf{B}E(y).$$

(v) This rule is a special case of Rule (v) with $\mathbf{B} = \mathbf{0}$.

(vi) Let $\mathbf{a}_l$ and $\mathbf{x}_l$, $l = 1, \dots, k$, denote the column vectors of $\mathbf{A}'$ and X, respectively. Then Equations (7.27), (7.29) and Rule (i) imply

$$E(X) = [E(\mathbf{x}_1), \dots, E(\mathbf{x}_k)] = [E(\mathbf{a}_1), \dots, E(\mathbf{a}_k)] = [\mathbf{a}_1, \dots, \mathbf{a}_k] = \mathbf{A}'.$$

(vii) Let $\mathbf{a}_l$ and $\mathbf{x}_l$, $l = 1, \dots, k$, denote the column vectors of $\mathbf{A}'$ and X, respectively. Then Equations (7.27), (7.29) and Rule (ii) imply

$$E(\mathbf{A}' + X) = E(\mathbf{a}_1 + \mathbf{x}_1, \dots, \mathbf{a}_k + \mathbf{x}_k) = [E(\mathbf{a}_1 + \mathbf{x}_1), \dots, E(\mathbf{a}_k + \mathbf{x}_k)]$$
$$= [\mathbf{a}_1 + E(\mathbf{x}_1), \dots, \mathbf{a}_k + E(\mathbf{x}_k)] = \mathbf{A}' + E(X).$$

(viii) Let $\mathbf{c}_i'$, $i = 1, \dots, l$, denote the row vectors of $\mathbf{C}$. Then Equations (7.27) and (7.29) imply

$$E(\mathbf{C}X) = E\begin{bmatrix} \mathbf{c}_1'X \\ \vdots \\ \mathbf{c}_l'X \end{bmatrix} = \begin{bmatrix} E(\mathbf{c}_1'X) \\ \vdots \\ E(\mathbf{c}_l'X) \end{bmatrix} = \begin{bmatrix} \mathbf{c}_1' E(X) \\ \vdots \\ \mathbf{c}_l' E(X) \end{bmatrix} = \mathbf{C} E(X).$$

(ix) Rule (viii), Equation (7.30), and the rules for the transpose of a matrix yield

$$\begin{aligned} E(\mathbf{C}XY'\mathbf{D}') &= \mathbf{C}E(XY'\mathbf{D}') &&[\text{(viii)}] \\ &= \mathbf{C}(E(\mathbf{D}YX'))' &&[(7.30)] \\ &= \mathbf{C}(\mathbf{D}E(YX'))' &&[\text{(viii)}] \\ &= \mathbf{C}E(YX')'\mathbf{D}' &&[(7.30)] \\ &= \mathbf{C}E(XY')\,\mathbf{D}'. &&[(7.30)] \end{aligned}$$

7.11 (i)

$$\begin{aligned} \Sigma_{xy} &= E([\mathbf{x} - E(\mathbf{x})]\,[\mathbf{y} - E(\mathbf{y})]') &&[(7.33)] \\ &= E(\mathbf{x}\,\mathbf{y}' - \mathbf{x}\,E(\mathbf{y})' - E(\mathbf{x})\,\mathbf{y}' + E(\mathbf{x})\,E(\mathbf{y})') \\ &= E(\mathbf{x}\,\mathbf{y}') - E(\mathbf{x})\,E(\mathbf{y})' - E(\mathbf{x})\,E(\mathbf{y}') + E(\mathbf{x})\,E(\mathbf{y})' &&[\text{Box 7.2 (iii)}] \\ &= E(\mathbf{x}\,\mathbf{y}') - E(\mathbf{x})\,E(\mathbf{y}'). \end{aligned}$$

(ii)

$$\begin{aligned} \Sigma_{\mathbf{a}+x,\,\mathbf{b}+y} &= E([\mathbf{a} + \mathbf{x} - E(\mathbf{a} + \mathbf{x})]\,[\mathbf{b} + \mathbf{y} - E(\mathbf{b} + \mathbf{y})]') &&[(7.33)] \\ &= E([\mathbf{a} + \mathbf{x} - \mathbf{a} - E(\mathbf{x})]\,[\mathbf{b} + \mathbf{y} - \mathbf{b} - E(\mathbf{y})]') &&[\text{Box 7.2 (ii)}] \\ &= E([\mathbf{x} - E(\mathbf{x})]\,[\mathbf{y} - E(\mathbf{y})]') \\ &= \Sigma_{xy}. &&[(7.33)] \end{aligned}$$

(iii)

$$\begin{aligned} \Sigma_{\mathbf{A}x,\,\mathbf{B}y} &= E([\mathbf{A}\mathbf{x} - E(\mathbf{A}\mathbf{x})]\,[\mathbf{B}\mathbf{y} - E(\mathbf{B}\mathbf{y})]') &&[(7.33)] \\ &= E([\mathbf{A}\mathbf{x} - \mathbf{A}E(\mathbf{x})][\mathbf{B}\mathbf{y} - \mathbf{B}E(\mathbf{y})]') &&[\text{Box 7.2 (iv)}] \\ &= E[\mathbf{A}[\mathbf{x} - E(\mathbf{x})](\mathbf{B}\,[\mathbf{y} - E(\mathbf{y})])'] \\ &= E(\mathbf{A}[\mathbf{x} - E(\mathbf{x})]\,[\mathbf{y} - E(\mathbf{y})]'\,\mathbf{B}') \\ &= \mathbf{A}E([\mathbf{x} - E(\mathbf{x})]\,[\mathbf{y} - E(\mathbf{y})]')\,\mathbf{B}' &&[\text{Box 7.2 (ix)}] \\ &= \mathbf{A}\,\Sigma_{xy}\,\mathbf{B}'. &&[(7.33)] \end{aligned}$$

(iv)

$$\Sigma_{xy} = E([x - E(x)] \, [y - E(y)]') \qquad [(7.33)]$$
$$= (E([x - E(x)] \, [y - E(y)]')')' \qquad [(A')' = A]$$
$$= E([[x - E(x)] \, [y - E(y)]']')' \qquad [(7.30)]$$
$$= E([y - E(y)] \, [x - E(x)]')' \qquad [(ab')' = ba']$$
$$= \Sigma'_{yx} . \qquad [(7.33)]$$

(v) Independence of the multivariate random variables X and Y implies $X_i \underset{P}{\perp\!\!\!\perp} Y_j$, for all $i = 1, \dots, n$ and $j = 1, \dots, m$. Therefore, Rule (vi) of Box 7.1 implies $Cov(X_i, Y_j) = 0$, for all $i = 1, \dots, n$ and $j = 1, \dots, m$. Equation (7.34) then implies $\Sigma_{xy} = 0$.

(vi) If $x \underset{P}{=} a$, then Rule (vii) of Box (7.1) yields $Cov(X_i, Y_j) = 0$, for all $i = 1, \dots, n$, $j = 1, \dots, m$. Equation (7.34) then implies $\Sigma_{xy} = 0$.

(vii)

$$\Sigma_{Ax + By, \, Cw + Dz}$$
$$= E([Ax + By - E(Ax + By)] \, [Cw + Dz - E(Cw + Dz)]') \qquad [(7.33)]$$
$$= E([Ax - E(Ax) + By - E(By)] \, [Cw - E(Cw) + Dz - E(Dz)]')$$
$$= E([Ax - AE(x)] \, [Cw - CE(w)]' + [Ax - AE(x)] \, [Dz - DE(z)]'$$
$$\quad + [By - BE(y)] \, [Cw - CE(w)]' + [By - BE(y)] \, [Dz - DE(z)]' \qquad [\text{Box } 7.2 \, (iv)]$$
$$= E(A[x - E(x)] \, [w - E(w)]'C') + E(A[x - E(x)] \, [z - E(z)]'D')$$
$$\quad + E(B[y - E(y)] \, [w - E(w)]'C') + E(B[y - E(y)] \, [z - E(z)]'D')$$
$$= AE([x - E(x)] \, [w - E(w)]')C' + AE([x - E(x)] \, [z - E(z)]')D'$$
$$\quad + BE([y - E(y)] \, [w - E(w)]')C' + BE([y - E(y)] \, [z - E(z)]')D' \qquad [\text{Box } 7.2 \, (ix)]$$
$$= A\Sigma_{xw} C' + A\Sigma_{xz} D' + B\Sigma_{yw} C' + B\Sigma_{yz} D'. \qquad [(7.33)]$$

(viii) If $x \underset{P}{=} z$, then Rule (x) of Box 7.1 implies that $Cov(X_i, Y_j) = Cov(Z_i, Y_j)$, for all $i = 1, \dots, n, j = 1, \dots, m$. Equation (7.34) then implies $\Sigma_{xy} = \Sigma_{zy}$.

7.12

$$Q^2_{Y|X} = \frac{Var[Q_{lin}(Y \mid X)]}{Var(Y)} \qquad [(7.40)]$$

$$= \frac{Cov[Y, Q_{lin}(Y \mid X)]}{Var(Y)} \qquad [(7.44)]$$

$$= \frac{Cov[Y, Q_{lin}(Y \mid X)] \cdot Var[Q_{lin}(Y \mid X)]}{Var[Q_{lin}(Y \mid X)] \cdot Var(Y)}$$

$$= \frac{Cov\,[Y, Q_{lin}(Y \mid X)] \cdot Cov\,[Y, Q_{lin}(Y \mid X)]}{Var\,[Q_{lin}(Y \mid X)] \cdot Var(Y)} \qquad [(7.44)]$$

$$= \frac{Cov\,[Y, Q_{lin}(Y \mid X)]^2}{SD[Q_{lin}(Y \mid X)]^2 \cdot SD(Y)^2}$$

$$= Corr\,[Y, Q_{lin}(Y \mid X)]^2.$$

Note that $Corr\,[Y, Q_{lin}(Y \mid X)] = 0$ if $Var\,[Q_{lin}(Y \mid X)] = 0$ [see Eq. (7.18)].

8

Some distributions

In chapter 5, we defined *random variables* as particular measurable mappings and their *distributions* as their image measures. There, we also extended the concept of independence of events and families of events to *independence of random variables* and *independence of families of random variables*. Furthermore, we introduced the concepts of a *probability function* and a *density* of real-valued random variables, which are useful for describing distributions. In chapters 6 and 7, we treated the expectation of a numerical random variable and related concepts such as variance, covariance, and correlation. In this chapter, we provide some examples illustrating how probability functions and densities describe the distribution of a random variable and how they can be used to compute expectations and variances of real-valued random variables.

8.1 Some distributions of discrete random variables

In this section, we treat some examples of distributions that are specified by probability functions of discrete random variables: the discrete uniform, binomial, Poisson, and geometric distributions.

8.1.1 Discrete uniform distribution

Reading the following definition, remember that p_X denotes the probability function assigning to each value x_i of a discrete random variable X its probability $p_X(x) = P(X=x_i) = P_X(\{x_i\})$ (see Def. 5.56 and Rem. 5.57). Furthermore, according to Remark 5.59, p_X uniquely determines the distribution P_X of X, which follows from σ-additivity of the measure P_X.

Definition 8.1 [Discrete uniform distribution on a finite set]
Let $\Omega'_X = \{x_1, \ldots, x_n\}$, $n \in \mathbb{N}$, and let $X \colon (\Omega, \mathcal{A}, P) \to (\Omega'_X, \mathcal{P}(\Omega'_X))$ be a random variable, where $\Omega'_X = \{x_1, \ldots, x_n\}$. Then X has a (discrete) uniform distribution on Ω'_X, if

$$p_X(x_i) = P(X=x_i) = \frac{1}{n}, \quad \forall\, i = 1, \ldots, n. \tag{8.1}$$

Probability and Conditional Expectation: Fundamentals for the Empirical Sciences, First Edition. Rolf Steyer and Werner Nagel.
© 2017 John Wiley & Sons, Ltd. Published 2017 by John Wiley & Sons, Ltd.
Companion website: http://www.probability-and-conditional-expectation.de

Let $\Omega'_X = \{x_1, \dots, x_n\}$ and let $\#A'$ denote the number elements of a finite set A'. According to Equation (5.46),

$$\left(\forall A' \subset \Omega'_X \colon P_X(A') = \frac{\#A'}{n} \right) \quad \Leftrightarrow \quad P_X \text{ is the uniform distribution on } \Omega'_X. \quad (8.2)$$

Hence, if X has a discrete uniform distribution, then the probability of an event $\{X \in A'\}$ only depends on $\#A'$, the number of elements of A', and not on the particular choice of these elements from Ω'_X. The uniform distribution on Ω'_X is the only distribution that has this property. A special case is treated in Example 8.6.

8.1.2 Bernoulli distribution

Now we consider n trials in which we observe whether or not an event A_i occurs at trial i, where $i = 1, \dots, n$. These events A_i will be indicated by independent identically distributed (i. i. d.) random variables $X_1, \dots, X_n$ with values $X_i(\omega) = 1$ if $\omega \in A_i$, and $X_i(\omega) = 0$ otherwise. Hence, $X_i = 1_{A_i}$, and $P(X_i = 1) = p$ for all $i = 1, \dots, n$.

Definition 8.2 [Finite sequence of Bernoulli variables]
Let $X_1, \dots, X_n \colon (\Omega, \mathcal{A}, P) \to (\{0, 1\}, \mathcal{P}(\{0, 1\}))$ be i. i. d. random variables with

$$P(X_i = 1) = 1 - P(X_i = 0) = p, \quad \forall \, i = 1, \dots, n, \quad (8.3)$$

where $0 \le p \le 1$. Then $X_1, \dots, X_n$ is called a sequence of n Bernoulli variables with parameter p, and $X = (X_1, \dots, X_n)$ is called an n-variate Bernoulli variable with parameter p.

Remark 8.3 [Bernoulli distribution]   Remember that $a^1 = a$ and $a^{1-1} = a^0 = 1$ for $a \in \mathbb{R}$. Hence, for all $i = 1, \dots, n$,

$$p_{X_i}(x) = p^x (1-p)^{1-x}, \quad \forall x \in \{0, 1\}. \quad (8.4)$$

The distribution of such a random variable X_i is called the *Bernoulli distribution*.   ◁

As we will show in Remark 8.5, a multivariate Bernoulli variable has a multivariate Bernoulli distribution that will now be defined.

Definition 8.4 [Multivariate Bernoulli distribution]
Let $X = (X_1, \dots, X_n) \colon (\Omega, \mathcal{A}, P) \to (\{0, 1\}^n, \mathcal{P}(\{0, 1\}^n))$ be an n-variate random variable and let $0 \le p \le 1$. If the probability function of X is

$$\forall (x_1, \dots, x_n) \in \{0, 1\}^n \colon \quad p_X(x_1, \dots, x_n) = \prod_{i=1}^{n} p^{x_i} (1-p)^{1-x_i}, \quad (8.5)$$

then P_X is called the multivariate Bernoulli distribution with parameters n and p.

Because $x_i \in \{0, 1\}$ for all $i = 1, \ldots, n$, the probability $p_X(x_1, \ldots, x_n)$ in (8.5) can also be written as

$$p_X(x_1, \ldots, x_n) = p^{\sum_{i=1}^n x_i} \cdot (1 - p)^{n - \sum_{i=1}^n x_i}, \tag{8.6}$$

where $\sum_{i=1}^n x_i$ is the number of ones in the sequence $x_1, \ldots, x_n$ (see Exercise 8.1).

Remark 8.5 [Probability function of a multivariate Bernoulli variable] Let $X = (X_1, \ldots, X_n)$ be an n-variate Bernoulli variable with parameter p. Then its probability function p_X is defined by (8.5) and its distribution, P_X, is the multivariate Bernoulli distribution with parameters n and p, which follows from (8.4), independence of $X_1, \ldots, X_n$, and Lemma 5.68.   ◁

Example 8.6 [A special case] For $p = .5$, Equation (8.6) implies $p_X(x_1, \ldots, x_n) = .5^n$, for all $(x_1, \ldots, x_n) \in \{0, 1\}^n$. Hence, this multivariate Bernoulli distribution is the discrete uniform distribution on $\{0, 1\}^n$; that is, in this case an n-variate Bernoulli variable $X = (X_1, \ldots, X_n)$ with parameter $p = .5$ is uniformly distributed on $\{0, 1\}^n$.   ◁

8.1.3 Binomial distribution

In the following definition, we use the binomial coefficient

$$\binom{n}{x} := \frac{n!}{x! \cdot (n - x)!}, \quad n \in \mathbb{N}_0, x = 0, 1, \ldots, n. \tag{8.7}$$

Furthermore, $a! := a \cdot (a - 1) \cdot \ldots \cdot 1$ denotes the *factorial of* $a \in \mathbb{N}_0$, where by convention $0! := 1$.

Definition 8.7 [Binomial distribution]
Let $X: (\Omega, \mathcal{A}, P) \to (\mathbb{N}_0, \mathcal{P}(\mathbb{N}_0))$ be a discrete random variable. If

$$\forall x \in \mathbb{N}_0: \quad p_X(x) = b_{n,p}(x) := \begin{cases} \binom{n}{x} p^x (1 - p)^{n-x}, & \text{if } x = 0, 1, \ldots, n \\ 0, & \text{if } x > n, \end{cases} \tag{8.8}$$

where $n \in \mathbb{N}$ and $0 \leq p \leq 1$, then we use the notation $X \sim B_{n,p}$ and say that X has a binomial distribution with parameters n and p.

Remark 8.8 [Distribution function] If $b_{n,p}$ is the probability function of X, that is, if $X \sim B_{n,p}$, then its *distribution function* is

$$\forall a \in \mathbb{R}: \quad F_X(a) = \begin{cases} 0, & \text{if } a < 0 \\ \sum_{x=0}^{k} b_{n,p}(x), & \text{if } k \leq a < k + 1 \quad \text{and} \quad k = 0, 1, \ldots, n - 1 \\ 1, & \text{if } a \geq n. \end{cases} \tag{8.9}$$

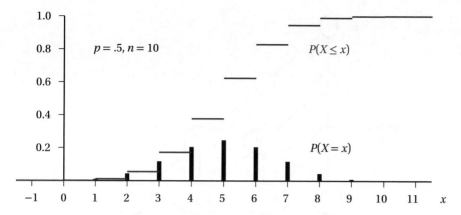

Figure 8.1   Probability function and distribution function of a binomial distribution.

Figure 8.1 shows the probability function and the distribution function of the binomial distribution for parameters $n = 10$ and $p = .5$. Figure 8.3 also shows the probability function for parameters $n = 10, p = .2$ and $n = 20, p = .1$. Note that probabilities close to zero may not be visible in this figure.   ◁

According to the following theorem, the sum of a sequence of n Bernoulli variables with parameter p has a binomial distribution with parameters n and p.

Theorem 8.9 [Sum of Bernoulli variables]
Let $X_1, \dots, X_n$ be a sequence of n Bernoulli variables with parameter p. Then

$$X := \sum_{i=1}^{n} X_i \sim \mathcal{B}_{n,p}. \tag{8.10}$$

(Proof p. 276)

Remark 8.10 [A reformulation using events]   Explicitly referring to events and their indicators, Theorem 8.9 may also be formulated as follows: If $(\Omega, \mathcal{A}, P)$ is a probability space and, for all $i = 1, \dots, n$, $A_i \in \mathcal{A}$ are independent with $P(A_i) = p \geq 0$, then $X := \sum_{i=1}^{n} 1_{A_i} \sim \mathcal{B}_{n,p}$.   ◁

Figure 8.2 illustrates the probability function p_X of $X = X_1 + \dots + X_4$, where X_i indicates 'success' in trial i, occurring with probability $p = P(X_i = 1)$ $(q := 1 - p)$. We consider $n = 0, 1, \dots, 4$ trials presented in the five 'rows' of the figure. The numbers in the circles and ellipses are the probabilities $p_X(x)$ of the number x of successes, which are computed by Equation (8.8) (see the 'columns' of the figure).

Example 8.11 [Flipping a coin n times]   A simple example is flipping a coin n times. The event A_i is *flipping heads at trial i*, $i = 1, \dots, n$. If $p = 1/2$, then we say that the coin is fair. If X is the number of flipping heads, that is, $X = \sum_{i=1}^{n} 1_{A_i}$, then, for $n = 2$, the values

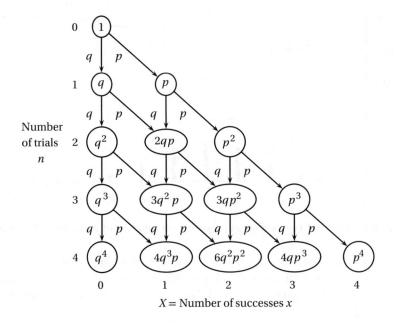

Figure 8.2   Probability function of the sum of i. i. d. Bernoulli variables.

of the binomial distribution are $b_{2,1/2}(0) = P(X=0) = 1/4$, $b_{2,1/2}(1) = P(X=1) = 1/2$, and $b_{2,1/2}(2) = P(X=2) = 1/4$, which have already been computed in Example 5.64. For the case $n = 4$, see Exercise 8.2. ◁

Example 8.12 [Tossing n dices]   Another example is tossing n dices. The events A_i could be *tossing a six with dice i*. If the probabilities are identical for all six possible outcomes of a single trial, then we say that the dice is fair. In this case, the parameter is $p = 1/6$. If X is the number of tossing a six, then, for $n=2$, the values of the binomial distribution are $b_{2,1/6}(0) = P(X=0) = 25/36$, $b_{2,1/6}(1) = P(X=1) = 10/36$, and $b_{2,1/6}(2) = P(X=2) = 1/36$. ◁

Corollary 8.13 shows how expectation and variance of a random variable X with a binomial distribution depend on the parameters n and p.

Corollary 8.13 [Expectation and variance]
If $X \sim B_{n,p}$, then $E(X) = np$ and $Var(X) = np\,(1-p)$.

(Proof p. 277)

8.1.4   Poisson distribution

Another discrete distribution is the Poisson distribution. According to Theorem 8.17, it is "close" to the binomial distribution if n is large and p small.

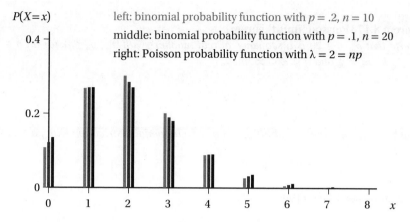

$P(X=x)$

left: binomial probability function with $p = .2$, $n = 10$
middle: binomial probability function with $p = .1$, $n = 20$
right: Poisson probability function with $\lambda = 2 = np$

Figure 8.3 Probability functions of binomial and Poisson distributions.

Definition 8.14 [Poisson distribution]
Let $X: (\Omega, \mathcal{A}, P) \to (\mathbb{N}_0, \mathscr{P}(\mathbb{N}_0))$ be a discrete random variable. If

$$p_X(x) = \frac{\lambda^x}{x!} \cdot e^{-\lambda}, \quad \forall\, x \in \mathbb{N}_0, \tag{8.11}$$

where $\lambda > 0$, then we use the notation $X \sim P_\lambda$ and say that X has a Poisson distribution with parameter λ.

Figure 8.3 displays this probability function for the parameter $\lambda = 2$.

Remark 8.15 [Distribution function] If $X \sim P_\lambda$, then its *distribution function* is obtained by sums of the probabilities $p_X(x)$ specified in Equation (8.11). More precisely,

$$\forall\, a \in \mathbb{R}: \quad F_X(a) = \begin{cases} 0, & \text{if } a < 0 \\ e^{-\lambda} \cdot \displaystyle\sum_{x=0}^{n} \frac{\lambda^x}{x!}, & \text{if } n \leq a < n+1 \text{ and } n \in \mathbb{N}_0. \end{cases} \tag{8.12}$$

◁

If X has a Poisson distribution, then, according to the following theorem, its expectation and variance are identical, and they are equal to the parameter λ.

Theorem 8.16 [Expectation and variance]
If $X \sim P_\lambda$, then $E(X) = Var(X) = \lambda$.

(Proof p. 277)

We can use the Poisson distribution for approximating the binomial distribution for large n and small p. The following theorem is the theoretical foundation.

Theorem 8.17 [Poisson limit theorem]
Suppose that p_n, $n \in \mathbb{N}$, is a sequence of real numbers with $0 \leq p_n \leq 1$ for all $n \in \mathbb{N}$. If $\lim_{n \to \infty} n \cdot p_n = \lambda$ and $0 < \lambda < \infty$, then,

$$\lim_{n \to \infty} b_{n, p_n}(x) = \frac{\lambda^x}{x!} \cdot e^{-\lambda}, \quad \forall \, x \in \mathbb{N}_0, \tag{8.13}$$

where b_{n, p_n} denotes the probability function of a binomial distribution with parameters n and p_n.

For a proof, see Bauer (1996).

Remark 8.18 [Poisson distribution as an approximation] Assuming $\lim_{n \to \infty} n \cdot p_n = \lambda > 0$ implies that the sequence $p_1, p_2, \ldots$ converges to 0. Hence, p_n will be small for a large n. Therefore, Equation (8.13) can be applied to a binomially distributed random variable X for large n and small p. In this case, p takes the role of p_n and $\lambda = n \cdot p$ takes the role of $\lim_{n \to \infty} n \cdot p_n$ in Theorem 8.17. Then,

$$P(X=x) = \binom{n}{x} p^x (1 - p)^{n-x} \approx \frac{\lambda^x}{x!} \cdot e^{-\lambda}, \quad \forall \, x = 0, 1, \ldots, n,$$

that is, the Poisson probability function approximates the binomial probability function (see Fig. 8.3). Obviously, the approximation is better for the parameter $n = 20, p = .1$ than for $n = 10, p = .2$ (i.e., it is better for larger n and smaller p).   ◁

8.1.5   Geometric distribution

Now we turn to an infinite sequence of Bernoulli trials, which is useful if the number of trials cannot be fixed in advance. An example is the random experiment of flipping a coin until the first *heads* occurs (see Example 8.24). In contrast to the last section, in which we considered a finite sequence of Bernoulli trials, now we will assume $p \neq 0$ and $p \neq 1$.

Definition 8.19 [Infinite sequence of Bernoulli variables]
Let $X_1, X_2, \ldots : (\Omega, \mathcal{A}, P) \to (\{0, 1\}, \mathcal{P}(\{0, 1\}))$ be i. i. d. random variables with

$$P(X_i=1) = 1 - P(X_i=0) = p, \quad \forall \, i = 1, 2, \ldots, \tag{8.14}$$

where $0 < p < 1$. Then $X_1, X_2, \ldots$ is called an infinite sequence of Bernoulli variables with parameter p.

The set Ω occurring in this definition can be specified as the set of all infinite sequences of elements of $\{0, 1\}$. Note that it is not obvious that there is an *infinite* sequence of *independent* random variables X_i satisfying (8.14). A proof that such an infinite sequence actually exists can be found in Klenke (2013, Th. 2.19 and Example 2.18).

According to Theorem 8.23 and Example 8.24, such an infinite sequence of Bernoulli variables is closely related to the geometric distribution that is defined as follows.

Definition 8.20 [Geometric distribution]
Let $X: (\Omega, \mathcal{A}, P) \to (\mathbb{N}, \mathcal{P}(\mathbb{N}))$ be a discrete random variable. If

$$p_X(x) = (1-p)^{x-1} \cdot p, \quad \forall x \in \mathbb{N}, \tag{8.15}$$

where $0 < p < 1$, then we use the notation $X \sim \mathcal{G}_p$ and say that X has a geometric distribution with parameter p.

Note that, for $0 < p < 1$,

$$\sum_{x=1}^{\infty} (1-p)^{x-1} = \frac{1}{1-(1-p)} = \frac{1}{p},$$

which implies $\sum_{x=1}^{\infty} p_X(x) = 1$.

Remark 8.21 [An alternative definition]   Sometimes the geometric distribution is alternatively defined for a random variable Y with values $y \in \mathbb{N}_0$. In this case, the probability function is $p_Y(y) = (1-p)^y \cdot p$. ◁

Theorem 8.22 [Expectation and variance]
If $X \sim \mathcal{G}_p$, then $E(X) = 1/p$ and $Var(X) = (1/p^2) - 1/p$.

(Proof p. 278)

In Theorem 8.23, we consider the probability function of the random variable X, the number of the trial in which the first *one* occurs in an infinite sequence of Bernoulli trials.

Theorem 8.23 [A class of random variables with geometric distributions]
Let $X_1, X_2, \ldots: (\Omega, \mathcal{A}, P) \to (\{0, 1\}, \mathcal{P}(\{0, 1\}))$ be an infinite sequence of Bernoulli variables with parameter p. If the random variable $X: (\Omega, \mathcal{A}, P) \to (\mathbb{N}, \mathcal{P}(\mathbb{N}))$ is defined by

$$X(\omega) = \min\{n \in \mathbb{N}: X_n(\omega) = 1\}, \quad \forall \omega \in \Omega, \tag{8.16}$$

then X has a geometric distribution with parameter p.

(Proof p. 279)

Example 8.24 [Flipping a coin until *heads* occurs]   Consider repeatedly flipping a coin and define the sequence of random variables $X_1, X_2, \ldots$ with $X_i(\omega) = 1$ if *heads* occurs at the ith flip and $X_i(\omega) = 0$, otherwise. This specifies an infinite sequence of Bernoulli variables with

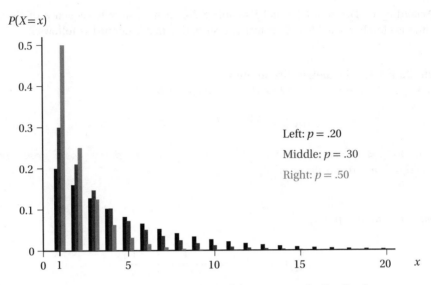

Figure 8.4   Probability functions of three geometric distributions.

parameter $p = .5$. Now define the random variable X by Equation (8.16). Hence, X is the index of the first of the variables X_i taking on the value 1, that is,

$$X(\omega) = n, \quad \text{if } X_1(\omega) = \ldots = X_{n-1}(\omega) = 0 \text{ and } X_n(\omega) = 1, \quad \forall\, \omega \in \Omega, \quad n \in \mathbb{N}.$$

Then X has a geometric distribution with parameter $p = .5$.   ◁

Remark 8.25 [Distribution function]   If $X \sim \mathcal{G}_p$, then,

$$\forall\, x \in \mathbb{R}: \quad F_X(x) = P(X \le x) = \begin{cases} 0, & \text{if } x < 1 \\ 1 - (1-p)^i, & \text{if } i \le x < i+1,\ i \in \mathbb{N}, \end{cases} \tag{8.17}$$

is the distribution function of X (see Exercise 8.3).   ◁

Another discrete distribution based on an infinite sequence $X_1, X_2, \ldots$ of Bernoulli variables is the negative binomial (or Pascal) distribution (see, e.g., Johnson *et al.*, 2005).

8.2   Some distributions of continuous random variables

8.2.1   Continuous uniform distribution

We begin with some examples using the densities of various random variables with continuous uniform distributions.

Example 8.26 [Uniform distribution on an interval] Let $X: (\Omega, \mathcal{A}, P) \to (\mathbb{R}, \mathcal{B})$ be a real-valued random variable. Then X has a *uniform distribution on the interval* $[a, b]$, $a < b$, a, $b \in \mathbb{R}$, if

$$f_X(x) = \frac{1}{b - a} \cdot 1_{[a,b]}(x). \tag{8.18}$$

◁

Example 8.27 [Uniform distribution on a rectangle] The random variable $X = (X_1, X_2): (\Omega, \mathcal{A}, P) \to (\mathbb{R}^2, \mathcal{B}_2)$ has a *uniform distribution on the rectangle* $[a, b] \times [c, d]$, $a < b$, $c < d$, $a, b, c, d \in \mathbb{R}$, if

$$f_X(x_1, x_2) = \frac{1}{(b - a) \cdot (d - c)} \cdot 1_{[a,b]}(x_1) \cdot 1_{[c,d]}(x_2) \tag{8.19}$$

(see Fig. 8.5). Note that X has a uniform distribution on the rectangle $[a, b] \times [c, d]$ if and only if the following three conditions hold:

(a) X_1 and X_2 are independent.

(b) X_1 has a uniform distribution on $[a, b]$.

(c) X_2 has a uniform distribution on $[c, d]$.

(See Cor. 5.100.)

◁

Example 8.28 [Uniform distribution on a circle] Let $X = (X_1, X_2): (\Omega, \mathcal{A}, P) \to (\mathbb{R}^2, \mathcal{B}_2)$ be a random variable. Then X has a *uniform distribution on the circle* $B_r = \{(x_1, x_2) \in \mathbb{R}^2 : x_1^2 + x_2^2 \leq r\}$, $0 < r \in \mathbb{R}$, if

$$f_X(x_1, x_2) = \frac{1}{\pi \cdot r^2} \, 1_{B_r}(x_1, x_2). \tag{8.20}$$

In this case, X_1 and X_2 are *not* independent.

◁

Definition 8.29 [Continuous uniform distribution on a bounded set]
Let $X: (\Omega, \mathcal{A}, P) \to (\mathbb{R}^n, \mathcal{B}_n)$ be a random variable and let $B \in \mathcal{B}_n$ such that $0 < \lambda_n(B) < \infty$. Then X has a (continuous) uniform distribution on the set B, denoted $X \sim \mathcal{U}_B$, if X has a density satisfying

$$f_X(x) = \frac{1}{\lambda_n(B)} \, 1_B(x), \quad \forall\, x \in \mathbb{R}^n. \tag{8.21}$$

Equation (5.40) and Theorem 3.68 imply: For $B \in \mathcal{B}_n$ with $0 < \lambda_n(B) < \infty$,

$$\left(\forall\, A' \in \mathcal{B}_n \colon P(X \in A') = \frac{\lambda_n(A')}{\lambda_n(B)} \right) \Leftrightarrow X \sim \mathcal{U}_B. \tag{8.22}$$

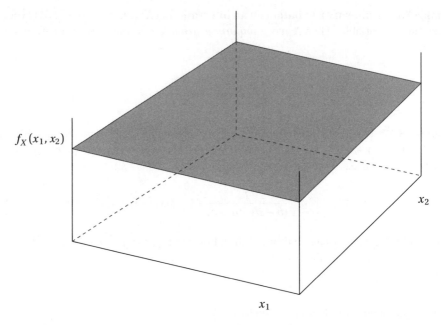

$f_X(x_1, x_2)$

x_2

x_1

Figure 8.5   Density of a bivariate uniform distribution on a rectangle.

Hence, the probability of an event $\{X \in A'\}$, $A' \in \mathcal{B}_n|_B$, only depends on the value $\lambda_n(A')$ of the Lebesgue measure λ_n. This is an invariance property that characterizes a continuous uniform distribution, that is, there is no other distribution on a bounded set $B \in \mathcal{B}_n$ that shares this property with the uniform distribution.

In the special case $n = 2$, the probability $P(X \in A')$ only depends on the area of $A' \in \mathcal{B}_2$; it does *not depend* on the location or the shape (triangle, ellipse, ...) of the set A'. Figure 8.5 shows the density of a bivariate uniform distribution on a rectangle.

8.2.2   Normal distribution

Definition 8.30 [Normal distribution]
A continuous random variable $X: (\Omega, \mathcal{A}, P) \to (\mathbb{R}, \mathcal{B})$ *has a normal distribution with parameters* $\mu \in \mathbb{R}$ *and* $\sigma^2 > 0$, *denoted* $X \sim \mathcal{N}_{\mu,\sigma^2}$, *if it has a density satisfying*

$$f_X(x) = \frac{1}{\sqrt{2\pi\sigma^2}} \cdot \exp\left(-\frac{(x - \mu)^2}{2\sigma^2}\right), \quad \forall\, x \in \mathbb{R}. \tag{8.23}$$

Figure 8.6 displays the densities of the normal distributions for three different parameter pairs of μ and σ^2. Comparing the three densities to each other illustrates that μ is a location parameter and σ^2 a scale parameter.

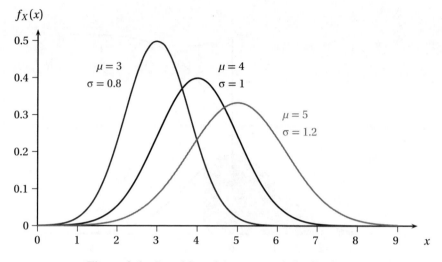

Figure 8.6   Densities of three normal distributions.

Theorem 8.31 [Expectation and variance]
If $X \sim \mathcal{N}_{\mu,\sigma^2}$, then $E(X) = \mu$ and $Var(X) = \sigma^2$.

For a proof, see Georgii (2008, Example 4.28).

The normal distribution with expectation $\mu = 0$ and variance $\sigma^2 = 1$ is called the *standard normal distribution*. In this case, Equation (8.23) simplifies to

$$f_X(x) = \frac{1}{\sqrt{2\pi}} \cdot \exp\left(-\frac{x^2}{2}\right), \quad \forall x \in \mathbb{R} \tag{8.24}$$

(see Fig. 8.7 and Exercise 8.4).

Remark 8.32 [Distribution function]   If $X \sim \mathcal{N}_{\mu,\sigma^2}$, then its *distribution function* is

$$F_X(x) = \frac{1}{\sqrt{2\pi\sigma^2}} \cdot \int_{-\infty}^{x} \exp\left(-\frac{(t-\mu)^2}{2\sigma^2}\right) dt, \quad \forall x \in \mathbb{R} \tag{8.25}$$

(see Cor. 5.95). If $\mu = 0$ and $\sigma^2 = 1$, then this distribution function is often denoted by Φ, and in this case Equation (8.25) simplifies to

$$\Phi(x) = \frac{1}{\sqrt{2\pi}} \cdot \int_{-\infty}^{x} \exp\left(-\frac{t^2}{2}\right) dt, \quad \forall x \in \mathbb{R}. \tag{8.26}$$

Figure 8.7 displays the graphs of the density and the distribution function of the standard normal distribution. The shaded area is the Riemann integral of the density from $-\infty$ to .6. Its value is $\Phi(.6) \approx 0.7257$, the value of the distribution function.   ◁

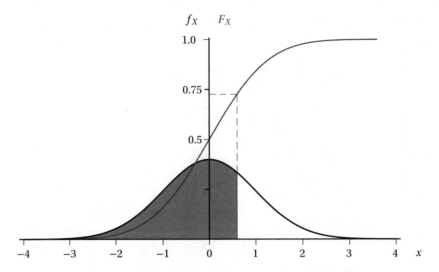

Figure 8.7   Density and distribution function of the standard normal distribution.

Remark 8.33 [Linear functions of X]   If $X \sim \mathcal{N}_{\mu,\sigma^2}$, then $\alpha + \beta\,X \sim \mathcal{N}_{\alpha+\beta\mu,\,\beta^2\sigma^2}$ (see Klenke, 2013, Exercise 1.5.3).   ◁

According to the central limit theorem, the limit (for $n \to \infty$) of the sum of n independent identically distributed (i. i. d.) random variables $X_1, \ldots, X_n$ with positive and finite variances has a normal distribution.

Theorem 8.34 [Central limit theorem]
Let $X_i \colon (\Omega, \mathcal{A}, P) \to (\mathbb{R}, \mathcal{B})$, $i = 1, 2, \ldots$, be a sequence of real-valued i. i. d. random variables with finite expectations $E(X_i) = \mu$ and finite variances $Var(X_i) = \sigma^2 > 0$. Furthermore, let $\overline{X}_n := S_n/n$, where $S_n := \sum_{i=1}^{n} X_i$, and

$$Z_n := \frac{S_n - n\,\mu}{\sigma \cdot \sqrt{n}} = \frac{(\overline{X}_n - \mu) \cdot \sqrt{n}}{\sigma}. \tag{8.27}$$

Then,

$$\lim_{n \to \infty} P(Z_n \leq z) = \Phi(z), \quad \forall\, z \in \mathbb{R}, \tag{8.28}$$

where $\Phi(z)$ denotes the distribution function of the standard normal distribution.

For a proof, see Georgii (2008, section 5.3). The second equation of (8.27) reveals that Z_n is the Z-transformation (see Rem. 6.33) of the sample mean $\overline{X}_n = \frac{1}{n}\sum_{i=1}^{n} X_i$, because $\sigma/\sqrt{n}$ is the standard deviation of $\overline{X}_n$ (see Exercise 6.6).

Remark 8.35 [Application of the central limit theorem]   The central limit theorem can be applied for the approximation of the distributions of sums of i. i. d. random variables. For large n and $x \in \mathbb{R}$,

$$P\left(\sum_{i=1}^{n} X_i \leq x\right) \approx \Phi\left(\frac{x - n\mu}{\sigma \cdot \sqrt{n}}\right). \qquad (8.29)$$

Note, if the X_i are discrete and integer-valued, then for all integers k,

$$P\left(\sum_{i=1}^{n} X_i \leq k\right) = P\left(\sum_{i=1}^{n} X_i < k+1\right), \qquad (8.30)$$

and therefore one may use the 'correction'

$$P\left(\sum_{i=1}^{n} X_i \leq k\right) \approx \Phi\left(\frac{k + \frac{1}{2} - n\mu}{\sigma \cdot \sqrt{n}}\right). \qquad (8.31)$$

Whether or not n is large enough for a good approximation can be dealt with by the Berry-Esséen bound [see Georgii, 2008, Remark 5.30 (c)]. A rough summary is: The more symmetric the distribution of the X_i, the faster the convergence of the limit in (8.28), that is, small n already yield good approximations. For example, for $n \geq 12$, good approximations of the distribution function of Z_n are already obtained, for the symmetric distributions of $X_i \sim \mathcal{U}_{[0,1]}$ and $X_i \sim \mathcal{B}_{n,1/2}$. In contrast, the approximation is still bad for the skewed (asymmetric) distribution of $X_i \sim \mathcal{B}_{n,1/20}$ even if $n = 50$. In the latter case, the approximation of a binomial distribution by a Poisson distribution (see Remark 8.18) is better than by a normal distribution. ◁

8.2.3   Multivariate normal distribution

Now we present a generalization of the univariate normal distribution considering an n-dimensional real-valued random variable $X = (X_1, \ldots, X_n)$. Reading the following definition, remember that an $n \times n$-matrix $\mathbf{A}$ is called *symmetric* if it is identical to its transpose (i.e., if $\mathbf{A} = \mathbf{A}'$), and that it is *positive definite* if $\mathbf{x}'\mathbf{A}\,\mathbf{x} > 0$ for all column vectors $\mathbf{x} \in \mathbb{R}^n$, $\mathbf{x} \neq \mathbf{0}$. Furthermore, $\det \boldsymbol{\Sigma}$ denotes the determinant of $\boldsymbol{\Sigma}$.

Definition 8.36 [Multivariate normal distribution]
Let $\boldsymbol{\mu}$ be an n-dimensional vector of real numbers and $\boldsymbol{\Sigma}$ a symmetric and positive definite $n \times n$-matrix of real numbers. Furthermore, let $X = (X_1, \ldots, X_n)$: $(\Omega, \mathcal{A}, P) \to (\mathbb{R}^n, \mathcal{B}_n)$ be an n-dimensional random variable. If the function $f_X \colon \mathbb{R}^n \to \mathbb{R}$ defined by

$$f_X(\mathbf{x}) = \frac{1}{\sqrt{(2\pi)^n \det \boldsymbol{\Sigma}}} \cdot \exp\left(-\frac{1}{2}(\mathbf{x} - \boldsymbol{\mu})'\boldsymbol{\Sigma}^{-1}(\mathbf{x} - \boldsymbol{\mu})\right), \quad \forall\, \mathbf{x} \in \mathbb{R}^n \qquad (8.32)$$

is a density of X, then we say that X has an n-variate (or n-dimensional) normal distribution with parameters $\boldsymbol{\mu}$ and $\boldsymbol{\Sigma}$, and we denote it by $X \sim \mathcal{N}_{\boldsymbol{\mu}, \boldsymbol{\Sigma}}$.

In section 7.4.2 we introduced the expectation of an n-dimensional random vector, and in section 7.4.3 the variance-covariance matrix of an n-variate random variable $X = (X_1, \dots, X_n)$. These concepts are used in the following theorem showing that the parameters $\boldsymbol{\mu}$ and $\boldsymbol{\Sigma}$ have a stochastic interpretation.

Theorem 8.37 [Expectation vector and covariance matrix]
Let $E(x)$ denote the expectation of the random vector $x = [X_1, \dots, X_n]'$ and $\boldsymbol{\Sigma}_{xx}$ the variance-covariance matrix of the n-variate random variable $X = (X_1, \dots, X_n)$. If $X \sim \mathcal{N}_{\boldsymbol{\mu}, \boldsymbol{\Sigma}}$, then $E(x) = \boldsymbol{\mu}$ and $\boldsymbol{\Sigma}_{xx} = \boldsymbol{\Sigma}$.

For a proof, see Georgii (2008, Th. 9.2).

Remark 8.38 [Univariate normal distribution] The univariate normal distribution (see Def. 8.30) is in fact a special case of Definition 8.36. This can be seen as follows: If $n = 1$ and $X \sim \mathcal{N}_{\mu,\sigma^2}$, then $\boldsymbol{\mu} = [\mu] = [E(X)]$, $\boldsymbol{\Sigma} = [\sigma^2]$, $\det \boldsymbol{\Sigma} = \sigma^2$, and $\boldsymbol{\Sigma}^{-1} = \left[\frac{1}{\sigma^2} \right]$. Hence, the density (8.32) simplifies to (8.23).   ◁

Remark 8.39 [Bivariate normal distribution]   If $n = 2$ and $X = (X_1, X_2) \sim \mathcal{N}_{\boldsymbol{\mu}, \boldsymbol{\Sigma}}$, then,

$$\boldsymbol{\mu} = [\mu_1, \mu_2]' = E([X_1, X_2]') = [E(X_1), E(X_2)]' \tag{8.33}$$

and

$$\boldsymbol{\Sigma} = \boldsymbol{\Sigma}_{xx} = \begin{bmatrix} \sigma_1^2 & \sigma_{12} \\ \sigma_{21} & \sigma_2^2 \end{bmatrix} = \begin{bmatrix} \sigma_1^2 & \rho \cdot \sigma_1 \sigma_2 \\ \rho \cdot \sigma_1 \sigma_2 & \sigma_2^2 \end{bmatrix}, \tag{8.34}$$

where $\rho = Corr(X_1, X_2)$, and the density (8.32) can also be written as:

$$f_X(x_1, x_2) = \frac{1}{2 \pi \sigma_1 \sigma_2 \sqrt{1 - \rho^2}} \cdot \exp\left(-\frac{z_1^2 - 2 \rho z_1 z_2 + z_2^2}{2(1 - \rho^2)} \right), \tag{8.35}$$

where $z_i := (x_i - \mu_i)/\sigma_i$ and $\sigma_i = \sqrt{\sigma_i^2}$, for $i = 1, 2$ (see Fisz, 1963, section 5.11). Figure 8.8 displays the density function of a bivariate normal distribution. The volume above the rectangle in the (x_1, x_2)-plane under the graph of the density is the probability that the bivariate random variable $X = (X_1, X_2)$ takes on a value in that rectangle.   ◁

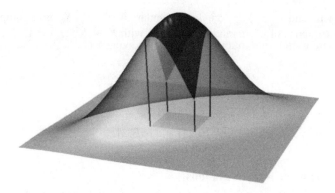

Figure 8.8 Density of a bivariate normal distribution.

Remark 8.40 [Correlation and independence] If (X_1, X_2) has a bivariate normal distribution and $\rho = 0$, then Equation (8.35) simplifies to

$$f_X(x_1, x_2) = \frac{1}{2\pi\sigma_1\sigma_2} \cdot \exp\left[-\frac{1}{2}\left(z_1^2 + z_2^2\right)\right]$$

$$= \frac{1}{\sqrt{2\pi\sigma_1^2}} \cdot \exp\left[-\frac{z_1^2}{2}\right] \cdot \frac{1}{\sqrt{2\pi\sigma_2^2}} \cdot \exp\left[-\frac{z_2^2}{2}\right] \qquad (8.36)$$

$$= f_{X_1}(x_1) \cdot f_{X_2}(x_2).$$

According to this equation, the density of (X_1, X_2) is the product of the univariate normal densities of X_1 and X_2, respectively [see Eq. (8.23)]. As shown in Exercise 8.5, this implies that X_1 and X_2 are independent if and only if $\rho = 0$. Hence, under bivariate normality, X_1 and X_2 are independent if and only if they are uncorrelated.   ◁

Theorem 8.41 [Linear combinations]
Let $X = (X_1, \ldots, X_n)\colon (\Omega, \mathcal{A}, P) \to (\mathbb{R}^n, \mathcal{B}_n)$ be an n-variate real-valued random variable and $x = [X_1, \ldots, X_n]'$. Then $X \sim \mathcal{N}_{\mu, \Sigma}$ if and only if

$$\forall\, a = [a_1, \ldots, a_n]' \in \mathbb{R}^n\colon\ a'x \sim \mathcal{N}_{a'\mu,\, a'\Sigma a}. \qquad (8.37)$$

For a proof and other characterizations of the multivariate normal distribution, see Tong (1990, ch. 3).

Remark 8.42 [Special cases] If $X_1, \ldots, X_n$ are independent random variables with $X_i \sim \mathcal{N}_{\mu_i, \sigma_i^2}$, $\mu_i \in \mathbb{R}$ and $\sigma_i^2 > 0$, $i = 1, \ldots, n$, then $\Sigma = \Sigma_{xx}$ [see Eq. (7.36)] is diagonal, and proposition (8.37) with $a = [1, \ldots, 1]'$ yields

$$\sum_{i=1}^{n} X_i \sim \mathcal{N}_{\mu_S, \sigma_S^2}$$

with $\mu_S = \sum_{i=1}^{n} \mu_i$ and $\sigma_S^2 = \sum_{i=1}^{n} \sigma_i^2$. In particular, if $X_1, \ldots, X_n$ is a sample, that is, if $X_1, \ldots, X_n$ is a sequence of i.i.d. random variables with $X_i \sim \mathcal{N}_{\mu, \sigma^2}$, $i = 1, \ldots, n$, $\mu \in \mathbb{R}$, and $\sigma^2 > 0$, then,

$$\frac{(\overline{X}_n - \mu) \cdot \sqrt{n}}{\sigma} \sim \mathcal{N}_{0,1}, \tag{8.38}$$

where $\overline{X}_n$ is the sample mean [see Eqs. (6.23) and (6.32)].   ◁

Theorem 8.43 [Linear transformations]
Let $X = (X_1, \ldots, X_n) \sim \mathcal{N}_{\mu, \Sigma}$ and $x = [X_1, \ldots, X_n]'$. Furthermore, for $m \leq n$, let $\mathbf{A}$ be an $m \times n$-matrix of real numbers of rank m and let $\mathbf{c}$ be a column m-vector of real numbers. Then,

$$\mathbf{A}x + \mathbf{c} \sim \mathcal{N}_{\mathbf{A}\mu + \mathbf{c},\, \mathbf{A}\Sigma\mathbf{A}'}. \tag{8.39}$$

For a proof, see Georgii (2008, Th. 9.5).

Example 8.44 [Univariate marginal distribution] For $m = 1$, $\mathbf{A} = [1, 0, \ldots, 0]$, and $\mathbf{c} = [0]$, Theorem 8.43 implies

$$X_1 \sim \mathcal{N}_{\mu_1, \sigma_1^2}, \tag{8.40}$$

where μ_1 is the first coordinate of μ and σ_1^2 the first diagonal element of Σ.   ◁

Example 8.45 [Bivariate marginal distribution] For $m = 2$,

$$\mathbf{A} = \begin{bmatrix} 1, 0, 0, \ldots, 0 \\ 0, 1, 0, \ldots, 0 \end{bmatrix},$$

and $\mathbf{c} = [0, 0]'$, Theorem 8.43 implies

$$\mathbf{A}x = [X_1, X_2]' \sim \mathcal{N}_{\mu_{12},\, \Sigma_{12}}, \tag{8.41}$$

where

$$\mu_{12} := [\mu_1, \mu_2]' = E([X_1, X_2]') = [E(X_1), E(X_2)]'$$

and

$$\Sigma_{12} := \Sigma_{\mathbf{A}x,\, \mathbf{A}x} = \begin{bmatrix} \sigma_1^2 & \sigma_{12} \\ \sigma_{21} & \sigma_2^2 \end{bmatrix} = \begin{bmatrix} \sigma_1^2 & \rho \cdot \sigma_1 \sigma_2 \\ \rho \cdot \sigma_1 \sigma_2 & \sigma_2^2 \end{bmatrix},$$

which is the variance-covariance matrix of (X_1, X_2) [cf. Eqs. (8.33) and (8.34)].   ◁

Examples 8.44 and 8.45 show that the appropriate choices of m, $\mathbf{A}$, and $\mathbf{c}$ immediately yield the following corollary:

Corollary 8.46 [Marginal distributions]
If $X = (X_1, \ldots, X_n) \sim \mathcal{N}_{\mu, \Sigma}$, then all marginal distributions are normal. In particular, $X_i \sim \mathcal{N}_{\mu_i, \sigma_i^2}$, $i = 1, \ldots, n$, where μ_i is the ith coordinate of μ and σ_i^2 the ith diagonal element of Σ.

8.2.4   Central χ^2-distribution

In the following definition, $\Gamma: \mathbb{R} \to \mathbb{R}$ denotes the *gamma function* defined by

$$\Gamma(a) := \int_0^\infty t^{a-1} e^{-t} dt, \quad \forall\, a \in \mathbb{R}, \quad a > 0. \tag{8.42}$$

Note that

$$\Gamma(a) = (a-1) \cdot \Gamma(a-1), \quad \text{for } a > 1. \tag{8.43}$$

Furthermore,

$$\Gamma(a) = (a-1)! \quad \text{for } a \in \mathbb{N}, \quad \text{and} \quad \Gamma\left(\frac{1}{2}\right) = \sqrt{\pi}. \tag{8.44}$$

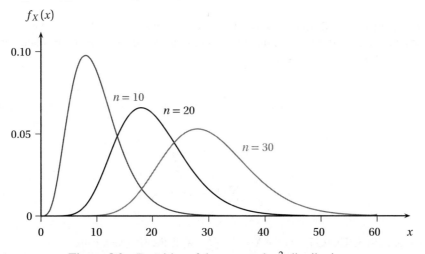

Figure 8.9   Densities of three central χ^2-distributions.

Definition 8.47 [Central χ^2-distribution]
Let $n \in \mathbb{N}$. A continuous nonnegative random variable $X: (\Omega, \mathcal{A}, P) \to (\mathbb{R}, \mathcal{B})$ has a central χ^2-distribution with n degrees of freedom, abbreviated $X \sim \chi_n^2$, if X has a density satisfying

$$f_X(x) = \begin{cases} \dfrac{x^{n/2-1} \cdot e^{-x/2}}{2^{n/2} \cdot \Gamma(n/2)}, & \text{if } x \geq 0 \\ 0, & \text{if } x < 0 \end{cases} \quad \forall x \in \mathbb{R}. \tag{8.45}$$

Normal and χ^2-distributions are related to each other as follows:

Theorem 8.48 [Relationship between normal and χ^2-distributions]
If $X_1, \ldots, X_n$ are i. i. d. random variables with standard normal distribution, then,

$$X := \sum_{i=1}^{n} X_i^2 \tag{8.46}$$

has a central χ^2-distribution with n degrees of freedom.

For a proof, see Fisz (1963, section 9.4).

Theorem 8.49 [Expectation and variance]
If $X \sim \chi_n^2$, then $E(X) = n$ and $Var(X) = 2n$.

(Proof p. 279)

Remark 8.50 [χ^2-distribution in statistics] Suppose that $X_1, \ldots, X_n$ is a sample of independent and normally distributed random variables with expectation $E(X_i) = \mu$ and variance $Var(X_i) = \sigma^2$, for $i = 1, \ldots, n$. Then,

$$X := \frac{1}{\sigma^2} \sum_{i=1}^{n} (X_i - \mu)^2 \sim \chi_n^2, \tag{8.47}$$

and

$$Y := \frac{1}{\sigma^2} \sum_{i=1}^{n} (X_i - \overline{X})^2 \sim \chi_{n-1}^2, \tag{8.48}$$

(see Georgii, 2008, Th. 9.17), where $\overline{X} = \frac{1}{n} \sum_{i=1}^{n} X_i$ is the *sample mean*. ◁

8.2.5 Central *t*-distribution

In the following definition, we again use the gamma function Γ defined by Equation (8.42).

Definition 8.51 [Central *t*-distribution]
Let $n \in \mathbb{N}$. A continuous random variable $X: (\Omega, \mathcal{A}, P) \to (\mathbb{R}, \mathcal{B})$ has a central *t*-distribution with *n* degrees of freedom, denoted $X \sim t_n$, if *X* has a density satisfying

$$f_X(x) = \frac{\Gamma((n+1)/2)}{\sqrt{n\pi} \cdot \Gamma(n/2)} \left(1 + \frac{x^2}{n}\right)^{-(n+1)/2}, \quad \forall x \in \mathbb{R}. \tag{8.49}$$

Figure 8.10 displays densities of three *t*-distributions with 1, 5, and 10 degrees of freedom, respectively.

Theorem 8.52 [Expectation and variance]
If $X \sim t_n$ and $n > 1$, then $E(X) = 0$, and if $n > 2$, then $Var(X) = n/(n-2)$.

For a proof, see Johnson *et al.* (1995).

Remark 8.53 [Cauchy density] If $X: (\Omega, \mathcal{A}, P) \to (\mathbb{R}, \mathcal{B})$ has a *t*-distribution with $n = 1$ degree of freedom, then the density f_X of *X* is also called the *(standard) Cauchy density*. In this case, the expectation of *X* does not exist (see Example 6.12). ◁

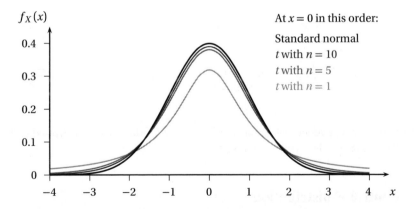

Figure 8.10 Densities of the standard normal and three *t*-distributions.

Theorem 8.54 [Relationship between t-, normal, and χ^2-distributions]
If $Z \sim \mathcal{N}_{0,1}$, $Y \sim \chi_n^2$, and Z and Y are independent, then,

$$X := \frac{Z}{\sqrt{\dfrac{Y}{n}}} \tag{8.50}$$

has a t-distribution with n degrees of freedom.

For a proof, see Fisz (1963, section 9.6) or Johnson *et al.* (1995, chapter 28).

Remark 8.55 [Convergence of densities]   The sequence f_{X_n}, $n = 1, 2, \ldots$, of densities of the t-distributions with n degrees of freedom converges to the density of a standard normal distribution for $n \to \infty$. That is, if f_{X_n}, $n = 1, 2, \ldots$, satisfy Equation (8.49), then,

$$\lim_{n \to \infty} f_{X_n}(x) = \frac{1}{\sqrt{2\pi}} \cdot \exp\left(\frac{-x^2}{2}\right), \quad \forall\, x \in \mathbb{R} \tag{8.51}$$

(see Fig. 8.10 and Exercise 8.6).   ◁

Remark 8.56 [t-distribution in statistics]   Suppose that $X_1, \ldots, X_n$ are independent and normally distributed random variables with expectation $E(X_i) = \mu$ and variance $Var(X_i) = \sigma^2$, for $i = 1, \ldots, n$. Then,

$$X := \frac{\overline{X} - \mu}{\frac{S}{\sqrt{n}}} = \sqrt{n} \cdot \frac{\overline{X} - \mu}{S} \sim t_{n-1}, \tag{8.52}$$

where

$$\overline{X} = \frac{1}{n} \sum_{i=1}^{n} X_i, \quad S^2 := \frac{1}{n-1} \sum_{i=1}^{n} (X_i - \overline{X})^2, \quad \text{and} \quad S := \sqrt{S^2}. \tag{8.53}$$

For the proof of independence of $\overline{X}$ and S, see Georgii (2008, Th. 9.17). Applying Theorem 8.54, (8.48), and (8.38) then yields the result.   ◁

8.2.6   Central F-distribution

In the following definition, we again use the gamma function Γ defined by Equation (8.42).

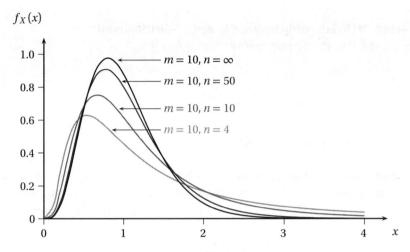

Figure 8.11 Densities of three F-distributions and the limit of the density of the F-distribution for $m = 10$ and n to ∞.

Definition 8.57 [Central F-distribution]
Let $m, n \in \mathbb{N}$ and let $X\colon (\Omega, \mathcal{A}, P) \to (\mathbb{R}, \mathcal{B})$ be a continuous nonnegative random variable. Then X has a central F-distribution with m and n degrees of freedom, abbreviated $X \sim F_{m,n}$, if it has a density satisfying

$$f_X(x) = \begin{cases} \dfrac{\Gamma((m+n)/2) \cdot m^{m/2} \cdot n^{n/2} \cdot x^{m/2-1}}{\Gamma(m/2) \cdot \Gamma(n/2) \cdot (n+mx)^{(m+n)/2}}, & \text{if } x \geq 0 \\ 0, & \text{if } x < 0 \end{cases} \quad \forall\, x \in \mathbb{R}. \quad (8.54)$$

Figure 8.11 displays the densities of three F-distributions and the limit of the density of the F-distribution for $m = 10$ and n to ∞.

Theorem 8.58 [Expectation and variance]
If $X \sim F_{m,n}$, then, for $n \geq 3$, the expectation of X is $E(X) = n/(n-2)$. For $n \leq 2$, the expectation of X is ∞. If $n \geq 5$, the variance of X is

$$Var(X) = \frac{2n^2 \cdot (m+n-2)}{m \cdot (n-2)^2 \cdot (n-4)}. \quad (8.55)$$

For $n \leq 2$, the variance does not exist and for $3 \leq n \leq 4$, the variance of X is infinite.

For a proof, see Johnson *et al.* (1995, chapter 27).

> **Theorem 8.59 [Relationship between F- and χ^2-distributions]**
> If $Z \sim \chi_m^2$ and $Y \sim \chi_n^2$ are independent, then,
>
> $$X := \frac{Z/m}{Y/n} \sim F_{m,n}. \tag{8.56}$$

For a proof, see Fisz (1963, section 9.7).

Remark 8.60 [F-distribution in statistics]   Let $Z_1, \ldots, Z_{n_1}, Y_1, \ldots, Y_{n_2}$ be independent normally distributed random variables with expectations $E(Z_i) = \mu_Z$, $E(Y_j) = \mu_Y$, and variances $Var(Z_i) = \sigma_Z^2$, $Var(Y_j) = \sigma_Y^2$, $i = 1, \ldots, n_1, j = 1, \ldots, n_2$, respectively. If $\sigma_Z^2 = \sigma_Y^2$, then,

$$X := \frac{S_Z^2}{S_Y^2} := \frac{\dfrac{1}{n_1 - 1} \displaystyle\sum_{i=1}^{n_1} (Z_i - \overline{Z})^2}{\dfrac{1}{n_2 - 1} \displaystyle\sum_{i=1}^{n_2} (Y_i - \overline{Y})^2} \sim F_{n_1-1, n_2-1}, \tag{8.57}$$

where

$$\overline{Z} := \frac{1}{n_1} \sum_{i=1}^{n_1} Z_i \quad \text{and} \quad \overline{Y} := \frac{1}{n_2} \sum_{j=1}^{n_2} Y_j \tag{8.58}$$

are the sample means [see (8.48) and Th. 8.59].   ◁

8.3   Proofs

Proof of Theorem 8.9

Let $x \in \{0, 1, \ldots, n\}$. Then, for any subset I_x of $\{1, \ldots, n\}$ that has x elements, the assumptions of independence and identical distributions of the $X_1, \ldots, X_n$ imply

$$P(\forall\, i \in I_x \colon X_i = 1, \forall\, i \in \{1, \ldots, n\} \setminus I_x \colon X_i = 0) = \prod_{i \in I_x} P(X_i = 1) \cdot \prod_{i \in \{1, \ldots, n\} \setminus I_x} [1 - P(X_i = 1)]$$

$$= p^x (1 - p)^{n-x}.$$

Note that different subsets of $\{1, \ldots, n\}$ represent disjoint events, even if the subsets have a nonempty intersection. Hence, as there are $\binom{n}{x}$ subsets I_x of $\{1, \ldots, n\}$ with x elements, additivity of P implies

$$P(X = x) = \binom{n}{x} p^x (1 - p)^{n-x}.$$

Proof of Corollary 8.13

According to Rule (x) of Box 4.1, the distribution P_X of X, which is a probability measure on $(\Omega', \mathcal{A}')$, is uniquely defined by the probability function p_X (see Def. 5.56). Furthermore, considering $g(X) := X$ or $g(X) := [X - E(X)]^2$, Corollary 6.17 shows that the expectation and the variance of X solely depend on its distribution P_X. Therefore, it suffices to derive the expectation and the variance of the random variable $X := \sum_{i=1}^{n} X_i$ on $(\Omega, \mathcal{A}, P)$ defined in Theorem 8.9 with independent and identically distributed $X_1, \ldots, X_n$. Hence, the expectation of X is

$$E(X) = E\left(\sum_{i=1}^{n} X_i\right) \qquad \text{[def. of } X]$$

$$= \sum_{i=1}^{n} E(X_i) \qquad \text{[Box 6.1 (vi)]}$$

$$= \sum_{i=1}^{n} P(X_i = 1) \qquad \text{[(6.4), (6.5)]}$$

$$= \sum_{i=1}^{n} p = np. \qquad \text{[}P(X_i = 1) = p]$$

Similarly, the variance of X is

$$Var(X) = Var\left(\sum_{i=1}^{n} X_i\right) \qquad \text{[def. of } X]$$

$$= \sum_{i=1}^{n} Var(X_i) \qquad \text{[Box 6.2 (vi), } \underset{P}{\perp\!\!\!\perp} X_i]$$

$$= \sum_{i=1}^{n} p(1-p) = np(1-p). \qquad \text{[Example 6.30, } P(X_i = 1) = p]$$

Proof of Theorem 8.16

If X has a Poisson distribution with parameter λ, then,

$$E(X) = \sum_{x=0}^{\infty} x \cdot p_X(x) = \sum_{x=1}^{\infty} x \cdot p_X(x) \qquad \text{[(6.6)]}$$

$$= \sum_{x=1}^{\infty} x \frac{\lambda^x}{x!} \cdot e^{-\lambda} \qquad \text{[(8.11)]}$$

$$= e^{-\lambda} \sum_{x=1}^{\infty} x \cdot \frac{\lambda \cdot \lambda^{x-1}}{x \cdot (x-1)!}$$

$$= e^{-\lambda} \lambda \cdot \sum_{x=1}^{\infty} \frac{\lambda^{x-1}}{(x-1)!}$$

$$= e^{-\lambda} \lambda \cdot \sum_{x=0}^{\infty} \frac{\lambda^x}{x!}$$

$$= e^{-\lambda} \lambda e^{\lambda} \qquad\qquad \left[e^{\lambda} = \sum_{x=0}^{\infty} \frac{\lambda^x}{x!} \right]$$

$$= \lambda. \qquad\qquad [e^{-\lambda} \cdot e^{\lambda} = e^{\lambda-\lambda}]$$

Because $E(X^2) = E[X \cdot (X - 1) + X] = E[X \cdot (X - 1)] + E(X)$ and $Var(X) = E(X^2) - E(X)^2$ [see Box 6.2 (i)], we consider

$$E[X \cdot (X - 1)] = \sum_{x=0}^{\infty} x(x - 1) \cdot p_X(x) \qquad\qquad [(6.16)]$$

$$= \sum_{x=0}^{\infty} x(x - 1) \frac{\lambda^x}{x!} \cdot e^{-\lambda} \qquad\qquad [(8.11)]$$

$$= e^{-\lambda} \sum_{x=2}^{\infty} x(x - 1) \cdot \frac{\lambda^2 \cdot \lambda^{x-2}}{x \cdot (x - 1)(x - 2)!}$$

$$= e^{-\lambda} \lambda^2 \cdot \sum_{x=2}^{\infty} \frac{\lambda^{x-2}}{(x - 2)!}$$

$$= e^{-\lambda} \lambda^2 \cdot \sum_{x=0}^{\infty} \frac{\lambda^x}{x!}$$

$$= e^{-\lambda} \lambda^2 e^{\lambda} \qquad\qquad \left[e^{\lambda} = \sum_{x=0}^{\infty} \frac{\lambda^x}{x!} \right]$$

$$= \lambda^2. \qquad\qquad [e^{-\lambda} \cdot e^{\lambda} = e^{\lambda-\lambda}]$$

Because $E(X) = \lambda$, we receive $Var(X) = E(X^2) - E(X)^2 = \lambda^2 + \lambda - \lambda^2 = \lambda$.

Proof of Theorem 8.22

If X has a geometric distribution with parameter p and we define $q := 1 - p$, then,

$$E(X) = \sum_{x=1}^{\infty} x \, p_X(x) \qquad\qquad [(6.6)]$$

$$= \sum_{x=1}^{\infty} x \, q^{x-1} \cdot p \qquad\qquad [(8.15)]$$

$$= p \cdot \sum_{x=1}^{\infty} x \, q^{x-1}$$

$$= p \cdot \sum_{x=1}^{\infty} \frac{d}{dq} q^x \qquad \left[\frac{d}{dq} q^x = x q^{x-1} \right]$$

$$= p \cdot \frac{d}{dq} \sum_{x=1}^{\infty} q^x$$

$$= p \cdot \frac{d}{dq} \left(\frac{1}{1-q} - 1 \right) \qquad \left[\sum_{x=1}^{\infty} q^x = \frac{1}{1-q} - 1 \right]$$

$$= p \cdot \frac{1}{(1-q)^2}$$

$$= p \cdot \frac{1}{p^2} = \frac{1}{p}. \qquad [p = 1 - q]$$

In the fifth equation, we used the fact that power series can be differentiated term-by-term within their radius of convergence.

The second moment $E(X^2) = \sum_{x=1}^{\infty} x^2 p_X(x)$ can be calculated analogously to $E(X^2)$, except for using the second derivative with respect to q. The variance is then obtained by $Var(X) = E(X^2) - E(X)^2$.

Proof of Theorem 8.23

Let $X_1, X_2, \ldots$ be an infinite sequence of Bernoulli variables and X be defined by Equation (8.16). Then, for all $x \in \mathbb{N}$:

$$p_X(x) = P(X=x) = P(X_1=0, X_2=0, X_{x-1}=0, \ldots, X_x=1)$$

$$= \left[\prod_{i=0}^{x-1} P(X_i=0) \right] \cdot P(X_x=1) \qquad [\underset{P}{\perp\!\!\!\perp} X_i, (5.28)]$$

$$= (1-p)^{x-1} \cdot p. \qquad [(8.14)]$$

According to Definition 8.20, this is the probability function of the geometric distribution.

Proof of Theorem 8.49

Let $X_1, \ldots, X_n$ be i.i.d. and $X_i \sim \mathcal{N}_{0,1}$, $i = 1, \ldots, n$, then $E(X_1) = 0$, $Var(X_1) = E(X_1^2) - E(X_1)^2 = E(X_1^2) = 1$. We start calculating:

$$E(X_1^4) = \int_{-\infty}^{\infty} x^4 f_{X_1}(x) \, dx \qquad [(6.13)]$$

$$= \frac{1}{\sqrt{2\pi}} \int_{-\infty}^{\infty} x^4 e^{-x^2/2} \, dx \qquad [(8.24)]$$

$$= \frac{1}{\sqrt{2\pi}} \left(-x^3 e^{-x^2/2} \Big|_{\infty}^{-\infty} + 3 \int_{\infty}^{-\infty} x^2 e^{-x^2/2} \, dx \right) \qquad \text{[integration by parts]}$$

$$= \frac{1}{\sqrt{2\pi}} \cdot \left(0 + 3 \cdot \sqrt{2\pi} \cdot E\left(X_1^2\right) \right) \qquad \text{[(6.22)]}$$

$$= \frac{1}{\sqrt{2\pi}} \cdot \left(3 \cdot \sqrt{2\pi} \right) = 3.$$

Using this result, we obtain

$$Var\left(X_1^2\right) = E\left(X_1^4\right) - E\left(X_1^2\right)^2 = 3 - 1 = 2.$$

If $X \sim \chi_n^2$, then this equation, $P_X = P_{\sum_{i=1}^n X_i^2}$ [see (8.46)], Corollary 6.17, $\underset{P}{\amalg} X_1, \ldots, X_n$, Box 6.1 (vii), and Box 6.2 (vi) yield

$$E(X) = E\left(\sum_{i=1}^n X_i^2 \right) = n \cdot E\left(X_1^2\right) = n$$

and

$$Var(X) = Var\left(\sum_{i=1}^n X_i^2 \right) = n \cdot Var\left(X_1^2\right) = 2n.$$

Exercises

8.1 Consider Definition 8.4 and show that (8.6) holds.

8.2 Consider the random variable X defined in Example 8.11 for $n = 4$ trials. Determine the elements of $\Omega = \{h, t\}^4$ that yield the value $X = 3$. Determine the probabilities for each $\{\omega\}$, $\omega \in \{X = 3\}$. Identify the four paths in Figure 8.2 that lead from knot 1 to knot 4 $p^3 q$. Also use this figure in order to determine $P(X=3)$.

8.3 Show that Equation (8.17) specifies the distribution function of X if $X \sim \mathcal{G}_p$.

8.4 Show that the expectation of a random variable Z that has a standard normal distribution is 0.

8.5 Use Corollary 5.100 to show that X_1 and X_2 are independent if and only if $\rho = 0$, provided that (X_1, X_2) has a bivariate normal distribution with density (8.35).

8.6 Prove the proposition of Remark 8.55.

Solutions

8.1
$$p_X(x_1, \ldots, x_n) = \prod_{i=1}^{n} p^{x_i}(1-p)^{1-x_i}$$

$$= p^{\sum_{i=1}^{n} x_i} \cdot (1-p)^{\sum_{i=1}^{n}(1-x_i)}$$

$$= p^{\sum_{i=1}^{n} x_i} \cdot (1-p)^{n-\sum_{i=1}^{n} x_i}.$$

8.2 If we consider $n = 4$ trials, then the value 3 of X is obtained by the $\binom{4}{3} = 4$ outcomes: $\omega_1 = (h, h, h, t)$, $\omega_2 = (h, h, t, h)$, $\omega_3 = (h, t, h, h)$, and $\omega_4 = (t, h, h, h)$. The corresponding probabilities are $P(\{\omega_1\}) = p^3 q$, $P(\{\omega_2\}) = p^2 q p$, $P(\{\omega_3\}) = p q p^2$, and $P(\{\omega_4\}) = q p^3$. Summing these four probabilities yields $P(X=3) = 4 \cdot p^3 q$. The four paths from knot 1 to knot $4\,p^3 q$ are: $1 \rightarrow p \rightarrow p^2 \rightarrow p^3 \rightarrow 4\,p^3 q$, $1 \rightarrow p \rightarrow p^2 \rightarrow 3\,q\,p^2$ $\rightarrow 4\,p^3 q$, $1 \rightarrow p \rightarrow 2\,q\,p \rightarrow 3\,q\,p^2 \rightarrow 4\,p^3 q$, and $1 \rightarrow q \rightarrow 2\,q\,p \rightarrow 3\,q\,p^2 \rightarrow 4\,p^3 q$. Using Figure 8.2 yields $P(X=3) = 4\,p^3 q = \binom{4}{3} p^3 q^{4-3}$. Because, in this example $p = q = .5$, this yields $P(X=3) = 4\,p^3 q = 4 \cdot .5^3 \cdot .5 = .25$.

8.3 For $x \in \mathbb{N}$,

$$1 - F_X(x) = P(X > x) \qquad\qquad [(5.51)]$$

$$= P(\min\{n \in \mathbb{N}: X_n(\omega) = 1\} > x) \qquad [\text{Th. 8.23}]$$

$$= P(X_1 = \ldots = X_x = 0)$$

$$= P(X_1 = 0) \cdot \ldots \cdot P(X_x = 0) \qquad [(5.28)]$$

$$= (1 - p)^x. \qquad\qquad [\text{Def. 8.19}]$$

Because X is a discrete random variable with values in $\mathbb{N}$, its distribution function is a right-continuous step function with jumps at $x \in \mathbb{N}$, which yields Equation (8.17).

8.4 According to Theorem 6.11 and Equation (8.26), we have to show that the function

$$h(z) := z \cdot \frac{1}{\sqrt{2\pi}} \cdot e^{-z^2/2} \qquad\qquad (8.59)$$

is integrable and that its integral is 0. For these purposes, we consider the positive and the negative parts of this function. The positive part is

$$h^+(z) := \begin{cases} 0, & \text{if } z < 0 \\ z \cdot \dfrac{1}{\sqrt{2\pi}} \cdot e^{-z^2/2}, & \text{if } z \geq 0, \end{cases}$$

and its integral is

$$\int_{-\infty}^{\infty} h^+(z)\, dz = \frac{1}{\sqrt{2\pi}} \cdot \int_0^{\infty} \left(z \cdot e^{-z^2/2} \right) dz$$

$$= \frac{1}{\sqrt{2\pi}} \cdot \left(-e^{-z^2/2} \Big|_0^{\infty} \right)$$

$$= \frac{1}{\sqrt{2\pi}} \cdot [0 - (-1)] = \frac{1}{\sqrt{2\pi}}.$$

The negative part is

$$h^-(z) := \begin{cases} -z \cdot \dfrac{1}{\sqrt{2\pi}} \cdot e^{-z^2/2}, & \text{if } z \leq 0 \\ 0 & \text{if } z > 0, \end{cases}$$

and its integral is

$$\int_{-\infty}^{\infty} h^-(z)\, dz = \frac{1}{\sqrt{2\pi}} \cdot \int_{-\infty}^{0} \left(-z \cdot e^{-z^2/2} \right) dz$$

$$= \frac{1}{\sqrt{2\pi}} \cdot \left(-e^{-z^2/2} \Big|_{-\infty}^{0} \right)$$

$$= \frac{1}{\sqrt{2\pi}} \cdot (1 - 0) = \frac{1}{\sqrt{2\pi}}.$$

Hence, according to Definition 3.28, the function $h(z)$ is integrable, and Equations (3.27) and (3.68) imply

$$\int_{-\infty}^{\infty} z \cdot \frac{1}{\sqrt{2\pi}} \cdot e^{-z^2/2}\, dz = \frac{1}{\sqrt{2\pi}} - \frac{1}{\sqrt{2\pi}} = 0.$$

8.5 If $\rho = 0$, then Equation (8.36) and Corollary 5.100 imply that X_1 and X_2 are independent. If $\rho \neq 0$, then $Cov(X_1, X_2) \neq 0$ [see (7.18)]. Finally, by contraposition, Rule (vi) of Box 7.1 implies that X_1 and X_2 are not independent.

8.6

$$\lim_{n \to \infty} \left(1 + \frac{x^2}{n} \right)^{-\frac{n+1}{2}} = \lim_{n \to \infty} \left[\left(1 + \frac{x^2}{n} \right)^n \right]^{-\frac{1}{2}} \left[1 + \frac{x^2}{n} \right]^{-\frac{1}{2}}$$

$$= \left[\lim_{n \to \infty} \left(1 + \frac{x^2}{n} \right)^n \right]^{-\frac{1}{2}} \left[\lim_{n \to \infty} \left(1 + \frac{x^2}{n} \right) \right]^{-\frac{1}{2}}$$

$$= \left(e^{x^2} \right)^{-1/2} \cdot 1^{-1/2}$$

$$= e^{-x^2/2}$$

(see Harris & Stocker, 1998).

Using the Stirling formula

$$\lim_{n \to \infty} \frac{n!}{\sqrt{2\pi n}\, n^n e^{-n}} = 1$$

and considering the subsequence of all even $n \in \mathbb{N}$ (for simplicity writing $\lim_{n \to \infty}$ instead of $\lim_{n \to \infty, n \text{ even}}$),

$$\lim_{n \to \infty} \frac{\Gamma\left(\frac{n}{2} + \frac{1}{2}\right)}{\sqrt{\pi}\sqrt{n}\,\Gamma\left(\frac{n}{2}\right)}$$

$$= \lim_{n \to \infty} \frac{1 \cdot 3 \cdot 5 \cdot \ldots \cdot (n-1)\sqrt{\pi}}{\sqrt{\pi}\sqrt{n}\left(\frac{n}{2}-1\right)!\, 2^{n/2}} \qquad\qquad [(8.43), (8.44)]$$

$$= \lim_{n \to \infty} \frac{\frac{n}{2} \cdot n!}{\sqrt{n}\left(\frac{n}{2}\right)!\, 2^{n/2} 2^{n/2}\left(\frac{n}{2}\right)!} \qquad\qquad \left[1 \cdot 3 \cdot 5 \cdot \ldots \cdot (n-1) = \frac{n!}{2^{n/2}\left(\frac{n}{2}\right)!}\right]$$

$$= \lim_{n \to \infty} \frac{n\sqrt{2\pi n}\, n^n e^{-n}}{2\sqrt{n}\, 2^n \sqrt{\pi n}\left(\frac{n}{2}\right)^{(n/2)} e^{-(n/2)} \sqrt{\pi n}\left(\frac{n}{2}\right)^{(n/2)} e^{-(n/2)}} \qquad [\text{Stirling for } n!,\, (n/2)!]$$

$$= \lim_{n \to \infty} \frac{n\sqrt{2\pi n}\, n^n e^{-n}}{2\sqrt{n}\, 2^n \pi n\, n^n\, 2^{-n} e^{-n}}$$

$$= \lim_{n \to \infty} \frac{\sqrt{2\pi}}{2\pi} = \frac{1}{\sqrt{2\pi}}.$$

The proof for the subsequence of all odd $n \in \mathbb{N}$ is analogous. Combining the two limits yields the proposition.

Part III

CONDITIONAL EXPECTATION AND REGRESSION

9

Conditional expectation value and discrete conditional expectation

In chapter 6, we introduced the concepts of covariance and correlation, which quantify the strength of the kind of dependence that can be described by a linear quasi-regression. In the next five chapters, we introduce the concept of a *conditional expectation* and a 'genuine' regression. These concepts can be used to describe how the $(X=x)$-conditional expectation values of a numerical random variable Y depend on the values of a (numerical, non-numerical, multivariate) random variable X. In this chapter, we start with the concepts $(X=x)$-*conditional expectation value* and *discrete conditional expectation*, presuming that X is a discrete random variable. In this case, the conditional expectation $E(Y \mid X)$ is easily defined as that random variable whose values are the conditional expectation values $E(Y \mid X=x)$. In chapter 10, we introduce the general concept of a conditional expectation, dropping the assumption that X is discrete. Chapter 11 is devoted to the concepts of *residual with respect to a conditional expectation*, *conditional variance*, *conditional covariance*, and *partial correlation*, and chapters 12 and 13 deal with parametrizations of a conditional expectation.

9.1 Conditional expectation value

Remember, the expectation of a numerical random variable $Y: (\Omega, \mathscr{A}, P) \to (\overline{\mathbb{R}}, \overline{\mathscr{B}})$ is defined by $E(Y) = \int Y \, dP$, using the probability measure P. Now we choose an event $B \in \mathscr{A}$ with $P(B) > 0$ and, instead of P, we consider the *B-conditional-probability measure* $P^B: \mathscr{A} \to [0, 1]$ defined by

$$P^B(A) := P(A \mid B), \quad \forall A \in \mathscr{A}, \tag{9.1}$$

(see Def. 4.29). Referring to this measure,

$$E^B(Y) := \int Y \, d P^B, \tag{9.2}$$

Probability and Conditional Expectation: Fundamentals for the Empirical Sciences, First Edition. Rolf Steyer and Werner Nagel.
© 2017 John Wiley & Sons, Ltd. Published 2017 by John Wiley & Sons, Ltd.
Companion website: http://www.probability-and-conditional-expectation.de

defines the P^B-*expectation of Y*, that is, the expectation of Y with respect to the measure P^B. Reading the following definition, also remember that

$$\{X=x\} = X^{-1}(\{x\}) = \{\omega \in \Omega: X(\omega) = x\}$$

denotes the event that the random variable $X: (\Omega, \mathcal{A}, P) \to (\Omega'_X, \mathcal{A}'_X)$ takes on the value x and that we defined $P(X=x) := P(\{X=x\})$ (see Rem. 5.4). Assuming $P(X=x) > 0$ and $A \in \mathcal{A}$, we denote

$$P(A \mid X=x) := P(A \mid \{X=x\}), \tag{9.3}$$

Furthermore, we denote $P^{X=x} := P^{\{X=x\}}$. Hence, according to Equation (9.1),

$$P^{X=x}(A) := P(A \mid X=x), \quad \forall A \in \mathcal{A}. \tag{9.4}$$

Remark 9.1 [$P^{X=x}$**-equivalence of** $f(X)$ **and** $f(x)$] Let $X: (\Omega, \mathcal{A}, P) \to (\Omega'_X, \mathcal{A}'_X)$ be a random variable, $x \in \Omega'_X$, $\{x\} \in \mathcal{A}'_X$, and $P(X=x) > 0$. Furthermore, let $f: (\Omega'_X, \mathcal{A}'_X) \to (\Omega', \mathcal{A}')$ be a measurable mapping, then $f(X) \underset{P^{X=x}}{=} f(x)$ (see Exercise 9.1). ◁

The probability measures P^B and $P^{X=x}$ are now used to define the conditional expectation value.

Definition 9.2 [Conditional expectation value]
Let $Y: (\Omega, \mathcal{A}, P) \to (\overline{\mathbb{R}}, \overline{\mathcal{B}})$ *be a random variable.*

(i) *If* $B \in \mathcal{A}$ *with* $P(B) > 0$ *and* Y *is quasi-integrable with respect to* P^B, *then we define*

$$E(Y \mid B) := E^B(Y) = \int Y \, dP^B, \tag{9.5}$$

call it the conditional expectation value of Y given the event B *(or the* B-conditional expectation value of Y), *and say that it exists.*

(ii) *If* $X: (\Omega, \mathcal{A}, P) \to (\Omega'_X, \mathcal{A}'_X)$ *is a random variable,* $x \in \Omega'_X$ *with* $\{x\} \in \mathcal{A}'_X$ *and* $P(X=x) > 0$, *and* Y *is quasi-integrable with respect to* $P^{X=x}$, *then we define*

$$E(Y \mid X=x) := E(Y \mid \{X=x\}) \tag{9.6}$$

and call it the conditional expectation value of Y given $X=x$ *(or the* $(X=x)$-conditional expectation value of Y), *and say that it exists.*

Note that $E(Y \mid B)$ can be infinite. The only restriction is that $B \in \mathcal{A}$ with $P(B) > 0$ and that Y is quasi-integrable with respect to P^B. Otherwise the integral $\int Y \, dP^B$ is not defined.

Remark 9.3 [Multivariate numerical or qualitative X]   Also note that the random variable X in Definition 9.2 (ii) can be numerical, that is, $\Omega'_X \subset \overline{\mathbb{R}}$, multivariate (see section 5.3), or non-numerical. If $X = (X_1, \dots, X_n)$ is a multivariate random variable, then we also use the notation $E(Y \mid X_1 = x_1, \dots, X_n = x_n)$ instead of $E(Y \mid X = x)$ for the conditional expectation value of Y given $(X_1, \dots, X_n) = (x_1, \dots, x_n)$ and call it the $(X_1 = x_1, \dots, X_n = x_n)$-*conditional expectation value of Y* or the *conditional expectation value of Y given $X_1 = x_1, \dots, X_n = x_n$.*   ◁

The following theorem addresses the relationship between the B-conditional expectation value of Y and the expectation of $1_B \cdot Y$ with respect to P.

Theorem 9.4 [B-conditional expectation value and the P-expectation]
Let the assumptions of Definition 9.2 (i) hold. Then,

(i) $E(Y)$ *exists* $\Rightarrow$ $E(Y \mid B)$ *exists.*

(ii) $E(Y)$ *is finite* $\Rightarrow$ $E(Y \mid B)$ *is finite.*

(iii) *Furthermore, if $E(Y^2)$ is finite, then*

$$E(Y \mid B) = \frac{1}{P(B)} \cdot \int 1_B \cdot Y \, dP = \frac{1}{P(B)} \cdot E(1_B \cdot Y) \tag{9.7}$$

$$= \frac{1}{P(B)} \cdot Cov(Y, 1_B) + E(Y). \tag{9.8}$$

(Proof p. 301)

Recall the following notation:

$$1_{X=x} := 1_{\{X=x\}}, \qquad P\,(X=x) := P(\{X=x\}), \tag{9.9}$$

and

$$E^{X=x}(Y) := E^{\{X=x\}}(Y). \tag{9.10}$$

Using this notation, Equations (9.5), (9.6), and (9.7) yield the following corollary.

Corollary 9.5 [$(X=x)$-conditional expectation value and P-expectation]
If the assumptions of Definition 9.2 (ii) hold, then,

$$E(Y \mid X=x) = E^{X=x}(Y) = \int Y \, dP^{X=x}$$

$$= \frac{1}{P(X=x)} \cdot \int 1_{X=x} \cdot Y \, dP = \frac{1}{P(X=x)} \cdot E(1_{X=x} \cdot Y). \tag{9.11}$$

Remark 9.6 [B-conditional probability] If $A \in \mathscr{A}$ and $P(B) > 0$, then,

$$E(1_A \mid B) = \frac{1}{P(B)} \cdot E(1_B \cdot 1_A) \qquad \text{[(9.7)]}$$

$$= \frac{1}{P(B)} \cdot P(A \cap B) \qquad \text{[Box 6.1 (iv)]} \qquad (9.12)$$

$$= P(A \mid B) = P^B(A). \qquad \text{[(4.2), (9.1)]}$$

Because P^B is a probability measure, these equations imply $0 \le E(1_A \mid B) \le 1$. ◁

Remark 9.7 [$(X=x)$-conditional probability] For $B = \{X=x\}$, using the notation (9.3) and (9.4), Equation (9.12) implies

$$E(1_A \mid X=x) = \frac{1}{P(X=x)} \cdot E(1_{X=x} \cdot 1_A)$$

$$= \frac{1}{P(X=x)} \cdot P(A \cap \{X=x\}) \qquad (9.13)$$

$$= P(A \mid X=x) = P^{X=x}(A),$$

provided that $A \in \mathscr{A}$ and $P(X=x) > 0$. Equations (9.13) show that $E(1_A \mid X=x)$ is identical to the conditional probability $P(A \mid \{X=x\})$ of A given the event $\{X=x\}$ (see Def. 4.12). The term $P(A \mid X=x)$ is also called the *conditional probability of A given $X=x$* or the *$(X=x)$-conditional probability of A.* ◁

Remark 9.8 [$(X=x)$-conditional probability of $\{Y=y\}$] If $Y: (\Omega, \mathscr{A}, P) \to (\Omega'_Y, \mathscr{A}'_Y)$ is a random variable and the assumptions of Definition 9.2 (ii) hold, then we also use the notation

$$P(Y=y \mid X=x) := P(\{Y=y\} \mid X=x) = P(1_{Y=y}=1 \mid X=x) = E(1_{Y=y} \mid X=x), \qquad (9.14)$$

and call it the *$(X=x)$-conditional probability of $\{Y=y\}$* [see Eqs. (9.9)]. Hence, Equations (9.3), (9.14), and (4.2) yield

$$P(Y=y \mid X=x) = \frac{P(Y=y, X=x)}{P(X=x)}. \qquad (9.15)$$

◁

9.2 Transformation theorem

If $P_Y^{X=x}: \mathscr{A}'_Y \to [0, 1]$ denotes the distribution of Y with respect to the $(X=x)$-conditional-probability measure $P^{X=x}$ and $E_Y^{X=x}(g)$ the expectation of g with respect to the distribution $P_Y^{X=x}$, then the transformation theorem (cf. Th. 6.13) for the conditional expectation value $E(Y \mid X=x)$ can be formulated as follows:

Theorem 9.9 [Transformation theorem for $E(Y \mid X=x)$]
Let $Y\colon (\Omega, \mathcal{A}, P) \to (\Omega'_Y, \mathcal{A}'_Y)$ and $X\colon (\Omega, \mathcal{A}, P) \to (\Omega'_X, \mathcal{A}'_X)$ be random variables, and $g\colon (\Omega'_Y, \mathcal{A}'_Y) \to (\overline{\mathbb{R}}, \overline{\mathcal{B}})$ be a measurable function. Furthermore, let $x \in \Omega'_X$ with $\{x\} \in \mathcal{A}'_X$ and $P(X=x) > 0$.

(i) If g is nonnegative or with finite expectation $E_Y^{X=x}(g)$, then,

$$E_Y^{X=x}(g) = \int g\, dP_Y^{X=x} = \int g(y)\, P_Y^{X=x}(dy)$$

$$= \int g(Y)\, dP^{X=x} = E^{X=x}[g(Y)] = E[g(Y) \mid X=x]. \tag{9.16}$$

(ii) $E_Y^{X=x}(g)$ is finite if and only if $E^{X=x}[g(Y)]$ is finite.

There are two important points in Equations (9.16). *First*, these equations show the relationship between integrals of the composition $g(Y)$ with respect to the conditional-probability measure $P^{X=x}$ on $(\Omega, \mathcal{A})$ on one side, and the distribution $P_Y^{X=x}$ of Y with respect to $P^{X=x}$ on the other side. *Second*, $E[g(Y) \mid X=x]$ is identical to the expectation of g with respect to the distribution $P_Y^{X=x}$ (i.e., the distribution of Y with respect to the probability measure $P^{X=x}$). Thus, using the distribution $P_{g(Y)}^{X=x}$ of $g(Y)$ with respect to $P^{X=x}$ is not necessary.

Remark 9.10 [$(X=x)$-conditional expectation value of $g(Y)$ and P-expectation] Equations (9.16) and (9.11) imply

$$E[g(Y) \mid X=x] = \frac{1}{P(X=x)} \cdot \int 1_{X=x} \cdot g(Y)\, dP = \frac{1}{P(X=x)} \cdot E[1_{X=x} \cdot g(Y)]. \tag{9.17}$$

◁

Remark 9.11 [A special case of the transformation theorem] Let $(\Omega'_Y, \mathcal{A}'_Y) = (\overline{\mathbb{R}}, \overline{\mathcal{B}})$ and g be the identity function $id\colon \overline{\mathbb{R}} \to \overline{\mathbb{R}}$, defined by $id(y) = y$ for all $y \in \overline{\mathbb{R}}$, which implies $id(Y) = Y$. If we assume that $x \in \Omega'_X$ with $\{x\} \in \mathcal{A}'_X$ and $P(X=x) > 0$, and that Y is nonnegative or with finite expectation $E^{X=x}(Y)$, then Equations (9.16) yield

$$E(Y \mid X=x) = E^{X=x}(Y) = \int Y\, dP^{X=x} = \int y\, P_Y^{X=x}(dy) = \int id\, dP_Y^{X=x}. \tag{9.18}$$

◁

Using the notation introduced in Equations (9.14) and (9.16), Theorem 9.9 (i), and Equations (6.3) and (6.6) imply the following corollary.

Corollary 9.12 [Y discrete, $g(Y)$ real-valued]
Let the assumptions of Theorem 9.9 (i) hold.

(i) If Y is discrete and we assume that there is a finite set $\{y_1, \dots, y_n\} \subset \Omega'_Y$, $\{y_1\}, \dots, \{y_n\} \in \mathscr{A}'_Y$ with $P_Y(\{y_1, \dots, y_n\}) = 1$, then,

$$E[g(Y) \mid X=x] = \sum_{i=1}^{n} g(y_i) \cdot P(Y=y_i \mid X=x). \qquad (9.19)$$

(ii) If Y is discrete and we assume that there is a countably infinite set $\{y_1, y_2, \dots\} \subset \Omega'_Y$, $\{y_i\} \in \mathscr{A}'_Y$, $i \in \mathbb{N}$, with $P_Y(\{y_1, y_2, \dots\}) = 1$, then,

$$E[g(Y) \mid X=x] = \sum_{i=1}^{\infty} g(y_i) \cdot P(Y=y_i \mid X=x). \qquad (9.20)$$

Note that, in this corollary, Y does not have to be real-valued or numerical. We only assume that $g(Y)$ is real-valued. In contrast, in the following theorem we have to assume that Y itself is real-valued. Remember that $\{y\} \in \overline{\mathscr{B}}$ if $y \in \mathbb{R}$ [see (1.22)].

Corollary 9.13 [Discrete and real-valued Y]
Let the assumptions of Definition 9.2 (ii) hold.

(i) If Y is discrete and there is a finite set $\{y_1, \dots, y_n\} \subset \mathbb{R}$ of real numbers with $P_Y(\{y_1, \dots, y_n\}) = 1$, then,

$$E(Y \mid X=x) = \sum_{i=1}^{n} y_i \cdot P(Y=y_i \mid X=x). \qquad (9.21)$$

(ii) If Y is discrete and there is a countably infinite set $\{y_1, y_2, \dots\} \subset \mathbb{R}$ of real numbers with $P_Y(\{y_1, y_2, \dots\}) = 1$, then,

$$E(Y \mid X=x) = \sum_{i=1}^{\infty} y_i \cdot P(Y=y_i \mid X=x). \qquad (9.22)$$

9.3   Other properties

Because $E(Y \mid B)$ is defined as the expectation $E^B(Y)$ of Y with respect to the probability measure P^B, all properties of the expectation with respect to P can be translated to $E(Y \mid B)$, simply by replacing P by P^B and $E(Y)$ by $E^B(Y) = E(Y \mid B)$. Box 9.1 is such a translation of Box 6.1. Note that, according to Theorem 9.4 (i), the conditional expectation value $E(Y \mid B)$ exists if $E(Y)$ exists, provided that $P(B) > 0$. Of course, the rules for $E(Y \mid B)$ also apply to the $(X=x)$-conditional expectation value $E(Y \mid X=x)$ [see Def. 9.2 (ii)].

Box 9.1   Rules of computation for B-conditional expectation values.

Let $Y: (\Omega, \mathscr{A}, P) \to (\overline{\mathbb{R}}, \overline{\mathscr{B}})$ be a random variable, let $A, B, C \in \mathscr{A}$ with $P(B) > 0$, let the conditional expectation value $E(Y \mid B)$ exist, and let $\alpha \in \mathbb{R}$. Then,

$$Y \underset{P^B}{=} \alpha \ \Rightarrow \ E(Y \mid B) = \alpha. \tag{i}$$

$$E(\alpha + Y \mid B) = \alpha + E(Y \mid B). \tag{ii}$$

$$E(\alpha \cdot Y \mid B) = \alpha \cdot E(Y \mid B). \tag{iii}$$

$$E(1_A \cdot 1_C \mid B) = P(A \cap C \mid B). \tag{iv}$$

For $i = 1, \dots, n$, let $Y_i: (\Omega, \mathscr{A}, P) \to (\mathbb{R}, \mathscr{B})$ be random variables with finite B-conditional expectation values $E(Y_i \mid B)$ and $\alpha_i \in \mathbb{R}$. Then,

$$E\left(\sum_{i=1}^{n} \alpha_i \cdot Y_i \ \middle| \ B \right) = \sum_{i=1}^{n} \alpha_i \cdot E(Y_i \mid B). \tag{v}$$

Let $X, Y: (\Omega, \mathscr{A}, P) \to (\overline{\mathbb{R}}, \overline{\mathscr{B}})$ be random variables that are nonnegative or with finite B-conditional expectation values. Then,

$$X \underset{P^B}{=} Y \ \Rightarrow \ E(X \mid B) = E(Y \mid B). \tag{vi}$$

$$X \underset{P^B}{=} Y \ \Leftrightarrow \ \forall A \in \mathscr{A}: E(1_A X \mid B) = E(1_A Y \mid B). \tag{vii}$$

$$X \underset{P^B}{\perp\!\!\!\perp} Y \ \Rightarrow \ E(X \cdot Y \mid B) = E(X \mid B) \cdot E(Y \mid B). \tag{viii}$$

However, there are additional properties when dealing with the relationship between the expectation and the conditional expectation value. Some of these have already been formulated in Theorem 9.4. Other additional properties are summarized in Box 9.2 and proved in Exercise 9.2.

Rule (ii) shows how the $(X=x)$-conditional expectation values $E(Y \mid X=x)$ can be computed from the conditional expectation values $E(Y \mid X=x, Z=z_i)$ and the conditional probabilities $P(Z=z_i \mid X=x)$. Hence, considering Equation (9.21) and Rule (ii) in Box 9.2 shows that we have two different equations for computing the conditional expectation value $E(Y \mid X=x)$. Finally, note that a special case of Rule (ii) is

$$E(Y) = \sum_{i=1}^{m} E(Y \mid Z=z_i) \cdot P(Z=z_i) \tag{9.23}$$

(see Exercise 9.3). According to this equation, we can also compute the expectation of Y from the conditional expectations $E(Y \mid Z=z_i)$ and the probabilities $P(Z=z_i)$.

Box 9.2   Rules of computation for $(X=x)$-conditional expectation values.

Let $Y: (\Omega, \mathscr{A}, P) \to (\overline{\mathbb{R}}, \overline{\mathscr{B}})$ and $X: (\Omega, \mathscr{A}, P) \to (\Omega'_X, \mathscr{A}'_X)$ be random variables, and let $x \in \Omega'_X$ with $\{x\} \in \mathscr{A}'_X$ and $P(X=x) > 0$. If $E(Y \mid X=x)$ exists, $f: (\Omega'_X, \mathscr{A}'_X) \to (\mathbb{R}, \mathscr{B})$ is a measurable function, and $E(Y^2), E[f(X)^2] < \infty$, then,

$$E[f(X) \cdot Y \mid X=x] = f(x) \cdot E(Y \mid X=x) = E[f(x) \cdot Y \mid X=x]. \tag{i}$$

If $Z: (\Omega, \mathscr{A}, P) \to (\Omega'_Z, \mathscr{A}'_Z)$ is a random variable and $z_1, \ldots, z_m \in \Omega'_Z$ such that $P_Z(\{z_1, \ldots, z_m\}) = 1$ and, for all $i = 1, \ldots, m$, $\{z_i\} \in \mathscr{A}'_Z$ and $P(X=x, Z=z_i) > 0$, then,

$$E(Y \mid X=x) = \sum_{i=1}^{m} E(Y \mid X=x, Z=z_i) \cdot P(Z=z_i \mid X=x). \tag{ii}$$

Correspondingly, if $z_1, z_2, \ldots \in \Omega'_Z$ such that $P_Z(\{z_1, z_2, \ldots\}) = 1$ and, for all $i = 1, 2 \ldots,$ $\{z_i\} \in \mathscr{A}'_Z$ and $P(X=x, Z=z_i) > 0$, then,

$$E(Y \mid X=x) = \sum_{i=1}^{\infty} E(Y \mid X=x, Z=z_i) \cdot P(Z=z_i \mid X=x). \tag{iii}$$

9.4   Discrete conditional expectation

The *discrete conditional expectation* $E(Y \mid X)$ of a numerical random variable $Y: (\Omega, \mathscr{A}, P) \to (\overline{\mathbb{R}}, \overline{\mathscr{B}})$ given a random variable $X: (\Omega, \mathscr{A}, P) \to (\Omega'_X, \mathscr{A}'_X)$ is now defined as that random variable on $(\Omega, \mathscr{A}, P)$ whose values are identical to the conditional expectation values $E(Y \mid X=x)$. In this definition, we have to assume that X is *discrete*, that is, we assume that there is a finite or countable set $\Omega'_0 \subset \Omega'_X$ such that $P_X(\Omega'_0) = 1$ and $P(X=x) > 0$ for all $x \in \Omega'_0$ (see Def. 5.56). In chapter 10, this limitation is dropped.

Definition 9.14 [Discrete conditional expectation]

Let $Y: (\Omega, \mathscr{A}, P) \to (\overline{\mathbb{R}}, \overline{\mathscr{B}})$ be a numerical random variable that is nonnegative or has a finite expectation, and let the random variable $X: (\Omega, \mathscr{A}, P) \to (\Omega'_X, \mathscr{A}'_X)$ be discrete.

(i) If $\{x_1, \ldots, x_m\} = \Omega'_0 \subset \Omega'_X$ such that $P_X(\Omega'_0) = 1$ and, for all $i = 1, \ldots, m$, $\{x_i\} \in \mathscr{A}'_X$ and $P(X=x_i) > 0$, then the discrete conditional expectation of Y given X is defined by

$$E(Y \mid X) := \sum_{i=1}^{m} E(Y \mid X=x_i) \cdot 1_{X=x_i}. \tag{9.24}$$

(ii) If $\{x_1, x_2, \ldots\} = \Omega'_0 \subset \Omega'_X$ such that $P_X(\Omega'_0) = 1$ and, for all $i = 1, 2, \ldots, \{x_i\} \in \mathscr{A}'_X$ and $P(X=x_i) > 0$, then the discrete conditional expectation of Y given X is defined by

$$E(Y \mid X) := \sum_{i=1}^{\infty} E(Y \mid X=x_i) \cdot 1_{X=x_i}. \tag{9.25}$$

Hence, in contrast to a conditional expectation value $E(Y \mid X=x)$, which is a real number, a discrete conditional expectation $E(Y \mid X)$ is a discrete *random variable* (see Def. 5.56) on $(\Omega, \mathscr{A}, P)$. Note that $E(Y \mid X)$ is a random variable taking a numerical value for each $\omega \in \Omega$. This means that we might look at its expectation, variance, covariance, and correlation with other random variables (see, e.g., Box 10.2 and section 11.2).

Remark 9.15 [X-conditional probability]   If $A \in \mathscr{A}$, then we use the notation

$$P(A \mid X) := E(1_A \mid X) \tag{9.26}$$

and call it the discrete *X-conditional probability of A*. If Y is dichotomous with values 0 and 1, we also use the notation $P(Y=1 \mid X)$ for the discrete *X-conditional probability of the event* $\{Y=1\}$. If Y is dichotomous with values 0 and 1, Equations (9.21) and (9.26) then yield

$$P(Y=1 \mid X) = E(Y \mid X). \tag{9.27}$$
◁

Remark 9.16 [Uniqueness and values of the conditional expectation]   An alternative way to write Equations (9.24) and (9.25) is

$$E(Y \mid X)(\omega) = \begin{cases} E(Y \mid X=x), & \text{if } \omega \in X^{-1}(\{x\}), \quad \forall\, x \in \Omega'_0 \\ 0, & \text{otherwise.} \end{cases} \tag{9.28}$$

Hence, the values of the conditional expectation $E(Y \mid X)$ are uniquely defined by Equations (9.24) and (9.25) for all $\omega \in \Omega$ (see Example 9.22). Assigning the value $E(Y \mid X)(\omega) = 0$ if $\omega \in \Omega \setminus X^{-1}(\Omega'_0)$ is arbitrary, but note that $P(\Omega \setminus X^{-1}(\Omega'_0)) = 0$. Hence, this arbitrary assignment is innocuous; it only occurs with probability 0. According to Equation (9.28) and Definition 5.56, this arbitrary assignment does not occur if $P(X=x) > 0$ for all $x \in X(\Omega)$, that is, if $\Omega'_0 = X(\Omega)$ is the image of Ω under X.   ◁

9.5   Discrete regression

Remark 9.17 [Measurability and factorization]   Definition 9.14 implies that the discrete conditional expectation $E(Y \mid X)$ is a random variable on $(\Omega, \mathscr{A}, P)$ that is measurable with respect to X. In more formal terms, $E(Y \mid X): (\Omega, \mathscr{A}, P) \to (\overline{\mathbb{R}}, \overline{\mathscr{B}})$ and $\sigma[E(Y \mid X)] \subset \sigma(X)$. The reason is that there is a measurable function $g: (\Omega'_X, \mathscr{A}'_X) \to (\overline{\mathbb{R}}, \overline{\mathscr{B}})$ that is defined by

$$g(x) = \begin{cases} E(Y \mid X=x), & \forall\, x \in \Omega'_0 \\ 0, & \text{otherwise.} \end{cases} \tag{9.29}$$

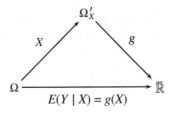

Figure 9.1   The conditional expectation $E(Y \mid X)$ as the composition of X and its factorization g.

Hence, $E(Y \mid X) = g(X)$ (see Fig. 9.1), and Lemma 2.52 implies that $E(Y \mid X)$ is measurable with respect to X. The function g is called the *factorization of $E(Y \mid X)$* or, if $\Omega_X' = \mathbb{R}^n$, the *discrete regression of Y on X*.   ◁

Definition 9.18 [Discrete regression]
Under the assumptions specified in Definition 9.14, the function $g \colon \Omega_X' \to \overline{\mathbb{R}}$ defined by Equation (9.29) is called the discrete regression of Y on X, if $(\Omega_X', \mathscr{A}_X') = (\mathbb{R}^n, \mathscr{B}_n), n \in \mathbb{N}$.

Remark 9.19 [Regressand and regressor] Considering the conditional expectation $E(Y \mid X)$, we call Y the *regressand* and X the *regressor*, provided that $(\Omega_X', \mathscr{A}_X') = (\mathbb{R}^n, \mathscr{B}_n)$, $n \in \mathbb{N}$.   ◁

Remark 9.20 [Multivariate numerical or qualitative regressors] In general, the codomain Ω_X' of X can be *any* set as long as there is a subset $\Omega_0' \subset \Omega_X'$ such that Ω_0' is finite, or countable with $P(X \in \Omega_0') = 1$ and $P(X = x) > 0$ for all $x \in \Omega_0'$. Hence, X can be uni- or multivariate (see Examples 9.21 and 9.22). If $X = (X_1, \dots, X_n)$ is a discrete multivariate random variable, then we also use the notation $E(Y \mid X_1, \dots, X_n)$ instead of $E(Y \mid X)$ for the conditional expectation of Y given X.   ◁

9.6   Examples

We treat two examples in some detail. Example 9.21 is straightforward, whereas Example 9.22 exemplifies that the values of a conditional expectation are uniquely defined by Equation (9.24) for all $\omega \in \Omega$.

Example 9.21 [Joe and Ann with randomized assignment – continued] Table 9.1 contains three discrete conditional expectations we may consider in the example already used in Example 1.9. All of them are random variables taking a numerical value for each $\omega \in \Omega$. According to Remark 9.16, the values of $E(Y \mid X)$ are the conditional expectation values $E(Y \mid X = x)$ for all $x \in \Omega_0'$, and they are 0 for all $x \in \Omega_X' \setminus \Omega_0'$.

We start by illustrating the discrete conditional expectation of Y given X. Both random variables, X and Y, are specified in Table 9.1. We consider the random variable $X \colon (\Omega, \mathscr{A}, P) \to (\mathbb{R}, \mathscr{B})$. In this case, $\Omega_X' = \mathbb{R}$ and $\Omega_0' = \{0, 1\}$. Because X takes on a value in $\Omega_0' = \{0, 1\}$ for

Table 9.1   Joe and Ann with randomized assignment: conditional expectations.

Elements of Ω			Observable random variables			Conditional expectations		
Unit	Treatment	Success	$P(\{\omega\})$	Person variable U	Treatment variable X	Outcome variable Y	$E(Y \mid X, U)$	$E(Y \mid X)$	$P(X=1 \mid U)$
(Joe, no, −)			.09	Joe	0	0	.7	.45	.4
(Joe, no, +)			.21	Joe	0	1	.7	.45	.4
(Joe, yes, −)			.04	Joe	1	0	.8	.6	.4
(Joe, yes, +)			.16	Joe	1	1	.8	.6	.4
(Ann, no, −)			.24	Ann	0	0	.2	.45	.4
(Ann, no, +)			.06	Ann	0	1	.2	.45	.4
(Ann, yes, −)			.12	Ann	1	0	.4	.6	.4
(Ann, yes, +)			.08	Ann	1	1	.4	.6	.4

all $\omega \in \Omega$, the discrete conditional expectation $E(Y \mid X)$ takes on either the value $E(Y \mid X=0)$ or the value $E(Y \mid X=1)$. It does not take on the value 0, because $\{X=x\} = \emptyset$ for all $x \in \mathbb{R} \setminus \{0, 1\}$ [see Eq. (9.28)].

Because Y is an indicator, $E(Y \mid X=x) = P(Y=1 \mid X=x)$ [see Eqs. (9.14) and (9.27)]. Hence, if we want to compute the values of $E(Y \mid X) = P(Y=1 \mid X)$, then we have to compute the conditional probabilities $P(Y=1 \mid X=x)$. For $x=0$, we receive

$$P(Y=1 \mid X=0) = \frac{P(Y=1, X=0)}{P(X=0)} = \frac{.21 + .06}{.09 + .21 + .24 + .06} = \frac{.27}{.6} = .45,$$

and for $x=1$,

$$P(Y=1 \mid X=1) = \frac{P(Y=1, X=1)}{P(X=1)} = \frac{.16 + .08}{.04 + .16 + .12 + .08} = \frac{.24}{.4} = .6.$$

Now we consider the conditional expectation $E(Y \mid X, U)$, where we condition on the random variable (X, U): $(\Omega, \mathscr{A}, P) \rightarrow [\mathbb{R} \times \Omega_U, \mathscr{B} \otimes \mathscr{P}(\Omega_U)]$, where $\Omega_U = \{Joe, Ann\}$, and $\Omega_0' = \{0, 1\} \times \Omega_U$. Note that

$$\forall \, (x, u) \in (\mathbb{R} \times \Omega_U) \setminus (\{0, 1\} \times \Omega_U): \{(X=x, U=u)\} = \emptyset.$$

Furthermore, because Y is an indicator, $E(Y \mid X, U) = P(Y=1 \mid X, U)$, and this conditional expectation has only four different values: the conditional probabilities $P(Y=1 \mid X=x,$

$U=u$). For $x=0$, $u=Joe$, we receive

$$P(Y=1 \mid X=0, U=Joe) = \frac{P(Y=1, X=0, U=Joe)}{P(X=0, U=Joe)} = \frac{.21}{.09 + .21} = .7,$$

for $x=1$, $u = Joe$,

$$P(Y=1 \mid X=1, U=Joe) = \frac{P(Y=1, X=1, U=Joe)}{P(X=1, U=Joe)} = \frac{.16}{.04 + .16} = .8,$$

for $x=0$, $u=Ann$,

$$P(Y=1 \mid X=0, U=Ann) = \frac{P(Y=1, X=0, U=Ann)}{P(X=0, U=Ann)} = \frac{.06}{.24 + .06} = .2,$$

and for $x=1$, $u=Ann$,

$$P(Y=1 \mid X=1, U=Ann) = \frac{P(Y=1, X=1, U=Ann)}{P(X=1, U=Ann)} = \frac{.08}{.12 + .08} = .4.$$

◁

Example 9.22 [No treatment for Joe] Let us use a second example in order to illustrate the concepts introduced above. Again the random experiment consists of sampling a person, observing the value x of the treatment variable X, that is, whether or not the sampled person receives a treatment ($x=1$ vs. $x=0$), and observing whether or not a success criterion is reached some time after treatment. In this new example, we fixed new probabilities of the elementary events. For instance, now the probability that Joe receives treatment is zero. This is useful to illustrate some general properties of discrete conditional expectations. Also note that the probabilities of the other elementary events have been changed as well. The only restriction on the probabilities of the elementary events in such a hypothetical example is that they sum up to one.

Using the probabilities displayed in Table 9.2, Equation (9.21) yields:

$$E(Y \mid X=1) = P(Y=1 \mid X=1) = \frac{P(Y=1, X=1)}{P(X=1)}$$

$$= \frac{0 + .152}{0 + 0 + .228 + .152} = .4$$

for the treatment condition $x = 1$. Applying the corresponding formula to the control condition $x = 0$ yields $E(Y \mid X=0) = (.348 + .024)/(.152 + .348 + .096 + .024) = .6$. Note that the conditional probabilities $P(Y=1 \mid X=1)$ and $P(Y=1 \mid X=0)$ do *not* necessarily add up to 1. In contrast, the sum of $P(Y=1 \mid X=1)$ and $P(Y=0 \mid X=1)$ and the sum of $P(Y=1 \mid X=0)$ and $P(Y=0 \mid X=0)$ are always equal to 1, provided that Y is dichotomous with values 0 and 1.

Table 9.2   No treatment for Joe with discrete conditional expectations.

Unit	Treatment	Success	$P(\{\omega\})$	Person variable U	Treatment variable X	Outcome variable Y	$E(Y \mid X, U)$	$E(Y \mid X)$	$P(X=1 \mid U)$
(Joe, no, −)			.152	*Joe*	0	0	.696	.6	0
(Joe, no, +)			.348	*Joe*	0	1	.696	.6	0
(Joe, yes, −)			0	*Joe*	1	0	0	.4	0
(Joe, yes, +)			0	*Joe*	1	1	0	.4	0
(Ann, no, −)			.096	*Ann*	0	0	.2	.6	.76
(Ann, no, +)			.024	*Ann*	0	1	.2	.6	.76
(Ann, yes, −)			.228	*Ann*	1	0	.4	.4	.76
(Ann, yes, +)			.152	*Ann*	1	1	.4	.4	.76

The column groups are: *Elements of Ω* (Unit, Treatment, Success), then $P(\{\omega\})$; *Observable random variables* (Person variable U, Treatment variable X, Outcome variable Y); *Conditional expectations* ($E(Y \mid X, U)$, $E(Y \mid X)$, $P(X=1 \mid U)$).

Table 9.2 also displays the conditional probability $P(X=1 \mid U)$, whose values are the treatment probabilities of Joe and Ann. For Joe, the treatment probability is $P(X=1 \mid U=Joe) = 0$, and for Ann it is

$$P(X=1 \mid U=Ann) = (.228 + .152)/(.096 + .024 + .228 + .152) = .76.$$

Finally, we compute the conditional probabilities $P(Y=1 \mid X=x, U=u)$ [see Eq. (9.14)]:

$$P(Y=1 \mid X=0, U=Joe) = \frac{P(Y=1, X=0, U=Joe)}{P(X=0, U=Joe)} = \frac{.348}{.152 + .348} = .696,$$

$$P(Y=1 \mid X=0, U=Ann) = \frac{P(Y=1, X=0, U=Ann)}{P(X=0, U=Ann)} = \frac{.024}{.096 + .024} = .2,$$

$$P(Y=1 \mid X=1, U=Ann) = \frac{P(Y=1, X=1, U=Ann)}{P(X=1, U=Ann)} = \frac{.152}{.228 + .152} = .4.$$

Note that $E(Y \mid X=1, U=Joe)$ is not defined, because $P(X=1, U=Joe) = 0$ (see Def. 9.2). However, according to Definition 9.14 (i), $E(Y \mid X, U)(\omega) = 0$ for $\omega \in \Omega \setminus \Omega_0' = \{(Joe, yes, −), (Joe, yes, +)\}$. Thus $E(Y \mid X, U)$ is uniquely defined, that is, the values $E(Y \mid X, U)(\omega)$ are uniquely defined for all $\omega \in \Omega$.   ◁

Example 9.23 [No treatment for Joe – continued] Using the results obtained in Example 9.22, Equation (9.24) yields:

$$E(Y \mid X) = \sum_{x=0}^{1} E(Y \mid X=x) \cdot 1_{X=x}$$
$$= E(Y \mid X=0) \cdot 1_{X=0} + E(Y \mid X=1) \cdot 1_{X=1}$$
$$= .6 \cdot 1_{X=0} + .4 \cdot 1_{X=1}.$$

Hence, the values $E(Y \mid X)(\omega)$ of the discrete X-conditional expectation of Y are

$$E(Y \mid X)(\omega) = .6 \cdot 1_{X=0}(\omega) + .4 \cdot 1_{X=1}(\omega) = .6, \quad \text{if } X(\omega) = 0,$$

and

$$E(Y \mid X)(\omega) = .6 \cdot 1_{X=0}(\omega) + .4 \cdot 1_{X=1}(\omega) = .4, \quad \text{if } X(\omega) = 1.$$

These are the only two values that $E(Y \mid X)$ takes on depending on the outcome $\omega \in \Omega$ of the random experiment considered (see the first column of Table 9.2). This example illustrates that $E(Y \mid X)$ is a random variable on $(\Omega, \mathcal{A}, P)$ just like X, Y, and U. Note again that the two values of $E(Y \mid X) = P(Y=1 \mid X)$ add up to 1 only by coincidence.

Table 9.2 shows two additional discrete conditional expectations, $E(X \mid U) = P(X=1 \mid U)$ as well as $E(Y \mid X, U) = P(Y=1 \mid X, U)$. Again using Equation (9.24) and the results obtained in Example 9.22, the discrete conditional probability $P(X=1 \mid U)$ is

$$P(X=1 \mid U) = \sum_{u} P(X=1 \mid U=u) \cdot 1_{U=u}$$
$$= P(X=1 \mid U=Joe) \cdot 1_{U=Joe} + P(X=1 \mid U=Ann) \cdot 1_{U=Ann}$$
$$= 0 \cdot 1_{U=Joe} + .76 \cdot 1_{U=Ann} = .76 \cdot 1_{U=Ann}.$$

Hence, the values $P(X=1 \mid U)(\omega)$ of the discrete U-conditional probability of the event $\{X=1\}$ are

$$P(X=1 \mid U)(\omega) = 0 \cdot 1_{U=Joe}(\omega) + .76 \cdot 1_{U=Ann}(\omega) = 0, \quad \text{if } U(\omega) = Joe$$

and

$$P(X=1 \mid U)(\omega) = 0 \cdot 1_{U=Joe}(\omega) + .76 \cdot 1_{U=Ann}(\omega) = .76, \quad \text{if } U(\omega) = Ann.$$

These are the only two values that $P(X=1 \mid U)$ takes on. Again, Table 9.2 shows how the values of $P(X=1 \mid U)$ are assigned to the outcomes $\omega \in \Omega$.

Finally, let us turn to the discrete conditional expectation $E(Y \mid X, U)$. Using the results obtained in Example 9.22, its values are

$$E(Y \mid X, U)(\omega) = E(Y \mid X=0, U=Joe) = .696, \quad \text{if } X(\omega) = 0 \text{ and } U(\omega) = Joe$$

$$E(Y \mid X, U)(\omega) = E(Y \mid X=0, U=Ann) = .2, \quad \text{if } X(\omega) = 0 \text{ and } U(\omega) = Ann$$

and

$$E(Y \mid X, U)(\omega) = E(Y \mid X=1, U=Ann) = .4, \quad \text{if } X(\omega) = 1 \text{ and } U(\omega) = Ann,$$

whereas

$$E(Y \mid X, U)(\omega) = 0, \quad \text{if } X(\omega) = 1 \text{ and } U(\omega) = Joe.$$

Note that the value $E(Y \mid X, U)(\omega)$ *is* defined for $\omega \in \{X=1, U=Joe\}$, although the conditional expectation value $E(Y \mid X=1, U=Joe)$ *is not* defined. Also note that in this case the value $E(Y \mid X, U)(\omega) = 0$ is arbitrarily fixed. However, because $P(X=1, U=Joe) = 0$, this has no disadvantageous consequences. [In chapter 10, we show that the values of a conditional expectation can arbitrarily be fixed for all elements ω of a subset A of Ω for which $P(A) = 0$.]

According to Equation (9.24), the discrete conditional expectation $E(Y \mid X, U)$ is

$$E(Y \mid X, U) = \sum_{(x,u)} E(Y \mid X=x, U=u) \cdot 1_{X=x, U=u}$$

$$= .696 \cdot 1_{X=0, U=Joe} + .2 \cdot 1_{X=0, U=Ann} + .4 \cdot 1_{X=1, U=Ann}. \tag{9.30}$$

The pair $(1, Joe)$ is not an element of the set Ω_0' (see Def. 9.14), and therefore the corresponding indicator $1_{X=1, U=Joe}$ does not occur in this sum. Hence, if

$$\omega \in \{X=1, U=Joe\} = \{(Joe, yes, -), (Joe, yes, +)\},$$

then all three indicators occurring in Equation (9.30) take on the value 0, implying that $E(Y \mid X, U)(\omega) = 0$ for these two elements of Ω. ◁

9.7 Proofs

Proof of Theorem 9.4

$$E(Y \mid B) = E^B(Y) = \int Y \, dP^B \qquad [(9.5)]$$

$$= \int Y \, d\left(\frac{1}{P(B)} \cdot 1_B\right) \odot P \qquad [\text{Lem. 4.32 (ii)}]$$

$$= \frac{1}{P(B)} \cdot \int 1_B \cdot Y \, dP \qquad [(3.72), (3.32)]$$

$$= \frac{1}{P(B)} \cdot E(1_B \cdot Y), \qquad [(6.1)]$$

which yields Equation (9.7). Using Lemma 3.33 (i) yields Theorem 9.4 (i), and Lemma 3.33 (ii) implies Theorem 9.4 (ii).

If $E(Y^2)$ is finite, then $E(Y)$ and $E(1_B \cdot Y)$ are finite as well [see Rem. 6.25 (iii) and Lemma 3.33 (ii)]. Therefore, according to Box 7.1 (i), $Cov(1_B, Y) = E(1_B \cdot Y) - E(1_B) \cdot E(Y)$ is finite. Hence,

$$E(Y \mid B) = \frac{1}{P(B)} \cdot E(1_B \cdot Y) \qquad\qquad [(9.7)]$$

$$= \frac{1}{P(B)} \cdot [Cov(1_B, Y) + E(1_B) \cdot E(Y)] \qquad [\text{Box 7.1 (i)}]$$

$$= \frac{1}{P(B)} \cdot Cov(1_B, Y) + E(Y). \qquad\qquad [(6.4)]$$

Exercises

9.1 Prove the proposition of Remark 9.1.

9.2 Prove the rules of computation of conditional expectation values displayed in Box 9.2.

9.3 Show that Equation (9.23) is a special case of Rule (ii) of Box 9.2.

Solutions

9.1 It is sufficient to prove $P^{X=x}(\{f(X) = f(x)\}^c) = 0$, which is equivalent to $P^{X=x}(\{f(X) = f(x)\}) = 1$.

$$P^{X=x}(f(X) = f(x)) = \frac{P(f(X) = f(x), X = x)}{P(X = x)} \qquad [(9.4), (9.15)]$$

$$= \frac{P(X = x)}{P(X = x)} \qquad [\{X = x\} \subset \{f(X) = f(x)\}, (9.9)]$$

$$= 1.$$

9.2 (i)

$$E[f(X) \cdot Y \mid X = x]$$

$$= \frac{1}{P(X = x)} \cdot E[1_{X=x} \cdot f(X) \cdot Y] \qquad [(9.11)]$$

$$= \frac{1}{P(X = x)} \cdot E[1_{X=x} \cdot f(x) \cdot Y] \qquad [1_{X=x} \cdot f(X) = 1_{X=x} \cdot f(x)]$$

$$= f(x) \cdot \frac{1}{P(X = x)} \cdot E(1_{X=x} \cdot Y) \qquad [\text{Box 6.1 (iii)}]$$

$$= f(x) \cdot E(Y \mid X = x) \qquad [(9.7)]$$

$$= E[f(x) \cdot Y \mid X = x]. \qquad [\text{Box 9.1 (iii)}]$$

(ii) $E(Y \mid X=x)$

$$= \frac{1}{P(X=x)} \cdot E(1_{X=x} \cdot Y) \hspace{4em} [(9.11)]$$

$$= \frac{1}{P(X=x)} \cdot E\left(1_{X=x} \cdot \sum_{i=1}^{m} 1_{Z=z_i} \cdot Y\right) \left[1 \underset{P}{=} \sum_{i=1}^{m} 1_{Z=z_i}, (5.33), \text{Rem. } 6.27\right]$$

$$= \sum_{i=1}^{m} \frac{1}{P(X=x)} \cdot E(1_{X=x} \cdot 1_{Z=z_i} \cdot Y) \hspace{4em} [\text{Box } 6.1 \text{ (vii)}]$$

$$= \sum_{i=1}^{m} \frac{P(X=x, Z=z_i)}{P(X=x)} \cdot \frac{1}{P(X=x, Z=z_i)} \cdot E(1_{X=x, Z=z_i} \cdot Y)$$

$$= \sum_{i=1}^{m} E(Y \mid X=x, Z=z_i) \cdot P(Z=z_i \mid X=x). \hspace{3em} [(9.11), (9.15)]$$

(iii)

$$\sum_{i=1}^{\infty} E(Y \mid X=x, Z=z_i) \cdot P(Z=z_i \mid X=x)$$

$$= \sum_{i=1}^{\infty} P(Z=z_i \mid X=x) \cdot \frac{1}{P(X=x, Z=z_i)} \cdot E(1_{X=x, Z=z_i} \cdot Y) \hspace{2em} [(9.11)]$$

$$= \sum_{i=1}^{\infty} \frac{1}{P(X=x)} \cdot E(1_{X=x, Z=z_i} \cdot Y) \hspace{4em} [(9.15)]$$

$$= \frac{1}{P(X=x)} \sum_{i=1}^{\infty} \int 1_{X=x, Z=z_i} \cdot Y \, dP \hspace{4em} [\text{Def. } 6.1]$$

$$= \frac{1}{P(X=x)} \sum_{i=1}^{\infty} \left[\int 1_{Z=z_i} \cdot 1_{X=x} \cdot Y^+ dP - \int 1_{Z=z_i} \cdot 1_{X=x} \cdot Y^- dP\right] \hspace{1em} [\text{Def. } 3.28]$$

$$= \frac{1}{P(X=x)} \left[\sum_{i=1}^{\infty} \int 1_{Z=z_i} \cdot 1_{X=x} \cdot Y^+ dP - \sum_{i=1}^{\infty} \int 1_{Z=z_i} \cdot 1_{X=x} \cdot Y^- dP\right]$$

$$[1_{X=x} \cdot Y \text{ is quasi-integrable}]$$

$$= \frac{1}{P(X=x)} \left[\int 1_{X=x} \cdot Y^+ dP - \int 1_{X=x} \cdot Y^- dP\right] \left[(3.65), 1 \underset{P}{=} \sum_{i=1}^{\infty} 1_{Z=z_i}, (5.33), (3.44)\right]$$

$$= \frac{1}{P(X=x)} \int 1_{X=x} \cdot Y \, dP \hspace{4em} [\text{Def. } 3.28]$$

$$= E(Y \mid X=x). \hspace{4em} [(9.11)]$$

9.3 This is easily seen considering the special case $X=\alpha$, $\alpha \in \Omega'_X$, that is, the case in which X is a constant. Then $\{X=\alpha\} = \Omega$, and Equation (9.11) yields $E(Y \mid X=\alpha) = E(Y)$, and $E(Y \mid X=\alpha, Z=z_i) = E(Y \mid Z=z_i)$. Hence, Rule (ii) of Box 9.2 yields

$$E(Y) = \sum_{i=1}^{m} E(Y \mid Z=z_i) \cdot P(Z=z_i).$$

10

Conditional expectation

In chapter 9, we treated the conditional expectation value given an event and the discrete conditional expectation $E(Y \mid X)$. In this chapter, we introduce the general concept of a conditional expectation given a σ-algebra $\mathscr{C}$. The price of this generalization is that a $\mathscr{C}$-conditional expectation is uniquely defined only up to P-equivalence; in other words, if there are two versions of such a $\mathscr{C}$-conditional expectation of a numerical random variable Y, then they are not necessarily identical, but they are P-equivalent. Furthermore, if $\mathscr{C}$ is generated by a random variable X, then a $\mathscr{C}$-conditional expectation is also called an X-*conditional expectation of* Y. This definition also applies if X is continuous. Hence, it even applies if $P(X=x) = 0$ for all values x of X. In this chapter, we also define the general concept of a regression as a factorization g of a conditional expectation $E(Y \mid X) = g(X)$, provided that X is real-valued. Furthermore, we define an $(X=x)$-conditional expectation value $E(Y \mid X=x)$ as a value $g(x)$ of the factorization g. This means that $E(Y \mid X=x)$ is defined even if $P(X=x) = 0$. However, $E(Y \mid X=x)$ is not uniquely defined. Nevertheless, we can formulate propositions about the conditional expectation values $E(Y \mid X=x)$ for P_X-almost all values x of X. Finally, we introduce the concept of *mean independence* and study its relationship to stochastic independence and uncorrelatedness.

10.1   Assumptions and definitions

Throughout this chapter, we will make the following assumptions and use the following notation.

Notation and assumptions 10.1
$Y: (\Omega, \mathscr{A}, P) \to (\overline{\mathbb{R}}, \overline{\mathscr{B}})$ *is a numerical random variable that is nonnegative or has a finite expectation* $E(Y)$. *Furthermore,* $\mathscr{C} \subset \mathscr{A}$ *is a* σ-*algebra, and* $X: (\Omega, \mathscr{A}, P) \to (\Omega'_X, \mathscr{A}'_X)$ *is a random variable.*

Probability and Conditional Expectation: Fundamentals for the Empirical Sciences, First Edition. Rolf Steyer and Werner Nagel.
© 2017 John Wiley & Sons, Ltd. Published 2017 by John Wiley & Sons, Ltd.
Companion website: http://www.probability-and-conditional-expectation.de

The definition of a conditional expectation given a σ-algebra is already found in Kolmogorov (1933/1977) (see also Kolmogorov, 1956). Reading the following definition, remember that, for a random variable $V: (\Omega, \mathscr{A}, P) \to (\overline{\mathbb{R}}, \overline{\mathscr{B}})$, we use $\sigma(V) = V^{-1}(\overline{\mathscr{B}})$ to denote the σ-algebra generated by V, and $\sigma(V) \subset \mathscr{C}$ means that V is $(\mathscr{C}, \overline{\mathscr{B}})$-measurable (see Def. 2.26 and Cor. 2.28).

Definition 10.2 [Conditional expectation given a σ-algebra]
Let the assumptions 10.1 hold. A random variable $V: (\Omega, \mathscr{A}, P) \to (\overline{\mathbb{R}}, \overline{\mathscr{B}})$ is called a (version of the) $\mathscr{C}$-conditional expectation of Y with respect to P, if the following two conditions hold:

(a) $\sigma(V) \subset \mathscr{C}$.

(b) $E(1_C \cdot V) = E(1_C \cdot Y), \quad \forall\, C \in \mathscr{C}$.

If V satisfies (a) and (b), then we also use the notation $E(Y \mid \mathscr{C}) := V$.

Remark 10.3 [X-conditional expectation] If the assumptions 10.1 hold, then we define

$$E(Y \mid X) := E[Y \mid \sigma(X)] \tag{10.1}$$

and call it a version of the *X-conditional expectation* of Y with respect to P. If there is no ambiguity, we will omit the reference to the measure P. Otherwise, we also use the term *X-conditional P-expectation of Y*. ◁

Remark 10.4 [Conditional probability given a σ-algebra] Let the assumptions 10.1 hold, and let $A \in \mathscr{A}$. Then we define

$$P(A \mid \mathscr{C}) := E(1_A \mid \mathscr{C}) \tag{10.2}$$

and call it a version of the *$\mathscr{C}$-conditional probability* of A with respect to P. Similarly, we define

$$P(A \mid X) := E[1_A \mid \sigma(X)] \tag{10.3}$$

and call it a version of the *X-conditional probability* of A with respect to P.
Furthermore, considering the event $\{Y=y\}$, we also use the notation

$$P(Y=y \mid X) := P(\{Y=y\} \mid X) = E(1_{Y=y} \mid X). \tag{10.4}$$
◁

Remark 10.5 [Conditioning on the smallest σ-algebra] If $\mathscr{C} = \{\Omega, \emptyset\}$, then Definition 10.2 (a) implies that $E(Y \mid \mathscr{C})$ is a constant (see Example 2.14), and in this case

$$E(Y \mid \mathscr{C}) = E(Y),$$

because $E[1_\Omega \cdot E(Y)] = E(Y) = E(1_\Omega \cdot Y)$ and $E[1_\emptyset \cdot E(Y)] = 0 = E(1_\emptyset \cdot Y)$ [see Def. 10.2(b)]. In fact, if $\mathscr{C} = \{\Omega, \emptyset\}$, then $E(Y)$ is the only version of the $\mathscr{C}$-conditional expectation of Y. Correspondingly, if X is a constant (i.e., if $X = \alpha$, $\alpha \in \Omega_X'$), then

$$E(Y \mid X) = E(Y).$$

◁

Remark 10.6 [C-conditional expectation value of a $\mathscr{C}$-conditional expectation] Consider an event $C \in \mathscr{C}$ with $P(C) > 0$ and a version V of the $\mathscr{C}$-conditional expectation of Y defined in Definition 10.2. Then the C-conditional expectation value of V (see Def. 9.2) is

$$E(V \mid C) = \frac{1}{P(C)} \cdot E(1_C \cdot V) \qquad [(9.7)]$$

$$= \frac{1}{P(C)} \cdot E(1_C \cdot Y) \qquad [\text{Def. 10.2 (b)}] \tag{10.5}$$

$$= E(Y \mid C). \qquad [(9.7)]$$

Inserting $E(Y \mid \mathscr{C}) = V$ into this equation shows that with condition (b) of Definition 10.2, we implicitly require

$$E[E(Y \mid \mathscr{C}) \mid C] = E(Y \mid C), \quad \forall\, C \in \mathscr{C} \text{ with } P(C) > 0. \tag{10.6}$$

◁

Remark 10.7 [Multivariate X] If $X = (X_1, \dots, X_n)$ is an n-variate random variable on $(\Omega, \mathscr{A}, P)$ (see section 5.3), then $E(Y \mid X)$ is also denoted by $E(Y \mid X_1, \dots, X_n)$.   ◁

Remark 10.8 [$\mathscr{C}$-conditional and X-conditional expectation] If $X \colon (\Omega, \mathscr{A}, P) \to (\mathbb{R}, \mathscr{B})$ is a *nonnegative* real-valued random variable, then $\sigma(X^2) = \sigma(X)$ [see Example 2.56 (i)]. Therefore, in this case $E(Y \mid X)$ and $E(Y \mid X^2)$ are just two different notations of the conditional expectation $E[Y \mid \sigma(X)]$. If X takes on also negative real numbers, then $\sigma(X^2) \subset \sigma(X)$, but $\sigma(X^2) = \sigma(X)$ does not necessarily hold [see Example 2.56 (ii)].   ◁

10.2   Existence and uniqueness

By its definition, it is not obvious that a conditional expectation exists and that it is well-defined. These issues are addressed in the following theorem.

Theorem 10.9 [Existence and uniqueness of a $\mathscr{C}$-conditional expectation]
Let the assumptions 10.1 hold. Then the following two propositions hold:

(i) *There is a $\mathscr{C}$-measurable random variable $V \colon (\Omega, \mathscr{A}, P) \to (\overline{\mathbb{R}}, \overline{\mathscr{B}})$ that is non-negative (if Y is nonnegative) or has a finite expectation $E(V)$ (if $E(Y)$ is finite) satisfying*

$$E(1_C \cdot V) = E(1_C \cdot Y), \quad \forall\, C \in \mathscr{C}. \tag{10.7}$$

(ii) *If $V, V^* \colon (\Omega, \mathscr{A}, P) \to (\overline{\mathbb{R}}, \overline{\mathscr{B}})$ satisfy (10.7) and $\sigma(V), \sigma(V^*) \subset \mathscr{C}$, then $V \underset{P}{=} V^*$.*

For a proof, see Bauer (1996, Theorem 15.1). Using the term *version of a conditional expectation* (see Def. 10.2) already hints at the fact that a conditional expectation, even if it exists, is not necessarily uniquely defined. However, according to Theorem 10.9 (ii), different versions of a conditional expectation are P-equivalent (see Remark 5.17).

Remark 10.10 [The sets $\mathscr{E}(Y \mid \mathscr{C})$ and $\mathscr{E}(Y \mid X)$]   We define $\mathscr{E}(Y \mid \mathscr{C})$ to be the set of all $\mathscr{C}$-measurable random variables satisfying Equation (10.7). Hence, $\mathscr{E}(Y \mid \mathscr{C})$ is the set of all versions of the $\mathscr{C}$-conditional expectation of Y with respect to the measure P. Similarly, $\mathscr{E}(Y \mid X)$ denotes the set of all versions of the X-conditional expectation of Y. The sets $\mathscr{P}(A \mid \mathscr{C})$ and $\mathscr{P}(A \mid X)$ are defined correspondingly for conditional probabilities of an event A.   ◁

Remark 10.11 [Consistency of definitions]   If X is discrete such that there is a finite or countable set $\Omega'_0 \subset \Omega'_X$ with $P_X(\Omega'_0) = 1$ and $P(X{=}x) > 0$ for all $x \in \Omega'_0$, then the discrete conditional expectation introduced in Definition 9.14 is a version of the conditional expectation of Y given X defined in Remark 10.3 (see Exercises 10.1 and 10.2). Hence, if X is discrete, then for all versions $E(Y \mid X) \in \mathscr{E}(Y \mid X)$,

$$E(Y \mid X) \underset{P}{=} \sum_{x \in \Omega'_0} E(Y \mid X{=}x) \cdot 1_{X=x} \tag{10.8}$$

and

$$\forall\, x \in \Omega'_0\colon \quad E(Y \mid X)(\omega) = E(Y \mid X{=}x), \quad \text{if } \omega \in \{X{=}x\}. \tag{10.9}$$

This equation shows that the conditional expectation $E(Y \mid X)$ describes how the $(X{=}x)$-conditional expectation values of Y depend on the values $x \in \Omega'_0$.   ◁

Remark 10.12 [Uniqueness of $E(Y \mid X)$]   If we add $\Omega'_0 = X(\Omega)$ to the assumptions of Remark 10.11, then,

$$E(Y \mid X) = \sum_{x \in X(\Omega)} E(Y \mid X{=}x) \cdot 1_{X=x}. \tag{10.10}$$

Hence, under these assumptions, $V = V^*$ for all $V, V^* \in \mathscr{E}(Y \mid X)$, that is, under these assumptions, there is only one single version of the X-conditional expectation of Y. The uniqueness properties of $E(Y \mid \mathscr{C})$ in the general case are formulated in the following section.   ◁

10.2.1   Uniqueness with respect to a probability measure

Remark 10.13 [Uniqueness of $E(Y \mid \mathscr{C})$ with respect to a probability measure]   Let the assumptions 10.1 hold and let Q be a probability measure on $(\Omega, \mathscr{A})$. Then we define

$$E(Y \mid \mathscr{C}) \text{ is } Q\text{-unique} \iff \forall\, V, V^* \in \mathscr{E}(Y \mid \mathscr{C})\colon V \underset{Q}{=} V^*. \tag{10.11}$$

This term is convenient not only for $Q = P$. According to the following remark, $E(Y \mid \mathscr{C})$ is P-unique.   ◁

Remark 10.14 [$\mathscr{E}(Y \mid \mathscr{C})$ is a *P*-equivalence class] Let $V \in \mathscr{E}(Y \mid \mathscr{C})$ and suppose that the random variable $V^*: (\Omega, \mathscr{A}, P) \rightarrow (\overline{\mathbb{R}}, \overline{\mathscr{B}})$ is $\mathscr{C}$-measurable with $V \underset{P}{=} V^*$. Then, according to Theorem 3.68 (i), Equation (10.7) also holds for V^*, and this implies $V^* \in \mathscr{E}(Y \mid \mathscr{C})$. Hence, if $V \in \mathscr{E}(Y \mid \mathscr{C})$, then Theorem 10.9 (ii) implies

$$V^* \in \mathscr{E}(Y \mid \mathscr{C}) \;\Leftrightarrow\; V^* \underset{P}{=} V \text{ and } V^* \text{ is } \mathscr{C}\text{-measurable.} \tag{10.12}$$

Therefore, $\mathscr{E}(Y \mid \mathscr{C})$ is the *P*-equivalence class of V in the set of all $\mathscr{C}$-measurable random variables (see Def. 2.74).   ◁

Remark 10.15 [*P*-equivalence and $\mathscr{C}$-measurability] Suppose that V is a version of the $\mathscr{C}$-conditional expectation, (i.e., $V \in \mathscr{E}(Y \mid \mathscr{C})$) and that $V^*: (\Omega, \mathscr{A}, P) \rightarrow (\overline{\mathbb{R}}, \overline{\mathscr{B}})$ is a random variable. Then (10.12) implies

$$V^* \in \mathscr{E}(Y \mid \mathscr{C}) \;\Rightarrow\; V^* \underset{P}{=} V.$$

However, $V^* \underset{P}{=} E(Y \mid \mathscr{C})$ may be true and yet $V^* \notin \mathscr{E}(Y \mid \mathscr{C})$, because $V^* \underset{P}{=} E(Y \mid \mathscr{C})$ does not imply that V^* is $\mathscr{C}$-measurable.   ◁

Remark 10.16 [Versions of $E(Y \mid \mathscr{C})$] For simplicity, we also say that V is a *version of* $E(Y \mid \mathscr{C})$, meaning $V \in \mathscr{E}(Y \mid \mathscr{C})$.   ◁

10.2.2   A necessary and sufficient condition of uniqueness

Now we present a necessary and sufficient condition for uniqueness of a conditional expectation for special σ-algebras $\mathscr{C}$. We consider a finite or countable partition of Ω, that is, a finite or countable set $\mathscr{E}$ of pairwise disjoint nonempty subsets A_i of Ω with $\bigcup_{A_i \in \mathscr{E}} A_i = \Omega$.

Theorem 10.17 [Uniqueness of $E(Y \mid \mathscr{C})$]
Let the assumptions 10.1 hold and let $\mathscr{C} = \sigma(\mathscr{E})$, where $\mathscr{E} = \{A_1, A_2, \ldots\}$ is a finite or countable partition of Ω. Then, $V = V^*$ for all $V, V^* \in \mathscr{E}(Y \mid \mathscr{C})$ if and only if

$$P(A_i) > 0, \quad \forall A_i \in \mathscr{E}. \tag{10.13}$$

(Proof p. 328)

Remark 10.18 [Values of a $\mathscr{C}$-conditional expectation] Under the assumptions of Theorem 10.17, the term $E(Y \mid \mathscr{C})$ is uniquely defined if and only if (10.13) holds, and in this case we can write

$$E(Y \mid \mathscr{C}) = \sum_{A_i \in \mathscr{E}} E(Y \mid A_i) \cdot 1_{A_i} \tag{10.14}$$

and

$$\forall A_i \in \mathscr{C}: \quad E(Y \mid \mathscr{C})(\omega) = E(Y \mid A_i), \quad \text{if } \omega \in A_i. \tag{10.15}$$

This equation shows that the conditional expectation $E(Y \mid \mathscr{C})$ describes how the conditional expectation values $E(Y \mid A_i)$ depend on the events $A_i \in \mathscr{C}$ (see Exercise 10.3).   ◁

10.2.3   Examples

Example 10.19 [No treatment for Joe – continued]   In Table 9.2, the conditional expectation $E(Y \mid X)$ of the outcome variable Y given the treatment variable X has only two different values, the conditional expectation values

$$E(Y \mid X=0) = .6 \quad \text{and} \quad E(Y \mid X=1) = .4.$$

The last but one column of Table 9.2 shows how these values are assigned to the eight possible outcomes $\omega \in \Omega$. These values and $E(Y \mid X)$ itself are uniquely defined.

In contrast, the conditional expectation $E(Y \mid X, U)$ of Y given the treatment variable X and the person variable U has four different values, .696, .2, .4, and 0 (see Example 9.23). Note that these four values define only one element, say V, of $\mathscr{E}(Y \mid X, U)$. If, instead of $E(Y \mid X, U)(\omega) = 0$ for $\omega \in \{(Joe, yes, -), (Joe, yes, +)\}$, we define

$$E(Y \mid X, U)(\omega) = \alpha, \quad \alpha \neq 0, \quad \alpha \in \mathbb{R}, \quad \text{for } \omega \in \{(Joe, yes, -), (Joe, yes, +)\},$$

then we have a new element, say V^*, of $\mathscr{E}(Y \mid X, U)$. Because α can be any nonzero real number, in this example, the set $\mathscr{E}(Y \mid X, U)$ is uncountably infinite. However, because $P(X=1, U=Joe) = 0$, two elements V and V^* of $\mathscr{E}(Y \mid X, U)$ are always identical with probability 1 (i.e., V and V^* are P-equivalent). Also note that $E(Y \mid X)$ and $E(Y \mid X, U)$ are random variables on the same probability space as the other random variables such as Y, X, and U.   ◁

Example 10.20 [No treatment for Joe – continued]   In Example 9.23, we specified the discrete conditional expectations $E(Y \mid X)$, $E(Y \mid X, U)$, and $P(X=1 \mid U)$. Now we check whether $E(Y \mid X)$ satisfies conditions (a) and (b) of Definition 10.2. First of all,

$$\mathscr{C} = \{\Omega, \emptyset, \{X=0\}, \{X=1\}\} = \sigma(X)$$

is the σ-algebra generated by X, where, in this example, $X: (\Omega, \mathscr{A}, P) \to (\Omega'_X, \mathscr{A}'_X)$ with $\Omega'_X = \{0, 1\}$ and $\mathscr{A}'_X = \{\Omega'_X, \emptyset, \{0\}, \{1\}\}$. If $V \in \mathscr{E}(Y \mid X)$ and $\overline{\mathscr{B}}$ is the Borel σ-algebra on $\overline{\mathbb{R}}$, then,

$$V^{-1}(B) = \begin{cases} \Omega, & \text{if } .4 \in B \text{ and } .6 \in B \\ \emptyset, & \text{if } .4 \notin B \text{ and } .6 \notin B \\ \{X=0\}, & \text{if } .4 \notin B \text{ and } .6 \in B \\ \{X=1\}, & \text{if } .4 \in B \text{ and } .6 \notin B, \end{cases} \quad \forall B \in \overline{\mathscr{B}}.$$

Hence, in this example, $V^{-1}(\overline{\mathscr{B}}) = \mathscr{C} = \sigma(X)$. Therefore, condition (a) of Definition 10.2 is satisfied for $V = E(Y \mid X)$ specified in Example 9.23.

Now we check condition (b) of Definition 10.2. For $C = \Omega$, this condition requires $E(1_\Omega \cdot V) = E(1_\Omega \cdot Y)$. The expectation of $1_\Omega \cdot Y$ is

$$E(1_\Omega \cdot Y) = E(Y) = P(Y=1) = .348 + 0 + .024 + .152 = .524,$$

and the expectation of $1_\Omega \cdot V$ is

$$E(1_\Omega \cdot V) = E(V) = .6 \cdot P(X=0) + .4 \cdot P(X=1)$$
$$= .6 \cdot (.152 + .348 + .096 + .024) + .4 \cdot (0 + 0 + .228 + .152) = .524$$

(see Table 9.2). For $C = \varnothing$, condition (b) of Definition 10.2 requires that $E(1_\varnothing \cdot V) = E(1_\varnothing \cdot Y)$, which is always true [see Box 6.1 (v)]. For $C = \{X=0\}$, condition (b) of Definition 10.2 requires $E(1_{X=0} \cdot V) = E(1_{X=0} \cdot Y)$, and the expectations of $1_{X=0} \cdot Y$ and $1_{X=0} \cdot V$ are

$$E(1_{X=0} \cdot Y) = E(1_{X=0} \cdot 1_{Y=1}) = P(X=0, Y=1) = .348 + .024 = .372$$

and

$$E(1_{X=0} \cdot V) = .6 \cdot P(1_{X=0}=1, X=0) + 0 \cdot P(1_{X=0}=1, X=1)$$
$$= .6 \cdot P(X=0) = .6 \cdot (.152 + .348 + .096 + .024) = .372.$$

Note that the random variable $1_{X=0} \cdot E(Y \mid X)$ has two values, .6 and 0 (see Table 9.2).

Finally, for $C = \{X=1\}$, condition (b) of Definition 10.2 requires that $E(1_{X=1} \cdot V) = E(1_{X=1} \cdot Y)$ and the expectations of $1_{X=1} \cdot Y$ and $1_{X=1} \cdot V$ are

$$E(1_{X=1} \cdot Y) = E(1_{X=1} \cdot 1_{Y=1}) = P(X=1, Y=1) = 0 + .152 = .152$$

and

$$E(1_{X=1} \cdot V) = 0 \cdot P(1_{X=1}=1, X=0) + .4 \cdot P(1_{X=1}=1, X=1)$$
$$= .4 \cdot P(X=1) = .4 \cdot (0 + 0 + .228 + .152) = .152$$

(see Table 9.2). Hence, $V = E(Y \mid X)$ satisfies conditions (a) and (b) of Definition 10.2. Therefore, $V = E(Y \mid X)$ is in fact an element of $\mathscr{E}(Y \mid X) = \mathscr{E}[Y \mid \sigma(X)]$. As mentioned before, in this example, $E(Y \mid X)$ is uniquely defined. This means that it is the only element of the set $\mathscr{E}(Y \mid X)$.   ◁

10.3   Rules of computation and other properties

10.3.1   Rules of computation

Some rules of computation for $\mathscr{C}$-conditional expectations are presented in Box 10.1, some of which are analog to the rules for expectations (see Box 6.1) and to the rules for B-conditional expectation values (see Box 9.1). In Rule (iv), the term $E[E(Y \mid \mathscr{C})]$ denotes the expectation

---

Box 10.1   Rules of computation for $\mathscr{C}$-conditional expectations.

Let $Y, Y_1, Y_2: (\Omega, \mathscr{A}, P) \to (\overline{\mathbb{R}}, \overline{\mathscr{B}})$ be numerical random variables that are nonnegative or have a finite expectation, $\mathscr{C} \subset \mathscr{A}$ a σ-algebra, and $\alpha \in \mathbb{R}$. Then,

$$E(\alpha \mid \mathscr{C}) \underset{P}{=} \alpha. \tag{i}$$

$$E(\alpha + Y \mid \mathscr{C}) \underset{P}{=} \alpha + E(Y \mid \mathscr{C}). \tag{ii}$$

$$E(\alpha \cdot Y \mid \mathscr{C}) \underset{P}{=} \alpha \cdot E(Y \mid \mathscr{C}). \tag{iii}$$

$$E[E(Y \mid \mathscr{C})] = E(Y). \tag{iv}$$

$$E\big[E(Y \mid \mathscr{C}) \mid \mathscr{C}_0\big] \underset{P}{=} E(Y \mid \mathscr{C}_0), \quad \text{if } \mathscr{C}_0 \subset \mathscr{C} \text{ is a } \sigma\text{-algebra.} \tag{v}$$

$$E(Y \mid \mathscr{C}) \underset{P}{=} E(Y), \quad \text{if } Y \underset{P}{\perp\!\!\!\perp} \mathscr{C}. \tag{vi}$$

$$E(Y \mid \mathscr{C}) \underset{P}{=} Y, \quad \text{if } Y \text{ is } \mathscr{C}\text{-measurable.} \tag{vii}$$

$$E(Y \mid \mathscr{C}) \underset{P}{=} E\big[Y \mid E(Y \mid \mathscr{C})\big]. \tag{viii}$$

$$E(Y_1 \mid \mathscr{C}) \underset{P}{=} E(Y_2 \mid \mathscr{C}), \quad \text{if } Y_1 \underset{P}{=} Y_2. \tag{ix}$$

$$E(Y) \text{ is finite} \quad \Rightarrow \quad \exists\, V \in \mathscr{E}(Y \mid \mathscr{C}): V \text{ is real-valued.} \tag{x}$$

$$E(Y^2) < \infty \quad \Rightarrow \quad E[E(Y \mid \mathscr{C})^2] < \infty. \tag{xi}$$

$$Cov[Y, E(Y \mid \mathscr{C})] = Var[E(Y \mid \mathscr{C})], \quad \text{if } E(Y^2) < \infty. \tag{xii}$$

$$Cov[Y_1, E(Y_2 \mid \mathscr{C})] = Cov(Y_1, Y_2), \quad \text{if } \sigma(Y_1) \subset \mathscr{C}, E(Y_1^2), E(Y_2^2) < \infty. \tag{xiii}$$

Let $E(Y_1^2), E(Y_2^2) < \infty$ or Y_1, Y_2 be nonnegative. Then, $\sigma(Y_1) \subset \mathscr{C}$ implies

$$E(Y_1 \cdot Y_2 \mid \mathscr{C}) \underset{P}{=} Y_1 \cdot E(Y_2 \mid \mathscr{C}). \tag{xiv}$$

If Y_1, Y_2 are nonnegative or real-valued with finite expectations, then there is a nonnegative (if Y_1 is nonnegative) or real-valued (if Y_1 is real-valued) version $E(Y_1 \mid \mathscr{C}) \in \mathscr{E}(Y_1 \mid \mathscr{C})$ and a nonnegative (if Y_2 is nonnegative) or real-valued (if Y_2 is real-valued) version $E(Y_2 \mid \mathscr{C}) \in \mathscr{E}(Y_2 \mid \mathscr{C})$ such that

$$E(Y_1 + Y_2 \mid \mathscr{C}) \underset{P}{=} E(Y_1 \mid \mathscr{C}) + E(Y_2 \mid \mathscr{C}). \tag{xv}$$

If $Y_1, \ldots, Y_n$ are real-valued with finite expectations and $\alpha_1, \ldots, \alpha_n \in \mathbb{R}$, then,

$$E\left(\sum_{i=1}^{n} \alpha_i \cdot Y_i \,\middle|\, \mathscr{C}\right) \underset{P}{=} \sum_{i=1}^{n} \alpha_i \cdot E(Y_i \mid \mathscr{C}). \tag{xvi}$$

(with respect to the measure P) of a $\mathscr{C}$-conditional expectation of Y. Similarly, in Rule (v), the term $E[E(Y \mid \mathscr{C}) \mid \mathscr{C}_0]$ denotes the $\mathscr{C}_0$-conditional expectation of the $\mathscr{C}$-conditional expectation of Y, where we presume $\mathscr{C}_0 \subset \mathscr{C}$. (For a proof of these rules, see Exercise 10.4).

For convenience, in Box 10.2 we translate these rules to X-conditional expectations, that is, to the case in which $\mathscr{C} = \sigma(X)$. Hence, these properties are special cases of those listed in Box 10.1, and they do not need proofs of their own.

Box 10.2   Rules of computation for X-conditional expectations.

Let $Y, Y_1, Y_2: (\Omega, \mathcal{A}, P) \to (\overline{\mathbb{R}}, \overline{\mathcal{B}})$ be numerical random variables that are nonnegative or have a finite expectation, let $X: (\Omega, \mathcal{A}, P) \to (\Omega'_X, \mathcal{A}'_X)$ be a random variable, let $f: (\Omega'_X, \mathcal{A}'_X) \to (\overline{\mathbb{R}}, \overline{\mathcal{B}})$ be measurable, and let $\alpha \in \mathbb{R}$. Then,

$$E(\alpha \mid X) \underset{P}{=} \alpha. \tag{i}$$

$$E(\alpha + Y \mid X) \underset{P}{=} \alpha + E(Y \mid X). \tag{ii}$$

$$E(\alpha \cdot Y \mid X) \underset{P}{=} \alpha \cdot E(Y \mid X). \tag{iii}$$

$$E[E(Y \mid X)] = E(Y). \tag{iv}$$

$$E\big[E(Y \mid X) \mid f(X)\big] \underset{P}{=} E[Y \mid f(X)]. \tag{v}$$

$$E(Y \mid X) \underset{P}{=} E(Y), \quad \text{if } Y \underset{P}{\perp\!\!\!\perp} X. \tag{vi}$$

$$E[f(X) \mid X] \underset{P}{=} f(X), \quad \text{if } f(X) \geq 0 \text{ or } E[f(X)] < \infty. \tag{vii}$$

$$E(Y \mid X) \underset{P}{=} E\big[Y \mid E(Y \mid X)\big]. \tag{viii}$$

$$E(Y_1 \mid X) \underset{P}{=} E(Y_2 \mid X), \quad \text{if } Y_1 \underset{P}{=} Y_2. \tag{ix}$$

$$E(Y) \text{ is finite} \quad \Rightarrow \quad \exists\, V \in \mathscr{E}(Y \mid X): V \text{ is real-valued.} \tag{x}$$

$$E(Y^2) < \infty \quad \Rightarrow \quad E[E(Y \mid X)^2] < \infty. \tag{xi}$$

$$Cov[Y, E(Y \mid X)] = Var[E(Y \mid X)], \quad \text{if } E(Y^2) < \infty. \tag{xii}$$

$$Cov[f(X), E(Y \mid X)] = Cov[f(X), Y], \quad \text{if } E[f(X)^2],\ E(Y^2) < \infty. \tag{xiii}$$

Let $E(Y^2), E[f(X)^2] < \infty$ or $Y, f(X)$ be nonnegative. Then,

$$E[f(X) \cdot Y \mid X] \underset{P}{=} f(X) \cdot E(Y \mid X). \tag{xiv}$$

If Y_1, Y_2 are nonnegative or real-valued with finite expectations, then there is a nonnegative (if Y_1 is nonnegative) or real-valued (if Y_1 is real-valued) version $E(Y_1 \mid X) \in \mathscr{E}(Y_1 \mid X)$ and a nonnegative (if Y_2 is nonnegative) or real-valued (if Y_2 is real-valued) version $E(Y_2 \mid X) \in \mathscr{E}(Y_2 \mid X)$ such that

$$E(Y_1 + Y_2 \mid X) \underset{P}{=} E(Y_1 \mid X) + E(Y_2 \mid X). \tag{xv}$$

If $Y_1, \dots, Y_n$ are real-valued with finite expectations and $\alpha_1, \dots, \alpha_n \in \mathbb{R}$, then,

$$E\left(\sum_{i=1}^{n} \alpha_i \cdot Y_i \,\Big|\, X \right) \underset{P}{=} \sum_{i=1}^{n} \alpha_i \cdot E(Y_i \mid X). \tag{xvi}$$

In some of these rules, we refer to the composition $f(X) = f \circ X$, where the function $f\colon (\Omega'_X, \mathcal{A}'_X) \to (\overline{\mathbb{R}}, \overline{\mathcal{B}})$ is assumed to be $(\mathcal{A}'_X, \overline{\mathcal{B}})$-measurable. Remember, according to Lemma 2.52, the composition $f(X)$ is measurable with respect to the σ-algebra $\mathcal{C} = X^{-1}(\mathcal{A}'_X) = \sigma(X)$ if $f\colon \Omega'_X \to \overline{\mathbb{R}}$ is $(\mathcal{A}'_X, \overline{\mathcal{B}})$-measurable. Furthermore, according to Corollary 2.53, the composition $f(X)$ is measurable with respect to the σ-algebra $\sigma(X)$ if $f\colon \Omega'_X \to \Omega'$, where Ω' is finite or countable but not necessarily a subset of $\overline{\mathbb{R}}$ and f is $(\mathcal{A}'_X, \mathcal{P}(\Omega'))$-measurable. According to Equation (2.29),

$$\sigma[f(X)] = [f(X)]^{-1}(\overline{\mathcal{B}}) = X^{-1}[f^{-1}(\overline{\mathcal{B}})] \subset X^{-1}(\mathcal{A}'_X) = \sigma(X).$$

In other words, we assume that the composition $f(X)$ is measurable with respect to the σ-algebra $\mathcal{C} = X^{-1}(\mathcal{A}'_X) = \sigma(X)$. Therefore, $\sigma[f(X)]$ can take the role of $\mathcal{C}_0$ in Rule (v) of Box 10.1. Furthermore, $f(X)$ takes the role of Y in Rule (vii) of Box 10.1, and the role of Y_1 in Rules (xiii) and (xiv) of Box 10.1.

10.3.2   Monotonicity

Box 10.3 displays some monotonicity properties that are proved in Exercise 10.5. Of course, if $X\colon (\Omega, \mathcal{A}, P) \to (\Omega'_X, \mathcal{A}'_X)$ is a random variable, then these properties also hold for $\mathcal{C} = \sigma(X)$ and $E(Y \mid X) = E[Y \mid \sigma(X)]$. For example, Rule (ix) can then be written as:

$$Y \underset{P}{\leq} Z \;\Rightarrow\; E(Y \mid X) \underset{P}{\leq} E(Z \mid X) \tag{10.16}$$

and Rule (v) yields

$$Y \underset{P}{\geq} \alpha \;\Rightarrow\; \exists\, V \in \mathcal{E}(Y \mid X)\colon V \geq \alpha. \tag{10.17}$$

10.3.3   Convergence theorems

Now we turn to convergence of $\mathcal{C}$-conditional expectations. Theorems 10.21 and 10.22 provide sufficient conditions that allow one to exchange taking the limit and taking the conditional expectation, for example,

$$\lim_{i \to \infty} E(Y_i \mid \mathcal{C}) \underset{P}{=} E(\lim_{i \to \infty} Y_i \mid \mathcal{C}).$$

This is not only of technical interest for many proofs; it also describes a continuity property of the conditional expectation: If Y_i is a good approximation to Y, then $E(Y_i \mid \mathcal{C})$ is a good approximation to $E(Y \mid \mathcal{C})$.

The first theorem deals with *monotone convergence*, and the second with *dominated convergence*. Reading these theorems, note that $\lim_{i \to \infty} Y_i \underset{P}{=} Y$ means

$$P(\{\omega \in \Omega\colon \lim_{i \to \infty} Y_i(\omega) = Y(\omega)\}) = 1, \tag{10.18}$$

that is, the sequence $(Y_i, i \in \mathbb{N})$ converges P-almost surely pointwise to Y.

Box 10.3   Monotonicity of conditional expectations.

Let $Y: (\Omega, \mathscr{A}, P) \to (\overline{\mathbb{R}}, \overline{\mathscr{B}})$ be a numerical random variable that is nonnegative or with finite expectation, $\mathscr{C} \subset \mathscr{A}$ be a σ-algebra, and $\alpha \in \mathbb{R}$. Then,

$$Y \underset{P}{\geq} 0 \text{ and } E(Y) = 0 \; \Rightarrow \; E(Y \mid \mathscr{C}) \underset{P}{=} 0. \tag{i}$$

$$Y \underset{P}{\geq} \alpha \; \Rightarrow \; E(Y \mid \mathscr{C}) \underset{P}{\geq} \alpha. \tag{ii}$$

$$Y \underset{P}{\leq} \alpha \; \Rightarrow \; E(Y \mid \mathscr{C}) \underset{P}{\leq} \alpha. \tag{iii}$$

$$Y \underset{P}{=} \alpha \; \Rightarrow \; E(Y \mid \mathscr{C}) \underset{P}{=} \alpha. \tag{iv}$$

$$Y \underset{P}{\geq} \alpha \; \Rightarrow \; \exists \, V \in \mathscr{E}(Y \mid \mathscr{C}): V \geq \alpha. \tag{v}$$

$$Y \underset{P}{=} \alpha \; \Rightarrow \; \exists \, V \in \mathscr{E}(Y \mid \mathscr{C}): V = \alpha. \tag{vi}$$

$$Y \underset{P}{\leq} \alpha \; \Rightarrow \; \exists \, V \in \mathscr{E}(Y \mid \mathscr{C}): V \leq \alpha. \tag{vii}$$

$$Y \underset{P}{=} \alpha \; \Rightarrow \; E(1_{Y=\alpha} \mid \mathscr{C}) \underset{P}{=} P(Y=\alpha \mid \mathscr{C}) \underset{P}{=} 1. \tag{viii}$$

Let $Y, Z: (\Omega, \mathscr{A}, P) \to (\overline{\mathbb{R}}, \overline{\mathscr{B}})$ be numerical random variables that are both nonnegative or such that both are real-valued with finite expectations and let $\mathscr{C} \subset \mathscr{A}$ be a σ-algebra. Then,

$$Y \underset{P}{\leq} Z \; \Rightarrow \; E(Y \mid \mathscr{C}) \underset{P}{\leq} E(Z \mid \mathscr{C}). \tag{ix}$$

If $A \in \mathscr{A}$ with $P(A) = 0$, then,

$$E(1_A \mid \mathscr{C}) \underset{P}{=} P(A \mid \mathscr{C}) \underset{P}{=} 0. \tag{x}$$

$$E(1_{A^c} \mid \mathscr{C}) \underset{P}{=} 1 - E(1_A \mid \mathscr{C}) \underset{P}{=} P(A^c \mid \mathscr{C}) \underset{P}{=} 1. \tag{xi}$$

Theorem 10.21 [Monotone convergence]
Let $Y, Y_i: (\Omega, \mathscr{A}, P) \to (\overline{\mathbb{R}}, \overline{\mathscr{B}})$, $i \in \mathbb{N}$, be random variables.

(i) *If the sequence $(Y_i, \; i \in \mathbb{N})$ is increasing with $\lim_{i \to \infty} Y_i \underset{P}{=} Y$, and $Y_i \geq 0, \forall \, i \in \mathbb{N}$, then,*

$$\lim_{i \to \infty} E(Y_i \mid \mathscr{C}) \underset{P}{=} E(Y \mid \mathscr{C}). \tag{10.19}$$

(ii) *If $Y_i \geq 0, \forall \, i \in \mathbb{N}$, then,*

$$E\left(\sum_{i=1}^{\infty} Y_i \,\middle|\, \mathscr{C} \right) \underset{P}{=} \sum_{i=1}^{\infty} E(Y_i \mid \mathscr{C}). \tag{10.20}$$

(Proof p. 329)

If the sequence $(Y_i, i \in \mathbb{N})$ is not increasing, then we need an additional assumption in order to guarantee convergence of the conditional expectations. A sufficient condition is that all $|Y_i|$ are dominated by the same P-integrable function Z.

Theorem 10.22 [Dominated convergence]
Let $Y_i \colon (\Omega, \mathscr{A}, P) \to (\overline{\mathbb{R}}, \overline{\mathscr{B}})$, $i \in \mathbb{N}$, *and* $Z \colon (\Omega, \mathscr{A}, P) \to (\overline{\mathbb{R}}, \overline{\mathscr{B}})$ *be random variables. If*

(a) Z *has a finite expectation and* $|Y_i| \leq Z$, $\forall\, i \in \mathbb{N}$, *and*

(b) $Y \colon (\Omega, \mathscr{A}, P) \to (\overline{\mathbb{R}}, \overline{\mathscr{B}})$ *is a random variable such that* $\lim_{i \to \infty} Y_i \underset{P}{=} Y$,

then,

$$\lim_{i \to \infty} E(Y_i \mid \mathscr{C}) \underset{P}{=} E(Y \mid \mathscr{C}). \qquad (10.21)$$

For a proof, see Klenke [2013, Theorem 8.14 (viii)].

10.4   Factorization, regression, and conditional expectation value

As mentioned in the Preface, much empirical research uses some kind of regression in order to investigate how the expectation of one random variable depends on the values of one or more other random variables. This applies to analysis of variance, regression analysis, the general linear model, the generalized linear model, factor analysis, structural equation models, hierarchical linear models, and analysis of qualitative data. Using these methods, we aim at learning about specific regressions. A regression is a special case of a factorization of an X-conditional expectation, in which X is real-valued. The concept of a factorization is also used for a general definition of a conditional expectation value.

10.4.1   Existence of a factorization

Lemma 2.52 can be applied to $E(Y \mid X)$, which, by definition, is measurable with respect to X. This immediately implies the following corollary:

Corollary 10.23 [Existence of a factorization of a conditional expectation]
Let $X \colon (\Omega, \mathscr{A}, P) \to (\Omega_X', \mathscr{A}_X')$ *and* $Y \colon (\Omega, \mathscr{A}, P) \to (\overline{\mathbb{R}}, \overline{\mathscr{B}})$ *be random variables, where* Y *is nonnegative or with finite expectation. If* $E(Y \mid X) \in \mathscr{E}(Y \mid X)$, *then there is a measurable function* $g \colon (\Omega_X', \mathscr{A}_X') \to (\overline{\mathbb{R}}, \overline{\mathscr{B}})$ *such that*

$$E(Y \mid X) = g \circ X. \qquad (10.22)$$

Remark 10.24 [Notation]   Instead of $g \circ X$, we also use the notation $g(X)$. Figure 9.1 displays the random variable X, a factorization g, and a version $E(Y \mid X) = g(X) \in \mathscr{E}(Y \mid X)$.   ◁

Definition 10.25 [Factorization and regression]
Under the assumptions of Corollary 10.23, the function $g: \Omega'_X \rightarrow \overline{\mathbb{R}}$ is called a factorization of $E(Y \mid X)$. If $(\Omega'_X, \mathscr{A}'_X) = (\mathbb{R}^n, \mathscr{B}_n)$, for $n \in \mathbb{N}$ and $g: \mathbb{R}^n \rightarrow \mathbb{R}$, then g is also called a regression of Y on X. In this case, Y is called the regressand and X the regressor.

Note that the definition of a regression does not refer to any parametric function. In contrast, this requirement is made in the definition of a linear quasi-regression (see section 7.1). The relationship between the regression and the linear quasi-regression is considered in the following section.

10.4.2 Conditional expectation and mean squared error

In Definition 7.2, we introduced the linear quasi-regression as that linear function $f: \mathbb{R} \rightarrow \mathbb{R}$ defined by $f(x) = \alpha_0 + \alpha_1 x, x \in \mathbb{R}$, that minimizes $MSE(a_0, a_1) = E([Y - (a_0 + a_1 X)]^2)$, where X is a real-valued random variable. In a sense, the linear quasi-regression is a function f such that the composition $Q_{lin}(Y \mid X) = f(X)$ is the best approximation of Y by a *linear function* of X. Now consider the approximation (with respect to the mean squared error) of Y by a more general function that is $\mathscr{C}$-measurable or X-measurable, respectively. Intuitively speaking, we ask for the best approximation of Y based on the information contained in $\mathscr{C}$ or in X.

Reading the following theorem, note that the right-hand sides of (10.23) and (10.24) do not depend on the particular choice of a version of $E(Y \mid X) \in \mathscr{E}(Y \mid X)$ [see Th. 10.9 (ii) and Rule (viii) of Box 6.1].

Theorem 10.26 [Conditional expectation and mean squared error]
Let $Y, Z: (\Omega, \mathscr{A}, P) \rightarrow (\mathbb{R}, \mathscr{B})$ be real-valued random variables with $E(Y^2), E(Z^2) < \infty$, let $\mathscr{C} \subset \mathscr{A}$ be a σ-algebra, and suppose that Z is $\mathscr{C}$-measurable. Then,

$$E[(Y - Z)^2] \geq E([Y - E(Y \mid \mathscr{C})]^2). \tag{10.23}$$

and

$$Z \underset{P}{=} E(Y \mid \mathscr{C}) \Leftrightarrow E[(Y - Z)^2] = E([Y - E(Y \mid \mathscr{C})]^2). \tag{10.24}$$

For a proof, see Klenke (2013, Cor. 8.17).

Remark 10.27 [Regression versus linear quasi-regression] If $E(Y^2) < \infty$, then Theorem 10.26 implies that $V \in \mathscr{E}(Y \mid X)$ is an X-measurable random variable with $E(V^2) < \infty$ minimizing the mean squared error $E[(Y - Z)^2]$ for all X-measurable random variables Z with $E(Z^2) < \infty$. Vice versa, if V is an X-measurable random variable with a finite second moment minimizing $E[(Y - Z)^2]$ for all X-measurable random variables Z with $E(Z^2) < \infty$, then $V \in \mathscr{E}(Y \mid X)$, provided that $E(Y^2) < \infty$.

In contrast, $Q_{lin}(Y \mid X)$ is a *linear function* of X minimizing $E([Y - (a_0 + a_1 X)]^2), a_0, a_1 \in \mathbb{R}$, provided that $E(Y^2) < \infty$. Hence, $Q_{lin}(Y \mid X)$ is the best (with respect to the mean squared error) approximation of Y in the set of all linear functions of X, whereas $E(Y \mid X)$ is the best approximation of Y in the set of all X-measurable functions, provided that the second moment of Y is finite.

According to Definition 7.2, $Q_{lin}(Y \mid X) = f(X)$, where $f \colon \mathbb{R} \to \mathbb{R}$ defined by $f(x) = \alpha_0 + \alpha_1 x$, $x \in \mathbb{R}$, is the linear quasi-regression of Y on X. In contrast, a regression of Y on X is a function $g \colon \mathbb{R}^n \to \mathbb{R}$ such that there is a version $E(Y \mid X) \in \mathscr{E}(Y \mid X)$ with $E(Y \mid X) = g(X)$. Even if we consider a random variable X with $\Omega'_X = \mathbb{R}^n$, then a regression g does not require that it is a linear function as specified in Equation (7.39). If g is a linear function with domain $\Omega'_X = \mathbb{R}^n$ and there is a version $E(Y \mid X) \in \mathscr{E}(Y \mid X)$ with $E(Y \mid X) = g(X)$, then $f = g$ and $Q_{lin}(Y \mid X) = E(Y \mid X)$.   ◁

10.4.3   Uniqueness of a factorization

A factorization of $E(Y \mid X)$ is not necessarily uniquely defined. This even applies if we consider a fixed version $E(Y \mid X)$ of the conditional expectation.

Remark 10.28 [Uniqueness of a factorization]   For two elements V and V^* of $\mathscr{E}(Y \mid X)$, there can be different factorizations g and g^* with $V = g(X)$ and $V^* = g^*(X)$. This is true even if $V = V^*$. Hence, there can be different factorizations of a single element $V \in \mathscr{E}(Y \mid X)$ (see Example 10.32). In other words, $V = g(X) = g^*(X)$, with $g \neq g^*$, is not necessarily contradictory. In this case $g(x) = g^*(x)$ for all $x \in X(\Omega)$, whereas $g(x) = g^*(x)$ does *not* hold for all $x \in \Omega'_X$. However, Theorem 10.9 (ii) and Corollary 5.25 (i) imply the following corollary: ◁

Corollary 10.29 [P_X-equivalence of factorizations]
Let the assumptions 10.1 hold and let $g, g^* \colon (\Omega'_X, \mathscr{A}'_X) \to (\mathbb{R}, \mathscr{B})$ be $(\mathscr{A}'_X, \mathscr{B})$-measurable functions. If $g(X), g^*(X) \in \mathscr{E}(Y \mid X)$, then,

$$g \underset{P_X}{=} g^*. \tag{10.25}$$

Remark 10.30 [P_X-equivalence]   Note that, according to (5.12), Equation (10.25) is equivalent to

$$g(x) = g^*(x), \quad \text{for } P_X\text{-a.a. } x \in \Omega'_X. \tag{10.26}$$

◁

According to Remark 10.12, $P(X = x) > 0$ for all $x \in X(\Omega)$ implies that $E(Y \mid X)$ is uniquely defined. According to the following corollary, this also applies to the factorization of $E(Y \mid X)$ if we additionally assume $\Omega'_X = X(\Omega)$.

Corollary 10.31 [Uniqueness of the factorization]
Let the assumptions 10.1 hold and assume $P(X = x) > 0$ for all $x \in \Omega'_X$. Then the function $g \colon \Omega'_X \to \overline{\mathbb{R}}$ satisfying $E(Y \mid X) = g(X)$ is uniquely defined.

Example 10.32 [No treatment for Joe – continued]   In Example 9.23, we specified $E(Y \mid X)$ with its two values

$$E(Y \mid X)(\omega) = E(Y \mid X = 0) = .6, \quad \forall\, \omega \in \Omega \text{ with } X(\omega) = 0,$$

and

$$E(Y \mid X)(\omega) = E(Y \mid X = 1) = .4, \quad \forall\, \omega \in \Omega \text{ with } X(\omega) = 1.$$

If we consider the treatment variable $X \colon (\Omega, \mathcal{A}, P) \to (\mathbb{R}, \mathcal{B})$, then $g \colon \mathbb{R} \to \mathbb{R}$ defined by

$$g(x) = \begin{cases} .6, & \text{if } x = 0 \\ .4, & \text{if } x = 1 \\ \alpha, & \text{otherwise} \end{cases}$$

is a factorization of the conditional expectation of Y given X for any choice of $\alpha \in \mathbb{R}$. This implies that there are different factorizations g, g^* with $g(X), g^*(X) \in \mathcal{E}(Y \mid X)$. However, g and g^* are P_X-equivalent. In contrast, if we consider $X \colon (\Omega, \mathcal{A}, P) \to (\Omega'_X, \mathcal{A}'_X)$, with $\Omega'_X = X(\Omega) = \{0, 1\}$ and $\mathcal{A}'_X = \mathcal{P}(\Omega'_X)$, then $P(X = x) > 0$ for all $x \in \Omega'_X$ and there is only one single factorization $g \colon \Omega'_X \to \mathbb{R}$ with the two values $g(0) = .6$ and $g(1) = .4$ (see Cor. 10.31).   ◁

10.4.4   Conditional expectation value

The concepts of a conditional expectation value $E(Y \mid X = x)$ and a conditional probability $P(A \mid X = x)$ have been introduced in Definition 9.2 and Remark 9.7 only for $P(X = x) > 0$. Now we drop this assumption and define these concepts more generally, again using the factorization of a conditional expectation.

Definition 10.33 [$(X = x)$-conditional expectation value]
Let $X \colon (\Omega, \mathcal{A}, P) \to (\Omega'_X, \mathcal{A}'_X)$ be a random variable, let $Y \colon (\Omega, \mathcal{A}, P) \to (\overline{\mathbb{R}}, \overline{\mathcal{B}})$ be nonnegative or with finite expectation $E(Y)$, and let $g \colon (\Omega'_X, \mathcal{A}'_X) \to (\overline{\mathbb{R}}, \overline{\mathcal{B}})$ be a function satisfying (10.22). Then the value $g(x)$ of g is called an $(X = x)$-conditional expectation value of Y and is denoted by $E(Y \mid X = x)$, that is,

$$E(Y \mid X = x) := g(x). \tag{10.27}$$

Remark 10.34 [$(X = x)$-conditional probability]   If 1_A is the indicator of $A \in \mathcal{A}$, then $E(1_A \mid X = x)$ is also called an $(X = x)$-*conditional probability* of A, and it is denoted by $P(A \mid X = x)$, that is,

$$P(A \mid X = x) := E(1_A \mid X = x). \tag{10.28}$$

Furthermore, considering the event $\{Y = y\}$, we also use the notation

$$P(Y = y \mid X = x) := P(\{Y = y\} \mid X = x) = E(1_{Y = y} \mid X = x). \tag{10.29}$$

◁

Remark 10.35 [Uniqueness and consistency of definitions] If $P(X=x) > 0$, then the $(X=x)$-conditional expectation value of Y is uniquely defined, and it is identical to the term introduced in Definition 9.2, that is,

$$E(Y \mid X=x) = E^{X=x}(Y), \quad \text{if } P(X=x) > 0 \tag{10.30}$$

(see Exercise 10.6). In the general case, $E(Y \mid X=x)$ is not uniquely defined. However, g is uniquely defined up to P_X-equivalence (see Cor. 10.29). ◁

Remark 10.36 [Versions of a conditional expectation with a discrete X] Let $Y: (\Omega, \mathcal{A}, P) \to (\overline{\mathbb{R}}, \overline{\mathcal{B}})$ be a random variable that is nonnegative or with finite expectation and suppose that $X: (\Omega, \mathcal{A}, P) \to (\Omega'_X, \mathcal{A}'_X)$ is discrete, implying that $\Omega'_0 \subset \Omega'_X$ is finite or countable with $P_X(\Omega'_0) = 1$ and $\{x\} \in \mathcal{A}'_X$ for all $x \in \Omega'_0$. Then, for all $\alpha \in \mathbb{R}$, the function $E(Y \mid X): \Omega \to \mathbb{R}$ defined by

$$E(Y \mid X)(\omega) := \begin{cases} E(Y \mid X=x), & \text{if } X(\omega) = x \text{ and } P(X=x) > 0 \\ \alpha, & \text{otherwise.} \end{cases} \tag{10.31}$$

is a version of $E(Y \mid X) \in \mathcal{E}(Y \mid X)$. This proposition follows from Remark 10.35 and proposition (10.12). ◁

Remark 10.37 [Values of the conditional expectation] Assume that $E(Y \mid X) = g(X) \in \mathcal{E}(Y \mid X)$. Then,

$$E(Y \mid X)(\omega) = g(x) = E(Y \mid X=x), \quad \forall \, \omega \in \Omega \text{ with } X(\omega) = x \tag{10.32}$$

(see Exercise 10.7). This also implies that the value of $E(Y \mid X)$ is constant on all sets $\{X=x\} = \{\omega \in \Omega: X(\omega) = x\}$. Note that this also holds if Ω is finite or countable and some $\omega \in \{X=x\}$ have probability $P(\{\omega\}) = 0$. As an example, see $E(Y \mid X)(\omega)$ for $\omega \in \{(Joe, yes, -), (Joe, yes, +)\}$ in Table 9.2. These two values are equal to $E(Y \mid X=1) = .4$, although $P(\{(Joe, yes, -)\}) = P(\{(Joe, yes, +)\}) = 0$. ◁

Remark 10.38 [Equivalent propositions] Let the assumptions 10.1 hold, and let $g(X) \in \mathcal{E}(Y \mid X)$ and $g^*: (\Omega'_X, \mathcal{A}'_X) \to (\overline{\mathbb{R}}, \overline{\mathcal{B}})$ be an $(\mathcal{A}'_X, \overline{\mathcal{B}})$-measurable function. Then proposition (10.12) and Theorem 2.49 yield

$$g(X) \underset{P}{=} g^*(X) \; \Leftrightarrow \; g^*(X) \in \mathcal{E}(Y \mid X). \tag{10.33}$$
  ◁

Definition 10.33, Corollary 5.25, Remark 5.26, Corollary 10.29, and Equation (10.22) imply the following corollary, according to which we may formulate propositions either in terms of $(X=x)$-conditional expectation values or, equivalently, in terms of the corresponding conditional expectations.

Corollary 10.39 [Equivalent propositions]
Let $Y_1, Y_2: (\Omega, \mathcal{A}, P) \rightarrow (\overline{\mathbb{R}}, \overline{\mathcal{B}})$ be two numerical random variables that are nonnegative or with finite expectations, and let $X: (\Omega, \mathcal{A}, P) \rightarrow (\Omega'_X, \mathcal{A}'_X)$ be a random variable. Then,

(i) $E(Y_1 \mid X) \underset{P}{=} E(Y_2 \mid X)$ is equivalent to

$$E(Y_1 \mid X=x) = E(Y_2 \mid X=x), \quad for\ P_X\text{-}a.a.\ x \in \Omega'_X. \qquad (10.34)$$

(ii) $E(Y_1 \mid X) \underset{P}{>} E(Y_2 \mid X)$ is equivalent to

$$E(Y_1 \mid X=x) > E(Y_2 \mid X=x), \quad for\ P_X\text{-}a.a.\ x \in \Omega'_X. \qquad (10.35)$$

(iii) $E(Y_1 \mid X) \underset{P}{\geq} E(Y_2 \mid X)$ is equivalent to

$$E(Y_1 \mid X=x) \geq E(Y_2 \mid X=x), \quad for\ P_X\text{-}a.a.\ x \in \Omega'_X. \qquad (10.36)$$

Remark 10.40 [$(X=x)$**-conditional expectation value of** $E(Y \mid X)$] Suppose that $f: (\Omega'_X, \mathcal{A}'_X) \rightarrow (\Omega'', \mathcal{A}'')$ is a measurable mapping, $w \in \Omega''$, $\{w\} \in \mathcal{A}''$, and $C = \{f(X)=w\}$. Then Equation (10.6) implies

$$E[E(Y \mid X) \mid f(X)=w] = E[Y \mid f(X)=w],$$
$$\text{if } w \in \Omega'' \text{ and } P[f(X)=w] > 0. \qquad (10.37)$$

As a special case, this equation yields

$$E[E(Y \mid X, Z) \mid X=x] = E(Y \mid X=x), \quad \text{if } x \in \Omega'_X \text{ with } P(X=x) > 0, \qquad (10.38)$$

provided that $Z: (\Omega, \mathcal{A}, P) \rightarrow (\Omega'_Z, \mathcal{A}'_Z)$ is a random variable, too. If Z takes on only a finite number of values $z_1, \dots, z_m$ and $P(X=x, Z=z_i) > 0$, for all $i = 1, \dots, m$, then Rule (ii) of Box 9.2 follows from Equations (10.38) and (9.21). Similarly, applying (9.20), Equation (10.38) yields Rule (iii) of Box 9.2 (see Exercise 10.8). ◁

Proposition (i) of Corollary 10.39 and Rule (v) of Box 10.2 yield

$$E[E(Y \mid X) \mid f(X)=w] = E[Y \mid f(X)=w], \quad for\ P_{f(X)}\text{-a.a}\ w \in \Omega'', \qquad (10.39)$$

which generalizes Equation (10.37). A special case of Equation (10.39) is

$$E[E(Y \mid X, Z) \mid X=x] = E(Y \mid X=x), \quad for\ P_X\text{-a.a}\ x \in \Omega'_X. \qquad (10.40)$$

Remark 10.41 [Expectation of a conditional expectation] Applying Definition 10.33 and Equation (6.13), Rule (iv) of Box 10.2 is equivalent to

$$E(Y) = \int E(Y \mid X=x) \, P_X(dx). \qquad (10.41)$$

Furthermore, if X only takes on a finite number of values $x_1, \ldots, x_n$, then Equation (10.41) can be written as:

$$E(Y) = \sum_{i=1}^{n} E(Y \mid X = x_i) \cdot P(X = x_i) \tag{10.42}$$

[see Eq. (6.3)].   ◁

10.5   Characterizing a conditional expectation by the joint distribution

Using the factorization and Equation (3.59) yields two conditions that are equivalent to those occurring in Definition 10.2. In these conditions, we refer to the joint distribution $P_{X,Y}$ of X and Y.

Theorem 10.42 [Conditional expectation and the joint distribution]
Let the assumptions 10.1 hold and assume that $g: (\Omega'_X, \mathscr{A}'_X) \to (\overline{\mathbb{R}}, \overline{\mathscr{B}})$ is measurable. Then, $g(X) \in \mathscr{E}(Y \mid X)$ if and only if

$$\int 1_{C'}(x) \cdot g(x) \; P_X(dx) = \int 1_{C'}(x) \cdot y \; P_{X,Y}[d(x, y)], \quad \forall \, C' \in \mathscr{A}'_X. \tag{10.43}$$

(Proof p. 329)

Remark 10.43 [Two alternative formulations]   According to (3.28) and (3.59), Equation (10.43) is equivalent to

$$\int_{C'} g \, dP_X = \int_{\{X \in C'\}} Y \, dP, \quad \forall \, C' \in \mathscr{A}'_X. \tag{10.44}$$

If $E_X(\cdot)$ denotes the expectation with respect to the distribution P_X and $1_{X \in C'}$ denotes the indicator of $\{X \in C'\}$, then Equation (10.44) may also be written as:

$$E_X(1_{C'} \cdot g) = E(1_{X \in C'} \cdot Y), \quad \forall \, C' \in \mathscr{A}'_X. \tag{10.45}$$

◁

Remark 10.44 [Conditional expectation with respect to a joint distribution]   Note that in Equation (10.43), we do not explicitly refer to the measure P. Instead we refer to $P_{X,Y}$, the joint distribution of X and Y. Therefore, $g(X) \in \mathscr{E}(Y \mid X)$ may also be called a version of the *conditional expectation* of Y on X with respect to $P_{X,Y}$. This can be used, for example, to consider a conditional expectation with respect to the $(Z = z)$-conditional distribution $P_{X,Y \mid Z = z}$ (see Def. 17.7).   ◁

10.6   Conditional mean independence

In section 5.4, we defined *independence* of two random variables X and Y. Furthermore, in chapter 7, introducing the covariance $Cov(X, Y)$ and the correlation $Corr(X, Y)$ of two numerical random variables, we also defined *uncorrelatedness* by $Cov(X, Y) = 0$. Now we add two other related concepts: *mean independence* and *conditional mean independence*.

Definition 10.45 [$\mathscr{C}$-conditional mean independence]
Let $Y: (\Omega, \mathscr{A}, P) \to (\overline{\mathbb{R}}, \overline{\mathscr{B}})$ be a numerical random variable that is nonnegative or has a finite expectation $E(Y)$ and let $\mathscr{D} \subset \mathscr{A}$ be a σ-algebra.

(i) Then Y is called mean independent from $\mathscr{D}$, *if*

$$E(Y \mid \mathscr{D}) \underset{P}{=} E(Y). \tag{10.46}$$

(ii) Let also $\mathscr{C} \subset \mathscr{A}$ be a σ-algebra and let $E(Y \mid \mathscr{C}, \mathscr{D})$ denote the conditional expectation of Y given $\sigma(\mathscr{C} \cup \mathscr{D})$. Then Y is called $\mathscr{C}$-conditionally mean independent from $\mathscr{D}$, if

$$E(Y \mid \mathscr{C}, \mathscr{D}) \underset{P}{=} E(Y \mid \mathscr{C}). \tag{10.47}$$

Analogously to $E(Y \mid \mathscr{C}, \mathscr{D}) := E[Y \mid \sigma(\mathscr{C} \cup \mathscr{D})]$, we use the notation $P(A \mid \mathscr{C}, \mathscr{D}) := P[A \mid \sigma(\mathscr{C} \cup \mathscr{D})]$ for the $\sigma(\mathscr{C} \cup \mathscr{D})$-conditional probability of $A \in \mathscr{A}$.

Remark 10.46 [X-conditional mean independence] Let $Y: (\Omega, \mathscr{A}, P) \to (\overline{\mathbb{R}}, \overline{\mathscr{B}})$ be a numerical random variable that is nonnegative or has a finite expectation $E(Y)$, and let $Z: (\Omega, \mathscr{A}, P) \to (\Omega'_Z, \mathscr{A}'_Z)$ be a random variable.

(i) Then Y is called *mean independent from Z* if

$$E(Y \mid Z) \underset{P}{=} E(Y). \tag{10.48}$$

(ii) Let also $X: (\Omega, \mathscr{A}, P) \to (\Omega'_X, \mathscr{A}'_X)$ be a random variable. Then Y is called *X-conditionally mean independent from Z* if

$$E(Y \mid X, Z) \underset{P}{=} E(Y \mid X). \tag{10.49}$$

◁

Remark 10.47 [A special case] Of course, if $\mathscr{D} \subset \mathscr{C}$, then Y is $\mathscr{C}$-conditionally mean independent from $\mathscr{D}$. In this case, $\sigma(\mathscr{C} \cup \mathscr{D}) = \sigma(\mathscr{C}) = \mathscr{C}$ and $E(Y \mid \mathscr{C}, \mathscr{D})$ is just a different notation of $E(Y \mid \mathscr{C})$. Correspondingly, assume that Z is measurable with respect to X. Then $\sigma(X, Z) = \sigma(X)$ and, therefore,

$$E(Y \mid X, Z) \underset{P}{=} E(Y \mid X), \quad \text{if } \sigma(Z) \subset \sigma(X). \tag{10.50}$$

Table 10.1   Joe and Ann with no treatment effects.

	$P(\{\omega\})$	Person variable U	Treatment variable X	Outcome variable Y	$E(Y \mid X, U)$	$E(Y \mid U)$	$E(Y \mid X)$	$P(X=1 \mid U)$
(*Joe, no, −*)	.16	*Joe*	0	0	.2	.2	.56	.6
(*Joe, no, +*)	.04	*Joe*	0	1	.2	.2	.56	.6
(*Joe, yes, −*)	.24	*Joe*	1	0	.2	.2	.44	.6
(*Joe, yes, +*)	.06	*Joe*	1	1	.2	.2	.44	.6
(*Ann, no, −*)	.06	*Ann*	0	0	.8	.8	.56	.4
(*Ann, no, +*)	.24	*Ann*	0	1	.8	.8	.56	.4
(*Ann, yes, −*)	.04	*Ann*	1	0	.8	.8	.44	.4
(*Ann, yes,+*)	.16	*Ann*	1	1	.8	.8	.44	.4

Hence, Y is X-conditionally mean independent from all random variables Z that are measurable with respect to X. In these cases, Z does not carry any information additional to X. In more formal terms, Z does not represent any event that is not already represented by X, that is, $\{Z \in A'\} \in \sigma(X)$, for all $A' \in \mathscr{A}'_Z$.   ◁

Example 10.48 [Joe and Ann with no treatment effect]   Table 10.1 displays an example for U-conditional mean independence of Y from X, that is,

$$E(Y \mid X, U) \underset{P}{=} E(Y \mid U).$$

The values of the conditional expectations $E(Y \mid X, U)$ and $E(Y \mid U)$ can be computed in the same way as in Example 9.21. This new example shows that $E(Y \mid X, U) \underset{P}{=} E(Y \mid U)$ does not imply $E(Y \mid X) \underset{P}{=} E(Y)$. Hence, although $E(Y \mid X) \underset{P}{=} E(Y)$ *does not hold* and the conditional expectation values $E(Y \mid X=x)$ *do depend* on the values x of X, in a sense, the treatment variable X is irrelevant once we condition on U. In other words, for *Joe*, success does not depend on whether or not he receives treatment, and the same is true for *Ann* [see the column headed $E(Y \mid X, U)$]. This example shows that the conditional expectation $E(Y \mid X)$ can be completely misleading if used for the evaluation of the effect of the treatment variable X on the outcome variable Y.   ◁

Remark 10.49 [Implication structure among different kinds of independence]   According-ing to Rule (vi) of Box 10.1, independence of Y and $\mathscr{C}$ implies that Y is mean independent from $\mathscr{C}$. Analogously, according to Rule (vi) of Box 10.2, independence of Y and X implies that Y is mean independent from X. Furthermore, mean independence of Y from X implies that

X and Y are uncorrelated, provided that X and Y are numerical and $E(X^2)$, $E(Y^2) < \infty$ (see Exercise 10.9). Hence, if $E(X^2)$, $E(Y^2) < \infty$, then,

$$Y \underset{P}{\perp\!\!\!\perp} X \Rightarrow E(Y \mid X) \underset{P}{=} E(Y) \tag{10.51}$$

and

$$E(Y \mid X) \underset{P}{=} E(Y) \Rightarrow Corr(X, Y) = Cov(X, Y) = 0. \tag{10.52}$$

◁

Remark 10.50 [Dichotomous Y] If Y is dichotomous with values 0 and 1 (see Example 5.10), then,

$$Y \underset{P}{\perp\!\!\!\perp} X \Leftrightarrow E(Y \mid X) \underset{P}{=} E(Y), \tag{10.53}$$

(see Exercise 10.10), which can equivalently be written as:

$$Y \underset{P}{\perp\!\!\!\perp} X \Leftrightarrow P(Y{=}1 \mid X) \underset{P}{=} P(Y{=}1) \tag{10.54}$$

[see (10.4)].
   If Y is dichotomous with values y_1 and y_2, then,

$$Y \underset{P}{\perp\!\!\!\perp} X \Leftrightarrow P(Y{=}y_1 \mid X) \underset{P}{=} P(Y{=}y_1), \tag{10.55}$$

because $P(1_{Y=y_1} = 1 \mid X) = P(Y{=}y_1 \mid X)$ and $P(1_{Y=y_1} = 1) = P(Y{=}y_1)$ [see again (10.4) and Def. 1.49].

◁

Now we turn to conditions that are equivalent to conditional mean independence. We start with a theorem that only applies to a *nonnegative* numerical random variable Y that *also has a finite expectation*. A second theorem also applies to a numerical random variable Y *with a finite second moment*.

Theorem 10.51 [$\mathscr{C}$-conditional mean independence if $E(Y)$ is finite]
Let $Y : (\Omega, \mathscr{A}, P) \to (\overline{\mathbb{R}}, \overline{\mathscr{B}})$ be a nonnegative random variable that has a finite expectation $E(Y)$ and let $\mathscr{C}, \mathscr{D} \subset \mathscr{A}$ be σ-algebras. Then the following two propositions are equivalent to each other:

(a) $E(Y \mid \mathscr{C}, \mathscr{D}) \underset{P}{=} E(Y \mid \mathscr{C})$.

(b) For all random variables $W : (\Omega, \mathscr{A}, P) \to (\overline{\mathbb{R}}, \overline{\mathscr{B}})$ that are nonnegative and $\mathscr{D}$-measurable,

$$E(W{\cdot}Y \mid \mathscr{C}) \underset{P}{=} E(W \mid \mathscr{C}) \cdot E(Y \mid \mathscr{C}). \tag{10.56}$$

(Proof p. 330)

326   PROBABILITY AND CONDITIONAL EXPECTATION

Remark 10.52 [Mean independence from a σ-algebra] For $\mathscr{C} = \{\Omega, \emptyset\}$, Theorem 10.51 and Remark 10.5 immediately yield the following proposition: If $Y\colon (\Omega, \mathscr{A}, P) \to (\overline{\mathbb{R}}, \overline{\mathscr{B}})$ is a nonnegative random variable that has a finite expectation $E(Y)$ and $\mathscr{D} \subset \mathscr{A}$ is a σ-algebra, then the following two propositions are equivalent to each other:

(a) $E(Y \mid \mathscr{D}) \underset{P}{=} E(Y)$.

(b) For all random variables $W\colon (\Omega, \mathscr{A}, P) \to (\mathbb{R}, \mathscr{B})$ that are nonnegative and $\mathscr{D}$-measurable,

$$E(W{\cdot}Y) = E(W) \cdot E(Y). \tag{10.57}$$

◁

In Theorem 10.51, we required that Y is nonnegative and has a finite expectation. The implication of conditional mean independence formulated in proposition (i) of the following theorem is *not* restricted to nonnegative random variables Y. Instead, we assume that Y has a finite second moment.

Theorem 10.53 [$\mathscr{C}$-conditional mean independence if $E(Y^2)$ is finite]
Let $Y\colon (\Omega, \mathscr{A}, P) \to (\overline{\mathbb{R}}, \overline{\mathscr{B}})$ be a random variable that has a finite second moment $E(Y^2)$, let $\mathscr{C}, \mathscr{D} \subset \mathscr{A}$ be σ-algebras, and consider:

(a) $E(Y \mid \mathscr{C}, \mathscr{D}) \underset{P}{=} E(Y \mid \mathscr{C})$.

(b) For all random variables $W\colon (\Omega, \mathscr{A}, P) \to (\mathbb{R}, \mathscr{B})$ that are $\mathscr{D}$-measurable and have a finite second moment $E(W^2)$,

$$E(W{\cdot}Y \mid \mathscr{C}) \underset{P}{=} E(W \mid \mathscr{C}) \cdot E(Y \mid \mathscr{C}). \tag{10.58}$$

Then,

(i) (a) implies (b).

(ii) If, additionally, Y is nonnegative, then (a) and (b) are equivalent to each other.
(Proof p. 330)

Remark 10.54 [Mean independence if $E(Y^2)$ is finite] For $\mathscr{C} = \{\Omega, \emptyset\}$, Theorem 10.53 and Remark 10.5 immediately yield the following proposition. Let $Y\colon (\Omega, \mathscr{A}, P) \to (\overline{\mathbb{R}}, \overline{\mathscr{B}})$ be a random variable that has a finite second moment $E(Y^2)$, let $\mathscr{D} \subset \mathscr{A}$ be a σ-algebra, and consider:

(a) $E(Y \mid \mathscr{D}) \underset{P}{=} E(Y)$.

(b) For all random variables $W\colon (\Omega, \mathscr{A}, P) \to (\mathbb{R}, \mathscr{B})$ that are $\mathscr{D}$-measurable and have a finite second moment $E(W^2)$,

$$E(W{\cdot}Y) = E(W) \cdot E(Y). \tag{10.59}$$

Then,

(i) (a) implies (b).

(ii) If Y is also nonnegative, then (a) and (b) are equivalent to each other.   ◁

For $\mathscr{C} = \sigma(X)$ and $\mathscr{D} = \sigma(Z)$, Theorem 10.51 immediately implies the following corollary.

Corollary 10.55 [X-conditional mean independence if $E(Y)$ is finite]
Let $Y: (\Omega, \mathscr{A}, P) \to (\overline{\mathbb{R}}, \overline{\mathscr{B}})$ be a nonnegative random variable that has a finite expectation $E(Y)$, and let $X: (\Omega, \mathscr{A}, P) \to (\Omega'_X, \mathscr{A}'_X)$ and $Z: (\Omega, \mathscr{A}, P) \to (\Omega'_Z, \mathscr{A}'_Z)$ be random variables. Then the following two propositions are equivalent to each other:

(a) $E(Y \mid X, Z) \underset{P}{=} E(Y \mid X)$.

(b) For all random variables $W: (\Omega, \mathscr{A}, P) \to (\mathbb{R}, \mathscr{B})$ that are nonnegative and Z-measurable,

$$E(W \cdot Y \mid X) \underset{P}{=} E(W \mid X) \cdot E(Y \mid X). \tag{10.60}$$

Similarly, for $\mathscr{C} = \sigma(X)$ and $\mathscr{D} = \sigma(Z)$, Theorem 10.53 immediately implies the following corollary.

Corollary 10.56 [X-conditional mean independence if $E(Y^2)$ is finite]
Let $Y: (\Omega, \mathscr{A}, P) \to (\overline{\mathbb{R}}, \overline{\mathscr{B}})$ be a random variable that has a finite second moment $E(Y^2)$, let $X: (\Omega, \mathscr{A}, P) \to (\Omega'_X, \mathscr{A}'_X)$ and $Z: (\Omega, \mathscr{A}, P) \to (\Omega'_Z, \mathscr{A}'_Z)$ be random variables, and consider:

(a) $E(Y \mid X, Z) \underset{P}{=} E(Y \mid X)$.

(b) For all random variables $W: (\Omega, \mathscr{A}, P) \to (\mathbb{R}, \mathscr{B})$ that are Z-measurable and have a finite second moment $E(W^2)$,

$$E(W \cdot Y \mid X) \underset{P}{=} E(W \mid X) \cdot E(Y \mid X). \tag{10.61}$$

Then,

(i) (a) implies (b).

(ii) If Y is also nonnegative, then (a) and (b) are equivalent to each other.

Remark 10.57 [Mean independence from a random variable]   For $X = \alpha$, $\alpha \in \Omega'_X$, this corollary and Remark 10.5 immediately yield the following proposition. Let $Y: (\Omega, \mathscr{A}, P) \to (\overline{\mathbb{R}}, \overline{\mathscr{B}})$ and $Z: (\Omega, \mathscr{A}, P) \to (\Omega'_Z, \mathscr{A}'_Z)$ be random variables with $E(Y^2) < \infty$, and consider:

(a) $E(Y \mid Z) \underset{P}{=} E(Y)$.

(b) For all random variables $W: (\Omega, \mathscr{A}, P) \to (\mathbb{R}, \mathscr{B})$ that are Z-measurable and have a finite second moment $E(W^2)$,

$$E(W \cdot Y) = E(W) \cdot E(Y). \tag{10.62}$$

Then,

(i) (a) implies (b).

(ii) If Y is also nonnegative, then (a) and (b) are equivalent to each other.   ◁

Remark 10.58 [Mean independence and uncorrelatedness] For $Z = W$, Proposition (i) of Remark 10.57 yields

$$E(Y \mid Z) \underset{P}{=} E(Y) \;\Rightarrow\; E(Z \cdot Y) = E(Z) \cdot E(Y), \tag{10.63}$$

provided that $E(Z^2)$, $E(Y^2) < \infty$. Proposition (10.63) is equivalent to (10.52), because $E(Z \cdot Y) = E(Z) \cdot E(Y)$ is equivalent to $Cov(Z, Y) = 0$, provided that $E(Z^2), E(Y^2) < \infty$ [see Box 7.1 (i)]. However, $Cov(Z, Y) = 0$ does *not* imply $E(Y \mid Z) \underset{P}{=} E(Y)$. In other words, uncorrelatedness does not imply mean independence.   ◁

Remark 10.59 [Dichotomous variables] If Y and Z are dichotomous with values 0 and 1, then $Y \underset{P}{=} 1_{Y=1}$ (see Example 5.10), and hence Proposition (10.63) can equivalently be written:

$$P(Y=1 \mid Z) \underset{P}{=} P(Y=1) \;\Rightarrow\; P(Y=1, Z=1) = P(Y=1) \cdot P(Z=1) \tag{10.64}$$

[see Eqs. (10.2), (6.5), and (1.33)]. Applying Remark 5.46 yields

$$P(Y=1 \mid Z) \underset{P}{=} P(Y=1) \;\Rightarrow\; Y \underset{P}{\perp\!\!\!\perp} Z. \tag{10.65}$$
◁

Further properties of conditional mean independence are treated in section 16.3, in particular the relationship between conditional independence and conditional mean independence.

10.7   Proofs

Proof of Theorem 10.17

Under the assumptions about $\mathscr{E} = \{A_1, A_2, \ldots\}$ and $\mathscr{C}$, a function $V: (\Omega, \mathscr{A}) \to (\overline{\mathbb{R}}, \overline{\mathscr{B}})$ is $\mathscr{C}$-measurable if and only if there are $\alpha_i \in \overline{\mathbb{R}}$, $i = 1, 2, \ldots$, such that $V = \sum_{i=1}^{\infty} \alpha_i 1_{A_i}$ (see Lemma 2.19). Hence, if $V, V^* \in \mathscr{E}(Y \mid \mathscr{C})$, then,

$$V = \sum_{i=1}^{\infty} \alpha_i 1_{A_i} \quad \text{and} \quad V^* = \sum_{i=1}^{\infty} \alpha_i^* 1_{A_i}, \qquad \alpha_i, \alpha_i^* \in \overline{\mathbb{R}}. \tag{10.66}$$

This implies

$$P(\{V \neq V^*\}) = \sum_{i:\ \alpha_i \neq \alpha_i^*} P(A_i).$$

Because $P(\{V \neq V^*\}) = 0$ (see Th. 16 and Def. 2.68), we can conclude

$$\sum_{i:\ \alpha_i \neq \alpha_i^*} P(A_i) = 0. \tag{10.67}$$

Hence, if (10.13) holds, then Equation (10.67) implies that there is no i with $\alpha_i \neq \alpha_i^*$, which implies $\{V \neq V^*\} = \emptyset$.

Now assume that there is an $A_j \in \mathscr{E}$ with $P(A_j) = 0$. Then define $V \in \mathscr{E}(Y \mid \mathscr{C})$ as in Equation (10.66) and $V^* = \sum_{i=1}^{\infty} \alpha_i^* 1_{A_i}$, with $\alpha_i^* = \alpha_i$ for $i \neq j$ and $\alpha_j^* := \alpha_j + 1$. This implies $V^* \underset{P}{=} V$ and $V^* \in \mathscr{E}(Y \mid \mathscr{C})$, but $\emptyset \neq A_j \subset \{V \neq V^*\}$. By contraposition, this proves that $\{V \neq V^*\} = \emptyset$ implies Equation (10.13). For $\mathscr{E} = \{A_1, \ldots, A_n\}$, the proof is analogous.

Proof of Theorem 10.21

(i) This proof is found in Bauer [1996, (15.13)]. Because the sequence $Y_i, i \in \mathbb{N}$, is increasing and the conditional expectation is monotone [see Rule (ix) of Box 10.3], we can conclude: $\lim_{i \to \infty} Y_i = \sup_{i \in \mathbb{N}} Y_i$ and $\lim_{i \to \infty} E(Y_i \mid \mathscr{C}) = \sup_{i \in \mathbb{N}} E(Y_i \mid \mathscr{C})$.

(ii) If $Y_i \geq 0$, for all $i \in \mathbb{N}$, then $\tilde{Y}_n := \sum_{i=1}^{n} Y_i$, $n \in \mathbb{N}$, is increasing and $\lim_{n \to \infty} \tilde{Y}_n = \sum_{i=1}^{\infty} Y_i$. Hence,

$$\begin{aligned}
E\left(\sum_{i=1}^{\infty} Y_i \,\middle|\, \mathscr{C} \right) &\underset{P}{=} E\left(\lim_{n \to \infty} \tilde{Y}_n \,\middle|\, \mathscr{C} \right) \\
&\underset{P}{=} \lim_{n \to \infty} E\left(\tilde{Y}_n \,\middle|\, \mathscr{C} \right) && [(10.19)] \\
&\underset{P}{=} \lim_{n \to \infty} E\left(\sum_{i=1}^{n} Y_i \,\middle|\, \mathscr{C} \right) \\
&\underset{P}{=} \lim_{n \to \infty} \sum_{i=1}^{n} E(Y_i \mid \mathscr{C}) && [\text{Box 10.1 (xvi)}] \\
&\underset{P}{=} \sum_{i=1}^{\infty} E(Y_i \mid \mathscr{C}).
\end{aligned}$$

Proof of Theorem 10.42

According to Lemma 2.52, $g(X)$ is measurable with respect to $\sigma(X)$. Therefore, according to Definition 10.2 (b), we only have to show that

$$E[1_C \cdot g(X)] = E(1_C \cdot Y), \quad \forall C \in \sigma(X), \tag{10.68}$$

and Equation (10.43) are equivalent to each other.

(10.43) $\Leftrightarrow$ (10.68) By definition, $\sigma(X) = \{X^{-1}(C'): C' \in \mathcal{A}'_X\}$. Therefore,

$$\int 1_{C'}(x) \cdot g(x)\, P_X(dx) = \int 1_{C'}(x) \cdot y\, P_{X,Y}[d(x,y)], \quad \forall\, C' \in \mathcal{A}'_X,$$

$$\Leftrightarrow \int 1_C \cdot g(X)\, dP = \int 1_C \cdot Y\, dP, \quad \forall\, C = X^{-1}(C') \in \sigma(X), \qquad [(3.59), (3.28)]$$

$$\Leftrightarrow E[1_C \cdot g(X)] = E(1_C \cdot Y), \quad \forall\, C \in \sigma(X). \qquad [(6.1)]$$

Proof of Theorem 10.51

(a) $\Rightarrow$ (b) If Y and W are nonnegative, then $Y \cdot W$ is nonnegative as well and, for $\mathcal{D}$-measurable W,

$$
\begin{aligned}
E(W \cdot Y \mid \mathcal{C}) &\underset{P}{=} E[E(W \cdot Y \mid \mathcal{C}, \mathcal{D}) \mid \mathcal{C}] && \text{[Box 10.1 (v)]} \\
&\underset{P}{=} E[W \cdot E(Y \mid \mathcal{C}, \mathcal{D}) \mid \mathcal{C}] && \text{[Box 10.1 (xiv)]} \\
&\underset{P}{=} E[W \cdot E(Y \mid \mathcal{C}) \mid \mathcal{C}] && \text{[(a), Box 10.1 (ix)]} \\
&\underset{P}{=} E(Y \mid \mathcal{C}) \cdot E(W \mid \mathcal{C}). && \text{[Box 10.1 (xiv)]}
\end{aligned}
\tag{10.69}
$$

(b) $\Rightarrow$ (a)

$$
\begin{aligned}
E[W \cdot E(Y \mid \mathcal{C}, \mathcal{D}) \mid \mathcal{C}] &\underset{P}{=} E[E(W \cdot Y \mid \mathcal{C}, \mathcal{D}) \mid \mathcal{C}] && \text{[Box 10.1 (xiv)]} \\
&\underset{P}{=} E(W \cdot Y \mid \mathcal{C}) && \text{[Box 10.1 (v)]} \\
&\underset{P}{=} E(Y \mid \mathcal{C}) \cdot E(W \mid \mathcal{C}) && \text{[(b)]} \\
&\underset{P}{=} E[W \cdot E(Y \mid \mathcal{C}) \mid \mathcal{C}]. && \text{[Box 10.1 (xiv)]}
\end{aligned}
\tag{10.70}
$$

Choosing in this equation $W = 1_D$, $D \in \mathcal{D}$, Definition 10.2 (b) yields

$$E[1_C \cdot 1_D \cdot E(Y \mid \mathcal{C}, \mathcal{D})] = E[1_C \cdot 1_D \cdot E(Y \mid \mathcal{C})], \quad \forall\, C \in \mathcal{C},$$

which is equivalent to

$$\int 1_{C \cap D} \cdot E(Y \mid \mathcal{C}, \mathcal{D})\, dP = \int 1_{C \cap D} \cdot E(Y \mid \mathcal{C})\, dP, \quad \forall\, C \in \mathcal{C}. \tag{10.71}$$

The set $\{C \cap D: C \in \mathcal{C}, D \in \mathcal{D}\}$ is $\cap$-stable and generates $\sigma(\mathcal{C} \cup \mathcal{D})$. Furthermore, finiteness of $E(Y) = E[E(Y \mid \mathcal{C}, \mathcal{D})] = E[E(Y \mid \mathcal{C})]$ [see Box 10.1, (iv)] implies that $E(Y \mid \mathcal{C}, \mathcal{D})$ and $E(Y \mid \mathcal{C})$ are integrable with respect to P. Hence, according to Theorem 3.68 (iv), we can conclude $E(Y \mid \mathcal{C}, \mathcal{D}) \underset{P}{=} E(Y \mid \mathcal{C})$.

Proof of Theorem 10.53

(a) $\Rightarrow$ (b) If $E(Y^2)$, $E(W^2) < \infty$, then $E(Y)$ and $E(Y \cdot W)$ are finite as well. Then, for $\mathcal{D}$-measurable W, (10.69) also applies to this case.

(b) $\Rightarrow$ (a) Choosing $W = 1_D$, $D \in \mathcal{D}$ in Equation (10.70), Definition 10.2 (b) yields

$$E[1_C \cdot 1_D \cdot E(Y \mid \mathcal{C}, \mathcal{D})] = E[1_C \cdot 1_D \cdot E(Y \mid \mathcal{C})], \quad \forall\, C \in \mathcal{C},$$

which is equivalent to

$$\int 1_{C \cap D} \cdot E(Y \mid \mathscr{C}, \mathscr{D}) \, dP = \int 1_{C \cap D} \cdot E(Y \mid \mathscr{C}) \, dP, \quad \forall \, C \in \mathscr{C}. \tag{10.72}$$

The set $\{C \cap D : C \in \mathscr{C}, D \in \mathscr{D}\}$ is $\cap$-stable and generates $\sigma(\mathscr{C} \cup \mathscr{D})$. Furthermore, finiteness of $E(Y) = E[E(Y \mid \mathscr{C}, \mathscr{D})] = E[E(Y \mid \mathscr{C})]$ [see Box 10.1, (iv)] implies that $E(Y \mid \mathscr{C}, \mathscr{D})$ and $E(Y \mid \mathscr{C})$ are integrable with respect to P. Hence, if additionally $Y \geq 0$, then there are versions of $E(Y \mid \mathscr{C}), E(Y \mid \mathscr{C}, \mathscr{D}) \geq 0$ [see Box 10.3 (v)] and, according to Theorem 3.68 (iv), we can conclude $E(Y \mid \mathscr{C}, \mathscr{D}) \underset{P}{=} E(Y \mid \mathscr{C})$.

Exercises

10.1 Show that, according to Definition 10.2, Equations (9.24) and (9.25) define an element of $\mathscr{E}(Y \mid X)$ provided that the assumptions of Definition 9.14 hold.

10.2 Table 9.2 presents an element, say V, of $\mathscr{E}(Y \mid X, U)$. Define an alternative element $V^* \in \mathscr{E}(Y \mid X, U)$, and show that the two elements are P-equivalent.

10.3 Under the assumptions of Theorem 10.17 and Equation (10.13), prove Equation (10.14).

10.4 Prove the propositions of Box 10.1.

10.5 Prove the propositions of Box 10.3.

10.6 Show that $P(X = x) > 0$ implies that the $(X = x)$-conditional expectation value of Y defined by Equation (10.27) is uniquely defined and identical to the term introduced in Definition 9.2.

10.7 Prove Equation (10.32).

10.8 Show that Equation (10.38) implies Rule (iii) of Box 9.2.

10.9 Show that mean independence of Y from X implies that X and Y are uncorrelated, provided that the second moments of X and Y are finite [see (10.52)].

10.10 Prove: If Y is dichotomous with values 0 and 1, then,

$$Y \underset{P}{\perp\!\!\!\perp} X \iff E(Y \mid X) \underset{P}{=} E(Y)$$

(see Rem. 10.50).

Solutions

10.1 Let $\Omega_0' \subset \Omega_X'$ denote the finite or countable set introduced in Definitions 9.14 (i) or (ii). Then $\{X = x\} \in \sigma(X)$ for all $x \in \Omega_0'$. This implies:
(a) For all $x \in \Omega_0'$, the indicator $1_{X=x}$ is X-measurable (see Example 2.12), which implies that $\sum_{x \in \Omega_0'} E(Y \mid X = x) \cdot 1_{X=x}$ is X-measurable as well (see Lemma 2.19).

(b) For $C \in \sigma(X)$, define $C_0 := C \cap X^{-1}(\Omega_0')$. Because, by definition, $P[X^{-1}(\Omega_0')] = P_X(\Omega_0') = 1$, this implies $P(C \setminus C_0) = 0$ [see Box 4.1 (v) and (vii)] and

$$E\left(1_C \cdot \sum_{x \in \Omega_0'} E(Y \mid X=x) \cdot 1_{X=x} \right)$$

$$= E\left(1_{C_0} \cdot \sum_{x \in \Omega_0'} E(Y \mid X=x) \cdot 1_{X=x} \right)$$

$$+ E\left(1_{C \setminus C_0} \cdot \sum_{x \in \Omega_0'} E(Y \mid X=x) \cdot 1_{X=x} \right) \qquad [(6.21)]$$

$$= E\left(\sum_{x \in \Omega_0'} E(Y \mid X=x) \cdot 1_{C_0} \cdot 1_{X=x} \right). \qquad [\text{Box } 6.1 \text{ (iii), (v)}]$$

Furthermore, denote $C_0' := X(C_0) = \{x \in \Omega_0' : X^{-1}(\{x\}) \subset C\}$. Then,

$$E\left(1_C \cdot \sum_{x \in \Omega_0'} E(Y \mid X=x) \cdot 1_{X=x} \right) = E\left(\sum_{x \in \Omega_0'} E(Y \mid X=x) \cdot 1_{C_0} \cdot 1_{X=x} \right)$$

$$= \sum_{x \in \Omega_0'} E\left[E(Y \mid X=x) \cdot 1_{C_0} \cdot 1_{X=x} \right] \qquad [\text{Box } 6.1 \text{ (vii), (3.65)}]$$

$$= \sum_{x \in C_0'} E\left[E(Y \mid X=x) \cdot 1_{X=x} \right] \qquad [\text{def. of } C_0]$$

$$= \sum_{x \in C_0'} E(Y \mid X=x) \cdot E(1_{X=x}) \qquad [\text{Box } 6.1 \text{ (iii)}]$$

$$= \sum_{x \in C_0'} E(Y \mid X=x) \cdot P(X=x) \qquad [(6.4)]$$

$$= \sum_{x \in C_0'} \left(\frac{1}{P(X=x)} \cdot E(1_{X=x} \cdot Y) \right) \cdot P(X=x) \qquad [(9.11)]$$

$$= \sum_{x \in C_0'} E(1_{X=x} \cdot Y)$$

$$= E\left(\sum_{x \in C_0'} 1_{X=x} \cdot Y \right) \qquad [\text{Box } 6.1 \text{ (vii), (3.65)}]$$

$$= E(1_{C_0} Y) \qquad [1_{C_0} = \sum_{x \in C_0'} 1_{X=x}]$$

$$= E(1_{C_0} Y) + E(1_{C \setminus C_0} Y) \qquad [(6.20)]$$

$$= E(1_C Y). \qquad [(6.21)]$$

10.2   Another element $V^* \in \mathcal{E}(Y \mid X, U)$ is obtained by defining

$$V^*(\omega) = \begin{cases} 9, & \text{if } \omega = \omega_3 \text{ or } \omega = \omega_4 \\ V(\omega), & \text{if } \omega \in \Omega, \omega \neq \omega_3, \omega \neq \omega_4, \end{cases}$$

where $\omega_3 = (Joe, yes, -)$ and $\omega_4 = (Joe, yes, +)$. For V and V^*, $P(A_1) = 1$, where $A_1 = \{\omega \in \Omega : V(\omega) = V^*(\omega)\}$. The probability $P(A_1) = 1$ is obtained from adding the probabilities of all six outcomes ω for which $P(\{\omega\}) > 0$ (see the second column of Table 9.2).

10.3   (a) If $\mathcal{E}$ is a finite or countable partition of Ω and $\sigma(\mathcal{E}) = \mathcal{C}$, then, for all $A_i \in \mathcal{E}$, the indicator 1_{A_i} is $\mathcal{C}$-measurable. This implies that $\sum_{A_i \in \mathcal{E}} E(Y \mid A_i) \cdot 1_{A_i}$ is $\mathcal{C}$-measurable as well (see Lemma 2.19). Hence, condition (a) of Definition 10.2 is satisfied.

(b) According to Lemma 1.20 and Equations (1.36) and (1.37), for all $C \in \mathcal{C}$,

$$1_C = \sum_{A_i \in \mathcal{E}, A_i \subset C} 1_{A_i}. \tag{10.73}$$

Hence,

$$
\begin{aligned}
E\left(1_C \cdot \sum_{A_i \in \mathcal{E}} E(Y \mid A_i) \cdot 1_{A_i}\right) &= E\left(\sum_{A_i \in \mathcal{E}} E(Y \mid A_i) \cdot 1_{A_i} \cdot 1_C\right) \\
&= \sum_{A_i \in \mathcal{E}} E\left[E(Y \mid A_i) \cdot 1_{A_i} \cdot 1_C\right] && \text{[Box 6.1 (vii), (3.65)]} \\
&= \sum_{A_i \in \mathcal{E}, A_i \subset C} E\left[E(Y \mid A_i) \cdot 1_{A_i}\right] && \text{[(10.73)]} \\
&= \sum_{A_i \in \mathcal{E}, A_i \subset C} E(Y \mid A_i) \cdot E(1_{A_i}) && \text{[Box 6.1 (iii)]} \\
&= \sum_{A_i \in \mathcal{E}, A_i \subset C} \frac{1}{P(A_i)} \cdot E(1_{A_i} \cdot Y) \cdot P(A_i) && \text{[(9.7), (6.4)]} \\
&= \sum_{A_i \in \mathcal{E}, A_i \subset C} E(1_{A_i} \cdot Y) \\
&= E\left(\sum_{A_i \in \mathcal{E}, A_i \subset C} 1_{A_i} \cdot Y\right) && \text{[Box 6.1 (vii)]} \\
&= E(1_C \cdot Y). && \text{[(10.73)]}
\end{aligned}
$$

This shows that condition (b) of Definition 10.2 is satisfied and that $\sum_{A_i \in \mathcal{E}} E(Y \mid A_i) \cdot 1_{A_i}$ is a version of $E(Y \mid \mathcal{C})$. Equation (10.14) then follows from Proposition (10.12) and the assumption that $\mathcal{E}$ is a countable partition of Ω and $P(A_i) > 0$ for all $A_i \in \mathcal{E}$ (see Theorem 10.17).

10.4 (ii) Both sides are $\mathscr{C}$-measurable. Furthermore, for all $C \in \mathscr{C}$,

$$
\begin{aligned}
E[1_C \cdot (\alpha + Y)] = E(1_C \alpha + 1_C Y) &= E(1_C \alpha) + E(1_C Y) && \text{[Box 6.1 (vi)]} \\
&= E(1_C \alpha) + E[1_C E(Y \mid \mathscr{C})] && \text{[Def. 10.2 (b)]} \\
&= E(1_C \alpha + 1_C E(Y \mid \mathscr{C})]) && \text{[Box 6.1 (vi)]} \\
&= E(1_C \cdot [\alpha + E(Y \mid \mathscr{C})]).
\end{aligned}
$$

Hence, according to conditions (a) and (b) of Definition 10.2, $\alpha + E(Y \mid \mathscr{C}) \in \mathscr{E}(\alpha + Y \mid \mathscr{C})$, and Proposition (10.12) yields (ii).

(iii) Both sides are $\mathscr{C}$-measurable. For all $C \in \mathscr{C}$,

$$
\begin{aligned}
E(1_C \cdot \alpha Y) = \alpha E(1_C \cdot Y) &= \alpha E(1_C \cdot Y) && \text{[Box 6.1 (iii)]} \\
&= \alpha E[1_C \cdot E(Y \mid \mathscr{C})]. && \text{[Def. 10.2 (b)]}
\end{aligned}
$$

Hence, according to Definition 10.2, $\alpha E(Y \mid \mathscr{C}) \in \mathscr{E}(\alpha Y \mid \mathscr{C})$, and Proposition (10.12) yields (iii).

(iv) This rule immediately follows from condition (b) of Definition 10.2 for $C = \Omega$, because

$$
\begin{aligned}
E[E(Y \mid \mathscr{C})] = E[1_\Omega E(Y \mid \mathscr{C})] &= E[1_\Omega E(Y \mid \mathscr{C})] && \text{[(3.31)]} \\
&= E(1_\Omega Y) && \text{[Def. 10.2 (b)]} \\
&= E(Y). && \text{[(3.31)]}
\end{aligned}
$$

(v) The terms on both sides of this equation are $\mathscr{C}_0$-measurable because of Definition 10.2 (a). Furthermore, for all $C \in \mathscr{C}_0 \subset \mathscr{C}$,

$$
\begin{aligned}
E(1_C E[E(Y \mid \mathscr{C}) \mid \mathscr{C}_0]) &= E(1_C E(Y \mid \mathscr{C})) && \text{[Def. 10.2 (b)]} \\
&= E(1_C Y). && \text{[Def. 10.2 (b)]}
\end{aligned}
$$

In the first equation, we apply Definition 10.2 (b) to $E[E(Y \mid \mathscr{C}) \mid \mathscr{C}_0]$ and $\mathscr{C}_0$, whereas in the second, we apply it to $E(Y \mid \mathscr{C})$ and $\mathscr{C}$. The last equation shows that $E[E(Y \mid \mathscr{C}) \mid \mathscr{C}_0] \in \mathscr{E}(Y \mid \mathscr{C}_0)$, and Proposition (10.12) yields (v).

(vi) The constant $E(Y)$ is measurable with respect to any σ-algebra $\mathscr{C}$ on Ω (see Example 2.10). Furthermore, if Y and $\mathscr{C}$ are independent, then Y and 1_C are independent for all $C \in \mathscr{C}$ (see Rem. 5.46). Hence, for $C \in \mathscr{C}$, $E(1_C Y) = E(1_C) E(Y)$ for all $C \in \mathscr{C}$ (see Th. 6.24). Therefore, Rules (ii) and (iii) of Box 6.1 yield

$$
E(1_C Y) = E(1_C) E(Y) = E[1_C E(Y)], \quad \forall\, C \in \mathscr{C}.
$$

(vii) We assume that Y is $\mathscr{C}$-measurable. Furthermore,

$$
E(1_C Y) = E(1_C Y), \quad \forall\, C \in \mathscr{C}
$$

obviously holds. Hence, according to Definition 10.2 (b) and Proposition (10.12), this implies that $Y \in \mathscr{E}(Y \mid \mathscr{C})$ and $E(Y \mid \mathscr{C}) \underset{P}{=} Y$.

(i) The σ-algebra generated by α is $\{\Omega, \varnothing\}$, which is a subset of every σ-algebra on Ω. Hence, $\sigma(\alpha) \subset \mathscr{C}$, which shows that Rule (i) is a special case of Rule (vii).

(viii)

$$
\begin{aligned}
E[Y \mid E(Y \mid \mathscr{C})] &\underset{P}{=} E[E(Y \mid \mathscr{C}) \mid E(Y \mid \mathscr{C})] \quad [\text{(v) with } \mathscr{C}_0 = \sigma[E(Y \mid \mathscr{C})]\,] \\
&\underset{P}{=} E(Y \mid \mathscr{C}). \qquad\qquad [\sigma[E(Y \mid \mathscr{C})] \subset \sigma[E(Y \mid \mathscr{C})], \text{(vii)}]
\end{aligned}
$$

(ix)

$$
\begin{aligned}
&Y_1 \underset{P}{=} Y_2 \\
&\Rightarrow \forall\, C \in \mathscr{C}: E(1_C \cdot Y_1) = E(1_C \cdot Y_2) \qquad\qquad [\text{Th. 3.48, (6.1), } \mathscr{C} \subset \mathscr{A}] \\
&\Rightarrow \forall\, C \in \mathscr{C}: E[1_C \cdot E(Y_1 \mid \mathscr{C})] = E(1_C \cdot Y_1) \\
&\quad = E(1_C \cdot Y_2) = E[1_C \cdot E(Y_2 \mid \mathscr{C})] \qquad\qquad [\text{Def. 10.2 (b)}] \\
&\Rightarrow E(Y_2 \mid \mathscr{C}) \in \mathscr{E}(Y_1 \mid \mathscr{C}). \qquad\qquad\qquad [\text{Def. 10.2}]
\end{aligned}
$$

According to Proposition (10.12), this implies (ix).

(x)

$$
\begin{aligned}
E(Y) \text{ finite } &\Rightarrow \forall\, V \in \mathscr{E}(Y \mid \mathscr{C}): E(V) \text{ finite} \qquad\qquad [E(V) = E(Y), \text{(iv)}] \\
&\Rightarrow \forall\, V \in \mathscr{E}(Y \mid \mathscr{C}): V \text{ is real-valued } P\text{-a.s.} \qquad [\text{Lemma 3.41}]
\end{aligned}
$$

Now let $V^* \in \mathscr{E}(Y \mid \mathscr{C})$ and $A := \{\omega \in \Omega: V(\omega) \notin \mathbb{R}\}$. Then $A \in \mathscr{C}$ and $P(A) = 0$. Define $V := 1_{\Omega \setminus A} \cdot V^*$. Then V is real-valued, $\mathscr{C}$-measurable, and $V \underset{P}{=} V^*$, which implies $V \in \mathscr{E}(Y \mid \mathscr{C})$.

(xv) If Y_1 (or Y_2) is real-valued and with finite expectation, then there is a real-valued version $E(Y_1 \mid \mathscr{C}) \in \mathscr{E}(Y_1 \mid \mathscr{C})$ [or $E(Y_2 \mid \mathscr{C}) \in \mathscr{E}(Y_2 \mid \mathscr{C})$] [see (x)]. If Y_1 (or Y_2) is nonnegative, then there is a nonnegative version $E(Y_1 \mid \mathscr{C}) \in \mathscr{E}(Y_1 \mid \mathscr{C})$ [or $E(Y_2 \mid \mathscr{C}) \in \mathscr{E}(Y_2 \mid \mathscr{C})$] [see Box 10.3 (v) for $\alpha = 0$]. [Note that the proof of Box 10.3 (v) uses Box 10.1 (iv) and (ii).]

For versions $E(Y_1 \mid \mathscr{C}) \in \mathscr{E}(Y_1 \mid \mathscr{C}), E(Y_2 \mid \mathscr{C}) \in \mathscr{E}(Y_2 \mid \mathscr{C})$ that are nonnegative or with finite expectations, the sum $E(Y_1 \mid \mathscr{C}) + E(Y_2 \mid \mathscr{C})$ is $\mathscr{C}$-measurable (see Th. 2.57), and for all $C \in \mathscr{C}$,

$$
\begin{aligned}
E\big(1_C\,[E(Y_1 \mid \mathscr{C}) + E(Y_2 \mid \mathscr{C})]\big) & \\
= E\big(1_C\,E(Y_1 \mid \mathscr{C})\big) + E\big(1_C\,E(Y_2 \mid \mathscr{C})\big) \qquad & [\text{Box 6.1 (vi)}] \\
= E(1_C\,Y_1) + E(1_C\,Y_2) \qquad & [\text{Def. 10.2 (b)}] \\
= E[1_C\,(Y_1 + Y_2)]. \qquad & [\text{Box 6.1 (vi)}]
\end{aligned}
$$

(xvi)

$$
\begin{aligned}
E(\alpha_1 Y_1 + \alpha_2 Y_2 \mid \mathscr{C}) & \\
\underset{P}{=} E(\alpha_1 Y_1 \mid \mathscr{C}) + E(\alpha_2 Y_2 \mid \mathscr{C}) \qquad & [\text{(xv)}] \\
\underset{P}{=} \alpha_1 E(Y_1 \mid \mathscr{C}) + \alpha_2 E(Y_2 \mid \mathscr{C}). \qquad & [\text{(iii)}]
\end{aligned}
$$

The equation for n summands follows by induction.

(xi)   Cor. 8.17 of Klenke (2013).

(xiv) If Y_1 is $\mathscr{C}$-measurable, then $Y_1 \cdot E(Y_2 \mid \mathscr{C})$ is $\mathscr{C}$-measurable as well [see Def. 10.2 (a), Th. 2.57]. First, consider the case $Y_1 = 1_{C^*}$, for $C^* \in \mathscr{C}$. Then, for all $C \in \mathscr{C}$,

$$\int 1_C \cdot E(1_{C^*} \cdot Y_2 \mid \mathscr{C})\, dP$$

$$= \int 1_C \cdot 1_{C^*} \cdot Y_2\, dP \qquad\qquad \text{[Def. 10.2 (b)]}$$

$$= \int 1_{C \cap C^*} \cdot Y_2\, dP \qquad\qquad \text{[(1.33)]}$$

$$= \int 1_{C \cap C^*} \cdot E(Y_2 \mid \mathscr{C})\, dP \quad [C \cap C^* \in \mathscr{C}, \text{Def. 10.2 (b)}]$$

$$= \int 1_C \cdot 1_{C^*} \cdot E(Y_2 \mid \mathscr{C})\, dP. \qquad\qquad \text{[(1.33)]}$$

Hence, according to Definition 10.2 (b) and Proposition (10.12),

$$E(1_{C^*} \cdot Y_2 \mid \mathscr{C}) \underset{P}{=} 1_{C^*} \cdot E(Y_2 \mid \mathscr{C}). \qquad\qquad (10.74)$$

If $E(Y_1^2)$, $E(Y_2^2) < \infty$ or Y_1, Y_2 nonnegative, then Remark 3.30, Box 10.1 (xi) and (xvi), and Theorem 10.21 imply, for all $\mathscr{C}$-measurable Y_1,

$$\int 1_C \cdot E(Y_1 \cdot Y_2 \mid \mathscr{C})\, dP = \int 1_C \cdot Y_1 \cdot E(Y_2 \mid \mathscr{C})\, dP, \quad \forall\, C \in \mathscr{C}.$$

Now, according to Definition 10.2 (b), $Y_1 \cdot E(Y_2 \mid \mathscr{C}) \in \mathscr{E}(Y_1 \cdot Y_2 \mid \mathscr{C})$, and Proposition (10.12) yields

$$E(Y_1 \cdot Y_2 \mid \mathscr{C}) \underset{P}{=} Y_1 \cdot E(Y_2 \mid \mathscr{C}).$$

(xii) Note that $E(Y^2) < \infty$ implies $E[E(Y \mid \mathscr{C})^2] < \infty$ [see Box 10.1 (xi)].

$$\begin{aligned}
&Cov\,[Y, E(Y \mid \mathscr{C})] \\
&= E[Y \cdot E(Y \mid \mathscr{C})] - E(Y) \cdot E[E(Y \mid \mathscr{C})] && \text{[Box 7.1 (i)]} \\
&= E(E[Y \cdot E(Y \mid \mathscr{C}) \mid \mathscr{C}]) - E(Y) \cdot E[E(Y \mid \mathscr{C})] && \text{[(iv)]} \\
&= E[E(Y \mid \mathscr{C}) \cdot E(Y \mid \mathscr{C})] - E[E(Y \mid \mathscr{C})] \cdot E[E(Y \mid \mathscr{C})] && \text{[(xiv), (iv)]} \\
&= Var\,[E(Y \mid \mathscr{C})]. && \text{[Box 6.2 (i)]}
\end{aligned}$$

(xiii) Note that $E(Y_2{}^2) < \infty$ implies $E[E(Y_2 \mid \mathscr{C})^2] < \infty$ [see Box 10.1 (xi)]. Hence,

$$\begin{aligned}
Cov\,(Y_1, Y_2) &= E(Y_1 \cdot Y_2) - E(Y_1) \cdot E(Y_2) && \text{[Box 7.1 (i)]} \\
&= E[E(Y_1 \cdot Y_2 \mid \mathscr{C})] - E(Y_1) \cdot E[E(Y_2 \mid \mathscr{C})] && \text{[(iv)]} \\
&= E[Y_1 \cdot E(Y_2 \mid \mathscr{C})] - E(Y_1) \cdot E[E(Y_2 \mid \mathscr{C})] && [\sigma(Y_1) \subset \mathscr{C}, \text{(xiv)}] \\
&= Cov\,[Y_1, E(Y_2 \mid \mathscr{C})]. && \text{[Box 7.1 (i)]}
\end{aligned}$$

10.5   (v) First, we prove

$$Y \underset{P}{\geq} 0 \; \Rightarrow \; \exists \, V \in \mathcal{E}(Y \mid \mathcal{C}): V \geq 0. \tag{10.75}$$

Let $V^* \in \mathcal{E}(Y \mid \mathcal{C})$. Define $A := \{\omega \in \Omega: V^*(\omega) < 0\} \in \mathcal{C}$ [see Rem. 2.67 (a)]. Applying Definition 10.2 (b) and Equation (3.50) to $1_A \cdot Y$ yields

$$\int 1_A \cdot V^* \, dP = \int 1_A \cdot Y \, dP \geq 0, \quad \text{if } Y \underset{P}{\geq} 0. \tag{10.76}$$

If we assume $P(A) > 0$, then Lemma 3.44 with $f := -V^*$ and Equation (3.32) yield

$$-\int 1_A \cdot V^* \, dP = \int 1_A \cdot (-V^*) \, dP > 0$$

and hence $\int 1_A \cdot V^* \, dP < 0$, which contradicts (10.76). Therefore, $P(A) = 0$ and thus $V^* \underset{P}{\geq} 0$.

If we define $V := 1_{\Omega \setminus A} \cdot V^* \geq 0$, then V is $\mathcal{C}$-measurable (see Th. 2.57), $V \underset{P}{=} V^*$, and $V \in \mathcal{E}(Y \mid \mathcal{C})$ [see (10.12)]. For $\alpha \in \mathbb{R}$, applying (10.75),

$$Y \underset{P}{\geq} \alpha \; \Rightarrow \; Y - \alpha \underset{P}{\geq} 0 \; \Rightarrow \; \exists \, V_\alpha \in \mathcal{E}(Y - \alpha \mid \mathcal{C}): V_\alpha \geq 0.$$

Now Rule (ii) of Box 10.1 implies that there is a $V := V_\alpha + \alpha$ such that $V \in \mathcal{E}(Y \mid \mathcal{C})$ and $V \geq \alpha$.

(vii) If Y is nonnegative and there is an $\alpha \in \mathbb{R}$ such that $Y \underset{P}{\leq} \alpha$, then (3.25) and (3.44) imply $0 \leq E(Y) \leq \alpha$. If $E(Y)$ is finite, then $E(-Y) = -E(Y)$ is finite as well. Furthermore,

$$\begin{aligned}
Y \underset{P}{\leq} \alpha \; &\Rightarrow \; -Y \underset{P}{\geq} -\alpha \\
&\Rightarrow \; \exists \, V^* \in \mathcal{E}(-Y \mid \mathcal{C}): V^* \geq -\alpha && \text{[(v)]} \\
&\Rightarrow \; \exists \, V \in \mathcal{E}(Y \mid \mathcal{C}): V \leq \alpha. && \text{[Box 10.1 (iii), } V := -V^*\text{]}
\end{aligned}$$

(vi)

$$\begin{aligned}
& Y \underset{P}{=} \alpha \\
\Rightarrow \; & Y \underset{P}{\geq} \alpha \wedge Y \underset{P}{\leq} \alpha \\
\Rightarrow \; & \exists \, V_1 \in \mathcal{E}(Y \mid \mathcal{C}): V_1 \geq \alpha \wedge \exists \, V_2 \in \mathcal{E}(Y \mid \mathcal{C}): V_2 \leq \alpha && \text{[(v), (vii)]} \\
\Rightarrow \; & \exists \, V_1, V_2 \in \mathcal{E}(Y \mid \mathcal{C}): \exists \, A \in \mathcal{C}: P(A) = 0 \\
& \quad \wedge \forall \, \omega \in \Omega \setminus A: V_1(\omega) = V_2(\omega) = \alpha && \text{[(10.12)]} \\
\Rightarrow \; & \exists \, V \in \mathcal{E}(Y \mid \mathcal{C}): V = \alpha. && \text{[}V := \alpha \cdot 1_A + V_1 \cdot 1_{\Omega \setminus A}\text{]}
\end{aligned}$$

(ii), (iv), and (iii) are direct implications of (v), (vi), and (vii).
(x), (xi) follow from (iv) and (5.11).
(i) is a straightforward implication of Theorem 3.43 and (iv).
(ix) is proved in Bauer (1996).

10.6   Let $g(X) \in \mathscr{E}(Y \mid X)$. Then, according to Equation (10.26), for all $g^*(X) \in \mathscr{E}(Y \mid X)$,

$$g(x) = g^*(x), \quad \text{for } P_X\text{-a.a. } x \in \Omega'_X.$$

Hence, if $P(X=x) > 0$ for an $x \in \Omega'_X$, then according to Remark 2.71, $g(x) = g^*(x)$, that is, $g(x)$ is uniquely defined. Furthermore, Equation (9.6) yields

$$E(Y \mid X=x) = \int Y \, dP^{X=x} \qquad\qquad [(9.5)]$$

$$= \frac{1}{P(X=x)} \cdot \int 1_{X=x} \cdot Y \, dP \qquad\qquad [(9.7)]$$

$$= \frac{1}{P(X=x)} \cdot \int 1_{X=x} \cdot g(X) \, dP \qquad\qquad [\text{Def. 10.2 (b), (10.22)}]$$

$$= \frac{1}{P(X=x)} \cdot \int 1_{X=x} \cdot g(x) \, dP \qquad\qquad [1_{X=x} \cdot g(X) = 1_{X=x} \cdot g(x)]$$

$$= \frac{1}{P(X=x)} \cdot g(x) \cdot \int 1_{X=x} \, dP \qquad\qquad [(3.32)]$$

$$= \frac{1}{P(X=x)} \cdot g(x) \cdot P(X=x) = g(x). \qquad\qquad [(3.8)]$$

10.7   If g is a factorization of $E(Y \mid X) \in \mathscr{E}(Y \mid X)$, then, for all $\omega \in \{X=x\}$,

$$\begin{aligned}
E(Y \mid X)(\omega) &= (g \circ X)(\omega) & [(10.22)] \\
&= g[X(\omega)] = g(x) & [\omega \in \{X=x\}] \\
&= E(Y \mid X=x). & [(10.27)]
\end{aligned}$$

10.8   Assume that Z is a discrete random variable with values $z_1, z_2, \ldots \in \Omega'_Z$ such that $P_Z(\{z_1, z_2, \ldots\}) = 1$ and, for all $i = 1, 2 \ldots$, $\{z_i\} \in \mathscr{A}'_Z$, and let g be a factorization of $E(Y \mid X, Z) = g(X, Z)$. Then, for all $x \in \Omega'_X$ with $P(X=x, Z=z_i) > 0$ for all $i \in \mathbb{N}$,

$$E(Y \mid X=x)$$

$$= E[E(Y \mid X, Z) \mid X=x] \qquad\qquad [(10.38)]$$

$$= E[g(X, Z) \mid X=x] \qquad\qquad [(10.22)]$$

$$= \sum_{i=1}^{\infty} g(x, z_i) \cdot P(X=x, Z=z_i \mid X=x) \qquad\qquad [\text{Rem. 10.35, (9.16), (9.22)}]$$

$$= \sum_{i=1}^{\infty} E(Y \mid X=x, Z=z_i) \cdot \frac{P(X=x, Z=z_i, X=x)}{P(X=x)} \qquad\qquad [(10.27), (4.2)]$$

$$= \sum_{i=1}^{\infty} E(Y \mid X=x, Z=z_i) \cdot \frac{P(X=x, Z=z_i)}{P(X=x)}$$

$$= \sum_{i=1}^{\infty} E(Y \mid X=x, Z=z_i) \cdot P(Z=z_i \mid X=x). \qquad\qquad [(4.2)]$$

10.9   If $E(Y \mid X) \underset{P}{=} E(Y)$, then,

$$
\begin{aligned}
Cov\,(X, Y) &= Cov\,[X, E(Y \mid X)] && \text{[Box 10.2 (xiii)]}\\
&= Cov\,[X, E(Y)] && [E(Y \mid X) \underset{P}{=} E(Y),\ \text{Box 7.1 (x)}]\\
&= 0. && \text{[Box 7.1 (vii)]}
\end{aligned}
$$

10.10   If Y is dichotomous with values 0 and 1, then $E(Y) = P(Y=1)$ [see Example 5.10 and Eq. (6.5)]. Hence, $E(Y)$ is finite, and Box 10.2 (vi) yields

$$
Y \underset{P}{\perp\!\!\!\perp} X \;\Rightarrow\; E(Y \mid X) \underset{P}{=} E(Y).
$$

For all $C' \in \mathscr{A}'_X$, the event $\{X \in C'\} = X^{-1}(C') \in \sigma(X)$ [see Eq. (2.14)]. Hence, if $E(Y \mid X) \underset{P}{=} E(Y)$, and therefore $P(Y=1 \mid X) \underset{P}{=} P(Y=1)$ [see Eq. (10.4)], then for all $C' \in \mathscr{A}'_X$,

$$
\begin{aligned}
P(Y=1, X \in C') &= \int 1_{X \in C'} \cdot 1_{Y=1}\, dP && \text{[(3.9), (1.33)]}\\[6pt]
&= \int 1_{X \in C'} \cdot P(Y=1 \mid X)\ dP && \text{[Def. 10.2 (b)]}\\[6pt]
&= \int 1_{X \in C'} \cdot P(Y=1)\, dP && [P(Y=1 \mid X) \underset{P}{=} P(Y=1)]\\[6pt]
&= P(Y=1) \cdot P(X \in C'), && \text{[(3.9)]}
\end{aligned}
$$

which, according to Remark 5.46, implies $Y \underset{P}{\perp\!\!\!\perp} X$.

11

Residual, conditional variance, and conditional covariance

In chapters 9 and 10, we introduced the concepts *conditional expectation* and *regression*. In this chapter, we turn to the *residual* of a conditional expectation. Its properties supplement the properties of conditional expectations. Oftentimes a residual is what econometricians call a *disturbance*, applied statisticians call an *error term*, and psychometricians call a *measurement error*. Furthermore, we define the *coefficient of determination*, which represents the proportion of variance of a regressand explained by the regressor. It appears under different names in special areas of applied statistics, ranging from *intra-class correlation* to *reliability* in psychometrics. The square root of the coefficient of determination is known as the *multiple correlation*. Next, we will define the concepts of a *conditional variance* and a *conditional covariance* given a σ-algebra and given a random variable, as well as the *partial correlation*. Just like the expectation has been used to define variance, covariance, and correlation, the conditional expectation can be used to define conditional variance, conditional covariance, and the partial correlation.

11.1 Residual with respect to a conditional expectation

In section 10.4.2, we showed that a conditional expectation $E(Y \mid \mathscr{C})$ is the best approximation of Y in the sense of minimizing the mean-squared error function. Now we study the properties of the deviation of Y from $E(Y \mid \mathscr{C})$. Defining this deviation, we refer to the following assumptions.

Assumptions 11.1
Let $Y: (\Omega, \mathscr{A}, P) \to (\mathbb{R}, \mathscr{B})$ be a real-valued random variable with finite expectation, $\mathscr{C} \subset \mathscr{A}$ a σ-algebra, and $E(Y \mid \mathscr{C})$ a real-valued version of the $\mathscr{C}$-conditional expectation of Y.

Probability and Conditional Expectation: Fundamentals for the Empirical Sciences, First Edition. Rolf Steyer and Werner Nagel.
© 2017 John Wiley & Sons, Ltd. Published 2017 by John Wiley & Sons, Ltd.
Companion website: http://www.probability-and-conditional-expectation.de

Note that, according to Rule (x) of Box 10.1, finiteness of $E(Y)$ implies that there is a real-valued version $E(Y \mid \mathscr{C})$. Hence, assuming that $E(Y \mid \mathscr{C})$ is real-valued is no substantial loss of generality. Referring to a real-valued version $E(Y \mid \mathscr{C})$ avoids the subtraction of ∞ and ∞ for values of Y and $E(Y \mid \mathscr{C})$, respectively.

Definition 11.2 [Residual with respect to a conditional expectation]
Under the assumptions 11.1,

$$\varepsilon := Y - E(Y \mid \mathscr{C}) \tag{11.1}$$

is called a version of the residual of Y with respect to $E(Y \mid \mathscr{C})$.

Remark 11.3 [Versions of the residual] If $E(Y \mid \mathscr{C})$, $E(Y \mid \mathscr{C})^* \in \mathscr{E}(Y \mid \mathscr{C})$ are real-valued and ε, ε^* are the respective residuals, then $\varepsilon \underset{P}{=} \varepsilon^*$.   ◁

Box 11.1 summarizes some properties of the residual, which are proved in Exercise 11.1. All these properties follow from the definition of a residual and the assumption that $E(Y \mid \mathscr{C})$ is a real-valued version of the $\mathscr{C}$-conditional expectation of Y.

Box 11.1   Rules of computation for a residual.

Let the assumptions 11.1 hold. Then the following properties hold for all real-valued versions of $E(Y \mid \mathscr{C})$ and all versions of the residual ε defined in (11.1):

$$\varepsilon \underset{P}{=} Y - E(Y \mid \mathscr{C}). \tag{i}$$

$$Y \underset{P}{=} E(Y \mid \mathscr{C}) + \varepsilon. \tag{ii}$$

$$E(\varepsilon) = 0. \tag{iii}$$

$$Var(Y) = Var[E(Y \mid \mathscr{C})] + Var(\varepsilon), \quad \text{if } E(Y^2) < \infty. \tag{iv}$$

$$\varepsilon \underset{P}{=} 0, \quad \text{if } Y \underset{P}{=} E(Y \mid \mathscr{C}). \tag{v}$$

Additionally, let $\mathscr{C}_0$ be a σ-algebra and $W: (\Omega, \mathscr{A}, P) \to (\Omega'_W, \mathscr{A}'_W)$ be a random variable. Then,

$$E(\varepsilon \mid \mathscr{C}_0) \underset{P}{=} 0, \quad \text{if } \mathscr{C}_0 \subset \mathscr{C}. \tag{vi}$$

$$E(\varepsilon \mid W) \underset{P}{=} 0, \quad \text{if } \sigma(W) \subset \mathscr{C}. \tag{vii}$$

If W is real-valued, $\sigma(W) \subset \mathscr{C}$, and $E(W^2)$, $E(Y^2) < \infty$, then,

$$Cov(\varepsilon, W) = E(\varepsilon \cdot W) = 0. \tag{viii}$$

$$Cov[W, E(Y \mid \mathscr{C})] = Cov[W, E(Y \mid \mathscr{C}) + \varepsilon] = Cov(W, Y). \tag{ix}$$

Remark 11.4 [Some special cases]   Because $E(Y \mid \mathscr{C})$ is $\mathscr{C}$-measurable, the following equations are special cases of Rules (vii) and (viii) of Box 11.1, respectively.

$$E\big[\varepsilon \mid E(Y \mid \mathscr{C})\big] \underset{P}{=} 0, \tag{11.2}$$

$$Cov\,[\varepsilon, E(Y \mid \mathscr{C})] = 0, \quad \text{if } E(Y^2) < \infty. \tag{11.3}$$

According to Equation (11.2), the conditional expectation of the residual ε given $E(Y \mid \mathscr{C})$ is 0 with probability 1. According to the second equation, the residual $\varepsilon = Y - E(Y \mid \mathscr{C})$ is uncorrelated with $E(Y \mid \mathscr{C})$ if $E(Y^2) < \infty$. [Note that finiteness of $E(E(Y \mid \mathscr{C})^2)$ follows from $E(Y^2) < \infty$; see Box 10.1 (xi).]

Now consider a random variable $X: (\Omega, \mathscr{A}, P) \to (\Omega'_X, \mathscr{A}'_X)$ with $\sigma(X) = \mathscr{C}$ and the residual $\varepsilon := Y - E(Y \mid X)$. Then a special case of Rule (vii) is

$$E(\varepsilon \mid X) \underset{P}{=} 0. \tag{11.4}$$

This property is illustrated in Figure 11.1. In this figure, the black points represent the values of the conditional expectation $E(\varepsilon \mid X)$, whereas the circles are possible values of ε. If $f(X)$ denotes the composition of X and a function $f: (\Omega'_X, \mathscr{A}'_X) \to (\mathbb{R}, \mathscr{B})$ that is $(\mathscr{A}'_X, \mathscr{B})$-measurable, then,

$$E[\varepsilon \mid f(X)] \underset{P}{=} 0 \tag{11.5}$$

with the special case

$$E\big[\varepsilon \mid E(Y \mid X)\big] \underset{P}{=} 0. \tag{11.6}$$

Furthermore, if $E(Y^2), E[f(X)^2] < \infty$, then,

$$Cov\,[\varepsilon, f(X)] = 0 \tag{11.7}$$

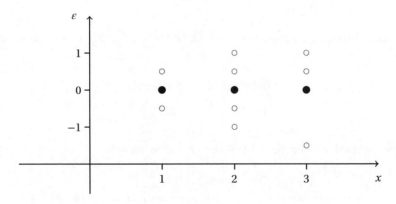

Figure 11.1   Conditional expectation of the residual on its regressor.

is a special case of Rule (viii). Hence, if $(\Omega'_X, \mathcal{A}'_X) = (\overline{\mathbb{R}}, \overline{\mathcal{B}})$ and $E(X^2)$, $E(Y^2) < \infty$, then,

$$Cov(\varepsilon, X) = 0 \tag{11.8}$$

is another special case of Rule (viii).

Now consider the residual $\varepsilon := Y - E(Y \mid X, Z)$, where $X: (\Omega, \mathcal{A}, P) \to (\Omega'_X, \mathcal{A}'_X)$ and $Z: (\Omega, \mathcal{A}, P) \to (\Omega'_Z, \mathcal{A}'_Z)$ are not necessarily real-valued. In this case,

$$E(\varepsilon \mid X, Z) \underset{P}{=} E(\varepsilon \mid X) \underset{P}{=} E(\varepsilon \mid Z) \underset{P}{=} 0 \tag{11.9}$$

are special cases of Rule (vii), where $\mathscr{C} = \sigma(X, Z)$. If we additionally assume X and Z to be numerical and $E(Y^2), E(X^2), E(Z^2) < \infty$, then,

$$Cov(X, \varepsilon) = E(X \cdot \varepsilon) = Cov(Z, \varepsilon) = E(Z \cdot \varepsilon) = 0 \tag{11.10}$$

are special cases of Rule (viii) (see Exercise 11.2).   ◁

Example 11.5 [No treatment for Joe – continued] Table 11.1 displays the conditional expectations $E(Y \mid X)$, $E(Y \mid X, U)$, and $P(X{=}1 \mid U)$, which have been computed in Examples 9.22 and 9.23. Additionally, it contains the residuals of Y with respect to these conditional expectations. First, we illustrate the property $E(\varepsilon) = 0$ for $\varepsilon = Y - E(Y \mid X)$. Looking at

Table 11.1   No treatment for Joe with conditional expectations and residuals.

Elements of Ω		Observables			Conditional expectations			Residuals		
Unit Treatment Success	$P(\{\omega\})$	Person variable U	Treatment variable X	Outcome variable Y	$E(Y \mid X, U)$	$E(Y \mid X)$	$P(X{=}1 \mid U)$	$Y - E(Y \mid X, U)$	$Y - E(Y \mid X)$	$X - P(X{=}1 \mid U)$
(*Joe, no,* −)	.152	*Joe*	0	0	.696	.6	0	−.696	−.6	0
(*Joe, no,* +)	.348	*Joe*	0	1	.696	.6	0	.304	.4	0
(*Joe, yes,* −)	0	*Joe*	1	0	0	.4	0	0	−.4	1
(*Joe, yes,* +)	0	*Joe*	1	1	0	.4	0	1	.6	1
(*Ann, no,* −)	.096	*Ann*	0	0	.2	.6	.76	−.2	−.6	−.76
(*Ann, no,* +)	.024	*Ann*	0	1	.2	.6	.76	.8	.4	−.76
(*Ann, yes,* −)	.228	*Ann*	1	0	.4	.4	.76	−.4	−.4	.24
(*Ann, yes,* +)	.152	*Ann*	1	1	.4	.4	.76	.6	.6	.24

the table reveals that $\varepsilon = Y - E(Y \mid X)$ has four different values: $-.6, .4, .6$, and $-.4$. Hence, according to Equation (6.3),

$$E(\varepsilon) = -.6 \cdot (.152 + .096) + .4 \cdot (.348 + .024) + .6 \cdot (0 + .152) - .4 \cdot (0 + .228)$$
$$= 0.$$

Second, we illustrate the property

$$E(\varepsilon \mid X) = 0$$

[see Eq. (11.4)]. Because X is an indicator variable with values 0 and 1, according to Equation (9.24) and Remark 10.35, it suffices to show that $E(\varepsilon \mid X=0) = 0$ and $E(\varepsilon \mid X=1) = 0$. The four values of $\varepsilon = Y - E(Y \mid X)$ occur with $(X=0)$-conditional probabilities

$$P(\varepsilon=-.6 \mid X=0) = \frac{.152 + .096}{.152 + .348 + .096 + .024} = .4,$$

$$P(\varepsilon=.4 \mid X=0) = \frac{.348 + .024}{.152 + .348 + .096 + .024} = .6,$$

$$P(\varepsilon=.6 \mid X=0) = 0, \qquad P(\varepsilon=-.4 \mid X=0) = 0,$$

and with $(X=1)$-conditional probabilities

$$P(\varepsilon=-.6 \mid X=1) = 0, \qquad P(\varepsilon=.4 \mid X=1) = 0$$

$$P(\varepsilon=.6 \mid X=1) = \frac{0 + .152}{0 + 0 + .152 + .228} = .4,$$

$$P(\varepsilon=-.4 \mid X=1) = \frac{0 + .228}{0 + 0 + .152 + .228} = .6,$$

respectively. Hence, according to Equation (9.21),

$$E(\varepsilon \mid X=0) = -.6 \cdot .4 + .4 \cdot .6 + .6 \cdot 0 - .4 \cdot 0 = 0$$

and

$$E(\varepsilon \mid X=1) = -.6 \cdot 0 + .4 \cdot 0 + .6 \cdot .4 - .4 \cdot .6 = 0.$$

Because X is dichotomous with values 0 and 1, and $P(X=0)$, $P(X=1) > 0$, we can conclude:

$$E(\varepsilon \mid X) = E(\varepsilon \mid X=0) \cdot 1_{X=0} + E(\varepsilon \mid X=1) \cdot 1_{X=1} = 0 \cdot 1_{X=0} + 0 \cdot 1_{X=1} = 0$$

[see Eq. (9.24) and Rem. 10.35]. ◁

11.2   Coefficient of determination and multiple correlation

The *coefficient of determination* quantifies the strength of the dependence of a numerical random variable Y on a σ-algebra $\mathscr{C}$, where we refer to the dependence described by the conditional expectation $E(Y \mid \mathscr{C})$. The *multiple correlation* is a closely related concept. Reading the following definition, remember that $E(Y^2) < \infty$ implies $Var(Y) < \infty$ and $E(Y) < \infty$ (see Rem. 6.25). It also implies that the conditional expectation $E(Y \mid \mathscr{C})$ is defined.

Definition 11.6 [Coefficient of determination]
Let $Y: (\Omega, \mathscr{A}, P) \to (\mathbb{R}, \mathscr{B})$ be a real-valued random variable with $E(Y^2) < \infty$ and $\mathscr{C} \subset \mathscr{A}$ be a σ-algebra. Then,

$$R^2_{Y|\mathscr{C}} := \begin{cases} \dfrac{Var[E(Y \mid \mathscr{C})]}{Var(Y)}, & \text{if } Var(Y) > 0 \\ 0, & \text{if } Var(Y) = 0, \end{cases} \tag{11.11}$$

is called the coefficient of determination of $E(Y \mid \mathscr{C})$.

Remark 11.7 [The case $Var(Y) = 0$] If $E(Y \mid \mathscr{C}) \underset{P}{=} E(Y)$ and $Var(Y) > 0$, then $R^2_{Y|\mathscr{C}} = Var[E(Y)]/Var(Y) = 0$ [see Eq. (11.11) and Box 6.2 (iv)]. Defining $R^2_{Y|\mathscr{C}} = 0$ if $Var(Y) = 0$ is arbitrary. However, $Var(Y) = 0$ if and only if there is an $\alpha \in \mathbb{R}$ such that $Y \underset{P}{=} \alpha$ [see again Box 6.2 (iv)], and $Y \underset{P}{=} \alpha$ implies $E(Y \mid \mathscr{C}) \underset{P}{=} E(Y)$ [see Box 6.1 (i) and Box 10.1 (i)]. Hence, Definition (11.11) implies $R^2_{Y|\mathscr{C}} = 0$ whenever $E(Y \mid \mathscr{C}) \underset{P}{=} E(Y)$, that is, whenever Y is mean independent from $\mathscr{C}$. Because $\mathscr{C} \underset{P}{\perp\!\!\!\perp} Y$ implies $E(Y \mid \mathscr{C}) \underset{P}{=} E(Y)$ [see Box 10.1 (vi)], *independence of $\mathscr{C}$ and Y* implies *mean independence* of Y from $\mathscr{C}$, which itself implies $R^2_{Y|\mathscr{C}} = 0$, and this implication holds irrespective of whether or not $Var(Y) > 0$.   ◁

Remark 11.8 [Range of the coefficient of determination]   Using Rule (iv) of Box 11.1 yields

$$R^2_{Y|\mathscr{C}} = \frac{Var[E(Y \mid \mathscr{C})]}{Var[E(Y \mid \mathscr{C})] + Var(\varepsilon)}, \tag{11.12}$$

provided that $Var(Y) > 0$. Because $Var(\varepsilon)$ is nonnegative, $0 \le R^2_{Y|\mathscr{C}} \le 1$. The number $R^2_{Y|\mathscr{C}}$ is close to 1 if the variance of the residual $\varepsilon = Y - E(Y \mid \mathscr{C})$ is small compared to the variance of the conditional expectation $E(Y \mid \mathscr{C})$. In contrast, $R^2_{Y|\mathscr{C}}$ is close to 0 if the variance of the residual is large compared to the variance of $E(Y \mid \mathscr{C})$.   ◁

Remark 11.9 [Conditions implying $R^2_{Y|\mathscr{C}} = 1$]   If $Var(Y) > 0$ and we assume that $Y \underset{P}{=} E(Y \mid \mathscr{C})$, then $R^2_{Y|\mathscr{C}} = 1$ [see Eq. (11.11) and Box 6.2 (v)]. Note that this does not necessarily mean that Y is $\mathscr{C}$-measurable. However, if Y is $\mathscr{C}$-measurable, then $Y \underset{P}{=} E(Y \mid \mathscr{C})$ already follows from Box 10.1 (vii).   ◁

Remark 11.10 [Alternative notation]   Suppose that $X: (\Omega, \mathcal{A}, P) \to (\Omega'_X, \mathcal{A}'_X)$ is a random variable and $\mathcal{C} = \sigma(X)$. Then we also use the notation $R^2_{Y|X}$ instead of $R^2_{Y|\mathcal{C}}$, that is,

$$R^2_{Y|X} := R^2_{Y|\sigma(X)} \,. \tag{11.13}$$

Equations (10.1) and (11.11) yield

$$R^2_{Y|X} = \begin{cases} \dfrac{Var[E(Y \mid X)]}{Var(Y)}, & \text{if } Var(Y) > 0 \\ 0, & \text{if } Var(Y) = 0. \end{cases} \tag{11.14}$$

If we consider the multivariate regressor $X = (X_1, \dots, X_n)$, then we also use the notation

$$R^2_{Y|X_1, \dots, X_n} := R^2_{Y|X} \,. \tag{11.15}$$

◁

Remark 11.11 [Correlation and the coefficient of determination]   Assume that $X: (\Omega, \mathcal{A}, P) \to (\mathbb{R}, \mathcal{B})$ is a real-valued random variable, $E(X^2) < \infty$, and that there is a version $E(Y \mid X) \in \mathcal{E}(Y \mid X)$ with

$$E(Y \mid X) = Q_{lin}(Y \mid X) = \beta_0 + \beta_1 X \tag{11.16}$$

(see Def. 7.2). Then,

$$\beta_1 = \frac{Cov(X, Y)}{Var(X)} \tag{11.17}$$

[see Th. 7.14 (ii)] and

$$R^2_{Y|X} = Corr(X, Y)^2, \tag{11.18}$$

which implies

$$R^2_{Y|X} = 0 \quad \Leftrightarrow \quad Corr(X, Y) = 0 \tag{11.19}$$

(see Exercise 11.3). Hence, under these assumptions, the correlation $Corr(X, Y)$ also quantifies the strength of the dependence of Y on X described by $E(Y \mid X)$. Both, $R^2_{Y|X}$ and $Corr(X, Y)$ are normed quantities. The first takes on its values in the interval $[0, 1]$, and the latter in the interval $[-1, 1]$. In contrast, the slope β_1 as well as $Cov(X, Y)$ quantify the strength of the dependence described by $Q_{lin}(Y \mid X)$ by real numbers without bounds.   ◁

Remark 11.12 [Quantifying the strength of dependence]   The term $R^2_{Y|X}$ quantifies the strength of the dependence of Y on X described by $E(Y \mid X)$, irrespective of whether or not Equation (11.16) holds. While $E(Y \mid X) = g(X)$ describes how the conditional expectation values $E(Y \mid X=x)$ of Y depend on the values x of X, the coefficient of determination $R^2_{Y|X}$

quantifies the strength of this dependence by a single real number between 0 and 1. Similarly, $R^2_{Y|\mathscr{C}}$ quantifies the strength of the dependence of Y on $\mathscr{C}$ described by $E(Y \mid \mathscr{C})$ (see Rem. 10.18).   ◁

Remark 11.13 [Uniqueness of $R^2_{Y|\mathscr{C}}$] If $V, V^* \in \mathscr{E}(Y \mid \mathscr{C})$, then V and V^* are P-equivalent, and, according to Rule (v) of Box 6.2, this implies $Var(V) = Var(V^*)$. Hence, Equation (11.11) implies that $R^2_{Y|\mathscr{C}}$ is identical for all versions $V \in \mathscr{E}(Y \mid \mathscr{C})$.   ◁

Remark 11.14 [Correlation of Y and the conditional expectation] The coefficient of determination $R^2_{Y|\mathscr{C}}$ is identical to the squared correlation of Y and $E(Y \mid \mathscr{C})$, that is,

$$R^2_{Y|\mathscr{C}} = Corr\,[Y,\ E(Y \mid \mathscr{C})]^2 \tag{11.20}$$

(see Exercise 11.4). Correspondingly,

$$R^2_{Y|X} = Corr\,[Y,\ E(Y \mid X)]^2. \tag{11.21}$$

Note that this equation does not rely on any parameterization of $E(Y \mid X)$.   ◁

Definition 11.15 [Multiple correlation]
Let $X: (\Omega, \mathscr{A}, P) \to (\Omega'_X, \mathscr{A}'_X)$ and $Y: (\Omega, \mathscr{A}, P) \to (\mathbb{R}, \mathscr{B})$ be random variables and assume $E(Y^2) < \infty$. Then,

$$R_{Y|X} := \sqrt{R^2_{Y|X}} \tag{11.22}$$

is called the multiple correlation of Y and X.

Equations (11.21) and (11.22) immediately imply

$$R_{Y|X} = Corr\,[Y,\ E(Y \mid X)]. \tag{11.23}$$

Remark 11.16 [Multivariate X] If $X = (X_1, \ldots, X_n)$ is a multivariate random variable, then we also use the notation

$$R_{Y|X_1, \ldots, X_n} := R_{Y|X}\,. \tag{11.24}$$
◁

Remark 11.17 [The multiple correlation is not symmetric] Note that, in contrast to a correlation of two numerical random variables, the multiple correlation of Y and X is not symmetric. That is, $R_{Y|X}$ can differ from $R_{X|Y}$ even if X is real-valued.   ◁

Example 11.18 [No treatment for Joe – continued] In Table 9.2, we displayed the conditional expectations $E(Y \mid X)$, $E(Y \mid X, U)$, and $P(X{=}1 \mid U)$. Now we compute $R^2_{Y|X}$ for the conditional expectation $E(Y \mid X)$. Looking at the table reveals that $E(Y \mid X)$ has two different values: .6, which occurs with probability

$$P[E(Y \mid X) = .6] = .152 + .348 + .096 + .024 = .62,$$

and .4, occurring with probability

$$P[E(Y \mid X) = .4] = 0 + 0 + .228 + .152 = .38.$$

Furthermore, the expectation of Y is

$$E(Y) = P(Y = 1) = .348 + 0 + .024 + .152 = .524$$

Hence, according to Equation (i) of Box 6.2,

$$\begin{aligned}
Var[E(Y \mid X)] &= E[E(Y \mid X)^2] - E[E(Y \mid X)]^2 \\
&= E[E(Y \mid X)^2] - E(Y)^2 \qquad \text{[Box 10.2 (iv)]} \\
&= (.6^2 \cdot .62 + .4^2 \cdot .38) - .524^2 \\
&\approx .284 - .2746 = .0094.
\end{aligned}$$

According to Equation (6.29), the variance of Y is $Var(Y) = P(Y = 1) \cdot [1 - P(Y = 1)] = .524 \cdot (1 - .524) \approx 0.2494$. This yields

$$R^2_{Y \mid X} = \frac{Var[E(Y \mid X)]}{Var(Y)} \approx \frac{.0094}{.2494} \approx .0377 \quad \text{and} \quad R_{Y \mid X} \approx .1941.$$

Similarly, the conditional expectation $E(Y \mid U)$ takes on each of the two values .696 and .352 with probability .5, that is,

$$P[E(Y \mid U) = .696] = P[E(Y \mid U) = .352] = .5.$$

Hence,

$$\begin{aligned}
Var[E(Y \mid U)] &= E[E(Y \mid U)^2] - E[E(Y \mid U)]^2 \qquad \text{[Box 6.2 (i)]} \\
&= E[E(Y \mid U)^2] - E(Y)^2 \qquad \text{[Box 10.2 (iv)]} \\
&= (.696^2 \cdot .5 + .352^2 \cdot .5) - .524^2 \\
&\approx .3042 - .2746 = .0296,
\end{aligned}$$

and this yields,

$$R^2_{Y \mid U} = \frac{Var[E(Y \mid U)]}{Var(Y)} \approx \frac{.0296}{.2494} \approx .1187 \quad \text{and} \quad R_{Y \mid U} \approx .3445.$$

Finally,

$$Var[E(Y \mid X, U)] = E[E(Y \mid X, U)^2] - E[E(Y \mid X, U)]^2 \qquad \text{[Box 6.2 (i)]}$$
$$= E[E(Y \mid X, U)^2] - E(Y)^2 \qquad \text{[Box 10.2 (iv)]}$$
$$= (.696^2 \cdot .5 + .2^2 \cdot .12 + .4^2 \cdot .38) - .524^2$$
$$\approx .3078 - .2746 = .0332.$$

Hence,

$$R^2_{Y|X,U} = \frac{Var[E(Y \mid X, U)]}{Var(Y)} \approx \frac{.0332}{.2494} \approx .1331 \quad \text{and} \quad R_{Y|X,U} \approx .3649.$$

Note that, in this example, $R^2_{Y|X,U}$ is smaller than the sum of $R^2_{Y|X}$ and $R^2_{Y|U}$. ◁

In the following theorem, we present a condition under which the coefficients of determination are additive [see Eq. (11.31)]. This theorem also contains a condition under which the coefficient α_1 of X in the equation $E(Y \mid X) = \alpha_0 + \alpha_1 X$ is identical to the coefficient β_1 of X in the equation $E(Y \mid X, Z) = \beta_0 + \beta_1 X + \beta_2 Z$ (see section 12.8 for a generalization).

Theorem 11.19 [Additivity of the coefficients of determination]
Let $X, Y, Z: (\Omega, \mathcal{A}, P) \to (\mathbb{R}, \mathcal{B})$ be three real-valued random variables with finite second moments and positive variances, and assume that there are $\beta_0, \beta_1, \beta_2, \gamma_0, \gamma_1 \in \mathbb{R}$, $E(Y \mid X, Z) \in \mathcal{E}(Y \mid X, Z)$, and $E(Z \mid X) \in \mathcal{E}(Z \mid X)$ such that

$$E(Y \mid X, Z) = \beta_0 + \beta_1 X + \beta_2 Z, \qquad (11.25)$$
$$E(Z \mid X) = \gamma_0 + \gamma_1 X. \qquad (11.26)$$

(i) *Then there are $\alpha_0, \alpha_1 \in \mathbb{R}$ such that*

$$E(Y \mid X) = \alpha_0 + \alpha_1 X. \qquad (11.27)$$

(ii) *If $E(Z \mid X) \underset{P}{=} E(Z)$ or $\beta_2 = 0$, then for α_0, α_1 occurring in (i),*

$$\alpha_0 = \beta_0 + \beta_2 E(Z) \qquad (11.28)$$

and

$$\alpha_1 = \beta_1. \qquad (11.29)$$

(iii) *If*

$$(E(Z \mid X) \underset{P}{=} E(Z) \text{ or } \beta_2 = 0) \quad \text{and} \quad (E(X \mid Z) \underset{P}{=} E(X) \text{ or } \beta_1 = 0),$$

then,

$$Var[E(Y \mid X, Z)] = Var[E(Y \mid X)] + Var[E(Y \mid Z)] \qquad (11.30)$$

and

$$R^2_{Y \mid X, Z} = R^2_{Y \mid X} + R^2_{Y \mid Z}. \qquad (11.31)$$

(Proof p. 359)

Remark 11.20 [Independence of X and Z]   Note that the condition specified in proposition (iii) of Theorem 11.19 is satisfied, for example, if X and Z are independent [see Rule (vi) of Box 10.2]. ◁

11.3   Conditional variance and covariance given a σ-algebra

The covariance $Cov(Y_1, Y_2)$ has been defined as the expectation of the product of the mean centered random variables $Y_1 - E(Y_1)$ and $Y_2 - E(Y_2)$, that is,

$$Cov(Y_1, Y_2) = E([Y_1 - E(Y_1)] \cdot [Y_2 - E(Y_2)]) \qquad (11.32)$$

(cf. Def. 7.8). Similarly, we define the $\mathscr{C}$-conditional covariance $Cov(Y_1, Y_2 \mid \mathscr{C})$ as the $\mathscr{C}$-*conditional expectation* of the product $[Y_1 - E(Y_1 \mid \mathscr{C})] \cdot [Y_2 - E(Y_2 \mid \mathscr{C})] = \varepsilon_1 \cdot \varepsilon_2$ of the residuals of Y_1 and Y_2 with respect to their $\mathscr{C}$-conditional expectations.

Definition 11.21 [Conditional covariance given a σ-algebra]
For $i = 1, 2$, let $Y_i \colon (\Omega, \mathscr{A}, P) \to (\mathbb{R}, \mathscr{B})$ be real-valued random variables with finite second moments, let $\mathscr{C} \subset \mathscr{A}$ be a σ-algebra, and define $\varepsilon_i := Y_i - E(Y_i \mid \mathscr{C})$. Then,

$$Cov(Y_1, Y_2 \mid \mathscr{C}) := E(\varepsilon_1 \cdot \varepsilon_2 \mid \mathscr{C}) \qquad (11.33)$$

is called a version of the $\mathscr{C}$-conditional covariance of Y_1 and Y_2.

Remark 11.22 [X-conditional covariance]   Let the assumptions of Definition 11.21 hold and let $X \colon (\Omega, \mathscr{A}, P) \to (\Omega'_X, \mathscr{A}'_X)$ be a random variable. Then,

$$Cov(Y_1, Y_2 \mid X) := Cov(Y_1, Y_2 \mid \sigma(X)) \qquad (11.34)$$

is called a version of the X-*conditional covariance* of Y_1 and Y_2. ◁

The $\mathscr{C}$-conditional variance is defined analogously.

Definition 11.23 [Conditional variance given a σ-algebra]
Let $Y: (\Omega, \mathcal{A}, P) \to (\mathbb{R}, \mathcal{B})$ be a real-valued random variable with $E(Y^2) < \infty$, let $\mathcal{C} \subset \mathcal{A}$ be a σ-algebra, and define $\varepsilon := Y - E(Y \mid \mathcal{C})$.

(i) Then,

$$Var(Y \mid \mathcal{C}) := E(\varepsilon^2 \mid \mathcal{C}) \tag{11.35}$$

is called a version of the $\mathcal{C}$-conditional variance of Y.

(ii) Let $Var(Y \mid \mathcal{C})$ be a nonnegative version of the $\mathcal{C}$-conditional variance of Y. Then,

$$SD\,(Y \mid \mathcal{C}) := \sqrt{Var(Y \mid \mathcal{C})} \tag{11.36}$$

is called a version of the $\mathcal{C}$-conditional standard deviation of Y.

Remark 11.24 [Conditional variance given X]   Let the assumptions of Definition 11.23 hold and let $X: (\Omega, \mathcal{A}, P) \to (\Omega_X', \mathcal{A}_X')$ be a random variable. Then,

$$Var(Y \mid X) := Var(Y \mid \sigma(X)) \tag{11.37}$$

is called a version of the *X-conditional variance* of Y. Correspondingly, let $Var(Y \mid X)$ be a nonnegative version of the X-conditional variance of Y. Then we call

$$SD\,(Y \mid X) := \sqrt{Var(Y \mid X)} \tag{11.38}$$

a version of the *X-conditional standard deviation* of Y.   ◁

11.4   Conditional variance and covariance given a value of a random variable

While the concepts defined above are random variables, the $(X=x)$-conditional covariance is a number. It is defined using the $(X=x)$-conditional expectation value $E(\varepsilon_1 \cdot \varepsilon_2 \mid X=x)$ that has been introduced as a value $g(x)$ of a factorization g of an X-conditional expectation $E(\varepsilon_1 \cdot \varepsilon_2 \mid X) = g(X)$ (see section 10.4.4).

Definition 11.25 [$(X=x)$-conditional variance and covariance]
Let $X: (\Omega, \mathcal{A}, P) \to (\Omega_X', \mathcal{A}_X')$ be a random variable.

(i) For $i = 1, 2$, let $Y_i: (\Omega, \mathcal{A}, P) \to (\mathbb{R}, \mathcal{B})$ be real-valued random variables with $E(Y_i^2) < \infty$, and let $\varepsilon_i := Y_i - E(Y_i \mid X)$. Then we call

$$Cov\,(Y_1, Y_2 \mid X=x) := E(\varepsilon_1 \cdot \varepsilon_2 \mid X=x) \tag{11.39}$$

an $(X=x)$-conditional covariance of Y_1 and Y_2.

(ii) *Let $Y: (\Omega, \mathcal{A}, P) \to (\mathbb{R}, \mathcal{B})$ be a real-valued random variable with $E(Y^2) < \infty$ and let $\varepsilon := Y - E(Y \mid X)$. Then we call*

$$Var(Y \mid X=x) := E(\varepsilon^2 \mid X=x) \qquad (11.40)$$

an $(X=x)$-conditional variance of Y.

(iii) *If, under the assumptions of (ii), $Var(Y \mid X)$ is a nonnegative version of the X-conditional variance of Y, then we call*

$$SD(Y \mid X=x) := \sqrt{Var(Y \mid X=x)} \qquad (11.41)$$

an $(X=x)$-conditional standard deviation of Y.

Remark 11.26 [Equivalent propositions]   Note that $Cov(Y_1, Y_2 \mid X=x)$ is uniquely defined only if $P(X=x) > 0$. However, even if $P(X=x) = 0$ for all $x \in \Omega_X'$, then we can still make propositions such as:

$$Cov(Y_1, Y_2 \mid X=x) = Cov(Z_1, Z_2 \mid X=x), \quad \text{for } P_X\text{-almost } x \in \Omega_X', \qquad (11.42)$$

provided that Z_1, Z_2 are real-valued random variables on $(\Omega, \mathcal{A}, P)$ with finite second moments. According to Corollary 10.39, this proposition is equivalent to

$$Cov(Y_1, Y_2 \mid X) \underset{P}{=} Cov(Z_1, Z_2 \mid X). \qquad (11.43)$$

Of course, the same applies to the X-conditional variance.   ◁

Remark 11.27 [Values of the conditional covariance]   As mentioned, the term defined in Remark 11.22 is a *random variable*. Its values are

$$Cov(Y_1, Y_2 \mid X)(\omega) = Cov(Y_1, Y_2 \mid X=x), \quad \text{if } X(\omega) = x \qquad (11.44)$$

(see Rem. 10.37). This also implies that the value of $Cov(Y_1, Y_2 \mid X)$ is constant on all sets $\{X=x\}$. Similarly,

$$Var(Y \mid X)(\omega) = Var(Y \mid X=x), \quad \text{if } X(\omega) = x. \qquad (11.45)$$
                                                                                ◁

Example 11.28 [Joe and Ann with self-selection]   Table 11.2 presents a new example with Joe and Ann. In this example, the probabilities of the elementary events differ from the example with randomized assignment (see Table 4.1). The values of the conditional expectations are computed analogously as in Example 9.21. In this new example, all individual treatment effects

$$E(Y \mid X=1, U=Joe) - E(Y \mid X=0, U=Joe)$$

and

$$E(Y \mid X=1, U=Ann) - E(Y \mid X=0, U=Ann)$$

Table 11.2   Joe and Ann with self-selection and residuals.

Outcomes ω		Observables			Conditional expectations				Residuals	
Unit Treatment Success	$P(\{\omega\})$	Person variable U	Treatment variable X	Outcome variable Y	$E(Y\mid X,U)$	$E(Y\mid X)$	$E(Y\mid U)$	$P(X=1\mid U)$	$\varepsilon_Y = Y - E(Y\mid U)$	$\varepsilon_X = X - P(X=1\mid U)$
(Joe, no, −)	.144	Joe	0	0	.7	.6	.704	.04	−.704	−.04
(Joe, no, +)	.336	Joe	0	1	.7	.6	.704	.04	.296	−.04
(Joe, yes, −)	.004	Joe	1	0	.8	.42	.704	.04	−.704	.96
(Joe, yes, +)	.016	Joe	1	1	.8	.42	.704	.04	.296	.96
(Ann, no, −)	.096	Ann	0	0	.2	.6	.352	.76	−.352	−.76
(Ann, no, +)	.024	Ann	0	1	.2	.6	.352	.76	.648	−.76
(Ann, yes, −)	.228	Ann	1	0	.4	.42	.352	.76	−.352	.24
(Ann, yes, +)	.152	Ann	1	1	.4	.42	.352	.76	.648	.24

are *positive*, whereas the difference $E(Y \mid X=1) - E(Y \mid X=0)$ is *negative*. Hence, in this example, this difference *cannot* be used to evaluate the treatment effect.

Let us consider the (unconditional) covariance of the treatment variable X and the outcome variable Y. Note that X and Y are indicator variables with values 0 and 1. Therefore, $E(X) = P(X=1)$, $E(Y) = P(Y=1)$, $E(X \cdot Y) = P(X=1, Y=1)$, and

$$Cov(X, Y) = E(X \cdot Y) - E(X) \cdot E(Y) = P(X=1, Y=1) - P(X=1) \cdot P(Y=1)$$

$$= (.016 + .152) - (.004 + .016 + .228 + .152) \cdot (.336 + .016 + .024 + .152)$$

$$= .168 - .4 \cdot .528 = -0.0432.$$

Hence, the treatment variable and the outcome variable have a *negative* covariance.

Now let us compute the $(U=u)$-conditional covariances of X and Y for $u=Joe$ and for $u=Ann$. First of all, note that $P(U=Joe) = .144 + .336 + .004 + .016 = .5$ and $P(U=Ann) = .096 + .024 + .228 + .152 = .5$. According to Equation (9.21), we have to sum the values of the product variable $\varepsilon_X \cdot \varepsilon_Y$ weighted by their $(U=u)$-conditional probabilities. Hence,

$$Cov(Y, X \mid U=Joe) = E(\varepsilon_X \cdot \varepsilon_Y \mid U=Joe)$$

$$= .04 \cdot .704 \cdot \frac{.144}{.5} - .04 \cdot .296 \cdot \frac{.336}{.5} - .96 \cdot .704 \cdot \frac{.004}{.5} + .96 \cdot .296 \cdot \frac{.016}{.5}$$

$$= .00384$$

and

$$Cov(Y, X \mid U=Ann) = E(\varepsilon_X \cdot \varepsilon_Y \mid U=Ann)$$

$$= .76 \cdot .352 \cdot \frac{.096}{.5} - .76 \cdot .648 \cdot \frac{.024}{.5} - .24 \cdot .352 \cdot \frac{.228}{.5} + .24 \cdot .648 \cdot \frac{.152}{.5}$$

$$= .03648.$$

Both conditional covariances are *positive*. Hence, in this example, the (unconditional) covariance of X and Y (which is negative) is highly misleading if used to evaluate the effects of the treatment on success, because for both persons the ($U=u$)-conditional (or person-specific) covariances of X and Y are positive. ◁

11.5   Properties of conditional variances and covariances

Boxes 11.2 and 11.3 summarize some important properties of conditional covariances and conditional variances. The rules for conditional variances are special cases of the corresponding rules for conditional covariances with $Y_1 = Y_2 = Y$ and $A = B$, respectively [see Box 11.2 (xiv)]. Hence, we only have to prove the rules for the conditional covariances (see Exercise 11.5).

For $n = 2$ variables Y_i and $m = 2$ variables Z_j, Equation (xiii) of Box 11.2 can be written as:

$$Cov(\alpha_1 Y_1 + \alpha_2 Y_2, \beta_1 Z_1 + \beta_2 Z_2 \mid \mathcal{C})$$
$$\underset{P}{=} \alpha_1 \beta_1 \, Cov(Y_1, Z_1 \mid \mathcal{C}) + \alpha_1 \beta_2 \, Cov(Y_1, Z_2 \mid \mathcal{C}) \tag{11.46}$$
$$+ \alpha_2 \beta_1 \, Cov(Y_2, Z_1 \mid \mathcal{C}) + \alpha_2 \beta_2 \, Cov(Y_2, Z_2 \mid \mathcal{C}).$$

Similarly, for two random variables Y_1 and Y_2, Rule (xiii) of Box 11.3 can also be written as:

$$Var(\alpha_1 Y_1 + \alpha_2 Y_2 \mid \mathcal{C})$$
$$\underset{P}{=} \alpha_1^2 \, Var(Y_1 \mid \mathcal{C}) + \alpha_2^2 \, Var(Y_2 \mid \mathcal{C}) + 2 \, \alpha_1 \, \alpha_2 \, Cov(Y_1, Y_2 \mid \mathcal{C}). \tag{11.47}$$

Example 11.29 [Conditional variance of an indicator]   Let $(\Omega, \mathcal{A}, P)$ be a probability space, let 1_A denote the indicator variable of $A \in \mathcal{A}$, and consider the random variable $X: (\Omega, \mathcal{A}, P) \to (\Omega'_X, \mathcal{A}'_X)$. Then, according to Rule (xiv) of Box 11.3,

$$Var(1_A \mid X) \underset{P}{=} P(A \mid X) \cdot [1 - P(A \mid X)]. \tag{11.48}$$

Hence, the X-conditional variance of an indicator variable does not contain any information additional to the X-conditional expectation $E(1_A \mid X) \underset{P}{=} P(A \mid X)$ (cf. Exercise 11.7). According to Corollaries 10.29 and 10.26, Equation (11.48) is equivalent to

$$Var(Y \mid X=x) = P(Y=1 \mid X=x) \cdot [1 - P(Y=1 \mid X=x)], \quad \text{for } P_X\text{-a.a. } x \in \Omega'_X. \tag{11.49}$$

◁

Box 11.2   Rules of computation for $\mathscr{C}$-conditional covariances.

For $i = 1, 2$, let $Y_i \colon (\Omega, \mathscr{A}, P) \to (\mathbb{R}, \mathscr{B})$ be real-valued random variables with $E(Y_i^2) < \infty$. Furthermore, let $\alpha, \beta \in \mathbb{R}$ and let $\mathscr{C}_0, \mathscr{C} \subset \mathscr{A}$ be σ-algebras. Then, the following properties hold for all real-valued versions of $E(Y_i \mid \mathscr{C})$ and all versions of the residuals $\varepsilon_i := Y_i - E(Y_i \mid \mathscr{C})$, $i = 1, 2$:

$$Cov\,(Y_1, Y_2 \mid \mathscr{C}) \underset{P}{=} E(Y_1 \cdot Y_2 \mid \mathscr{C}) - E(Y_1 \mid \mathscr{C}) \cdot E(Y_2 \mid \mathscr{C}). \tag{i}$$

$$Cov\,(Y_1, Y_2 \mid \mathscr{C}) \underset{P}{=} Cov\,(\varepsilon_1, \varepsilon_2 \mid \mathscr{C}). \tag{ii}$$

$$Cov\,(Y_1, Y_2 \mid \mathscr{C}) \underset{P}{=} 0, \quad \text{if } Y_1 \underset{P}{=} \alpha. \tag{iii}$$

$$Cov\,(\alpha + Y_1,\ \beta + Y_2 \mid \mathscr{C}) \underset{P}{=} Cov\,(Y_1, Y_2 \mid \mathscr{C}). \tag{iv}$$

$$Cov\,(\alpha\,Y_1,\ \beta\,Y_2 \mid \mathscr{C}) \underset{P}{=} \alpha\,\beta\,Cov\,(Y_1, Y_2 \mid \mathscr{C}). \tag{v}$$

$$E\big[Cov\,(Y_1, Y_2 \mid \mathscr{C}) \mid \mathscr{C}_0\big] \underset{P}{=} E(\varepsilon_1 \cdot \varepsilon_2 \mid \mathscr{C}_0) \underset{P}{=} Cov\,(Y_1, Y_2 \mid \mathscr{C}_0), \quad \text{if } \mathscr{C}_0 \subset \mathscr{C}. \tag{vi}$$

$$E[Cov\,(Y_1, Y_2 \mid \mathscr{C})] = E(\varepsilon_1 \cdot \varepsilon_2) = Cov\,(\varepsilon_1, \varepsilon_2). \tag{vii}$$

$$Cov\,(Y_1, Y_2) = Cov\,[E(Y_1 \mid \mathscr{C}), E(Y_2 \mid \mathscr{C})] + E[Cov\,(Y_1, Y_2 \mid \mathscr{C})] \tag{viii}$$

$$= Cov\,[E(Y_1 \mid \mathscr{C}), E(Y_2 \mid \mathscr{C})] + Cov\,(\varepsilon_1, \varepsilon_2). \tag{ix}$$

If we additionally assume $\mathscr{C}_0 \subset \mathscr{C}$, then,

$$Cov\,(Y_1, Y_2, \mid \mathscr{C}_0) \underset{P}{=} Cov\,\big[E(Y_1 \mid \mathscr{C}), E(Y_2 \mid \mathscr{C}) \mid \mathscr{C}_0\big] + E\big[Cov\,(Y_1, Y_2 \mid \mathscr{C}) \mid \mathscr{C}_0\big] \tag{x}$$

$$\underset{P}{=} Cov\,\big[E(Y_1 \mid \mathscr{C}), E(Y_2 \mid \mathscr{C}) \mid \mathscr{C}_0\big] + Cov\,(\varepsilon_1, \varepsilon_2 \mid \mathscr{C}_0). \tag{xi}$$

Let $W_1, W_2, Y_1, Y_2 \colon (\Omega, \mathscr{A}, P) \to (\mathbb{R}, \mathscr{B})$ be random variables such that $E(W_1^4)$, $E(W_2^4), E(Y_1^4)$, and $E(Y_2^4) < \infty$. If W_1 and W_2 are $\mathscr{C}$-measurable, then,

$$Cov\,(W_1 \cdot Y_1,\ W_2 \cdot Y_2 \mid \mathscr{C}) \underset{P}{=} W_1 \cdot W_2 \cdot Cov\,(Y_1, Y_2 \mid \mathscr{C}). \tag{xii}$$

For $i = 1, \ldots, n, j = 1, \ldots, m$, let $Y_i, Z_j \colon (\Omega, \mathscr{A}, P) \to (\mathbb{R}, \mathscr{B})$ be random variables with $E(Y_i^2), E(Z_j^2) < \infty$, and let $\alpha_i, \beta_j \in \mathbb{R}$. Then,

$$Cov\left(\sum_{i=1}^{n} \alpha_i\,Y_i,\ \sum_{j=1}^{m} \beta_j\,Z_j \,\middle|\, \mathscr{C}\right) \underset{P}{=} \sum_{i=1}^{n} \sum_{j=1}^{m} \alpha_i\,\beta_j\,Cov\,(Y_i, Z_j \mid \mathscr{C}). \tag{xiii}$$

If $A, B \in \mathscr{A}$, then,

$$Cov\,(1_A, 1_B \mid \mathscr{C}) \underset{P}{=} P(A \cap B \mid \mathscr{C}) - P(A \mid \mathscr{C}) \cdot P(B \mid \mathscr{C}). \tag{xiv}$$

Box 11.3   Rules of computation for conditional variances.

Let $Y\colon (\Omega, \mathcal{A}, P) \to (\mathbb{R}, \mathcal{B})$ denote a real-valued random variable with $E(Y^2) < \infty$. Furthermore, let $\alpha \in \mathbb{R}$ and let $\mathcal{C} \subset \mathcal{A}$ be a σ-algebra. Then the following properties hold for all real-valued versions of $E(Y \mid \mathcal{C})$ and all versions of the residual $\varepsilon := Y - E(Y \mid \mathcal{C})$:

$$Var(Y \mid \mathcal{C}) \underset{P}{=} E(Y^2 \mid \mathcal{C}) - E(Y \mid \mathcal{C})^2. \tag{i}$$

$$Var(Y \mid \mathcal{C}) \underset{P}{=} Var(\varepsilon \mid \mathcal{C}). \tag{ii}$$

$$Var(Y \mid \mathcal{C}) \underset{P}{=} 0, \quad \text{if } Y \underset{P}{=} \alpha. \tag{iii}$$

$$Var(\alpha + Y \mid \mathcal{C}) \underset{P}{=} Var(Y \mid \mathcal{C}). \tag{iv}$$

$$Var(\alpha\, Y \mid \mathcal{C}) \underset{P}{=} \alpha^2\, Var(Y \mid \mathcal{C}). \tag{v}$$

$$E\big[Var(Y \mid \mathcal{C}) \mid \mathcal{C}_0\big] \underset{P}{=} E(\varepsilon^2 \mid \mathcal{C}_0) \underset{P}{=} Var(\varepsilon \mid \mathcal{C}_0), \quad \text{if } \mathcal{C}_0 \subset \mathcal{C}. \tag{vi}$$

$$E[Var(Y \mid \mathcal{C})] = E(\varepsilon^2) = Var(\varepsilon). \tag{vii}$$

$$Var(Y) = Var[E(Y \mid \mathcal{C})] + E[Var(Y \mid \mathcal{C})] \tag{viii}$$

$$= Var[E(Y \mid \mathcal{C})] + Var(\varepsilon). \tag{ix}$$

If we additionally assume that the random variable $X\colon (\Omega, \mathcal{A}, P) \to (\Omega'_X, \mathcal{A}'_X)$ is $\mathcal{C}$-measurable, then,

$$Var(Y \mid X) \underset{P}{=} Var\big[E(Y \mid \mathcal{C}) \mid X\big] + E\big[Var(Y \mid \mathcal{C}) \mid X\big] \tag{x}$$

$$\underset{P}{=} Var\big[E(Y \mid \mathcal{C}) \mid X\big] + Var(\varepsilon \mid X). \tag{xi}$$

Let $X, Y\colon (\Omega, \mathcal{A}, P) \to (\mathbb{R}, \mathcal{B})$ be random variables with $E(X^4), E(Y^4) < \infty$. If X is $\mathcal{C}$-measurable, then,

$$Var(X \cdot Y \mid \mathcal{C}) \underset{P}{=} X^2 \cdot Var(Y \mid \mathcal{C}). \tag{xii}$$

Let $Y_i\colon (\Omega, \mathcal{A}, P) \to (\mathbb{R}, \mathcal{B})$ be random variables with $E(Y_i^2) < \infty$ and $\alpha_i \in \mathbb{R}$, where $i = 1, \dots, n$. Then,

$$Var\left(\sum_{i=1}^{n} \alpha_i Y_i \,\Big|\, \mathcal{C}\right) \underset{P}{=} \sum_{i=1}^{n} \alpha_i^2\, Var(Y_i \mid \mathcal{C}) + \sum_{i=1}^{n} \sum_{j=1, j \neq i}^{n} \alpha_i\, \alpha_j\, Cov(Y_i, Y_j \mid \mathcal{C}). \tag{xiii}$$

If $A \in \mathcal{A}$, then,

$$Var(\mathbb{1}_A \mid \mathcal{C}) \underset{P}{=} P(A \mid \mathcal{C}) \cdot [1 - P(A \mid \mathcal{C})]. \tag{xiv}$$

Remark 11.30 [Two more properties of conditional covariances] If $\varepsilon_1 = Y_1 - E(Y_1 \mid \mathscr{C})$
and $W \colon (\Omega, \mathscr{A}, P) \to (\mathbb{R}, \mathscr{B})$ is a random variable and $E(W^2) < \infty$, then,

$$Cov(\varepsilon_1, W \mid \mathscr{C}_0) \underset{P}{=} E(\varepsilon_1 \cdot W \mid \mathscr{C}_0), \quad \text{if } \mathscr{C}_0 \subset \mathscr{C}, \tag{11.50}$$

and

$$Cov(Y_1, W \mid \mathscr{C}) \underset{P}{=} 0, \quad \text{if } \sigma(W) \subset \mathscr{C}. \tag{11.51}$$

For a proof, see Exercise 11.6.   ◁

11.6 Partial correlation

Another concept used to describe a certain kind of dependence between two random variables
Y_1 and Y_2 is the *partial correlation*, which is the correlation of the residuals of Y_1 and Y_2 with
respect to the conditional expectations $E(Y_1 \mid \mathscr{C})$ and $E(Y_2 \mid \mathscr{C})$, respectively.

Definition 11.31 [Partial correlation]
*For $i = 1, 2$, let $Y_i \colon (\Omega, \mathscr{A}, P) \to (\mathbb{R}, \mathscr{B})$ be random variables with $E(Y_i^2) < \infty$, let $\mathscr{C} \subset \mathscr{A}$
be a σ-algebra, and define $\varepsilon_i := Y_i - E(Y_i \mid \mathscr{C})$ for real-valued versions $E(Y_i \mid \mathscr{C})$. Then
we call*

$$Corr(Y_1, Y_2; \mathscr{C}) := Corr(\varepsilon_1, \varepsilon_2) \tag{11.52}$$

*the partial correlation of Y_1 and Y_2 given $\mathscr{C}$. If $X \colon (\Omega, \mathscr{A}, P) \to (\Omega'_X, \mathscr{A}'_X)$ is a random
variable, then we call*

$$Corr(Y_1, Y_2; X) := Corr[Y_1, Y_2; \sigma(X)] \tag{11.53}$$

the partial correlation of Y_1 and Y_2 given X.

Remark 11.32 [Formulas for the partial correlation] If $Var(Y_1)$, $Var(Y_2) > 0$, and $R^2_{Y_1 \mid \mathscr{C}}$,
$R^2_{Y_2 \mid \mathscr{C}} < 1$, then,

$$Corr(Y_1, Y_2; \mathscr{C}) = \frac{Corr(Y_1, Y_2) - R_{Y_1 \mid \mathscr{C}} \cdot R_{Y_2 \mid \mathscr{C}} \cdot Corr[E(Y_1 \mid \mathscr{C}), E(Y_2 \mid \mathscr{C})]}{\sqrt{1 - R^2_{Y_1 \mid \mathscr{C}}} \cdot \sqrt{1 - R^2_{Y_2 \mid \mathscr{C}}}}, \tag{11.54}$$

where $R^2_{Y_i \mid \mathscr{C}} = Var[E(Y_i \mid \mathscr{C})]/Var(Y_i)$, $i = 1, 2$, denotes the coefficient of determination (see
Def. 11.6 and Exercise 11.8). Similarly, if, for $i = 1, 2$, there are versions $E(Y_i \mid X)$ with

$E(Y_i \mid X) = \beta_{i0} + \beta_{i1}X$ with slopes $\beta_{i1} \neq 0$ and $Corr(Y_i, X)^2 < 1$, then Equation (11.54) simplifies to

$$Corr(Y_1, Y_2; X) = \frac{Corr(Y_1, Y_2) - Corr(Y_1, X) \cdot Corr(Y_2, X)}{\sqrt{1 - Corr(Y_1, X)^2} \cdot \sqrt{1 - Corr(Y_2, X)^2}} \tag{11.55}$$

(see Exercise 11.9).   ◁

Remark 11.33 [$(X=x)$-conditional correlation]   If $P(X=x) > 0$ and, for $i = 1, 2$, $Var(Y_i \mid X=x) > 0$, then we define

$$Corr(Y_1, Y_2 \mid X=x) = \frac{Cov(Y_1, Y_2 \mid X=x)}{\sqrt{Var(Y_1 \mid X=x)} \cdot \sqrt{Var(Y_2 \mid X=x)}}, \tag{11.56}$$

and call it the $(X=x)$-conditional correlation of Y_1 and Y_2. If $Var(Y_1 \mid X=x) = 0$ or $Var(Y_2 \mid X=x) = 0$, then we define $Corr(Y_1, Y_2 \mid X=x) = 0$.   ◁

Remark 11.34 [Interpretation of the partial correlation]   The definition of the partial correlation $Corr(Y_1, Y_2; X)$, Rule (vii) of Box 11.2, and Rule (vii) of Box 11.3 imply

$$Corr(Y_1, Y_2; X) = \frac{E[Cov(Y_1, Y_2 \mid X)]}{\sqrt{E[Var(Y_1 \mid X)]} \cdot \sqrt{E[Var(Y_2 \mid X)]}}. \tag{11.57}$$
                                                                                   ◁

Example 11.35 [Joe and Ann with self-selection – continued]   We compute the partial correlation $Corr(Y, X; U)$ in the example presented in Table 11.2. For this purpose, we use Equations (11.52), (11.53), and (7.18). The covariance of the two residuals is

$$\begin{aligned}
Cov(\varepsilon_Y, \varepsilon_X) &= E(\varepsilon_Y \cdot \varepsilon_X) \\
&= (-.704 \cdot (-.04)) \cdot .144 + (.296 \cdot (-.04)) \cdot .336 + (-.704 \cdot .96) \cdot .004 \\
&\quad + (.296 \cdot .96) \cdot .016 + (-.352 \cdot (-.76)) \cdot .096 + (.648 \cdot (-.76)) \cdot .024 \\
&\quad + (-.352 \cdot .24) \cdot .228 + (.648 \cdot .24) \cdot .152 = .02016,
\end{aligned}$$

the variance of ε_Y is

$$\begin{aligned}
Var(\varepsilon_Y) = E\left(\varepsilon_Y^2\right) &= (-.704)^2 \cdot .144 + .296^2 \cdot .336 + (-.704)^2 \cdot .004 \\
&\quad + .296^2 \cdot .016 + (-.352)^2 \cdot .096 + .648^2 \cdot .024 \\
&\quad + (-.352)^2 \cdot .228 + .648^2 \cdot .152 = .21824,
\end{aligned}$$

and the variance of ε_X is

$$\begin{aligned}
Var(\varepsilon_X) = E\left(\varepsilon_X^2\right) &= (-.04)^2 \cdot .144 + (-.04)^2 \cdot .336 + .96^2 \cdot .004 + .96^2 \cdot .016 \\
&\quad + (-.76)^2 \cdot .096 + (-.76)^2 \cdot .024 + .24^2 \cdot .228 + .24^2 \cdot .152 = .1104.
\end{aligned}$$

Hence,

$$Corr\,(Y, X; U) = \frac{Cov\,(\varepsilon_Y, \varepsilon_X)}{SD\,(\varepsilon_Y) \cdot SD\,(\varepsilon_X)} = \frac{.02016}{\sqrt{.21824} \cdot \sqrt{.1104}} \approx .1299,$$

which is a *positive* number. Again, this indicates that using the (unconditional) covariance of X and Y or the (unconditional) correlation for the evaluation of the effects of the treatment on success would be highly misleading, because in Example 11.28 we showed that $Cov\,(X, Y)$ is negative although the treatment effects are positive for both persons. For Joe it is $.8 - .7 = .1$, and for Ann $.4 - .2 = .2$ (see the column headed $E(Y \mid X, U)$ in Table 11.2).   ◁

11.7   Proofs

Proof of Theorem 11.19

(i) Equation (11.27) can be derived as follows:

$$
\begin{aligned}
E(Y \mid X) &\underset{P}{=} E[E(Y \mid X, Z) \mid X] && \text{[Box 10.2 (v)]}\\
&\underset{P}{=} E(\beta_0 + \beta_1\,X + \beta_2\,Z \mid X) && \text{[(11.25)]}\\
&\underset{P}{=} \beta_0 + \beta_1\,X + \beta_2\,E(Z \mid X) && \text{[Box 10.2 (xvi), (vii)]}\\
&\underset{P}{=} \beta_0 + \beta_1\,X + \beta_2\,(\gamma_0 + \gamma_1\,X) && \text{[(11.26)]}\\
&\underset{P}{=} (\beta_0 + \beta_2\,\gamma_0) + (\beta_1 + \gamma_1)\,X\\
&\underset{P}{=} \alpha_0 + \alpha_1\,X,
\end{aligned}
$$

with $\alpha_0 := \beta_0 + \beta_2\,\gamma_0$ and $\alpha_1 := \beta_1 + \gamma_1$.

(ii) If $E(Z \mid X) \underset{P}{=} E(Z)$, then the third line of the equations above yields Equations (11.28) and (11.29). If $\beta_2 = 0$, then the proof of (i) shows that $E(Y \mid X) \underset{P}{=} \beta_0 + \beta_1 X \underset{P}{=} \alpha_0 + \alpha_1 X$, which proves the proposition.

(iii) If $E(X \mid Z) \underset{P}{=} E(X)$ or $\beta_1 = 0$, then,

$$
\begin{aligned}
E(Y \mid Z) &\underset{P}{=} E[E(Y \mid X, Z) \mid Z] && \text{[Box 10.2 (v)]}\\
&\underset{P}{=} E(\beta_0 + \beta_1\,X + \beta_2\,Z \mid Z) && \text{[(11.25)]}\\
&\underset{P}{=} \beta_0 + \beta_1\,E(X \mid Z) + \beta_2\,Z && \text{[Box 10.2 (i), (xvi), (vii)]}\\
&\underset{P}{=} \beta_0 + \beta_1\,E(X) + \beta_2\,Z. && \text{[}E(X \mid Z) \underset{P}{=} E(X) \text{ or } \beta_1 = 0\text{]}
\end{aligned}
$$

Hence, our assumption implies

$$Var\,[E(Y \mid Z)] = \beta_2^2\,Var\,(Z) \tag{11.58}$$

[see Box 6.2 (ii), (iii)]. Now consider

$$Var[E(Y \mid X, Z)]$$
$$= Var(\beta_0 + \beta_1 X + \beta_2 Z) \qquad\qquad\qquad\qquad [(11.25)]$$
$$= \beta_1^2 Var(X) + \beta_2^2 Var(Z) + 2\,\beta_1\beta_2 Cov(X, Z). \qquad [\text{Box 6.2 (ii), Box 7.1 (viii)}]$$

Assuming

$$[E(Z \mid X) \underset{P}{=} E(Z) \text{ or } \beta_2 = 0] \quad \text{and} \quad [E(X \mid Z) \underset{P}{=} E(X) \text{ or } \beta_1 = 0]$$

yields $2\,\beta_1\beta_2 Cov(X, Z) = 0$, because $E(Z \mid X) \underset{P}{=} E(Z)$ and $E(X \mid Z) \underset{P}{=} E(X)$ both imply $Cov(X, Z) = 0$ [see Eq. (10.52)]. Hence,

$$Var[E(Y \mid X, Z)] = \beta_1^2 Var(X) + \beta_2^2 Var(Z)$$
$$= \alpha_1^2 Var(X) + \beta_2^2 Var(Z) \qquad\qquad [(11.29)]$$
$$= Var[E(Y \mid X)] + Var[E(Y \mid Z)],$$

because $Var[E(Y \mid X)] = \alpha_1^2 Var(X)$ and $Var[E(Y \mid Z)] = \beta_2^2 Var(Z)$ [see (i), Eq. (11.58), and Box 6.2 (ii) and (iii)]. Now Equation (11.31) follows, dividing both sides by $Var(Y)$ and using the definition of the coefficient of determination.

Exercises

11.1 Prove the rules of computation for the residual $\varepsilon = Y - E(Y \mid \mathscr{C})$ summarized in Box 11.1.

11.2 Show that for $\varepsilon = Y - E(Y \mid X, Z)$, the equations $Cov(X, \varepsilon) = Cov(Z, \varepsilon) = 0$ are special cases of Rule (viii) of Box 11.1 if we consider the conditional expectation $E(Y \mid X, Z)$, assume X and Z to be numerical, and $E(Y^2)$, $E(X^2)$, $E(Z^2) < \infty$.

11.3 Prove Equations (11.17) and (11.18).

11.4 Show that $R_{Y \mid \mathscr{C}}^2 = Corr\,[Y, E(Y \mid \mathscr{C})]^2$.

11.5 Prove the rules of Box 11.2.

11.6 Prove the propositions of Remark 11.30.

11.7 Show: If Y is a dichotomous random variable on $(\Omega, \mathscr{A}, P)$ with values 0 and 1, and $P(X=x) > 0$, then $Var(Y \mid X=x) = P(Y=1 \mid X=x) \cdot [1 - P(Y=1 \mid X=x)]$. Furthermore, if $P(X=x_1), P(X=x_2) > 0, P(Y=1 \mid X=x_1) \neq P(Y=1 \mid X=x_2)$ and $P(Y=1 \mid X=x_1) \neq 1 - P(Y=1 \mid X=x_2)$, then $Var(Y \mid X=x_1) \neq Var(Y \mid X=x_2)$.

11.8 Show that Equation (11.54) holds for $Corr(Y_1, Y_2; \mathscr{C})$.

11.9 Prove Equation (11.55).

Solutions

11.1 (i) This rule directly follows from Theorem 10.9 (ii) and Proposition (2.36).

(ii) This rule directly follows from (i) and the assumption that Y and $E(Y \mid \mathscr{C})$ are real-valued.

(v)

$$\varepsilon \underset{P}{=} Y - E(Y \mid \mathscr{C}) \qquad\qquad\qquad [(\text{i})]$$

$$\underset{P}{=} Y - Y = 0. \qquad\qquad\qquad [Y \underset{P}{=} E(Y \mid \mathscr{C})]$$

(vi) This rule can be derived as follows:

$$E(\varepsilon \mid \mathscr{C}_0) \underset{P}{=} E[Y - E(Y \mid \mathscr{C}) \mid \mathscr{C}_0] \qquad\qquad [\text{def. of } \varepsilon]$$

$$\underset{P}{=} E(Y \mid \mathscr{C}_0) - E[E(Y \mid \mathscr{C}) \mid \mathscr{C}_0] \qquad [\text{Box 10.1 (xvi)}]$$

$$\underset{P}{=} E(Y \mid \mathscr{C}_0) - E(Y \mid \mathscr{C}_0) \qquad\qquad [\text{Box 10.1 (v)}]$$

$$\underset{P}{=} 0.$$

(iii) This rule is a special case of Rule (vi) for $\mathscr{C}_0 = \{\Omega, \varnothing\}$ (see Rem. 10.5).

(vii) This rule is a special case of Rule (vi) for $\mathscr{C}_0 := \sigma(W) \subset \mathscr{C}$.

(viii) Note that $E(Y^2) < \infty$ implies $E[E(Y \mid \mathscr{C})^2] < \infty$ [see Box 10.1 (xi)], which in turn implies $E(\varepsilon^2) = E(Y^2) + E[E(Y \mid \mathscr{C})^2] - 2E[Y \cdot E(Y \mid \mathscr{C})] < \infty$ (see Rem. 7.1).

$$Cov(\varepsilon, W) = E(\varepsilon \cdot W) - E(\varepsilon) \cdot E(W) \qquad [\text{Box 7.1 (i)}]$$

$$= E(\varepsilon \cdot W) - 0 \qquad\qquad\qquad [(\text{iii})]$$

$$= E[E(\varepsilon \cdot W \mid \mathscr{C})] \qquad\qquad [\text{Box 10.1 (iv)}]$$

$$= E[W \cdot E(\varepsilon \mid \mathscr{C})] \qquad\qquad [\text{Box 10.1 (xiv)}]$$

$$= E(W \cdot 0) \qquad\qquad\qquad\qquad [(\text{vi})]$$

$$= E(0) = 0. \qquad\qquad\qquad [\text{Box 6.1 (i)}]$$

(iv)

$$Var(Y) = Var[E(Y \mid \mathscr{C}) + \varepsilon] \qquad\qquad\qquad\qquad [(\text{ii})]$$

$$= Var[E(Y \mid \mathscr{C})] + Var(\varepsilon) + 2 \cdot Cov[E(Y \mid \mathscr{C}), \varepsilon] \qquad [\text{Box 7.1 (viii)}]$$

$$= Var[E(Y \mid \mathscr{C})] + Var(\varepsilon). \qquad\qquad\qquad\qquad [(\text{viii})]$$

In the last equation, we used the fact that $E(Y \mid \mathscr{C})$ is $\mathscr{C}$-measurable, thus taking the role of W in Rule (viii) of Box 11.1.

(ix)   Note that $E(Y^2) < \infty$ implies $E[E(Y \mid \mathscr{C})^2] < \infty$ [see Box 10.1 (xi)]. Hence,

$$
\begin{aligned}
Cov(W, Y) &= Cov[W, E(Y \mid \mathscr{C}) + \varepsilon] && [(\text{ii})] \\
&= Cov[W, E(Y \mid \mathscr{C})] + Cov(W, \varepsilon) && [\text{Box 7.1 (ix)}] \\
&= Cov[W, E(Y \mid \mathscr{C})] + 0. && [(\text{viii})]
\end{aligned}
$$

11.2 $E(Y \mid X, Z) \underset{P}{=} E(Y \mid \mathscr{C})$ with $\mathscr{C} := \sigma(X, Z)$. Hence, X and Z are both $\mathscr{C}$-measurable [see Eq. (2.17)], and Rule (viii) of Box 11.1 applies if we assume $E(Y^2)$, $E(X^2)$, and $E(Z^2) < \infty$.

11.3   Equation (11.17) follows from Theorem 7.14 (ii). Now assume $Var(X), Var(Y) > 0$. Then,

$$
\begin{aligned}
R^2_{Y|X} &= \frac{Var[E(Y \mid X)]}{Var(Y)} && [(11.14)] \\[2mm]
&= \frac{Var(\beta_0 + \beta_1 X)}{Var(Y)} && [(11.16)] \\[2mm]
&= \frac{\beta_1^2 \, Var(X)}{Var(Y)} && [\text{Box 6.2 (ii), (iii)}] \\[2mm]
&= \frac{\left(\dfrac{Cov\,(X, Y)}{Var(X)}\right)^2 Var(X)}{Var(Y)} && [(11.17)] \\[2mm]
&= \frac{Cov(X, Y)^2}{Var(X) \cdot Var(Y)} \\[2mm]
&= Corr(X, Y)^2. && [(7.18)]
\end{aligned}
$$

If $Var(Y) = 0$, then $R^2_{Y|X} = Corr(X, Y) = 0$ by the definitions of the two terms. If $Var(X) = 0$, $Var(Y) > 0$, then $Corr(X, Y) = 0$ by definition and $Var[E(Y \mid X)] = Var(\beta_0 + \beta_1 X) = \beta_1^2 \, Var(X) = 0$, and Equation (11.14) implies $R^2_{Y|X} = 0$ as well. Hence, in both cases Equation (11.18) holds.

11.4   Assume that $Var(Y), Var[E(Y \mid \mathscr{C})] > 0$. Then,

$$
\begin{aligned}
Corr[Y, E(Y \mid \mathscr{C})] &= \frac{Cov[Y, E(Y \mid \mathscr{C})]}{SD(Y) \cdot SD[E(Y \mid \mathscr{C})]} && [(7.18)] \\[2mm]
&= \frac{Var[E(Y \mid \mathscr{C})]}{SD(Y) \cdot SD[E(Y \mid \mathscr{C})]} && [\text{Box 10.1 (xii)}] \\[2mm]
&= \frac{SD[E(Y \mid \mathscr{C})] \cdot SD[E(Y \mid \mathscr{C})]}{SD(Y) \cdot SD[E(Y \mid \mathscr{C})]} \\[2mm]
&= \frac{SD[E(Y \mid \mathscr{C})]}{SD(Y)}.
\end{aligned}
$$

Squaring both sides and inserting the definition of $R^2_{Y|\mathscr{C}}$ yield

$$R^2_{Y|\mathscr{C}} = Corr\,[Y,\,E(Y\mid\mathscr{C})]^2.$$

If $Var(Y) = 0$, then $Corr\,[Y,E(Y\mid\mathscr{C})]^2 = 0 = R^2_{Y|\mathscr{C}}$, which follows from the definitions of $Corr\,[Y,E(Y\mid\mathscr{C})]$ and $R^2_{Y|\mathscr{C}}$. If $Var[E(Y\mid\mathscr{C})] = 0$, then $Corr\,[Y,E(Y\mid\mathscr{C})]^2 = 0$ (see the definition of $Corr\,[Y,E(Y\mid\mathscr{C})]$), and $R^2_{Y|\mathscr{C}} = 0$, either because $Var[E(Y\mid\mathscr{C})] = 0$ or, if $Var(Y) = 0$, by definition of $R^2_{Y|\mathscr{C}}$.

11.5 (i)

$$Cov\,(Y_1, Y_2\mid\mathscr{C})$$
$$\underset{P}{=} E(\varepsilon_1 \cdot \varepsilon_2\mid\mathscr{C}) \qquad\qquad\qquad\qquad\qquad [(11.33)]$$
$$\underset{P}{=} E\left([Y_1 - E(Y_1\mid\mathscr{C})]\cdot[Y_2 - E(Y_2\mid\mathscr{C})]\mid\mathscr{C}\right) \qquad [(11.1)]$$
$$\underset{P}{=} E\big[Y_1\cdot Y_2 - Y_1\cdot E(Y_2\mid\mathscr{C}) - E(Y_1\mid\mathscr{C})\cdot Y_2 + E(Y_1\mid\mathscr{C})\cdot E(Y_2\mid\mathscr{C})\mid\mathscr{C}\big]$$
$$\underset{P}{=} E(Y_1\cdot Y_2\mid\mathscr{C}) - E[Y_1\cdot E(Y_2\mid\mathscr{C})\mid\mathscr{C}] - E[E(Y_1\mid\mathscr{C})\cdot Y_2\mid\mathscr{C}]$$
$$\quad + E[E(Y_1\mid\mathscr{C})\cdot E(Y_2\mid\mathscr{C})\mid\mathscr{C}] \qquad\qquad [\text{Box } 10.1\ (\text{xvi})]$$
$$\underset{P}{=} E(Y_1\cdot Y_2\mid\mathscr{C}) - E(Y_1\mid\mathscr{C})\cdot E(Y_2\mid\mathscr{C}). \qquad [\text{Box } 10.1\ (\text{xiv})]$$

(ii)

$$Cov\,(\varepsilon_1, \varepsilon_2\mid\mathscr{C}) \underset{P}{=} E(\varepsilon_1\cdot\varepsilon_2\mid\mathscr{C}) - E(\varepsilon_1\mid\mathscr{C})\cdot E(\varepsilon_2\mid\mathscr{C}) \qquad [(\text{i})]$$
$$\underset{P}{=} E(\varepsilon_1\cdot\varepsilon_2\mid\mathscr{C}) \qquad\qquad\qquad\qquad [\text{Box } 11.1\ (\text{vi})]$$
$$\underset{P}{=} Cov\,(Y_1, Y_2\mid\mathscr{C}). \qquad\qquad\qquad\quad [(11.33)]$$

(iii)

$$Cov\,(Y_1, Y_2\mid\mathscr{C}) \underset{P}{=} E(Y_1\cdot Y_2\mid\mathscr{C}) - E(Y_1\mid\mathscr{C})\cdot E(Y_2\mid\mathscr{C}) \qquad [(\text{i})]$$
$$\underset{P}{=} E(\alpha\,Y_2\mid\mathscr{C}) - \alpha\,E(Y_2\mid\mathscr{C}) \qquad [Y_1 \underset{P}{=} \alpha,\ \text{Box } 10.3\ (\text{iv})]$$
$$\underset{P}{=} \alpha\,E(Y_2\mid\mathscr{C}) - \alpha\,E(Y_2\mid\mathscr{C}) \underset{P}{=} 0. \qquad [\text{Box } 10.1\ (\text{iii})]$$

(iv)

$$Cov\,(\alpha + Y_1,\ \beta + Y_2\mid\mathscr{C})$$
$$\underset{P}{=} E\left([\alpha + Y_1 - E(\alpha + Y_1\mid\mathscr{C})]\cdot[\beta + Y_2 - E(\beta + Y_2\mid\mathscr{C})]\mid\mathscr{C}\right) \qquad [(11.33)]$$
$$\underset{P}{=} E\left([Y_1 - E(Y_1\mid\mathscr{C})]\cdot[Y_2 - E(Y_2\mid\mathscr{C})]\mid\mathscr{C}\right) \qquad [\text{Box } 10.1\ (\text{ii})]$$
$$\underset{P}{=} E(\varepsilon_1\cdot\varepsilon_2\mid\mathscr{C}) \qquad\qquad\qquad\qquad\qquad [(11.1)]$$
$$\underset{P}{=} Cov\,(Y_1, Y_2\mid\mathscr{C}). \qquad\qquad\qquad\qquad [(11.33)]$$

(v)

$$Cov(\alpha\, Y_1,\, \beta\, Y_2 \mid \mathscr{C})$$

$$\underset{P}{=} E\Big(\big[\alpha\, Y_1 - E(\alpha\, Y_1 \mid \mathscr{C})\big] \cdot \big[\beta\, Y_2 - E(\beta\, Y_2 \mid \mathscr{C})\big] \,\Big|\, \mathscr{C}\Big) \qquad\qquad [(11.33)]$$

$$\underset{P}{=} E\Big(\big[\alpha \cdot [Y_1 - E(Y_1 \mid \mathscr{C})]\big] \cdot \big[\beta \cdot [Y_2 - E(Y_2 \mid \mathscr{C})]\big] \,\Big|\, \mathscr{C}\Big) \qquad [\text{Box 10.1 (iii)}]$$

$$\underset{P}{=} E(\alpha\, \varepsilon_1 \cdot \beta\, \varepsilon_2 \mid \mathscr{C}) \qquad\qquad\qquad\qquad\qquad\qquad [(11.1)]$$

$$\underset{P}{=} \alpha\, \beta\, E(\varepsilon_1 \cdot \varepsilon_2 \mid \mathscr{C}) \qquad\qquad\qquad\qquad\qquad\quad [\text{Box 10.1 (iii)}]$$

$$\underset{P}{=} \alpha\, \beta\, Cov(Y_1, Y_2 \mid \mathscr{C}). \qquad\qquad\qquad\qquad\qquad [(11.33)]$$

(vi)

$$Cov(Y_1, Y_2 \mid \mathscr{C}_0) \underset{P}{=} E(\varepsilon_1 \cdot \varepsilon_2 \mid \mathscr{C}_0) \qquad\qquad [(11.33)]$$

$$\underset{P}{=} E\big[E(\varepsilon_1 \cdot \varepsilon_2 \mid \mathscr{C}) \mid \mathscr{C}_0\big] \qquad [\text{Box 10.1 (v)}]$$

$$\underset{P}{=} E\big[Cov(Y_1, Y_2 \mid \mathscr{C}) \mid \mathscr{C}_0\big]. \qquad [(11.33)]$$

(vii)

$$E[Cov(Y_1,\, Y_2 \mid \mathscr{C})] = E[E(\varepsilon_1 \cdot \varepsilon_2 \mid \mathscr{C})] \qquad\qquad [(11.33)]$$

$$= E(\varepsilon_1 \cdot \varepsilon_2) \qquad\qquad\quad [\text{Box 10.1 (iv)}]$$

$$= Cov(\varepsilon_1, \varepsilon_2). \qquad [\text{Box 7.1 (i), Box 11.1 (iii)}]$$

(ix),(viii)

$$Cov(Y_1,\, Y_2) = Cov(E(Y_1 \mid \mathscr{C}) + \varepsilon_1,\, E(Y_2 \mid \mathscr{C}) + \varepsilon_2) \qquad [\text{Box 11.1 (ii)}]$$

$$= Cov(E(Y_1 \mid \mathscr{C}), E(Y_2 \mid \mathscr{C})) + Cov(E(Y_1 \mid \mathscr{C}), \varepsilon_2)$$

$$+ Cov(\varepsilon_1,\, E(Y_2 \mid \mathscr{C})) + Cov(\varepsilon_1, \varepsilon_2) \qquad [\text{Box 7.1 (ix)}]$$

$$= Cov(E(Y_1 \mid \mathscr{C}), E(Y_2 \mid \mathscr{C})) + Cov(\varepsilon_1, \varepsilon_2). \qquad [\text{Box 11.1 (viii)}]$$

$$= Cov(E(Y_1 \mid \mathscr{C}), E(Y_2 \mid \mathscr{C})) + E[Cov(Y_1,\, Y_2 \mid \mathscr{C})]. \qquad [(\text{vii})]$$

(xii)   For $i = 1, 2$, consider the residual of $W_i \cdot Y_i$ with respect to its $\mathscr{C}$-conditional expectation,

$$W_i \cdot Y_i - E(W_i \cdot Y_i \mid \mathscr{C}) \underset{P}{=} W_i \cdot Y_i - W_i \cdot E(Y_i \mid \mathscr{C}) \qquad [\text{Box 10.1 (xiv)}]$$

$$\underset{P}{=} W_i \cdot [Y_i - E(Y_i \mid \mathscr{C})] \underset{P}{=} W_i \cdot \varepsilon_i.$$

This equation implies

$$Cov(W_1 \cdot Y_1,\, W_2 \cdot Y_2 \mid \mathscr{C}) \underset{P}{=} E(W_1 \cdot \varepsilon_1 \cdot W_2 \cdot \varepsilon_2 \mid \mathscr{C}) \qquad\qquad [(11.33)]$$

$$\underset{P}{=} W_1 \cdot W_2 \cdot E(\varepsilon_1 \cdot \varepsilon_2 \mid \mathscr{C}) \qquad [\text{Box 10.1 (xiv)}]$$

$$\underset{P}{=} W_1 \cdot W_2 \cdot Cov(Y_1, Y_2 \mid \mathscr{C}). \qquad\qquad [(11.33)]$$

(xiii) Define

$$\varepsilon_i := Y_i - E(Y_i \mid \mathscr{C}), \quad i = 1, \dots, n, \quad \text{and} \quad \delta_j := Z_j - E(Z_j \mid \mathscr{C}), \quad j = 1, \dots, m.$$

Then,

$$\sum_{i=1}^{n} \alpha_i Y_i - E\left(\sum_{i=1}^{n} \alpha_i Y_i \mid \mathscr{C} \right) \underset{P}{=} \sum_{i=1}^{n} \alpha_i [Y_i - E(Y_i \mid \mathscr{C})] \qquad \text{[Box 10.1 (xvi)]}$$

$$\underset{P}{=} \sum_{i=1}^{n} \alpha_i \, \varepsilon_i,$$

and, analogously,

$$\sum_{j=1}^{m} \beta_j Z_j - E\left(\sum_{j=1}^{m} \beta_j Z_j \mid \mathscr{C} \right) = \sum_{j=1}^{m} \beta_j \, \delta_j.$$

Hence,

$$Cov\left(\sum_{i=1}^{n} \alpha_i Y_i, \sum_{j=1}^{m} \beta_j Z_j \mid \mathscr{C} \right) \underset{P}{=} E\left[\left(\sum_{i=1}^{n} \alpha_i \, \varepsilon_i \right) \cdot \left(\sum_{j=1}^{m} \beta_j \, \delta_j \right) \mid \mathscr{C} \right] \qquad \text{[(11.33)]}$$

$$\underset{P}{=} E\left(\sum_{i=1}^{n} \sum_{j=1}^{m} \alpha_i \, \beta_j \cdot \varepsilon_i \cdot \delta_j \mid \mathscr{C} \right)$$

$$\underset{P}{=} \sum_{i=1}^{n} \sum_{j=1}^{m} \alpha_i \, \beta_j \, E(\varepsilon_i \cdot \delta_j \mid \mathscr{C}) \qquad \text{[Box 10.1 (xvi)]}$$

$$\underset{P}{=} \sum_{i=1}^{n} \sum_{j=1}^{m} \alpha_i \, \beta_j \, Cov\,(Y_i, Z_j \mid \mathscr{C}). \qquad \text{[(11.33)]}$$

(x),(xi)

$$Cov\,(Y_1, \ Y_2 \mid \mathscr{C}_0)$$
$$= Cov\,(E(Y_1 \mid \mathscr{C}) + \varepsilon_1, E(Y_2 \mid \mathscr{C}) + \varepsilon_2 \mid \mathscr{C}_0) \qquad \text{[Box 11.1 (ii)]}$$
$$= Cov\,(E(Y_1 \mid \mathscr{C}), E(Y_2 \mid \mathscr{C}) \mid \mathscr{C}_0) + Cov\,(E(Y_1 \mid \mathscr{C}), \varepsilon_2 \mid \mathscr{C}_0)$$
$$\quad + Cov\,(\varepsilon_1, \ E(Y_2 \mid \mathscr{C}) \mid \mathscr{C}_0) + Cov\,(\varepsilon_1, \varepsilon_2 \mid \mathscr{C}_0) \qquad \text{[(xiii)]}$$
$$= Cov\,(E(Y_1 \mid \mathscr{C}), E(Y_2 \mid \mathscr{C}) \mid \mathscr{C}_0) + Cov\,(\varepsilon_1, \varepsilon_2 \mid \mathscr{C}_0) \qquad \text{[Box 11.1 (viii)]}$$
$$= Cov\,(E(Y_1 \mid \mathscr{C}), E(Y_2 \mid \mathscr{C}) \mid \mathscr{C}_0) + E[Cov\,(Y_1, Y_2 \mid \mathscr{C}) \mid \mathscr{C}_0]. \qquad \text{[(vii)]}$$

Hence,

$$Cov\,(Y_1, Y_2, \mid \mathscr{C}_0)$$
$$\underset{P}{=} Cov\,[E(Y_1 \mid \mathscr{C}), E(Y_2 \mid \mathscr{C}) \mid \mathscr{C}_0] + E[Cov\,(Y_1, Y_2 \mid \mathscr{C}) \mid \mathscr{C}_0]$$
$$\underset{P}{=} Cov\,[E(Y_1 \mid \mathscr{C}), E(Y_2 \mid \mathscr{C}) \mid \mathscr{C}_0] + Cov\,(\varepsilon_1, \varepsilon_2 \mid \mathscr{C}_0). \qquad \text{[(ii), (vi), (ii)]}$$

(xiv)

$$Cov\,(1_A,\,1_B \mid \mathscr{C}) \underset{P}{=} E(1_A \cdot 1_B \mid \mathscr{C}) - E(1_A \mid \mathscr{C}) \cdot E(1_B \mid \mathscr{C}) \qquad [(i)]$$

$$\underset{P}{=} P(A \cap B \mid \mathscr{C}) - P(A \mid \mathscr{C}) \cdot P(B \mid \mathscr{C}). \qquad [(1.33),\ (10.2)]$$

11.6

(11.50)

$$Cov\,(\varepsilon_1,\,W \mid \mathscr{C}_0) \underset{P}{=} E\Big(\big[\varepsilon_1 - E(\varepsilon_1 \mid \mathscr{C}_0)\big] \cdot \big[W - E(W \mid \mathscr{C}_0)\big] \mid \mathscr{C}_0 \Big) \qquad [(11.33)]$$

$$\underset{P}{=} E\big(\varepsilon_1 \cdot [W - E(W \mid \mathscr{C}_0)] \mid \mathscr{C}_0 \big) \qquad [\mathscr{C}_0 \subset \mathscr{C},\ \text{Box 11.1 (vi)}]$$

$$\underset{P}{=} E\big(\varepsilon_1 \cdot W - \varepsilon_1 \cdot E(W \mid \mathscr{C}_0) \mid \mathscr{C}_0 \big)$$

$$\underset{P}{=} E(\varepsilon_1 \cdot W \mid \mathscr{C}_0) - E\big(\varepsilon_1 \cdot E(W \mid \mathscr{C}_0) \mid \mathscr{C}_0 \big) \qquad [\text{Box 10.1 (xvi)}]$$

$$\underset{P}{=} E(\varepsilon_1 \cdot W \mid \mathscr{C}_0) - E(\varepsilon_1 \mid \mathscr{C}_0) \cdot E(W \mid \mathscr{C}_0) \qquad [\text{Box 10.1 (xiv)}]$$

$$\underset{P}{=} E(\varepsilon_1 \cdot W \mid \mathscr{C}_0). \qquad [\text{Box 11.1 (vi)}]$$

(11.51)

$$Cov\,(Y_1,\,W \mid \mathscr{C})$$

$$\underset{P}{=} E\Big(\big[Y_1 - E(Y_1 \mid \mathscr{C})\big] \cdot \big[W - E(W \mid \mathscr{C})\big] \mid \mathscr{C} \Big) \qquad [(11.33)]$$

$$\underset{P}{=} E\big(Y_1 - E(Y_1 \mid \mathscr{C}) \mid \mathscr{C} \big) \cdot \big(W - E(W \mid \mathscr{C}) \big) \qquad [\sigma(W) \subset \mathscr{C},\ \text{Box 10.1 (xiv)}]$$

$$\underset{P}{=} 0. \qquad [\text{Box 11.1 (vi)}]$$

11.7   If $P\,(X{=}x) > 0$, then (11.48) and Remark 10.35 imply

$$Var(Y \mid X{=}x) = P(Y{=}1 \mid X{=}x) \cdot [1 - P(Y{=}1 \mid X{=}x)].$$

Now, define $a := P(Y{=}1 \mid X{=}x_1)$ and $b := P(Y{=}1 \mid X{=}x_2)$. Then,

$$a \cdot (1 - a) = b \cdot (1 - b)$$
$$\Rightarrow a - a^2 = b - b^2$$
$$\Rightarrow a - b = (a - b) \cdot (a + b).$$

If $a - b = 0$, then $a = b$. Furthermore, if $a - b \neq 0$, then $a + b = 1$ and hence $a = 1 - b$. Thus, we have shown that $a \cdot (1 - a) = b \cdot (1 - b)$ implies $a = b$ or $a = 1 - b$. The contraposition of this implication and substituting a and b by the conditional probabilities prove the proposition.

11.8   Because $E(Y_1 \mid \mathscr{C})$ and $E(Y_2 \mid \mathscr{C})$ are $\mathscr{C}$-measurable, Rule (ix) of Box 11.1 implies

$$Cov\,[Y_1,\,E(Y_2 \mid \mathscr{C})] = Cov\,[E(Y_1 \mid \mathscr{C}),\,E(Y_2 \mid \mathscr{C})] = Cov\,[Y_2,\,E(Y_1 \mid \mathscr{C})].$$

Therefore,

$$Cov(\varepsilon_1, \varepsilon_2)$$

$$= Cov[Y_1 - E(Y_1 \mid \mathscr{C}), Y_2 - E(Y_2 \mid \mathscr{C})]$$

$$= Cov(Y_1, Y_2) + Cov[E(Y_1 \mid \mathscr{C}), E(Y_2 \mid \mathscr{C})]$$
$$\quad - Cov[Y_1, E(Y_2 \mid \mathscr{C})] - Cov[Y_2, E(Y_1 \mid \mathscr{C})] \qquad \text{[Box 7.1 (ix)]}$$

$$= Cov(Y_1, Y_2) - Cov[E(Y_1 \mid \mathscr{C}), E(Y_2 \mid \mathscr{C})]$$

$$= SD(Y_1) \cdot SD(Y_2) \cdot Corr(Y_1, Y_2) - Cov[E(Y_1 \mid \mathscr{C}), E(Y_2 \mid \mathscr{C})] \qquad \text{[(7.18)]}$$

$$= SD(Y_1) \cdot SD(Y_2) \cdot \left[Corr(Y_1, Y_2) - \frac{Cov[E(Y_1 \mid \mathscr{C}), E(Y_2 \mid \mathscr{C})]}{SD(Y_1) \cdot SD(Y_2)} \right]$$

$$= SD(Y_1) \cdot SD(Y_2) \cdot \left[Corr(Y_1, Y_2) - R_{Y_1 \mid \mathscr{C}} R_{Y_2 \mid \mathscr{C}} \cdot \frac{Cov[E(Y_1 \mid \mathscr{C}), E(Y_2 \mid \mathscr{C})]}{SD[E(Y_1 \mid \mathscr{C})] \cdot SD[E(Y_2 \mid \mathscr{C})]} \right]$$
$$\text{[(11.11), (11.22)]}$$

$$= SD(Y_1) \cdot SD(Y_2) \cdot \left[Corr(Y_1, Y_2) - R_{Y_1 \mid \mathscr{C}} R_{Y_2 \mid \mathscr{C}} \cdot Corr[E(Y_1 \mid \mathscr{C}), E(Y_2 \mid \mathscr{C})] \right].$$

Furthermore, for $i = 1, 2$,

$$SD(\varepsilon_i) = \sqrt{Var(\varepsilon_i)} = \sqrt{Var(Y_i) - Var[E(Y_i \mid \mathscr{C})]} \qquad \text{[Box 11.1 (iv)]}$$

$$= \sqrt{Var(Y_i) - Var(Y_i) \cdot R^2_{Y_i \mid \mathscr{C}}} = \sqrt{Var(Y_i) \cdot (1 - R^2_{Y_i \mid \mathscr{C}})} \qquad \text{[(11.11)]}$$

$$= SD(Y_i) \cdot \sqrt{1 - R^2_{Y_i \mid \mathscr{C}}},$$

which implies

$$SD(\varepsilon_1) \cdot SD(\varepsilon_2) = SD(Y_1) \cdot SD(Y_2) \cdot \sqrt{1 - R^2_{Y_1 \mid \mathscr{C}}} \cdot \sqrt{1 - R^2_{Y_2 \mid \mathscr{C}}}.$$

Using these results, Definition (11.52) yields

$$Corr(Y_1, Y_2; \mathscr{C})$$

$$:= Corr(\varepsilon_1, \varepsilon_2) = \frac{Cov(\varepsilon_1, \varepsilon_2)}{SD(\varepsilon_1) \cdot SD(\varepsilon_2)}$$

$$= \frac{SD(Y_1) \cdot SD(Y_2) \cdot \left[Corr(Y_1, Y_2) - R_{Y_1 \mid \mathscr{C}} \cdot R_{Y_2 \mid \mathscr{C}} \cdot Corr[E(Y_1 \mid \mathscr{C}), E(Y_2 \mid \mathscr{C})] \right]}{SD(Y_1) \cdot SD(Y_2) \cdot \sqrt{1 - R^2_{Y_1 \mid \mathscr{C}}} \cdot \sqrt{1 - R^2_{Y_2 \mid \mathscr{C}}}}$$

$$= \frac{Corr(Y_1, Y_2) - R_{Y_1 \mid \mathscr{C}} \cdot R_{Y_2 \mid \mathscr{C}} \cdot Corr[E(Y_1 \mid \mathscr{C}), E(Y_2 \mid \mathscr{C})]}{\sqrt{1 - R^2_{Y_1 \mid \mathscr{C}}} \cdot \sqrt{1 - R^2_{Y_2 \mid \mathscr{C}}}},$$

which is Equation (11.54).

11.9 If there are β_{i0}, $\beta_{i1} \in \mathbb{R}$, $i = 1, 2$, such that

$$E(Y_1 \mid X) \underset{P}{=} \beta_{10} + \beta_{11}X \quad \text{and} \quad E(Y_2 \mid X) \underset{P}{=} \beta_{20} + \beta_{21}X, \qquad \beta_{11}, \beta_{21} \neq 0,$$

then,

$$E(Y_1 \mid X) \underset{P}{=} a_0 + a_1 E(Y_2 \mid X), \quad \text{with} \quad a_0 = \beta_{10} - \beta_{20} \cdot \frac{\beta_{11}}{\beta_{21}} \quad \text{and} \quad a_1 = \frac{\beta_{11}}{\beta_{21}}.$$

Furthermore, $\beta_{i1} \neq 0$, $Corr(Y_i, X)^2 < 1$, $i = 1, 2$, and Equations (7.18) and (11.17) imply $0 < Corr(Y_i, X)^2 < 1$. Now we consider two cases.

Case 1: The slopes β_{11} and β_{21} have *identical signs*. Then $a_1 > 0$ and the correlations $Corr(Y_1, X)$, $Corr(Y_2, X)$ have identical signs as well [see (11.17)], and

$$R_{Y_1 \mid X} \cdot R_{Y_2 \mid X} = Corr(Y_1, X) \cdot Corr(Y_2, X)$$

[see Eqs. (11.18) and (11.22)] and $Corr[E(Y_1 \mid X), E(Y_2 \mid X)] = 1$ [see Cor. 7.23]. This implies that Equation (11.54) simplifies to

$$Corr(Y_1, Y_2; X) = \frac{Corr(Y_1, Y_2) - Corr(Y_1, X) \cdot Corr(Y_2, X)}{\sqrt{1 - Corr(Y_1, X)^2} \cdot \sqrt{1 - Corr(Y_2, X)^2}}.$$

Case 2: The slopes β_{11} and β_{21} have *different signs*. Then $a_1 < 0$ and the correlations $Corr(Y_1, X)$, $Corr(Y_2, X)$ have different signs as well [see (11.17)]. In this case, the same equation holds, because $R_{Y_1 \mid X} \cdot R_{Y_2 \mid X} = -Corr(Y_1, X) \cdot Corr(Y_2, X)$ [see Eqs. (11.18) and (11.22)] and, according to Corollary 7.23, $Corr[E(Y_1 \mid X), E(Y_2 \mid X)] = -1$.

12

Linear regression

In chapter 10 we introduced the general concepts of a conditional expectation and a regression, and in chapter 11 we treated the *residual* with respect to a conditional expectation, the concepts *conditional variance*, *conditional covariance*, and *partial correlation*. Now we turn to *parameterizations* of a conditional expectation. A parameterization serves to describe a conditional expectation with a few parameters (real numbers). Oftentimes, these parameters have important meanings that differ between different parameterizations. We treat a *linear parameterization* of a conditional expectation, which is also called the *linear regression*. We start with the basic ideas, present the definitions, treat some examples, consider the relationship between a linear regression and a linear quasi-regression, and deal with uniqueness of a linear parameterization and the identification of the regression coefficients. Finally, we present a theorem on the invariance of regression coefficients and a theorem on the existence of a linear regression if the regressand and the regressors have a joint multivariate normal distribution.

12.1 Basic ideas

Consider the random variables $X: (\Omega, \mathcal{A}, P) \to (\mathbb{R}, \mathcal{B})$ and $Y: (\Omega, \mathcal{A}, P) \to (\overline{\mathbb{R}}, \overline{\mathcal{B}})$, and let Y be nonnegative or with finite expectation. Furthermore, let $\mathcal{C} \subset \mathcal{A}$ be a σ-algebra and assume that X is $\mathcal{C}$-measurable. Now assume that there are a real-valued version $E(Y \mid \mathcal{C})$ of the $\mathcal{C}$-conditional expectation of Y and coefficients $\beta_0, \beta_1 \in \mathbb{R}$ such that

$$E(Y \mid \mathcal{C}) = \beta_0 + \beta_1 X. \tag{12.1}$$

Then we call the function $g: \mathbb{R} \to \mathbb{R}$ defined by

$$g(x) := \beta_0 + \beta_1 x, \quad \forall\, x \in \mathbb{R}, \tag{12.2}$$

a *linear parameterization of $E(Y \mid \mathcal{C})$ in X*. This definition implies $g(X) \in \mathcal{E}(Y \mid \mathcal{C})$, where $g(X)$ denotes the composition of X and g.

Probability and Conditional Expectation: Fundamentals for the Empirical Sciences, First Edition. Rolf Steyer and Werner Nagel.
© 2017 John Wiley & Sons, Ltd. Published 2017 by John Wiley & Sons, Ltd.
Companion website: http://www.probability-and-conditional-expectation.de

If $\mathscr{C} = \sigma(X)$, then a function $g: \mathbb{R} \to \overline{\mathbb{R}}$ satisfying $E(Y \mid \mathscr{C}) = E(Y \mid X) = g(X)$ always exists (see Rem. 10.3 and Cor. 10.23). However, g is not necessarily a linear function. Hence, even if $\mathscr{C} = \sigma(X)$, a *linear parameterization* of $E(Y \mid \mathscr{C}) = E(Y \mid X)$ in X does *not necessarily exist*. Yet, if we assume that

(a) Equations (12.1) and (12.2) hold, and

(b) the variance of X is positive and finite,

then $E(Y \mid X)$ and g, and therefore the coefficients β_0 and β_1, are uniquely defined (see section 12.5 and cf. Rem. 11.11). Under the assumptions (a) and (b), the function g is also called the *linear regression* of Y on X and the numbers β_0 and β_1 are called *regression coefficients*.

Remark 12.1 [Composition of X and a linear function] Because the conditional expectation $E(Y \mid \mathscr{C})$ is a function with domain Ω, strictly speaking, it is not a linear function itself. Assuming that Equation (12.1) holds and saying that $E(Y \mid \mathscr{C})$ is a linear function of X, we mean that $E(Y \mid \mathscr{C})$ *is the composition* of the random variable $X: (\Omega, \mathscr{A}, P) \to (\mathbb{R}, \mathscr{B})$ and the linear function $g: \mathbb{R} \to \mathbb{R}$ satisfying (12.2). This is why g and not $E(Y \mid \mathscr{C})$ itself is called a linear parameterization and a linear regression if the assumptions (a) and (b) hold.   ◁

Remark 12.2 [Estimation] Although estimation is beyond the scope of this book, it is worthwhile noting that estimation is one of the reasons why a parameterization is useful. The definition of a concrete version of a conditional expectation $E(Y \mid X)$ requires that we know for all $\omega \in \Omega$ which values $E(Y \mid X)(\omega)$ are assigned to ω. In empirical applications, these values are often *unknown*, that is, we do not know which concrete number $E(Y \mid X)(\omega)$ is assigned to a concrete ω. In these cases, estimating the values of the conditional expectation may be an issue. In particular, if Equations (12.1) and (12.2) hold and the variance of X is positive and finite, then the coefficients β_0 and β_1 – and with them $E(Y \mid X) = \beta_0 + \beta_1 X$ – can be computed from estimable quantities such as the variance of X and the expectations and the covariance of X and Y. In this case, estimation of the values of $E(Y \mid X)$ is relatively simple because the variance of X as well as the expectations and the covariance of X and Y can be estimated in a data sample.   ◁

Example 12.3 [Joe and Ann with self-selection – continued] Table 12.1 (cf. also Table 11.2) shows nine random variables, the first five of which may be called *observable* (or *manifest*), whereas the last four are *unobservable* (or *latent*). The difference between the two kinds of random variables is that, in empirical applications, the values of the conditional expectations, the unobservable random variables, are unknown parameters that we might wish to estimate in a sample. These parameters can be computed from the joint distributions of the random variables involved. In this fictitious example, the information about the joint distribution of the random variables U, X, and Y is contained in the second column of the table, whereas in empirical applications these parameters usually have to be estimated using a data sample. Examples in case are the conditional expectation values $E(Y \mid U=Joe, X=0) = .7$ and $E(Y \mid X=0) = .6$. In contrast to the values of the conditional expectations, the values of the five observables are known for all eight possible outcomes $\omega \in \Omega$ of the random experiment. For example, if $\omega = (Joe, no, -)$, then $U(\omega) = Joe$, $X(\omega) = 0$, and $Y(\omega) = 0$, and these values are known, because the definitions of these observables do not involve unknown parameters that depend on the joint distribution of the random variables involved.

Table 12.1 Joe and Ann with self-selection: conditional expectations.

Unit Treatment Success	$P(\{\omega\})$	Person variable U	Indicator for Joe $1_{U=Joe}$	Indicator for Ann $1_{U=Ann}$	Treatment variable X	Outcome variable Y	$E(Y\mid X,U)$	$E(Y\mid X)$	$E(Y\mid U)$	$P(X=1\mid U)$
$(Joe, no, -)$	.144	Joe	1	0	0	0	.7	.6	.704	.04
$(Joe, no, +)$	.336	Joe	1	0	0	1	.7	.6	.704	.04
$(Joe, yes, -)$	.004	Joe	1	0	1	0	.8	.42	.704	.04
$(Joe, yes, +)$	.016	Joe	1	0	1	1	.8	.42	.704	.04
$(Ann, no, -)$	.096	Ann	0	1	0	0	.2	.6	.352	.76
$(Ann, no, +)$	.024	Ann	0	1	0	1	.2	.6	.352	.76
$(Ann, yes, -)$	.228	Ann	0	1	1	0	.4	.42	.352	.76
$(Ann, yes, +)$	.152	Ann	0	1	1	1	.4	.42	.352	.76

In this example, we may consider, for instance, the conditional expectations

$$E(Y \mid X) = .6 - .18 \cdot X, \tag{12.3}$$
$$E(Y \mid U) = .352 + .352 \cdot 1_{U=Joe}, \tag{12.4}$$

and

$$E(Y \mid X, U) = .2 + .2 \cdot X + .5 \cdot 1_{U=Joe} - .1 \cdot X \cdot 1_{U=Joe}. \tag{12.5}$$

The computation of the parameters in Equations (12.3) and (12.5) is illustrated in Examples 12.16 and 12.24. ◁

12.2 Assumptions and definitions

In this section, we often refer to the following assumptions and the following notation.

Notation and assumptions 12.4
$Y: (\Omega, \mathscr{A}, P) \to (\overline{\mathbb{R}}, \overline{\mathscr{B}})$ and $X := (X_1, \dots, X_n): (\Omega, \mathscr{A}, P) \to (\mathbb{R}^n, \mathscr{B}_n)$ are random variables, where Y is nonnegative or has a finite expectation $E(Y)$. Furthermore, $\mathscr{C} \subset \mathscr{A}$ is a σ-algebra and X is $\mathscr{C}$-measurable.

Referring to these assumptions, we define a linear parameterization as follows:

Definition 12.5 [Linear parameterization]
Let the assumptions 12.4 hold and let $\beta_0, \beta_1, \ldots, \beta_n \in \mathbb{R}$. *If there is a real-valued version* $E(Y \mid \mathcal{C}) \in \mathcal{E}(Y \mid \mathcal{C})$ *such that*

$$E(Y \mid \mathcal{C}) = \beta_0 + \sum_{i=1}^{n} \beta_i X_i, \qquad (12.6)$$

then the function $g \colon \mathbb{R}^n \to \mathbb{R}$ *defined by*

$$g(x) := \beta_0 + \sum_{i=1}^{n} \beta_i x_i, \quad \forall\, x = (x_1, \ldots, x_n) \in \mathbb{R}^n, \qquad (12.7)$$

is called a linear parameterization of $E(Y \mid \mathcal{C})$ *in X.*

Note that, even if it exists, a linear parameterization of $E(Y \mid \mathcal{C})$ is not uniquely defined unless additional assumptions hold (see Example 12.9). Uniqueness of a linear parameterization is treated in Corollary 12.31.

Remark 12.6 [Another notation]   If $\sigma(X) = \mathcal{C}$, then Equation (12.6) is equivalent to

$$E(Y \mid X) = \beta_0 + \sum_{i=1}^{n} \beta_i X_i. \qquad (12.8)$$

◁

Remark 12.7 [X-conditional mean independence]   Equation (12.6) implies that $E(Y \mid \mathcal{C}) \in \mathcal{E}(Y \mid X)$, which in turn implies that Y is X-conditionally mean independent from $\mathcal{C}$ (see Def. 10.45). If $\sigma(X) \neq \mathcal{C}$, then this conditional mean independence does not necessarily hold. ◁

Remark 12.8 [Other versions and other factorizations]   Note that a linear parameterization g of $E(Y \mid \mathcal{C})$ in X is a factorization of $E(Y \mid X)$, that is, $E(Y \mid X) = g(X)$ is the composition of X and g (see section 10.4). Also note that there may be other factorizations g^* of $E(Y \mid X)$ and versions in $V^* \in \mathcal{E}(Y \mid \mathcal{C})$ that do not satisfy Equations (12.6) and (12.7), respectively. However, according to Theorem 10.9 (ii), Equation (12.6) implies

$$V^* \underset{P}{=} \beta_0 + \sum_{i=1}^{n} \beta_i X_i, \quad \forall\, V^* \in \mathcal{E}(Y \mid \mathcal{C}). \qquad (12.9)$$

Furthermore, if g, g^* are factorizations of versions $V, V^* \in \mathcal{E}(Y \mid X)$, respectively, then,

$$g^* \underset{P_X}{=} g \qquad (12.10)$$

(see Cor. 10.29). ◁

Example 12.9 [Constant regressor]   Suppose that $E(Y \mid \mathscr{C}) = g(X) = \beta_0 + \beta_1 X$ and X is a constant function, that is, there is an $\alpha \in \mathbb{R}$ such that, for all $\omega \in \Omega$, $X(\omega) = \alpha$. Then, $Var(X) = 0$ and $g, g^*\colon \mathbb{R} \to \mathbb{R}$ defined by

$$\forall\, x \in \mathbb{R}\colon g(x) = \beta_0 + \beta_1 x \qquad \text{and} \qquad g^*(x) = (\beta_0 - \gamma\alpha) + (\beta_1 + \gamma)\, x, \quad 0 \neq \gamma \in \mathbb{R},$$

are two linear parameterizations in X that differ from each other but satisfy $g(X) = g^*(X) = E(Y \mid \mathscr{C})$. This example shows that a linear parameterization of $E(Y \mid \mathscr{C})$ in X is not uniquely defined unless additional assumptions hold. ◁

Remark 12.10 [Linear parameterizations in different random variables]   Consider Equation (12.6), which involves a linear parameterization g of $E(Y \mid \mathscr{C})$ in X. If $Z = (Z_1, \dots, Z_m)\colon \Omega \to \mathbb{R}^m$ is an m-variate random variable on $(\Omega, \mathscr{A}, P)$, and $\sigma(Z) \subset \mathscr{C}$, then for one and the same version $E(Y \mid \mathscr{C}) \in \mathscr{E}(Y \mid \mathscr{C})$, there may also be a linear parameterization $f\colon \mathbb{R}^m \to \mathbb{R}$ of $E(Y \mid \mathscr{C})$ in Z with coefficients $\gamma_0, \gamma_1, \dots, \gamma_m \in \mathbb{R}$ satisfying

$$f(z) = \gamma_0 + \sum_{i=1}^{m} \gamma_i\, z_i, \quad \forall\, z = (z_1, \dots, z_m) \in \mathbb{R}^m. \tag{12.11}$$

In other words, one and the same version $E(Y \mid \mathscr{C}) \in \mathscr{E}(Y \mid \mathscr{C})$ may have several parameterizations such as f and g that are linear in $Z = (Z_1, \dots, Z_m)$ and $X = (X_1, \dots, X_n)$, respectively. Note that the regression coefficients $\gamma_0, \gamma_1, \dots, \gamma_m \in \mathbb{R}$ and $\beta_0, \beta_1, \dots, \beta_n \in \mathbb{R}$ pertaining to the two parameterizations f and g may differ from each other. ◁

Remark 12.11 [Conditional expectation values]   If g is a linear parameterization of $E(Y \mid X)$ in X satisfying Equation (12.7), then, according to Definition 10.33,

$$\begin{aligned} E(Y \mid X{=}x) &= E(Y \mid X_1{=}x_1, \dots, X_n{=}x_n) \\ &= g(x) = \beta_0 + \beta_1 x_1 + \cdots + \beta_n x_n, \quad \forall\, x = (x_1, \dots, x_n) \in \mathbb{R}^n. \end{aligned} \tag{12.12}$$

Note that another factorization g^* of $E(Y \mid X)$ might yield another conditional expectation value $E(Y \mid X{=}x)$ if $P(X{=}x) = 0$. ◁

Remark 12.12 [P_X-equivalence of different parameterizations]   If g, g^* are factorizations of versions $V, V^* \in \mathscr{E}(Y \mid X)$, then, according to Equation (12.10),

$$g(x) = g^*(x), \quad \text{for } P_X\text{-a.a. } x \in \mathbb{R}^n, \tag{12.13}$$

[see Eq. (10.26)]. ◁

12.3   Examples

Example 12.13 [Univariate real-valued X]   If $X\colon (\Omega, \mathscr{A}, P) \to (\mathbb{R}, \mathscr{B})$ and

$$E(Y \mid X) = \beta_0 + \beta_1 X, \tag{12.14}$$

then the function $g\colon \mathbb{R} \to \mathbb{R}$ defined by

$$g(x) = \beta_0 + \beta_1 x, \quad x \in \mathbb{R}, \tag{12.15}$$

is a linear parameterization of $E(Y \mid X)$ in X. If $0 < Var(X) < \infty$, then g is uniquely defined and also called the *simple linear regression* of Y on X (see Cor. 12.31 and Rem. 12.34).   ◁

Example 12.14 [Intercept and slope]   If $E(Y \mid X) = \beta_0 + \beta_1 X$, then,

$$\beta_0 = E(Y \mid X=0). \tag{12.16}$$

Furthermore, if $x_1, x_2 \in \mathbb{R}$ and $x_2 > x_1$, then,

$$\beta_1 = \frac{1}{x_2 - x_1} \cdot [E(Y \mid X=x_2) - E(Y \mid X=x_1)] \tag{12.17}$$

[see Exercise 7.2 and Eq. (12.12)]. Equation (12.17) yields

$$\beta_1 = E(Y \mid X=x_2) - E(Y \mid X=x_1), \quad \text{if } x_2 - x_1 = 1. \tag{12.18}$$

This justifies calling β_0 the *intercept* and β_1 the *slope* of $E(Y \mid X)$, respectively (see Fig. 7.3). Note that these equations also apply if $P(X=0) = P(X=x_1) = P(X=x_2) = 0$. They even apply if $0, x_1, x_2 \notin X(\Omega)$.   ◁

Example 12.15 [Dichotomous regressor]   If X is dichotomous with values 0 and 1 (see Example 5.10), then there is always a version $E(Y \mid X) \in \mathscr{E}(Y \mid X)$ such that

$$E(Y \mid X) = \beta_0 + \beta_1 X \tag{12.19}$$

with

$$\beta_0 = E(Y \mid X=0), \tag{12.20}$$

and

$$\beta_1 = E(Y \mid X=1) - E(Y \mid X=0) \tag{12.21}$$

(for a proof, see Th. 12.37).   ◁

Example 12.16 [Joe and Ann with self-selection – continued]   In Table 12.1, X is dichotomous. According to Example 12.15, this implies that there is a linear parameterization of $E(Y \mid X)$ in X satisfying Equation (12.19). In this example,

$$E(Y \mid X) = .6 - .18\, X, \tag{12.22}$$

and the function $g\colon \mathbb{R} \to \mathbb{R}$ defined by $g(x) = .6 - .18\, x, x \in \mathbb{R}$, is a linear parameterization of $E(Y \mid X)$ in X.

The intercept is most easily obtained via Equation (12.20):

$$\beta_0 = E(Y \mid X=0) = P(Y=1 \mid X=0)$$
$$= \frac{P(Y=1, X=0)}{P(X=0)} = \frac{.336 + .024}{.144 + .336 + .096 + .024} = .6. \tag{12.23}$$

The slope is obtained via Equation (12.21):

$$\beta_1 = E(Y \mid X=1) - E(Y \mid X=0) = P(Y=1 \mid X=1) - P(Y=1 \mid X=0)$$
$$= \frac{P(Y=1, X=1)}{P(X=1)} - P(Y=1 \mid X=0) \tag{12.24}$$
$$= \frac{.016 + .152}{.004 + .016 + .228 + .152} - .6 = .42 - .6 = -.18.$$

◁

Example 12.17 [Dichotomous regressor – continued] Continue Example 12.15 and define the random variable $Z: (\Omega, \mathcal{A}, P) \to (\mathbb{R}, \mathcal{B})$ by $Z := 2X - 1$. Then Z is dichotomous with values -1 and 1, and

$$\forall \omega \in \Omega: \quad X(\omega) = 0 \Leftrightarrow Z(\omega) = -1 \quad \text{and} \quad X(\omega) = 1 \Leftrightarrow Z(\omega) = 1.$$

Note that $\sigma(Z) = \sigma(X)$ holds for the σ-algebras generated by X and Z. Because $X = \frac{1}{2}(Z+1)$,

$$E(Y \mid X) = \beta_0 + \beta_1 X = \beta_0 + \frac{\beta_1}{2}(Z+1) = \beta_0 + \frac{\beta_1}{2} + \frac{\beta_1}{2}Z = E(Y \mid Z). \tag{12.25}$$

the function $g^*: \mathbb{R} \to \mathbb{R}$ defined by $g^*(z) = \alpha_0 + \alpha_1 z$, $z \in \mathbb{R}$, is a linear parameterization of $E(Y \mid X)$ in Z, where

$$\alpha_0 = \beta_0 + \frac{\beta_1}{2} = E(Y \mid X=0) + \frac{E(Y \mid X=1) - E(Y \mid X=0)}{2}$$
$$= \frac{E(Y \mid X=1) + E(Y \mid X=0)}{2} \tag{12.26}$$

and

$$\alpha_1 = \frac{\beta_1}{2} = \frac{E(Y \mid X=1) - E(Y \mid X=0)}{2}. \tag{12.27}$$

Note that

$$\{X=1\} = \left\{\frac{1}{2}(Z+1) = 1\right\} = \{Z=1\},$$
$$\{X=0\} = \left\{\frac{1}{2}(Z+1) = 0\right\} = \{Z=-1\}. \tag{12.28}$$

Because X and Z are dichotomous with positive probabilities for both of their values, Equation (12.28) and $E(Y \mid X=x) = E(Y \mid \{X=x\})$ [see Eq. (9.6)] imply

$$\alpha_0 = \frac{E(Y \mid Z=1) + E(Y \mid Z=-1)}{2}, \tag{12.29}$$

$$\alpha_1 = \frac{E(Y \mid Z=1) - E(Y \mid Z=-1)}{2}. \tag{12.30}$$

Comparing Equations (12.26) and (12.27) to Equations (12.20) and (12.21) shows that the meaning of the regression coefficients depends on the choice of the random variable, here X or Z, with respect to which we consider a linear parameterization. ◁

Example 12.18 [Joe and Ann with self-selection – continued]   Using the results of Example 12.16 as well as Equations (12.25) to (12.27) yields

$$E(Y \mid X) = E(Y \mid Z) = \frac{.42 + .6}{2} + \frac{.42 - .6}{2} Z = .51 - .09 Z. \tag{12.31}$$

Note again that $E(Y \mid X)$ and $E(Y \mid Z)$ are only different notations for a version of $\mathscr{E}(Y \mid \mathscr{C})$, where $\mathscr{C} = \sigma(X) = \sigma(Z)$ and that $g \colon \mathbb{R} \to \mathbb{R}$ defined by $g(x) = .6 - .18\,x$, $x \in \mathbb{R}$, and $g^* \colon \mathbb{R} \to \mathbb{R}$ defined by $g^*(z) = .51 - .09\,z$, $z \in \mathbb{R}$, are two different linear parameterizations of one and the same version $E(Y \mid \mathscr{C}) \in \mathscr{E}(Y \mid \mathscr{C})$, one is linear in X, the other one linear in Z. ◁

Example 12.19 [Quadratic function]   Let $X_1 \colon (\Omega, \mathscr{A}, P) \to (\mathbb{R}, \mathscr{B})$ be a real-valued random variable, let $X_2 := X_1^2$, $X := (X_1, X_2)$, and assume that there is a version $E(Y \mid X) \in \mathscr{E}(Y \mid X)$ with

$$E(Y \mid X) = \beta_0 + \beta_1 X_1 + \beta_2 X_1^2. \tag{12.32}$$

Then the function $g \colon \mathbb{R}^2 \to \mathbb{R}$ defined by

$$g(x) = \beta_0 + \beta_1 x_1 + \beta_2 x_2, \quad x = (x_1, x_2) \in \mathbb{R}^2, \tag{12.33}$$

is a linear parameterization of $E(Y \mid X)$ in $X = (X_1, X_1^2)$. ◁

Example 12.20 [Logarithmic function]   Consider $Z \colon (\Omega, \mathscr{A}, P) \to (\mathbb{R}, \mathscr{B})$, a real-valued and positive random variable; define $X := \ln Z$; and assume that there is a version $E(Y \mid X) \in \mathscr{E}(Y \mid X)$ with

$$E(Y \mid X) = \beta_0 + \beta_1 \ln Z = \beta_0 + \beta_1 X. \tag{12.34}$$

Then the function $g \colon \mathbb{R} \to \mathbb{R}$ defined by

$$g(x) = \beta_0 + \beta_1 x, \quad x \in \mathbb{R}, \tag{12.35}$$

is a linear parameterization of $E(Y \mid X)$ in $X = \ln Z$. Note that $\mathscr{E}(Y \mid Z) = \mathscr{E}(Y \mid X)$, because $\sigma(X) = \sigma(Z)$ (see Exercise 12.1). ◁

Example 12.21 [Two regressors] If $X_i: (\Omega, \mathcal{A}, P) \to (\mathbb{R}, \mathcal{B})$, $i = 1, 2$, are univariate real-valued random variables, $X := (X_1, X_2)$, and there is a version $E(Y \mid X) \in \mathcal{E}(Y \mid X)$ with

$$E(Y \mid X) = \beta_0 + \beta_1 X_1 + \beta_2 X_2, \tag{12.36}$$

then the function $g: \mathbb{R}^2 \to \mathbb{R}$ defined by

$$g(x) = \beta_0 + \beta_1 x_1 + \beta_2 x_2, \quad x = (x_1, x_2) \in \mathbb{R}^2, \tag{12.37}$$

is a linear parameterization of $E(Y \mid X)$ in $X = (X_1, X_2)$. ◁

Example 12.22 [Two regressors and their product] Let $X_i: (\Omega, \mathcal{A}, P) \to (\mathbb{R}, \mathcal{B})$, $i = 1, 2$, be univariate real-valued random variables; define $X_3 := X_1 \cdot X_2$ and $X := (X_1, X_2, X_3)$; and assume that there is a version $E(Y \mid X) \in \mathcal{E}(Y \mid X)$ with

$$E(Y \mid X) = \beta_0 + \beta_1 X_1 + \beta_2 X_2 + \beta_3 X_1 \cdot X_2. \tag{12.38}$$

Then the function $g: \mathbb{R}^3 \to \mathbb{R}$ defined by

$$g(x) = \beta_0 + \beta_1 x_1 + \beta_2 x_2 + \beta_3 x_3, \quad \forall\, x = (x_1, x_2, x_3) \in \mathbb{R}^3, \tag{12.39}$$

is a linear parameterization of $E(Y \mid X)$ in $X = (X_1, X_2, X_1 \cdot X_2)$. ◁

Remark 12.23 [Dichotomous random variables] Note that the linear parameterization of $E(Y \mid X)$ specified by Equations (12.38) and (12.39) always exists, if X_1 and X_2 are dichotomous with values 0 and 1 (see Example 5.10 and Exercise 12.2). ◁

Example 12.24 [Joe and Ann with self-selection – continued] Consider the random variables X and $1_{U=Joe}$ specified in Table 12.1, and define $Z := (Z_1, Z_2, Z_3) := (X, 1_{U=Joe}, X \cdot 1_{U=Joe})$. Then, according to Example 12.22 and Remark 12.23,

$$E(Y \mid Z) = \beta_0 + \beta_1 X + \beta_2 1_{U=Joe} + \beta_3 X \cdot 1_{U=Joe}, \tag{12.40}$$

and the function $g: \mathbb{R}^3 \to \mathbb{R}$ defined by

$$g(z) = .2 + .2\, z_1 + .5\, z_2 - .1\, z_3, \quad \forall\, z = (z_1, z_2, z_3) \in \mathbb{R}^3,$$

is a linear parameterization of $E(Y \mid Z)$ in $Z = (X, 1_{U=Joe}, X \cdot 1_{U=Joe})$.

The coefficients β_0 to β_3 in Equation (12.40) can also be obtained as follows (cf. Exercise 12.2). Table 12.1 and Equation (12.40) yield:

$$\beta_0 = E(Y \mid Z_1=0, Z_2=0, Z_3=0) = P(Y=1 \mid X=0, U=Ann) = .2,$$
$$\beta_0 + \beta_1 = E(Y \mid Z_1=1, Z_2=0, Z_3=0) = P(Y=1 \mid X=1, U=Ann) = .4,$$
$$\beta_0 + \beta_2 = E(Y \mid Z_1=0, Z_2=1, Z_3=0) = P(Y=1 \mid X=0, U=Joe) = .7,$$
$$\beta_0 + \beta_1 + \beta_2 + \beta_3 = E(Y \mid Z_1=1, Z_2=1, Z_3=1) = P(Y=1 \mid X=1, U=Joe) = .8.$$

Solving these equations for the four coefficients and inserting them into Equation (12.40) yield

$$E(Y \mid Z) = .2 + .2\,X + .5\,1_{U=Joe} - .1\,X \cdot 1_{U=Joe}. \tag{12.41}$$

In this example, $\sigma(Z) = \sigma(X, U)$ and $Z(\Omega) = \{0, 1\}^3$. According to Remark 10.12, this implies $E(Y \mid Z) = E(Y \mid X, U)$ and that the function g is also a linear parameterization of $E(Y \mid X, U)$ in $Z = (X, 1_{U=Joe}, X \cdot 1_{U=Joe})$.   ◁

Remark 12.25 [Generalizing the examples]   Generalizing the Examples 12.19 to 12.24, let $Z: (\Omega, \mathcal{A}, P) \to (\Omega'_Z, \mathcal{A}'_Z)$ be a (univariate or multivariate) random variable. For $i = 1, \dots, n$, let $h_i: (\Omega'_Z, \mathcal{A}'_Z) \to (\mathbb{R}, \mathcal{B})$ be measurable functions, define $X_i := h_i(Z)$, and $X := (X_1, \dots, X_n)$. If there is a version $E(Y \mid X) \in \mathcal{E}(Y \mid X)$ such that

$$E(Y \mid X) = \beta_0 + \sum_{i=1}^{n} \beta_i\, h_i(Z), \tag{12.42}$$

then $g: \mathbb{R}^n \to \mathbb{R}$ defined by

$$g(x) = \beta_0 + \sum_{i=1}^{n} \beta_i\, x_i, \quad \forall\, x = (x_1, \dots, x_n) \in \mathbb{R}^n, \tag{12.43}$$

is a linear parameterization of $E(Y \mid X)$ in $X = (h_1(Z), \dots, h_n(Z))$. Remember that $\sigma(X) \subset \sigma(Z)$, but note that $\sigma(X) = \sigma(Z)$ does not necessarily hold. If $\sigma(X) \neq \sigma(Z)$, then there is not necessarily a version $E(Y \mid Z) \in \mathcal{E}(Y \mid Z)$ with $E(Y \mid Z) = E(Y \mid X)$. However, if we assume that there is a version $E(Y \mid Z) \in \mathcal{E}(Y \mid Z)$ with $E(Y \mid Z) = E(Y \mid X)$, then

$$E(Y \mid Z) = \beta_0 + \sum_{i=1}^{n} \beta_i\, h_i(Z). \tag{12.44}$$

In this case, the function g is also a linear parameterization of $E(Y \mid Z)$ in $X = (h_1(Z), \dots, h_n(Z))$.   ◁

12.4   Linear quasi-regression

In the following corollary, $Q_{lin}(Y \mid X_1, \dots, X_n)$ denotes the function that has been introduced in Definition 7.28. This corollary immediately follows from Theorem 10.26 and Definition 12.5.

Corollary 12.26 [Linear regression and linear quasi-regression]
Let the assumptions 12.4 hold, and suppose that $E(Y^2)$, $E(X_i^2) < \infty$, $i = 1, \dots, n$. *If there is a version* $E(Y \mid \mathcal{C}) \in \mathcal{E}(Y \mid \mathcal{C})$ *with* $E(Y \mid \mathcal{C}) = \beta_0 + \sum_{i=1}^{n} \beta_i\, X_i$, *where* $\beta_0, \beta_1, \dots, \beta_n \in \mathbb{R}$, *then,*

$$E(Y \mid \mathcal{C}) = Q_{lin}(Y \mid X_1, \dots, X_n) = \beta_0 + \sum_{i=1}^{n} \beta_i\, X_i. \tag{12.45}$$

Hence, if $E(Y \mid \mathcal{C}) = \beta_0 + \sum_{i=1}^{n} \beta_i X_i$, then $E(Y \mid \mathcal{C})$ and $Q_{lin}(Y \mid X_1, \ldots, X_n) = f(X)$ are identical, where $f(X)$ is the composition of X and the linear quasi-regression $f: \mathbb{R}^n \to \mathbb{R}$ (see Def. 7.28).

Remark 12.27 [Dichotomous regressor]   If X is dichotomous with values 0 and 1 (see Example 5.10), then $E(Y \mid X) = Q_{lin}(Y \mid X)$ (see Example 12.15 and Th. 12.37). If $P(X=x_i) > 0$ for at least three different $x_i \in \mathbb{R}$, then it is not necessarily true that $Q_{lin}(Y \mid X) \in \mathcal{E}(Y \mid X)$. If $Q_{lin}(Y \mid X) \notin \mathcal{E}(Y \mid X)$, then there are no $\beta_0, \beta_1 \in \mathbb{R}$ such that the function g defined by $g(x) = \beta_0 + \beta_1 x, x \in \mathbb{R}$, is a linear parameterization in X of a version $E(Y \mid X) \in \mathcal{E}(Y \mid X)$ (see Example 12.41).   ◁

Remark 12.28 [Unbounded regressor, dichotomous regressand]   Suppose that X and Y are real-valued random variables on $(\Omega, \mathcal{A}, P)$, and that Y is dichotomous with values 0 and 1. Because $P(0 \leq Y \leq 1) = 1$, Rules (ii) and (iii) of Box 10.3 imply

$$0 \underset{P}{\leq} E(Y \mid X) = P(Y=1 \mid X) \underset{P}{\leq} 1. \tag{12.46}$$

Suppose that the regressor X is *not P-almost surely bounded*, that is, suppose

$$\forall c \in \mathbb{R}, \ c > 0: \ P(X < -c) + P(X > c) > 0. \tag{12.47}$$

Then there is no linear parameterization of $E(Y \mid X)$ in X with slope $\beta_1 \neq 0$ (see Exercise 12.3 and cf. ch. 13). Note that the premise (12.47) holds, for example, if X has a normal distribution.   ◁

Example 12.29 [No treatment for Joe – continued]   In the example presented in Table 9.2,

$$E(Y \mid X) = \beta_0 + \beta_1 X = .6 - .2 X \tag{12.48}$$

(see Example 12.15), and

$$g(x) = \beta_0 + \beta_1 x = .6 - .2 x, \quad \forall x \in \mathbb{R}, \tag{12.49}$$

defines a linear parameterization $g: \mathbb{R} \to \mathbb{R}$ of $E(Y \mid X)$ in X. Hence, according to Remark 12.11, we may define

$$E(Y \mid X=x) = g(x) = .6 - .2 x, \quad \forall x \in \mathbb{R}, \tag{12.50}$$

(see Def. 10.33). For $x=0$, Equation (12.50) yields $E(Y \mid X=0) = .6$, and it yields $E(Y \mid X=1) = .4$ for $x=1$. Note that the definition of the conditional expectation values $E(Y \mid X=x)$ for $x \in \mathbb{R} \setminus \{0, 1\}$ via Equation (12.50) is arbitrary, because $P(X \in \mathbb{R} \setminus \{0, 1\}) = 0$. Using any other factorization of $E(Y \mid X)$ for the definition of the conditional expectation values $E(Y \mid X=x)$ for $x \in \mathbb{R} \setminus \{0, 1\}$ would do as well.

Remark 12.27 and Definition 7.28 imply that the function $MSE: \mathbb{R}^2 \to [0, \infty[$ defined by

$$MSE(a_0, a_1) = E([Y - (a_0 + a_1 X)]^2), \quad \forall (a_0, a_1) \in \mathbb{R}^2, \tag{12.51}$$

has its minimum for $(a_0, a_1) = (.6, -.2)$. Hence, in this example,

$$E(Y \mid X) = Q_{lin}(Y \mid X) = .6 - .2\,X,$$

and the function $g: \mathbb{R} \to \mathbb{R}$ defined by Equation (12.49) is a linear parameterization of $E(Y \mid X)$ in X. ◁

12.5   Uniqueness and identification of regression coefficients

In Example 12.9, we showed that a linear parameterization is not uniquely defined unless additional assumptions hold. Such assumptions are specified in Corollary 12.31, which uses the following notation and general assumptions:

Notation and assumptions 12.30
Let the assumptions 12.4 hold. Furthermore, $x := [X_1, \ldots, X_n]'$ is the column vector of $X = (X_1, \ldots, X_n)$, $\mu := [E(X_1), \ldots, E(X_n)]'$ the column vector of the expectations of $X_1, \ldots, X_n$, and $\beta := [\beta_1, \ldots, \beta_n]'$ a column vector of n real numbers.

Assuming finite second moments of Y and $X_1, \ldots, X_n$,

$$\Sigma_{xx} = E\big([x - \mu]\,[x - \mu]'\big) = \begin{bmatrix} \sigma^2_{X_1} & \sigma_{X_1 X_2} & \cdots & \sigma_{X_1 X_n} \\ \sigma_{X_2 X_1} & \sigma^2_{X_2} & \cdots & \sigma_{X_2 X_n} \\ \vdots & \vdots & \ddots & \vdots \\ \sigma_{X_n X_1} & \sigma_{X_n X_2} & \cdots & \sigma^2_{X_n} \end{bmatrix}$$

denotes the variance-covariance matrix of $X = (X_1, \ldots, X_n)$ (see section 7.4.3). Furthermore,

$$\Sigma_{yx} = [\sigma_{YX_1}, \ldots, \sigma_{YX_n}]$$

denotes the row vector of the covariances $Cov\,(Y, X_i) = \sigma_{YX_i}$, $i = 1, \ldots, n$, and $\Sigma_{xy} := \Sigma'_{yx}$ the column vector of these covariances. Remember that the notation $X = (X_1, \ldots, X_n)$ refers to an *n-variate random variable*, whereas $x = [X_1, \ldots, X_n]$ denotes the *row vector* of the random variables $X_1, \ldots, X_n$.

The following corollary immediately follows from Theorem 7.30 and Corollary 12.26. It shows how to compute (identify) the regression coefficients of a linear parameterization of $E(Y \mid \mathscr{C})$ in X, and it specifies sufficient conditions under which such a linear parameterization of $E(Y \mid \mathscr{C})$ is uniquely defined.

Corollary 12.31 [Identification of parameters]
Let the assumptions 12.30 hold. If there is a version $E(Y \mid \mathscr{C}) \in \mathscr{E}\,(Y \mid \mathscr{C})$ with

$$E(Y \mid \mathscr{C}) = \beta_0 + \beta'x = \beta_0 + \sum_{i=1}^{n} \beta_i\, X_i, \qquad (12.52)$$

then,

$$\beta_0 = E(Y) - \beta' \mu. \tag{12.53}$$

If, in addition, Y and $X_1, \ldots, X_n$ have finite second moments and the inverse Σ_{xx}^{-1} exists, then,

$$\beta = \Sigma_{xx}^{-1} \Sigma_{xy}, \tag{12.54}$$

and the linear parameterization $g \colon \mathbb{R}^n \to \mathbb{R}$ of $E(Y \mid \mathscr{C})$ in X with

$$g(x) = \beta_0 + \sum_{i=1}^{n} \beta_i x_i, \quad \forall x = (x_1, \ldots, x_n) \in \mathbb{R}^n, \tag{12.55}$$

is uniquely defined.

12.6 Linear regression

As already mentioned in the Preface, much empirical research uses some kind of regression in order to investigate how the conditional expectation values of one random variable depend on the values of one or more other random variables. In Definition 10.25, we introduced the concept of a regression as a special case of a factorization of a conditional expectation in which the regressor X is numerical. In Definition 12.5, we defined the concept of a parameterization of $E(Y \mid \mathscr{C})$ that is linear in X. Such parameterizations of $E(Y \mid \mathscr{C})$ are not necessarily uniquely defined. If there is a parameterization g of $E(Y \mid \mathscr{C})$ that is linear in X and satisfies Equation (12.55), then we call it *the linear regression of Y on X*. According to Corollary 12.31, the linear regression is uniquely defined.

Definition 12.32 [Linear regression]
Let the assumptions 12.30 hold, and suppose that there is a version $E(Y \mid \mathscr{C}) \in \mathscr{E}(Y \mid \mathscr{C})$ such that Equation (12.52) holds. Furthermore, assume that Y and $X_1, \ldots, X_n$ have finite second moments and that the inverse Σ_{xx}^{-1} exists. Then the function $g \colon \mathbb{R}^n \to \mathbb{R}$ defined by Equation (12.55) is called the linear regression of Y on X *(with respect to P).*

Figure 12.1 shows the conditional expectation $E(Y \mid X)$ as the composition of X and the linear regression g. Hence, while the conditional expectation $E(Y \mid X)$ is a function with domain Ω and codomain $\mathbb{R}$, the linear regression g is a function with domain $\mathbb{R}^n$ and codomain $\mathbb{R}$.

Remark 12.33 [Simple and multiple linear regression] If $n \geq 2$, then the linear regression of Y on X is also called the *multiple linear regression* of Y on X. The coefficients $\beta_0, \beta_1, \ldots, \beta_n$ are called *regression coefficients*, and β_0 the *intercept*. If $n = 1$, then a linear regression of Y on X is also called a *simple linear regression* of Y on X with *slope* β_1.   ◁

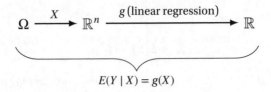

$$E(Y \mid X) = g(X)$$

Figure 12.1 $E(Y \mid X)$ as the composition of X and the linear regression g.

Remark 12.34 [Simple regression as a special case] If $n = 1$ and we define $X := X_1$, then Equation (12.52) can be written as:

$$E(Y \mid \mathscr{C}) = \beta_0 + \beta_1 X. \tag{12.56}$$

If Equation (12.56) holds, then there is a version $E(Y \mid X) \in \mathscr{E}(Y \mid X)$ such that

$$E(Y \mid \mathscr{C}) = E(Y \mid X) = \beta_0 + \beta_1 X = Q_{lin}(Y \mid X). \tag{12.57}$$

Therefore, we obtain the same results for the regression coefficients that have already been described in Theorem 7.14. In particular, Equation (12.53) yields

$$\beta_0 = E(Y) - \beta_1 E(X), \tag{12.58}$$

and (12.54) implies

$$\beta_1 = \frac{Cov(X, Y)}{Var(X)}. \tag{12.59}$$

If $n = 1$ and $X = X_1$, then $\Sigma_{xx}^{-1} = \left[\frac{1}{Var(X)} \right]$ and the existence of the inverse Σ_{xx}^{-1} is equivalent to $Var(X) > 0$ (see Th. 7.14 and Rem. 7.25).   ◁

In the following theorem, we consider the special case that $Z = (Y, X_1, \ldots, X_n)$ has an $(n + 1)$-variate normal distribution (see Def. 8.36).

Theorem 12.35 [Linear parameterization and normal distribution]
Let the assumptions 12.30 hold and let $Z := (Y, X_1, \ldots, X_n)$ be an $(n + 1)$-variate real-valued random variable on $(\Omega, \mathscr{A}, P)$ with $Z \sim \mathcal{N}_{\mu_z, \Sigma_{zz}}$. Furthermore, assume that the inverse Σ_{xx}^{-1} exists. Then, there is a version $E(Y \mid X) \in \mathscr{E}(Y \mid X)$ with

$$E(Y \mid X) = \beta_0 + \boldsymbol{\beta}' x$$

such that Equations (12.53) and (12.54) hold for β_0 and $\boldsymbol{\beta}$, respectively.

Existence is proved by Rao (1973, Eq. 8a.2.16), and Corollary 12.31 implies that Equations (12.53) and (12.54) hold for β_0 and $\boldsymbol{\beta} = [\beta_1, \ldots, \beta_n]'$.

12.7   Parameterizations of a discrete conditional expectation

Now we consider two parameterizations of a discrete conditional expectation $E(Y \mid Z)$. In both cases, there are only a finite number of values of Z. In the first case, the possible values of Z are denoted $z_1, \ldots, z_n$. In the second case, the notation is changed for didactic reasons.

Theorem 12.36 [Means as coefficients]
Let the assumptions 12.30 hold, assume that $Z: (\Omega, \mathscr{A}, P) \to (\Omega'_Z, \mathscr{A}'_Z)$ is a discrete random variable, and $P(Z \in \{z_1, \ldots, z_n\}) = 1$ with $P(Z=z_i) > 0$, for all $i = 1, \ldots, n$. Furthermore, define $X_i := 1_{Z=z_i}$, $i = 1, \ldots, n$, and $X := (X_1, \ldots, X_n)$. Then there is a version $E(Y \mid Z) \in \mathscr{E}(Y \mid Z)$ with

$$E(Y \mid Z) = \sum_{i=1}^{n} \beta_i \, 1_{Z=z_i} = \sum_{i=1}^{n} \beta_i \, X_i, \qquad (12.60)$$

where

$$\beta_i = E(Y \mid Z=z_i), \quad \forall \, i = 1, \ldots, n. \qquad (12.61)$$

The function $g: \mathbb{R}^n \to \mathbb{R}$ defined by

$$g(x) = 0 + \sum_{i=1}^{n} \beta_i \, x_i, \quad \forall \, x = (x_1, \ldots, x_n) \in \mathbb{R}^n, \qquad (12.62)$$

is a linear parameterization of $E(Y \mid Z)$ in $X = (1_{Z=z_1}, \ldots, 1_{Z=z_n})$. If $Z(\Omega) = \{z_1, \ldots, z_n\}$, then $V = V^*$ for all $V, V^* \in \mathscr{E}(Y \mid Z)$.

(Proof p. 388)

Hence, in this parameterization, the coefficients of Equation (12.7) are

$$\beta_0 = 0, \quad \beta_1 = E(Y \mid Z=z_1), \quad \ldots, \quad \beta_n = E(Y \mid Z=z_n), \qquad (12.63)$$

that is, in this parameterization, the coefficients β_i, $i = 1, \ldots, n$, are the $(Z=z_i)$-conditional expectation values of Y.

In the following theorem, we present another linear parameterization of $E(Y \mid Z)$, generalizing Example 12.15. For convenience, the possible values of Z are now denoted $z_0, z_1, \ldots, z_n$. Aside from this change in the notation, the assumptions in Theorems 12.36 and 12.37 are identical.

Theorem 12.37 [Differences between means as coefficients]
Let the assumptions 12.4 hold, and assume that $Z: (\Omega, \mathscr{A}, P) \to (\Omega'_Z, \mathscr{A}'_Z)$ is discrete, $P(Z \in \{z_0, z_1, \ldots, z_n\}) = 1$ with $P(Z=z_i) > 0$, for all $i = 0, 1, \ldots, n$. Furthermore, define

$X_i := 1_{Z=z_i}$, $i = 1, \dots, n$, and $X := (X_1, \dots, X_n)$. *Then there is a version* $E(Y \mid Z) \in$ $\mathscr{E}(Y \mid Z)$ *with*

$$E(Y \mid Z) = \beta_0 + \sum_{i=1}^{n} \beta_i\, 1_{Z=z_i} = \beta_0 + \sum_{i=1}^{n} \beta_i\, X_i, \tag{12.64}$$

where

$$\beta_0 = E(Y \mid Z=z_0) \tag{12.65}$$

and

$$\beta_i = E(Y \mid Z=z_i) - E(Y \mid Z=z_0), \quad \forall\, i = 1, \dots, n. \tag{12.66}$$

The function $g\colon \mathbb{R}^n \to \mathbb{R}$ *defined by*

$$g(x) = \beta_0 + \sum_{i=1}^{n} \beta_i\, x_i, \quad \forall\, x = (x_1, \dots, x_n) \in \mathbb{R}^n, \tag{12.67}$$

is a linear parameterization of $E(Y \mid Z)$ *in* $X = (1_{Z=z_1}, \dots, 1_{Z=z_n})$. *If* $Z(\Omega) = \{z_0, z_1, \dots, z_n\}$, *then* $V = V^*$ *for all* $V, V^* \in \mathscr{E}(Y \mid Z)$.

(Proof p. 388)

Hence, in contrast to Equation (12.63), in this parameterization the coefficients are

$$\begin{aligned} \beta_0 &= E(Y \mid Z=z_0), \\ \beta_1 &= E(Y \mid Z=z_1) - E(Y \mid Z=z_0), \\ &\ \ \vdots \\ \beta_n &= E(Y \mid Z=z_n) - E(Y \mid Z=z_0). \end{aligned} \tag{12.68}$$

Now we present a lemma on the covariance matrix of the indicators $1_{Z=z_i}$ for the values $z_0, z_1, \dots, z_n$ of a discrete random variable Z. In particular, this lemma helps to prove that the covariance matrix of the indicators $1_{Z=z_1}, \dots, 1_{Z=z_n}$ is regular. This implies that the inverse of this covariance matrix exists so that Corollary 12.31 can be applied.

Lemma 12.38 [Covariance matrix of indicators]
Let $Z\colon (\Omega, \mathcal{A}, P) \to (\Omega_Z', \mathcal{A}_Z')$ *be discrete with* $P(Z \in \{z_0, z_1, \dots, z_n\}) = 1$ *and* $p_i := P(Z=z_i) > 0$, *for all* $i = 0, 1, \dots, n$. *Then the following two propositions hold:*

(i) *The second moments of the indicators* $1_{Z=z_1}, \ldots, 1_{Z=z_n}$ *are finite, and for all* $i, j = 0, 1, \ldots, n$,

$$\sigma_{ij} := Cov(1_{Z=z_i}, 1_{Z=z_j}) = \begin{cases} p_i \cdot (1 - p_i), & if \ i = j \\ -p_i \cdot p_j, & if \ i \neq j. \end{cases} \tag{12.69}$$

(ii) *For all* $i = 1, \ldots, n$,

$$\sigma_{ii} > \sum_{j=1, j \neq i}^{n} |\sigma_{ij}|. \tag{12.70}$$

(Proof p. 389)

Remark 12.39 [Strict diagonal dominance]   Proposition (ii) of Lemma 12.38 implies that the covariance matrix of the indicators $1_{Z=z_1}, \ldots, 1_{Z=z_n}$ is *strictly diagonally dominant*, that is, it satisfies

$$|\sigma_{ii}| > \sum_{j=1, j \neq i}^{n} |\sigma_{ij}|, \quad \forall \, i = 1, \ldots, n. \tag{12.71}$$

◁

Remark 12.40 [Regularity of the covariance matrix of indicators]   According to Corollary 5.6.17 of Horn and Johnson (1991), (12.71) implies that the covariance matrix of the indicators $1_{Z=z_1}, \ldots, 1_{Z=z_n}$ is regular. In contrast, the covariance matrix of the indicators $1_{Z=z_0}, 1_{Z=z_1}, \ldots, 1_{Z=z_n}$ *is not* strictly diagonally dominant, and it is *not regular*. The reason is that $1_{Z=z_0} = 1 - \sum_{i=1}^{n} 1_{Z=z_i}$, that is, $1_{Z=z_0}$ is a linear combination of $1_{Z=z_1}, \ldots, 1_{Z=z_n}$, which implies that the covariance matrix of $1_{Z=z_0}, 1_{Z=z_1}, \ldots, 1_{Z=z_n}$ is not regular (see Exercise 12.4). ◁

Example 12.41 [Tom, Jim, and Kate – continued]   Table 5.1 displays an example in which the treatment variable X has three values. The conditional expectation $E(Y \mid X) = P(Y = 1 \mid X)$ has three values as well, namely the following conditional probabilities:

$$P(Y = 1 \mid X = 0) = \frac{(10 + 15 + 8) / 99}{(10 + 10 + 5 + 15 + 12 + 8) / 99} = .55,$$

$$P(Y = 1 \mid X = 1) = \frac{(6 + 5 + 3) / 99}{(2 + 6 + 3 + 5 + 5 + 3) / 99} = .58\overline{3},$$

$$P(Y = 1 \mid X = 2) = \frac{(4 + 3 + 1) / 99}{(1 + 4 + 2 + 3 + 4 + 1) / 99} = .5\overline{3}.$$

There are several linear parameterizations of the conditional expectation $E(Y \mid X)$. For example, we can use the linear parameterization in (X, X^2) specified in Equation (12.33),

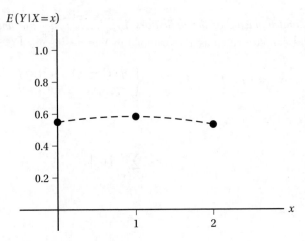

Figure 12.2   A regression with a parameterization that is linear in (X, X^2) but not in X.

which yields

$$P(Y=1 \mid X=0) = \beta_0,$$
$$P(Y=1 \mid X=1) = \beta_0 + \beta_1 + \beta_2,$$
$$P(Y=1 \mid X=2) = \beta_0 + 2 \cdot \beta_1 + 4 \cdot \beta_2.$$

Solving this equation system results in

$$\beta_0 = P(Y=1 \mid X=0) = .55,$$
$$\beta_1 = \frac{1}{2}\Big[- P(Y=1 \mid X=2) + P(Y=1 \mid X=1) - 3 \cdot P(Y=1 \mid X=0)\Big] = -.8,$$
$$\beta_2 = \frac{1}{2}\Big[P(Y=1 \mid X=2) - 2 \cdot P(Y=1 \mid X=1) + P(Y=1 \mid X=0)\Big] = -0.041\overline{6}.$$

In this example, $\beta_2 \neq 0$, and there is no linear parameterization of $E(Y \mid X)$ in X. This is illustrated in Figure 12.2 (see Exercise 12.5). Note that X and the conditional expectation $E(Y \mid X) = P(Y=1 \mid X)$ have only three different values.

We could also use the linear parameterization specified in Equation (12.62). In this case, according to (12.63), the coefficients are

$$\beta_0 = 0, \quad \beta_1 = P(Y=1 \mid X=0) = .55, \quad \beta_2 = P(Y=1 \mid X=1) = .58\overline{3},$$
$$\beta_3 = P(Y=1 \mid X=2) = .5\overline{3}.$$

If we use the linear parameterization specified in Equation (12.67), then, according to (12.68), the coefficients are

$$\beta_0 = P(Y=1 \mid X=0) = .55,$$
$$\beta_1 = P(Y=1 \mid X=1) - P(Y=1 \mid X=0) = .0\overline{3},$$
$$\beta_2 = P(Y=1 \mid X=2) - P(Y=1 \mid X=0) = -.01\overline{6}.$$

◁

Remark 12.42 [*n*-valued regressor] If X is a discrete random variable with n different values, then there is a version $P(Y=1 \mid X)$ with a linear parameterization in $(X, X^2, \ldots, X^{n-1})$, that is, $P(Y=1 \mid X)$ can be written as a polynomial in X of degree $n - 1$ (see Süli & Mayers, 2003, Theorem 6.1, Lagrange's interpolation theorem).   ◁

12.8   Invariance of regression coefficients

Remark 12.43 [Simple versus multiple regression] Let the assumptions 12.4 hold with $n = 2$, and let the second moments of Y, X_1, and X_2 be finite. Assume that the inverse of the covariance matrix of (X_1, X_2) exists and that there is a version $E(Y \mid X_1, X_2) \in \mathscr{E}(Y \mid X_1, X_2)$ such that

$$E(Y \mid X_1, X_2) = \beta_0 + \beta_1 X_1 + \beta_2 X_2. \tag{12.72}$$

Furthermore, assume that there is a version $E(Y \mid X_1) \in \mathscr{E}(Y \mid X_1)$ with

$$E(Y \mid X_1) = \alpha_0 + \alpha_1 X_1. \tag{12.73}$$

Then, $\alpha_1 = \beta_1$ does not necessarily hold.   ◁

In the following theorem, we formulate a sufficient condition for $\alpha_1 = \beta_1$ and generalize it to the *n*-variate case, where $n \geq 2$. That is, instead of $E(Y \mid X_1)$, we consider $E(Y \mid X_1, \ldots, X_m)$ and replace $E(Y \mid X_1, X_2)$ by $E(Y \mid X_1, \ldots, X_n)$, presuming $m < n$.

Theorem 12.44 [Invariance of regression coefficients]
Let the assumptions 12.4 hold, let $m < n$, and suppose that there is a version
$E(Y \mid X_1, \ldots, X_n) \in \mathscr{E}(Y \mid X_1, \ldots, X_n)$ such that

$$E(Y \mid X_1, \ldots, X_n) = \beta_0 + \sum_{i=1}^{m} \beta_i X_i + \sum_{i=m+1}^{n} \beta_i X_i. \tag{12.74}$$

If

$$\forall\, i = m+1, \ldots, n: \left(\beta_i = 0 \text{ or } E(X_i \mid X_1, \ldots, X_m) \underset{P}{=} E(X_i) \right), \tag{12.75}$$

then there is a version $E(Y \mid X_1, \ldots, X_m) \in \mathscr{E}(Y \mid X_1, \ldots, X_m)$ such that

$$E(Y \mid X_1, \ldots, X_m) = \alpha_0 + \sum_{i=1}^{m} \alpha_i X_i \tag{12.76}$$

with

$$\alpha_0 = \beta_0 + \sum_{i=m+1}^{n} \beta_i E(X_i) \tag{12.77}$$

and

$$\alpha_i = \beta_i, \quad \forall\, i = 1, \dots, m. \tag{12.78}$$

(Proof p. 390)

Note that

$$X_1, \dots, X_m \perp\!\!\!\perp_P X_i, \quad \forall\, i = m+1, \dots, n$$

is a sufficient condition for

$$E(X_i \mid X_1, \dots, X_m) \underset{P}{=} E(X_i), \quad \forall\, i = m+1, \dots, n$$

[see Eq. (12.75) and Box 10.2 (vi)].

12.9 Proofs

Proof of Theorem 12.36

The existence of a version $E(Y \mid Z)$ satisfying (12.60) and (12.61) immediately follows from Definition 9.14 and Remark 10.11. Note that

$$\sigma(X) = \sigma(1_{Z=z_1}, \dots, 1_{Z=z_n}) = \sigma\big(\{\{Z=z_i\}: i = 1, \dots, n\,\}\big) \subset \sigma(Z),$$

where $X := (X_1, \dots, X_n) = (1_{Z=z_1}, \dots, 1_{Z=z_n})$. This implies that the function g defined by Equation (12.62) is a linear parameterization of $E(Y \mid Z)$ in $X = (1_{Z=z_1}, \dots, 1_{Z=z_n})$ with coefficients

$$\beta_0 = 0, \ \beta_1 = E(Y \mid Z=z_1), \ \dots, \ \beta_n = E(Y \mid Z=z_n)$$

(see Def. 12.5). Uniqueness is an immediate implication of Remark 10.12, provided that $Z(\Omega) = \{z_1, \dots, z_n\}$.

Proof of Theorem 12.37

Let $h: \Omega'_Z \to \mathbb{R}$ be a function such that $h(Z) \in \mathscr{E}(Y \mid Z)$. Then,

$$h(Z) \underset{P}{=} \sum_{i=0}^{n} h(z_i) \cdot 1_{Z=z_i} \qquad\qquad [(5.34)]$$

$$\underset{P}{=} \sum_{i=0}^{n} E(Y \mid Z=z_i) \cdot 1_{Z=z_i} \qquad\qquad [(10.27)]$$

$$\underset{P}{=} E(Y \mid Z=z_0) \cdot 1_{Z=z_0} + \sum_{i=1}^{n} E(Y \mid Z=z_i) \cdot 1_{Z=z_i}$$

$$\underset{P}{=} E(Y \mid Z=z_0) - \sum_{i=1}^{n} E(Y \mid Z=z_0) \cdot 1_{Z=z_i} + \sum_{i=1}^{n} E(Y \mid Z=z_i) \cdot 1_{Z=z_i} \qquad [(5.33)]$$

$$\underset{P}{=} E(Y \mid Z=z_0) + \sum_{i=1}^{n} [E(Y \mid Z=z_i) - E(Y \mid Z=z_0)] \cdot 1_{Z=z_i}$$

$$\underset{P}{=} \beta_0 + \sum_{i=1}^{n} \beta_i \cdot 1_{Z=z_i},$$

where $\beta_0 := E(Y \mid Z=z_0)$ and $\beta_i := E(Y \mid Z=z_i) - E(Y \mid Z=z_0)$, for $i = 1, \ldots, n$. Because the function on the right-hand side of the last equation is Z-measurable (see Rem. 2.17), it is an element of $\mathscr{E}(Y \mid Z)$, that is,

$$\beta_0 + \sum_{i=1}^{n} \beta_i \cdot 1_{Z=z_i} \in \mathscr{E}(Y \mid Z).$$

This proves equations (12.64), (12.65), and (12.66). Note that

$$\sigma(X) = \sigma(\{\{Z=z_i\}: i = 1, \ldots, n\}) \subset \sigma(\{\{Z=z_i\}: i = 0, 1, \ldots, n\}) \subset \sigma(Z),$$

where $X := (X_1, \ldots, X_n) = (1_{Z=z_1}, \ldots, 1_{Z=z_n})$. This implies that g [see Eq. (12.67)] is a linear parameterization of $E(Y \mid Z)$ in $X = (1_{Z=z_1}, \ldots, 1_{Z=z_n})$ (see Def. 12.5). If $Z(\Omega) = \{z_0, z_1, \ldots, z_n\}$, then Remark 10.12 implies that $E(Y \mid Z)$ is the only version in $\mathscr{E}(Y \mid Z)$.

Proof of Lemma 12.38

(i) For $i = j$, Equation (12.69) immediately follows from Equation (6.29) for the event $A = \{Z=z_i\}$. For $i \neq j$, Equation (12.69) follows from Equation (7.14) and the fact that $P(Z=z_i, Z=z_j) = 0$ if $i \neq j$. Equation (12.69) also shows that the variances and covariances of the indicators $1_{Z=z_1}, \ldots, 1_{Z=z_n}$ are finite, which implies that their second moments are finite as well [see Box 7.1 (i)].

(ii) This proposition can be derived as follows: For all $i = 1, \ldots, n$,

$$\sigma_{ii} = p_i \cdot (1 - p_i) \qquad\qquad [(12.69) \text{ for } i = j]$$

$$= p_i \cdot \sum_{j=0, j \neq i}^{n} p_j \qquad\qquad \left[1 - p_i = \Sigma_{j=0, j \neq i}^{n} p_j\right]$$

$$= \sum_{j=1, j \neq i}^{n} p_i \, p_j + p_i \cdot p_0$$

$$> \sum_{j=1, j \neq i}^{n} p_i p_j \qquad\qquad [p_i > 0, \forall\, i = 0, 1, \ldots, n,\ \ p_i \cdot p_0 > 0]$$

$$= \sum_{j=1, j \neq i}^{n} |-p_i p_j| \qquad\qquad [p_i p_j = |-p_i p_j|\,]$$

$$= \sum_{j=1, j \neq i}^{n} |\sigma_{ij}|\,. \qquad\qquad [(12.69)\ \text{for}\ i \neq j]$$

Proof of Theorem 12.44

$$E(Y \mid X_1, \ldots, X_m)$$

$$\underset{P}{=} E\big(E(Y \mid X_1, \ldots, X_n) \mid X_1, \ldots, X_m\big) \qquad [m < n,\ \text{Box 10.2 (v)}]$$

$$\underset{P}{=} E\left(\beta_0 + \sum_{i=1}^{m} \beta_i X_i + \sum_{i=m+1}^{n} \beta_i X_i \,\Big|\, X_1, \ldots, X_m\right) \qquad [(12.74)]$$

$$\underset{P}{=} \beta_0 + E\left(\sum_{i=1}^{m} \beta_i X_i \,\Big|\, X_1, \ldots, X_m\right) + E\left(\sum_{i=m+1}^{n} \beta_i X_i \,\Big|\, X_1, \ldots, X_m\right) \quad [\text{Box 10.2 (xvi), (i)}]$$

$$\underset{P}{=} \beta_0 + \sum_{i=1}^{m} \beta_i X_i + \sum_{i=m+1}^{n} \beta_i E(X_i \mid X_1, \ldots, X_m) \qquad [\text{Box 10.2 (xiv), (xvi)}]$$

$$\underset{P}{=} \beta_0 + \sum_{i=1}^{m} \beta_i X_i + \sum_{i=m+1}^{n} \beta_i E(X_i) \qquad [(12.75)]$$

$$\underset{P}{=} \alpha_0 + \sum_{i=1}^{m} \beta_i X_i\,. \qquad [(12.77)]$$

Exercises

12.1   Consider Example 12.20 and show that $\sigma(X) = \sigma(Z)$.

12.2   Let X and Z be dichotomous random variables on $(\Omega, \mathcal{A}, P)$ with values x_1, x_2, and z_1, z_2, respectively (see Example 5.10), and $k\colon \{0, 1\}^2 \to \mathbb{R}$ a (not necessarily measurable) function. Show that there are $\beta_0, \ldots, \beta_3 \in \mathbb{R}$ such that

$$k(X, Z) \underset{P}{=} \beta_0 + \beta_1 \cdot 1_{X = x_2} + \beta_2 \cdot 1_{Z = z_2} + \beta_3 \cdot 1_{X = x_2} \cdot 1_{Z = z_2},$$

and determine these coefficients in terms of the values of $k(X, Z)$. [Special cases are $k(X, Z) = E(Y \mid X, Z)$ (see Example 12.22) and $k(X, Z) = \text{logit}[P(Y = 1 \mid X,\ Z)]$ (see Example 13.24 with $Z = U$).]

12.3   Show that under the assumptions made in Remark 12.28, there is no linear parameterization g of $E(Y \mid X)$ in X with $g(x) = \beta_0 + \beta_1 x$, $x \in \mathbb{R}$, and slope $\beta_1 \neq 0$.

12.4 Assume $X_0, X_1, \ldots, X_n\colon (\Omega, \mathscr{A}, P) \to (\mathbb{R}, \mathscr{B})$ are random variables with finite second moments. Show: If $X_0 = \mathbf{a}'\mathbf{x} + b$, where $\mathbf{x} := [X_1, \ldots, X_n]'$, $\mathbf{a}' := [a_1, \ldots, a_n] \in \mathbb{R}^n$, and $b \in \mathbb{R}$, then the variance-covariance matrix of $X_0, X_1, \ldots, X_n$ is singular.

12.5 Compute the parameters of the quadratic function displayed in Figure 12.2 using the three conditional probabilities computed in Example 12.41.

Solutions

12.1 Because the function $\ln\colon\ ]0, \infty[\ \to \mathbb{R}$ is continuous, it is $(\mathscr{B}|_{]0,\infty[}, \mathscr{B})$-measurable (see Klenke, 2013, Th. 1.88). Hence, for $X = \ln Z$, Lemma 2.52 implies $\sigma(X) \subset \sigma(Z)$. Furthermore, the exponential function $\exp\colon \mathbb{R} \to \mathbb{R}$ is continuous as well, and therefore it is $(\mathscr{B}, \mathscr{B})$-measurable (see again Klenke, 2013, Th. 1.88). Because $Z = \exp(X)$, Lemma 2.52 implies $\sigma(Z) \subset \sigma(X)$, and this yields $\sigma(Z) = \sigma(X)$.

12.2 Note that $1_{X=x_2} = (1 - 1_{X=x_1})$, $1_{Z=z_2} = (1 - 1_{Z=z_1})$, and $1_{X=x_i,Z=z_j} = 1_{X=x_i} \cdot 1_{Z=z_j}$, $i, j = 1, 2$ [see Eq. (1.33)]. Hence,

$$
\begin{aligned}
k(X, Z) \underset{P}{=}\ & k(x_1, z_1) \cdot 1_{X=x_1} \cdot 1_{Z=z_1} + k(x_2, z_1) \cdot 1_{X=x_2} \cdot 1_{Z=z_1} \\
& + k(x_1, z_2) \cdot 1_{X=x_1} \cdot 1_{Z=z_2} + k(x_2, z_2) \cdot 1_{X=x_2} \cdot 1_{Z=z_2} \qquad\text{[Cor. 5.63]} \\
=\ & k(x_1, z_1) \cdot (1 - 1_{X=x_2}) \cdot (1 - 1_{Z=z_2}) + k(x_2, z_1) \cdot 1_{X=x_2} \cdot (1 - 1_{Z=z_2}) \\
& + k(x_1, z_2) \cdot (1 - 1_{X=x_2}) \cdot 1_{Z=z_2} + k(x_2, z_2) \cdot 1_{X=x_2} \cdot 1_{Z=z_2} \\
=\ & k(x_1, z_1) + \big(k(x_2, z_1) - k(x_1, z_1)\big) \cdot 1_{X=x_2} + \big(k(x_1, z_2) - k(x_1, z_1)\big) \cdot 1_{Z=z_2} \\
& + \big(k(x_2, z_2) - k(x_2, z_1) - k(x_1, z_2) + k(x_1, z_1)\big) \cdot 1_{X=x_2} \cdot 1_{Z=z_2},
\end{aligned}
$$

where the last equation is obtained by multiplying out the parentheses and rearranging terms. Hence,

$$
\begin{aligned}
\beta_0 &= k(x_1, z_1), \\
\beta_1 &= k(x_2, z_1) - k(x_1, z_1), \\
\beta_2 &= k(x_1, z_2) - k(x_1, z_1), \\
\beta_3 &= k(x_2, z_2) - k(x_2, z_1) - k(x_1, z_2) + k(x_1, z_1).
\end{aligned}
$$

12.3 This proposition follows from

$$
P[g(X) < 0] + P[g(X) > 1] =
\begin{cases}
P\!\left(X < -\dfrac{\beta_0}{\beta_1}\right) + P\!\left(X > \dfrac{1 - \beta_0}{\beta_1}\right) > 0, & \text{if } \beta_1 > 0 \\[2ex]
P\!\left(X > -\dfrac{\beta_0}{\beta_1}\right) + P\!\left(X < \dfrac{1 - \beta_0}{\beta_1}\right) > 0, & \text{if } \beta_1 < 0,
\end{cases}
$$

because $P[g(X) < 0] + P[g(X) > 1] > 0$ is a contradiction to (12.46).

12.4 Let Σ denote the variance-covariance matrix of $X_0, X_1, \ldots, X_n$. Then,

$$
\Sigma = \begin{bmatrix}
\sigma_{X_0}^2 & \sigma_{X_0 X_1} & \cdots & \sigma_{X_0 X_n} \\
\sigma_{X_1 X_0} & \sigma_{X_1}^2 & \cdots & \sigma_{X_1 X_n} \\
\vdots & \vdots & \ddots & \vdots \\
\sigma_{X_n X_0} & \sigma_{X_n X_1} & \cdots & \sigma_{X_n}^2
\end{bmatrix}
$$

$$
= E\left[\begin{pmatrix} \mathbf{a'} x - E(\mathbf{a'} x) \\ x - E(x) \end{pmatrix} \left(\mathbf{a'} x - E(\mathbf{a'} x), x' - E(x') \right) \right] \qquad [X_0 = \mathbf{a'} x + b, \, (7.35)]
$$

$$
= E\begin{bmatrix} \mathbf{a'}[x - E(x)][x' - E(x')]\,\mathbf{a} & \mathbf{a'}[x - E(x)][x' - E(x')] \\ [x - E(x)][x' - E(x')]\,\mathbf{a} & [x - E(x)][x' - E(x')] \end{bmatrix} \qquad \text{[Box 7.2 (iii), (v)]}
$$

$$
= \begin{bmatrix} \mathbf{a'}\, E\big([x - E(x)][x' - E(x')]\big)\,\mathbf{a} & \mathbf{a'}\, E\big([x - E(x)][x' - E(x')]\big) \\ E\big([x - E(x)][x' - E(x')]\big)\,\mathbf{a} & E\big([x - E(x)][x' - E(x')]\big) \end{bmatrix}. \qquad [(7.29)]
$$

The first row of this matrix is obtained by multiplying the lower two submatrices by $\mathbf{a'}$ from the left. Hence, the first row of this variance-covariance matrix is a linear combination of its other rows. This implies that Σ is singular.

12.5 For $x_1 = x$ and $x_2 = x^2$, the parameterization $g(x) = \beta_0 + \beta_1\, x_1 + \beta_2\, x_2$ [see Eq. (12.33)] yields

$$
.55 = \beta_0,
$$

if $x = 0$,

$$
.58\overline{3} = \beta_0 + \beta_1 + \beta_2,
$$

if $x = 1$, and

$$
.53\overline{3} = \beta_0 + \beta_1 \cdot 2 + \beta_2 \cdot 4,
$$

if $x = 2$. Hence, $\beta_0 = .55$, and solving the last two equations for the remaining two unknowns yields $\beta_1 \approx .0750$ and $\beta_2 \approx -.041\overline{6}$.

13

Linear logistic regression

In chapter 12, we treated the notions of a linear parameterization of a conditional expectation and of a *linear regression*. In this chapter, we turn to the *linear logit regression*, presuming that the regressand, say Y, is an indicator variable. In Remark 10.4, we noted that in this case a version $E(Y \mid \mathscr{C})$ of the $\mathscr{C}$-conditional expectation of Y is also called a version of the $\mathscr{C}$-conditional probability of the event $\{Y = 1\}$ and that it is also denoted by $P(Y = 1 \mid \mathscr{C})$. As noted in Remark 12.28, if X is a $\mathscr{C}$-measurable real-valued random variable on $(\Omega, \mathscr{A}, P)$ and X is P-almost surely unbounded, then there is no linear parameterization g in X of $E(Y \mid \mathscr{C}) = P(Y = 1 \mid \mathscr{C}) = g(X)$ with a nonzero slope. However, in this case, there might be a linear logistic parameterization.

We begin with the logit transformation, define the logit of a $\mathscr{C}$-conditional probability $P(Y = 1 \mid \mathscr{C})$, a linear logistic parameterization, and then present a theorem on uniqueness and the identification of the parameters. Finally, the concept of a linear logit regression is defined.

13.1 Logit transformation of a conditional probability

The general assumptions and notation are as follows:

> **Notation and assumptions 13.1**
> Let $Y \colon (\Omega, \mathscr{A}, P) \to (\mathbb{R}, \mathscr{B})$ be a *dichotomous random variable with values* 0 *and* 1*, let* $\mathscr{C} \subset \mathscr{A}$ *be a σ-algebra, and assume that there is a version* $P(Y = 1 \mid \mathscr{C}) \in \mathscr{P}(Y = 1 \mid \mathscr{C})$ *with* $0 < P(Y = 1 \mid \mathscr{C}) < 1$.

Remark 13.2 [Necessary condition] If $\{Y = 1\} \in \mathscr{C}$ or $\{Y = 0\} \in \mathscr{C}$, then there is no version $P(Y = 1 \mid \mathscr{C}) \in \mathscr{P}(Y = 1 \mid \mathscr{C})$ with $0 < P(Y = 1 \mid \mathscr{C}) < 1$. Furthermore, there is no version $P(Y = 1 \mid \mathscr{C}) \in \mathscr{P}(Y = 1 \mid \mathscr{C})$ with $0 < P(Y = 1 \mid \mathscr{C}) < 1$ if $\sigma(Y) \subset \mathscr{C}$ (see Exercise 13.1). ◁

Probability and Conditional Expectation: Fundamentals for the Empirical Sciences, First Edition. Rolf Steyer and Werner Nagel.
© 2017 John Wiley & Sons, Ltd. Published 2017 by John Wiley & Sons, Ltd.
Companion website: http://www.probability-and-conditional-expectation.de

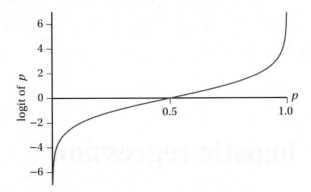

Figure 13.1   Graph of the logit transformation of p.

Remark 13.3 [The logit transformation]   Each version $P(Y=1 \mid \mathscr{C})$ of the conditional probability is P-almost surely bounded, because

$$P(\{\omega \in \Omega : 0 \leq P(Y=1 \mid \mathscr{C})(\omega) \leq 1\}) = 1 \qquad (13.1)$$

[see Box 10.3 (ii), (iii)]. If there is a version $P(Y=1 \mid \mathscr{C})$ with $0 < P(Y=1 \mid \mathscr{C}) < 1$, then this version can be transformed using the logit transformation, that is, using the function logit: $]0, 1[\rightarrow \mathbb{R}$ defined by

$$\text{logit}(p) := \ln\left(\frac{p}{1-p}\right) = \ln(p) - \ln(1-p), \quad \forall\, p \in\,]0, 1[, \qquad (13.2)$$

where ln denotes the natural logarithm. If p represents a probability, then $\ln(\frac{p}{1-p})$ is also called the *log-odds* of p. Figure 13.1 shows the graph of such a logit transformation. In the context of generalized linear models (see, e.g., Agresti, 2015; McCullagh & Nelder, 1989), this function is an example of a *link function*. Some algebra yields

$$p = \frac{\exp[\text{logit}(p)]}{1 + \exp[\text{logit}(p)]}, \quad \forall\, p \in\,]0, 1[, \qquad (13.3)$$

(see Exercise 13.2).   ◁

Remark 13.4 [Logistic function]   The function $h: \mathbb{R} \rightarrow\,]0, 1[$ specified by

$$h(x) = \frac{\exp(x)}{1 + \exp(x)}, \quad \forall\, x \in \mathbb{R}, \qquad (13.4)$$

is called the *logistic function*. Equation (13.3) implies that the logistic function is the inverse of the logit function.   ◁

Definition 13.5 [Logit of $P(Y=1 \mid \mathscr{C})$]
Let the assumptions 13.1 hold. Then,

$$\text{logit}\,[P(Y=1 \mid \mathscr{C})] := \ln\left(\frac{P(Y=1 \mid \mathscr{C})}{1 - P(Y=1 \mid \mathscr{C})}\right) \tag{13.5}$$

is called the logit of $P(Y=1 \mid \mathscr{C})$.

Remark 13.6 [One-to-one transformation] Note that $\text{logit}\,[P(Y=1 \mid \mathscr{C})]$ denotes the composition of $P(Y=1 \mid \mathscr{C})$ and the function logit defined by Equation (13.2). Hence, $\text{logit}\,[P(Y=1 \mid \mathscr{C})]$ is a random variable on $(\Omega, \mathscr{A}, P)$. Also note that $P(Y=1 \mid \mathscr{C})$ and its logit contain the same information, that is, the σ-algebras they generate are identical (see Lemma 13.7). While $P(Y=1 \mid \mathscr{C})$ informs us about the $\mathscr{C}$-conditional probability of the event $\{Y=1\}$ *on the probability scale* with values between 0 and 1, a logit of $P(Y=1 \mid \mathscr{C})$ informs us about this conditional probability *on the log-odds scale* with values between $-\infty$ and ∞. Applying (13.3) yields

$$P(Y=1 \mid \mathscr{C}) = \frac{\exp(\text{logit}[P(Y=1 \mid \mathscr{C})])}{1 + \exp(\text{logit}[P(Y=1 \mid \mathscr{C})])}. \tag{13.6}$$

Hence, $P(Y=1 \mid \mathscr{C})$ is uniquely determined by $\text{logit}[P(Y=1 \mid \mathscr{C})]$, and vice versa [see Eq. (13.5)]. ◁

Lemma 13.7 [σ-Algebra generated by $P(Y=1 \mid \mathscr{C})$ and its logit]
Let the assumptions 13.1 hold. Then

$$\sigma[P(Y=1 \mid \mathscr{C})] = \sigma(\text{logit}\,[P(Y=1 \mid \mathscr{C})]). \tag{13.7}$$

(Proof p. 407)

Remark 13.8 [Motivation for considering the logit of $P(Y=1 \mid \mathscr{C})$] Suppose that X is a real-valued $\mathscr{C}$-measurable random variable on $(\Omega, \mathscr{A}, P)$ and that X is not P-almost surely bounded [see (12.47)]. Then it is still possible to assume that there are a version $P(Y=1 \mid \mathscr{C}) \in \mathscr{P}(Y=1 \mid \mathscr{C})$ and numbers $\lambda_0, \lambda_1 \in \mathbb{R}$ such that

$$\text{logit}\,[P(Y=1 \mid \mathscr{C})] = \lambda_0 + \lambda_1 X. \tag{13.8}$$

In contrast, assuming $P(Y=1 \mid \mathscr{C}) = \lambda_0 + \lambda_1 X$ for real numbers $\lambda_0, \lambda_1, \lambda_1 \neq 0$, would be contradictory if X is not P-almost surely bounded. ◁

13.2   Linear logistic parameterization

Notation and assumptions 13.9
Let $Y: (\Omega, \mathcal{A}, P) \to (\mathbb{R}, \mathcal{B})$ be a dichotomous random variable with values 0 and 1, let $\mathcal{C} \subset \mathcal{A}$ be a σ-algebra, and assume that there is a version $P(Y=1 \mid \mathcal{C}) \in \mathcal{P}(Y=1 \mid \mathcal{C})$ with $0 < P(Y=1 \mid \mathcal{C}) < 1$. Furthermore, let $X_i: (\Omega, \mathcal{A}, P) \to (\mathbb{R}, \mathcal{B}), i = 1, \ldots, n$, be real-valued random variables, define $X := (X_1, \ldots, X_n): (\Omega, \mathcal{A}, P) \to (\mathbb{R}^n, \mathcal{B}_n)$, and assume that X is $\mathcal{C}$-measurable.

Using this notation and these assumptions, a linear logistic parameterization of a conditional probability $P(Y=1 \mid \mathcal{C})$ is now defined as follows:

Definition 13.10 [Linear logistic parameterization]
Let the assumptions 13.9 hold. If there are $\lambda_0, \lambda_1, \ldots, \lambda_n \in \mathbb{R}$ and a version $P(Y=1 \mid \mathcal{C}) \in \mathcal{P}(Y=1 \mid \mathcal{C})$ such that

$$P(Y=1 \mid \mathcal{C}) = \frac{\exp(\lambda_0 + \sum_{i=1}^n \lambda_i X_i)}{1 + \exp(\lambda_0 + \sum_{i=1}^n \lambda_i X_i)}, \tag{13.9}$$

then the function g: $\mathbb{R}^n \to [0, 1]$ satisfying

$$g(x) = \frac{\exp(\lambda_0 + \sum_{i=1}^n \lambda_i x_i)}{1 + \exp(\lambda_0 + \sum_{i=1}^n \lambda_i x_i)}, \quad \forall x = (x_1, \ldots, x_n) \in \mathbb{R}^n, \tag{13.10}$$

is called a linear logistic parameterization of $P(Y=1 \mid \mathcal{C})$ in X.

If $\mathcal{C} = \sigma(X)$, then Equation (13.9) is equivalent to

$$P(Y=1 \mid X) = \frac{\exp(\lambda_0 + \sum_{i=1}^n \lambda_i X_i)}{1 + \exp(\lambda_0 + \sum_{i=1}^n \lambda_i X_i)}. \tag{13.11}$$

Remark 13.11 [Univariate real-valued X]   If Equation (13.9) holds for $n = 1$, then there is a version $P(Y=1 \mid X) \in \mathcal{P}(Y=1 \mid X)$ such that

$$P(Y=1 \mid \mathcal{C}) = P(Y=1 \mid X) = \frac{\exp(\lambda_0 + \lambda_1 X)}{1 + \exp(\lambda_0 + \beta_1 X)}. \tag{13.12}$$

In Example 17.82, we present a sufficient condition of Equation (13.12) related to the normal distribution.   ◁

Remark 13.12 [Conditional probabilities] If g is a linear logistic parameterization of $P(Y=1 \mid X)$ in X satisfying Equation (13.10), then, according to Definition 10.33, we can define

$$P(Y=1 \mid X=x) = P(Y=1 \mid X_1=x_1, \ldots, X_n=x_n)$$

$$:= g(x) = \frac{\exp(\lambda_0 + \sum_{i=1}^{n} \lambda_i x_i)}{1 + \exp(\lambda_0 + \sum_{i=1}^{n} \lambda_i x_i)}, \quad \forall x \in \mathbb{R}^n. \tag{13.13}$$

This definition is convenient, but note that another factorization g^* of $P(Y=1 \mid X)$ might yield other conditional probabilities $P(Y=1 \mid X=x)$ for values x of X with $P(X=x) = 0$. However, according to Equation (12.10), if g, g^* are factorizations of two versions $V, V^* \in \mathcal{P}(Y=1 \mid X)$, then $g(x) = g^*(x)$, for P_X-almost all $x \in \mathbb{R}^n$ [see Eq. (12.13)].   ◁

Remark 13.13 [Meaning of coefficients] Figure 13.2 displays the graphs of logistic transformations in which the logits of $P(Y=1 \mid X)$ are linear functions $\lambda_0 + \lambda_1 X$. As is easily seen,

$$P\left(Y=1 \mid X = -\frac{\lambda_0}{\lambda_1}\right) = \frac{\exp(0)}{1 + \exp(0)} = \frac{1}{1+1} = \frac{1}{2}.$$

This equation shows that $x = -\dfrac{\lambda_0}{\lambda_1}$ is the point on the x-axis at which the conditional probability $P(Y=1 \mid X=x)$ is .5. Furthermore, the derivative of the linear parameterization g with respect to x is

$$\frac{d}{dx} g(x) = \frac{d}{dx} \frac{\exp(\lambda_0 + \lambda_1 x)}{1 + \exp(\lambda_0 + \lambda_1 x)} = \frac{\lambda_1 \exp(\lambda_0 + \lambda_1 x)}{(1 + \exp(\lambda_0 + \lambda_1 x))^2}. \tag{13.14}$$

Hence, the derivative (i.e., the slope) of g at $x = -\dfrac{\lambda_0}{\lambda_1}$ is $\dfrac{\lambda_1}{4}$ (see Exercise 13.3).   ◁

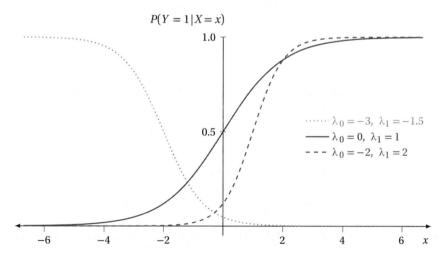

Figure 13.2 Graphs of three logistic functions.

13.3   A parameterization of a discrete conditional probability

In Theorem 12.37, we already considered a parameterization of a discrete conditional expectation $E(Y \mid Z)$ in which the parameters $\beta_1, \ldots, \beta_n$ are the differences $E(Y \mid Z=z_i) - E(Y \mid Z=z_0)$, $i = 1, \ldots, n$. If Y is dichotomous, then there is also a logistic parameterization of the conditional probability $E(Y \mid Z) = P(Y=1 \mid Z)$.

Theorem 13.14 [Existence of the logit effects]
Let $Y: (\Omega, \mathcal{A}, P) \to (\mathbb{R}, \mathcal{B})$ and $Z: (\Omega, \mathcal{A}, P) \to (\Omega_Z', \mathcal{A}_Z')$ be random variables, where Y is dichotomous with values 0 and 1, and Z is discrete with $P(Z \in \{z_0, z_1, \ldots, z_n\}) = 1$, and assume $0 < P(Z=z_i) < 1$ and $0 < P(Y=1 \mid Z=z_i) < 1$, for all $i = 0, 1, \ldots, n$. Then $P(Y=1 \mid Z)$ is uniquely defined, and there are coefficients $\beta_0, \beta_1, \ldots, \beta_n, \lambda_0, \lambda_1, \ldots, \lambda_n \in \mathbb{R}$ such that

$$P(Y=1 \mid Z) = \beta_0 + \sum_{i=1}^{n} \beta_i \cdot 1_{Z=z_i} \tag{13.15}$$

$$= \frac{\exp\left[\lambda_0 + \sum_{i=1}^{n} \lambda_i \cdot 1_{Z=z_i}\right]}{1 + \exp\left[\lambda_0 + \sum_{i=1}^{n} \lambda_i \cdot 1_{X=x_i}\right]} \tag{13.16}$$

with

$$\beta_0 = P(Y=1 \mid Z=z_0) \tag{13.17}$$

$$= \frac{\exp(\lambda_0)}{1 + \exp(\lambda_0)} \tag{13.18}$$

and

$$\beta_i = P(Y=1 \mid Z=z_0) - P(Y=1 \mid Z=z_i) \tag{13.19}$$

$$= \frac{\exp(\lambda_0 + \lambda_i)}{1 + \exp(\lambda_0 + \lambda_i)} - \frac{\exp(\lambda_0)}{1 + \exp(\lambda_0)}. \tag{13.20}$$

(Proof p. 408)

Remark 13.15 [Log odds]   In terms of conditional probabilities, the logit intercept can be written as:

$$\lambda_0 = \ln\left[\frac{P(Y=1 \mid Z=z_0)}{1 - P(Y=1 \mid Z=z_0)}\right]. \tag{13.21}$$

Hence, λ_0 is the log odds of $P(Y=1 \mid Z=z_0)$. Similarly,

$$\lambda_0 + \lambda_i = \ln\left[\frac{P(Y=1 \mid Z=z_i)}{1 - P(Y=1 \mid Z=z_i)}\right], \quad \forall\, i = 1, \ldots, n, \tag{13.22}$$

[see Eqs. (13.17) to (13.20) and (13.2)]. This equation shows that $\lambda_0 + \lambda_i$ is the log odds of $P(Y=1 \mid Z=z_i)$.   ◁

Remark 13.16 [Log odds ratio]   Equations (13.21) and (13.22) immediately imply

$$\lambda_i = \ln\left[\frac{P(Y=1 \mid Z=z_i)}{1 - P(Y=1 \mid Z=z_i)}\right] - \ln\left[\frac{P(Y=1 \mid Z=z_0)}{1 - P(Y=1 \mid Z=z_0)}\right] \tag{13.23}$$

$$= \ln\left[\left(\frac{P(Y=1 \mid Z=z_i)}{1 - P(Y=1 \mid Z=z_i)}\right) \middle/ \left(\frac{P(Y=1 \mid Z=z_0)}{1 - P(Y=1 \mid Z=z_0)}\right)\right] \tag{13.24}$$

$$= \ln\left[\frac{P(Y=1 \mid Z=z_i) \cdot (1 - P(Y=1 \mid Z=z_0))}{(1 - P(Y=1 \mid Z=z_i)) \cdot P(Y=1 \mid Z=z_0)}\right], \quad \forall\, i = 1, \dots, n.$$

Hence, the logit effect λ_i, $i = 1, \dots, n$, is the difference between the log odds of $P(Y=1 \mid Z=z_i)$ and $P(Y=1 \mid Z=z_0)$, respectively [see Eq. (13.23)]. Equation (13.24) shows that λ_i is the log odds ratio of $P(Y=1 \mid Z=z_i)$ and $P(Y=1 \mid Z=z_0)$.   ◁

Remark 13.17 [Odds ratio]   The value of the exponential function for the argument λ_i is

$$\exp(\lambda_i) = \left(\frac{P(Y=1 \mid Z=z_i)}{1 - P(Y=1 \mid Z=z_i)}\right) \middle/ \left(\frac{P(Y=1 \mid Z=z_0)}{1 - P(Y=1 \mid Z=z_0)}\right)$$

$$= \frac{P(Y=1 \mid Z=z_i) \cdot (1 - P(Y=1 \mid Z=z_0))}{(1 - P(Y=1 \mid Z=z_i)) \cdot P(Y=1 \mid Z=z_0)}, \quad \forall\, i = 1, \dots, n. \tag{13.25}$$

This equation shows that the number $\exp(\lambda_i)$ is the odds ratio of $P(Y=1 \mid Z=z_i)$ and $P(Y=1 \mid Z=z_0)$.   ◁

Remark 13.18 [Risk ratio]   Another closely related parameter is

$$\kappa_i := \frac{P(Y=1 \mid Z=z_i)}{P(Y=1 \mid Z=z_0)}, \quad \forall\, i = 1, \dots, n. \tag{13.26}$$

This parameter is called the *risk ratio of $P(Y=1 \mid Z=z_i)$ and $P(Y=1 \mid Z=z_0)$*.   ◁

Remark 13.19 [Four kinds of effect parameters]   Hence, under the assumptions of Theorem 13.14, we may consider four different kinds of effect parameters: β_i, λ_i, $\exp(\lambda_i)$, and κ_i. They all quantify the effect of x_i compared to x_0 on Y, each one on a different scale.   ◁

13.4   Identification of coefficients of a linear logistic parameterization

The following theorem specifies sufficient conditions under which a linear logit parameterization of $P(Y=1 \mid X)$ is uniquely defined.

Theorem 13.20 [Identification of coefficients and uniqueness]
Let the assumptions 13.9 hold and, let $\lambda_0 \in \mathbb{R}$, and $\boldsymbol{\lambda} \in \mathbb{R}^n$. Furthermore, define $\boldsymbol{x} = [X_1, \ldots, X_n]$ and assume

(a) *There is a version $P(Y=1 \mid \mathscr{C}) \in \mathscr{P}(Y=1 \mid \mathscr{C})$ such that Equation (13.9) holds.*

(b) *$X_1, \ldots, X_n$ have finite second moments.*

(c) *The inverse $\boldsymbol{\Sigma}_{xx}^{-1}$ of the covariance matrix of $X = (X_1, \ldots, X_n)$ exists.*

Then, using

$$L := \text{logit}[P(Y=1 \mid \mathscr{C})] = \lambda_0 + \boldsymbol{\lambda}' \boldsymbol{x}, \tag{13.27}$$

the following two equations hold:

$$\lambda_0 = E(L) - \boldsymbol{\lambda}' \boldsymbol{\mu}, \tag{13.28}$$

$$\boldsymbol{\lambda} = \boldsymbol{\Sigma}_{xx}^{-1} \boldsymbol{\Sigma}_{xl}, \tag{13.29}$$

where $\boldsymbol{\mu} := [E(X_1), \ldots, E(X_n)]'$; and $\boldsymbol{\Sigma}_{xl}$ denotes the column vector of the covariances $\text{Cov}(X_i, L)$. The coefficient λ_0 and $\boldsymbol{\lambda}$ are uniquely determined, and hence, the linear logistic parameterization $g \colon \mathbb{R}^n \to [0, 1]$ of $P(Y=1 \mid \mathscr{C})$ satisfying

$$g(x_1, \ldots, x_n) = \frac{\exp(\lambda_0 + \sum_{i=1}^{n} \lambda_i x_i)}{1 + \exp(\lambda_0 + \sum_{i=1}^{n} \lambda_i x_i)}, \quad \forall\, (x_1, \ldots, x_n) \in \mathbb{R}^n, \tag{13.30}$$

is uniquely defined.

(Proof p. 408)

Remark 13.21 [Identification versus estimation]   Note that, for a version $P(Y=1 \mid \mathscr{C}) \in \mathscr{P}(Y=1 \mid \mathscr{C})$ satisfying assumptions 13.1, the logit of $P(Y=1 \mid \mathscr{C})$ is uniquely defined. Because the expectation vector $\boldsymbol{\mu} = [E(X_1), \ldots, E(X_n)]$, the covariance matrix $\boldsymbol{\Sigma}_{xx}$, and the covariance vector $\boldsymbol{\Sigma}_{xl}$ in Equation (13.29) are also uniquely defined, we can conclude that the coefficients of the linear logistic parameterization are uniquely defined, (or 'identified') as well. Estimation in the logistic case is more difficult as compared to the linear regression, because, in contrast to Y (see Rem. 12.2), the random variable $L = \text{logit}[P(Y=1 \mid \mathscr{C})]$ is nonobservable. For methods of estimation, see, for example, Agresti (2015) or McCullagh and Nelder (1989).   ◁

13.5   Linear logistic regression and linear logit regression

In Definition 13.10, we defined the concept of a linear logistic parameterization of $P(Y=1 \mid \mathscr{C})$ in X. Such parameterizations of $P(Y=1 \mid \mathscr{C})$ are not necessarily uniquely defined. If there is a linear logistic parameterization g of $P(Y=1 \mid \mathscr{C})$ in X that satisfies Equation (13.30), then we call it *the linear logistic regression of Y on X*. According to Theorem 13.20, the linear logistic regression is uniquely defined.

Definition 13.22 [Linear logistic regression and linear logit regression]
*Let the assumptions 13.9 hold and suppose there are an $P(Y=1 \mid \mathscr{C}) \in \mathscr{P}(Y=1 \mid \mathscr{C})$
and $\lambda_0, \lambda_1, \dots, \lambda_n \in \mathbb{R}$ such that Equation (13.9) holds. Furthermore, assume that Y and
$X_1, \dots, X_n$ have finite second moments and that the inverse Σ_{xx}^{-1} of the covariance matrix
of $X = (X_1, \dots, X_n)$ exists. Then the function $g \colon \mathbb{R}^n \to [0, 1]$ defined by Equation (13.30)
is called the* **linear logistic regression** *or the* **linear inverse logit regression**,
and the function $f \colon \mathbb{R}^n \to \mathbb{R}$ defined by

$$f(x_1, \dots, x_n) = \lambda_0 + \sum_{i=1}^{n} \lambda_i x_i, \quad \forall\, (x_1, \dots, x_n) \in \mathbb{R}^n, \tag{13.31}$$

is called the **linear logit regression** *of Y on X.*

Figure 13.3 shows $P(Y=1 \mid X)$ as the composition of the functions introduced above.
According to this figure, $P(Y=1 \mid X)$ is the composition of X and the linear logistic regression g, which itself is the composition of the linear logit regression f and the logistic function
h [see Eq. (13.4)] that transforms a logit into a probability.

Remark 13.23 [Simple and multiple linear logistic regression]   If $n \geq 2$, then a linear logistic regression is also called a *multiple linear logistic regression*. If $n = 1$, then it is also called
a *simple linear logistic regression*.   ◁

Example 13.24 [Joe and Ann with randomized assignment – continued]   Table 9.1 shows
the random variables U, X, and Y as well as the conditional expectations $E(Y \mid X, U)$ and
$E(Y \mid X)$. Because X is dichotomous with values 0 and 1, the conditional probability $E(Y \mid X) =
P(Y=1 \mid X)$ can always be written as a linear function of X. Applying the equations of
Example 12.15 yields

$$P(Y=1 \mid X) = .45 + .15 \cdot X \tag{13.32}$$

(see also Example 12.16).

Figure 13.3 $P(Y=1 \mid X)$ as the composition of X, the linear logit regression f, and the logistic
function h.

Alternatively, we can also write the *logit* of $E(Y \mid X) = P(Y=1 \mid X)$ as a linear function of X. This yields

$$P(Y=1 \mid X) = \frac{\exp(\alpha_0 + \alpha_1 \cdot X)}{1 + \exp(\alpha_0 + \alpha_1 \cdot X)} \approx \frac{\exp(-.201 + .606 \cdot X)}{1 + \exp(-.201 + .606 \cdot X)} \qquad (13.33)$$

(see Exercise 13.4). Inserting the two values of X, this equation yields the probabilities

$$P(Y=1 \mid X)(\omega) = P(Y=1 \mid X=0) = .45, \quad \text{for } \omega \in \{X=0\}$$

and

$$P(Y=1 \mid X)(\omega) = P(Y=1 \mid X=1) = .60, \quad \text{for } \omega \in \{X=1\},$$

which is consistent with Equation (13.32). The function $g \colon \mathbb{R} \to [0, 1]$ defined by

$$g(x) \approx \frac{\exp(-.201 + .606 \cdot x)}{1 + \exp(-.201 + .606 \cdot x)}, \quad \forall x \in \mathbb{R},$$

is the linear logistic regression of Y on X.   ◁

Remark 13.25 [Linear versus linear logistic regression]   In this particular example, there is no compelling reason to prefer the logistic linear regression over the (ordinary) linear regression. However, this does not apply any more if X is unbounded (see Rem. 13.8).

Furthermore, note that the standard computer programs (and the underlying statistical models) for linear regressions assume equality (homogeneity) of the conditional variances $Var(Y \mid X=x)$ for different values x of X. In contrast, computer programs (and the underlying statistical models) for the analysis of linear logistic regressions allow for heterogeneous $(X=x)$-conditional variances of Y.

Remember that

$$Var(Y \mid X=x) = P(Y=1 \mid X=x) \cdot [1 - P(Y=1 \mid X=x)], \quad \text{for } P_X\text{-a.a. } x \in \Omega'_X \qquad (13.34)$$

[see Eq. (11.49)]. Equation (13.34) shows that $Var(Y \mid X=x)$ depends on $P(Y=1 \mid X=x)$. In Example 13.24, the $(X=0)$-conditional variance is

$$Var(Y \mid X=0) = P(Y=1 \mid X=0) \cdot [1 - P(Y=1 \mid X=0)] = .45 \cdot (1 - .45) = .2475,$$

whereas the $(X=1)$-conditional variance is

$$Var(Y \mid X=1) = P(Y=1 \mid X=1) \cdot [1 - P(Y=1 \mid X=1)] = .6 \cdot (1 - .6) = .24.$$

◁

Example 13.26 [Joe and Ann with randomized assignment – continued]   According to Remark 12.23, the logit of $E(Y \mid X, U) = P(Y=1 \mid X, U)$ can be written as a linear function in $X, 1_{U=Ann}, X \cdot 1_{U=Ann}$. Hence, we can write

$$P(Y=1 \mid X, U) = \frac{\exp(\lambda_0 + \lambda_1 X + \lambda_2 \cdot 1_{U=Ann} + \lambda_3 X \cdot 1_{U=Ann})}{1 + \exp(\lambda_0 + \lambda_1 X + \lambda_2 \cdot 1_{U=Ann} + \lambda_3 X \cdot 1_{U=Ann})}$$

$$\approx \frac{\exp(.847 + .539 X - 2.234 \cdot 1_{U=Ann} + .442 X \cdot 1_{U=Ann})}{1 + \exp(.847 + .539 X - 2.234 \cdot 1_{U=Ann} + .442 X \cdot 1_{U=Ann})}$$

(see Exercise 13.5) Inserting the two values of X and the two values of $1_{U=Ann}$, this equation yields the four probabilities $P(Y=1 \mid X=x, U=u)$ listed in Table 9.1. The function $g \colon \mathbb{R}^3 \to [0, 1]$ defined by

$$g(x) \approx \frac{\exp(.847 + .539 \cdot x_1 - 2.234 \cdot x_2 + .442 \cdot x_3)}{1 + \exp(.847 + .539 \cdot x_1 - 2.234 \cdot x_2 + .442 \cdot x_3)}, \quad \forall\, x \in \mathbb{R}^3,$$

is the linear logistic regression of Y on $X = (X_1, X_2, X_3) = (X, 1_{U=Ann}, X \cdot 1_{U=Ann})$.
   Rearranging the equation for $P(Y=1 \mid X, U)$ yields

$$P(Y=1 \mid X, U) \underset{P}{=} \frac{\exp((\lambda_0 + \lambda_2 \cdot 1_{U=Ann}) + (\lambda_1 + \lambda_3 \cdot 1_{U=Ann}) X)}{1 + \exp((\lambda_0 + \lambda_2 \cdot 1_{U=Ann}) + (\lambda_1 + \lambda_3 \cdot 1_{U=Ann}) X)}$$

$$\approx \frac{\exp((.847 - 2.234 \cdot 1_{U=Ann}) + (.539 + .442 \cdot 1_{U=Ann}) X)}{1 + \exp((.847 - 2.234 \cdot 1_{U=Ann}) + (.539 + .442 \cdot 1_{U=Ann}) X)},$$

showing that the logit is $f_0(U) + f_1(U) \cdot X$ with

$$f_0(U) = \lambda_0 + \lambda_2 \cdot 1_{U=Ann} \approx .847 - 2.234 \cdot 1_{U=Ann}$$

and

$$f_1(U) = \lambda_1 + \lambda_3 \cdot 1_{U=Ann} = .539 + .442 \cdot 1_{U=Ann}.$$

The function $f_0(U)$ is called the *logit intercept function* and $f_1(U)$ the *logit effect function*. Note that

$$\alpha_1 \approx .606$$
$$\neq E[f_1(U)] \approx .539 + .442 \cdot E(1_{U=Ann}) \approx .7599.$$

Hence, although X and U are independent (see Example 5.37), the slope α_1 of the logit in the logistic parameterization of $E(Y \mid X) = P(Y=1 \mid X)$ is *not* equal to the expectation of the logit effect function $f_1(U)$ of the logit in the logistic parameterization of $E(Y \mid X, U) = P(Y=1 \mid X, U)$.
   From a methodological point of view, this means that randomized assignment of a unit to one of two treatment conditions – which creates independence of a treatment variable X and the person variable U – does not imply that the slope α_1 of the logit in the logistic parameterization of $E(Y \mid X) = P(Y=1 \mid X)$ can be interpreted as an average effect of treatment variable on Y

[or on the logit of the linear logit parameterization of $E(Y \mid X)$]. In examples in which $f_1(U) = \lambda_1$ is a constant, this implies that $\alpha_1 = \lambda_1$ does not follow from independence of X and U. In contrast, compare the corresponding invariance property formulated in Theorem 12.44 for a linear parameterization. ◁

Remark 13.27 [Normal distribution and linear logistic regression] Let X be a real-valued continuous random variable, let $A \in \mathscr{A}$ with $0 < P(A) < 1$, and let $Y = 1_A$. Furthermore, assume that X has a normal distribution with respect to each of the two conditional-probability measures $P^{Y=y}$, $y = 0, 1$, and that $Var(X \mid Y=0) = Var(X \mid Y=1)$. Then $P(Y=1 \mid X)$ has a linear logistic parameterization (for a proof see Examples 17.80 to 17.82). ◁

Example 13.28 [Joe and Ann with latent abilities] Table 13.1 describes a random experiment that consists of sampling a person from the set $\Omega_U = \{Joe, Ann\}$ of persons and observing whether (+) or not (−) problem 1, 2, and 3, respectively, are solved. The probabilities of the elementary events are displayed in the second column, which can be used to compute the values of the conditional probabilities $P(Y_i=1 \mid U)$ for Joe and for Ann, or more precisely, the values $P(Y_i=1 \mid U)(\omega)$ if $\omega \in \{U=Joe\}$ and if $\omega \in \{U=Ann\}$. The last three columns present the logit transformations of these conditional probabilities, that is,

$$\mathrm{logit}_i := \mathrm{logit}[P(Y_i=1 \mid U)] = \ln \left(\frac{P(Y_i=1 \mid U)}{1 - P(Y_i=1 \mid U)} \right), \quad \forall\, i = 1, \dots, m, \quad (13.35)$$

(see Def. 13.5), where $m = 3$. The probability space $(\Omega, \mathscr{A}, P)$ is completely specified by $\Omega := \{\omega_1, \dots, \omega_{16}\}$, $\mathscr{A} := \mathscr{P}(\Omega)$, and the 16 probabilities $P(\{\omega_i\})$ of the outcomes $\omega_i \in \Omega$. Also note that the conditional probabilities $P(Y_i=1 \mid U)$ and their logits are completely determined by the 16 probabilities $P(\{\omega_i\})$.

The probability measure P is such that it satisfies the following two conditions for $m = 3$, the number of items considered in Table 13.1:

$$\forall\, i, j \in \{1, \dots, m\} \; \exists\, \beta_{ij} \in \mathbb{R}: \quad \mathrm{logit}_i - \mathrm{logit}_j = \beta_{ij}, \quad (13.36)$$

$$P(Y_i=1 \mid U, Y_1, \dots, Y_{i-1}, Y_{i+1}, \dots, Y_m) = P(Y_i=1 \mid U), \quad \forall\, i = 1, \dots, m. \quad (13.37)$$

While (13.36) is easily checked by inspecting the last three columns of Table 13.1, we delay showing that (13.37) holds to Example 16.51. If (13.36) and (13.37) hold, then we say that Y_1, Y_2, and Y_3 satisfy the (two assumptions of the) Rasch model (cf. Rasch, 1960/1980). Assumption (13.36) is called *Rasch homogeneity*. According to this assumption, the graphs of the logits of each pair of items are translations of each other. Assumption (13.37) postulates U-conditional mean independence of each item Y_i from the other items $Y_1, \dots, Y_{i-1}, Y_{i+1}, \dots, Y_m$ [see Def. 10.45 (ii) and Rem. 46 (ii)]. In chapter 16, it is shown that this assumption is equivalent to U-*conditional independence* of the random variables Y_i (see Example 16.51 for more details).

Now define

$$\xi := \mathrm{logit}_1 \quad \text{and} \quad \beta_i := \beta_{1i}, \quad \forall\, i = 1, \dots, m. \quad (13.38)$$

Table 13.1 Joe and Ann with a latent ability variable.

Outcomes ω (Unit, Problem 1, Problem 2, Problem 3)	$P(\{\omega\})$	U	Y_1	Y_2	Y_3	$P(Y_1=1\mid U)$	$P(Y_2=1\mid U)$	$P(Y_3=1\mid U)$	logit_1	logit_2	logit_3
$\omega_1 = (Joe, -, -, -)$	.0492	Joe	0	0	0	.5000	.7311	.2689	0	1	−1
$\omega_2 = (Joe, -, -, +)$	.0181	Joe	0	0	1	.5000	.7311	.2689	0	1	−1
$\omega_3 = (Joe, -, +, -)$	.1336	Joe	0	1	0	.5000	.7311	.2689	0	1	−1
$\omega_4 = (Joe, +, -, -)$	.0492	Joe	1	0	0	.5000	.7311	.2689	0	1	−1
$\omega_5 = (Joe, -, +, +)$	.0492	Joe	0	1	1	.5000	.7311	.2689	0	1	−1
$\omega_6 = (Joe, +, -, +)$	.0181	Joe	1	0	1	.5000	.7311	.2689	0	1	−1
$\omega_7 = (Joe, +, +, -)$	.1336	Joe	1	1	0	.5000	.7311	.2689	0	1	−1
$\omega_8 = (Joe, +, +, +)$	.0492	Joe	1	1	1	.5000	.7311	.2689	0	1	−1
$\omega_9 = (Ann, -, -, -)$	.0080	Ann	0	0	0	.7311	.8808	.5000	1	2	0
$\omega_{10} = (Ann, -, -, +)$	.0080	Ann	0	0	1	.7311	.8808	.5000	1	2	0
$\omega_{11} = (Ann, -, +, -)$	.0592	Ann	0	1	0	.7311	.8808	.5000	1	2	0
$\omega_{12} = (Ann, +, -, -)$	.0218	Ann	1	0	0	.7311	.8808	.5000	1	2	0
$\omega_{13} = (Ann, -, +, +)$	.0592	Ann	0	1	1	.7311	.8808	.5000	1	2	0
$\omega_{14} = (Ann, +, -, +)$	.0218	Ann	1	0	1	.7311	.8808	.5000	1	2	0
$\omega_{15} = (Ann, +, +, -)$	.1610	Ann	1	1	0	.7311	.8808	.5000	1	2	0
$\omega_{16} = (Ann, +, +, +)$	.1610	Ann	1	1	1	.7311	.8808	.5000	1	2	0

Note: Logits are exact, whereas the probabilities are rounded.

In applications in which the Y_i indicate whether or not a problem is solved, the random variable ξ is called the *latent ability variable*, and β_i the *difficulty parameter of problem i*, respectively. Inserting the definitions of ξ and β_i into (13.36), we receive

$$\text{logit}_i = \xi - \beta_i, \quad \forall\, i = 1, \dots, m. \tag{13.39}$$

Furthermore, inserting Equation (13.39) into (13.35) and solving it for $P(Y_i = 1 \mid U)$ yield

$$P(Y_i = 1 \mid U) = \frac{\exp(\xi - \beta_i)}{1 + \exp(\xi - \beta_i)} \tag{13.40}$$

$$= P(Y_i = 1 \mid \xi), \quad \forall\, i = 1, \dots, m. \tag{13.41}$$

According to Equation (13.41), the graphs of the logits of the conditional probabilities $P(Y_i = 1 \mid \xi)$ and $P(Y_j = 1 \mid \xi)$ are translations of each other. Figure 13.4 shows the graphs of the factorizations $g \colon \mathbb{R} \to [0, 1]$ of the conditional probabilities $P(Y_i = 1 \mid \xi) = g(\xi)$, that is, the graphs of the regressions of Y_i on ξ (see Def. 10.25). This figure illustrates that these graphs are translations (parallel to the ξ axis) of each other.

The difficulty parameters β_1, β_2, and β_3 can be computed as follows: Using Equations (13.36) and (13.38), and the logits displayed in Table 13.1, we receive

$$\begin{aligned}
\beta_1 &= \beta_{11} = \text{logit}_1 - \text{logit}_1 = 0, \\
\beta_2 &= \beta_{12} = \text{logit}_1 - \text{logit}_2 = -1, \\
\beta_3 &= \beta_{13} = \text{logit}_1 - \text{logit}_3 = 1.
\end{aligned} \tag{13.42}$$

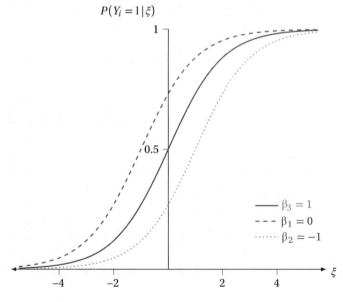

Figure 13.4   Graphs of the ξ-conditional probabilities of items Y_i satisfying the Rasch model.

Note that the definitions of the latent variable ξ and the difficulty parameters β_i are arbitrary to some degree. Defining $\xi^* := \xi + a$ and $\beta_i^* := \beta_i + a$ for any $a \in \mathbb{R}$ would do as well, that is, ξ^* and β_i^* also satisfy Equation (13.40).

Also note that the definition of ξ and Corollary 2.53 imply that there is a function $f: \Omega_U \to \mathbb{R}$ such that ξ is the composition $f \circ U = f(U)$ of U and f. The function f assigns to each $u \in \Omega_U$ its value $f(u)$, where

$$\forall\, u \in \Omega_U \;\; \forall\, \omega \in \{U=u\}\!: \quad f(u) = f[U(\omega)] = \xi(\omega) \qquad\qquad [(2.25)]$$
$$= \text{logit}[P(Y_1 = 1 \mid U)(\omega)] \quad [(13.35),\,(13.38)] \qquad (13.43)$$
$$= \text{logit}[P(Y_1 = 1 \mid U = u)]. \qquad\qquad [(10.9)]$$

Hence, the values of ξ are attributes of the persons in the set Ω_U. This justifies calling such a value the *ability of person u*, if the items Y_i indicate whether or not problem i is solved.

Finally, note that the values of the latent variable ξ are unknown in empirical applications. However, in Example 16.51 we treat some details on estimating the value $f(u)$ from the response pattern $(y_1, \ldots, y_m)$ of the items $Y_1, \ldots, Y_m$ that is observed for a person $u \in \Omega_U$ if the random experiment is actually conducted, that is, if we actually sample a person from a set of persons and observe, for $i = 1, \ldots, m$, whether or not problem i is solved.   ◁

13.6 Proofs

Proof of Lemma 13.7

Let $V: (\Omega, \mathscr{A}, P) \to (\mathbb{R}, \mathscr{B})$ be a version in $\mathscr{P}(Y=1 \mid \mathscr{C})$ with values in $]0, 1[$. Furthermore, let $g:]0, 1[\to \mathbb{R}$ be a continuous and strictly monotone function with $g(]0, 1[) = \mathbb{R}$. Then the inverse function $g^{-1}: \mathbb{R} \to]0, 1[$ exists (see Ellis & Gulick, 2006, section 7.1), and it is continuous (Ellis & Gulick, 2006, Th. 7.4) and strictly monotone with $g^{-1}(\mathbb{R}) =]0, 1[$. Hence,

$$\sigma[g(V)]$$
$$= (g \circ V)^{-1}(\mathscr{B}) \qquad\qquad\qquad\qquad\qquad [(2.14)]$$
$$= V^{-1} \circ g^{-1}(\mathscr{B}) \qquad\qquad\qquad\qquad\qquad [(2.29)]$$
$$= V^{-1} \circ g^{-1}[\sigma(\{]\!-\!\infty, b]: b \in \mathbb{R}\})] \qquad\qquad [(1.19)]$$
$$= \sigma[V^{-1} \circ g^{-1}(\{]\!-\!\infty, b]: b \in \mathbb{R}\})] \qquad\qquad [(2.12)]$$
$$= \sigma[V^{-1}(\{]\!-\!\infty, g^{-1}(b)]: b \in \mathbb{R}\})] \qquad [\text{monotonicity, continuity}]$$
$$= \sigma[V^{-1}(]\!-\!\infty, c]: c \in\,]0, 1[\})] \qquad\qquad [\text{domain of } g]$$
$$= \sigma[V^{-1}(\{]\!-\!\infty, b]: b \in \mathbb{R}\})]$$
$$\qquad\qquad [b \leq 0\!: V^{-1}(]\!-\!\infty, b]) = \varnothing,\; b \geq 1\!: V^{-1}(]\!-\!\infty, b]) = \mathbb{R}]$$
$$= V^{-1}[\sigma(\{]\!-\!\infty, b]: b \in \mathbb{R}\})] \qquad\qquad [(2.12)]$$
$$= V^{-1}(\mathscr{B}) \qquad\qquad\qquad\qquad\qquad\qquad [(1.19)]$$
$$= \sigma(V)\,. \qquad\qquad\qquad\qquad\qquad\qquad\quad [(2.14)]$$

The result $\sigma[g(V)] = \sigma(V)$ can now be applied to $g = \text{logit}$ [see Eq. (13.2)].

Proof of Theorem 13.14

By definition, $P(Y=1 \mid Z) = E(1_{Y=1} \mid Z)$. Hence, the existence of coefficients $\beta_0, \beta_1, \dots, \beta_n$ and a version $P(Y=1 \mid Z) \in \mathcal{P}(Y=1 \mid Z)$ satisfying Equations (13.15), (13.17), and (13.19) has already been proved in Theorem 12.37. In order to show that there are $\lambda_0, \lambda_1, \dots, \lambda_n$ satisfying Equation (13.16), we define

$$\lambda_0 := \mathrm{logit}[P(Y=1 \mid Z=z_0)], \tag{13.44}$$

[see Eq. (13.2)] and

$$\lambda_i := \mathrm{logit}[P(Y=1 \mid Z=z_i)] - \mathrm{logit}[P(Y=1 \mid Z=z_0)], \quad \forall\, i = 1, \dots, n. \tag{13.45}$$

These definitions and Equation (13.3) then yield

$$P(Y=1 \mid Z=z_0) = \frac{\exp(\mathrm{logit}[P(Y=1 \mid Z=z_0)])}{1 + \exp(\mathrm{logit}[P(Y=1 \mid Z=z_0)])} = \frac{\exp(\lambda_0)}{1 + \exp(\lambda_0)},$$

and

$$P(Y=1 \mid Z=z_i) = \frac{\exp(\mathrm{logit}[P(Y=1 \mid Z=z_i)])}{1 + \exp(\mathrm{logit}[P(Y=1 \mid Z=z_i)])} = \frac{\exp(\lambda_0 + \lambda_i)}{1 + \exp(\lambda_0 + \lambda_i)}, \quad \forall\, i = 1, \dots, n.$$

Hence, Equation (13.17) implies

$$\beta_0 = P(Y=1 \mid Z=z_0) = \frac{\exp(\lambda_0)}{1 + \exp(\lambda_0)},$$

and Equation (13.19) yields

$$\begin{aligned}
\beta_i &= P(Y=1 \mid Z=z_i) - P(Y=1 \mid Z=z_0) \\
&= \frac{\exp(\lambda_0 + \lambda_i)}{1 + \exp(\lambda_0 + \lambda_i)} - \frac{\exp(\lambda_0)}{1 + \exp(\lambda_0)}, \quad \forall\, i = 1, \dots, n.
\end{aligned}$$

Proof of Theorem 13.20

Denote $x := [X_1, \dots, X_n]'$, $\mu := [E(X_1), \dots, E(X_n)]'$, $\lambda = [\lambda_1, \dots, \lambda_n]'$, as well as

$$L := \mathrm{logit}[P(Y=1 \mid \mathscr{C})] = \lambda_0 + \lambda' x$$

[see Eqs. (13.27) and (13.9)]. Taking the expectation on both sides, using the definition of μ, and rearranging yield

$$\lambda_0 = E(L) - \lambda' \mu.$$

Furthermore, consider the n-dimensional covariance vector

$$\Sigma_{xl} = \Sigma_{x,\lambda_0 + \lambda'x} \qquad\qquad [(13.27)]$$
$$= \Sigma_{xx}\,\lambda. \qquad\qquad \text{[Box 7.3 (ii), (iii)]}$$

Multiplying both sides by Σ_{xx}^{-1} yields

$$\lambda = \Sigma_{xx}^{-1}\,\Sigma_{xl}.$$

This equation also shows that the vector λ is uniquely defined, and this implies that λ_0 is uniquely defined as well. Uniqueness of λ_0 and λ implies that the linear logit parameterization g satisfying Equation (13.30) is uniquely defined as well.

Exercises

13.1 Prove the propositions of Remark 13.2.

13.2 Prove Equation (13.3).

13.3 Calculate the derivative (13.14).

13.4 Consider Example 13.24 and compute the coefficients of the linear logit parameterization of $E(Y \mid X)$.

13.5 Consider Example 13.24 and compute the coefficients of the linear logit parameterization of $E(Y \mid X, U)$.

Solutions

13.1 Note that $1_{Y=0} \cdot Y = 1_{Y=0} \cdot 0 = 0$ [see Eq. (2.49)], Hence, if $\{Y=0\} \in \mathcal{C}$, then Definition 10.2 (b) yields

$$\int 1_{Y=0} \cdot P(Y=1 \mid \mathcal{C})\ dP = \int 1_{Y=0} \cdot Y\ dP = \int 0\ dP = 0. \qquad (13.46)$$

Let $P(Y=1 \mid \mathcal{C}) \in \mathcal{P}(Y=1 \mid \mathcal{C})$ be a version with $P(Y=1 \mid \mathcal{C}) \geq 0$ [for its existence, see Box 10.3 (v)]. Then Theorem 3.43 yields $1_{Y=0} \cdot P(Y=1 \mid \mathcal{C}) \underset{P}{=} 0$, that is, $P(A) = 0$ for $A := \{\omega \in \Omega : 1_{Y=0}(\omega) \cdot P(Y=1 \mid \mathcal{C})(\omega) \neq 0\}$. This implies $P(A^c) = 1$ and

$$\{Y=0\} \cap A^c \subset \{P(Y=1 \mid \mathcal{C}) = 0\} \cap A^c,$$

which in turn yields

$$
\begin{aligned}
&P(P(Y=1 \mid \mathcal{C}) = 0) \\
&= P(\{P(Y=1 \mid \mathcal{C}) = 0\} \cap A^c) \qquad\qquad &&\text{[Box 4.1 (viii)]} \\
&\geq P(\{Y=0\} \cap A^c) \qquad\qquad &&\text{[Box 4.1 (v)]} \\
&= P(Y=0) \qquad\qquad &&\text{[Box 4.1 (viii)]} \\
&> 0. \qquad\qquad &&\text{[Y dichotomous, Example 5.10]}
\end{aligned}
$$

Hence, we have shown that $\{Y=0\} \in \mathscr{C}$ implies $P(P(Y=1 \mid \mathscr{C}) = 0) > 0$, which in turn implies that there is no version $P(Y=1 \mid \mathscr{C}) > 0$, that is, there is no version $P(Y=1 \mid \mathscr{C})$ with $\{\omega \in \Omega \colon P(Y=1 \mid \mathscr{C})(\omega) \leq 0\} = \emptyset$ [see Rem. 2.77 and (10.12)]. Now,

$$\{Y=1\} \in \mathscr{C}$$

$$\Rightarrow \int 1_{Y=1} \cdot Y \, dP = \int 1_{Y=1} \cdot P(Y=1 \mid \mathscr{C}) \, dP \qquad \text{[Def. 10.2 (b)]}$$

$$\Rightarrow \int (1 - 1_{Y=0}) \cdot Y \, dP = \int (1 - 1_{Y=0}) \cdot P(Y=1 \mid \mathscr{C}) \, dP$$

$$\text{[Y dichotomous, (5.33)]}$$

$$\Rightarrow E(Y) - \int 1_{Y=0} \cdot Y \, dP = E(Y) - \int 1_{Y=0} \cdot P(Y=1 \mid \mathscr{C}) \, dP \qquad \text{[(3.33), (6.1)]}$$

$$\Rightarrow \int 1_{Y=0} \cdot Y \, dP = \int 1_{Y=0} \cdot P(Y=1 \mid \mathscr{C}) \, dP. \qquad \text{[$E(Y)$ is finite]}$$

The last equation is identical to (13.46). Hence, we can also conclude that there is no version $P(Y=1 \mid \mathscr{C}) > 0$ if $\{Y=1\} \in \mathscr{C}$.

13.2

$$\exp[\text{logit}(p)] = \exp\left[\ln\left(\frac{p}{1-p}\right)\right] = \frac{p}{1-p}.$$

Hence,

$$p = \frac{\dfrac{p}{1-p}}{\dfrac{1}{1-p}} = \frac{\dfrac{p}{1-p}}{\dfrac{1-p+p}{1-p}} = \frac{\dfrac{p}{1-p}}{\dfrac{1-p}{1-p} + \dfrac{p}{1-p}} = \frac{\dfrac{p}{1-p}}{1 + \dfrac{p}{1-p}} = \frac{\exp[\text{logit}(p)]}{1 + \exp[\text{logit}(p)]}.$$

13.3   The chain rule and the quotient rule of differential calculus yield

$$\frac{d}{dx} g(x) = \frac{d}{dx} \frac{\exp(\lambda_0 + \lambda_1 x)}{1 + \exp(\lambda_0 + \lambda_1 x)}$$

$$= \frac{\lambda_1 \exp(\lambda_0 + \lambda_1 x)(1 + \exp(\lambda_0 + \lambda_1 x)) - \exp(\lambda_0 + \lambda_1 x) \cdot \lambda_1 \exp(\lambda_0 + \lambda_1 x)}{(1 + \exp(\lambda_0 + \lambda_1 x))^2}$$

$$= \frac{\lambda_1 \exp(\lambda_0 + \lambda_1 x)(1 + \exp(\lambda_0 + \lambda_1 x) - \exp(\lambda_0 + \lambda_1 x))}{(1 + \exp(\lambda_0 + \lambda_1 x))^2}$$

$$= \frac{\lambda_1 \exp(\lambda_0 + \lambda_1 x)}{(1 + \exp(\lambda_0 + \lambda_1 x))^2}.$$

13.4   Inserting the value $x = 0$ in the equation

$$\text{logit}[P(Y=1 \mid X=x)] = \ln\left(\frac{P(Y=1 \mid X=x)}{1 - P(Y=1 \mid X=x)}\right) = \alpha_0 + \alpha_1 x$$

[see Eqs. (13.2) and (13.13)] yields

$$\text{logit}[P(Y=1 \mid X=0)] = \ln\left(\frac{.45}{1-.45}\right) \approx -.201 \approx \alpha_0,$$

and inserting the value $x = 1$ yields

$$\text{logit}[P(Y=1 \mid X=1)] = \ln\left(\frac{.6}{1-.6}\right) \approx .406 \approx \alpha_0 + \alpha_1.$$

[In R, these values are obtained by $\texttt{qlogis}(.45)$ and $\texttt{qlogis}(.60)$, respectively.] Solving the last equation yields $\alpha_1 \approx .406 - (-.201) = .606$.

13.5   For $x_1 = 0$, $x_2 = 0$, and $x_3 = 0$, the equation

$$\text{logit}[P(Y=1 \mid X=0, U=Ann)] = \ln\left(\frac{P(Y=1 \mid X=0, U=Ann)}{1 - P(Y=1 \mid X=0, U=Ann)}\right)$$
$$= \lambda_0 + \lambda_1 x_1 + \lambda_2 x_2 + \lambda_3 x_3$$

[see Eqs. (13.2) and (13.13)] yields

$$\ln\left(\frac{P(Y=1 \mid X=0, U=Joe)}{1 - P(Y=1 \mid X=0, U=Joe)}\right) = \ln\left(\frac{.7}{1-.7}\right) \approx -.847 = \lambda_0,$$

for $x_1 = 1$, $x_2 = 0$, and $x_3 = 0$ it yields

$$\ln\left(\frac{P(Y=1 \mid X=1, U=Joe)}{1 - P(Y=1 \mid X=1, U=Joe)}\right) = \ln\left(\frac{.8}{1-.8}\right) \approx 1.386 \approx \lambda_0 + \lambda_1,$$

for $x_1 = 0$, $x_2 = 1$, and $x_3 = 0$ it yields

$$\ln\left(\frac{P(Y=1 \mid X=0, U=Ann)}{1 - P(Y=1 \mid X=0, U=Ann)}\right) = \ln\left(\frac{.2}{1-.2}\right) \approx -1.386 \approx \lambda_0 + \lambda_2,$$

and for $x_1 = 1$, $x_2 = 1$, and $x_3 = 1$ it yields

$$\ln\left(\frac{P(Y=1 \mid X=1, U=Ann)}{1 - P(Y=1 \mid X=1, U=Ann)}\right) = \ln\left(\frac{.4}{1-.4}\right) \approx -.406 \approx \lambda_0 + \lambda_1 + \lambda_2 + \lambda_3.$$

[In R, these values are obtained by $\texttt{qlogis}(.7)$ to $\texttt{qlogis}(.4)$.] Solving the second equation yields $\lambda_1 \approx .539$, solving the third equation yields $\lambda_2 \approx -2.234$, and solving the last one yields $\lambda_3 \approx .442$.

14

Conditional expectation with respect to a conditional-probability measure

In chapter 10, we treated the concept of a $\mathscr{C}$-conditional expectation $E(Y \mid \mathscr{C})$ with respect to a probability measure P on a measurable space $(\Omega, \mathscr{A})$. In this chapter, we introduce the concept of a $\mathscr{C}$-conditional expectation $E^B(Y \mid \mathscr{C})$ of Y with respect to the *conditional-probability measure P^B* on $(\Omega, \mathscr{A})$ (see Def. 4.29). A special case with $\mathscr{C} = \sigma(X)$ is the X-conditional expectation of Y with respect to P^B, which is also denoted by $E^B(Y \mid X)$. If $B = \{Z=z\}$ is the event that a random variable Z on $(\Omega, \mathscr{A}, P)$ takes on the value z and $P(Z=z) > 0$, then we use the notation $E^{Z=z}(Y \mid X)$ and call it a version of the *X-conditional expectation of Y with respect to $P^{Z=z}$*.

In empirical applications, the conditional expectation $E^{Z=z}(Y \mid X)$ can be used to describe how the conditional expectation values of Y depend on the values x of X given that Z takes on the value z. The dependency of Y on X described by $E^{Z=z}(Y \mid X)$ may not only differ for different values z_1 and z_2 of Z, but also differ from the dependency described by the X-conditional expectation $E(Y \mid X)$ of Y with respect to P. If, for instance, X denotes a treatment variable and $Z = sex$ with values m (male) and f (female), then $E^{Z=m}(Y \mid X)$ and $E^{Z=f}(Y \mid X)$ refer to the X-conditional expectation of Y for males and females, respectively. In a data sample, these are the conditional expectations estimated using only the y-values and x-values obtained within the male and female subsamples, respectively. In contrast, $E^{X=x}(Y \mid Z)$ refers to the Z-conditional expectation of Y given treatment x, and this is the conditional expectation estimated in the analysis of experimental or quasi-experimental data using only the y-values and z-values obtained in treatment condition x. If the treatment variable X is dichotomous with values 0 (control) and 1 (treatment), then $g_1(Z) := E^{X=1}(Y \mid Z) - E^{X=0}(Y \mid Z)$ and g_1 is the Z-conditional-effect function of X. The values $g_1(z)$ are the effects of X on Y given the value z of Z.

Probability and Conditional Expectation: Fundamentals for the Empirical Sciences, First Edition. Rolf Steyer and Werner Nagel.
© 2017 John Wiley & Sons, Ltd. Published 2017 by John Wiley & Sons, Ltd.
Companion website: http://www.probability-and-conditional-expectation.de

Table 14.1 Joe and Ann with randomized assignment: conditional expectations with respect to $P^{X=x}$.

Outcomes ω (Unit, Treatment, Success)	$P(\{\omega\})$	Person variable U	Treatment variable X	Outcome variable Y	$E(Y\mid X, U)$	$E(Y\mid X)$	$P(X=1\mid U)$	$E^{X=0}(Y\mid U)$	$E^{X=1}(Y\mid U)$	$g_1(U)$	$P^{X=0}(\{\omega\})$	$P^{X=1}(\{\omega\})$
(Joe, no, −)	.09	Joe	0	0	.7	.45	.4	.7	.8	.1	.15	0
(Joe, no, +)	.21	Joe	0	1	.7	.45	.4	.7	.8	.1	.35	0
(Joe, yes, −)	.04	Joe	1	0	.8	.6	.4	.7	.8	.1	0	.1
(Joe, yes, +)	.16	Joe	1	1	.8	.6	.4	.7	.8	.1	0	.4
(Ann, no, −)	.24	Ann	0	0	.2	.45	.4	.2	.4	.2	.4	0
(Ann, no, +)	.06	Ann	0	1	.2	.45	.4	.2	.4	.2	.1	0
(Ann, yes, −)	.12	Ann	1	0	.4	.6	.4	.2	.4	.2	0	.3
(Ann, yes, +)	.08	Ann	1	1	.4	.6	.4	.2	.4	.2	0	.2

14.1 Introductory examples

Example 14.1 [Joe and Ann with randomized assignment – continued] Table 14.1 displays the random variables U, X, and Y, and, among other things, the conditional expectations $E(Y\mid X)$ and $E(Y\mid X, U)$, the values of which have already been computed in Example 9.21. Here, the conditional expectation $E(Y\mid X, U)$ is uniquely defined. According to Remark 12.23, it has a linear parameterization in $(1_{U=Joe}, X, 1_{U=Joe}\cdot X)$. The coefficient of this linear parameterization can be computed analogously to Example 12.24, which yields

$$E(Y\mid X, U) = .2 + .5\cdot 1_{U=Joe} + .2\cdot X - .1\cdot 1_{U=Joe}\cdot X$$
$$= (.2 + .5\cdot 1_{U=Joe}) + (.2 - .1\cdot 1_{U=Joe})\cdot X \qquad (14.1)$$
$$= g_0(U) + g_1(U)\cdot X.$$

In this equation, g_0 is the U-conditional-*intercept function* that assigns the person-specific intercept to each value u of the person variable U, and g_1 is the U-conditional-*effect function* that assigns the person-specific effect of X on Y to each value u of U. The function $g_1(U)$ is a random variable on $(\Omega, \mathscr{A}, P)$. It is the composition of the person variable U and the effect function g_1. In this example, the treatment effect for *Joe* is $g_1(Joe) = .2 - .1\cdot 1 = .1$, and for *Ann* it is $g_1(Ann) = .2 - .1\cdot 0 = .2$. Hence,

$$g_1(U)(\omega) = .2 - .1\cdot 1_{U=Joe}(\omega) = \begin{cases} .1, & \text{if } \omega \in \{U=Joe\} \\ .2, & \text{if } \omega \in \{U=Ann\}. \end{cases}$$

In Theorem 15.3, we show that Equation (14.1) implies

$$g_0(U) = E^{X=0}(Y \mid U) \tag{14.2}$$

and

$$g_1(U) = E^{X=1}(Y \mid U) - E^{X=0}(Y \mid U), \tag{14.3}$$

where $E^{X=x}(Y \mid U)$, $x = 0, 1$, denotes the U-conditional expectation of Y with respect to the conditional-probability measure $P^{X=x}$.

Furthermore, because $E(1_{U=Joe}) = P(U=Joe) = .5$ [see Eq. (6.4)], the expectation of the effect function is

$$E_U(g_1) = E[g_1(U)]$$
$$= E(.2 - .1 \cdot 1_{U=Joe}) = .2 - .1 \cdot E(1_{U=Joe}) = .2 - .1 \cdot .5 = .15 \tag{14.4}$$

[see Eq. (6.13)]. For simplicity, this expectation is also called the *average treatment effect*. In this example, $E[g_1(U)]$ is also the slope of X in the equation

$$E(Y \mid X) = .45 + .15 \cdot X, \tag{14.5}$$

which can be computed by applying the equations of Example 12.15 (see also Example 12.16).

From a methodological point of view, note that in this example the function g_1 and its expectation have a causal interpretation as a U-conditional-effect function and the average effect of the treatment. Furthermore, the slope of X in Equation (14.5) is also identical to the average causal effect of the treatment variable X. As shown in Corollary 15.18, this follows from independence of X and U [see the column headed $P(X=1 \mid U)$ in Table 14.1, Example 5.37, and Rem. 10.59]. In empirical applications, in which we draw one person and then assign him or her to one of two treatment conditions (see Table 14.1), independence of X and U is created by assigning the drawn person to treatment x *with identical probabilities for all persons* such that $P(X=1 \mid U) = P(X=1)$ (see Rem. 10.50).   ◁

Example 14.2 [Joe and Ann with self-selection – continued]   Table 14.2 shows another example with Joe and Ann. Some of the conditional expectations displayed have already been presented in Table 11.2. In Table 14.2, the treatment probability *does* depend on the person drawn, that is, $P(X=1 \mid U) \neq P(X=1)$ [see the column headed $P(X=1 \mid U)$]. Therefore, according to Remark 10.50, X and U are not independent. In applied statistics, such a dependence between X and U is often created by 'self-selection' into one of the treatment conditions.

Inspecting the columns headed by $E(Y \mid X, U)$ in Tables 14.1 and 14.2 shows that these columns are identical. Therefore, Equation (14.1) still holds for $E(Y \mid X, U)$, and this implies that the conditional treatment effect for *Joe* is again $g_1(Joe) = .1$, and for *Ann* it is $g_1(Ann) = .2$. Furthermore, because $P(U=Joe) = .5$ still holds, the average treatment effect is again $E[g_1(U)] = .15$ [see also Eq. (14.4)]. In contrast to Example 14.1, $E[g_1(U)]$ is *not identical* to the slope of X in the equation

$$E(Y \mid X) = .60 - .18 \cdot X, \tag{14.6}$$

which can be computed by applying the equations of Example 12.15 (see also Example 12.16). Now the slope $-.18$ has no causal interpretation, because all individual treatment effects are

Table 14.2   Joe and Ann with self-selection: conditional expectations with respect to $P^{X=x}$.

Unit	Treatment	Success	$P(\{\omega\})$	Person variable U	Treatment variable X	Outcome variable Y	$E(Y \mid X, U)$	$E(Y \mid X)$	$P(X = 1 \mid U)$	$E^{X=0}(Y \mid U)$	$E^{X=1}(Y \mid U)$	$g_1(U)$	$P^{X=0}(\{\omega\})$	$P^{X=1}(\{\omega\})$
(Joe, no, −)			.144	Joe	0	0	.7	.6	.04	.7	.8	.1	.24	0
(Joe, no, +)			.336	Joe	0	1	.7	.6	.04	.7	.8	.1	.56	0
(Joe, yes, −)			.004	Joe	1	0	.8	.42	.04	.7	.8	.1	0	.01
(Joe, yes, +)			.016	Joe	1	1	.8	.42	.04	.7	.8	.1	0	.04
(Ann, no, −)			.096	Ann	0	0	.2	.6	.76	.2	.4	.2	.16	0
(Ann, no, +)			.024	Ann	0	1	.2	.6	.76	.2	.4	.2	.04	0
(Ann, yes, −)			.228	Ann	1	0	.4	.42	.76	.2	.4	.2	0	.57
(Ann, yes, +)			.152	Ann	1	1	.4	.42	.76	.2	.4	.2	0	.38

positive (.1 for Joe, .2 for Ann), whereas this slope is negative. Hence, this slope would be extremely misleading if used for the evaluation of the treatment. Note that the phenomenon that $E[g_1(U)]$ and the slope of X in Equation (14.6) are *not identical* can only occur if X and U are not independent (see Theorem 15.14).

Hence, while the function g_1 and its expectation $E_U(g_1) = E[g_1(U)]$ can still be causally interpreted as a U-conditional-effect function and the average effect of the treatment, respectively, the slope of X in Equation (14.6) does not have a causal meaning.   ◁

Example 14.3 [No treatment for Joe – continued]   Table 14.3 displays a third example with Joe and Ann. Note that the values of the versions $E(Y \mid X)$ and $E(Y \mid X, U)$ in this table are specified for all $\omega \in \Omega$ (see Examples 9.22, 9.23, and Exercise 14.1). However, in this example, there are infinitely many versions of the conditional expectation $E(Y \mid X, U)$. In Table 9.2, we already specified a version $V \in \mathscr{E}(Y \mid X, U)$ with

$$V(\omega) = 0, \quad \text{if } \omega \in \{(Joe, yes, -), (Joe, yes, +)\},$$

and in Example 10.19, we noted that assigning any other real number instead would do as well. For instance, assigning

$$V^*(\omega) = 9, \quad \text{if } \omega \in \{(Joe, yes, -), (Joe, yes, +)\},$$

we define a second version $V^* \in \mathscr{E}(Y \mid X, U)$, provided, of course, that the other values are assigned as in Table 9.2. The version $V^* = E(Y \mid X, U)$ is displayed in Table 14.3. In Example 10.19, we also noted that two versions V and V^* of $E(Y \mid X, U)$ are P-equivalent.

Table 14.3 No treatment for Joe: conditional expectations with respect to $P^{X=x}$.

Unit	Treatment	Success	$P(\{\omega\})$	Person variable U	Treatment variable X	Outcome variable Y	$E(Y\mid X,U)$	$E(Y\mid X)$	$P(X=1\mid U)$	$E^{X=0}(Y\mid U)$	$E^{X=1}(Y\mid U)$	$P^{X=0}(\{\omega\})$ (rounded)	$P^{X=1}(\{\omega\})$
	Outcomes ω			Observables				Conditional expectations					
(Joe, no, −)			.152	Joe	0	0	.696	.6	0	.696	9	.245	0
(Joe, no, +)			.348	Joe	0	1	.696	.6	0	.696	9	.561	0
(Joe, yes, −)			0	Joe	1	0	9	.4	0	.696	9	0	0
(Joe, yes, +)			0	Joe	1	1	9	.4	0	.696	9	0	0
(Ann, no, −)			.096	Ann	0	0	.2	.6	.76	.2	.4	.155	0
(Ann, no, +)			.024	Ann	0	1	.2	.6	.76	.2	.4	.039	0
(Ann, yes, −)			.228	Ann	1	0	.4	.4	.76	.2	.4	0	.6
(Ann, yes, +)			.152	Ann	1	1	.4	.4	.76	.2	.4	0	.4

For the version V [see the column headed $E(Y\mid X, U)$ in Table 9.2], we can write

$$V = .2 + .496 \cdot 1_{U=Joe} + .2 \cdot X - .896 \cdot 1_{U=Joe} \cdot X$$
$$= (.2 + .496 \cdot 1_{U=Joe}) + (.2 - .896 \cdot 1_{U=Joe}) \cdot X \qquad (14.7)$$
$$= g_0(U) + g_1(U) \cdot X.$$

(The coefficients of this equation are obtained analogously as in Example 12.24.) In this equation, $g_0(U) := .2 + .496 \cdot 1_{U=Joe}$ holds for the U-conditional-*intercept function*, and

$$g_1(U) := .2 - .896 \cdot 1_{U=Joe}$$

holds for the U-conditional-*effect function* g_1. For the version V, the value of g_1 for *Joe* is $g_1(Joe) = .2 - .896 \cdot 1 = -.696$, and for *Ann* it is $g_1(Ann) = .2 - .896 \cdot 0 = .2$.

Furthermore, the version V^* satisfies

$$V^* = .2 + .496 \cdot 1_{U=Joe} + .2 \cdot X + 8.104 \cdot 1_{U=Joe} \cdot X$$
$$= (.2 + .496 \cdot 1_{U=Joe}) + (.2 + 8.104 \cdot 1_{U=Joe}) \cdot X \qquad (14.8)$$
$$= g_0^*(U) + g_1^*(U) \cdot X.$$

In contrast to the two different versions of $E(Y\mid X, U)$, different versions of the U-conditional-effect function are not necessarily P_U-equivalent. For instance, if we consider the version $V \in \mathscr{E}(Y\mid X, U)$ specified above, then the associated effect function g_1 has the two values

$$g_1(Joe) = .2 - .896 \cdot 1 = -.696 \quad \text{and} \quad g_1(Ann) = .2 - .896 \cdot 0 = .2,$$

whereas considering the version $V^* \in \mathscr{E}(Y \mid X, U)$ specified above, the associated effect function $g_1^*(U)$ has the two values

$$g_1^*(Joe) = .2 + 8.104 \cdot 1 = 8.304 \quad \text{and} \quad g_1^*(Ann) = .2 + 8.104 \cdot 0 = .2.$$

Consequently, the expectations of different versions of the effect function also differ from
each other. Because $P(U=Joe) = .152 + .348 + 0 + 0 = .5$ and $P(U=Ann) = .096 + .024 + .228 + .152 = .5$, the expectation of g_1 is

$$E_U(g_1) = E[g_1(U)] = -.696 \cdot P(U=Joe) + .2 \cdot P(U=Ann) = -.696 \cdot .5 + .2 \cdot .5 = -.248,$$

whereas the expectation of g_1^* is

$$E_U(g_1^*) = E[g_1^*(U)] = 8.304 \cdot P(U=Joe) + .2 \cdot P(U=Ann) = 8.304 \cdot .5 + .2 \cdot .5 = 4.252.$$

For both computations, we used Equation (6.15).   ◁

Remark 14.4 [Methodological conclusions] Examples 14.1 to 14.3 show that the notions
effect function, conditional effects, and average effects are crucial for the evaluation of treatments, interventions, and expositions. This applies not only to U-conditional effects but also to
effects conditioning on other variables, say Z, such as $Z = gender$, $Z = severity\ of\ symptoms$,
$Z = educational\ status$, and so on. Our examples with a dichotomous treatment variable X
with values 0 and 1 show: Although a conditional expectation $E(Y \mid X, U)$ is uniquely defined
up to P-equivalence, this guarantees neither that the effect function is uniquely defined up
to P_U-equivalence nor that the expectations of g_1 and g_1^* are identical for different versions
g_1, g_1^* of the effect function. This suggests that we need to learn more about the effect function and the conditional expectations $E^{X=0}(Y \mid U)$ and $E^{X=1}(Y \mid U)$ [see Eq. (14.3)]. In more
general terms, we need to learn more about a conditional expectation $E^B(Y \mid Z)$ with respect
to a conditional-probability measure P^B.   ◁

14.2 Assumptions and definitions

In section 4.2, we considered a probability space $(\Omega, \mathscr{A}, P)$ and an event $B \in \mathscr{A}$ with $P(B) > 0$.
According to Theorem 4.28, the function $P^B: \mathscr{A} \rightarrow [0, 1]$ defined by

$$P^B(A) = P(A \mid B) = \frac{P(A \cap B)}{P(B)}, \quad \forall A \in \mathscr{A}, \tag{14.9}$$

is a probability measure on $(\Omega, \mathscr{A})$ called the *B-conditional-probability measure*. We also
noted that $(\Omega, \mathscr{A}, P^B)$ is a probability space sharing the measurable space $(\Omega, \mathscr{A})$ with the
original probability space $(\Omega, \mathscr{A}, P)$.

In section 9.1, we used the conditional-probability measure P^B in order to introduce the
B-conditional expectation value

$$E(Y \mid B) = E^B(Y) = \int Y \, dP^B, \tag{14.10}$$

assuming $P(B) > 0$ [see Eq. (9.2)] and that the expectation of Y with respect to P^B exists (see
Def. 6.1), that is, that Y is quasi-integrable with respect to P^B (see Def. 3.28). Hence, $E(Y \mid B)$

denotes the B-conditional expectation of Y and, by definition, it is identical to the expectation $E^B(Y)$ of Y with respect to P^B. If Z is a random variable on $(\Omega, \mathcal{A}, P)$ and $B = \{Z=z\} = \{\omega \in \Omega: Z(\omega) = z\}$ with $P(Z=z) > 0$, then we use the notation $E^{Z=z}(Y)$ and

$$E(Y \mid Z=z) = E^{Z=z}(Y) \tag{14.11}$$

instead of $E^B(Y)$ as well as $P^{Z=z}$ instead of P^B. Note, however, that $E(Y \mid Z=z)$ is also defined if $P(Z=z) = 0$. [For the definition see Eq. (10.27), and for uniqueness see Rem. 10.28.]

In this section, we often refer to the following assumptions and notation:

Notation and assumptions 14.5
Let $Y: (\Omega, \mathcal{A}, P) \to (\overline{\mathbb{R}}, \overline{\mathcal{B}})$ be a random variable, $B \in \mathcal{A}$ with $P(B) > 0$, $\mathcal{C} \subset \mathcal{A}$ a σ-algebra, and P^B defined by Equation (14.9). Furthermore, assume that Y is nonnegative or such that $E^B(Y)$ is finite.

Remark 14.6 [Finite expectation with respect to P^B] If Y is a random variable with finite expectation $E(Y)$, then $E^B(Y)$ is finite, too (see Exercise 14.2). In contrast, finiteness of $E^B(Y)$ does not imply that $E(Y)$ is finite. ◁

In chapter 10, we defined a version of a $\mathcal{C}$-conditional expectation $E(Y \mid \mathcal{C})$ with respect to the measure P (see Def. 10.2). For convenience, we repeat this definition of a $\mathcal{C}$-conditional expectation, but now it is for a conditional-probability measure P^B defined by Equation (14.9).

Reading the following definition, note that V, X, Y are random variables on the probability space $(\Omega, \mathcal{A}, P)$ if and only if they are random variables on $(\Omega, \mathcal{A}, P^B)$, provided, of course, that $P(B) > 0$ so that P^B is defined (see Exercise 14.3). Also remember that $\sigma(V) = V^{-1}(\overline{\mathcal{B}})$ denotes the σ-algebra generated by the random variable $V: (\Omega, \mathcal{A}, P) \to (\overline{\mathbb{R}}, \overline{\mathcal{B}})$ (see section 2.3.2).

Definition 14.7 [$\mathcal{C}$-conditional expectation with respect to P^B]
Let the assumptions 14.5 hold. A random variable $V: (\Omega, \mathcal{A}, P) \to (\overline{\mathbb{R}}, \overline{\mathcal{B}})$ is called a version of the $\mathcal{C}$-conditional expectation of Y with respect to P^B, if the following two conditions hold:

(a) $\sigma(V) \subset \mathcal{C}$.

(b) $E^B(1_C \cdot V) = E^B(1_C \cdot Y), \quad \forall C \in \mathcal{C}$.

If V satisfies (a) and (b), then we also use the notation $E^B(Y \mid \mathcal{C}) := V$.

According to Equation (6.1), condition (b) of this definition is equivalent to

$$\int 1_C \cdot V \, dP^B = \int 1_C \cdot Y \, dP^B, \quad \forall C \in \mathcal{C}. \tag{14.12}$$

This equation shows more clearly how the measure P^B is involved in the definition of $E^B(Y \mid \mathcal{C})$.

Remark 14.8 [Existence] The only difference between Definitions 10.2 and 14.7 is that, instead of referring to P, now we refer to the conditional-probability measure P^B defined by Equation (14.9). Hence, Theorem 10.9 (i) implies that, under the assumptions of Definition 14.7, there exists at least one version $E^B(Y \mid \mathscr{C})$.   ◁

Remark 14.9 [Uniqueness] Theorem 10.9 (ii) implies that two versions V and V^* of the $\mathscr{C}$-conditional expectation of Y with respect to P^B are P^B-equivalent. In other words, $E^B(Y \mid \mathscr{C})$ is P^B-unique (see Rem. 10.13).   ◁

Remark 14.10 [Set of all versions of the $\mathscr{C}$-conditional P^B-expectation] The notation $E^B(Y \mid \mathscr{C})$ refers to a *version* of the $\mathscr{C}$-conditional expectation of Y with respect to P^B. In contrast, the *set* of all random variables V on $(\Omega, \mathscr{A}, P)$ that satisfy conditions (a) and (b) of Definition 14.7 is denoted by $\mathscr{E}^B(Y \mid \mathscr{C})$. Hence, we can write $E^B(Y \mid \mathscr{C}) \in \mathscr{E}^B(Y \mid \mathscr{C})$ (see also Exercise 14.4).   ◁

Remark 14.11 [$E^B(Y \mid \mathscr{C})$ is not necessarily P-unique] Although $E^B(Y \mid \mathscr{C})$ is P^B-unique, it is *not necessarily P-unique* (see Rem. 10.13). In section 14.6 we present necessary and sufficient conditions for P-uniqueness of $E^B(Y \mid \mathscr{C})$, a property that has important implications (see, e.g., section 14.6.4 and Box 14.1).   ◁

Remark 14.12 [Properties of $E^B(Y \mid \mathscr{C})$] Because $E^B(Y \mid \mathscr{C})$ is a conditional expectation, the properties that have been treated in detail in chapters 10 and 11 analogously also apply to $E^B(Y \mid \mathscr{C})$. We simply have to exchange the probability measure P by P^B, the expectation $E(\cdot)$ by $E^B(\cdot)$, the variance $Var(\cdot)$ by $Var^B(\cdot)$, and the covariance $Cov(\cdot, \cdot)$ by $Cov^B(\cdot, \cdot)$.   ◁

Remark 14.13 [Existence of a real-valued version $E^B(Y \mid \mathscr{C})$] Box 10.1 (x) immediately yields: If the assumptions 14.5 hold and $E^B(Y)$ is finite, then there is a *real-valued* version $V \in \mathscr{E}^B(Y \mid \mathscr{C})$.   ◁

Remark 14.14 [$\mathscr{C}$-conditional probability with respect to P^B] Let the assumptions 14.5 hold and let $A \in \mathscr{A}$. Then we call

$$P^B(A \mid \mathscr{C}) := E^B(1_A \mid \mathscr{C}) \tag{14.13}$$

a *version of the $\mathscr{C}$-conditional probability of A with respect to P^B*. Correspondingly, $\mathscr{P}^B(A \mid \mathscr{C})$ denotes the set of all these versions $P^B(A \mid \mathscr{C})$.   ◁

Now we adapt notation and terminology of a $\mathscr{C}$-conditional expectation with respect to a conditional-probability measure to the case in which the σ-algebra $\mathscr{C}$ is generated by a random variable.

Remark 14.15 [X-conditional expectation with respect to P^B] Let the assumptions 14.5 hold and assume that $X: (\Omega, \mathscr{A}, P) \to (\Omega'_X, \mathscr{A}'_X)$ is a random variable. Then we define

$$E^B(Y \mid X) := E^B(Y \mid \sigma(X)) \tag{14.14}$$

and call it a *version of the X-conditional expectation of Y with respect to P^B* or a *version of the X-conditional P^B-expectation of Y*.   ◁

Correspondingly, we use $\mathscr{E}^B(Y \mid X)$ to denote the set of all versions of the X-conditional expectation of Y with respect to P^B.

Remark 14.16 [X-conditional probability of an event with respect to P^B] If $A \in \mathscr{A}$, then,

$$P^B(A \mid X) := E^B(1_A \mid X) \tag{14.15}$$

is called *a version of the X-conditional probability of A with respect to P^B*, and $\mathscr{P}^B(A \mid X)$ denotes the set of all these versions.   ◁

If B is the event $\{Z=z\} = \{\omega \in \Omega: Z(\omega) = z\}$ that a random variable Z takes on the value z, then we adapt the notation and the terminology correspondingly.

Notation and assumptions 14.17
Let $Y: (\Omega, \mathscr{A}, P) \to (\overline{\mathbb{R}}, \overline{\mathscr{B}})$ and $Z: (\Omega, \mathscr{A}, P) \to (\Omega'_Z, \mathscr{A}'_Z)$ be random variables and $\mathscr{C} \subset \mathscr{A}$ a σ-algebra. Furthermore, let $z \in \Omega'_Z$ with $P(Z=z) > 0$ and assume that Y is nonnegative or such that $E^{Z=z}(Y)$ is finite.

Remark 14.18 [Conditional expectation with respect to $P^{Z=z}$] Let the assumptions 14.17 hold. Then we use the notation

$$E^{Z=z}(Y \mid \mathscr{C}) := E^{\{Z=z\}}(Y \mid \mathscr{C}) \tag{14.16}$$

and call it a version of the $\mathscr{C}$-conditional expectation of Y with respect to $P^{Z=z}$. The measure $P^{Z=z}$ is defined by Equation (14.9) with $B = \{Z=z\}$. Correspondingly, $\mathscr{E}^{Z=z}(Y \mid \mathscr{C})$ denotes the set of all versions of the $\mathscr{C}$-conditional expectation of Y with respect to $P^{Z=z}$.   ◁

Remark 14.19 [$\mathscr{C}$-conditional probability with respect to $P^{Z=z}$] Correspondingly, for $A \in \mathscr{A}$ we define $P^{Z=z}(A \mid \mathscr{C}) := E^{Z=z}(1_A \mid \mathscr{C})$, a version of the $\mathscr{C}$-conditional probability of the event A with respect to the measure $P^{Z=z}$, and we use $\mathscr{P}^{Z=z}(A \mid \mathscr{C})$ to denote the family of all versions of the $\mathscr{C}$-conditional probability of the event A with respect to the measure $P^{Z=z}$.   ◁

In the next definition, we additionally consider a random variable X and use it such that $\sigma(X)$ takes the role of the σ-algebra $\mathscr{C}$.

Notation and assumptions 14.20
Let $X: (\Omega, \mathscr{A}, P) \to (\Omega'_X, \mathscr{A}'_X)$, $Y: (\Omega, \mathscr{A}, P) \to (\overline{\mathbb{R}}, \overline{\mathscr{B}})$, and $Z: (\Omega, \mathscr{A}, P) \to (\Omega'_Z, \mathscr{A}'_Z)$ be random variables. Furthermore, let $z \in \Omega'_Z$ with $P(Z=z) > 0$, and assume that Y is nonnegative or such that $E^{Z=z}(Y)$ is finite.

Under these assumptions, we define an X-conditional expectation with respect to a $(Z=z)$-conditional probability measure as follows:

Remark 14.21 [X-conditional expectation with respect to $P^{Z=z}$] Let the assumptions 14.20 hold. Then,

$$E^{Z=z}(Y \mid X) := E^{\{Z=z\}}(Y \mid \sigma(X)), \tag{14.17}$$

is called a *version of the X-conditional expectation of Y with respect to $P^{Z=z}$.*   ◁

Again, note the distinction between a version $E^{Z=z}(Y \mid X)$ and $\mathscr{E}^{Z=z}(Y \mid X)$, the family of all versions of the X-conditional expectation of Y with respect to $P^{Z=z}$. Of course, what has been said in Remark 14.12 about the properties of a $\mathscr{C}$-conditional expectation with respect to P^B applies to $E^{Z=z}(Y \mid X)$ as well.

Remark 14.22 [X-conditional probability with respect to $P^{Z=z}$] Correspondingly, for $A \in \mathscr{A}$ we define $P^{Z=z}(A \mid X) := E^{Z=z}(1_A \mid X)$, a version of the X-conditional probability of the event A with respect to the measure $P^{Z=z}$, and we use $\mathscr{P}^{Z=z}(A \mid X)$ to denote the family of all these versions.   ◁

Example 14.23 [Joe and Ann with self-selection – continued] We continue Example 14.2, illustrating how to compute $E^{X=0}(Y \mid U)$ and $E^{X=1}(Y \mid U)$, which, in this example, are both uniquely defined. First we compute the probabilities of the elementary events with respect to the measures $P^{X=0}$ and $P^{X=1}$, and then specify $E^{X=0}(Y \mid U)$ and $E^{X=1}(Y \mid U)$.

Enumerating the eight elementary events $\{\omega_1\}$ to $\{\omega_8\}$ from top to bottom of the first column of Table 14.2, the probabilities of these elementary events with respect to $P^{X=0}$ can be computed as follows:

$$P^{X=0}(\{\omega_1\}) = P^{X=0}[\{(Joe, no, -)\}] = \frac{P[\{(Joe, no, -)\} \cap \{X=0\}]}{P(X=0)}$$

$$= \frac{P[\{(Joe, no, -)\}]}{P(X=0)} = \frac{.144}{.144 + .336 + .096 + .024} = .24.$$

For the elementary event $\{\omega_2\}$, we obtain

$$P^{X=0}(\{\omega_2\}) = P^{X=0}[\{(Joe, no, +)\}] = \frac{P[\{(Joe, no, +)\} \cap \{X=0\}]}{P(X=0)}$$

$$= \frac{P[\{(Joe, no, +)\}]}{P(X=0)} = \frac{.336}{.144 + .336 + .096 + .024} = .56.$$

For $\{\omega_5\}$, we obtain

$$P^{X=0}(\{\omega_5\}) = P^{X=0}[\{(Ann, no, -)\}] = \frac{P[\{(Ann, no, -)\} \cap \{X=0\}]}{P(X=0)}$$

$$= \frac{P[\{(Ann, no, -)\}]}{P(X=0)} = \frac{.096}{.144 + .336 + .096 + .024} = .16;$$

and, for $\{\omega_6\}$,

$$P^{X=0}(\{\omega_6\}) = P^{X=0}[\{(Ann, no, +)\}] = \frac{P[\{(Ann, no, +)\} \cap \{X=0\}]}{P(X=0)}$$

$$= \frac{P[\{(Ann, no, +)\}]}{P(X=0)} = \frac{.024}{.144 + .336 + .096 + .024} = .04.$$

The probabilities of the other four elementary events with respect to $P^{X=0}$ are 0 (see the last but one column of Table 14.2). The probabilities of the eight elementary events with respect to $P^{X=1}$ are computed analogously (see also the last column of Table 14.2).

Now we specify the U-conditional expectation of Y with respect to $P^{X=0}$. Because Y is an indicator variable with values 0 and 1, the conditional expectation $E^{X=0}(Y \mid U)$ can also be denoted by $P^{X=0}(Y=1 \mid U)$ [see Eq. (14.15)]. It has two different values, one for $\omega \in \{U=Joe\}$ and one for $\omega \in \{U=Ann\}$. These values can be computed as follows:

$$P^{X=0}(Y=1 \mid U=Joe) = \frac{P^{X=0}(Y=1, U=Joe)}{P^{X=0}(U=Joe)} = \frac{.56}{.24 + .56} = .7$$

and

$$P^{X=0}(Y=1 \mid U=Ann) = \frac{P^{X=0}(Y=1, U=Ann)}{P^{X=0}(U=Ann)} = \frac{.04}{.04 + .16} = .2.$$

The results of the corresponding computations for $P^{X=1}(Y=1 \mid U)$ are displayed in the third from the right column of Table 14.2.

Note that, in this example, there is only one single element in $\mathscr{E}^{X=0}(Y \mid U)$ and one single element in $\mathscr{E}^{X=1}(Y \mid U)$, that is, the conditional expectations $E^{X=0}(Y \mid U)$ and $E^{X=1}(Y \mid U)$ are uniquely defined. In contrast, in Example 14.24 there is one single element in $\mathscr{E}^{X=0}(Y \mid U)$, but an infinite number of different elements in $\mathscr{E}^{X=1}(Y \mid U)$.   ◁

Example 14.24 [No treatment for Joe – continued]   We continue Example 14.3. In this example, $E^{X=0}(Y \mid U)$ is uniquely defined (see Exercise 14.7). In contrast, this is not true for $E^{X=1}(Y \mid U)$. The probabilities of the elementary events with respect to the measures $P^{X=x}$ are computed analogously to Example 14.23. The results are displayed in the last two columns of Table 14.3.

Now we specify a version of the U-conditional expectation of Y with respect to the measure $P^{X=1}$. Because Y is an indicator variable with values 0 and 1, a version of this conditional expectation can also be denoted by $P^{X=1}(Y=1 \mid U)$ [see Eq. (14.15)]. It has two different values, one for $U(\omega) = Joe$ and one for $U(\omega) = Ann$. The latter is

$$P^{X=1}(Y=1 \mid U=Ann) = \frac{P^{X=1}(Y=1, U=Ann)}{P^{X=1}(U=Ann)} = \frac{.4}{.6 + .4} = .4.$$

In contrast, the fraction $P^{X=1}(Y=1, U=Joe)/P^{X=1}(U=Joe)$ is not defined, because $P^{X=1}(U=Joe) = 0$. Nevertheless, a value $P^{X=1}(Y=1 \mid U=Joe)$ of a factorization of $P^{X=1}(Y=1 \mid U)$ is defined [see Def. 10.33 and also Eq. (14.34)]. In this case, we can choose any real number as the value of $P^{X=1}(Y=1 \mid U)$ for $\omega \in \{U=Joe\} = \{\omega_1, \dots, \omega_4\}$. For example, ple,

$$V_1(\omega) = \begin{cases} 9, & \text{if } U(\omega) = Joe \\ .4, & \text{if } U(\omega) = Ann, \end{cases}$$

[see Eq. (10.31) and also Eq. (14.41)] defines a first element of $\mathscr{P}^{X=1}(Y=1 \mid U)$ (see Table 14.3), and

$$V_1^*(\omega) = \begin{cases} 0, & \text{if } U(\omega) = Joe \\ .4, & \text{if } U(\omega) = Ann, \end{cases}$$

is a second element of the set $\mathscr{P}^{X=1}(Y=1 \mid U)$. Obviously, V_1 and V_1^* are $P^{X=1}$-equivalent, because $P^{X=1}(A_1) = 0$, where

$$\begin{aligned} A_1 &:= \{\omega \in \Omega \colon V_1(\omega) \neq V_1^*(\omega)\} = \{\omega_1, \dots, \omega_4\} \\ &= \{(Joe, no, -), (Joe, no, +), (Joe, yes, -), (Joe, yes, +)\}. \end{aligned}$$

This probability can be computed by

$$P^{X=1}(A_1) = \sum_{\omega \in A_1} P^{X=1}(\{\omega\}) = 0 + 0 + 0 + 0 = 0$$

(see the last column of Table 14.3). In contrast,

$$P(A_1) = \sum_{\omega \in A_1} P(\{\omega\}) = .152 + .348 + 0 + 0 = .5.$$

Hence, the two versions V_1 and V_1^* of the U-conditional $P^{X=1}$-expectation of Y are $P^{X=1}$-equivalent, that is, they are equivalent with respect to the measure $P^{X=1}$. However, the two versions are *not equivalent* with respect to the measure P. (This issue will be treated in more detail in section 14.6.)

Note that the values of a version $E^{X=x}(Y \mid U)$ of the conditional expectation are defined for *all* elements $\omega \in \Omega$ and that these conditional expectations are random variables on all three probability spaces $(\Omega, \mathscr{A}, P)$, $(\Omega, \mathscr{A}, P^{X=0})$, and $(\Omega, \mathscr{A}, P^{X=1})$, because they share the same measurable space $(\Omega, \mathscr{A})$ (see Def. 5.1). Furthermore, the values of such a version $E^{X=x}(Y \mid U)$ only depend on the person drawn. This illustrates that they are measurable with respect to U [see Def. 14.7 (b)]. ◁

14.3 Properties

Remark 14.25 [Rules of computation] The rules of computation for conditional expectations $E(Y \mid \mathscr{C})$ with respect to a probability measure P analogously hold for conditional expectation $E^B(Y \mid \mathscr{C})$ with respect to a conditional-probability measure P^B. For example, according to Rule (iv) of Box 10.1,

$$E^B[E^B(Y \mid \mathscr{C})] = E^B(Y). \tag{14.18}$$

Similarly, according to Rule (iii) of Box 10.1,

$$E^B(\alpha \cdot Y \mid \mathscr{C}) \underset{P^B}{=} \alpha \cdot E^B(Y \mid \mathscr{C}), \quad \alpha \in \mathbb{R}. \tag{14.19}$$

We simply have to exchange the notation $E(\cdot)$, which refers to the measure P, by $E^B(\cdot)$ referring to the conditional-probability measure P^B and, of course, exchange P by P^B. ◁

In the following theorem we extend Equation (14.19), showing how to deal with a $\mathscr{C}$-conditional expectation with respect to $P^{Z=z}$ of $f(Z) \cdot Y$.

Theorem 14.26 [Regressand $f(Z) \cdot Y$]
Let the assumptions 14.17 hold. If $f : (\Omega'_Z, \mathscr{A}'_Z) \to (\overline{\mathbb{R}}, \overline{\mathscr{B}})$ is a measurable function and $f(z) \in \mathbb{R}$, then,

$$E^{Z=z}[f(Z) \cdot Y \mid \mathscr{C}] \underset{P^{Z=z}}{=} f(z) \cdot E^{Z=z}(Y \mid \mathscr{C}). \tag{14.20}$$

(Proof p. 439)

Remark 14.27 [Two special cases] For the constant 1 taking the role of Y, Rule (i) of Box 10.1 yields $E^{Z=z}(Y \mid \mathscr{C}) \underset{P^{Z=z}}{=} 1$. Therefore, Equation (14.20) implies

$$E^{Z=z}[f(Z) \mid \mathscr{C}] \underset{P^{Z=z}}{=} f(z). \tag{14.21}$$

Another special case of Equation (14.20) is

$$E^{Z=z}[f(Z)] = f(z), \tag{14.22}$$

which follows from Remark 10.5 and (14.21) for $\mathscr{C} = \{\Omega, \emptyset\}$. ◁

Remark 14.28 [Two probability spaces] There are also some properties of a conditional expectation $E^B(Y \mid \mathscr{C})$ that are related to the fact that two probability spaces, $(\Omega, \mathscr{A}, P)$ and $(\Omega, \mathscr{A}, P^B)$, are involved. By definition, a version of the conditional expectation $E^B(Y \mid \mathscr{C})$ with respect to P^B is a random variable on the probability space $(\Omega, \mathscr{A}, P^B)$. Therefore, it is also a random variable on $(\Omega, \mathscr{A}, P)$. However, different elements of $\mathscr{E}^B(Y \mid \mathscr{C})$ are not necessarily P-equivalent; they are necessarily equivalent only with respect to P^B. Hence, if $V, V^* \in \mathscr{E}^B(Y \mid \mathscr{C})$, then the expectations $E(V)$ and $E(V^*)$ with respect to P may differ from each other, whereas $E^B(V)$ and $E^B(V^*)$ are necessarily identical. These issues are treated in detail in section 14.6.2. ◁

14.4 Partial conditional expectation

Now we introduce the concept of a *partial conditional expectation* using a factorization of a version $g(X, Z) = E(Y \mid X, Z) \in \mathscr{E}(Y \mid X, Z)$. We show how this concept is related to a conditional expectation with respect to a conditional-probability measure. In Definition 14.29, we refer to the functions $g_z \colon \Omega_X' \to \overline{\mathbb{R}}$ that, for all $z \in \Omega_Z'$, are defined by

$$g_z(x) = g(x, z), \quad \forall\, x \in \Omega_X'. \tag{14.23}$$

Referring to the concept of an $(X = x, Z = z)$-conditional expectation value introduced in Definition 10.33, we can write

$$g_z(x) = g(x, z) = E(Y \mid X = x, Z = z), \quad \forall\, (x, z) \in \Omega_X' \times \Omega_Z'. \tag{14.24}$$

Note that, in Equations (14.23) to (14.25), we do not assume $P(Z = z) > 0$.

Definition 14.29 [Partial conditional expectation]
Let $X \colon (\Omega, \mathcal{A}, P) \to (\Omega_X', \mathcal{A}_X')$, $Y \colon (\Omega, \mathcal{A}, P) \to (\overline{\mathbb{R}}, \overline{\mathcal{B}})$, and $Z \colon (\Omega, \mathcal{A}, P) \to (\Omega_Z', \mathcal{A}_Z')$ be random variables and assume that Y is nonnegative or with finite expectation $E(Y)$. Furthermore, let $g(X, Z) = E(Y \mid X, Z) \in \mathscr{E}\,(Y \mid X, Z)$ and, for $z \in \Omega_Z'$, let the function g_z, be defined by Equation (14.23). Then the function $E(Y \mid X, Z = z) \colon \Omega \to \overline{\mathbb{R}}$ defined by

$$E(Y \mid X, Z = z) := g_z(X) \tag{14.25}$$

is called a version of the partial $(X, Z = z)$-conditional expectation of Y (with respect to P).

To emphasize, for each $z \in \Omega_Z'$, the function $E(Y \mid X, Z = z)$ denotes the composition of X and g_z. Hence, for each $z \in \Omega_Z'$ it is a random variable on $(\Omega, \mathcal{A}, P)$ that is X-measurable (see Lemma 2.52). In Theorem 14.33, we show that $E(Y \mid X, Z = z)$ is a version of the conditional expectation of Y on X with respect to $P^{Z=z}$, provided that $P(Z = z) > 0$.

Remark 14.30 [Partial conditional probability]   If $A \in \mathcal{A}$, then we also use the notation $P(A \mid X, Z = z) := E(1_A \mid X, Z = z)$ and call it the partial $(X, Z = z)$-conditional probability of the event A (with respect to P). Furthermore, if Y is dichotomous with values 0 and 1, then we also use the notation $P(Y = 1 \mid X, Z = z) := E(Y \mid X, Z = z)$ and call it the partial $(X, Z = z)$-conditional probability of the event $\{Y = 1\}$, or simply of $Y = 1$ (with respect to P).   ◁

Remark 14.31 [Factorization and partial conditional expectation]   If the assumptions of Definition 14.29 hold, then,

$$1_{Z=z} \cdot g(X, Z) = 1_{Z=z} \cdot g_z(X). \tag{14.26}$$

(For a proof, see Exercise 14.5.)   ◁

Remark 14.32 [Discrete Z]   Under the assumptions of Definition 14.29, suppose that $Z \colon (\Omega, \mathcal{A}, P) \to (\Omega_Z', \mathcal{A}_Z')$ is discrete and $Z(\Omega) \subset \Omega_Z'$ is finite or countable with $\{z\} \in \mathcal{A}_Z'$ for all $z \in Z(\Omega)$. Then,

$$E(Y \mid X, Z) = \sum_{z \in Z(\Omega)} E(Y \mid X, Z = z) \cdot 1_{Z=z} \tag{14.27}$$

holds *for the specific version* $E(Y \mid X, Z)$ that is used in Definition 14.29 (see Exercise 14.6). Furthermore,

$$V = \sum_{z \in Z(\Omega)} E(Y \mid X, Z=z) \cdot 1_{Z=z}, \quad \forall V \in \mathscr{E}(Y \mid X, Z). \tag{14.28}$$

$\triangleleft$

According to the following theorem, the partial conditional expectation $E(Y \mid X, Z=z)$ is also a version of the X-conditional expectation of Y with respect to $P^{Z=z}$, provided that $z \in \Omega'_Z$ with $P(Z=z) > 0$.

Theorem 14.33 [Relationship between $E(Y \mid X, Z=z)$ and $E^{Z=z}(Y \mid X)$]
Let the assumptions of Definition 14.29 hold and suppose that $z \in \Omega'_Z$ with $P(Z=z) > 0$. Then,

$$E(Y \mid X, Z=z) \in \mathscr{E}^{Z=z}(Y \mid X), \tag{14.29}$$

and therefore

$$E(Y \mid X, Z=z) = E^{Z=z}(Y \mid X), \quad \forall E^{Z=z}(Y \mid X) \in \mathscr{E}^{Z=z}(Y \mid X). \tag{14.30}$$

(Proof p. 439)

Remark 14.34 [An immediate implication]   If the assumptions of Theorem 14.33 hold and if Z is discrete with $P(Z \in \Omega'_0) = 1$ and $P(Z=z) > 0$ for all $z \in \Omega'_0$ (see Def. 5.56), then,

$$V = \sum_{z \in \Omega'_0} E^{Z=z}(Y \mid X) \cdot 1_{Z=z}, \quad \forall V \in \mathscr{E}(Y \mid X, Z). \tag{14.31}$$

$\triangleleft$

14.5   Factorization

Let the assumptions 14.5 hold and let $X \colon (\Omega, \mathscr{A}, P) \to (\Omega'_X, \mathscr{A}'_X)$ be a random variable. Because $E^B(Y \mid X)$ is measurable with respect to X, Lemma 2.52 implies that there is a measurable function $g_B \colon (\Omega'_X, \mathscr{A}'_X) \to (\mathbb{R}, \mathscr{B})$ such that

$$E^B(Y \mid X) = g_B(X) \tag{14.32}$$

is the composition of X and g_B. A function g_B satisfying Equation (14.32) is called a *factorization* of the version $E^B(Y \mid X) \in \mathscr{E}^B(Y \mid X)$ (see sections 10.4.1 and 10.4.4).

14.5.1   Conditional expectation value with respect to P^B

The values of a factorization g_B of $E^B(Y \mid X)$ are called *$(X=x)$-conditional expectation values with respect to P^B*, and they are denoted by

$$E^B(Y \mid X=x) := g_B(x), \quad \forall x \in \Omega'_X \tag{14.33}$$

(see Def. 10.33). Instead of $E^B(1_A \mid X=x)$, we also use the notation $P^B(A \mid X=x)$, provided that $A \in \mathscr{A}$.

Correspondingly, under the assumptions 14.5,

$$E^{Z=z}(Y \mid X=x) := g_{Z=z}(x), \quad \forall\, x \in \Omega'_X, \tag{14.34}$$

and $P^{Z=z}(A \mid X=x) := E^{Z=z}(1_A \mid X=x)$ for $A \in \mathscr{A}$. Note that $g_{Z=z}$ is not necessarily identical to the function g_z defined by Equation (14.23).

Remark 14.35 [Relationship between the functions g_z and $g_{Z=z}$] Let $g(X, Z) \in \mathscr{E}(Y \mid X, Z)$ and $g_{Z=z}(X) \in \mathscr{E}^{Z=z}(Y \mid X)$. Then the relationship between the function g_z defined by Equation (14.23) and the function $g_{Z=z}$ defined by Equation (14.34) is as follows:

(i) $g_z \underset{P_X^{Z=z}}{=} g_{Z=z}$ if $P(Z=z) > 0$.

(ii) $g_z = g_{Z=z}$ if $z \in \Omega'_Z$ with $P(Z=z, X=x) > 0$ for all $x \in \Omega'_X$.

(see Exercise 14.8). Note that $g_{Z=z}$ is only defined if $P(Z=z) > 0$. ◁

Remark 14.36 [Relationship between factorizations] Suppose that the assumptions 14.20 hold, where Y is nonnegative or with finite expectation $E(Y)$. Then Theorem 14.33 implies

$$\begin{aligned} E^{Z=z}(Y \mid X=x) &= E(Y \mid X=x, Z=z) \\ &= E(Y \mid X=x, 1_{Z=z}=1), \quad \text{for } P_X^{Z=z}\text{-a.a. } x \in \Omega'_X, \end{aligned} \tag{14.35}$$

where $E^{Z=z}(Y \mid X=x)$ and $E(Y \mid X=x, Z=z)$ are the conditional expectation values defined by (14.34) and (10.27), respectively (see Exercise 14.9). Note that $B = \{1_B = 1\}$. Therefore, according to (14.33), for $B \in \mathscr{A}$ with $P(B) > 0$, $Z = 1_B$, and $z = 1$, the first of these two equations yields

$$E^B(Y \mid X=x) = E(Y \mid X=x, 1_B=1), \quad \text{for } P_X^B\text{-a.a. } x \in \Omega'_X. \tag{14.36}$$
◁

Remark 14.37 [A sufficient condition for uniqueness] Applying Remark 2.71 to Equation (14.35) yields

$$E^{Z=z}(Y \mid X=x) = E(Y \mid X=x, Z=z) = E^{X=x, Z=z}(Y) \tag{14.37}$$

for all $x \in \Omega'_X$ for which $P^{Z=z}(X=x) > 0$, or equivalently, for which $P(X=x, Z=z) > 0$, where $E^{X=x, Z=z}(Y)$ denotes the expectation of Y with respect to the conditional-probability measure P^B with $B = \{X=x\} \cap \{Z=z\}$. Hence, $E^{Z=z}(Y \mid X=x)$ is uniquely defined if $P(X=x, Z=z) > 0$. ◁

14.5.2 Uniqueness of factorizations

Remark 14.38 [Uniqueness of factorizations] For a fixed version $E^B(Y \mid X)$, the factorization g_B of $E^B(Y \mid X)$ is uniquely defined, provided that Ω'_X is identical to the image $X(\Omega) = \{X(\omega): \omega \in \Omega\}$. If $\Omega'_X \neq X(\Omega)$, then there can be different factorizations of a single version $E^B(Y \mid X)$ (see Rem. 10.28).

If g_B and g_B^* are factorizations of two versions V and V^* of $\mathscr{E}^B(Y \mid X)$, respectively, then, according to Corollary 10.29,

$$g_B \underset{P_X^B}{=} g_B^*, \tag{14.38}$$

that is, g_B and g_B^* are P_X^B-equivalent, where P_X^B denotes the probability measure on $(\Omega_X', \mathscr{A}_X')$ defined by

$$P_X^B(A') = P^B(X \in A'), \quad \forall A' \in \mathscr{A}_X'. \tag{14.39}$$

Hence, because $P^B(X \in A')$ is just another notation for $P^B[X^{-1}(A')]$, P_X^B is the image measure of P^B under X (see Def. 5.3 and Rem. 5.5).   ◁

Remark 14.39 [Values of the X-conditional expectation with respect to P^B] If $P(B) > 0$ and $E^B(Y \mid X) \in \mathscr{E}^B(Y \mid X)$, then, for all $x \in \Omega_X'$,

$$E^B(Y \mid X)(\omega) = E^B(Y \mid X=x), \quad \forall \omega \in \{X=x\} \tag{14.40}$$

(see Rem. 10.37). This also implies that the value of $E^B(Y \mid X)$ is constant on all sets $\{X=x\}$, $x \in \Omega_X'$. Correspondingly, if $P(Z=z) > 0$, then, for all $x \in \Omega_X'$,

$$E^{Z=z}(Y \mid X)(\omega) = E^{Z=z}(Y \mid X=x), \quad \forall \omega \in \{X=x\}. \tag{14.41}$$

In other words, whenever the random variable X takes on the value x, and this is the case if $\omega \in \{X=x\}$, then the random variable $E^{Z=z}(Y \mid X)$ takes on the value $E^{Z=z}(Y \mid X=x)$. Note that Equation (14.41) also holds if Ω is finite or countable and some $\omega \in \{X=x\}$ have the probability $P^{Z=z}(\{\omega\}) = 0$. [As an example, consider the values of $E^{X=1}(Y \mid U)$ for $\omega \in \{(Joe, yes, -), (Joe, yes, +)\}$ in Example 14.24.]   ◁

14.6 Uniqueness

14.6.1 A necessary and sufficient condition of uniqueness

In Theorem 10.17, we presented a necessary and sufficient condition for uniqueness of a conditional expectation. In the following corollary, we translate this result to a conditional expectation with respect to P^B.

Corollary 14.40 [Uniqueness of $E^B(Y \mid \mathscr{C})$]
Let the assumptions 14.5 hold, let $\mathscr{C}$ be a finite or countable partition of Ω, and assume $\mathscr{C} = \sigma(\mathscr{C})$. Then $V = V^*$ for all $V, V^* \in \mathscr{E}^B(Y \mid \mathscr{C})$, if and only if

$$P^B(A) > 0, \quad \forall A \in \mathscr{C}. \tag{14.42}$$

Remark 14.41 [Values of $E^B(Y \mid \mathcal{C})$] In other words, under the assumptions of Corollary 14.40, the conditional expectation $E^B(Y \mid \mathcal{C})$ is uniquely defined if and only if (14.42) holds. Furthermore, if (14.42) holds, then,

$$E^B(Y \mid \mathcal{C}) = \sum_{A \in \mathcal{E}} E^B(Y \mid A) \cdot 1_A \tag{14.43}$$

[see Eq. (10.14)] and

$$\forall A \in \mathcal{E}: \quad E^B(Y \mid \mathcal{C})(\omega) = E^B(Y \mid A), \quad \text{if } \omega \in A \tag{14.44}$$

[see Eq. (10.15)]. The last equation shows that $E^B(Y \mid \mathcal{C})$ describes how the conditional expectation values $E^B(Y \mid A)$ depend on the events $A \in \mathcal{E}$. ◁

The corresponding result for the X-conditional expectation $E^B(Y \mid X)$ of Y with respect to P^B is as follows:

Corollary 14.42 [Uniqueness of $E^B(Y \mid X)$]
Let the assumptions 14.5 hold. Furthermore, let $X: (\Omega, \mathcal{A}, P) \to (\Omega'_X, \mathcal{A}'_X)$ be a random variable such that Ω'_X is finite or countable, $X(\Omega) = \Omega'_X$, and $\mathcal{A}'_X = \mathcal{P}(\Omega'_X)$. Then $V = V^*$ for all $V, V^* \in \mathcal{E}^B(Y \mid X)$, if and only if

$$P^B(X{=}x) > 0, \quad \forall x \in X(\Omega). \tag{14.45}$$

Remark 14.43 [Values of $E^B(Y \mid X)$] Hence, under the assumptions of Corollary 14.42, $E^B(Y \mid X)$ is uniquely defined if and only if (14.45) holds. And, if (14.45) holds, then,

$$E^B(Y \mid X) = \sum_{x \in X(\Omega)} E^B(Y \mid X{=}x) \cdot 1_{X=x} \tag{14.46}$$

and

$$\forall x \in X(\Omega): \quad E^B(Y \mid X)(\omega) = E^B(Y \mid X{=}x), \quad \text{if } \omega \in \{X{=}x\}. \tag{14.47}$$

This equation shows that the conditional expectation $E^B(Y \mid X)$ describes how the conditional expectation values $E^B(Y \mid X{=}x)$ depend on the values x of X. ◁

Example 14.44 [Joe and Ann] Applying Corollary 14.42 to the introductory Examples 14.1 and 14.2 implies that the conditional expectations $E^{X=0}(Y \mid U)$ and $E^{X=1}(Y \mid U)$ are uniquely defined. Furthermore, according to Remark 14.37, in these two examples, the conditional expectation values $E^{X=x}(Y \mid U{=}u)$ are also uniquely defined, which follows from the fact that $P^{X=x}(U{=}u) > 0$ for all pairs (x, u) of values of X and U. In contrast, in Example 14.3, only $E^{X=0}(Y \mid U)$ is uniquely defined, but $E^{X=1}(Y \mid U)$ is not. In fact, in this example, $E^{X=1}(Y \mid U)$ is even not P-unique, although it is $P^{X=1}$-unique. This issue is dealt with in section 14.6.2. ◁

14.6.2   Uniqueness with respect to P and other probability measures

A conditional expectation $E^B(Y \mid \mathscr{C})$ is always P^B-unique (see Rem. 14.9). However, we may also ask if $E^B(Y \mid \mathscr{C})$ is Q-unique, where Q is *any* probability measure on $(\Omega, \mathscr{A})$. This includes $Q = P$, but also $Q = P^C$, where $C \in \mathscr{A}$ and $C \neq B$ (see Remarks 5.17 and 5.20). If $E^B(Y \mid \mathscr{C})$ is Q-unique, then two versions $V, V^* \in \mathscr{E}^B(Y \mid \mathscr{C})$ have identical distributions, expectations, variances, and so on not only with respect to P^B, but also with respect to Q (i.e., $Q_V = Q^*_V$) (see Cors. 5.24 and 6.17). The following example shows why this is of interest.

Example 14.45 [Pre-post design]   Suppose that X is an indicator variable with values 0 (control) and 1 (treatment), Y represents *life satisfaction after treatment*, and Z *life satisfaction before treatment*. Then P-uniqueness of $E^{X=x}(Y \mid Z)$ is crucial if we consider

$$E(Y \mid X, Z) \underset{P}{=} g_0(Z) + g_1(Z) \cdot X$$

with

$$g_1(Z) \underset{P}{=} E^{X=1}(Y \mid Z) - E^{X=0}(Y \mid Z),$$

or the expectation

$$E[g_1(Z)] = E\big[E^{X=1}(Y \mid Z)\big] - E\big[E^{X=0}(Y \mid Z)\big].$$

Furthermore, for $x = 0$ and $x = 1$, we may also consider the $(X=x)$-conditional expectation values of $g_1(Z)$

$$E[g_1(Z) \mid X=x] = E\big[E^{X=1}(Y \mid Z) \mid X=x\big] - E\big[E^{X=0}(Y \mid Z) \mid X=x\big],$$

the *average effect of X on Y* given $x = 0$ (control) and $x = 1$ (treatment), respectively. Considering $E[g_1(Z) \mid X=x]$, where $x = 0$ or $x = 1$, it is crucial that $E^{X=0}(Y \mid Z)$ and $E^{X=1}(Y \mid Z)$ are unique with respect to the measure $P^{X=x}$ for the following reason: For $x = 0$ or $x = 1$, if $P(X=x) > 0$ and $E^{X=1}(Y \mid Z)$ is $P^{X=x}$-unique, then the conditional expectation value $E\big[E^{X=1}(Y \mid Z) \mid X=x\big]$ is identical for different versions $E^{X=1}(Y \mid Z) \in \mathscr{E}^{X=1}(Y \mid Z)$. Correspondingly, for $x = 0$ or $x = 1$, if $P(X=x) > 0$ and $E^{X=0}(Y \mid Z)$ is $P^{X=x}$-unique, then $E\big[E^{X=0}(Y \mid Z) \mid X=x\big]$ is identical for different versions $E^{X=0}(Y \mid Z) \in \mathscr{E}^{X=0}(Y \mid Z)$ (see Exercise 14.10).   ◁

14.6.3   Necessary and sufficient conditions of P-uniqueness

Now we present conditions that are equivalent to P-uniqueness of $E^B(Y \mid \mathscr{C})$. Note that, in this theorem, we do not refer to the expectation $E^B(V)$ of V with respect to the measure P^B, but to the expectation $E(V)$ of V with respect to P.

Theorem 14.46 [Conditions equivalent to P-uniqueness of $E^B(Y \mid \mathscr{C})$]
Let the assumptions 14.5 hold. Then the following propositions are equivalent to each other:

(a) $E^B(Y \mid \mathscr{C})$ *is P-unique.*

(b) $P \underset{\mathscr{C}}{\ll} P^B$.

(c) $P(B \mid \mathscr{C}) \underset{P}{>} 0.$

If there is a version $V \in \mathscr{E}^B(Y \mid \mathscr{C})$ *such that* $E(V)$ *is finite, then each of (a) to (c) is also equivalent to*

   (d) $\forall\, V, V^* \in \mathscr{E}^B(Y \mid \mathscr{C})$: $E(V) = E(V^*)$.

(Proof p. 440)

Remark 14.47 [Sufficient conditions for finiteness of $E[E^B(Y \mid \mathscr{C})]$]   Remember, the expectation of a random variable Y exists if $\int Y^+\, dP$ or $\int Y^-\, dP$ is finite (see Def. 3.28). Hence, the expectation $E[E^B(Y \mid \mathscr{C})]$ of any version $E^B(Y \mid \mathscr{C}) \in \mathscr{E}^B(Y \mid \mathscr{C})$ exists and is finite, for example if one of the following conditions holds:

   (a) $\mathscr{C}$ is a finite set and $E^B(Y)$ is finite (see Exercise 14.11).

   (b) $E^B(Y \mid \mathscr{C})$ has only a finite number of real values (see Rem. 6.5).

   (c) $E^B(Y \mid \mathscr{C})$ is P-almost surely bounded on both sides, that is,
   $\exists\, \alpha \in \mathbb{R}$: $-\alpha \underset{P}{\leq} E^B(Y \mid \mathscr{C}) \underset{P}{\leq} \alpha$ [see Eq. (3.50)].

   (d) Y is P-almost surely bounded on both sides, that is, $\exists\, \alpha \in \mathbb{R}$: $-\alpha \underset{P}{\leq} Y \underset{P}{\leq} \alpha$
   $[E^B(\cdot) = E^{1_B = 1}(\cdot)$, and see Eq. (14.30), and Box 10.3 (v) and (vii), and (c)].

A special case of (d) is $0 \underset{P}{\leq} Y \underset{P}{\leq} \alpha$, for $0 < \alpha \in \mathbb{R}$. Another one is $Y = 1_A$, if $A \in \mathscr{A}$.   ◁

   In the following corollary, we translate Theorem 14.46 to the special case of an X-conditional expectation $E^B(Y \mid X)$ with respect to P^B and apply Lemma 5.29.

Corollary 14.48 [P-uniqueness of $E^B(Y \mid X)$]
Let the assumptions 14.5 hold and let $X: (\Omega, \mathscr{A}, P) \to (\Omega'_X, \mathscr{A}'_X)$ be a random variable. Then the following propositions are equivalent to each other:

   (a) $E^B(Y \mid X)$ *is P-unique.*

   (b) $P \underset{\sigma(X)}{\ll} P^B$.

   (c) $P(B \mid X) \underset{P}{>} 0$.

   (d) $P_X \underset{\mathscr{A}'_X}{\ll} P_X^B$.

If there is a version $V \in \mathscr{E}^B(Y \mid X)$ *such that* $E(V)$ *is finite, then each of (a) to (d) is also equivalent to*

   (e) $\forall\, V, V^* \in \mathscr{E}^B(Y \mid X)$: $E(V) = E(V^*)$.

Remark 14.49 [Absolute continuity if X is discrete] Under the assumptions of Corollary 14.48: If X is discrete and $\{x\} \in \mathscr{A}'_X$ for all $x \in \Omega'_X$, then $P_X \underset{\mathscr{A}'_X}{\ll} P^B_X$ [see Cor. 14.48 (d)] is equivalent to

$$\forall x \in \Omega'_X: \quad P^B(X=x) = 0 \;\Rightarrow\; P(X=x) = 0 \tag{14.48}$$

(see Exercise 14.12).   ◁

Example 14.50 [Joe and Ann with randomized assignment – continued] The last two columns of Table 14.1 display the conditional-probability measures $P^{X=0}$ and $P^{X=1}$. The values of $P^{X=1} = P^B$ with $B = \{X=1\}$ were already computed in Example 4.34. The last two columns of Table 14.1 show that $P^{X=x}(U=u) > 0$ for all pairs of values of X and U. This implies that the conditional expectations $E^{X=x}(Y \mid U)$ are uniquely defined for both values of X (see Cor. 14.42), which in turn implies that they are P-unique. Furthermore, according to Equation (14.35), the expectations

$$E[E^{X=x}(Y \mid U)] = \sum_u E(Y \mid X=x, U=u) \cdot P(U=u), \quad x = 0, 1, \tag{14.49}$$

are finite. According to Remark 14.47 (b), this follows from the fact that the conditional expectation values $E(Y \mid X=x, U=u) = E^{X=x}(Y \mid U=u)$ are finite. Finally, $E(V) = E(V^*) = E[E^{X=x}(Y \mid U)]$, for all $V, V^* \in \mathscr{E}^{X=x}(Y \mid U)$ [see Cor. 14.48 (e)].   ◁

Example 14.51 [No treatment for Joe – continued] Continuing Example 14.24, consider the event

$$\{X=1\} = \{(Joe, yes, -), (Joe, yes, +), (Ann, yes, -), (Ann, yes, +)\},$$

that the *drawn person is treated*, and the event

$$\{U=Joe\} = \{(Joe, no, -), (Joe, no, +), (Joe, yes, -), (Joe, yes, +)\},$$

that *Joe is drawn*. In this example, we already computed the $(X=1)$-conditional probability $P^{X=1}(U=Joe) = 0$ and the (unconditional) probability $P(U=Joe) = .50$. Because $\{U=Joe\} \in \sigma(U)$, in this example, it is not true that $P^{X=1}(A) = 0$ implies $P(A) = 0$ for all $A \in \sigma(U)$. Therefore, $P \underset{\sigma(U)}{\ll} P^{X=1}$ does not hold (see Def. 3.70). Hence, Corollary 14.48 implies that the conditional expectation $E^{X=1}(Y \mid U)$ is not P-unique (see also Examples 14.3 and 14.24).

In Table 14.3, the values of $E^{X=1}(Y \mid U)$ are not uniquely defined for all four $\omega \in \{U=Joe\}$. Instead of 9, we could have assigned *any* real number to these four possible outcomes ω, because $P^{X=1}(U=Joe) = 0$. Nevertheless, $E^{X=1}(Y \mid U)$ is $P^{X=1}$-unique. However, because $E^{X=1}(Y \mid U)$ is not P-unique, in this example, $E(V_1) = E(V_1^*)$ does *not* hold for all $V_1, V_1^* \in \mathscr{E}^{X=1}(Y \mid U)$ (see Th. 14.46). This has already been illustrated in Example 14.3.   ◁

14.6.4   Properties related to *P*-uniqueness

Box 14.1 summarizes some important properties related to *P*-uniqueness (for proofs, see Exercise 14.13), some of which have already been treated and illustrated in section 14.6.3. In the following remarks, we comment on some of the implications of *P*-uniqueness.

Remark 14.52 [Implications of *P*-uniqueness]   Suppose that Y is nonnegative or with finite expectation $E(Y)$, that $E^B(Y \mid \mathscr{C})$ is *P*-unique, and $C \in \mathscr{A}$ with $P(C) > 0$. Then, according to property (v) of Box 14.1, $E^B(Y \mid \mathscr{C})$ is also P^C-unique, and according to property (vi) of Box 14.1, the distributions P_V^C with respect to P^C of all versions $V \in \mathscr{E}^B(Y \mid \mathscr{C})$ are identical (see Cor. 5.24). This implies, for example, that the expectation $E^C[E^B(Y \mid \mathscr{C})]$ of $E^B(Y \mid \mathscr{C})$ with respect to the conditional-probability measure P^C is identical for all versions $V \in \mathscr{E}^B(Y \mid \mathscr{C})$ [see Box 14.1 (vii)]. The same applies to its variance $Var^C[E^B(Y \mid \mathscr{C})]$ [see Rem. 6.27 and Box 6.2 (v)] as well as to its covariance $Cov^C[E^B(Y \mid \mathscr{C}), W]$ with another random variable $W \colon (\Omega, \mathscr{A}, P) \to (\mathbb{R}, \overline{\mathscr{B}})$ [see Box 7.1 (x)], provided that this variance and this covariance with respect to P^C exist.   ◁

Box 14.1   *P*-uniqueness of $E^B(Y \mid \mathscr{C})$.

Let $(\Omega, \mathscr{A}, P)$ be a probability space, let $\mathscr{C}, \mathscr{D} \subset \mathscr{A}$ be σ-algebras, let $B, C \in \mathscr{A}$ with $P(B), P(C) > 0$, and let $Y \colon (\Omega, \mathscr{A}, P) \to (\mathbb{R}, \overline{\mathscr{B}})$ be a random variable that is nonnegative or with finite expectation $E(Y)$. Then,

$$E^B(Y \mid \mathscr{C}) \text{ is } P\text{-unique} \;\Leftrightarrow\; \forall\, V, V^* \in \mathscr{E}^B(Y \mid \mathscr{C}) \colon V \underset{P}{=} V^* \tag{i}$$

$$E^B(Y \mid \mathscr{C}) \text{ is } P\text{-unique} \;\Leftrightarrow\; P \underset{\mathscr{C}}{\ll} P^B \tag{ii}$$

$$E^B(Y \mid \mathscr{C}) \text{ is } P\text{-unique} \;\Leftrightarrow\; P(B \mid \mathscr{C}) \underset{P}{>} 0 \tag{iii}$$

$$E^B(Y \mid \mathscr{C}) \text{ is } P\text{-unique} \;\Rightarrow\; E^B(Y \mid \mathscr{D}) \text{ is } P\text{-unique}, \quad \text{if } \mathscr{D} \subset \mathscr{C} \tag{iv}$$

$$E^B(Y \mid \mathscr{C}) \text{ is } P\text{-unique} \;\Rightarrow\; E^B(Y \mid \mathscr{C}) \text{ is } P^C\text{-unique} \tag{v}$$

$$E^B(Y \mid \mathscr{C}) \text{ is } P\text{-unique} \;\Rightarrow\; \forall\, V, V^* \in \mathscr{E}^B(Y \mid \mathscr{C}) \colon P_V^C = P_{V^*}^C \tag{vi}$$

$$E^B(Y \mid \mathscr{C}) \text{ is } P\text{-unique} \;\Rightarrow\; \forall\, V, V^* \in \mathscr{E}^B(Y \mid \mathscr{C}) \colon E^C(V) = E^C(V^*). \tag{vii}$$

If $E^B(Y \mid \mathscr{C})$ or $E^C(Y \mid \mathscr{C})$ is real-valued and $\alpha, \beta \in \mathbb{R}$, then,

$$E^B(Y \mid \mathscr{C}), E^C(Y \mid \mathscr{C}) \text{ are } P\text{-unique} \;\Rightarrow\; \alpha\, E^B(Y \mid \mathscr{C}) + \beta\, E^C(Y \mid \mathscr{C}) \text{ is } P\text{-unique}. \tag{viii}$$

If $X \colon (\Omega, \mathscr{A}, P) \to (\Omega_X', \mathscr{A}_X')$ is a random variable and Ω_X' finite or countable, then,

$$(\forall\, x \in \Omega_X' \colon P^B(X = x) > 0) \;\Rightarrow\; E^B(Y \mid X) \text{ is } P\text{-unique}. \tag{ix}$$

Remark 14.53 [The special case $C = \Omega$] A special case is $C = \Omega$. Hence, under P-uniqueness of $E^B(Y \mid \mathscr{C})$, the following equations hold for all $V, V^* \in \mathscr{E}^B(Y \mid \mathscr{C})$:

$$P_V = P_{V^*}, \tag{14.50}$$

$$E(V) = E(V^*), \tag{14.51}$$

$$Var(V) = Var(V^*), \tag{14.52}$$

$$Cov(V, W) = Cov(V^*, W), \tag{14.53}$$

provided that these expectations, variances, and covariances with respect to the measure P exist.   ◁

Remark 14.54 [Another implication of P-uniqueness] Let the assumptions 14.5 hold. Furthermore, let $\mathscr{D} \subset \mathscr{A}$ be a σ-algebra, $\mathscr{C} \subset \mathscr{D}$, and $C \in \mathscr{A}$ with $P(C) > 0$, and suppose that $E^B(Y \mid \mathscr{C})$ is P^C-unique. Then,

$$E^C[E^B(Y \mid \mathscr{C}) \mid \mathscr{D}] \underset{P^C}{=} E^B(Y \mid \mathscr{C}), \tag{14.54}$$

which follows from Rule (vii) of Box 10.1, because, by definition, $E^B(Y \mid \mathscr{C})$ is $\mathscr{C}$-measurable, and because we assume $\mathscr{C} \subset \mathscr{D}$. For the special case $C = \Omega$, this yields: If $\mathscr{C} \subset \mathscr{D}$ and $E^B(Y \mid \mathscr{C})$ is P-unique, then,

$$E[E^B(Y \mid \mathscr{C}) \mid \mathscr{D}] \underset{P}{=} E^B(Y \mid \mathscr{C}). \tag{14.55}$$

◁

Remark 14.55 [Expectation of a linear combination] An implication of Box 14.1 (viii) and Box 6.1 (vii) is:

$$E\big[\alpha \cdot E^B(Y \mid \mathscr{C}) + \beta \cdot E^C(Y \mid \mathscr{C})\big] = \alpha \cdot E\big[E^B(Y \mid \mathscr{C})\big] + \beta \cdot E\big[E^C(Y \mid \mathscr{C})\big], \tag{14.56}$$

provided that $E^B(Y \mid \mathscr{C})$ and $E^C(Y \mid \mathscr{C})$ are P-unique, that $E^B(Y \mid \mathscr{C})$ or $E^C(Y \mid \mathscr{C})$ is real-valued, and the expectation $E\big[E^B(Y \mid \mathscr{C})\big]$ or $E\big[E^C(Y \mid \mathscr{C})\big]$ is finite. Under these assumptions, for all real-valued versions $V_B \in \mathscr{E}^B(Y \mid \mathscr{C})$ and $V_C \in \mathscr{E}^C(Y \mid \mathscr{C})$,

$$E(\alpha \cdot V_B + \beta \cdot V_C) = \alpha \cdot E(V_B) + \beta \cdot E(V_C).$$

◁

Example 14.56 [Joe and Ann with self-selection – continued] Consider again Table 14.2. In this example, the function $g_1(U) = E^{X=1}(Y \mid U) - E^{X=0}(Y \mid U)$ is a uniquely defined random variable on $(\Omega, \mathscr{A}, P)$, because $E^{X=0}(Y \mid U)$ as well as $E^{X=1}(Y \mid U)$ are uniquely defined, that is, each of the sets $\mathscr{E}^{X=0}(Y \mid U)$ and $\mathscr{E}^{X=1}(Y \mid U)$ has only one single element. This implies that the average treatment effect

$$E[g_1(U)] = E\big[E^{X=1}(Y \mid U) - E^{X=0}(Y \mid U)\big]$$

is uniquely defined as well. Note that the assumptions of Box 14.1 (viii) are less restrictive, because they allow that each of the sets $\mathscr{E}^{X=0}(Y \mid U)$ and $\mathscr{E}^{X=1}(Y \mid U)$ has more than one element. The requirement of Box 14.1 (viii) is not uniqueness but only P-uniqueness.   ◁

Example 14.57 [No treatment for Joe – continued]   In the example presented in Table 14.3, the set $\mathscr{E}^{X=0}(Y \mid U)$ has only one single element. However, $\mathscr{E}^{X=1}(Y \mid U)$ has infinitely many elements: Replacing the value 9 by any other real number yields a new element of $\mathscr{E}^{X=1}(Y \mid U)$. More important, it is *not true* that all these elements are pairwise P-equivalent. Hence, in this example, the function $g_1(U) = E^{X=1}(Y \mid U) - E^{X=0}(Y \mid U)$ is *not P-unique* (see Example 14.51). In other words, it is *not true* that all elements of the set

$$\left\{ V_1 - V_0 \colon V_0 \in \mathscr{E}^{X=0}(Y \mid U), V_1 \in \mathscr{E}^{X=1}(Y \mid U) \right\}$$

are pairwise P-equivalent. Therefore, there can be versions $V_0, V_0^* \in \mathscr{E}^{X=0}(Y \mid U)$ and $V_1, V_1^* \in \mathscr{E}^{X=1}(Y \mid U)$ such that

$$E(V_1 - V_0) \neq E(V_1^* - V_0^*).$$

This means that there is no uniquely defined average treatment effect $E[g_1(U)]$ in this example.   ◁

In the following corollary we extend Theorem 14.33 by adding another assumption. Remember, if assumptions of Definition 14.29 hold and $P(Z=z) > 0$, then according to Theorem 14.33,

$$E^{Z=z}(Y \mid X) \underset{P^{Z=z}}{=} E(Y \mid X, Z=z), \tag{14.57}$$

referring to the partial conditional expectation $E(Y \mid X, Z=z)$ [see Def. 14.29].

Corollary 14.58 [Implications of P^C-uniqueness of $E^{Z=z}(Y \mid X)$]
Let the assumptions of Definition 14.29 hold, suppose $P(Z=z) > 0$, and let $C \in \mathscr{A}$ with $P(C) > 0$. If $E^{Z=z}(Y \mid X)$ is P^C-unique, then,

$$E^{Z=z}(Y \mid X) \underset{P^C}{=} E(Y \mid X, Z=z), \tag{14.58}$$

and

$$E^{Z=z}(Y \mid X=x) = E(Y \mid X=x, Z=z), \quad \text{for } P_X^C\text{-a.a. } x \in \Omega_X'. \tag{14.59}$$

(Proof p. 442)

Remark 14.59 [Implications of P-uniqueness of $E^{Z=z}(Y \mid X)$]   For $C = \Omega$, this corollary yields: If $E^{Z=z}(Y \mid X)$ is P-unique, then,

$$E^{Z=z}(Y \mid X) \underset{P}{=} E(Y \mid X, Z=z), \tag{14.60}$$

and

$$E^{Z=z}(Y \mid X=x) = E(Y \mid X=x, Z=z), \quad \text{for } P_X\text{-a.a. } x \in \Omega_X'. \tag{14.61}$$

◁

Now we consider the family of factorizations g_B of $E^B(Y \mid X)$, which are defined by Equation (14.32). Because each element of $\mathscr{E}^B(Y \mid X)$ has at least one factorization g_B, there is a family of factorizations, which are random variables on the probability space $(\Omega_X', \mathscr{A}_X', P_X)$.

The next corollary immediately follows from Corollary 5.25 (i) if P^C takes the role of P and P_X^C the role of P_X.

Corollary 14.60 [P^C-uniqueness and P_X^C-uniqueness]
Let the assumptions 14.5 hold, let $X: (\Omega, \mathscr{A}, P) \to (\Omega_X', \mathscr{A}_X')$ be a random variable, let $V = g_B(X)$ and $V^* = g_B^*(X)$ be two elements of $\mathscr{E}^B(Y \mid X)$, and let $C \in \mathscr{A}$ with $P(C) > 0$. Then,

$$V \underset{P^C}{=} V^* \Leftrightarrow g_B \underset{P_X^C}{=} g_B^*. \tag{14.62}$$

Remark 14.61 [P-uniqueness and P_X-uniqueness] For $C = \Omega$, Corollary 14.60 yields: If $V = g_B(X)$ and $V^* = g_B^*(X)$ are two elements of $\mathscr{E}^B(Y \mid X)$, then,

$$g_B(X) \underset{P}{=} g_B^*(X) \Leftrightarrow g_B \underset{P_X}{=} g_B^*. \tag{14.63}$$

Note that both sides of (14.63) are equivalent not only to each other but also to

$$g_B(x) = g_B^*(x), \quad \text{for } P_X\text{-a.a. } x \in \Omega_X'. \tag{14.64}$$
◁

Remark 14.62 [Some formulas for the expectation of $E^B(Y \mid X)$] Suppose that the assumptions 14.5 hold, where Y is nonnegative or with finite expectation $E(Y)$, and let $X: (\Omega, \mathscr{A}, P) \to (\Omega_X', \mathscr{A}_X')$ be a random variable. If $E^B(Y \mid X)$ is P-unique, then, according to Equations (6.13) and (14.36),

$$E[E^B(Y \mid X)] = \int E^B(Y \mid X{=}x) \, P_X(dx) = \int E(Y \mid X{=}x, 1_B{=}1) \, P_X(dx). \tag{14.65}$$

Furthermore, if X is discrete (see Def. 5.56) and $P_X(\Omega_0') = 1$, then,

$$
\begin{aligned}
E[E^B(Y \mid X)] &= \int E^B(Y \mid X{=}x) \, P_X(dx) && [(6.13)] \\
&= \sum_{x \in \Omega_0'} E^B(Y \mid X{=}x) \cdot P(X{=}x) && [(6.15)] \\
&= \sum_{x \in \Omega_0'} E(Y \mid X{=}x, 1_B{=}1) \cdot P(X{=}x). && [(14.36)]
\end{aligned}
\tag{14.66}
$$
◁

Remark 14.63 [Some formulas for the expectation of $E^{Z=z}(Y \mid X)$] Correspondingly, let the assumptions 14.20 hold, where Y is nonnegative or with finite expectation $E(Y)$. If $E^{Z=z}(Y \mid X)$ is P-unique, then for $B = \{Z{=}z\}$, Equations (14.65) and (14.61) yield

$$E[E^{Z=z}(Y \mid X)] = \int E^{Z=z}(Y \mid X{=}x) \, P_X(dx) = \int E(Y \mid X{=}x, Z{=}z) \, P_X(dx), \tag{14.67}$$

and, if X is discrete (see Def. 5.56) and $P_X(\Omega'_0) = 1$, then Equations (14.66) and (14.61) yield

$$\begin{aligned} E[E^{Z=z}(Y \mid X)] &= \int E^{Z=z}(Y \mid X=x)\, P_X(dx) \\ &= \sum_{x \in \Omega'_0} E(Y \mid X=x, Z=z) \cdot P(X=x). \end{aligned} \qquad (14.68)$$

◁

14.7   Conditional mean independence with respect to $P^{Z=z}$

According to the following theorem, a numerical random variable Y on $(\Omega, \mathcal{A}, P)$ that is non-negative or with finite expectation $E(Y)$ is $\mathcal{C}$-conditionally mean independent from Z with respect to $P^{Z=z}$. For simplicity, we use the following notation:

$$\sigma(\mathcal{C}, \mathcal{D}) := \sigma(\mathcal{C} \cup \mathcal{D}), \qquad \sigma(\mathcal{C}, Z) := \sigma(\mathcal{C} \cup \sigma(Z)), \qquad (14.69)$$
$$E^{Z=z}(Y \mid \mathcal{C}, \mathcal{D}) := E^{Z=z}(Y \mid \sigma(\mathcal{C} \cup \mathcal{D})), \qquad (14.70)$$
$$E^{Z=z}(Y \mid \mathcal{C}, Z) := E^{Z=z}(Y \mid \sigma[\mathcal{C} \cup \sigma(Z)]). \qquad (14.71)$$

Theorem 14.64 [Conditional mean independence]
Let the assumptions 14.17 hold, where Y is nonnegative or with finite expectation $E(Y)$. Furthermore, let $Z \colon (\Omega, \mathcal{A}, P) \to (\Omega'_Z, \mathcal{A}'_Z)$ be a random variable, and suppose that $z \in \Omega'_Z$ with $P(Z=z) > 0$. Then,

$$E^{Z=z}(Y \mid \mathcal{C}), E(Y \mid \mathcal{C}, Z) \in \mathscr{E}^{Z=z}(Y \mid \mathcal{C}, Z),$$

which implies

$$E^{Z=z}(Y \mid \mathcal{C}, Z) \underset{P^{Z=z}}{=} E^{Z=z}(Y \mid \mathcal{C}) \underset{P^{Z=z}}{=} E(Y \mid \mathcal{C}, Z). \qquad (14.72)$$

(Proof p. 442)

Remark 14.65 [A caveat]   If $\sigma(\mathcal{C}, Z) \neq \mathcal{C}$, then $E^{Z=z}(Y \mid \mathcal{C}, Z)$ and $E(Y \mid \mathcal{C}, Z)$ are not necessarily elements of $\mathscr{E}^{Z=z}(Y \mid \mathcal{C})$ [see Def. 14.7 (a)]. Nevertheless,

$$E^{Z=z}(Y \mid \mathcal{C}) \underset{P^{Z=z}}{=} E(Y \mid \mathcal{C}, Z).$$

According to Box 6.1 (viii) and Box 10.2 (iv), this implies

$$E^{Z=z}(Y) = E^{Z=z}\big[E^{Z=z}(Y \mid \mathcal{C})\big] = E^{Z=z}\big[E(Y \mid \mathcal{C}, Z)\big], \qquad (14.73)$$

and, for $\mathcal{C} = \sigma(X)$, using Equation (14.14),

$$E^{Z=z}(Y) = E^{Z=z}\big[E^{Z=z}(Y \mid X)\big] = E^{Z=z}\big[E(Y \mid X, Z)\big]. \qquad (14.74)$$

Furthermore, using (14.30), this equation yields

$$E^{Z=z}(Y) = E^{Z=z}[E^{Z=z}(Y \mid X)] = E^{Z=z}[E(Y \mid X, Z=z)] \tag{14.75}$$

Note that we still presume that the assumptions of Theorem 14.64 hold. ◁

Remark 14.66 [Two implications concerning mean independence] Let the assumptions 14.20 hold. Then,

$$E^{Z=z}(Y \mid \mathcal{C}, X) \underset{p^{Z=z}}{=} E^{Z=z}(Y \mid \mathcal{C}), \quad \text{if } \sigma(X) \subset \sigma(Z) \tag{14.76}$$

(see Exercise 14.14). Furthermore, considering a σ-algebra $\mathcal{D}$ and assuming that Y is nonnegative or with finite expectation $E(Y)$,

$$E^{Z=z}[E(Y \mid \mathcal{C}, Z) \mid \mathcal{D}] \underset{p^{Z=z}}{=} E^{Z=z}(Y \mid \mathcal{D}), \quad \text{if } \mathcal{D} \subset \sigma(\mathcal{C}, Z) \tag{14.77}$$

(see Exercise 14.15). ◁

In the following theorem, we generalize the propositions of Remark 14.63.

Theorem 14.67 [Expectation of $E^{Z=z}(Y \mid X, W)$ with respect to $P^{W=w}$]
Let the assumptions 14.20 hold, where Y is nonnegative or with finite expectation $E(Y)$, let $W: (\Omega, \mathcal{A}, P) \to (\Omega'_W, \mathcal{A}'_W)$ be a random variable, let $w \in \Omega'_W$ with $P(W=w) > 0$, and assume that $E^{Z=z}(Y \mid X, W)$ is $P^{W=w}$-unique. Then,

$$E^{W=w}[E^{Z=z}(Y \mid X, W)] = \int E^{Z=z}(Y \mid X=x, W=w) \, P_X^{W=w}(dx)$$

$$= \int E(Y \mid X=x, W=w, Z=z) \, P_X^{W=w}(dx). \tag{14.78}$$

(Proof p. 443)

Remark 14.68 [Discrete X] If the assumptions of Theorem 14.67 hold, if X is discrete (see Def. 5.56), and $P_X(\Omega'_0) = 1$, then Equations (14.78), (6.15), (14.59), and (9.13) yield

$$E^{W=w}[E^{Z=z}(Y \mid X, W)] = \int E^{Z=z}(Y \mid X=x, W=w) \, P_X^{W=w}(dx)$$

$$= \sum_{x \in \Omega'_0} E(Y \mid X=x, W=w, Z=z) \cdot P^{W=w}(X=x) \tag{14.79}$$

$$= \sum_{x \in \Omega'_0} E(Y \mid X=x, W=w, Z=z) \cdot P(X=x \mid W=w). $$

◁

In the following theorem, we study an implication of conditional mean independence [see Def. 10.45 (ii)] on conditional expectations with respect to $P^{Z=z}$.

Theorem 14.69 [An implication of conditional mean independence]
Let the assumptions 14.17 hold, where Y is nonnegative or with finite expectation $E(Y)$, and let $\mathscr{C}, \mathscr{D} \subset \mathscr{A}$ be σ-algebras. Then,

$$E(Y \mid \mathscr{C}, \mathscr{D}, Z) \underset{P}{=} E(Y \mid \mathscr{C}, Z) \;\Rightarrow\; E^{Z=z}(Y \mid \mathscr{C}, \mathscr{D}) \underset{P^{Z=z}}{=} E^{Z=z}(Y \mid \mathscr{C}). \qquad (14.80)$$

(Proof p. 444)

Remark 14.70 [An implication of conditional mean independence]   Let the assumptions of Theorem 14.69 hold. If X and W are two random variables on the probability space $(\Omega, \mathscr{A}, P)$, then,

$$E(Y \mid X, W, Z) \underset{P}{=} E(Y \mid X, Z) \;\Rightarrow\; E^{Z=z}(Y \mid X, W) \underset{P^{Z=z}}{=} E^{Z=z}(Y \mid X). \qquad (14.81)$$

Furthermore, for $X = \alpha$, $\alpha \in \Omega'_X$, we can conclude

$$E(Y \mid W, Z) \underset{P}{=} E(Y \mid Z) \;\Rightarrow\; E^{Z=z}(Y \mid W) \underset{P^{Z=z}}{=} E^{Z=z}(Y). \qquad (14.82)$$

◁

14.8   Proofs

Proof of Theorem 14.26

Because $f(Z) \underset{P^{Z=z}}{=} f(z)$ (see Rem. 9.1),

$$E^{Z=z}\big[f(Z) \cdot Y \mid \mathscr{C}\big] \underset{P^{Z=z}}{=} E^{Z=z}\big[f(z) \cdot Y \mid \mathscr{C}\big] \qquad \text{[Box 10.1 (ix)]}$$

$$\underset{P^{Z=z}}{=} f(z) \cdot E^{Z=z}(Y \mid \mathscr{C}). \qquad \text{[Box 10.1 (iii)]}$$

Proof of Theorem 14.33

Let the assumptions of Definition 14.29 hold. Because $g \colon \Omega'_X \times \Omega'_Z \to \overline{\mathbb{R}}$ is $(\mathscr{A}'_X \otimes \mathscr{A}'_Z, \overline{\mathscr{B}})$-measurable, the function g_z defined in (14.23) is $(\mathscr{A}'_X, \overline{\mathscr{B}})$-measurable (see Bauer, 2001, Lemma 23.5). Hence, condition (a) of Definition 14.7 holds. Furthermore, for all $C \in \sigma(X)$, we can conclude $C \cap \{Z=z\} \in \sigma(X, Z)$, and therefore, for all $C \in \sigma(X)$,

$$\int 1_C \, E(Y \mid X, Z=z) \, dP^{Z=z} = \frac{1}{P(Z=z)} \int 1_C \cdot 1_{Z=z} \cdot E(Y \mid X, Z=z) \, dP \qquad \text{[(9.16), (9.11)]}$$

$$= \frac{1}{P(Z=z)} \int 1_C \cdot 1_{Z=z} \cdot g_z(X) \, dP \qquad \text{[(14.25)]}$$

$$= \frac{1}{P(Z=z)} \int 1_C \cdot 1_{Z=z} \cdot g(X, Z) \, dP \qquad \text{[(14.26)]}$$

$$= \frac{1}{P(Z=z)} \int 1_C \cdot 1_{Z=z} \cdot E(Y \mid X, Z) \, dP \qquad \text{[Ass. of Def. 14.29]}$$

$$= \frac{1}{P(Z=z)} \int 1_C \cdot 1_{Z=z} \cdot Y \, dP \qquad \text{[Def. 10.2 (b)]}$$

$$= \int 1_C \cdot Y \, dP^{Z=z}. \qquad \text{[(9.16)]}$$

This shows that condition (b) of Definition 14.7 holds as well.

Proof of Theorem 14.46

(b) $\Rightarrow$ (a) Remember that (b) is defined by

$$\forall \, C \in \mathcal{C}: \quad P^B(C) = 0 \; \Rightarrow \; P(C) = 0.$$

If $V, V^* \in \mathcal{E}^B(Y \mid \mathcal{C})$, then $\{V \neq V^*\} \in \mathcal{C}$ [see Rem. 2.67 (c)], $P^B(\{V \neq V^*\}) = 0$, and (b) implies $P(\{V \neq V^*\}) = 0$.

(a) $\Rightarrow$ (b) This proposition is proved by contraposition, that is, we show $\neg$ (b) $\Rightarrow \neg$ (a). Assume that there is an $A \in \mathcal{C}$ with $P^B(A) = 0$ and $P(A) > 0$, and let $V \in \mathcal{E}^B(Y \mid \mathcal{C})$ with $V(\omega) = 0$ for all $\omega \in A$. [Note that if $V' \in \mathcal{E}^B(Y \mid \mathcal{C})$ and $A \in \mathcal{C}$ with $P^B(A) = 0$, then $V := V' \cdot 1_{A^c} \in \mathcal{E}^B(Y \mid \mathcal{C})$.] Then $V^* = V + 1_A$ is also $\mathcal{C}$-measurable and, for all $C \in \mathcal{C}$,

$$\int 1_C \cdot V^* \, dP^B = \int 1_C \cdot (V + 1_A) \, dP^B$$

$$= \int 1_C \cdot V \, dP^B + \int 1_C \cdot 1_A \, dP^B \qquad \text{[(3.33)]} \qquad (14.83)$$

$$= \int 1_C \cdot V \, dP^B, \qquad \text{[Lemma 3.45]}$$

because $\int 1_C \cdot 1_A \, dP^B = P^B(C \cap A) = 0$. Hence, according to Definition 14.7, $V^* = V + 1_A \in \mathcal{E}^B(Y \mid \mathcal{C})$. However, $P(\{V \neq V^*\}) = P(A) > 0$.

(c) $\Rightarrow$ (b) Let $P(B \mid \mathcal{C}) \underset{P}{>} 0$. This implies $P(A) = 0$, where

$$A = \{\omega \in \Omega \colon P(B \mid \mathcal{C})(\omega) \leq 0\},$$

and, according to Remark 2.67 (a), $A \in \mathcal{C}$. However, if $P(A) = 0$, then, according to Rule (ix) of Box 4.1,

$$P(A \cup C) = P(C), \quad \forall \, C \in \mathcal{C}. \qquad (14.84)$$

Now, let $C \in \mathcal{C}$ with $P^B(C) = 0$. This yields

$$\int P(B \mid \mathcal{C}) \cdot 1_C \, dP = \int 1_B \cdot 1_C \, dP \qquad \text{[Def. 10.2 (b)]}$$

$$= P(B \cap C) = P^B(C) \cdot P(B) = 0. \qquad \text{[(1.33), (3.9), (4.14)]}$$

Because $P(B \mid \mathscr{C}) \cdot 1_C \underset{P}{\geq} 0$ [see Box 10.3 (ii)], Lemma 3.44 yields $P(B \mid \mathscr{C}) \cdot 1_C \underset{P}{=} 0$, which is equivalent to

$$
\begin{aligned}
1 &= P(\{\omega \in \Omega \colon P(B \mid \mathscr{C})(\omega) \cdot 1_C(\omega) = 0\}) \\
&= P(\{\omega \in \Omega \colon \omega \in A \text{ or } \omega \in C^c\}) = P(A \cup C^c) \\
&= P(C^c).
\end{aligned}
\qquad\qquad [(14.84)]
$$

Hence, $P(C) = 0$, which shows that $P(B \mid \mathscr{C}) \underset{P}{>} 0$ implies (b).

(b) $\Rightarrow$ (c) This proposition is proved by contraposition, that is, we show $\neg$ (c) $\Rightarrow \neg$ (b). Again, let $A = \{\omega \in \Omega \colon P(B \mid \mathscr{C})(\omega) \leq 0\}$, and assume $P(A) > 0$. Now, $P(B \mid \mathscr{C})(\omega) \leq 0$ for all $\omega \in A$ implies $P(B \mid \mathscr{C}) \cdot 1_A \leq 0$. Therefore,

$$
\begin{aligned}
0 &\geq E[P(B \mid \mathscr{C}) \cdot 1_A] & [(3.50)] \\
&= E[E(1_B \mid \mathscr{C}) \cdot 1_A] & [(10.4)] \\
&= E[E(1_B \cdot 1_A \mid \mathscr{C})] & [A \in \mathscr{C}, \text{Box 10.1 (xiv)}] \\
&= E(1_B \cdot 1_A) & [10.1 \text{ (iv)}] \\
&= E(1_{A \cap B}) & [(1.33)] \\
&= P(A \cap B). & [(6.4)]
\end{aligned}
$$

Because $P(B) > 0$, the equation $P(A \cap B) = P^B(A) \cdot P(B) = 0$ implies $P^B(A) = 0$ [see (4.14)]. Hence, $A \in \mathscr{C}$ with $P^B(A) = P(A \cap B)/P(B) = 0$ and $P(A) > 0$.

(b) $\Rightarrow$ (d) As has been shown above, (b) is equivalent to P-uniqueness of $E^B(Y \mid \mathscr{C})$, and according to Box 6.1 (ix), P-uniqueness of $E^B(Y \mid \mathscr{C})$ implies (d).

(d) $\Rightarrow$ (b) This proposition is proved by contraposition, that is, we show $\neg$ (b) $\Rightarrow \neg$ (d). Assume that there is an $A \in \mathscr{C}$ with $P^B(A) = 0$ and $P(A) > 0$, and let $V \in \mathscr{E}^B(Y \mid \mathscr{C})$ be real-valued [see Box 10.1 (x)] with a finite expectation $E(V)$. Then $V^* = V + 1_A$ is also $\mathscr{C}$-measurable and, for all $C \in \mathscr{C}$,

$$
\int 1_C \cdot V^* \, dP^B = \int 1_C \cdot V \, dP^B. \qquad [(14.83)]
$$

Therefore, according to Definition 14.7, $V^* = V + 1_A \in \mathscr{E}^B(Y \mid \mathscr{C})$ and $P(\{V \neq V^*\}) = P(A) > 0$. Now,

$$
\begin{aligned}
E(V^*) &= \int V^* \, dP & [(6.1)] \\
&= \int (V + 1_A) \, dP & [V^* := V + 1_A] \\
&= \int V \, dP + \int 1_A \, dP & [(3.33)] \\
&= E(V) + P(A). & [(6.1), (6.4)]
\end{aligned}
$$

Because $E(V)$ is finite and $P(A) > 0$, it follows that $E(V) \neq E(V^*)$.

Proof of Corollary 14.58

If $E^{Z=z}(Y \mid X)$ is P^C-unique, then all pairs of elements of $\mathscr{E}^{Z=z}(Y \mid X)$ are P^C-equivalent. According to Theorem 14.33, $E(Y \mid X, Z=z)$ is an element of $\mathscr{E}^{Z=z}(Y \mid X)$, which implies Equation (14.58). Finally, according to Corollary 5.25, Equation (14.59) is equivalent to (14.58).

Proof of Theorem 14.64

First, we show that $E(Y \mid \mathscr{C}, Z) \in \mathscr{E}^{Z=z}(Y \mid \mathscr{C}, Z)$. By Definition 10.2 (a), $\sigma[E(Y \mid \mathscr{C}, Z)] \subset \sigma(\mathscr{C}, Z)$, which implies that condition (a) of Definition 14.7 holds. In order to show condition (b) of Definition 14.7, note that, for all $C \in \sigma(\mathscr{C}, Z)$, $\{Z=z\}$, $\{Z=z\} \cap C \in \sigma(\mathscr{C}, Z)$ and $1_{\{Z=z\} \cap C} = 1_{Z=z} \cdot 1_C$. Hence, for all $C \in \sigma(\mathscr{C}, Z)$,

$$E^{Z=z}[1_C \cdot E(Y \mid \mathscr{C}, Z)] = \frac{1}{P(Z=z)} E[1_{Z=z} \cdot 1_C \cdot E(Y \mid \mathscr{C}, Z)] \qquad [(9.11)]$$

$$= \frac{1}{P(Z=z)} E(1_{Z=z} \cdot 1_C \cdot Y) \qquad [\text{Def. 10.2 (b)}]$$

$$= E^{Z=z}(1_C \cdot Y). \qquad [(9.11)]$$

Now we show that $E^{Z=z}(Y \mid \mathscr{C}) \in \mathscr{E}^{Z=z}(Y \mid \mathscr{C}, Z)$. By Definition 14.7 (a) and the definition of $\sigma(\mathscr{C}, Z)$,

$$\sigma[E^{Z=z}(Y \mid \mathscr{C})] \subset \mathscr{C} \subset \sigma(\mathscr{C}, Z),$$

and hence, $\sigma[E^{Z=z}(Y \mid \mathscr{C})] \subset \sigma(\mathscr{C}, Z)$.

Now we show that condition (b) of Definition 14.7 holds. Note that $\sigma(Z)|_{\{Z=z\}} = \{\emptyset, \{Z=z\}\}$ (see the definition of the trace of a set system in Example 1.10). Therefore,

$$\mathscr{C} \cup \sigma(Z)|_{\{Z=z\}} = \left(\mathscr{C}|_{\{Z=z\}}\right) \cup \left(\sigma(Z)|_{\{Z=z\}}\right) \qquad [\text{Example 1.10}]$$

$$= \left(\mathscr{C}|_{\{Z=z\}}\right) \cup \{\emptyset, \{Z=z\}\} \qquad (14.85)$$

$$= \mathscr{C}|_{\{Z=z\}}. \qquad \left[\emptyset \in \mathscr{C}, \Omega \cap \{Z=z\} \in \mathscr{C}|_{\{Z=z\}}\right]$$

Now

$$\sigma(\mathscr{C}, Z)|_{\{Z=z\}} = \sigma\left([\mathscr{C} \cup \sigma(Z)]|_{\{Z=z\}}\right) \qquad [(14.69)]$$

$$= \sigma(\mathscr{C}|_{\{Z=z\}}) \qquad [(14.85)]$$

$$= \mathscr{C}|_{\{Z=z\}}. \qquad [(1.15)]$$

Hence, for all $C \in \sigma(\mathscr{C}, Z)$, there is an $A_C \in \mathscr{C}$ such that $\{Z=z\} \cap C = \{Z=z\} \cap A_C$, which implies

$$1_{Z=z} \cdot 1_C = 1_{Z=z} \cdot 1_{A_C}. \qquad (14.86)$$

Therefore, for all $C \in \sigma(\mathscr{C}, Z)$,

$$E^{Z=z}[1_C \cdot E^{Z=z}(Y \mid \mathscr{C})] = \frac{1}{P(Z=z)} E[1_{Z=z} \cdot 1_C \cdot E^{Z=z}(Y \mid \mathscr{C})] \qquad [(9.11)]$$

$$= \frac{1}{P(Z=z)} E[1_{Z=z} \cdot 1_{A_C} \cdot E^{Z=z}(Y \mid \mathscr{C})] \qquad [(14.86)]$$

$$= E^{Z=z}[1_{A_C} \cdot E^{Z=z}(Y \mid \mathscr{C})] \qquad [(9.11)]$$

$$= E^{Z=z}(1_{A_C} \cdot Y) \qquad [\text{Def. 14.7 (b)}]$$

$$= \frac{1}{P(Z=z)} E(1_{Z=z} \cdot 1_{A_C} \cdot Y) \qquad [(9.11)]$$

$$= \frac{1}{P(Z=z)} E(1_{Z=z} \cdot 1_C \cdot Y) \qquad [(14.86)]$$

$$= E^{Z=z}(1_C \cdot Y). \qquad [(9.11)]$$

Proof of Theorem 14.67

Let $g_{Z=z}(X, W) = E^{Z=z}(Y \mid X, W)$ [see (14.34)]. Then,

$$g_{Z=z}(X, W) \cdot 1_{W=w} = g_{Z=z}(X, w) \cdot 1_{W=w}. \qquad (14.87)$$

Furthermore,

$$E^{W=w}\left[E^{Z=z}(Y \mid X, W)\right] = \frac{1}{P(W=w)} \int E^{Z=z}(Y \mid X, W) \cdot 1_{W=w} \, dP \qquad [(9.11)]$$

$$= \frac{1}{P(W=w)} \int E^{Z=z}(Y \mid X, W=w) \cdot 1_{W=w} \, dP \qquad [(14.87)]$$

$$= \int E^{Z=z}(Y \mid X, W=w) \, dP^{W=w} \qquad [(9.11)]$$

$$= \int E^{Z=z}(Y \mid X=x, W=w) \, P_X^{W=w}(dx), \qquad [(6.13)]$$

which proves the first equation of (14.78).

Now let $g(X, W, Z) = E(Y \mid X, W, Z)$. Then, for $g_z(X, W)$ [see (14.23)],

$$g_z(X, W) \cdot 1_{W=w} = g_z(X, w) \cdot 1_{W=w}. \qquad (14.88)$$

Furthermore, because $E^{Z=z}(Y \mid X, W)$ is $P^{W=w}$-unique,

$$E^{W=w}\left[E^{Z=z}(Y \mid X, W)\right] = E^{W=w}\left[E(Y \mid X, W, Z=z)\right] \qquad [(14.58), \text{Box 6.1 (ix)}]$$

$$= \frac{1}{P(W=w)} \int E(Y \mid X, W, Z=z) \cdot 1_{W=w} \, dP \qquad [(9.11)]$$

$$= \frac{1}{P(W=w)} \int E(Y \mid X, W=w, Z=z) \cdot 1_{W=w} \, dP \qquad [(14.88)]$$

$$= \int E(Y \mid X, W=w, Z=z) \, dP^{W=w} \qquad [(9.11)]$$

$$= \int E(Y \mid X=x, W=w, Z=z) \, P_X^{W=w}(dx). \qquad [(6.13)]$$

This proves the second equation of (14.78).

Proof of Theorem 14.69

Note that $E^{Z=z}(Y \mid \mathscr{C})$ is $\mathscr{C}$-measurable and therefore also $\sigma(\mathscr{C}, \mathscr{D})$-measurable [see Eq. (14.69)]. Furthermore,

$$E^{Z=z}(Y \mid \mathscr{C}, \mathscr{D}) \underset{P^{Z=z}}{=} E(Y \mid \mathscr{C}, \mathscr{D}, Z) \qquad\qquad\qquad \text{[Eqs. (14.72)]}$$

$$\underset{P^{Z=z}}{=} E(Y \mid \mathscr{C}, Z) \qquad [E(Y \mid \mathscr{C}, \mathscr{D}, Z) \underset{P}{=} E(Y \mid \mathscr{C}, Z), \text{Cor. 5.22}]$$

$$\underset{P^{Z=z}}{=} E^{Z=z}(Y \mid \mathscr{C}). \qquad\qquad\qquad\qquad \text{[Eqs. (14.72)]}$$

Exercises

14.1 Consider Table 14.3. Why do the values of the conditional expectation $E(Y \mid X, U)$ have to be identical for $\omega_3 = (Joe, yes, -)$ and $\omega_4 = (Joe, yes, +)$?

14.2 Show that if $Y: (\Omega, \mathscr{A}, P) \to (\overline{\mathbb{R}}, \overline{\mathscr{B}})$ is a random variable with finite expectation $E(Y)$, then $E^B(Y)$ is finite.

14.3 Let $(\Omega, \mathscr{A}, P)$, $(\Omega, \mathscr{A}, Q)$ be probability spaces, and consider the random variable $X: (\Omega, \mathscr{A}, P) \to (\Omega'_X, \mathscr{A}'_X)$. Show that X is a random variable on $(\Omega, \mathscr{A}, P)$ if and only if it is also a random variable on $(\Omega, \mathscr{A}, Q)$.

14.4 Show that the assumptions 14.5 imply $\mathscr{E}^B(1_B \cdot Y \mid \mathscr{C}) = \mathscr{E}^B(Y \mid \mathscr{C})$.

14.5 Prove Equation (14.26).

14.6 Prove Equation (14.27).

14.7 Compute the values of the conditional expectation $E^{X=0}(Y \mid U)$ in the example presented in Table 14.3.

14.8 Prove propositions (i) and (ii) of Remark 14.35.

14.9 Show that Theorem 14.33 implies Equation (14.35).

14.10 Prove: For all $x \in X(\Omega) = \{0, 1\}$: If $P(X=x) > 0$ and $E^{X=0}(Y \mid Z)$ is $P^{X=x}$-unique, then $E(V_0 \mid X=x) = E(V_0^* \mid X=x)$ for all versions $V_0, V_0^* \in \mathscr{E}^{X=0}(Y \mid Z)$.

14.11 Show that the expectation $E(V)$ of a version $V \in \mathscr{E}^B(Y \mid \mathscr{C})$ exists and is finite if $\mathscr{C}$ is a finite set and $E^B(Y)$ is finite.

14.12 Prove Remark 14.49 for X being discrete.

14.13 Prove the propositions summarized in Box 14.1.

14.14 Prove Equation (14.76).

14.15 Prove Equation (14.77).

Solutions

14.1 If $E(Y \mid X, U)(\omega_3) \neq E(Y \mid X, U)(\omega_4)$, then $E(Y \mid X, U)$ would not be measurable with respect to (X, U). Measurability of a random variable V with respect to (X, U) requires that V takes on only one single value for all $\omega \in (X, U)^{-1}(\{(1, Joe)\})$ (see Cor. 2.53).

14.2 According to Equation (9.11),

$$E^B(Y) = \frac{1}{P(B)} E(1_B \cdot Y).$$

If Y has a finite expectation with respect to P, then $\int Y^+ \, dP < \infty$ and $\int Y^- \, dP < \infty$. Because $0 \leq 1_B \cdot Y^+ \leq Y^+$ and $0 \leq 1_B \cdot Y^- \leq Y^-$, Equation (3.24) yields $\int 1_B \cdot Y^+ \, dP < \infty$ and $\int 1_B \cdot Y^- \, dP < \infty$, and therefore $-\infty < \int 1_B \cdot Y \, dP < \infty$. Therefore, if $-\infty < E(Y) < \infty$, then $-\infty < E(1_B \cdot Y) < \infty$.

14.3 Definition 5.1 of a random variable $X: (\Omega, \mathscr{A}, P) \rightarrow (\Omega'_X, \mathscr{A}'_X)$ only requires that there is a probability measure, here denoted P, on a measurable space $(\Omega, \mathscr{A})$ and that X is measurable with respect to $\mathscr{A}$, that is, $X^{-1}(A') \in \mathscr{A}$, for all $A' \in \mathscr{A}'_X$. Hence, if X is a random variable on $(\Omega, \mathscr{A}, P)$, then it is also a random variable on $(\Omega, \mathscr{A}, Q)$, whenever P and Q are probability measures on $(\Omega, \mathscr{A})$.

14.4 If $V \in \mathscr{E}^B(1_B \cdot Y \mid \mathscr{C})$, then, according to Definition 14.7 (a), it is measurable with respect to $\mathscr{C}$. Furthermore, for all $C \in \mathscr{C}$,

$$
\begin{aligned}
E^B(1_C \cdot V) &= E^B(1_C \cdot 1_B \cdot Y) && \text{[Def. 14.7 (b)]} \\
&= \frac{1}{P(B)} \cdot E(1_C \cdot 1_B \cdot 1_B \cdot Y) && \text{[(9.7)]} \\
&= \frac{1}{P(B)} \cdot E(1_C \cdot 1_B \cdot Y) && [1_B \cdot 1_B = 1_B] \\
&= E^B(1_C \cdot Y). && \text{[(9.7)]}
\end{aligned}
$$

Hence, $V \in \mathscr{E}^B(Y \mid \mathscr{C})$. Vice versa, if $V^* \in \mathscr{E}^B(Y \mid \mathscr{C})$, then it is measurable with respect to $\mathscr{C}$ and, for all $C \in \mathscr{C}$,

$$
\begin{aligned}
E^B(1_C \cdot V^*) &= E^B(1_C \cdot Y) && \text{[Def. 14.7 (b)]} \\
&= \frac{1}{P(B)} \cdot E(1_C \cdot 1_B \cdot Y) && \text{[(9.7)]} \\
&= \frac{1}{P(B)} \cdot E(1_C \cdot 1_B \cdot 1_B \cdot Y) && [1_B \cdot 1_B = 1_B] \\
&= E^B(1_C \cdot 1_B \cdot Y). && \text{[(9.7)]}
\end{aligned}
$$

Hence, $V^* \in \mathscr{E}^B(1_B \cdot Y \mid \mathscr{C})$.

14.5 For all $\omega \in \Omega$,

$$
\begin{aligned}
1_{Z=z}(\omega) \cdot g[X(\omega), Z(\omega)] &= \begin{cases} 0, & \text{if } 1_{Z=z}(\omega) = 0 \\ g_z[X(\omega)], & \text{if } 1_{Z=z}(\omega) = 1 \end{cases} \\
&= \begin{cases} 1_{Z=z}(\omega) \cdot g_z[X(\omega)], & \text{if } 1_{Z=z}(\omega) = 0 \\ 1_{Z=z}(\omega) \cdot g_z[X(\omega)], & \text{if } 1_{Z=z}(\omega) = 1 \end{cases} \\
&= 1_{Z=z}(\omega) \cdot g_z[X(\omega)].
\end{aligned}
$$

14.6   For all $\omega \in \Omega$,

$$\left(\sum_{z \in Z(\Omega)} E(Y \mid X, Z=z) \cdot 1_{Z=z} \right)(\omega)$$

$$= \left(\sum_{z \in Z(\Omega)} g_z(X) \cdot 1_{Z=z} \right)(\omega) \qquad\qquad [(14.25)]$$

$$= \left(\sum_{z \in Z(\Omega)} g(X, Z) \cdot 1_{Z=z} \right)(\omega) \qquad\qquad [(14.26)]$$

$$= \sum_{z \in Z(\Omega)} g(X, Z)(\omega) \cdot 1_{Z=z}(\omega) \qquad\qquad [(2.31)]$$

$$= g(X, Z)(\omega) \qquad\qquad\qquad [(2.26)]$$

$$= E(Y \mid X, Z)(\omega). \qquad\qquad\qquad [\text{Def. 14.29}]$$

14.7   The values of the conditional expectation $E^{X=0}(Y \mid U)$ are the two conditional expectation values $E^{X=0}(Y \mid U=Joe)$ and $E^{X=0}(Y \mid U=Ann)$ with respect to the measure $P^{X=0}$. Because $E^{X=0}(Y \mid U=u) = P^{X=0}(Y=1 \mid U=u)$, they can be computed as follows:

$$P^{X=0}(Y=1 \mid U=Joe) = \frac{P^{X=0}(Y=1, U=Joe)}{P^{X=0}(U=Joe)} \approx \frac{.561}{.561 + .245} \approx .696$$

and

$$P^{X=0}(Y=1 \mid U=Ann) = \frac{P^{X=0}(Y=1, U=Ann)}{P^{X=0}(U=Ann)} \approx \frac{.039}{.039 + .155} \approx .2.$$

14.8   (i) If $g(X, Z) \in \mathscr{E}(Y \mid X, Z)$ and $g_{Z=z}(X) \in \mathscr{E}^{Z=z}(Y \mid X)$, then for all $C \in \sigma(X)$,

$$\int 1_C \cdot g_z(X) \, dP^{Z=z}$$

$$= \frac{1}{P(Z=z)} \int 1_C \cdot 1_{Z=z} \cdot g_z(X) \, dP \qquad\qquad [(9.11)]$$

$$= \frac{1}{P(Z=z)} \int 1_C \cdot 1_{Z=z} \cdot g(X, Z) \, dP \qquad\qquad [(14.26)]$$

$$= \frac{1}{P(Z=z)} \int 1_C \cdot 1_{Z=z} \cdot Y \, dP \qquad\qquad [C \cap \{Z=z\} \in \sigma(X, Z), \text{ Def. 10.2 (b)}]$$

$$= \int 1_C \cdot Y \, dP^{Z=z} \qquad\qquad\qquad [(9.11)]$$

$$= \int 1_C \cdot g_{Z=z}(X) \, dP^{Z=z}. \qquad\qquad\qquad [\text{Def. 14.7 (b)}]$$

Hence, $g_z(X) \underset{P^{Z=z}}{=} g_{Z=z}(X)$ (see Th. 3.48) or, equivalently, $g_z(x) = g_{Z=z}(x)$, for $P_X^{Z=z}$-a.a. $x \in \Omega_X'$ [see Cor. 5.25 (a) and (5.12)].

(ii) If $P(X=x, Z=z) > 0$ for all $x \in \Omega'_X$, then $P(Z=z) > 0$ [see Box 4.1 (v)] and

$$P_X^{Z=z}(\{x\}) = P^{Z=z}(X=x) = \frac{P(X=x, Z=z)}{P(Z=z)} > 0, \quad \forall x \in \Omega'_X.$$

Hence, if $P(X=x, Z=z) > 0$, then Remark 2.71 and $g_z(X) \underset{P^{Z=z}}{=} g_{Z=z}(X)$ imply $g_z(x) = g_{Z=z}(x)$ for all $x \in \Omega'_X$.

14.9

$$E(Y \mid X, Z=z) \underset{P^{Z=z}}{=} E^{Z=z}(Y \mid X) \qquad\qquad [(14.30)]$$

$$\underset{P^{Z=z}}{=} E^{1_{Z=z}=1}(Y \mid X) \qquad\qquad [\{Z=z\} = \{1_{Z=z} = 1\}]$$

$$\underset{P^{Z=z}}{=} E(Y \mid X, 1_{Z=z}=1). \qquad\qquad [(14.30)]$$

For $E(Y \mid X, Z) = g(X, Z) \in \mathscr{E}\,(Y \mid X, Z)$ and $P_X^{Z=z}$-a.a. $x \in \Omega'_X$,

$$E(Y \mid X=x, Z=z) = g(x, z) \qquad\qquad [(10.27)]$$

$$= g_z(x) \qquad\qquad [(14.23)]$$

$$= g_{Z=z}(x) \qquad\qquad [\text{Rem. 14.35 (i)}]$$

$$= E^{Z=z}(Y \mid X=x). \qquad\qquad [(14.34)]$$

14.10 This proposition follows from the definition of uniqueness of a conditional expectation with respect to a probability measure $P^{X=x}$, equivalence of two random variables with respect to a probability measure $P^{X=x}$, Corollaries 5.24 and 6.17, and Equations (9.5) and (9.6) with $B = \{X=x\}$.

14.11 If $\mathscr{C} = \{A_1, \dots, A_n\}$ is finite, then there is a finite partition $\{B_1, \dots, B_m\}$ of Ω with $\mathscr{C} = \sigma(\{B_1, \dots, B_m\})$ (see Rem. 1.21). Then, according to Lemma 2.19 and Definition 14.7 (b),

$$V = \sum_{j=1}^m \alpha_j 1_{B_j}, \quad \text{where} \quad \begin{cases} \alpha_j = \displaystyle\int 1_{B_j} \cdot Y \, dP^B, & \text{if } P^B(B_j) > 0 \\[2mm] \text{any } \alpha_j \in \mathbb{R}, & \text{if } P^B(B_j) = 0, \end{cases}$$

is a version $V \in \mathscr{E}^B(Y \mid \mathscr{C})$. Hence, if $E^B(Y) = \int Y \, dP^B$ is finite, then, for any such choice of the numbers α_j,

$$E(V) = \sum_{j=1}^m \alpha_j \cdot P(B_j)$$

[see Eq. (6.3)] is finite as well.

14.12   (d) ⇒ (14.48) By definition, condition (d) of Corollary 14.48 is equivalent to

$$\forall A' \in \mathscr{A}'_X: \quad P^B_X(A') = 0 \;\Rightarrow\; P_X(A') = 0.$$

If X is discrete and $\{x\} \in \mathscr{A}'_X$ for all $x \in \Omega'_X$, then we can choose $A' = \{x\}$ for all $x \in \Omega'_X$, and this yields

$$\forall x \in \Omega'_X: \quad P^B_X(\{x\}) = 0 \;\Rightarrow\; P_X(\{x\}) = 0,$$

which is (14.48) in a different notation [see Eqs. (5.2) and (5.4)].

(14.48) ⇒ (d) Assume that (14.48) holds and let $A' \in \mathscr{A}'_X$ with $P^B_X(A') = 0$. Then,

$$
\begin{aligned}
P^B_X(A') &= \sum_{\substack{x \in A' \\ P(X=x)>0}} P^B_X(\{x\}) &&\text{[Th. 4.28, Box 4.1 (i)]}\\
&= \sum_{\substack{x \in A' \\ P(X=x)>0}} P^B(X=x) &&\text{[(5.2), (5.4)]}\\
&= 0.
\end{aligned}
$$

Because a sum of nonnegative summands is 0 if and only if all summands are 0, the last equation implies $P^B(X=x) = 0$ for all $x \in A'$. Now (14.48) yields $P(X=x) = 0$ for all $x \in A'$. Therefore,

$$P(A') = \sum_{\substack{x \in A' \\ P(X=x)>0}} P(\{x\}) = \sum_{\substack{x \in A' \\ P(X=x)>0}} P(X=x) = 0$$

[see again Eqs. (5.2) and (5.4)].

14.13   (i) This is the definition of P-uniqueness of $E^B(Y \mid \mathscr{C})$ (see Rem. 10.13).

(ii), (iii) These propositions have been proved in Theorem 14.46.

(iv) This proposition follows from Theorem 14.46, because $P \underset{\mathscr{C}}{\ll} P^B$ is equivalent to

$$\forall C \in \mathscr{C}: \quad P^B(C) = 0 \;\Rightarrow\; P(C) = 0,$$

and $\mathscr{D} \subset \mathscr{C}$ implies that this implication also holds for all $C \in \mathscr{D}$, that is, $P \underset{\mathscr{D}}{\ll} P^B$.

(v) This proposition immediately follows from applying Corollary 5.22 to $E^B(Y \mid \mathscr{C})$.

(vi), (vii) According to (v), P-uniqueness of $E^B(Y \mid \mathscr{C})$ implies $V \underset{P^C}{=} V^*$. Theorem 2.85 then implies $P^C_V = P^C_{V^*}$. If the two distributions are identical, then the corresponding expectations are identical as well (see Cor. 6.17).

(viii) This proposition immediately follows from Remark 2.76 (iii).

(ix) Note that $P^B(X=x) > 0$, for all $x \in \Omega'_X$, implies $X(\Omega) = \Omega'_X$. Therefore, and because uniqueness of $E^B(Y \mid X)$ implies that $E^B(Y \mid X)$ is P-unique, the proposition is an immediate implication of Corollary 14.42.

14.14 If X is measurable with respect to Z, then,

$$E^{Z=z}(Y \mid \mathscr{C}, X) \underset{p^{Z=z}}{=} E^{Z=z}\left[E^{Z=z}(Y \mid \mathscr{C}, Z) \mid \mathscr{C}, X\right] \qquad \text{[Box 10.1 (v)]}$$

$$\underset{p^{Z=z}}{=} E^{Z=z}\left[E^{Z=z}(Y \mid \mathscr{C}) \mid \mathscr{C}, X\right] \qquad \text{[(14.72), Box 10.1 (ix)]}$$

$$\underset{p^{Z=z}}{=} E^{Z=z}(Y \mid \mathscr{C}). \qquad \text{[Box 10.1 (vii)]}$$

14.15 If $\mathscr{D} \subset \sigma(\mathscr{C}, Z)$, then,

$$E^{Z=z}(Y \mid \mathscr{D}) \underset{p^{Z=z}}{=} E^{Z=z}[E^{Z=z}(Y \mid \mathscr{C}, Z) \mid \mathscr{D}] \qquad \text{[Box 10.1 (v)]}$$

$$\underset{p^{Z=z}}{=} E^{Z=z}[E(Y \mid \mathscr{C}, Z) \mid \mathscr{D}]. \qquad \text{[(14.72), Box 10.1 (ix)]}$$

15

Effect functions of a discrete regressor

In chapter 14, we treated $E^{X=x}(Y \mid Z)$, the Z-conditional expectation of Y with respect to the $(X = x)$-conditional-probability measure $P^{X=x}$. There we already noted that, if the values of X represent treatment conditions, then $E^{X=x}(Y \mid Z)$ refers to the Z-conditional expectation of Y given treatment x. If X is dichotomous with values 0 and 1, then the values $g_1(z)$ of the function $g_1(Z) := E^{X=1}(Y \mid Z) - E^{X=0}(Y \mid Z)$ are the effects of X on Y given the value z of Z, and g_1 is called the *Z-conditional-effect function of X*.

From a methodological point of view, the values $g_1(z)$ of the Z-conditional-effect function of X are of interest for at least two reasons. First, the conditional effect $g_1(z)$ describes the effect of X on Y for a fixed value z of Z. Thus, the impact of Z on X and Y is controlled by keeping Z constant on one of its values. Second, the conditional effect $g_1(z)$ is *more specific* and therefore more informative than the unconditional effect $E(Y \mid X = 1) - E(Y \mid X = 0)$. Knowing such conditional effects, we can choose individualized treatments.

In this chapter, we introduce the concepts of conditional-intercept functions and conditional-effect functions and consider these functions for the parameterizations of the conditional expectation $E(Y \mid X, Z)$ that have been treated in chapters 12 and 13.

15.1 Assumptions and definitions

In section 14.1, we treated three examples that motivated introducing the conditional expectations $E^{X=x}(Y \mid Z)$. These examples, in which the person variable U takes the role of Z, also motivate the present chapter on conditional-effect functions. In Examples 14.1 and 14.2, the conditional-effect function g_1 and each of its two values $g_1(Joe)$ and $g_1(Ann)$ are uniquely defined, whereas in Example 14.3 this is not the case. While in the latter example, the value $g_1(Ann)$ is uniquely defined, the value $g_1(Joe)$ is not, and we can choose any real number as the value $g_1(Joe) = g_1(U)(\omega)$ if $U(\omega) = Joe$, and still $g_0(U)$ and $g_1(U)$ satisfy

$$E(Y \mid X, U) \underset{P}{=} g_0(U) + g_1(U) \cdot X, \tag{15.1}$$

Probability and Conditional Expectation: Fundamentals for the Empirical Sciences, First Edition. Rolf Steyer and Werner Nagel.
© 2017 John Wiley & Sons, Ltd. Published 2017 by John Wiley & Sons, Ltd.
Companion website: http://www.probability-and-conditional-expectation.de

for all versions $E(Y \mid X, U) \in \mathscr{E}(Y \mid X, U)$. In other words, the function $g_1(U)$ specified in Example 14.3 is only one out of infinitely many versions of such a conditional-effect function satisfying Equation (15.1), and even the expectations $E[g_1(U)]$ are not necessarily identical. Therefore, we have to introduce an assumption that guarantees that the values of the functions $g_0(U)$ and $g_1(U)$ are not arbitrary and that the expectations $E[g_0(U)]$ and $E[g_1(U)]$ are uniquely defined. Instead of the person variable U used in the examples with Joe and Ann, now we choose Z as a random variable with respect to which we consider intercept and effect functions.

Throughout this chapter, we refer to the following assumptions and notation.

Notation and assumptions 15.1
Let $X \colon (\Omega, \mathscr{A}, P) \to (\Omega'_X, \mathscr{A}'_X)$, $Y \colon (\Omega, \mathscr{A}, P) \to (\mathbb{R}, \mathscr{B})$, and $Z \colon (\Omega, \mathscr{A}, P) \to (\Omega'_Z, \mathscr{A}'_Z)$ be random variables, where $E(Y)$ is finite. Furthermore, assume that X is discrete with $P(X \in \{x_0, x_1, \ldots, x_n\}) = 1$ and $P(X = x_i) > 0$, for all $i = 0, 1, \ldots, n$.

Remark 15.2 [An additional assumption]   The following additional assumption is often used in this chapter:

$$P(X = x_i \mid Z) \underset{P}{>} 0, \quad \forall\, i = 0, 1, \ldots, n. \tag{15.2}$$

According to Corollary 14.48, this assumption is equivalent to each of the following conditions:

P-uniqueness of the conditional expectations $E^{X = x_i}(Y \mid Z)$, $\quad \forall\, i = 0, 1, \ldots, n,$ (15.3)

$$P \underset{\sigma(Z)}{\ll} P^{X = x_i}, \quad \forall\, i = 0, 1, \ldots, n, \qquad \text{(absolute continuity)} \tag{15.4}$$

$$E(V_i) = E(V_i^*), \quad \forall\, V_i, V_i^* \in \mathscr{E}^{X = x_i}(Y \mid Z), \quad \forall\, i = 0, 1, \ldots, n. \tag{15.5}$$

Also remember, P-uniqueness of the conditional expectations $E^{X = x_i}(Y \mid Z)$ implies that different versions of $E^{X = x_i}(Y \mid Z)$ do not only have identical distributions with respect to the conditional-probability measure $P^{X = x_i}$ [see Def. 4.29 and (9.4)] but also with respect to P. ◁

15.2   Intercept function and effect functions

Theorem 15.3 [Existence of intercept function and effect functions]
Let the assumptions 15.1 hold. Then there are an $E(Y \mid X, Z) \in \mathscr{E}(Y \mid X, Z)$ and, for all $i = 0, 1, \ldots, n$, real-valued $E^{X = x_i}(Y \mid Z) \in \mathscr{E}^{X = x_i}(Y \mid Z)$ such that

$$E(Y \mid X, Z) = g_0(Z) + \sum_{i=1}^{n} g_i(Z) \cdot 1_{X = x_i}, \tag{15.6}$$

with

$$g_0(Z) := E^{X=x_0}(Y \mid Z), \tag{15.7}$$

and

$$g_i(Z) := E^{X=x_i}(Y \mid Z) - E^{X=x_0}(Y \mid Z), \quad \forall\, i = 1, \ldots, n. \tag{15.8}$$

(Proof p. 466)

Referring to Equation (15.6), now we can define the intercept function and the effect functions as follows.

Definition 15.4 [Intercept function and effect functions]
Let the assumptions 15.1 hold as well as (15.2). Then the function $g_0 : (\Omega'_Z, \mathscr{A}'_Z) \to (\mathbb{R}, \mathscr{B})$ is called the Z-conditional-intercept function, and, for all $i = 1, \ldots, n$, the function $g_i : (\Omega'_Z, \mathscr{A}'_Z) \to (\mathbb{R}, \mathscr{B})$, the Z-conditional-effect function of x_i versus x_0 on Y which pertains to the version $E(Y \mid X, Z) \in \mathscr{E}(Y \mid X, Z)$ in Equation (15.6).

Remark 15.5 [The functions g_i versus the functions $g_i(Z)$] Note that the functions $g_i(Z)$, $i = 0, 1, \ldots, n$, denote the compositions of Z and g_i. Because Z is a random variable on $(\Omega, \mathscr{A}, P)$, the compositions $g_i(Z)$ are random variables on $(\Omega, \mathscr{A}, P)$ as well. In contrast, the functions $g_i : (\Omega'_Z, \mathscr{A}'_Z) \to (\mathbb{R}, \mathscr{B}), i = 0, 1, \ldots, n$, are not random variables on $(\Omega, \mathscr{A}, P)$. However, they are random variables on the probability space $(\Omega'_Z, \mathscr{A}'_Z, P_Z)$, where P_Z denotes the distribution of Z.   ◁

Remark 15.6 [P-uniqeness of the intercept function and effect functions] Suppose that the assumptions of Definition 15.4 hold. Then, (15.3) and (2.36) imply that all measurable functions $g_0^*, g_1^*, \ldots, g_n^* : (\Omega'_Z, \mathscr{A}'_Z) \to (\mathbb{R}, \mathscr{B})$ with

$$\left(g_0^*(Z) + \sum_{i=1}^{n} g_i^*(Z) \cdot 1_{X=x_i} \right) \in \mathscr{E}(Y \mid X, Z)$$

are P_Z-unique, and according to Corollary 5.25 (i), this implies

$$g_i^*(Z) \underset{P}{=} g_i(Z), \quad \forall\, i = 0, 1, \ldots, n. \tag{15.9}$$

Hence, under the assumptions mentioned above, the compositions of the intercept and effect functions are P-unique. This in turn implies that the expectations and the variances of the intercept and effect functions are uniquely defined, provided that they exist.   ◁

Remark 15.7 [Partial conditional expectation $E(Y \mid X, Z=z)$]   Let the assumptions of Definition 15.4 hold. Then Equation (15.6) implies

$$E(Y \mid X, Z=z) = g_0(z) + \sum_{i=1}^{n} g_i(z) \cdot 1_{X=x_i} \tag{15.10}$$

for the partial conditional expectation $E(Y \mid X, Z=z)$ (see Def. 14.29). This equation justifies the terminology introduced in Definition 15.4.   ◁

Remark 15.8 [Conditional intercepts and conditional effects]   Let the assumptions of Definition 15.4 hold and assume $z \in \Omega_Z'$ with $P(Z=z) > 0$. Then Equation (15.10) and (14.29) imply that there is an $E^{Z=z}(Y \mid X) \in \mathscr{E}^{Z=z}(Y \mid X)$ with

$$E^{Z=z}(Y \mid X) = E(Y \mid X, Z=z) = g_0(z) + \sum_{i=1}^{n} g_i(z) \cdot 1_{X=x_i}. \tag{15.11}$$

If $P(Z=z) > 0$, then, according to Remark 2.71, the coefficients $g_i(z)$, $i = 0, 1, \ldots, n$, are uniquely determined. The number $g_0(z)$ is called the $(Z=z)$-*conditional intercept* and $g_i(z)$ the $(Z=z)$-*conditional effect* of x_i vs. x_0 on Y, where $i = 1, \ldots, n$. Equation (15.11) and Theorem 12.37 imply that $g_0(z)$ is the intercept and $g_i(z)$, $i = 1, \ldots, n$, are the regression coefficients pertaining to a linear parameterization in $1_{X=x_1}, \ldots, 1_{X=x_n}$ of the X-conditional expectation of Y with respect to the measure $P^{Z=z}$.

Furthermore, if Y has finite second moments with respect to $P^{Z=z}$ and the matrix of the covariances of the indicators $1_{X=x_1}, \ldots, 1_{X=x_n}$ with respect to $P^{Z=z}$ is regular, then Corollary 12.31 can be applied. Note that regularity of the matrix of the covariances of indicators $1_{X=x_1}, \ldots, 1_{X=x_n}$ with respect to $P^{Z=z}$ holds, if (15.2) and $P(Z=z) > 0$, because then $P^{Z=z}(X=x_i) > 0$ for all $i = 1, \ldots, n$ (see Lemma 12.38 and Rem. 12.39).

Also note that the prerequisite of Corollary 12.31 that $E(Y^2)$ is finite can be neglected because finiteness of $E(Y)$ already implies that $E(Y \cdot 1_{X=x_i})$ is finite [see Lemma 3.33 (ii)] and that the covariance vector of $(1_{X=x_1}, \ldots, 1_{X=x_n})$ and Y exists. Therefore, Equations (12.53) and (12.54) can be applied as well.   ◁

Remark 15.9 [Versions of $E^{Z=z}(Y \mid X)$]   Under the assumptions of Remark 15.8, Equation (15.11) and Remark 14.9 immediately imply

$$V_z \underset{P^{Z=z}}{=} g_0(z) + \sum_{i=1}^{n} g_i(z) \cdot 1_{X=x_i}, \quad \forall\, V_z \in \mathscr{E}^{Z=z}(Y \mid X). \tag{15.12}$$
◁

Remark 15.10 [Partial conditional expectation $E(Y \mid X=x_i, Z)$]   Let the assumptions of Definition 15.4 hold and let $E(Y \mid X=x_i, Z)$ denote the partial conditional expectation (see Def. 14.29 and Eq. 14.23). Then Equation (15.6) implies that there is a version $E(Y \mid X=x_i, Z)$ such that

$$E(Y \mid X=x_i, Z) = g_0(Z) + g_i(Z), \quad \forall\, i = 1, \ldots, n. \tag{15.13}$$
◁

Remark 15.11 [Conditional expectation $E^{X=x_i}(Y \mid Z)$] If the assumptions of Definition 15.4 hold, then, for all $i = 1, \ldots, n$, there is an $E^{X=x_i}(Y \mid Z) \in \mathscr{E}^{X=x_i}(Y \mid Z)$ with

$$E^{X=x_i}(Y \mid Z) = E(Y \mid X=x_i, Z) = g_0(Z) + g_i(Z). \tag{15.14}$$

$\triangleleft$

Remark 15.12 [Versions of $E^{X=x_i}(Y \mid Z)$] Remark 14.9 and Equation (15.14) imply, for all $i = 1, \ldots, n$,

$$V_{x_i} \underset{P^{X=x_i}}{=} g_0(Z) + g_i(Z), \quad \forall \, V_{x_i} \in \mathscr{E}^{X=x_i}(Y \mid Z). \tag{15.15}$$

$\triangleleft$

Example 15.13 [No treatment for Joe – continued] In Example 14.3, the conditional expectation $E^{X=1}(Y \mid Z)$ is not P-unique. Although the U-conditional-intercept function g_0 and the conditional effect $g_1(Ann)$ are uniquely defined, the U-conditional-effect function g_1 is not defined, because $P(X=1 \mid U) \underset{P}{>} 0$ [see (15.2)] does not hold. In this example, there are (infinitely) many functions g_1 satisfying Equation (15.1). This example emphasizes the importance of (15.2), which, in the definition of a conditional-effect function (see Def. 15.4), is assumed to hold. $\triangleleft$

15.3   Implications of independence of X and Z for regression coefficients

In the following theorem, we presume that the assumptions of Definition 15.4 hold, which implies that there is a version $E(Y \mid X, Z)$ of the (X, Z)-conditional expectation of Y such that Equation (15.6) holds. Now we consider the implications for the conditional expectation $E(Y \mid X)$ if we additionally assume

$$E[E^{X=x_i}(Y \mid Z) \mid X] \underset{P}{=} E[E^{X=x_i}(Y \mid Z)], \quad \forall \, i = 0, 1, \ldots, n. \tag{15.16}$$

Note that this equation follows from independence of X and Z [see (14.32) with $B = \{X=x_i\}$, (14.17), Theorem 5.52, and Box 10.2 (vi)].

Remember, if the assumptions of Definition 15.4 hold, then, according to Theorem 12.37, there is a version $E(Y \mid X) \in \mathscr{E}(Y \mid X)$ such that

$$E(Y \mid X) = \beta_0 + \sum_{i=1}^{n} \beta_i \cdot 1_{X=x_i} \tag{15.17}$$

with

$$\beta_0 = E(Y \mid X=x_0) \tag{15.18}$$

and

$$\beta_i = E(Y \mid X=x_i) - E(Y \mid X=x_0). \tag{15.19}$$

Theorem 15.14 [Mean independence and average effects]
Let the assumptions of Definition 15.4 hold, and let $g_i(Z)$, $i = 1, \ldots, n$, be the functions defined in Theorem 15.3. If Equation (15.16) holds, then there is a version $E(Y \mid X) \in \mathscr{E}(Y \mid X)$ satisfying Equations (15.17) to (15.19) with

$$\beta_0 = E[g_0(Z)] \tag{15.20}$$

and

$$\beta_i = E[g_i(Z)], \quad \forall\, i = 1, \ldots, n. \tag{15.21}$$

(Proof p. 467)

Remark 15.15 [Uniqueness of regression coefficients]   According to Equations (15.20) and (15.21), the regression coefficients $\beta_0, \beta_1, \ldots, \beta_n$ are uniquely defined. The crucial assumption for uniqueness is $P(X \in \{x_0, x_1, \ldots, x_n\}) = 1$ and $P(X=x_i) > 0$, for all $i = 0, 1, \ldots, n$.   ◁

Remark 15.16 [Independence of X and Z]   Independence of X and Z implies (15.16), which in turn implies

$$E[g_i(Z) \mid X] \underset{P}{=} E[g_i(Z)], \quad \forall\, i = 0, 1, \ldots, n. \tag{15.22}$$

Note that Equation (15.22) also immediately follows from Theorem 5.52 and Box 10.2 (vi).   ◁

Remark 15.17 [The role of randomization]   From a methodological point of view it should be noted that independence of a treatment variable X and a variable Z can be created by randomized assignment of the observational unit (e.g., a person) to one of the treatment conditions, provided that Z represents a pretreatment variable. Randomized assignment creates independence of X and *all* pretreatment variables. Examples for such pretreatment variables are the person variable U (see Table 14.1 for a concrete example) as well as any function of U such as sex, race, and any other attribute of persons prior to treatment.

This means that we can ignore a pretreatment variable Z in a randomized experiment with treatment variable X and response variable Y and still interpret the coefficient β_i as the average effect $E[g_i(Z)]$ of x_i compared to x_0 on Y, where $i = 1, \ldots, n$ [see Eq. (15.21)].   ◁

According to Corollary 15.18, independence of X and Z implies P-uniqueness of the functions $g_i(Z)$, provided that we presume $P(X=x_i) > 0$ for all $i = 0, 1, \ldots, n$.

Corollary 15.18 [Independence and P-uniqeness]
Let the assumptions 15.1 hold. If $X \underset{P}{\perp\!\!\!\perp} Z$, then $E^{X=x_i}(Y \mid Z)$ and the functions $g_i(Z)$ in Equations (15.7) and (15.8) are P-unique, for all $i = 0, 1, \ldots, n$.

(Proof p. 467)

Example 15.19 [Joe and Ann with randomized assignment – continued] In Example 14.1, we already showed that the slope in

$$E(Y \mid X) = .45 + .15 \cdot X,$$

[see Eq. (14.5)] is identical to the expectation $E_U(g_1) = E[g_1(U)] = .15$ of the U-conditional-effect function g_1. Similarly, using Equation (14.1) and $E(1_{U=Joe}) = .5$ (see Table 14.1),

$$E[g_0(U)] = E(.2 + .5 \cdot 1_{U=Joe}) = .2 + .5 \cdot E(1_{U=Joe}) = .2 + .5 \cdot .5 = .45,$$

which is identical to the intercept in the equation for $E(Y \mid X)$. This illustrates Theorem 15.14. $\triangleleft$

Example 15.20 [Joe and Ann with self-selection – continued] In Example 14.2, we already showed that the slope in

$$E(Y \mid X) = .6 - .18 \cdot X,$$

[see Eq. (14.6)] is not identical to the expectation

$$E_U(g_1) = E[g_1(U)] = .15$$

of the U-conditional-effect function g_1. In this example, neither independence of U and X nor $E[E_U(g_1) \mid X] = E[g_1(U)]$ hold, and the slope $-.18$ in the equation for $E(Y \mid X)$ cannot be used for the evaluation of the treatment effect. This emphasizes the importance of assumption (15.16), which is the crucial assumption made in Theorem 15.14. $\triangleleft$

15.4 Adjusted effect functions

Remark 15.21 [Methodological background] In Definition 15.4, we introduced the Z-conditional-intercept function g_0 as well as the Z-conditional-effect functions g_i, $i = 1, \ldots, n$, where $Z = (Z_1, \ldots, Z_m)$ can be an m-dimensional random variable consisting of m unidimensional random variables. Examples of such random variables are *pretest* (Z_1), *sex* (Z_2), *educational status* (Z_3), *body mass index* (Z_4), and *blood type* (Z_5). As mentioned before, conditioning on a (possibly multidimensional) random variable $Z = (Z_1, \ldots, Z_m)$ also serves to obtain more specific effects that are more informative than unconditional effects.

However, the $(Z_1, \ldots, Z_m)$-conditional effects described by the effect functions g_i might be too fine-grained and one may wish to reaggregate them. The most radical reaggregation is to consider the (unconditional) expectation (the 'average') of the $(Z_1, \ldots, Z_m)$-conditional effects.

Aside from the average effects, that is, the (unconditional) expectations of the functions $g_i(Z_1, \ldots, Z_m)$, we might also be interested in the conditional expectation values of the functions $g_i(Z_1, \ldots, Z_m)$ given $Z_2 = male$ and given $Z_2 = female$, or in the conditional expectation values of the functions $g_i(Z_1, \ldots, Z_m)$ given various values of Z_1 *(pretest)*. This way of reaggregation may be called *coarsening the effects*. Knowing such coarsened effects is very important; for example, if, knowing the $(Z_1, \ldots, Z_m)$-conditional-effect functions (e.g., from a previous study), we want to conduct an as much as possible individualized treatment, but are not able to assess all components $Z_1, \ldots, Z_m$ of Z, but just Z_1 or just Z_1 and Z_2.

In Definition 15.22, we consider reaggregating the Z-conditional-intercept function and the Z-conditional-effect functions to a W-conditional-effect function that is adjusted for Z, where W is another random variable. If $Z = (Z_1, \ldots, Z_m)$ with $m \geq 2$, then $W = Z_1$ and $W = (Z_1, Z_2)$ are examples in case. In these two cases, W would be Z-measurable.

Note that reaggregating the Z-conditional-effect functions is not equivalent to ignoring Z and conditioning on W instead. More precisely, assume

(a) $Z = (Z_1, \ldots, Z_m)$ and there are real-valued functions g_i such that

$$E(Y \mid X, Z) \underset{P}{=} g_0(Z) + \sum_{i=1}^{n} g_i(Z) \cdot 1_{X=x_i}, \tag{15.23}$$

(b) $W = (Z_{i_1}, \ldots, Z_{i_k})$, for $\{i_1, \ldots, i_k\} \subset \{1, \ldots, m\}$, and

(c) there are real-valued functions f_i such that

$$E(Y \mid X, W) \underset{P}{=} f_0(W) + \sum_{i=1}^{n} f_i(W) \cdot 1_{X=x_i}. \tag{15.24}$$

Then $f_i(W) = E[g_i(Z) \mid W]$, for $i \in \{i_1, \ldots, i_k\}$, does not necessarily hold (see Example 15.24).

Considering $E(Y \mid X, W)$ instead of $E(Y \mid X, Z)$, we might miss the purpose of controlling and adjusting for important confounders contained in Z but not in W. In contrast, using the conditional expectation $E[g_i(Z) \mid W]$ (see Def. 15.22), we still control for Z, and with it we control for potential confounders contained in Z. Although, reaggregating the Z-conditional-effect functions in this way, that is, considering $E[g_i(Z) \mid W]$, we obtain less informative and less individualized conditional effects, the purpose of controlling and adjusting for important confounders is still fulfilled. ◁

Definition 15.22 [Adjusted effect function]
Let the assumptions of Definition 15.4 hold, let $E(Y \mid X, Z) \in \mathscr{E}(Y \mid X, Z)$, let g_i, $i = 1, \ldots, n$, be the functions satisfying Equation (15.6), and let $W: (\Omega, \mathscr{A}, P) \to (\Omega'_W, \mathscr{A}'_W)$ be a random variable.

   (i) *Then, for each $i = 1, \ldots, n$, the conditional expectation $E[g_i(Z) \mid W]$ is called a Z-adjusted W-conditional-effect function of x_i versus x_0 on Y.*

   (ii) *Furthermore, for each $i = 1, \ldots, n$, the expectation $E[g_i(Z)]$ is called the Z-adjusted effect of x_i versus x_0 on Y or the average of the Z-condition-al effects of x_i versus x_0 on Y.*

According to Remark 15.2, the conditional expectations $E[g_i(Z) \mid W]$, $i = 1, \ldots, n$, are P-unique and the expectations $E[g_i(Z)]$ are uniquely determined. The values of $E[g_i(Z) \mid W]$ are the $(W=w)$-conditional effects of x_i versus x_0 on Y that are adjusted for Z.

Remark 15.23 [Z-adjusted W-conditional expectation of Y given $X = x_i$]   Using the definition of the functions $g_i(Z)$ [see Eq. (15.8)] and Rule (xv) of Box 10.2 yield

$$
\begin{aligned}
&E[g_i(Z) \mid W] \\
&= E[E^{X=x_i}(Y \mid Z) - E^{X=x_0}(Y \mid Z) \mid W] \quad\quad\quad\quad\quad (15.25) \\
&= E[E^{X=x_i}(Y \mid Z) \mid W] - E[E^{X=x_0}(Y \mid Z) \mid W], \quad \forall\, i = 1, \dots, n.
\end{aligned}
$$

A conditional expectation $E[E^{X=x_i}(Y \mid Z) \mid W]$ is called a *Z-adjusted W-conditional expectation of Y given $X = x_i$*, where $i = 0, 1, \dots, n$.   ◁

Example 15.24 [Four persons with self-selection to treatment]   In Table 15.1 we present a new example displaying the conditional expectations $E(Y \mid X, U)$, $E(Y \mid X, Z)$, $E(Y \mid X)$, and $P(X = 1 \mid U)$ with dichotomous random variables X, Y, and Z, where X indicates with its values 0 and 1 whether or not the person is treated, Y indicates with its values 0 and 1 whether or not the person is successful, the values of Z are *male* and *female*, and U is the person variable with values *Joe*, *Jim*, *Sue*, and *Ann*.

Table 15.1   Four persons with self-selection to treatment.

Outcomes ω		Observables					Conditional expectations			
Unit / Treatment / Success	$P(\{\omega\})$	Person variable U	Sex Z	Treatment variable X	Outcome variable Y		$E(Y \mid X, U)$	$E(Y \mid X, Z)$	$E(Y \mid X)$	$P(X = 1 \mid U)$
(Joe, no, −)	.0675	Joe	m	0	0		.7	.66	.63	.1
(Joe, no, +)	.1575	Joe	m	0	1		.7	.66	.63	.1
(Joe, yes, −)	.0050	Joe	m	1	0		.8	.44	.44	.1
(Joe, yes, +)	.0200	Joe	m	1	1		.8	.44	.44	.1
(Jim, no, −)	.0175	Jim	m	0	0		.3	.66	.63	.9
(Jim, no, +)	.0075	Jim	m	0	1		.3	.66	.63	.9
(Jim, yes, −)	.1350	Jim	m	1	0		.4	.44	.44	.9
(Jim, yes, +)	.0900	Jim	m	1	1		.4	.44	.44	.9
(Sue, no, −)	.0600	Sue	f	0	0		.7	.60	.63	.2
(Sue, no, +)	.1400	Sue	f	0	1		.7	.60	.63	.2
(Sue, yes, −)	.0200	Sue	f	1	0		.6	.44	.44	.2
(Sue, yes, +)	.0300	Sue	f	1	1		.6	.44	.44	.2
(Ann, no, −)	.0400	Ann	f	0	0		.2	.60	.63	.8
(Ann, no, +)	.0100	Ann	f	0	1		.2	.60	.63	.8
(Ann, yes, −)	.1200	Ann	f	1	0		.4	.44	.44	.8
(Ann, yes, +)	.0800	Ann	f	1	1		.4	.44	.44	.8

We use the example of Table 15.1 to show that reaggregating the U-conditional-effect function is not equivalent to ignoring U and conditioning on Z instead. In order to formulate this more precisely, consider the equations

$$E(Y \mid X, U) = g_0(U) + g_1(U) \cdot X \tag{15.26}$$

and

$$E(Y \mid X, Z) = f_0(Z) + f_1(Z) \cdot X. \tag{15.27}$$

We show that, in this example, the function $f_1(Z)$ is not identical to the U-adjusted Z-conditional expectation $E[g_1(U) \mid Z]$.

Using the values $E(Y \mid X{=}x, U{=}u)$ of the conditional expectation $E(Y \mid X, U)$ displayed in Table 15.1 yields the following values of the functions g_i of Equation (15.26):

$$g_0(Joe) = .7, \qquad g_0(Jim) = .3, \qquad g_0(Sue) = .7, \qquad g_0(Ann) = .4, \tag{15.28}$$

and

$$g_1(Joe) = .1, \qquad g_1(Jim) = .1, \qquad g_1(Sue) = -.1, \qquad g_1(Ann) = .2. \tag{15.29}$$

Similarly, using the values $E(Y \mid X{=}x, Z{=}z)$ of the conditional expectation $E(Y \mid X, Z)$ displayed in Table 15.1 yields the following values of the functions f_i of Equation (15.27):

$$f_0(male) = .66, \qquad f_0(female) = .6 \tag{15.30}$$

and

$$f_1(male) = .22, \qquad f_1(female) = .16. \tag{15.31}$$

Hence, the $(Z{=}male)$-conditional effect of X on Y is .22 and the $(Z{=}female)$-conditional effect of X on Y is .16. In this example, these two numbers are misleading if used for the evaluation of the treatment effect, because the individual treatment effects for Joe and Jim are both .1, and for Sue and Ann they are $-.1$ and .2, respectively.

In contrast, computing the two values of the U-adjusted Z-conditional expectation $E[g_1(U) \mid Z]$ yields

$$E[g_1(U) \mid Z{=}male] = \sum_u g_1(u) \cdot P(U{=}u \mid Z{=}male) = .1 \cdot .5 + .1 \cdot .5 - .1 \cdot 0 + .2 \cdot 0 = .1,$$

and, correspondingly,

$$E[g_1(U) \mid Z{=}female] = \sum_u g_1(u) \cdot P(U{=}u \mid Z{=}female)$$
$$= .1 \cdot 0 + .1 \cdot 0 - .1 \cdot .5 + .2 \cdot .5 = .5.$$

Hence, the U-adjusted $(Z{=}male)$-conditional expectation value $E[g_1(U) \mid Z{=}male] = .1$ is the average effect of the treatment for males, and the U-adjusted $(Z{=}female)$-conditional expectation value $E[g_1(U) \mid Z=female] = .5$ is the average effect of the treatment for females. These effects are less informative than the person-specific effects $g_1(u)$ [see (15.29)]. Nevertheless, in this example they still inform us about the (conditional) expectations of these person-specific effects for the two sexes. In other words, in this example, they are the average treatment effects of the males and of the females, respectively.   ◁

15.5   Logit effect functions

In the previous sections of this chapter, we studied the conditional intercept and conditional-effect functions $g_0, g_1, \ldots, g_n$. Now we consider the special case, in which Y is dichotomous with values 0 and 1 (see Example 5.10). In this case, $E(Y \mid X, Z)$ is also called a conditional probability and is denoted by $P(Y=1 \mid X, Z)$ (see Rem. 10.4). If Y is dichotomous, then aside from the Z-conditional-intercept functions and the Z-conditional-effect functions $g_0, g_1, \ldots, g_n$, there are also Z-conditional logit intercept and logit effect functions (see Example 13.24), denoted f_i.

Theorem 15.25 [Existence of the logit effect functions]

Let the assumptions 15.1 hold, let Y be dichotomous with values 0 and 1, and suppose there is a $P(Y=1 \mid X, Z) \in \mathscr{P}(Y=1 \mid X, Z)$ with $0 < P(Y=1 \mid X, Z) < 1$. Then there are a version $P(Y=1 \mid X, Z) \in \mathscr{P}(Y=1 \mid X, Z)$, measurable functions $g_0, g_1, \ldots, g_n \colon (\Omega'_Z, \mathscr{A}'_Z) \to (\mathbb{R}, \mathscr{B})$, measurable functions $f_0, f_1, \ldots, f_n \colon (\Omega'_Z, \mathscr{A}'_Z) \to (\mathbb{R}, \mathscr{B})$, and, for all $i = 0, 1, \ldots, n$, a real-valued $P^{X=x_i}(Y=1 \mid Z) \in \mathscr{P}^{X=x_i}(Y=1 \mid Z)$ such that

$$P(Y=1 \mid X, Z) = g_0(Z) + \sum_{i=1}^{n} g_i(Z) \cdot 1_{X=x_i} \tag{15.32}$$

$$= \frac{\exp\left[f_0(Z) + \sum_{i=1}^{n} f_i(Z) \cdot 1_{X=x_i}\right]}{1 + \exp\left[f_0(Z) + \sum_{i=1}^{n} f_i(Z) \cdot 1_{X=x_i}\right]} \tag{15.33}$$

with

$$g_0(Z) := P^{X=x_0}(Y=1 \mid Z) \tag{15.34}$$

$$= \frac{\exp[f_0(Z)]}{1 + \exp[f_0(Z)]} \tag{15.35}$$

and, for $i = 1, \ldots, n$,

$$g_i(Z) := P^{X=x_i}(Y=1 \mid Z) - P^{X=x_0}(Y=1 \mid Z) \tag{15.36}$$

$$= \frac{\exp[f_0(Z) + f_i(Z)]}{1 + \exp[f_0(Z) + f_i(Z)]} - \frac{\exp[f_0(Z)]}{1 + \exp[f_0(Z)]}. \tag{15.37}$$

(Proof p. 467)

Remark 15.26 [P-uniqueness of the functions $f_i(Z)$] Let the assumptions of Theorem 15.25 and the additional assumption (15.2) hold. Then, for all measurable functions $f_0^*, f_1^*, \ldots, f_n^*: (\Omega_Z', \mathscr{A}_Z') \to (\mathbb{R}, \mathscr{B})$ satisfying Equations (15.34) to (15.37),

$$f_i^*(Z) \underset{P}{=} f_i(Z), \quad \forall\, i = 0, 1, \ldots, n, \tag{15.38}$$

(see Exercise 15.1). Hence, $E[f_i^*(Z)] = E[f_i(Z)], i = 1, \ldots, n$, provided that these expectations exist.   ◁

Definition 15.27 [Logit intercept function and logit effect function]
Let the assumptions of Theorem 15.25 as well as (15.2) hold. Then the function $f_0: (\Omega_Z', \mathscr{A}_Z') \to (\mathbb{R}, \mathscr{B})$ is called the Z-conditional logit intercept function, and, for all $i = 1, \ldots, n$, the function $f_i: (\Omega_Z', \mathscr{A}_Z') \to (\mathbb{R}, \mathscr{B})$, the Z-conditional logit effect function of x_i versus x_0 on Y which pertains to the version $P(Y=1 \mid X, Z) \in \mathscr{P}(Y=1 \mid X, Z)$ in Equation (15.33).

Remark 15.28 [Partial conditional probability $P(Y=1 \mid X, Z=z)$] Let the assumptions of Theorem 15.25 hold. Then Equations (15.33) and (14.25) imply

$$P(Y=1 \mid X, Z=z) = \frac{\exp\left(f_0(z) + \sum_{i=1}^{n} f_i(z) \cdot 1_{X=x_i}\right)}{1 + \exp\left(f_0(z) + \sum_{i=1}^{n} f_i(z) \cdot 1_{X=x_i}\right)} \tag{15.39}$$

for the partial conditional probability $P(Y=1 \mid X, Z=z)$ (see Rem. 14.30). This equation justifies the terminology introduced in Definition 15.27.   ◁

Remark 15.29 [$(Z=z)$-conditional logit intercept and effects] If the assumptions of Theorem 15.25 hold and $z \in \Omega_Z'$ with $P(Z=z) > 0$, then Equation (15.39) and (14.29) imply that there is a $P^{Z=z}(Y=1 \mid X) \in \mathscr{P}^{Z=z}(Y=1 \mid X)$ with

$$P^{Z=z}(Y=1 \mid X) = P(Y=1 \mid X, Z=z) = \frac{\exp\left[f_0(z) + \sum_{i=1}^{n} f_i(z) \cdot 1_{X=x_i}\right]}{1 + \exp\left[f_0(z) + \sum_{i=1}^{n} f_i(z) \cdot 1_{X=x_i}\right]}. \tag{15.40}$$

Equation (15.40) and Theorem 13.20 imply that $f_0(z)$ is the intercept and $f_i(z)$, $i = 1, \ldots, n$, are the coefficients pertaining to a linear logistic parameterization of $P^{Z=z}(Y=1 \mid X)$. According to Lemma 12.38 and Remark 12.39, $P^{Z=z}(X=x_i) > 0$ for all $i = 1, \ldots, n$ implies that the matrix of the covariances of the random variables $1_{X=x_1}, \ldots, 1_{X=x_n}$ with respect to the measure $P^{Z=z}$ is regular. Therefore, we can apply Theorem 13.20 for identifying the coefficients $f_0(z)$ and $f_i(z)$, $i = 1, \ldots, n$.

Under the assumptions of Theorem 15.25, Remark 14.9 and Equation (15.40) imply

$$V_z \underset{P^{Z=z}}{=} \frac{\exp\left(f_0(z) + \sum_{i=1}^n f_i(z) \cdot 1_{X=x_i}\right)}{1 + \exp\left(f_0(z) + \sum_{i=1}^n f_i(z) \cdot 1_{X=x_i}\right)}, \quad \forall\, V_z \in \mathscr{P}^{Z=z}(Y=1 \mid X). \tag{15.41}$$

◁

Remark 15.30 [Conditional probability $P^{X=x_i}(Y=1 \mid Z)$] Remark 14.9 and Equation (15.32) imply

$$V_0 \underset{P^{X=x}}{=} \frac{\exp(f_0(Z))}{1 + \exp(f_0(Z))}, \quad \forall\, V_0 \in \mathscr{P}^{X=0}(Y=1 \mid Z), \tag{15.42}$$

and, for all $i = 1, \dots, n$,

$$V_i \underset{P^{X=x_i}}{=} \frac{\exp(f_0(Z) + f_i(Z))}{1 + \exp(f_0(Z) + f_i(Z))}, \quad \forall\, V_i \in \mathscr{P}^{X=x_i}(Y=1 \mid Z). \tag{15.43}$$

◁

Remark 15.31 [Log odds functions] In terms of conditional probabilities, the Z-conditional logit intercept function f_0 satisfies

$$f_0(Z) = \ln\left[\frac{P^{X=x_0}(Y=1 \mid Z)}{1 - P^{X=x_0}(Y=1 \mid Z)}\right] \tag{15.44}$$

(cf. Rem. 13.15). Hence, f_0 may also be called the *Z-conditional log odds function* of $P^{X=x_0}(Y=1 \mid Z)$. Similarly,

$$f_0(Z) + f_i(Z) = \ln\left[\frac{P^{X=x_i}(Y=1 \mid Z)}{1 - P^{X=x_i}(Y=1 \mid Z)}\right], \quad i = 1, \dots, n \tag{15.45}$$

(see Exercise 15.2). The function $f_0 + f_i \colon (\Omega'_Z, \mathscr{A}'_Z) \to (\mathbb{R}, \mathscr{B})$ satisfying $(f_0 + f_i)(Z) = f_0(Z) + f_i(Z)$ is called the *Z-conditional log odds function* of $P^{X=x_i}(Y=1 \mid Z)$. ◁

Remark 15.32 [Log odds ratio functions] Equations (15.44) and (15.45) imply

$$f_i(Z) = \ln\left[\frac{P^{X=x_i}(Y=1 \mid Z)}{1 - P^{X=x_i}(Y=1 \mid Z)}\right] - \ln\left[\frac{P^{X=x_0}(Y=1 \mid Z)}{1 - P^{X=x_0}(Y=1 \mid Z)}\right] \tag{15.46}$$

$$= \ln\left[\frac{P^{X=x_i}(Y=1 \mid Z)}{1 - P^{X=x_i}(Y=1 \mid Z)} \middle/ \frac{P^{X=x_0}(Y=1 \mid Z)}{1 - P^{X=x_0}(Y=1 \mid Z)}\right], \quad i = 1, \dots, n, \tag{15.47}$$

for the Z-conditional logit effect functions f_i (cf. Rem. 13.16).

Referring to Equation (15.47), f_i, $i = 1, \dots, n$, is also called the *Z-conditional log odds ratio function of $P^{X=x_i}(Y=1 \mid Z)$ and $P^{X=x_0}(Y=1 \mid Z)$.* ◁

Remark 15.33 [Odds ratio functions] The exponential function of $f_i(Z)$ is

$$\exp[f_i(Z)] = \frac{P^{X=x_i}(Y=1\mid Z)}{1 - P^{X=x_i}(Y=1\mid Z)} \bigg/ \frac{P^{X=x_0}(Y=1\mid Z)}{1 - P^{X=x_0}(Y=1\mid Z)}, \quad i = 1,\dots,n. \qquad (15.48)$$

The composite function $\exp(f_i)\colon (\Omega'_Z, \mathscr{A}'_Z) \to (\mathbb{R}, \mathscr{B})$ is called the *Z-conditional odds ratio function of* $P^{X=x_i}(Y=1\mid Z)$ *and* $P^{X=x_0}(Y=1\mid Z)$ (cf. Rem. 13.17). ◁

Remark 15.34 [Risk ratio functions] Another closely related function describing conditional effects of $(X=x_i)$ compared to $(X=x_0)$ is

$$k_i(Z) := \frac{P^{X=x_i}(Y=1\mid Z)}{P^{X=x_0}(Y=1\mid Z)}, \quad i = 1,\dots,n. \qquad (15.49)$$

The function $k_i\colon (\Omega'_Z, \mathscr{A}'_Z) \to (\mathbb{R}, \mathscr{B})$ satisfying (15.49) is called the *Z-conditional risk ratio function of* $P^{X=x_i}(Y=1\mid Z)$ *and* $P^{X=x_0}(Y=1\mid Z)$ (cf. Rem. 13.18). ◁

Remark 15.35 [Four kinds of conditional-effect functions] Hence, under the assumptions of Theorem 15.25, we considered four kinds of different Z-conditional-effect functions: g_i, f_i, $\exp(f_i)$, and k_i. They all describe Z-conditional-effect functions of x_i compared to x_0 on Y on different scales. ◁

15.6   Implications of independence of X and Z for the logit regression coefficients

In section 15.3, we already treated the implication of

$$E[g_i(Z)\mid X] = E[g_i(Z)], \quad \forall\, i = 0, 1, \dots, n, \qquad (15.50)$$

[see Eq. (15.22)]. According to Theorem 15.14, this equation implies

$$E[g_i(Z)] = \beta_i, \quad i = 0, 1, \dots, n, \qquad (15.51)$$

where β_i are the parameters in the equation

$$P(Y=1\mid X) = \beta_0 + \sum_{i=1}^{n} \beta_i \cdot 1_{X=x_i}. \qquad (15.52)$$

Note that, under the assumptions of Theorem 15.25, there are coefficients $\beta_0, \beta_1, \dots, \beta_n \in \mathbb{R}$ such that Equation (15.52) holds. Remember, according to Remark 15.16, Equation (15.50) holds if X and Z are independent.

Remark 15.36 [Mean independence and logit effect functions] Note that the analog to Equation (15.51) *does not hold* for the expectation of the logit effect functions f_i specified in

Equation (15.33). That is, although, under the assumptions of Theorem 15.25, there are always coefficients $\alpha_0, \alpha_1, \ldots, \alpha_n \in \mathbb{R}$ such that

$$P(Y=1 \mid X) = \frac{\exp\left(\alpha_0 + \sum_{i=1}^{n} \alpha_i \cdot 1_{X=x_i}\right)}{1 + \exp\left(\alpha_0 + \sum_{i=1}^{n} \alpha_i \cdot 1_{X=x_i}\right)}, \tag{15.53}$$

neither

$$E[f_i(Z) \mid X] \underset{P}{=} E[f_i(Z)], \quad \forall\, i = 0, 1, \ldots, n, \tag{15.54}$$

nor does independence of X and Z imply that α_i is identical to $E[f_i(Z)]$ (see Example 15.38). In fact, it is even possible that α_i is negative and $E[f_i(Z)]$ is positive (see Example 15.39). ◁

Remark 15.37 [Implications for methodology]   From a methodological point of view, this means that randomized assignment of a unit to one of the treatment conditions – which creates independence of a treatment variable X and the person variable U – does not imply that the regression coefficients in the logistic parameterization of $P(Y=1 \mid X)$ can be interpreted as the expectation of the corresponding functions $f_i(U)$. As mentioned before, it is even possible that a coefficient α_i is negative and $E[f_i(Z)]$ is positive (see Example 15.39). In contrast, compare Equations (15.50) to (15.53) and the invariance property formulated in Theorem 12.44 for a linear parameterization of a conditional expectation. ◁

Example 15.38 [Joe and Ann with randomized assignment – continued]   Now we show by an example that independence of X and Z does *not imply* that the coefficients α_i of Equation (15.53) are identical to $E[f_i(Z)]$, where the functions f_i are specified in Theorem 15.25. [In contrast, compare Equations (14.4) and (14.5) for the effect function g_1.]

In Examples 13.24 and 13.26, we computed

$$P(Y=1 \mid X) = \frac{\exp(\alpha_0 + \alpha_1 \cdot X)}{1 + \exp(\alpha_0 + \alpha_1 \cdot X)} \approx \frac{\exp(-.201 + .606 \cdot X)}{1 + \exp(-.201 + .606 \cdot X)} \tag{15.55}$$

and

$$P(Y=1 \mid X, U) = \frac{\exp\left[(\lambda_0 + \lambda_2 \cdot 1_{U=Ann}) + (\lambda_1 + \lambda_3 \cdot 1_{U=Ann}) \cdot X\right]}{1 + \exp\left[(\lambda_0 + \lambda_2 \cdot 1_{U=Ann}) + (\lambda_1 + \lambda_3 \cdot 1_{U=Ann}) \cdot X\right]}$$

$$\approx \frac{\exp\left[(.847 - 2.234 \cdot 1_{U=Ann}) + (.539 + .442 \cdot 1_{U=Ann}) \cdot X\right]}{1 + \exp\left[(.847 - 2.234 \cdot 1_{U=Ann}) + (.539 + .442 \cdot 1_{U=Ann}) \cdot X\right]}, \tag{15.56}$$

showing that the logit of $P(Y=1 \mid X, U)$ is $f_0(U) + f_1(U) \cdot X$ with logit intercept function f_0 satisfying

$$f_0(U) = \lambda_0 + \lambda_2 \cdot 1_{U=Ann} \approx .847 - 2.234 \cdot 1_{U=Ann}$$

and logit effect function f_1 satisfying

$$f_1(U) = \lambda_1 + \lambda_3 \cdot 1_{U=Ann} = .539 + .442 \cdot 1_{U=Ann}.$$

Note that

$$\alpha_1 \approx .606 \neq E[f_1(U)] \approx .539 + .442 \cdot E(1_{U=Ann}) \approx .76.$$

Hence, although X and U are independent, the logit effect α_1 of X in the logistic parameterization of $P(Y=1 \mid X)$ is *not equal* to the expectation of $f_1(U)$ in the logistic parameterization of $P(Y=1 \mid X, U)$. ◁

Example 15.39 [Joe and Ann: reversed average logit effect]   Table 15.2 displays an example in which the coefficient α_1 in the equation

$$P(Y=1 \mid X) = \frac{\exp(\alpha_0 + \alpha_1 \cdot X)}{1 + \exp(\alpha_0 + \alpha_1 \cdot X)} \tag{15.57}$$

is negative, whereas the expectation $E[f_1(U)]$ of the function $f_1(U)$ in the equation

$$P(Y=1 \mid X, U) \underset{P}{=} \frac{\exp\big(f_0(U) + f_1(U) \cdot X\big)}{1 + \exp\big(f_0(U) + f_1(U) \cdot X\big)} \tag{15.58}$$

Table 15.2   Joe and Ann: reversed average logit effect.

| Outcomes ω | | Observables | | | Conditional expectations | | | | | | | |
Unit / Treatment / Success	$P(\{\omega\})$	Person variable U	Treatment variable X	Outcome variable Y	$P(Y=1 \mid X, U)$	$P(Y=1 \mid X)$	$P(X=1 \mid U)$	$P^{X=0}(Y=1 \mid U)$	$P^{X=1}(Y=1 \mid U)$	$\mathrm{logit}[P(Y=1 \mid X, U)]$	$g_1(U) = P^{X=1}(Y=1 \mid U) - P^{X=0}(Y=1 \mid U)$	$f_1(U)$
$(Joe, no, -)$	.067	Joe	0	0	.732	.732	.5	.732	.984	1.005	.252	3.114
$(Joe, no, +)$	.183	Joe	0	1	.732	.732	.5	.732	.984	1.005	.252	3.114
$(Joe, yes, -)$	.004	Joe	1	0	.984	.626	.5	.732	.984	4.119	.252	3.114
$(Joe, yes, +)$	.246	Joe	1	1	.984	.626	.5	.732	.984	4.119	.252	3.114
$(Ann, no, -)$	.067	Ann	0	0	.732	.732	.5	.732	.268	1.005	−.464	−2.010
$(Ann, no, +)$	.183	Ann	0	1	.732	.732	.5	.732	.268	1.005	−.464	−2.010
$(Ann, yes, -)$	.183	Ann	1	0	.268	.626	.5	.732	.268	−1.005	−.464	−2.010
$(Ann, yes, +)$	.067	Ann	1	1	.268	.626	.5	.732	.268	−1.005	−.464	−2.010

is positive, although X and U are independent. The coefficients of Equation (15.57) are $\alpha_0 \approx 1.005$ and $\alpha_1 \approx -.490$, whereas the expectation of the conditional logit effect function is $E_U(f_1) = E[f_1(U)] \approx .552$ (see Exercise 15.3).   ◁

15.7   Proofs

Proof of Theorem 15.3

For all $i = 0, 1, \ldots, n$,

$$E(Y) \text{ is finite} \;\Rightarrow\; E^{X=x_i}(Y) \text{ is finite} \qquad\qquad \text{[Rem. 14.6]}$$

$$\Rightarrow \quad \text{there is a real-valued } E^{X=x_i}(Y \mid Z) \in \mathscr{E}^{X=x_i}(Y \mid Z). \quad \text{[Box 10.2 (x)]}$$

According to Theorem 10.9 and Corollary 10.23, finiteness of $E(Y)$ also implies that there is a $g(X, Z) \in \mathscr{E}(Y \mid X, Z)$ such that, for real-valued versions $E^{X=x_i}(Y \mid Z)$,

$$g(X, Z) \underset{P}{=} \sum_{i=0}^{n} E^{X=x_i}(Y \mid Z) \cdot 1_{X=x_i} \qquad\qquad \text{[(14.31)]}$$

$$\underset{P}{=} E^{X=x_0}(Y \mid Z) \cdot 1_{X=x_0} + \sum_{i=1}^{n} E^{X=x_i}(Y \mid Z) \cdot 1_{X=x_i}$$

$$\underset{P}{=} E^{X=x_0}(Y \mid Z) - \sum_{i=1}^{n} E^{X=x_0}(Y \mid Z) \cdot 1_{X=x_i} + \sum_{i=1}^{n} E^{X=x_i}(Y \mid Z) \cdot 1_{X=x_i} \quad \text{[(5.33)]}$$

$$\underset{P}{=} E^{X=x_0}(Y \mid Z) + \sum_{i=1}^{n} [E^{X=x_i}(Y \mid Z) - E^{X=x_0}(Y \mid Z)] \cdot 1_{X=x_i}.$$

Because the function on the right-hand side of the last equation is (X, Z)-measurable and P-equivalent to $g(X, Z) \in \mathscr{E}(Y \mid X, Z)$ (see Th. 2.57), it is an element of $\mathscr{E}(Y \mid X, Z)$. Defining the specific version

$$E(Y \mid X, Z) := E^{X=x_0}(Y \mid Z) + \sum_{i=1}^{n} [E^{X=x_i}(Y \mid Z) - E^{X=x_0}(Y \mid Z)] \cdot 1_{X=x_i}$$

completes the proof.

Proof of Theorem 15.14

For all versions $E(Y \mid X) \in \mathscr{E}(Y \mid X)$,

$$E(Y \mid X) \underset{P}{=} E[E(Y \mid X, Z) \mid X] \qquad\qquad \text{[Box 10.2 (v)]}$$

$$\underset{P}{=} E\left[g_0(Z) + \sum_{i=1}^{n} g_i(Z) \cdot 1_{X=x_i} \;\middle|\; X \right] \qquad\qquad \text{[(15.6)]}$$

$$\underset{P}{=} E[g_0(Z) \mid X] + \sum_{i=1}^{n} E\left[g_i(Z) \cdot 1_{X=x_i} \mid X \right] \qquad\qquad \text{[Box 10.2 (xv)]}$$

$$\underset{P}{=} E[g_0(Z) \mid X] + \sum_{i=1}^{n} E[g_i(Z) \mid X] \cdot 1_{X=x_i} \qquad [\sigma(1_{X=x_i}) \subset \sigma(X), (10.74), (2.36)]$$

$$\underset{P}{=} E[g_0(Z)] + \sum_{i=1}^{n} E[g_i(Z)] \cdot 1_{X=x_i}. \qquad [(15.22)]$$

According to Remark 2.18, the right-hand side of the last equation above is X-measurable. Because it is P-equivalent to all versions $E(Y \mid X) \in \mathcal{E}(Y \mid X)$, it is an element of $\mathcal{E}(Y \mid X)$ [see (10.12)], and it satisfies Equations (15.17) to (15.21).

Proof of Corollary 15.18

According to Box 10.2 (vi) and Equations (6.5) and (10.4), independence of X and Z implies $P(X=x_i \mid Z) \underset{P}{=} P(X=x_i)$. Because $P(X=x_i) > 0$, we can conclude $P(X=x_i \mid Z) \underset{P}{>} 0$. Furthermore, $P(X=x_i \mid Z) \underset{P}{>} 0$ is equivalent to P-uniqueness of $E^{X=x_i}(Y \mid Z)$ [see Cor. 14.48 (a) and (c)]. According to Box 14.1 (viii), this implies P-uniqueness of the functions $g_i(Z)$, $i = 0, 1, \dots, n$.

Proof of Theorem 15.25

By definition, $P(Y=1 \mid X, Z) = E(1_{Y=1} \mid X, Z)$. Hence, the existence of measurable functions $g_0, g_1, \dots, g_n \colon (\Omega'_Z, \mathcal{A}'_Z) \to (\mathbb{R}, \mathcal{B})$ satisfying Equations (15.32), (15.34), and (15.36) has already been proved in Theorem 15.3. In order to show that there are measurable functions f_0, $f_1, \dots, f_n \colon (\Omega'_Z, \mathcal{A}'_Z) \to (\mathbb{R}, \mathcal{B})$ satisfying Equation (15.33), we define

$$l_i(Z) := \mathrm{logit}\left[P^{X=x_i}(Y=1 \mid Z)\right], \quad \forall\, i = 0, 1, \dots, n, \qquad (15.59)$$

using the logit of $P^{X=x_i}(Y=1 \mid Z)$ defined by Equation (13.5). Furthermore, we define

$$f_0(Z) := l_0(Z), \qquad (15.60)$$

and

$$f_i(Z) := l_i(Z) - l_0(Z), \quad \forall\, i = 1, \dots, n. \qquad (15.61)$$

These definitions and Equation (13.6) then yield

$$P^{X=x_i}(Y=1 \mid Z) = \frac{\exp\left(\mathrm{logit}\left[P^{X=x_i}(Y=1 \mid Z)\right]\right)}{1 + \exp\left(\mathrm{logit}\left[P^{X=x_i}(Y=1 \mid Z)\right]\right)} = \frac{\exp\left(l_i(Z)\right)}{1 + \exp\left(l_i(Z)\right)}, \quad \forall\, i = 0, 1, \dots, n.$$

Hence, for $i = 0$, Equation (15.34) implies

$$g_0 = P^{X=x_0}(Y=1 \mid Z) = \frac{\exp\left(f_0(Z)\right)}{1 + \exp\left(f_0(Z)\right)},$$

which proves (15.35). Furthermore, Equation (15.36) yields

$$g_i(Z) = P^{X=x_i}(Y=1 \mid Z) - P^{X=x_0}(Y=1 \mid Z)$$

$$= \frac{\exp\big(f_0(Z) + f_i(Z)\big)}{1 + \exp\big(f_0(Z) + f_i(Z)\big)} - \frac{\exp\big(f_0(Z)\big)}{1 + \exp\big(f_0(Z)\big)}, \quad \forall\, i = 1, \dots, n,$$

which proves (15.37).

Finally, for all $\omega \in \Omega$,

$$\Big(f_0(Z) + \sum_{i=1}^{n} f_i(Z) \cdot 1_{X=x_i}\Big)(\omega) = \begin{cases} \big(f_0(Z) + f_j(Z)\big)(\omega), & \text{if } X(\omega) = x_j, j = 1, \dots, n \\ f_0(Z)(\omega), & \text{otherwise.} \end{cases}$$

For all $\omega \in \{X=x_0\}$,

$$\frac{\exp\Big[\big(f_0(Z) + \sum_{i=1}^{n} f_i(Z) \cdot 1_{X=x_i}\big)(\omega)\Big]}{1 + \exp\Big[\big(f_0(Z) + \sum_{i=1}^{n} f_i(Z) \cdot 1_{X=x_i}\big)(\omega)\Big]}$$

$$= \frac{\exp[\, f_0(Z)(\omega)]}{1 + \exp[\, f_0(Z)(\omega)]} \qquad\qquad [1_{X=x_i}(\omega) = 0, \forall\, i = 1, \dots, n]$$

$$= g_0(Z)(\omega). \qquad\qquad\qquad\qquad\qquad\qquad\qquad [(15.35)]$$

Hence, for all $j = 1, \dots, n$ and all $\omega \in \{X=x_j\}$,

$$\frac{\exp\Big[\big(f_0(Z) + \sum_{i=1}^{n} f_i(Z) \cdot 1_{X=x_i}\big)(\omega)\Big]}{1 + \exp\Big[\big(f_0(Z) + \sum_{i=1}^{n} f_i(Z) \cdot 1_{X=x_i}\big)(\omega)\Big]}$$

$$= \frac{\exp[(f_0(Z) + f_j(Z))(\omega)]}{1 + \exp[(f_0(Z) + f_j(Z))(\omega)]}$$

$$= [g_0(Z) + g_j(Z)](\omega) \qquad\qquad\qquad [(15.36), (15.35)]$$

$$= \Big[g_0(Z) + \sum_{i=1}^{n} g_i(Z) \cdot 1_{X=x_i}\Big](\omega) \qquad\qquad [1_{X=x_i}(\omega) = 1 \Leftrightarrow x_i = x_j]$$

$$= P(Y=1 \mid X, Z)(\omega). \qquad\qquad\qquad\qquad [(15.32)]$$

Exercises

15.1 Prove Equation (15.38).

15.2 Prove Equations (15.44) and (15.45).

15.3 Using the results displayed in Table 15.2, compute the coefficients α_0 and α_1 of Equation (15.57) as well as the expectation of the conditional logit effect function f_1.

Solutions

15.1 According to Theorem 14.46, assuming (15.2) implies that the conditional probabilities $P^{X=x_i}(Y=1 \mid Z)$ are P-unique, which in turn implies that the functions g_i are P-unique [see (2.36)]. If, for all $i = 0, 1, \ldots, n$, the $P^{X=x_i}(Y=1 \mid Z)$ are P-unique, then the functions $l_i(Z)$ and their differences [see Eqs. (15.60) and (15.61)] are P-unique as well [see (2.34) and Rem. 2.76].

15.2 The composition of Z and the conditional logit intercept function can be written as:

$$f_0(z) = \mathrm{logit}\left[P^{X=x_0}(Y=1 \mid Z)\right] \qquad\qquad [(15.35),\ (13.3)]$$

$$= \ln\left[\frac{P^{X=x_0}(Y=1 \mid Z)}{1 - P^{X=x_0}(Y=1 \mid Z)}\right]. \qquad\qquad [(13.2)]$$

Furthermore,

$$f_0(Z) + f_i(Z) = \mathrm{logit}\left[P^{X=x_i}(Y=1 \mid Z)\right] \qquad\qquad [(15.37),\ (13.3)]$$

$$= \ln\left[\frac{P^{X=x_i}(Y=1 \mid Z)}{1 - P^{X=x_i}(Y=1 \mid Z)}\right]. \qquad\qquad [(13.2)]$$

15.3 Using Equations (13.6) (13.8) and the conditional probabilities $P(Y=1 \mid X=x)$ displayed in Table 15.2 yields

$$\alpha_0 = \mathrm{logit}[P(Y=1 \mid X=0)] \approx \ln(.732/(1 - .732)) \approx 1.005,$$
$$\alpha_0 + \alpha_1 = \mathrm{logit}[P(Y=1 \mid X=1)] \approx \ln(.626/(1 - .626)) \approx .515,$$

and $\alpha_1 \approx .515 - 1.005 = -.490$.
  Similarly,

$$\lambda_0 = \mathrm{logit}[P(Y=1 \mid X=0, U=Joe)] \approx \ln(.732/(1 - .732)) \approx 1.005,$$
$$\lambda_0 + \lambda_1 = \mathrm{logit}[P(Y=1 \mid X=1, U=Joe)] \approx \ln(.984/(1 - .984)) \approx 4.119,$$
$$\lambda_0 + \lambda_2 = \mathrm{logit}[P(Y=1 \mid X=0, U=Ann)] \approx \ln(.732/(1 - .732)) \approx 1.005,$$
$$\lambda_0 + \lambda_1 + \lambda_2 + \lambda_3 = \mathrm{logit}[P(Y=1 \mid X=1, U=Ann)] \approx \ln(.268/(1 - .268)) \approx -1.005,$$

and this yields

$$\lambda_1 \approx 4.119 - 1.005 = 3.114,$$
$$\lambda_2 \approx 1.005 - 1.005 = 0,$$
$$\lambda_3 \approx -1.005 - 1.005 - 3.114 - 0 = -5.124.$$

Hence, the composition of U and the logit effect function f_1 is

$$f_1(U) = \lambda_1 + \lambda_3 \cdot 1_{U=Ann} \approx 3.114 - 5.124 \cdot 1_{U=Ann},$$

and taking its expectation we receive

$$E_U(f_1) = E[f_1(U)] \qquad\qquad\qquad [(6.13)]$$
$$\approx 3.114 - 5.124 \cdot .5 = .552. \qquad \text{[Table 15.2, Box 6.1 (ii), (iii)]}$$

Part IV

CONDITIONAL INDEPENDENCE AND CONDITIONAL DISTRIBUTION

16

Conditional independence

In section 4.3 we introduced independence of events and of sets of events, and in section 4.4 we treated conditional independence of events given an event that has a positive probability. There, we noted that conditional independence given an event B is equivalent to independence with respect to the conditional-probability measure P^B. Furthermore, in section 5.4 we used these definitions in order to introduce independence of random variables. In this chapter we generalize these concepts and define conditional independence of events, of sets of events, and of random variables given a σ-*algebra* or given a *random variable*. Furthermore, we study the relationship between conditional independence and conditional mean independence (see section 10.6). For further implications of conditional independence, see section 17.6.

16.1 Assumptions and definitions

In this section, we make the following assumptions and use the following notation.

> **Notation and assumptions 16.1**
> *Let* $Z: (\Omega, \mathscr{A}, P) \to (\Omega'_Z, \mathscr{A}'_Z)$ *be a random variable, let* $\mathscr{C} \subset \mathscr{A}$ *be a σ-algebra, and let* $A, B \in \mathscr{A}$.

Also remember that $P(A \mid \mathscr{C}) := E(1_A \mid \mathscr{C})$, where $E(1_A \mid \mathscr{C})$ denotes the $\mathscr{C}$-conditional expectation of the indicator 1_A of $A \in \mathscr{A}$, and $P(A \mid Z) := E(1_A \mid Z)$ the Z-conditional probability of A (see Def. 10.2 and Rem. 10.4). Finally, see Remark 5.17 for the concept of P-equivalence and section 10.2 for propositions on P-equivalence of conditional expectations.

Probability and Conditional Expectation: Fundamentals for the Empirical Sciences, First Edition. Rolf Steyer and Werner Nagel.
© 2017 John Wiley & Sons, Ltd. Published 2017 by John Wiley & Sons, Ltd.
Companion website: http://www.probability-and-conditional-expectation.de

16.1.1   Two events

Definition 16.2 [$\mathscr{C}$-conditional independence of two events]
Let the assumptions 16.1 hold.

(i) *A and B are called $\mathscr{C}$-conditionally P-independent, denoted $A \underset{P}{\perp\!\!\!\perp} B \mid \mathscr{C}$, if*

$$P(A \cap B \mid \mathscr{C}) \underset{P}{=} P(A \mid \mathscr{C}) \cdot P(B \mid \mathscr{C}). \tag{16.1}$$

(ii) *A and B are called Z-conditionally P-independent, denoted $A \underset{P}{\perp\!\!\!\perp} B \mid Z$, if*

$$P(A \cap B \mid Z) \underset{P}{=} P(A \mid Z) \cdot P(B \mid Z). \tag{16.2}$$

A synonym for $\mathscr{C}$-conditional P-independence of A and B is *conditional independence of A and B given $\mathscr{C}$ with respect to the probability measure P.*

Remark 16.3 [Symmetry]   Obviously, $\mathscr{C}$-conditional independence of A and B with respect to P is *symmetric* in the following sense:

$$A \underset{P}{\perp\!\!\!\perp} B \mid \mathscr{C} \;\Leftrightarrow\; B \underset{P}{\perp\!\!\!\perp} A \mid \mathscr{C}. \tag{16.3}$$

$\lhd$

Note that we did not exclude that the events A or B are elements of the σ-algebra $\mathscr{C}$. This case is considered in the following lemma.

Lemma 16.4 [A sufficient condition]
Let the assumptions 16.1 hold. Then,

$$A \in \mathscr{C} \text{ or } B \in \mathscr{C} \;\Rightarrow\; A \underset{P}{\perp\!\!\!\perp} B \mid \mathscr{C} \tag{16.4}$$

and

$$A \in \sigma(Z) \text{ or } B \in \sigma(Z) \;\Rightarrow\; A \underset{P}{\perp\!\!\!\perp} B \mid Z. \tag{16.5}$$

(Proof p. 493)

Remark 16.5 [An immediate implication]   If $(\Omega, \mathscr{A}, P)$ is a probability space and $A, B \in \mathscr{A}$, then $A \underset{P}{\perp\!\!\!\perp} B \mid \mathscr{A}$. $\lhd$

Corollary 16.6 [Two formulations of Z-conditional P-independence]
Let the assumptions 16.1 hold. Then the following propositions are equivalent to each other:

(i) $A \underset{P}{\perp\!\!\!\perp} B \mid Z$.

(ii) $P(A \cap B \mid Z=z) = P(A \mid Z=z) \cdot P(B \mid Z=z), \quad$ *for P_Z-a.a. $z \in \Omega'_Z$.*

(Proof p. 493)

Remark 16.7 [The special case $P(Z=z) > 0$]   Corollary 16.6 and Remark 2.71 imply: If $A \underset{P}{\perp\!\!\!\perp} B \mid Z$ and $P(Z=z) > 0$, then,

$$P(A \cap B \mid Z=z) = P(A \mid Z=z) \cdot P(B \mid Z=z). \tag{16.6}$$

According to Equation (10.30), $P(A \mid Z=z) = P(A \mid \{Z=z\})$, if $P(Z=z) > 0$. Therefore, Equation (16.6) is consistent with Equation (4.27).   ◁

Definition 16.8 [$(Z=z)$-conditional independence of two events]
Let the assumptions 16.1 hold and assume that $z \in \Omega_Z'$ with $P(Z=z) > 0$. Then A and B are called $(Z=z)$-conditionally independent, denoted $A \underset{P}{\perp\!\!\!\perp} B \mid Z=z$, if Equation (16.6) holds.

Remark 16.9 [Conditioning on a constant]   Let the assumptions 16.1 hold. If $\mathscr{C} = \{\Omega, \varnothing\}$, then Remark 10.5 implies $P(A \mid \mathscr{C}) = P(A)$, $P(B \mid \mathscr{C}) = P(B)$, and $P(A \cap B \mid \mathscr{C}) = P(A \cap B)$. Hence, in this case, $\mathscr{C}$-*conditional independence* and (unconditional) *independence* of two events are equivalent, that is,

$$\mathscr{C} = \{\Omega, \varnothing\} \;\Rightarrow\; (A \underset{P}{\perp\!\!\!\perp} B \mid \mathscr{C} \;\Leftrightarrow\; A \underset{P}{\perp\!\!\!\perp} B). \tag{16.7}$$

If $Z=\alpha$, $\alpha \in \Omega_Z'$, that is, if Z is a constant random variable, then $\sigma(Z) = \{\Omega, \varnothing\}$. Therefore, (16.7) implies

$$(\exists\, \alpha \in \Omega_Z' : Z = \alpha) \;\Rightarrow\; (A \underset{P}{\perp\!\!\!\perp} B \mid Z \;\Leftrightarrow\; A \underset{P}{\perp\!\!\!\perp} B). \tag{16.8}$$

A more general proposition is that *independence* and Z-*conditional independence* of two events are equivalent if $Z \underset{P}{=} \alpha$, $\alpha \in \Omega_Z'$, that is,

$$(\exists\, \alpha \in \Omega_Z' : Z \underset{P}{=} \alpha) \;\Rightarrow\; (A \underset{P}{\perp\!\!\!\perp} B \mid Z \;\Leftrightarrow\; A \underset{P}{\perp\!\!\!\perp} B) \tag{16.9}$$

(see Exercise 16.1).   ◁

16.1.2   Two sets of events

Using the concept of conditional independence of two events, we define $\mathscr{C}$-conditional P-independence of two *sets of events*. One or both of these sets of events can be a σ-algebra, but this is not required.

Definition 16.10 [$\mathscr{C}$-conditional independence of two sets of events]
Let the assumptions 16.1 hold and let $\mathscr{D}, \mathscr{E} \subset \mathscr{A}$. Then:

*(i) $\mathscr{D}$ and $\mathscr{E}$ are called $\mathscr{C}$-conditionally P-independent, denoted $\mathscr{D} \underset{P}{\perp\!\!\!\perp} \mathscr{E} \mid \mathscr{C}$, if
for all pairs $(A, B) \in \mathscr{D} \times \mathscr{E}$, the events A and B are $\mathscr{C}$-conditionally P-independent.*

*(ii) $\mathscr{D}$ and $\mathscr{E}$ are called Z-conditionally P-independent, denoted $\mathscr{D} \underset{P}{\perp\!\!\!\perp} \mathscr{E} \mid Z$, if
for all pairs $(A, B) \in \mathscr{D} \times \mathscr{E}$, the events A and B are Z-conditionally P-independent.*

Remark 16.11 [Conditioning on a constant]   Remark 16.9 implies

$$\mathscr{C} = \{\Omega, \varnothing\} \;\Rightarrow\; (\mathscr{D} \underset{P}{\perp\!\!\!\perp} \mathscr{E} \mid \mathscr{C} \;\Leftrightarrow\; \mathscr{D} \underset{P}{\perp\!\!\!\perp} \mathscr{E}). \tag{16.10}$$

Hence, $\mathscr{D} \underset{P}{\perp\!\!\!\perp} \mathscr{E}$ (see Def. 4.40) is a special case of $\mathscr{D} \underset{P}{\perp\!\!\!\perp} \mathscr{E} \mid \mathscr{C}$ with $\mathscr{C} = \{\Omega, \varnothing\}$. Similarly,

$$(\exists\, \alpha \in \Omega'_Z \colon Z = \alpha) \;\Rightarrow\; (\mathscr{D} \underset{P}{\perp\!\!\!\perp} \mathscr{E} \mid Z \;\Leftrightarrow\; \mathscr{D} \underset{P}{\perp\!\!\!\perp} \mathscr{E}). \tag{16.11}$$

Again, a more general proposition, which follows from (16.9), is that *independence* and
Z-conditional independence of two sets of events are equivalent if $Z \underset{P}{=} \alpha$, $\alpha \in \Omega'_Z$, that is,

$$(\exists\, \alpha \in \Omega'_Z \colon Z \underset{P}{=} \alpha) \;\Rightarrow\; (\mathscr{D} \underset{P}{\perp\!\!\!\perp} \mathscr{E} \mid Z \;\Leftrightarrow\; \mathscr{D} \underset{P}{\perp\!\!\!\perp} \mathscr{E}) \tag{16.12}$$

(see Exercise 16.2). ◁

According to the following lemma, two σ-algebras are conditionally independent if
two $\cap$-stable set systems (see Def. 1.36) that generate these σ-algebras are conditionally
independent.

Lemma 16.12 [$\cap$-Stable generators]
*Let the assumptions 16.1 hold, let $\mathscr{D}, \mathscr{E} \subset \mathscr{A}$, and suppose that $\mathscr{D}$ and $\mathscr{E}$ are $\cap$-stable.
Then,*

$$\mathscr{D} \underset{P}{\perp\!\!\!\perp} \mathscr{E} \mid \mathscr{C} \;\Leftrightarrow\; \sigma(\mathscr{D}) \underset{P}{\perp\!\!\!\perp} \sigma(\mathscr{E}) \mid \mathscr{C}. \tag{16.13}$$

(Proof p. 493)

16.1.3   Two random variables

The sets of events $\mathscr{D}$ and $\mathscr{E}$ occurring in Definition 16.10 may also be two σ-algebras, such
as $\sigma(X)$ and $\sigma(Y)$, the σ-algebras generated by random variables X and Y, respectively (see
Def. 2.26). This case will now be used to define *conditional P-independence of two random
variables*.

Definition 16.13 [Conditional P-independence of two random variables]
Let $X: (\Omega, \mathcal{A}, P) \to (\Omega'_X, \mathcal{A}'_X)$, $Y: (\Omega, \mathcal{A}, P) \to (\Omega'_Y, \mathcal{A}'_Y)$, and $Z: (\Omega, \mathcal{A}, P) \to (\Omega'_Z, \mathcal{A}'_Z)$
be random variables and $\mathcal{C} \subset \mathcal{A}$ a σ-algebra. Then:

(i) X and Y are called $\mathcal{C}$-conditionally P-independent, denoted $X \underset{P}{\perp\!\!\!\perp} Y \mid \mathcal{C}$, if
$\sigma(X)$ and $\sigma(Y)$ are $\mathcal{C}$-conditionally P-independent.

(ii) X and Y are called Z-conditionally P-independent, denoted $X \underset{P}{\perp\!\!\!\perp} Y \mid Z$, if
$\sigma(X)$ and $\sigma(Y)$ are Z-conditionally P-independent.

Remark 16.14 [Three equivalent notations]   Let $\{X \in A'\} := \{\omega \in \Omega: X(\omega) \in A'\}$ and
$\{Y \in B'\} := \{\omega \in \Omega: Y(\omega) \in B'\}$ denote the events that X takes on a value in A' and Y takes on
a value in B', respectively. Then we use the following equivalent notations for $\mathcal{C}$-conditional
independence of two random variables:

(i) $X \underset{P}{\perp\!\!\!\perp} Y \mid \mathcal{C}$.

(ii) $\sigma(X) \underset{P}{\perp\!\!\!\perp} \sigma(Y) \mid \mathcal{C}$.

(iii) $\{X \in A'\} \underset{P}{\perp\!\!\!\perp} \{Y \in B'\} \mid \mathcal{C}, \quad \forall\, (A', B') \in \mathcal{A}'_X \times \mathcal{A}'_Y$.

Each of these notations is equivalent to

$$P(X \in A', Y \in B' \mid \mathcal{C}) \underset{P}{=} P(X \in A' \mid \mathcal{C}) \cdot P(Y \in B' \mid \mathcal{C}), \quad \forall\, (A', B') \in \mathcal{A}'_X \times \mathcal{A}'_Y. \quad (16.14)$$

Analogously, $X \underset{P}{\perp\!\!\!\perp} Y \mid Z$ is equivalent to

$$P(X \in A', Y \in B' \mid Z) \underset{P}{=} P(X \in A' \mid Z) \cdot P(Y \in B' \mid Z), \quad \forall\, (A', B') \in \mathcal{A}'_X \times \mathcal{A}'_Y. \quad (16.15)$$

◁

Remark 16.15 [An implication for random variables]   The following proposition follows
from Definition 16.13. If $\mathcal{D}, \mathcal{E} \subset \mathcal{A}$, $\sigma(X) \subset \mathcal{D}$, and $\sigma(Y) \subset \mathcal{E}$, then,

$$\mathcal{D} \underset{P}{\perp\!\!\!\perp} \mathcal{E} \mid \mathcal{C} \;\Rightarrow\; X \underset{P}{\perp\!\!\!\perp} Y \mid \mathcal{C}. \quad (16.16)$$

◁

16.2   Properties

Now we study some implications of conditional independence of two sets of events and
of two random variables. According to Example 1.37, the set $\{A\}$ is a $\cap$-stable genera-
tor of $\sigma(\{A\}) = \{A, A^c, \Omega, \emptyset\}$. Furthermore, $\sigma(1_A) = \sigma(\{A\})$ (see Example 2.31). Therefore,
Lemma 16.2, Definitions 16.2 and 16.13, and $1_{B^c} = 1 - 1_B$ imply the following corollary.

Corollary 16.16 [Equivalent propositions]
Let the assumptions 16.1 hold. Then the following propositions are equivalent to each other:

(i) $\{A\} \underset{P}{\perp\!\!\!\perp} \{B\} \mid \mathscr{C}$.

(ii) $\sigma(\{A\}) \underset{P}{\perp\!\!\!\perp} \sigma(\{B\}) \mid \mathscr{C}$.

(iii) $1_A \underset{P}{\perp\!\!\!\perp} 1_B \mid \mathscr{C}$.

(iv) $A \underset{P}{\perp\!\!\!\perp} B \mid \mathscr{C}$.

(v) $B \underset{P}{\perp\!\!\!\perp} A \mid \mathscr{C}$.

(vi) $A \underset{P}{\perp\!\!\!\perp} B^c \mid \mathscr{C}$.

Lemma 16.4 and Definition 16.13 imply the following corollary.

Corollary 16.17 [Sub-σ-algebras]
Let the assumptions 16.1 hold and let $\mathscr{D}, \mathscr{E} \subset \mathscr{A}$. Then,

$$(\mathscr{D} \subset \mathscr{C} \text{ or } \mathscr{E} \subset \mathscr{C}) \Rightarrow \mathscr{D} \underset{P}{\perp\!\!\!\perp} \mathscr{E} \mid \mathscr{C}. \tag{16.17}$$

Furthermore, if the assumptions of Definition 16.13 hold, then,

$$(\sigma(X) \subset \sigma(Z) \text{ or } \sigma(Y) \subset \sigma(Z)) \Rightarrow X \underset{P}{\perp\!\!\!\perp} Y \mid Z. \tag{16.18}$$

Remark 16.18 [An implication for compositions]   Corollary 16.17 and Lemma 2.52 imply: If $g(Z)$ is a composition of Z and an $(\mathscr{A}'_Z, \overline{\mathscr{B}})$-measurable function $g: (\Omega'_Z, \mathscr{A}'_Z) \to (\overline{\mathbb{R}}, \overline{\mathscr{B}})$, then $X \underset{P}{\perp\!\!\!\perp} g(Z) \mid Z$.   ◁

Now we consider conditional independence of two discrete random variables. Remember, a random variable $X: (\Omega, \mathscr{A}, P) \to (\Omega'_X, \mathscr{A}'_X)$ is called *discrete* if there is a finite or countable set $\Omega'_{X0} \subset \Omega'_X$ with $P_X(\Omega'_{X0}) = 1$ and $\{x\} \in \mathscr{A}'_X$ for all $x \in \Omega'_{X0}$ (see Def. 5.56). If we consider two discrete random variables X, Y and adopt the corresponding notation for Y, then Equation (16.14) can be replaced by a simpler equation specified in the following corollary.

Corollary 16.19 [Conditional independence of discrete random variables]
Let $X: (\Omega, \mathscr{A}, P) \to (\Omega'_X, \mathscr{A}'_X)$ and $Y: (\Omega, \mathscr{A}, P) \to (\Omega'_Y, \mathscr{A}'_Y)$ be discrete random variables and let $\mathscr{C} \subset \mathscr{A}$ be a σ-algebra. Then X and Y are $\mathscr{C}$-conditionally P-independent, if and only if

$$P(X{=}x, Y{=}y \mid \mathscr{C}) \underset{P}{=} P(X{=}x \mid \mathscr{C}) \cdot P(Y{=}y \mid \mathscr{C}), \quad \forall\, (x, y) \in \Omega'_{X0} \times \Omega'_{Y0}. \tag{16.19}$$

(For a proof, see the proof of the more general Corollary 16.47.)

Table 16.1   Joe and Ann with no individual treatment effects.

Unit	Treatment	Success	$P(\{\omega\})$	Person variable U	Treatment variable X	Outcome variable Y	$P(Y=1\mid X, U) = E(Y\mid X, U)$	$P(Y=1\mid X) = E(Y\mid X)$	$P(Y=1\mid U) = E(Y\mid U)$	$P(X=1\mid U) = E(X\mid U)$	$P(X=1, Y=1\mid U)$
(Joe, no, −)	.04			Joe	0	0	.6	.76	.6	3/4	9/20
(Joe, no, +)	.06			Joe	0	1	.6	.76	.6	3/4	9/20
(Joe, yes, −)	.12			Joe	1	0	.6	.68	.6	3/4	9/20
(Joe, yes, +)	.18			Joe	1	1	.6	.68	.6	3/4	9/20
(Ann, no, −)	.08			Ann	0	0	.8	.76	.8	1/3	8/30
(Ann, no, +)	.32			Ann	0	1	.8	.76	.8	1/3	8/30
(Ann, yes, −)	.04			Ann	1	0	.8	.68	.8	1/3	8/30
(Ann, yes, +)	.16			Ann	1	1	.8	.68	.8	1/3	8/30

The column groupings are: *Outcomes* ω (Unit, Treatment, Success), *Observables* ($P(\{\omega\})$, Person variable U, Treatment variable X, Outcome variable Y), *Conditional probabilities*.

Equation (16.19) only refers to all pairs $(x, y) \in \Omega'_{X0} \times \Omega'_{Y0}$, whereas (16.14) refers to all pairs (A', B') of elements in the σ-algebras $\mathscr{A}'_X$ and $\mathscr{A}'_Y$, respectively.

Remark 16.20 [Conditioning on a random variable] If $Z: (\Omega, \mathscr{A}, P) \to (\Omega'_Z, \mathscr{A}'_Z)$, then (16.19) can be written as:

$$P(X=x, Y=y \mid Z) \underset{P}{=} P(X=x \mid Z) \cdot P(Y=y \mid Z), \quad \forall\, (x, y) \in \Omega'_{X0} \times \Omega'_{Y0}. \qquad (16.20)$$

◁

Example 16.21 [Joe and Ann with no individual treatment effects] Table 16.1 displays a new example with Joe and Ann. As before, the random experiment consists of drawing Joe or Ann, each one with probability .5, observing whether $(X=1)$ or not $(X=0)$ the drawn person receives treatment, and whether $(Y=1)$ or not $(Y=0)$ a success criterion is reached. In this example, X and Y are U-conditionally independent with respect to P. In other words, conditioning on the events $\{U=Joe\}$ or $\{U=Ann\}$, X and Y are P-independent. Intuitively speaking, this means that there are no individual treatment effects.

According to Definition 16.13, we have to show that the two σ-algebras $\sigma(X)$ and $\sigma(Y)$ are U-conditionally P-independent, where $X: (\Omega, \mathscr{A}, P) \to (\Omega'_X, \mathscr{A}'_X)$ and $Y: (\Omega, \mathscr{A}, P) \to (\Omega'_Y, \mathscr{A}'_Y)$ with $\Omega'_X = \Omega'_Y = \{0, 1\}$ and $\mathscr{A}'_X = \mathscr{A}'_Y = \{\{0, 1\}, \emptyset, \{0\}, \{1\}\}$. In other words, we have to show

$$P(X \in A', Y \in B' \mid U) = P(X \in A' \mid U) \cdot P(Y \in B' \mid U), \quad \forall\, (A', B') \in \mathscr{A}'_X \times \mathscr{A}'_Y. \qquad (16.21)$$

According to Corollary 16.16, it suffices to consider $A' = B' = \{1\}$, using the notation $P(X \in A' \mid U) = P(X=1 \mid U)$, $P(Y \in B' \mid U) = P(Y=1 \mid U)$, and $P(X \in A', Y \in B' \mid U) = P(X=1, Y=1 \mid U)$. Looking at the last three columns of conditional probabilities in Table 16.1 shows that, in this example,

$$P(X=1, Y=1 \mid U) = P(X=1 \mid U) \cdot P(Y=1 \mid U). \tag{16.22}$$

◁

The following theorem adds another important property of conditional independence that is also useful in some proofs.

Theorem 16.22 [A property equivalent to conditional P-independence]
Let $(\Omega, \mathcal{A}, P)$ be a probability space, let $\mathscr{C}, \mathscr{D}, \mathscr{E} \subset \mathcal{A}$, and let $\mathscr{C}, \mathscr{D}$ be σ-algebras. Then,

$$\mathscr{E} \underset{P}{\perp\!\!\!\perp} \mathscr{D} \mid \mathscr{C} \;\Leftrightarrow\; \forall A \in \mathscr{E}: P(A \mid \mathscr{C}, \mathscr{D}) \underset{P}{=} P(A \mid \mathscr{C}). \tag{16.23}$$

(Proof p. 495)

In the following remark, we use the notation $E(Y \mid \mathscr{C}, Z) := E(Y \mid \sigma[\mathscr{C} \cup \sigma(Z)])$ for a version of the $\sigma[\mathscr{C} \cup \sigma(Z)]$-conditional expectation of Y and $P(A \mid \mathscr{C}, Z) := P(A \mid \sigma[\mathscr{C} \cup \sigma(Z)])$ for a version of the $\sigma[\mathscr{C} \cup \sigma(Z)]$-conditional probability of an event $A \in \mathcal{A}$.

Remark 16.23 [Sufficient condition for conditional independence of events]   An immediate implication of Theorem 16.22 is

$$\forall A, B \in \mathcal{A}: \left(P(A \mid 1_B, \mathscr{C}) \underset{P}{=} P(A \mid \mathscr{C}) \;\Rightarrow\; A \underset{P}{\perp\!\!\!\perp} B \mid \mathscr{C} \right). \tag{16.24}$$

This proposition may also be written as:

$$\forall A, B \in \mathcal{A}: \left(E(1_A \mid 1_B, \mathscr{C}) \underset{P}{=} E(1_A \mid \mathscr{C}) \;\Rightarrow\; A \underset{P}{\perp\!\!\!\perp} B \mid \mathscr{C} \right). \tag{16.25}$$

◁

The following corollary follows from Theorem 16.22 for $A := \{Y=y\}$.

Corollary 16.24 [An implication of conditional independence]
Let $Y: (\Omega, \mathcal{A}, P) \to (\Omega'_Y, \mathcal{A}'_Y)$ be a random variable with $\{y\} \in \mathcal{A}'_Y$ for all $y \in \Omega'_Y$, and let $\mathscr{C}, \mathscr{D} \subset \mathcal{A}$ be σ-algebras. Then,

$$Y \underset{P}{\perp\!\!\!\perp} \mathscr{D} \mid \mathscr{C} \;\Rightarrow\; \forall y \in \Omega'_Y: P(Y=y \mid \mathscr{C}, \mathscr{D}) \underset{P}{=} P(Y=y \mid \mathscr{C}). \tag{16.26}$$

Remark 16.25 [Rewriting the corollary for random variables]   In addition to the assumptions of Corollary 16.24, let $X: (\Omega, \mathcal{A}, P) \to (\Omega'_X, \mathcal{A}'_X)$ and $Z: (\Omega, \mathcal{A}, P) \to (\Omega'_Z, \mathcal{A}'_Z)$ be random variables. Then,

$$Y \underset{P}{\perp\!\!\!\perp} X \mid Z \;\Rightarrow\; \forall y \in \Omega'_Y: P(Y=y \mid X, Z) \underset{P}{=} P(Y=y \mid Z). \tag{16.27}$$

◁

Box 16.1   Notation.

Let $(\Omega, \mathcal{A}, P)$ be a probability space, let $A, B \in \mathcal{A}$, let $\mathcal{C}, \mathcal{D}, \mathcal{E} \subset \mathcal{A}$, where $\mathcal{C}$ is a σ-algebra, and let X, Y, Z be random variables on $(\Omega, \mathcal{A}, P)$. Then,

Symbol	Meaning
$A \underset{P}{\perp\!\!\!\perp} B \mid \mathcal{C}$	$\mathcal{C}$-conditional P-independence of A and B
$\mathcal{E} \underset{P}{\perp\!\!\!\perp} \mathcal{D} \mid \mathcal{C}$	$\mathcal{C}$-conditional P-independence of $\mathcal{E}$ and $\mathcal{D}$
$X \underset{P}{\perp\!\!\!\perp} Y \mid \mathcal{C}$	$\mathcal{C}$-conditional P-independence of X and Y
$X \underset{P}{\perp\!\!\!\perp} \mathcal{D} \mid \mathcal{C}$	$\mathcal{C}$-conditional P-independence of X and $\mathcal{D}$ [i.e., $\sigma(X) \underset{P}{\perp\!\!\!\perp} \mathcal{D} \mid \mathcal{C}$]
$A \underset{P}{\perp\!\!\!\perp} B \mid Z$	Z-conditional P-independence of A and B [i.e., $A \underset{P}{\perp\!\!\!\perp} B \mid \sigma(Z)$]
$\mathcal{E} \underset{P}{\perp\!\!\!\perp} \mathcal{D} \mid Z$	Z-conditional P-independence of $\mathcal{E}$ and $\mathcal{D}$ [i.e., $\mathcal{E} \underset{P}{\perp\!\!\!\perp} \mathcal{D} \mid \sigma(Z)$]

Now we consider a special case of Theorem 16.22 for a *discrete* random variable. This means that we consider a random variable $Y \colon (\Omega, \mathcal{A}, P) \to (\Omega'_Y, \mathcal{A}'_Y)$ such that there is a finite or countable set $\Omega'_{Y0} \subset \Omega'_Y$ with $P_Y(\Omega'_{Y0}) = 1$ and $\{y\} \in \mathcal{A}'_Y$ for all $y \in \Omega'_{Y0}$ (see Def. 5.56). In this theorem, we also use the notation $Y \underset{P}{\perp\!\!\!\perp} \mathcal{D} \mid \mathcal{C}$ introduced in Box 16.1.

Theorem 16.26 [A proposition equivalent to conditional independence]
Let $Y \colon (\Omega, \mathcal{A}, P) \to (\Omega'_Y, \mathcal{A}'_Y)$ be a discrete random variable and let $\mathcal{C}, \mathcal{D} \subset \mathcal{A}$ be σ-algebras. Then,

$$Y \underset{P}{\perp\!\!\!\perp} \mathcal{D} \mid \mathcal{C} \;\Leftrightarrow\; \forall\, y \in \Omega'_{Y0} \colon P(Y=y \mid \mathcal{C}, \mathcal{D}) \underset{P}{=} P(Y=y \mid \mathcal{C}). \tag{16.28}$$

(Proof p. 495)

Remark 16.27 [Rewriting the theorem for random variables]   Suppose that $X \colon (\Omega, \mathcal{A}, P) \to (\Omega'_X, \mathcal{A}'_X)$, $Y \colon (\Omega, \mathcal{A}, P) \to (\Omega'_Y, \mathcal{A}'_Y)$, and $Z \colon (\Omega, \mathcal{A}, P) \to (\Omega'_Z, \mathcal{A}'_Z)$ are random variables, and that Y is discrete. Then, for $\mathcal{D} = \sigma(X)$ and $\mathcal{C} = \sigma(Z)$, Theorem 16.26 immediately yields:

$$Y \underset{P}{\perp\!\!\!\perp} X \mid Z \;\Leftrightarrow\; \forall\, y \in \Omega'_{Y0} \colon P(Y=y \mid X, Z) \underset{P}{=} P(Y=y \mid Z). \tag{16.29}$$

◁

Box 16.2 summarizes some important propositions on conditional independence of σ-algebras, and Box 16.3 translates these propositions to conditional independence of random variables. Proofs are provided in Exercise 16.3.

Considering a value $z \in \Omega'_Z$ for which $P(Z=z) > 0$, some rules of Box 16.2 may also have implications for conditional independence with respect to a conditional-probability measure $P^{Z=z}$. For example, if X is a numerical Z-measurable function, then according to

Box 16.2   Conditional independence of set systems.

Let $(\Omega, \mathcal{A}, P)$ be a probability space; $\mathcal{C}, \mathcal{D}, \mathcal{D}_1, \mathcal{D}_2, \mathcal{D}_3 \subset \mathcal{A}$ be σ-algebras; and $\mathcal{E}, \mathcal{F}$, $\mathcal{G} \subset \mathcal{A}$. Then:

$$\mathcal{E} \underset{P}{\perp\!\!\!\perp} \mathcal{F} \mid \mathcal{C} \;\Leftrightarrow\; \mathcal{F} \underset{P}{\perp\!\!\!\perp} \mathcal{E} \mid \mathcal{C} \qquad\qquad\text{(i)}$$

$$\mathcal{E} \underset{P}{\perp\!\!\!\perp} \sigma(\mathcal{F}, \mathcal{G}) \mid \mathcal{C} \;\Rightarrow\; \mathcal{E} \underset{P}{\perp\!\!\!\perp} \mathcal{F} \mid \mathcal{C} \text{ and } \mathcal{E} \underset{P}{\perp\!\!\!\perp} \mathcal{G} \mid \mathcal{C} \qquad\text{(ii)}$$

$$\underset{P}{\perp\!\!\!\perp} \mathcal{D}_1, \mathcal{D}_2, \mathcal{D}_3 \;\Rightarrow\; \mathcal{D}_1 \underset{P}{\perp\!\!\!\perp} \mathcal{D}_2 \mid \mathcal{D}_3 \qquad\qquad\text{(iii)}$$

$$\mathcal{F} \subset \mathcal{C} \;\Rightarrow\; \mathcal{E} \underset{P}{\perp\!\!\!\perp} \mathcal{F} \mid \mathcal{C} \qquad\qquad\text{(iv)}$$

$$\mathcal{F} = \{\Omega, \varnothing\} \;\Rightarrow\; \mathcal{E} \underset{P}{\perp\!\!\!\perp} \mathcal{F} \mid \mathcal{C} \qquad\qquad\text{(v)}$$

$$\mathcal{E} \underset{P}{\perp\!\!\!\perp} \mathcal{F} \mid \mathcal{C} \text{ and } \mathcal{G} \subset \mathcal{F} \;\Rightarrow\; \mathcal{E} \underset{P}{\perp\!\!\!\perp} \mathcal{G} \mid \mathcal{C} \qquad\text{(vi)}$$

$$\mathcal{D}_1 \underset{P}{\perp\!\!\!\perp} \mathcal{D}_2 \mid \mathcal{C} \;\Leftrightarrow\; \mathcal{D}_1 \underset{P}{\perp\!\!\!\perp} \sigma(\mathcal{C}, \mathcal{D}_2) \mid \mathcal{C} \qquad\qquad\text{(vii)}$$

$$\mathcal{E} \underset{P}{\perp\!\!\!\perp} \sigma(\mathcal{D}_1, \mathcal{D}_2) \mid \mathcal{C} \;\Leftrightarrow\; \mathcal{E} \underset{P}{\perp\!\!\!\perp} \mathcal{D}_1 \mid \mathcal{C} \text{ and } \mathcal{E} \underset{P}{\perp\!\!\!\perp} \mathcal{D}_2 \mid \sigma(\mathcal{C}, \mathcal{D}_1) \qquad\text{(viii)}$$

$$\mathcal{E} \underset{P}{\perp\!\!\!\perp} \sigma(\mathcal{D}_1, \mathcal{D}_2) \;\Leftrightarrow\; \mathcal{E} \underset{P}{\perp\!\!\!\perp} \mathcal{D}_1 \text{ and } \mathcal{E} \underset{P}{\perp\!\!\!\perp} \mathcal{D}_2 \mid \mathcal{D}_1. \qquad\text{(ix)}$$

Let $\mathcal{F}_0 \subset \mathcal{F}, \mathcal{G}_0 \subset \mathcal{G}$ and $\mathcal{F}_0, \mathcal{F}, \mathcal{G}_0, \mathcal{G} \subset \mathcal{A}$ be σ-algebras. Then:

$$\mathcal{F} \underset{P}{\perp\!\!\!\perp} \mathcal{G} \mid \mathcal{C} \;\Rightarrow\; \mathcal{F} \underset{P}{\perp\!\!\!\perp} \mathcal{G} \mid \sigma(\mathcal{C}, \mathcal{F}_0, \mathcal{G}_0). \qquad\qquad\text{(x)}$$

Let $X_1, X_2, X: (\Omega, \mathcal{A}, P) \to (\Omega'_X, \mathcal{A}'_X)$ be random variables. Then:

$$X_1 \underset{P}{=} X_2 \text{ and } X_1 \underset{P}{\perp\!\!\!\perp} \mathcal{E} \mid \mathcal{C} \;\Rightarrow\; X_2 \underset{P}{\perp\!\!\!\perp} \mathcal{E} \mid \mathcal{C} \qquad\qquad\text{(xi)}$$

$$X \underset{P}{=} \alpha, \alpha \in \Omega'_X \;\Rightarrow\; X \underset{P}{\perp\!\!\!\perp} \mathcal{E} \mid \mathcal{C}. \qquad\qquad\text{(xii)}$$

Lemma 2.52, there is a measurable function $g: (\overline{\mathbb{R}}, \overline{\mathcal{B}}) \to (\overline{\mathbb{R}}, \overline{\mathcal{B}})$ such that $X = g(Z)$. This implies that $X \underset{P^{Z=z}}{=} g(z)$ (see Rem. 9.1). Therefore, for $\alpha = g(z)$, Box 16.2 (xii) yields the following corollary.

Corollary 16.28 [Measurability and conditional independence with respect to $P^{Z=z}$]

Let $X: (\Omega, \mathcal{A}, P) \to (\overline{\mathbb{R}}, \overline{\mathcal{B}})$ *and* $Z: (\Omega, \mathcal{A}, P) \to (\Omega'_Z, \mathcal{A}'_Z)$ *be random variables, let* $P(Z = z) > 0$, *and* $\mathcal{C}, \mathcal{E} \subset \mathcal{A}$, *where* $\mathcal{C}$ *is a* σ-algebra. *Then,*

$$X \text{ is } Z\text{-measurable} \;\Rightarrow\; X \underset{P^{Z=z}}{\perp\!\!\!\perp} \mathcal{E} \mid \mathcal{C}. \qquad\qquad(16.30)$$

Box 16.3   Conditional independence of random variables.

Let W, X, X_1, X_2, X_3, Y, and Z be random variables on the probability space $(\Omega, \mathcal{A}, P)$. Then:

$$X \underset{P}{\perp\!\!\!\perp} Y \mid Z \Leftrightarrow Y \underset{P}{\perp\!\!\!\perp} X \mid Z \tag{i}$$

$$X \underset{P}{\perp\!\!\!\perp} (W, Y) \mid Z \Rightarrow X \underset{P}{\perp\!\!\!\perp} W \mid Z \text{ and } X \underset{P}{\perp\!\!\!\perp} Y \mid Z \tag{ii}$$

$$\underset{i=1}{\overset{3}{\perp\!\!\!\perp}} X_i \Rightarrow X_1 \underset{P}{\perp\!\!\!\perp} X_2 \mid X_3 \tag{iii}$$

$$\sigma(Y) \subset \sigma(Z) \Rightarrow X \underset{P}{\perp\!\!\!\perp} Y \mid Z \tag{iv}$$

$$X = \alpha, \alpha \in \Omega'_X \Rightarrow X \underset{P}{\perp\!\!\!\perp} Y \mid Z \tag{v}$$

$$X \underset{P}{\perp\!\!\!\perp} Y \mid Z \text{ and } \sigma(W) \subset \sigma(Y) \Rightarrow X \underset{P}{\perp\!\!\!\perp} W \mid Z \tag{vi}$$

$$X \underset{P}{\perp\!\!\!\perp} Y \mid Z \Leftrightarrow X \underset{P}{\perp\!\!\!\perp} (Z, Y) \mid Z \tag{vii}$$

$$X \underset{P}{\perp\!\!\!\perp} (W, Y) \mid Z \Leftrightarrow X \underset{P}{\perp\!\!\!\perp} Y \mid Z \text{ and } X \underset{P}{\perp\!\!\!\perp} W \mid (Z, Y) \tag{viii}$$

$$X \underset{P}{\perp\!\!\!\perp} (W, Y) \Leftrightarrow X \underset{P}{\perp\!\!\!\perp} Y \text{ and } X \underset{P}{\perp\!\!\!\perp} W \mid Y. \tag{ix}$$

If Y_0 and W_0 are random variables on $(\Omega, \mathcal{A}, P)$ that are measurable with respect to Y and W, respectively, then,

$$W \underset{P}{\perp\!\!\!\perp} Y \mid Z \Rightarrow W \underset{P}{\perp\!\!\!\perp} Y \mid (Z, Y_0, W_0) \tag{x}$$

$$X_1 \underset{P}{=} X_2 \text{ and } X_1 \underset{P}{\perp\!\!\!\perp} Y \mid Z \Rightarrow X_2 \underset{P}{\perp\!\!\!\perp} Y \mid Z \tag{xi}$$

$$X \underset{P}{=} \alpha, \alpha \in \Omega'_X \Rightarrow X \underset{P}{\perp\!\!\!\perp} Y \mid Z. \tag{xii}$$

Remark 16.29 [Z-measurability and independence with respect to $P^{Z=z}$] For $\mathcal{C} = \{\Omega, \varnothing\}$, Corollary 16.28 implies that $\mathcal{E}$ and X are independent with respect to $P^{Z=z}$, that is,

$$X \text{ is Z-measurable} \Rightarrow X \underset{P^{Z=z}}{\perp\!\!\!\perp} \mathcal{E}. \tag{16.31}$$

This result is also an immediate implication of Lemmas 2.52 and 5.51.   ◁

Example 16.30 [Conditional independence of treatment and person variables] Table 16.2 displays some parameters of a new random experiment. In this experiment, a person (unit) is drawn from a set of six persons with a sampling probability of $P(U=u) = 1/6$ for each person. There are four males ($Z=m$) and two females ($Z=f$). If a male person is drawn, then he receives treatment ($X=1$) with probability $P(X=1 \mid U=u) = 3/4$; if a female person is drawn, then she gets treatment with probability $P(X=1 \mid U=u) = 1/4$. The table also displays the conditional expectation values $E(Y \mid X=0, U=u)$ and $E(Y \mid X=1, U=u)$ of a real-valued outcome variable Y.

Table 16.2 Z-conditional independence of X and U.

Persons u			Conditional expectation values		
Person variable U	$P(U=u)$	Sex Z	$E(Y \mid X=0, U=u)$	$E(Y \mid X=1, U=u)$	$P(X=1 \mid U=u)$
u_1	1/6	m	68	81	3/4
u_2	1/6	m	78	86	3/4
u_3	1/6	m	88	100	3/4
u_4	1/6	m	98	103	3/4
u_5	1/6	f	106	114	1/4
u_6	1/6	f	116	130	1/4

In this example,

$$P(X=1 \mid Z, U) = P(X=1 \mid Z) = \frac{3}{4} \cdot 1_{Z=m} + \frac{1}{4} \cdot 1_{Z=f}.$$

However, if the first of these two equations holds, then $P(X=0 \mid Z, U) = P(X=0 \mid Z)$ holds as well, because $P(X=0 \mid Z) = 1 - P(X=1 \mid Z)$ and $P(X=0 \mid Z, U) = 1 - P(X=1 \mid Z, U)$ (see again Exercise 16.4). Therefore,

$$P(X=x \mid Z, U) = P(X=x \mid Z), \quad x = 0, 1, \tag{16.32}$$

and, according to (16.29), the treatment variable X and the person variable U are conditionally independent given Z (sex). ◁

Example 16.31 [Joe and Ann with no individual treatment effects – continued]   In Example 16.21, we presented Table 16.1 and showed that X and Y are U-conditionally independent. According to Theorem 16.26,

$$P(Y=y \mid X, U) \underset{P}{=} P(Y=y \mid U), \quad y = 0, 1,$$

is equivalent to U-conditional independence of X and Y. Comparing the columns headed by $P(Y=1 \mid X, U)$ and $P(Y=1 \mid U)$ in Table 16.1 confirms that the displayed equation is satisfied for $y=1$. However, this implies that $P(Y=0 \mid X, U) \underset{P}{=} P(Y=0 \mid U)$ holds as well (see Exercise 16.4). ◁

16.3 Conditional independence and conditional mean independence

Now we generalize the product rule for the expectations of independent random variables [see Box 6.1 (x) and Proposition (16.7)] using the notation $X \underset{P}{\perp\!\!\!\perp} Y \mid \mathscr{C}$ introduced in Definition 16.13.

Theorem 16.32 [Product rule under conditional P-independence]
Let $X, Y\colon (\Omega, \mathscr{A}, P) \to (\mathbb{R}, \mathscr{B})$ be random variables that are both nonnegative or both with finite expectations, and let $\mathscr{C} \subset \mathscr{A}$ be a σ-algebra. Then,

$$X \underset{P}{\perp\!\!\!\perp} Y \mid \mathscr{C} \;\Rightarrow\; E(X \cdot Y \mid \mathscr{C}) \underset{P}{=} E(X \mid \mathscr{C}) \cdot E(Y \mid \mathscr{C}). \qquad (16.33)$$

(Proof p. 495)

Remark 16.33 [Conditioning on a random variable]   If $Z\colon (\Omega, \mathscr{A}, P) \to (\Omega'_Z, \mathscr{A}'_Z)$ is a random variable, then, under the assumptions of Theorem 16.32,

$$X \underset{P}{\perp\!\!\!\perp} Y \mid Z \;\Rightarrow\; E(X \cdot Y \mid Z) \underset{P}{=} E(X \mid Z) \cdot E(Y \mid Z). \qquad (16.34)$$

Hence, under the assumptions of Theorem 16.32, and if X and Y have finite second moments, then Z-conditional independence of X and Y also implies $Cov(X, Y \mid Z) \underset{P}{=} 0$ [see Rule (i) of Box 11.2].   ◁

In section 10.6, we introduced the concept of conditional mean independence. Now we consider its relationship to conditional independence.

Theorem 16.34 [Conditional mean independence]
Let $(\Omega, \mathscr{A}, P)$ be a probability space, let $\mathscr{C}, \mathscr{D}, \mathscr{E} \subset \mathscr{A}$ be σ-algebras, and let $Y\colon (\Omega, \mathscr{A}, P) \to (\mathbb{R}, \overline{\mathscr{B}})$ be a random variable that is nonnegative or has a finite expectation $E(Y)$. If $\sigma(Y) \subset \mathscr{E}$, then,

$$\mathscr{E} \underset{P}{\perp\!\!\!\perp} \mathscr{D} \mid \mathscr{C} \;\Rightarrow\; E(Y \mid \mathscr{C}, \mathscr{D}) \underset{P}{=} E(Y \mid \mathscr{C}). \qquad (16.35)$$

(Proof p. 496)

An immediate implication is

$$Y \underset{P}{\perp\!\!\!\perp} \mathscr{D} \mid \mathscr{C} \;\Rightarrow\; E(Y \mid \mathscr{C}, \mathscr{D}) \underset{P}{=} E(Y \mid \mathscr{C}). \qquad (16.36)$$

Remark 16.35 [Conditioning on random variables]   If $X\colon (\Omega, \mathscr{A}, P) \to (\Omega'_X, \mathscr{A}'_X)$ and $Z\colon (\Omega, \mathscr{A}, P) \to (\Omega'_Z, \mathscr{A}'_Z)$ are random variables, and $Y\colon (\Omega, \mathscr{A}, P) \to (\mathbb{R}, \overline{\mathscr{B}})$ is a random variable that is nonnegative or has a finite expectation $E(Y)$, then,

$$Y \underset{P}{\perp\!\!\!\perp} X \mid Z \;\Rightarrow\; E(Y \mid X, Z) \underset{P}{=} E(Y \mid Z) . \qquad (16.37)$$

◁

Remark 16.36 [Implications of independence] For Z being a constant, the last remark yields

$$Y \underset{P}{\perp\!\!\!\perp} X \;\Rightarrow\; E(Y \mid X) \underset{P}{=} E(Y) \qquad (16.38)$$

[cf. Box 10.2 (vi)].   ◁

While Theorem 16.34 deals with an implication of conditional independence on conditional mean independence, now we present some conditions that are equivalent to conditional independence.

Theorem 16.37 [Characterizations of conditional independence]
Let $(\Omega, \mathcal{A}, P)$ be a probability space, let $\mathcal{C}, \mathcal{D}, \mathcal{E} \subset \mathcal{A}$ be σ-algebras, and let $W, Y \colon (\Omega, \mathcal{A}, P) \to (\mathbb{R}, \mathcal{B})$ be real-valued random variables. Then the following propositions are equivalent to each other:

(i) $\mathcal{E} \underset{P}{\perp\!\!\!\perp} \mathcal{D} \mid \mathcal{C}$.

(ii) $E(Y \mid \mathcal{C}, \mathcal{D}) \underset{P}{=} E(Y \mid \mathcal{C})$, for all nonnegative Y with finite expectation and $\sigma(Y) \subset \mathcal{E}$.

(iii) $E(W \cdot Y \mid \mathcal{C}) \underset{P}{=} E(W \mid \mathcal{C}) \cdot E(Y \mid \mathcal{C})$, for all nonnegative Y with finite expectation and $\sigma(Y) \subset \mathcal{E}$ and all nonnegative W with $\sigma(W) \subset \mathcal{D}$.

(iv) $E\big[E(Y \mid \mathcal{C}) \mid \mathcal{D}\big] \underset{P}{=} E(Y \mid \mathcal{D})$, for all nonnegative Y with finite expectation and $\sigma(Y) \subset \sigma(\mathcal{C}, \mathcal{E})$.

(Proof p. 497)

In the next theorem, we assume $Z \underset{P}{\perp\!\!\!\perp} Y \mid \mathcal{C}$ and present an implication for the conditional expectation $E^{Z=z}(Y \mid \mathcal{C})$ (see ch. 14).

Theorem 16.38 [An implication of $\mathcal{C}$-conditional P-independence]
Let $Y \colon (\Omega, \mathcal{A}, P) \to (\overline{\mathbb{R}}, \overline{\mathcal{B}})$ and $Z \colon (\Omega, \mathcal{A}, P) \to (\Omega'_Z, \mathcal{A}'_Z)$ be random variables, let $z \in \Omega'_Z$ with $\{z\} \in \mathcal{A}'_Z$ and $P(Z=z) > 0$, and let Y be nonnegative or such that $E^{Z=z}(Y) =$

$\int Y \, dP^{Z=z}$ *is finite. Furthermore, let* $\mathscr{C} \subset \mathscr{A}$ *be a σ-algebra. If* $Z \perp\!\!\!\perp_P Y \mid \mathscr{C}$*, then* $\mathscr{E}(Y \mid \mathscr{C}) \subset \mathscr{E}^{Z=z}(Y \mid \mathscr{C})$*, and therefore,*

$$E^{Z=z}(Y \mid \mathscr{C}) \underset{P^{Z=z}}{=} E(Y \mid \mathscr{C}). \tag{16.39}$$

(Proof p. 498)

Note that, even if $Z \perp\!\!\!\perp_P Y \mid \mathscr{C}$, then $E^{Z=z}(Y \mid \mathscr{C})$ is not necessarily an element of $\mathscr{E}(Y \mid \mathscr{C})$, because Equation (16.39) does *not imply* $E^{Z=z}(Y \mid \mathscr{C}) \underset{P}{=} E(Y \mid \mathscr{C})$. For $\mathscr{C} = \{\Omega, \varnothing\}$, Theorem 16.38 implies the following corollary:

Corollary 16.39 [An implication of P-independence]
Let the assumptions of Theorem 16.38 hold. If $Z \perp\!\!\!\perp_P Y$*, then,*

$$E^{Z=z}(Y) = E(Y). \tag{16.40}$$

16.4   Families of events

Now we extend the concept of conditional independence to more than two events, more than two sets of events, and more than two random variables.

Three events A_1, A_2, A_3 are called $\mathscr{C}$*-conditionally P-independent*, if

$$P(A_i \cap A_j \mid \mathscr{C}) \underset{P}{=} P(A_i \mid \mathscr{C}) \cdot P(A_j \mid \mathscr{C}), \quad \forall \, i, j = 1, 2, 3, \ i \neq j, \tag{16.41}$$

(pairwise conditional independence), and

$$P(A_1 \cap A_2 \cap A_3 \mid \mathscr{C}) \underset{P}{=} P(A_1 \mid \mathscr{C}) \cdot P(A_2 \mid \mathscr{C}) \cdot P(A_3 \mid \mathscr{C}) \tag{16.42}$$

(triple-wise conditional independence). We use the notation

$$\perp\!\!\!\perp_P A_1, A_2, A_3 \mid \mathscr{C}$$

for $\mathscr{C}$-conditional P-independence of A_1, A_2, A_3. Four events are conditionally P-independent if the corresponding product rule holds for all pairs, all triples, and the quadruple. The general definition is as follows:

Definition 16.40 [Conditional independence of a family of events]
Let $(\Omega, \mathcal{A}, P)$ be a probability space, I be a nonempty set, $A_i \in \mathcal{A}$, for all $i \in I$, and $\mathcal{C} \subset \mathcal{A}$ be a σ-algebra. Then,

(i) $(A_i, i \in I)$ is called a family of $\mathcal{C}$-conditionally P-independent events, denoted $\underset{P}{\perp\!\!\!\perp} (A_i, i \in I) \mid \mathcal{C}$, if

$$P\left(\bigcap_{i \in J} A_i \,\Big|\, \mathcal{C}\right) \underset{P}{=} \prod_{i \in J} P(A_i \mid \mathcal{C}), \quad \text{for all finite sets } J \subset I. \qquad (16.43)$$

(ii) Let $Z: (\Omega, \mathcal{A}, P) \to (\Omega'_Z, \mathcal{A}'_Z)$ be a random variable. Then, $(A_i, i \in I)$ is called a family of Z-conditionally independent events, denoted $\underset{P}{\perp\!\!\!\perp} (A_i, i \in I) \mid Z$, if (16.43) holds for $\mathcal{C} = \sigma(Z)$.

16.5 Families of set systems

Using the concept of conditional P-independence of families of events, we can now define *conditional P-independence of families of set systems* or of sets of events. Three sets $\mathcal{E}_1, \mathcal{E}_2, \mathcal{E}_3$ of events are called $\mathcal{C}$-conditionally P-independent if, for all $i, j = 1, 2, 3, i \neq j$,

$$P(A_i \cap A_j \mid \mathcal{C}) \underset{P}{=} P(A_i \mid \mathcal{C}) \cdot P(A_j \mid \mathcal{C}), \quad \forall\, (A_i, A_j) \in \mathcal{E}_i \times \mathcal{E}_j, \qquad (16.44)$$

and

$$P(A_1 \cap A_2 \cap A_3 \mid \mathcal{C}) \underset{P}{=} P(A_1 \mid \mathcal{C}) \cdot P(A_2 \mid \mathcal{C}) \cdot P(A_3 \mid \mathcal{C}),$$
$$\forall\, (A_1, A_2, A_3) \in \mathcal{E}_1 \times \mathcal{E}_2 \times \mathcal{E}_3. \qquad (16.45)$$

The general definition for a family $(\mathcal{E}_i, i \in I)$ of sets of events is as follows:

Definition 16.41 [Conditional independence of a family of sets of events]
Let $(\Omega, \mathcal{A}, P)$ be a probability space, I be a nonempty set, and $\mathcal{E}_i \subset \mathcal{A}, i \in I$.

(i) Then, $(\mathcal{E}_i, i \in I)$ is called a family of $\mathcal{C}$-conditionally P-independent sets of events, denoted $\underset{P}{\perp\!\!\!\perp} (\mathcal{E}_i, i \in I) \mid \mathcal{C}$, if each family $(A_i, i \in I)$ of events $A_i \in \mathcal{E}_i, i \in I$, is $\mathcal{C}$-conditionally P-independent.

(ii) Furthermore, let $Z: (\Omega, \mathcal{A}, P) \to (\Omega'_Z, \mathcal{A}'_Z)$ be a random variable. Then, $(\mathcal{E}_i, i \in I)$ is called a family of Z-conditionally P-independent sets of events, denoted $\underset{P}{\perp\!\!\!\perp} (\mathcal{E}_i, i \in I) \mid Z$, if each family $(A_i, i \in I)$ of events $A_i \in \mathcal{E}_i$, $i \in I$, is Z-conditionally P-independent.

Remark 16.42 [Subfamilies]   This definition immediately implies

$$\underset{P}{\perp\!\!\!\perp} (\mathcal{E}_i, i \in I) \mid \mathcal{C} \iff \forall\, J \subset I: \underset{P}{\perp\!\!\!\perp} (\mathcal{E}_i, i \in J) \mid \mathcal{C}. \qquad (16.46)$$

$\triangleleft$

Remark 16.43 [Smallest σ-algebra] If $\mathscr{C} = \{\Omega, \emptyset\}$, then,

$$\underset{P}{\perp\!\!\!\perp} (\mathscr{E}_i, i \in I) \mid \mathscr{C} \quad \Leftrightarrow \quad \underset{P}{\perp\!\!\!\perp} (\mathscr{E}_i, i \in I) \tag{16.47}$$

[see Eq. (10.2) and Remark 10.5]. Hence, P-independence of families of events (see Def. 4.40) is a special case of $\mathscr{C}$-conditional P-independence of families of events.   ◁

Theorem 16.44 [Conditional independence of ∩-stable set systems]
If $(\Omega, \mathscr{A}, P)$ is a probability space, $\mathscr{C} \subset \mathscr{A}$ is a σ-algebra, and $(\mathscr{E}_i, i \in I)$ is a family of ∩-stable set systems with $\mathscr{E}_i \subset \mathscr{A}, i \in I$, then,

$$\underset{P}{\perp\!\!\!\perp} (\mathscr{E}_i, i \in I) \mid \mathscr{C} \quad \Leftrightarrow \quad \underset{P}{\perp\!\!\!\perp} (\sigma(\mathscr{E}_i), i \in I) \mid \mathscr{C}. \tag{16.48}$$

(Proof p. 499)

16.6   Families of random variables

Now we turn to the concept of conditional P-independence of (families of) random variables. Three random variables X_1, X_2, and X_3 are called $\mathscr{C}$-conditionally P-independent if Equations (16.44) and (16.45) hold for $\mathscr{E}_i = \sigma(X_i)$. Note that $\mathscr{C}$-conditional P-independence of X_1, X_2, and X_3 implies that X_1 and X_2, X_1 and X_3, as well as X_2 and X_3 are $\mathscr{C}$-conditionally P-independent. Of course, the same applies to Z-conditional P-independence.

Definition 16.45 [Conditional independence of random variables]
Let I be a nonempty set, let $X_i : (\Omega, \mathscr{A}, P) \to (\Omega_i', \mathscr{A}_i'), i \in I$, be random variables, and let $\mathscr{C} \subset \mathscr{A}$ be a σ-algebra.

(i) $(X_i, i \in I)$ is called a family of $\mathscr{C}$-conditionally P-independent random variables, denoted $\underset{P}{\perp\!\!\!\perp} (X_i, i \in I) \mid \mathscr{C}$, if

$$\underset{P}{\perp\!\!\!\perp} (\sigma(X_i), i \in I) \mid \mathscr{C}. \tag{16.49}$$

(ii) Let $Z : (\Omega, \mathscr{A}, P) \to (\Omega_Z', \mathscr{A}_Z')$ be a random variable. Then $(X_i, i \in I)$ is called a family of Z-conditionally P-independent random variables, denoted $\underset{P}{\perp\!\!\!\perp} (X_i, i \in I) \mid Z$, if

$$\underset{P}{\perp\!\!\!\perp} (\sigma(X_i), i \in I) \mid Z. \tag{16.50}$$

Remark 16.46 [Conditional independence of subfamilies] If $(X_i, i \in I)$ is a family of $\mathscr{C}$-conditionally P-independent random variables, that is, if $\underset{P}{\perp\!\!\!\perp} (X_i, i \in I) \mid \mathscr{C}$, then $\underset{P}{\perp\!\!\!\perp} (X_i, i \in J) \mid \mathscr{C}$, for all $J \subset I$, because in this case any finite subset of J is also a finite subset of I (cf. Def. 16.40). For example, if X_1, X_2, X_3 are $\mathscr{C}$-conditionally P-independent, then they are also pairwise $\mathscr{C}$-conditionally P-independent (cf. Rem. 5.40).   ◁

Now we consider conditional independence of n discrete random variables, generalizing Corollary 16.19. Remember that a random variable $X_i \colon (\Omega, \mathscr{A}, P) \to (\Omega_i', \mathscr{A}_i')$ is called *discrete* if there is a finite or countable set $\Omega_{i0}' \subset \Omega_i'$ with $P_{X_i}(\Omega_{i0}') = 1$ and $\{x_i\} \in \mathscr{A}_i'$ for all $x_i \in \Omega_{i0}'$, $i = 1, \dots, n$ (see Def. 5.56).

Corollary 16.47 [Conditional independence of n discrete random variables]
Let $X_i \colon (\Omega, \mathscr{A}, P) \to (\Omega_i', \mathscr{A}_i')$, $i = 1, \dots, n$, be discrete random variables and let $\mathscr{C} \subset \mathscr{A}$ be a σ-algebra. Then, $\underset{P}{\perp\!\!\!\perp} (X_i, i = 1, \dots, n) \mid \mathscr{C}$ if and only if

$$P(X_1 = x_1, \dots, X_n = x_n \mid \mathscr{C}) \underset{P}{=} P(X_1 = x_1 \mid \mathscr{C}) \cdot \ldots \cdot P(X_n = x_n \mid \mathscr{C}),$$

$$\forall\, (x_1, \dots, x_n) \in \Omega_{10}' \times \dots \times \Omega_{n0}' \,. \tag{16.51}$$

(Proof p. 499)

Remark 16.48 [Conditioning on a random variable]   If $Z \colon (\Omega, \mathscr{A}, P) \to (\Omega_Z', \mathscr{A}_Z')$, $n = 2$, $X_1 = X$, and $X_2 = Y$, then (16.51) yields:

$$P(X = x, Y = y \mid Z) \underset{P}{=} P(X = x \mid Z) \cdot P(Y = y \mid Z), \quad \forall\, (x, y) \in \Omega_{X0}' \times \Omega_{Y0}'. \tag{16.52}$$

◁

Lemma 16.49 [Unions of independent σ-algebras]
Under the assumptions of Definition 16.45, let $(X_i, i \in I)$ be a family of $\mathscr{C}$-conditionally P-independent random variables, let J be a nonempty set, and let $\{I_j, j \in J\}$ be a set of pairwise disjoint subsets of I. Then,

$$\underset{P}{\perp\!\!\!\perp} (\sigma(X_i, i \in I_j), j \in J) \mid \mathscr{C}. \tag{16.53}$$

(Proof p. 499)

If $Y_1, \dots, Y_m \colon (\Omega, \mathscr{A}, P) \to (\mathbb{R}, \mathscr{B})$ are discrete random variables and X is a random variable on $(\Omega, \mathscr{A}, P)$, then the following corollary presents a useful characterization of X-conditional independence of the $Y_1, \dots, Y_m$.

Corollary 16.50 [X-conditional independence of discrete random variables]
Let $X \colon (\Omega, \mathscr{A}, P) \to (\Omega_X', \mathscr{A}_X')$ be a random variable, let $(Y_1, \dots, Y_m) \colon (\Omega, \mathscr{A}, P) \to (\Omega_1' \times \dots \times \Omega_m', \mathscr{A}_1' \otimes \dots \otimes \mathscr{A}_m')$ be a discrete multivariate random variable, and let $\Omega_{i0}' \subset \Omega_i'$ be finite or countable with $P(Y_i \in \Omega_{i0}') = 1$, $i = 1, \dots, n$. Then the following propositions are equivalent to each other:

(i) $\underset{P}{\perp\!\!\!\perp} Y_1, \dots, Y_m \mid X.$

(ii) For all $i = 1, \dots, m$ and all $y \in \Omega'_{i0}$:

$$P(Y_i = y \mid X, Y_1, \dots, Y_{i-1}, Y_{i+1}, \dots, Y_m) \underset{P}{=} P(Y_i = y \mid X).$$

(iii) For all $i = 2, \dots, m$ and all $y \in \Omega'_{i0}$:

$$P(Y_i = y \mid X, Y_1, \dots, Y_{i-1}) \underset{P}{=} P(Y_i = y \mid X).$$

(iv) For all $i = 1, \dots, m-1$ and all $y \in \Omega'_{i0}$:

$$P(Y_i = y \mid X, Y_{i+1}, \dots, Y_m) \underset{P}{=} P(Y_i = y \mid X).$$

(Proof p. 500)

Example 16.51 [Joe and Ann with latent abilities – continued] Table 13.1 describes an example in which three dichotomous variables Y_1, Y_2, Y_3 satisfy the assumptions (13.36) and (13.37) that define a Rasch model. Because Y is dichotomous with values 0 and 1, the second of these assumptions,

$$P(Y_i = 1 \mid U, Y_1, \dots, Y_{i-1}, Y_{i+1}, \dots, Y_m) = P(Y_i = 1 \mid U), \quad \forall i = 1, \dots, m, \quad (16.54)$$

is equivalent to $\underset{P}{\perp\!\!\!\perp} (Y_1, \dots, Y_m) \mid U$ (see Cor. 16.50) and, because

$$P(Y_i = 0 \mid U, Y_1, \dots, Y_{i-1}, Y_{i+1}, \dots, Y_m) = 1 - P(Y_i = 1 \mid U, Y_1, \dots, Y_{i-1}, Y_{i+1}, \dots, Y_m),$$

also to

$$P(Y_i = 0 \mid U, Y_1, \dots, Y_{i-1}, Y_{i+1}, \dots, Y_m) = P(Y_i = 0 \mid U), \quad \forall i = 1, \dots, m.$$

Using the probabilities displayed in Table 13.1, Equation (4.2), and Notation (5.4), Equation (16.54) can be illustrated for $i = 1$ as follows:

$$P(Y_1 = 1 \mid U = Joe, Y_2 = 1, Y_3 = 1) = \frac{P(Y_1 = 1, U = Joe, Y_2 = 1, Y_3 = 1)}{P(U = Joe, Y_2 = 1, Y_3 = 1)}$$

$$= \frac{.0492}{.0492 + .0492} = .5$$

$$= P(Y_1 = 1 \mid U = Joe).$$

Analogously, we obtain

$$P(Y_1=1 \mid U=u, Y_2=y_2, Y_3=y_3) = \frac{P(Y_1=1, U=u, Y_2=y_2, Y_3=y_3)}{P(U=Joe, Y_2=y_2, Y_3=y_3)}$$
$$= P(Y_1=1 \mid U=u)$$

for all other combinations of values of U, Y_2, and Y_3. The corresponding equations for $P(Y_2=1 \mid U=Joe, Y_1, Y_3)$ and $P(Y_3=1 \mid U=Joe, Y_1, Y_2)$, which are also assumed with (16.54), can be checked analogously.

Assumptions (13.36) and (13.37) and the definition of ξ [see (13.38)] also imply ξ-conditional independence of the random variables $Y_1, \ldots, Y_m$, that is, they imply

$$P(Y_i=1 \mid \xi, Y_1, \ldots, Y_{i-1}, Y_{i+1}, \ldots, Y_m) = P(Y_i=1 \mid \xi), \quad \forall i = 1, \ldots, m \qquad (16.55)$$

(see Exercise 16.5), which is equivalent to

$$\forall J \subset \{1, \ldots, m\}, \forall y_i \in \{0, 1\}: \quad P\left(\bigcap_{i \in J} \{Y_i = y_i\} \,\middle|\, \xi \right) = \prod_{i \in J} P(Y_i = y_i \mid \xi) \qquad (16.56)$$

(see Exercise 16.6).

Applying Equation (16.56) for $J = \{1, 2, 3\}$ and Equation (13.40) to the conditional probabilities $P(Y_i=1 \mid U) = P(Y_i=1 \mid \xi)$ displayed in Table 13.1 yields the functions

$$P(Y_1=0, Y_2=1, Y_3=1 \mid \xi) = P(Y_1=0 \mid \xi) \cdot P(Y_2=1 \mid \xi) \cdot P(Y_3=1 \mid \xi)$$
$$= \left(1 - \frac{\exp(\xi - \beta_1)}{1 + \exp(\xi - \beta_1)} \right) \cdot \frac{\exp(\xi - \beta_2)}{1 + \exp(\xi - \beta_2)} \cdot \frac{\exp(\xi - \beta_3)}{1 + \exp(\xi - \beta_3)}$$

(see Fig. 16.1), where the difficulty parameters β_i have been computed in Equation (13.42). The conditional probabilities $P(Y_1=1, Y_2=0, Y_3=1 \mid \xi)$ and $P(Y_1=1, Y_2=1, Y_3=0 \mid \xi)$ are computed analogously.

These functions are called *likelihood functions*. The graphs of these functions in Figure 16.1 illustrate that the most likely ability parameter of a person solving exactly two items is the value of ξ at which the functions $P(Y_1=y_1, Y_2=y_2, Y_3=y_3 \mid \xi)$ have their maximum.

The likelihood functions can also be used for the estimation of the ability of a person, for which the score pattern $(0, 1, 1)$, $(1, 0, 1)$, or $(1, 1, 0)$ is observed. The value of ξ (ability) at which these functions have their maximum, the *maximum likelihood estimate*, is the most likely ability score of such a person. For each value of ξ other than this maximum likelihood estimate, the probability of such a score pattern is smaller. ◁

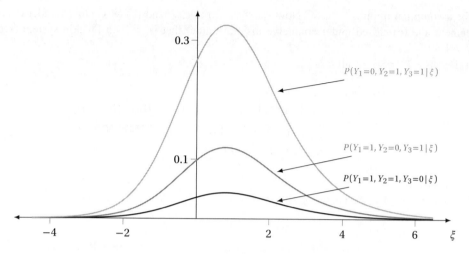

Figure 16.1 Likelihood functions for the response patterns of three variables Y_i satisfying the Rasch model.

16.7 Proofs

Proof of Lemma 16.4

If $B \in \mathscr{C}$, then 1_B is $(\mathscr{C}, \mathscr{B})$-measurable (see Example 2.12). Hence,

$$
\begin{aligned}
P(A \cap B \mid \mathscr{C}) &\underset{P}{=} E(1_{A \cap B} \mid \mathscr{C}) \\
&\underset{P}{=} E(1_A \cdot 1_B \mid \mathscr{C}) && [(1.33)] \\
&\underset{P}{=} E(1_A \mid \mathscr{C}) \cdot 1_B && [\text{Box } 10.1 \text{ (xiv)}] \\
&\underset{P}{=} E(1_A \mid \mathscr{C}) \cdot E(1_B \mid \mathscr{C}). && [\text{Box } 10.1 \text{ (vii)}]
\end{aligned}
$$

If we assume $A \in \mathscr{C}$, then the proof is analogous, exchanging the roles of A and B. Proposition (16.5) is an immediate implication of (16.4).

Proof of Corollary 16.6

$$
\begin{aligned}
& A \underset{P}{\perp\!\!\!\perp} B \mid Z \\
\Leftrightarrow\; & P(A \cap B \mid Z) \underset{P}{=} P(A \mid Z) \cdot P(B \mid Z) && [\text{Def. } 16.2 \text{ (ii)}] \\
\Leftrightarrow\; & P(A \cap B \mid Z=z) = P(A \mid Z=z) \cdot P(B \mid Z=z), \quad \text{for } P_Z\text{-a.a. } z \in \Omega'_Z. && [\text{Cor. } 10.39 \text{ (i)}]
\end{aligned}
$$

Proof of Lemma 16.12

(i) $\sigma(\mathscr{D}) \underset{P}{\perp\!\!\!\perp} \sigma(\mathscr{E}) \mid \mathscr{C} \Rightarrow \mathscr{D} \underset{P}{\perp\!\!\!\perp} \mathscr{E} \mid \mathscr{C}$ follows from Definition 16.10 (i), because $\mathscr{D} \subset \sigma(\mathscr{D})$ and $\mathscr{E} \subset \sigma(\mathscr{E})$.

(ii) $\mathscr{D} \underset{P}{\perp\!\!\!\perp} \mathscr{E} \mid \mathscr{C} \Rightarrow \sigma(\mathscr{D}) \underset{P}{\perp\!\!\!\perp} \sigma(\mathscr{E}) \mid \mathscr{C}$. Assume $\mathscr{D} \underset{P}{\perp\!\!\!\perp} \mathscr{E} \mid \mathscr{C}$, apply Equation (10.2), and define the set of events

$$
\mathscr{F} := \{A \in \mathscr{A} : \forall B \in \mathscr{E} : E(1_A \cdot 1_B \mid \mathscr{C}) \underset{P}{=} E(1_A \mid \mathscr{C}) \cdot E(1_B \mid \mathscr{C})\}.
$$

The assumption implies $\mathcal{D} \subset \mathcal{F}$. Now we show (a) $\Omega \in \mathcal{F}$ and (b) $\mathcal{F}$ is closed under complement and (c) closed under countable disjoint union, that is, $\mathcal{F}$ is a Dynkin system (see Def. 1.40).

(a) Because $1_\Omega = 1$, for all $B \in \mathscr{E}$,

$$
\begin{aligned}
E(1_\Omega \cdot 1_B \mid \mathscr{C}) &\underset{P}{=} E(1 \cdot 1_B \mid \mathscr{C}) \\
&\underset{P}{=} 1 \cdot E(1_B \mid \mathscr{C}) && \text{[Box 10.1 (iii)]} \\
&\underset{P}{=} E(1 \mid \mathscr{C}) \cdot E(1_B \mid \mathscr{C}) && \text{[Box 10.1 (i)]} \\
&\underset{P}{=} E(1_\Omega \mid \mathscr{C}) \cdot E(1_B \mid \mathscr{C}).
\end{aligned}
$$

Hence, $\Omega \in \mathcal{F}$.

(b) Assume that $A \in \mathcal{F}$. Then, for all $B \in \mathscr{E}$,

$$
\begin{aligned}
E(1_{A^c} \cdot 1_B \mid \mathscr{C}) &\underset{P}{=} E[(1 - 1_A) \cdot 1_B \mid \mathscr{C}] \\
&\underset{P}{=} E(1_B \mid \mathscr{C}) - E(1_A \cdot 1_B \mid \mathscr{C}) && \text{[Box 10.1 (xvi)]} \\
&\underset{P}{=} E(1_B \mid \mathscr{C}) - E(1_A \mid \mathscr{C}) \cdot E(1_B \mid \mathscr{C}) && \text{[def. of $\mathcal{F}$]} \\
&\underset{P}{=} [1 - E(1_A \mid \mathscr{C})] \cdot E(1_B \mid \mathscr{C}) \\
&\underset{P}{=} E(1 - 1_A \mid \mathscr{C}) \cdot E(1_B \mid \mathscr{C}) && \text{[Box 10.1 (i), (xvi)]} \\
&\underset{P}{=} E(1_{A^c} \mid \mathscr{C}) \cdot E(1_B \mid \mathscr{C}).
\end{aligned}
$$

Hence, $A \in \mathcal{F}$ implies $A^c \in \mathcal{F}$.

(c) Assume $A_1, A_2, \ldots \in \mathcal{F}$ and $A_i \cap A_j = \emptyset$ for $i \neq j$. Then, for all $B \in \mathscr{E}$,

$$
\begin{aligned}
E\left(1_{\bigcup_{i=1}^\infty A_i} \cdot 1_B \,\middle|\, \mathscr{C}\right) &\underset{P}{=} E\left[\left(\sum_{i=1}^\infty 1_{A_i}\right) \cdot 1_B \,\middle|\, \mathscr{C}\right] && \text{[(1.37)]} \\
&\underset{P}{=} E\left(\sum_{i=1}^\infty (1_{A_i} \cdot 1_B) \,\middle|\, \mathscr{C}\right) \\
&\underset{P}{=} \sum_{i=1}^\infty E(1_{A_i} \cdot 1_B \mid \mathscr{C}) && \text{[Th. 10.21 (ii)]} \\
&\underset{P}{=} \sum_{i=1}^\infty E(1_{A_i} \mid \mathscr{C}) \cdot E(1_B \mid \mathscr{C}) && \text{[def. of $\mathcal{F}$]} \\
&\underset{P}{=} E\left(\sum_{i=1}^\infty 1_{A_i} \,\middle|\, \mathscr{C}\right) \cdot E(1_B \mid \mathscr{C}) && \text{[Th. 10.21 (ii)]} \\
&\underset{P}{=} E\left(1_{\bigcup_{i=1}^\infty A_i} \,\middle|\, \mathscr{C}\right) \cdot E(1_B \mid \mathscr{C}). && \text{[(1.37)]}
\end{aligned}
$$

Hence, $\bigcup_{i=1}^\infty A_i \in \mathcal{F}$. This proves that $\mathcal{F}$ is a Dynkin system (see Def. 1.40). Because $\mathcal{D} \subset \mathcal{F}$, we can conclude $\delta(\mathcal{D}) \subset \mathcal{F}$, where $\delta(\mathcal{D})$ denotes the Dynkin system generated by $\mathcal{D}$, that is, the smallest Dynkin system that contains $\mathcal{D}$ (see Rem. 1.14). According to Theorem 1.41 (ii), $\cap$-stability of $\mathcal{D}$ implies $\delta(\mathcal{D}) = \sigma(\mathcal{D})$. Hence, $\sigma(\mathcal{D}) \subset \mathcal{F}$.

Now we have shown: $\mathcal{D} \underset{P}{\perp\!\!\!\perp} \mathscr{E} \mid \mathscr{C} \Rightarrow \sigma(\mathcal{D}) \underset{P}{\perp\!\!\!\perp} \mathscr{E} \mid \mathscr{C}$. Because $\sigma(\mathcal{D}) \underset{P}{\perp\!\!\!\perp} \mathscr{E} \mid \mathscr{C} \Leftrightarrow \mathscr{E} \underset{P}{\perp\!\!\!\perp} \sigma(\mathcal{D}) \mid \mathscr{C}$, this also implies $\sigma(\mathcal{D}) \underset{P}{\perp\!\!\!\perp} \mathscr{E} \mid \mathscr{C} \Rightarrow \sigma(\mathcal{D}) \underset{P}{\perp\!\!\!\perp} \sigma(\mathscr{E}) \mid \mathscr{C}$.

Proof of Theorem 16.22

$$\mathscr{E} \underset{P}{\perp\!\!\!\perp} \mathscr{D} \mid \mathscr{C}$$

$$\Leftrightarrow E(1_A \cdot 1_B \mid \mathscr{C}) \underset{P}{=} E(1_A \mid \mathscr{C}) \cdot E(1_B \mid \mathscr{C}), \quad \forall (A, B) \in \mathscr{E} \times \mathscr{D} \qquad [(10.2), \text{Def. } 16.10 \text{ (i)}]$$

$$\Leftrightarrow E(1_A \cdot W \mid \mathscr{C}) \underset{P}{=} E(1_A \mid \mathscr{C}) \cdot E(W \mid \mathscr{C}), \quad \forall A \in \mathscr{E} \text{ and all nonnegative } \mathscr{D}\text{-measurable } W$$

$$\Leftrightarrow E(1_A \mid \mathscr{C}, \mathscr{D}) \underset{P}{=} E(1_A \mid \mathscr{C}), \quad \forall A \in \mathscr{E} \qquad [\text{Th. } 10.51]$$

$$\Leftrightarrow P(A \mid \mathscr{C}, \mathscr{D}) \underset{P}{=} P(A \mid \mathscr{C}), \quad \forall A \in \mathscr{E}. \qquad [(10.2)]$$

The equivalence of the second and third propositions is obtained by using the proof of Theorem 16.32, with W taking the role of X, where $W = \sum_{i=1}^{\infty} \alpha_i 1_{A_i}$ and $Y = 1_A$.

Proof of Theorem 16.26

$\Rightarrow$ This is an implication of Corollary 16.24.

$\Leftarrow$ For $A'_Y \in \mathscr{A}'_Y$, define $A'_{Y0} := A'_Y \cap \Omega'_{Y0}$, which is finite or countable. Then, $1_{Y \in A'_Y} \underset{P}{=} 1_{Y \in A'_{Y0}}$. Furthermore,

$$P(Y = y \mid \mathscr{C}, \mathscr{D}) \underset{P}{=} P(Y = y \mid \mathscr{C}), \quad \forall y \in \Omega'_{Y0}$$

$$\Rightarrow E\left(\sum_{y \in A'_{Y0}} 1_{Y=y} \,\middle|\, \mathscr{C}, \mathscr{D} \right) \underset{P}{=} E\left(\sum_{y \in A'_{Y0}} 1_{Y=y} \,\middle|\, \mathscr{C} \right) \qquad [(10.2), \text{Th. } 10.21 \text{ (ii)}]$$

$$\Rightarrow E(1_{Y \in A'_{Y0}} \mid \mathscr{C}, \mathscr{D}) \underset{P}{=} E(1_{Y \in A'_{Y0}} \mid \mathscr{C}) \qquad [(1.36)\,(1.37)]$$

$$\Rightarrow E(1_{Y \in A'_Y} \mid \mathscr{C}, \mathscr{D}) \underset{P}{=} E(1_{Y \in A'_Y} \mid \mathscr{C}) \qquad [\text{Box } 10.1 \text{ (ix)}]$$

$$\Rightarrow Y \underset{P}{\perp\!\!\!\perp} \mathscr{D} \mid \mathscr{C}. \qquad [(10.2), \text{Th. } 16.22]$$

Proof of Theorem 16.32

Case 1: X and Y nonnegative. If $X \colon (\Omega, \mathscr{A}, P) \to (\mathbb{R}, \mathscr{B})$ is a nonnegative random variable, then it is also $(\sigma(X), \mathscr{B})$-measurable, and therefore, according to Theorem 3.39 (ii), there are a sequence $A_1, A_2, \ldots \in \sigma(X)$ and a sequence of nonnegative real numbers $\alpha_1, \alpha_2, \ldots$ such that $X = \sum_{i=1}^{\infty} \alpha_i 1_{A_i}$. Analogously, a nonnegative random variable Y can be represented as $Y = \sum_{j=1}^{\infty} \beta_j 1_{B_j}$, $\beta_j \geq 0$, $B_j \in \sigma(Y)$, $j \in \mathbb{N}$. Hence,

$$E(X \cdot Y \mid \mathscr{C}) \underset{P}{=} E\left(\sum_{i=1}^{\infty} \alpha_i 1_{A_i} \cdot \sum_{j=1}^{\infty} \beta_j 1_{B_j} \,\middle|\, \mathscr{C} \right)$$

$$\underset{P}{=} E\left(\sum_{i=1}^{\infty} \sum_{j=1}^{\infty} \alpha_i \beta_j 1_{A_i} \cdot 1_{B_j} \,\middle|\, \mathscr{C} \right)$$

$$\underset{P}{=} \sum_{i=1}^{\infty} \sum_{j=1}^{\infty} \alpha_i \beta_j \, E(1_{A_i} \cdot 1_{B_j} \mid \mathscr{C}) \qquad [\text{Th. } 10.21 \text{ (ii)}, \text{Box } 10.1 \text{ (iii)}]$$

$$\underset{P}{=} \sum_{i=1}^{\infty} \sum_{j=1}^{\infty} \alpha_i \beta_j \, E(1_{A_i} \mid \mathscr{C}) \cdot E(1_{B_j} \mid \mathscr{C}) \qquad [\sigma(X) \underset{P}{\perp\!\!\!\perp} \sigma(Y) \mid \mathscr{C}]$$

$$\underset{P}{=} E\left(\sum_{i=1}^{\infty} \alpha_i 1_{A_i} \,\middle|\, \mathscr{C} \right) \cdot E\left(\sum_{j=1}^{\infty} \beta_j 1_{B_j} \,\middle|\, \mathscr{C} \right) \qquad [\text{Th. } 10.21 \text{ (ii)}, \text{Box } 10.1 \text{ (iii)}]$$

$$\underset{P}{=} E(X \mid \mathscr{C}) \cdot E(Y \mid \mathscr{C}).$$

Case 2: Let X^+, Y^+ and X^-, Y^- be the *positive* and *negative parts* of X and Y (see Def. 2.62). These functions are nonnegative and measurable. Note that $\sigma(X^+)$, $\sigma(X^-) \subset \sigma(X)$ and $\sigma(Y^+)$, $\sigma(Y^-) \subset \sigma(Y)$ (see Th. 2.66). Hence, (16.33) holds for the four pairs of random variables (X^+, Y^+), (X^+, Y^-), (X^-, Y^+), and (X^-, Y^-). If $E(X)$ and $E(Y)$ are finite, then $E(X^+)$, $E(X^-)$, $E(Y^+)$, and $E(Y^-)$ are finite as well, and according to Box 10.1 (x), there are real-valued versions of the conditional expectations that are used in the following equations. Because $X = X^+ - X^-$ and $Y = Y^+ - Y^-$, it can be shown that (16.33) also holds if we assume finite expectations:

$$E(X \mid \mathcal{C}) \cdot E(Y \mid \mathcal{C})$$

$$\underset{P}{=} E(X^+ - X^- \mid \mathcal{C}) \cdot E(Y^+ - Y^- \mid \mathcal{C})$$

$$\underset{P}{=} [E(X^+ \mid \mathcal{C}) - E(X^- \mid \mathcal{C})] \cdot [E(Y^+ \mid \mathcal{C}) - E(Y^- \mid \mathcal{C})] \qquad \text{[Box 10.1 (xvi)]}$$

$$\underset{P}{=} E(X^+ \mid \mathcal{C}) \cdot E(Y^+ \mid \mathcal{C}) - E(X^+ \mid \mathcal{C}) \cdot E(Y^- \mid \mathcal{C})$$
$$\quad - E(X^- \mid \mathcal{C}) \cdot E(Y^+ \mid \mathcal{C}) + E(X^- \mid \mathcal{C}) \cdot E(Y^- \mid \mathcal{C})$$

$$\underset{P}{=} E(X^+ \cdot Y^+ \mid \mathcal{C}) - E(X^+ \cdot Y^- \mid \mathcal{C}) - E(X^- \cdot Y^+ \mid \mathcal{C}) + E(X^- \cdot Y^- \mid \mathcal{C}) \qquad \text{[(16.33)]}$$

$$\underset{P}{=} E(X^+ \cdot Y^+ + X^+ \cdot Y^- - X^- \cdot Y^+ + X^- \cdot Y^- \mid \mathcal{C}) \qquad \text{[Box 10.1 (xvi)]}$$

$$\underset{P}{=} E[(X^+ - X^-) \cdot (Y^+ - Y^-) \mid \mathcal{C}]$$

$$\underset{P}{=} E(X \cdot Y \mid \mathcal{C}).$$

Proof of Theorem 16.34

Part 1. If $Y: (\Omega, \mathcal{A}, P) \to (\overline{\mathbb{R}}, \overline{\mathcal{B}})$ is nonnegative, then it is also $(\sigma(Y), \overline{\mathcal{B}})$-measurable, and therefore, according to Theorem 3.19 (ii), there is a sequence $A_1, A_2, \ldots \in \sigma(Y)$ and a sequence of nonnegative real numbers $\alpha_1, \alpha_2, \ldots$ such that

$$Y = \sum_{i=1}^{\infty} \alpha_i 1_{A_i}, \quad \alpha_i \geq 0, \; A_i \in \sigma(Y).$$

Hence, if $\sigma(Y) \subset \mathcal{E}$, then

$$\mathcal{E} \underset{P}{\perp\!\!\!\perp} \mathcal{D} \mid \mathcal{C} \; \Rightarrow \; P(A \mid \mathcal{C}, \mathcal{D}) \underset{P}{=} P(A \mid \mathcal{C}), \quad \forall A \in \mathcal{E} \qquad \text{[(16.23)]}$$

$$\Rightarrow \; \sum_{i=1}^{\infty} \alpha_i E(1_{A_i} \mid \mathcal{C}, \mathcal{D}) \underset{P}{=} \sum_{i=1}^{\infty} \alpha_i E(1_{A_i} \mid \mathcal{C}) \qquad \text{[(10.2), (2.39)]}$$

$$\Rightarrow \; E\left(\sum_{i=1}^{\infty} \alpha_i 1_{A_i} \mid \mathcal{C}, \mathcal{D} \right) \underset{P}{=} E\left(\sum_{i=1}^{\infty} \alpha_i E(1_{A_i} \mid \mathcal{C}) \right) \qquad \text{[Th. 10.21 (ii)]}$$

$$\Rightarrow \; E(Y \mid \mathcal{C}, \mathcal{D}) \underset{P}{=} E(Y \mid \mathcal{C}). \qquad \text{[Th. 3.19 (ii)]}$$

Part 2. If Y has a finite expectation, we use the positive part Y^+ and the negative part Y^- of

$$Y = Y^+ - Y^-$$

(see Rem. 2.62). Because both Y^+ and Y^- are nonnegative and $\sigma(Y^+) \subset \sigma(Y) \subset \mathcal{E}$, $\sigma(Y^-) \subset \sigma(Y) \subset \mathcal{E}$, we can use Part 1 of this proof. Hence,

$$
\begin{aligned}
E(Y \mid \mathcal{C}, \mathcal{D}) &\underset{P}{=} E(Y^+ - Y^- \mid \mathcal{C}, \mathcal{D}) && \text{[Rem. 2.62]}\\
&\underset{P}{=} E(Y^+ \mid \mathcal{C}, \mathcal{D}) - E(Y^- \mid \mathcal{C}, \mathcal{D}) && \text{[Box 10.1 (xvi)]}\\
&\underset{P}{=} E(Y^+ \mid \mathcal{C}) - E(Y^- \mid \mathcal{C}) && \text{[Part 1]}\\
&\underset{P}{=} E(Y^+ - Y^- \mid \mathcal{C}) && \text{[Box 10.1 (xvi)]}\\
&\underset{P}{=} E(Y \mid \mathcal{C}). && \text{[Rem. 2.62]}
\end{aligned}
$$

Proof of Theorem 16.37

We thank Ernesto San Martin for hinting at proposition (iv). In this proof, we use the notation $Y \in \langle \mathcal{E} \rangle^+$ for 'Y is nonnegative and $\sigma(Y) \subset \mathcal{E}$'.

(i) $\Rightarrow$ (ii) This implication immediately follows from Theorem 16.34.
(ii) $\Rightarrow$ (i)

$$
\begin{aligned}
\text{(ii)} &\Rightarrow \forall\, B \in \mathcal{E}: E(1_B \mid \mathcal{C}, \mathcal{D}) \underset{P}{=} E(1_B \mid \mathcal{C}) && [1_B \in \langle \mathcal{E} \rangle^+]\\
&\Rightarrow \forall\, B \in \mathcal{E}: P(B \mid \mathcal{C}, \mathcal{D}) \underset{P}{=} P(B \mid \mathcal{C}) && [(10.2)]\\
&\Rightarrow \mathcal{E} \underset{P}{\perp\!\!\!\perp} \mathcal{D} \mid \mathcal{C}. && [(16.23)]
\end{aligned}
$$

(ii) $\Leftrightarrow$ (iii) This immediately follows from Theorem 10.51.
(iv) $\Rightarrow$ (iii) Let $W, X, Y \colon (\Omega, \mathcal{A}, P) \to (\mathbb{R}, \mathcal{B})$ be random variables. Then, for all $X \in \langle \mathcal{C} \rangle^+$, $W \in \langle \mathcal{D} \rangle^+$, $Y \in \langle \mathcal{E} \rangle^+$,

$$
\begin{aligned}
&E[X \cdot E(W \cdot Y \mid \mathcal{C})]\\
&= E[E(X \cdot W \cdot Y \mid \mathcal{C})] && \text{[Box 10.1 (xiv)]}\\
&= E(X \cdot W \cdot Y) && \text{[Box 10.1 (iv)]}\\
&= E[E(X \cdot W \cdot Y \mid \mathcal{D})] && \text{[Box 10.1 (iv)]}\\
&= E[W \cdot E(X \cdot Y \mid \mathcal{D})] && \text{[Box 10.1 (xiv)]}\\
&= E(W \cdot E[E(X \cdot Y \mid \mathcal{C}) \mid \mathcal{D}]) && [\text{(iv)}, \sigma(X \cdot Y) \subset \sigma(\mathcal{C}, \mathcal{E})]\\
&= E(E[W \cdot E(X \cdot Y \mid \mathcal{C}) \mid \mathcal{D}]) && [\sigma(W) \subset \sigma(\mathcal{D}), \text{Box 10.1 (xiv)}]\\
&= E(W \cdot E(X \cdot Y \mid \mathcal{C})) && \text{[Box 10.1 (iv)]}\\
&= E(E[W \cdot E(X \cdot Y \mid \mathcal{C}) \mid \mathcal{C}]) && \text{[Box 10.1 (iv)]}\\
&= E(X \cdot E(W \mid \mathcal{C}) \cdot E(Y \mid \mathcal{C})). && [\sigma(X), \sigma(E(Y \mid \mathcal{C})) \subset \mathcal{C}, \text{Box 10.1 (xiv)}]
\end{aligned}
$$

Because this equation holds for all $X \in \langle \mathcal{C} \rangle^+$, it also holds for all 1_C, $C \in \mathcal{C}$, and using Definition 10.2 (b) we conclude: $E(W \cdot Y \mid \mathcal{C}) \underset{P}{=} E(W \mid \mathcal{C}) \cdot E(Y \mid \mathcal{C})$.

(i) $\Rightarrow$ (iv) Note that we have already shown that (i) and (ii) are equivalent to each other. Therefore, we can use (ii) in this part of the proof. For all nonnegative random variables Y with finite expectation and $\sigma(Y) \subset \sigma(\mathscr{C}, \mathscr{E})$,

$$\mathscr{E} \underset{P}{\perp\!\!\!\perp} \mathscr{D} \mid \mathscr{C}$$

$$\Rightarrow \sigma(\mathscr{C}, \mathscr{E}) \underset{P}{\perp\!\!\!\perp} \mathscr{D} \mid \mathscr{C} \qquad\qquad \text{[Box 16.2 (vii)]}$$

$$\Rightarrow E(Y \mid \mathscr{C}) \underset{P}{=} E(Y \mid \mathscr{C}, \mathscr{D}) \qquad\qquad \text{[(ii)]}$$

$$\Rightarrow E[E(Y \mid \mathscr{C}) \mid \mathscr{D}] \underset{P}{=} E[E(Y \mid \mathscr{C}, \mathscr{D}) \mid \mathscr{D}] \qquad \text{[Box 10.1 (ix)]}$$

$$\Rightarrow E[E(Y \mid \mathscr{C}) \mid \mathscr{D}] \underset{P}{=} E(Y \mid \mathscr{D}). \qquad\qquad \text{[Box 10.1 (v)]}$$

Proof of Theorem 16.38

$E^{Z=z}(Y \mid \mathscr{C})$ and $E(Y \mid \mathscr{C})$ are both $\mathscr{C}$-measurable. Hence, it suffices to show that condition (b) of Definition 14.7 holds for $E(Y \mid \mathscr{C})$, that is, it suffices to show

$$E^{Z=z}[1_C \cdot E(Y \mid \mathscr{C})] = E^{Z=z}(1_C \cdot Y), \quad \forall\, C \in \mathscr{C}. \qquad (16.57)$$

Now,

$$E[1_A \cdot E(Y \mid \mathscr{C}, Z)] = E(1_A \cdot Y), \quad \forall\, A \in \sigma(\mathscr{C}, Z). \qquad \text{[Def. 10.2 (b)]}$$

Because, for $C \in \sigma(\mathscr{C}, Z)$, $\{Z=z\} \cap C \in \sigma(\mathscr{C}, Z)$, and $1_{\{Z=z\} \cap C} = 1_{Z=z} \cdot 1_C$, Definition 10.2 (b) implies

$$E[1_{Z=z} \cdot 1_C \cdot E(Y \mid \mathscr{C}, Z)] = E(1_{Z=z} \cdot 1_C \cdot Y), \quad \forall\, C \in \mathscr{C}.$$

Dividing both sides by $P(Z=z)$ yields

$$\frac{1}{P(Z=z)} E[1_{Z=z} \cdot 1_C \cdot E(Y \mid \mathscr{C}, Z)] = \frac{1}{P(Z=z)} E(1_{Z=z} \cdot 1_C \cdot Y), \quad \forall\, C \in \mathscr{C},$$

which, according to Equation (9.11), is equivalent to

$$E^{Z=z}[1_C \cdot E(Y \mid \mathscr{C}, Z)] = E^{Z=z}(1_C \cdot Y), \quad \forall\, C \in \mathscr{C}.$$

According to Theorem 16.34, assuming $Z \underset{P}{\perp\!\!\!\perp} Y \mid \mathscr{C}$ implies $E(Y \mid \mathscr{C}, Z) \underset{P}{=} E(Y \mid \mathscr{C})$ and Corollary 5.22 yields $E(Y \mid \mathscr{C}, Z) \underset{P^{Z=z}}{=} E(Y \mid \mathscr{C})$. Now, Equation (16.57) follows from Box 6.1 (viii).

Proof of Theorem 16.44

$\Leftarrow$ This proposition immediately follows from Definition 16.41 (i).

$\Rightarrow$ If $(\mathscr{E}_i, i \in I)$ is a family of $\cap$-stable set systems with $\mathscr{E}_i \subset \mathscr{A}, i \in I$, and $J \subset I$ is finite, then,

$$\underset{P}{\perp\!\!\!\perp} (\mathscr{E}_i, i \in I) \mid \mathscr{C} \Rightarrow \underset{P}{\perp\!\!\!\perp} (\mathscr{E}_i, i \in J) \mid \mathscr{C} \qquad [(16.46)]$$

$$\Rightarrow \underset{P}{\perp\!\!\!\perp} (\sigma(\mathscr{E}_i), i \in J) \mid \mathscr{C}.$$

This last implication follows from repeatedly applying (ii) (b) of the proof of Lemma 16.12. Because the implications above hold for all finite $J \subset I$, this yields $\underset{P}{\perp\!\!\!\perp} (\sigma(\mathscr{E}_i), i \in I) \mid \mathscr{C}$.

Proof of Corollary 16.47

$\underset{P}{\perp\!\!\!\perp} (X_i, i = 1, \ldots, n) \mid \mathscr{C} \Rightarrow (16.51)$ immediately follows from Definitions 16.45 and 16.41, because $\{X_i = x_i\} \in X_i^{-1}(\mathscr{A}_i')$ for all $x_i \in \Omega_{i0}', i = 1, \ldots, n$.

$(16.51) \Rightarrow \underset{P}{\perp\!\!\!\perp} (X_i, i = 1, \ldots, n) \mid \mathscr{C}$. For $i = 1, \ldots, n$, let $A_i' \in \mathscr{A}_i'$ and $A_{i0}' := A_i' \cap \Omega_{i0}'$, which implies that the A_{i0}' are finite or countable. Then,

$$1_{X_i \in A_i'} \underset{P}{=} 1_{X_i \in A_{i0}'} \quad \forall i = 1, \ldots, n. \qquad (16.58)$$

Furthermore,

$$P(X_1 \in A_1', \ldots, X_n \in A_n' \mid \mathscr{C})$$

$$\underset{P}{=} E(1_{X_1 \in A_{10}'} \cdot \ldots \cdot 1_{X_n \in A_{n0}'} \mid \mathscr{C}) \qquad [(16.58), \text{Box } 10.1 \text{ (ix)}]$$

$$\underset{P}{=} E\left(\sum_{x_1 \in A_{10}'} \cdots \sum_{x_n \in A_{n0}'} 1_{X_1 = x_1} \cdot \ldots \cdot 1_{X_n = x_n} \,\middle|\, \mathscr{C} \right) \qquad [(1.36)\ (1.37)]$$

$$\underset{P}{=} \sum_{x_1 \in A_{10}'} \cdots \sum_{x_n \in A_{n0}'} E(1_{X_1 = x_1} \cdot \ldots \cdot 1_{X_n = x_n} \mid \mathscr{C}) \qquad [\text{Th. } 10.21 \text{ (ii)}]$$

$$\underset{P}{=} \sum_{x_1 \in A_{10}'} \cdots \sum_{x_n \in A_{n0}'} E(1_{X_1 = x_1} \mid \mathscr{C}) \cdot \ldots \cdot E(1_{X_n = x_n} \mid \mathscr{C}) \qquad [(16.51)]$$

$$\underset{P}{=} \left(\sum_{x_1 \in A_{10}'} E(1_{X_1 = x_1} \mid \mathscr{C}) \right) \cdot \ldots \cdot \left(\sum_{x_n \in A_{n0}'} E(1_{X_n = x_n} \mid \mathscr{C}) \right)$$

$$\underset{P}{=} E\left(\sum_{x_1 \in A_{10}'} 1_{X_1 = x_1} \,\middle|\, \mathscr{C} \right) \cdot \ldots \cdot E\left(\sum_{x_n \in A_{n0}'} 1_{X_n = x_n} \,\middle|\, \mathscr{C} \right) \qquad [\text{Th. } 10.21 \text{ (ii)}]$$

$$\underset{P}{=} E(1_{X_1 \in A_{10}'} \mid \mathscr{C}) \cdot \ldots \cdot E(1_{X_n \in A_{n0}'} \mid \mathscr{C}) \qquad [(1.36)\ (1.37)]$$

$$\underset{P}{=} P(X_1 \in A_1' \mid \mathscr{C}) \cdot \ldots \cdot P(X_n \in A_n' \mid \mathscr{C}). \qquad [(16.58), \text{Box } 10.1 \text{ (ix)}]$$

Now, Remark 16.46 yields the result.

Proof of Lemma 16.49

This proof is analogous to the proof of Theorem 6.5 of Bauer (1996). For all $j \in J$, consider

$$\mathscr{E}_j := \left\{ \bigcap_{k=1}^{n} A_{i_k} : A_{i_k} \in \sigma(X_{i_k}), \{i_1, \ldots, i_n\} \subset I_j, n \in \mathbb{N} \right\}.$$

The set system $\mathscr{E}_j$ is ∩-stable because the intersection of two finite intersections of sets is again a finite intersection, and the σ-algebras $\sigma(X_{i_k})$ are ∩-stable as well. Furthermore, if $\underset{P}{\perp\!\!\!\perp}\,(\sigma(X_i), i \in I) \mid \mathscr{C}$, then according to Definitions 16.45 (i) and 16.41,

$$\underset{P}{\perp\!\!\!\perp}\,(\mathscr{E}_j, j \in J) \mid \mathscr{C},$$

because the intersection of finitely many sets that are finite intersections is again an intersection of finitely many sets. Because $\mathscr{E}_j \subset \sigma(X_i, i \in I_j)$ for all $j \in J$ and $\sigma(X_i) \subset \mathscr{E}_j$ for all $i \in I_j, j \in J$, we can conclude $\sigma(\mathscr{E}_j) = \sigma(X_i, i \in I_j)$ (see Rem. 2.39 and Def. 1.13). Therefore, applying Theorem 16.44 completes the proof.

Proof of Corollary 16.50

(i) ⇒ (ii) For all $i = 1, \ldots, m$ and all $y_i \in \Omega'_{i0}$,

$$\underset{P}{\perp\!\!\!\perp}\,Y_1, \ldots, Y_m \mid X$$
$$\Rightarrow Y_i \underset{P}{\perp\!\!\!\perp} \sigma(Y_1, \ldots, Y_{i-1}, Y_{i+1}, \ldots, Y_m) \mid X \qquad\qquad \text{[Def. 16.45, Lemma 16.49]}$$
$$\Rightarrow 1_{Y_i=y_i} \underset{P}{\perp\!\!\!\perp} \sigma(Y_1, \ldots, Y_{i-1}, Y_{i+1}, \ldots, Y_m) \mid X \qquad\qquad [\sigma(1_{Y_i=y_i}) \subset \sigma(Y_i)]$$
$$\Rightarrow P[Y_i=y_i \mid X, \sigma(Y_1, \ldots, Y_{i-1}, Y_{i+1}, \ldots, Y_m)] \underset{P}{=} P(Y_i=y_i \mid X) \qquad [(10.4), (16.37)]$$
$$\Rightarrow P(Y_i=y_i \mid X, Y_1, \ldots, Y_{i-1}, Y_{i+1}, \ldots, Y_m) \underset{P}{=} P(Y_i=y_i \mid X).$$

The last implication follows from

$$\sigma[\sigma(Y_1, \ldots, Y_{i-1}, Y_{i+1}, \ldots, Y_m) \cup \sigma(X)] = \sigma(Y_1, \ldots, Y_{i-1}, Y_{i+1}, \ldots, Y_m, X).$$

(ii) ⇒ (iv) For all $i = 1, \ldots, m$ and all $y_i \in \Omega'_{i0}$,

$$P(Y_i=y_i \mid X, Y_1, \ldots, Y_{i-1}, Y_{i+1}, \ldots, Y_m) \underset{P}{=} P(Y_i=y_i \mid X)$$
$$\Rightarrow E\big(E(1_{Y_i=y_i} \mid X, Y_1, \ldots, Y_{i-1}, Y_{i+1}, \ldots, Y_m) \mid X, Y_{i+1}, \ldots, Y_m\big)$$
$$\underset{P}{=} E\big(E(1_{Y_i=y_i} \mid X) \mid X, Y_{i+1}, \ldots, Y_m\big) \qquad\qquad \text{[(10.4), Box 10.2 (ix)]}$$
$$\Rightarrow E(1_{Y_i=y_i} \mid X, Y_{i+1}, \ldots, Y_m) \underset{P}{=} E(1_{Y_i=y_i} \mid X) \qquad\qquad \text{[Box 10.2 (v), (xiv), (i)]}$$
$$\Rightarrow P(Y_i=y_i \mid X, Y_{i+1}, \ldots, Y_m) \underset{P}{=} P(Y_i=y_i \mid X). \qquad\qquad \text{[(10.4)]}$$

(iv) ⇒ (i) If

$$P(Y_i=y_i \mid X, Y_{i+1}, \ldots, Y_m) \underset{P}{=} P(Y_i=y_i \mid X), \quad \forall\, y_i \in \Omega'_{i0}, \forall\, i = 1, \ldots, m-1,$$

then (10.4) and Th. 16.37 (ii) and (iii) [with $\mathscr{E} = \sigma(1_{Y_i=y_i})$, $\mathscr{C} = \sigma(X)$, $\mathscr{D} = \sigma(Y_{i+1}, \ldots, Y_m)$] imply

$$E(1_{Y_i=y_i} \cdot 1_{Y_{i+1}=y_{i+1}} \cdot \ldots \cdot 1_{Y_m=y_m} \mid X) \underset{P}{=} E(1_{Y_i=y_i} \mid X) \cdot E(1_{Y_{i+1}=y_{i+1}} \cdot \ldots \cdot 1_{Y_m=y_m} \mid X),$$
$$\forall\, y_j \in \Omega'_{j0}, j = i+1, \ldots m, \forall\, i = 1, \ldots, m-1.$$

Sequential application of this implication for $i = 1, \ldots, m - 1$ yields

$$E(1_{Y_1 = y_1} \cdots 1_{Y_m = y_m} \mid X) \underset{P}{=} E(1_{Y_1 = y_1} \mid X) \cdots E(1_{Y_m = y_m} \mid X), \ \forall \, y_j \in \Omega'_{j0}, \ j = 1, \ldots, m.$$

According to (10.4), this in turn implies

$$P(Y_1 = y_1, \ldots, Y_m = y_m \mid X) \underset{P}{=} \prod_{i=1}^{m} P(Y_i = y_i \mid X), \quad \forall \, y_j \in \Omega'_{j0}, \ j = 1, \ldots m.$$

Now Corollary 16.47 yields

$$\underset{P}{\perp\!\!\!\perp} \ Y_1, \ldots, Y_m \mid X.$$

(ii) $\Rightarrow$ (iii) This proof is analogous to the proof of (ii) $\Rightarrow$ (iv), where the role of $Y_{i+1}, \ldots, Y_m$ is taken by $Y_1, \ldots, Y_{i-1}$, and the role of $i = 1, \ldots, m - 1$ by $i = 2, \ldots, m$.

(iii) $\Rightarrow$ (i) This proof is analogous to the proof of (iv) $\Rightarrow$ (i), where again the role of $Y_{i+1}, \ldots, Y_m$ is taken by $Y_1, \ldots, Y_{i-1}$, and the role of $i = 1, \ldots, m - 1$ by $i = 2, \ldots, m$.

Exercises

16.1 Prove proposition (16.9).

16.2 Prove proposition (16.12).

16.3 Prove the propositions of Box 16.2.

16.4 Comparing the columns for $P(Y = 1 \mid X, U)$ and $P(Y = 1 \mid U)$ in Table 16.1 reveals that $P(Y = 1 \mid X, U) \underset{P}{=} P(Y = 1 \mid U)$. Show that this implies $P(Y = 0 \mid X, U) \underset{P}{=} P(Y = 0 \mid U)$.

16.5 Show that Assumptions (13.36) and (13.37) and the definition of ξ imply Equation (16.55).

16.6 Show that Equations (16.55) and (16.56) are equivalent to each other.

Solutions

16.1 If $Z \underset{P}{=} \alpha, \ \alpha \in \Omega'_Z$, then,

$$A \underset{P}{\perp\!\!\!\perp} B \mid Z$$
$$\Leftrightarrow P(A \cap B \mid Z = \alpha) = P(A \mid Z = \alpha) \cdot P(B \mid Z = \alpha) \quad [\text{Cor. 16.6}, P(Z = \alpha) = 1]$$
$$\Leftrightarrow P(A \cap B) = P(A) \cdot P(B) \quad\quad [\text{Rem. 10.35, Def. 4.12}, P(Z = \alpha) = 1]$$
$$\Leftrightarrow A \underset{P}{\perp\!\!\!\perp} B. \quad\quad\quad\quad\quad\quad\quad\quad\quad [\text{Def. 4.37 (i)}]$$

16.2
$$\mathcal{D} \underset{P}{\perp\!\!\!\perp} \mathcal{E} \mid Z \Leftrightarrow \forall \, (A, B) \in \mathcal{D} \times \mathcal{E} : A, B \underset{P}{\perp\!\!\!\perp} \mid Z \quad [\text{Def. 16.10 (ii)}]$$
$$\Leftrightarrow \forall \, (A, B) \in \mathcal{D} \times \mathcal{E} : A \underset{P}{\perp\!\!\!\perp} B \quad\quad [Z \underset{P}{=} \alpha, (16.9)]$$
$$\Leftrightarrow \mathcal{D} \underset{P}{\perp\!\!\!\perp} \mathcal{E}. \quad\quad\quad\quad\quad\quad\quad\quad [\text{Def. 4.40}]$$

16.3   (i)   This proposition follows from Definitions 16.2 (i) and 16.10 (i) of conditional independence and commutativity of $\cap$ and multiplication.

(ii)   This proposition follows from monotonicity of generated σ-algebras (see Def. 1.13 and Rem. 1.23).

(iii)   If $\underset{P}{\perp\!\!\!\perp}\, \mathcal{D}_1, \mathcal{D}_2, \mathcal{D}_3$, then, according to Equation (4.23),

$$\mathcal{D}_i \underset{P}{\perp\!\!\!\perp} \mathcal{D}_j, \quad i \neq j, \quad i,j = 1,2,3. \tag{16.59}$$

Furthermore, Equations (4.23) and Equation (4.24) yield

$$\left\{A \cap B \colon (A,B) \in \mathcal{D}_1 \times \mathcal{D}_2\right\} \underset{P}{\perp\!\!\!\perp} \mathcal{D}_3. \tag{16.60}$$

Hence, for all $(A,B) \in \mathcal{D}_1 \times \mathcal{D}_2$,

$$
\begin{aligned}
P(A \cap B \mid \mathcal{D}_3) &\underset{P}{=} E(1_{A \cap B} \mid \mathcal{D}_3) && [(10.2)] \\
&\underset{P}{=} E(1_{A \cap B}) && [(16.60),\ \text{Box } 10.1\ (\text{vi})] \\
&= E(1_A \cdot 1_B) && [(1.33)] \\
&= E(1_A) \cdot E(1_B) && [(16.59),\ \text{Box } 6.1\ (\text{x})] \\
&\underset{P}{=} E(1_A \mid \mathcal{D}_3) \cdot E(1_B \mid \mathcal{D}_3) && [(16.59),\ \text{Box } 10.1\ (\text{vi})] \\
&\underset{P}{=} P(A \mid \mathcal{D}_3) \cdot P(B \mid \mathcal{D}_3), && [(10.2)]
\end{aligned}
$$

that is, $\mathcal{D}_1 \underset{P}{\perp\!\!\!\perp} \mathcal{D}_2 \mid \mathcal{D}_3$.

(iv)   This is an implication of Corollary 16.4.

(v)   This is a special case of (iv).

(vi)

$$
\begin{aligned}
\mathcal{E} \underset{P}{\perp\!\!\!\perp} \mathcal{F} \mid \mathcal{C} \;&\Rightarrow\; A \underset{P}{\perp\!\!\!\perp} B \mid \mathcal{C}, \quad \forall\, (A,B) \in \mathcal{E} \times \mathcal{F} && [\text{Def. } 16.10\ (\text{i})] \\
&\Rightarrow\; A \underset{P}{\perp\!\!\!\perp} B \mid \mathcal{C}, \quad \forall\, (A,B) \in \mathcal{E} \times \mathcal{G} && [\mathcal{G} \subset \mathcal{F}] \\
&\Rightarrow\; \mathcal{E} \underset{P}{\perp\!\!\!\perp} \mathcal{G} \mid \mathcal{C}. && [\text{Def. } 16.10\ (\text{i})]
\end{aligned}
$$

(vii)   Using $\sigma[\mathcal{C}, \sigma(\mathcal{C}, \mathcal{D}_2)] = \sigma(\mathcal{C}, \mathcal{D}_2)$ [see Eq. (1.13)] yields

$$
\begin{aligned}
\mathcal{D}_1 \underset{P}{\perp\!\!\!\perp} \mathcal{D}_2 \mid \mathcal{C} \;&\Leftrightarrow\; P(A \mid \mathcal{C}, \mathcal{D}_2) \underset{P}{=} P(A \mid \mathcal{C}), \quad \forall\, A \in \mathcal{D}_1 && [\text{Th. } 16.22] \\
&\Leftrightarrow\; P[A \mid \mathcal{C}, \sigma(\mathcal{C}, \mathcal{D}_2)] \underset{P}{=} P(A \mid \mathcal{C}), \quad \forall\, A \in \mathcal{D}_1 \\
&\Leftrightarrow\; \mathcal{D}_1 \underset{P}{\perp\!\!\!\perp} \sigma(\mathcal{C}, \mathcal{D}_2) \mid \mathcal{C}. && [\text{Th. } 16.22]
\end{aligned}
$$

(viii) $\Rightarrow$

$$\mathcal{E} \underset{P}{\perp\!\!\!\perp} \sigma(\mathcal{D}_1, \mathcal{D}_2) \mid \mathcal{C} \;\Rightarrow\; \mathcal{E} \underset{P}{\perp\!\!\!\perp} \mathcal{D}_1 \mid \mathcal{C}, \quad [\text{Box } 16.2\ (\text{vi})]$$

and using $\sigma[\mathscr{C}, \sigma(\mathscr{D}_1, \mathscr{D}_2)] = \sigma[\mathscr{D}_2, \sigma(\mathscr{D}_1, \mathscr{C})]$ (see Rem. 1.16) yields

$$\mathscr{E} \underset{P}{\perp\!\!\!\perp} \sigma(\mathscr{D}_1, \mathscr{D}_2) \mid \mathscr{C} \quad \text{and} \quad \mathscr{E} \underset{P}{\perp\!\!\!\perp} \mathscr{D}_1 \mid \mathscr{C}$$

$$\Rightarrow P[A \mid \mathscr{C}, \sigma(\mathscr{D}_1, \mathscr{D}_2)] \underset{P}{=} P(A \mid \mathscr{C}) \quad \text{and} \quad P(A \mid \mathscr{C}, \mathscr{D}_1) \underset{P}{=} P(A \mid \mathscr{C}), \quad \forall A \in \mathscr{E}$$

[Th. 16.22]

$$\Rightarrow P[A \mid \mathscr{D}_2, \sigma(\mathscr{D}_1, \mathscr{C})] \underset{P}{=} P(A \mid \mathscr{C}, \mathscr{D}_1), \quad \forall A \in \mathscr{E} \qquad \text{[Rem. 2.73 (iii)]}$$

$$\Rightarrow \mathscr{E} \underset{P}{\perp\!\!\!\perp} \mathscr{D}_2 \mid \sigma(\mathscr{D}_1, \mathscr{C}). \qquad \text{[16.2 (ii)]}$$

$\Leftarrow$

$$\mathscr{E} \underset{P}{\perp\!\!\!\perp} \mathscr{D}_1 \mid \mathscr{C} \quad \text{and} \quad \mathscr{E} \underset{P}{\perp\!\!\!\perp} \mathscr{D}_2 \mid \sigma(\mathscr{C}, \mathscr{D}_1)$$

$$\Rightarrow P(A \mid \mathscr{C}, \mathscr{D}_1) \underset{P}{=} P(A \mid \mathscr{C}) \quad \text{and} \quad P[A \mid \mathscr{C}, \sigma(\mathscr{D}_1, \mathscr{D}_2)] \underset{P}{=} P(A \mid \mathscr{C}, \mathscr{D}_1), \forall A \in \mathscr{E}$$

[(1.13), Th. 16.22]

$$\Rightarrow P[A \mid \mathscr{C}, \sigma(\mathscr{D}_1, \mathscr{D}_2)] \underset{P}{=} P(A \mid \mathscr{C}), \quad \forall A \in \mathscr{E} \qquad \text{[Rem. 2.73 (iii)]}$$

$$\Rightarrow \mathscr{E} \underset{P}{\perp\!\!\!\perp} \sigma(\mathscr{D}_1, \mathscr{D}_2) \mid \mathscr{C}. \qquad \text{[Th. 16.22]}$$

(ix)   This proposition immediately follows from (viii) for $\mathscr{C} = \{\Omega, \emptyset\}$.

(x)

$$\mathscr{F} \underset{P}{\perp\!\!\!\perp} \mathscr{G} \mid \mathscr{C} \Rightarrow \sigma(\mathscr{F}, \mathscr{F}_0) \underset{P}{\perp\!\!\!\perp} \mathscr{G} \mid \mathscr{C} \qquad [\mathscr{F}_0 \subset \mathscr{F}, \sigma(\mathscr{F}_0, \mathscr{F}) = \mathscr{F}]$$

$$\Rightarrow \mathscr{F} \underset{P}{\perp\!\!\!\perp} \mathscr{G} \mid \sigma(\mathscr{C}, \mathscr{F}_0) \qquad \qquad [\text{(viii)}]$$

$$\Rightarrow \mathscr{F} \underset{P}{\perp\!\!\!\perp} \sigma(\mathscr{G}, \mathscr{G}_0) \mid \sigma(\mathscr{C}, \mathscr{F}_0) \qquad [\mathscr{G}_0 \subset \mathscr{G}, \sigma(\mathscr{G}_0, \mathscr{G}) = \mathscr{G}]$$

$$\Rightarrow \mathscr{F} \underset{P}{\perp\!\!\!\perp} \mathscr{G} \mid \sigma(\mathscr{C}, \mathscr{F}_0, \mathscr{G}_0).$$

[(viii), $\sigma[\sigma(\mathscr{C}, \mathscr{F}_0), \mathscr{G}_0] = \sigma(\mathscr{C}, \mathscr{F}_0, \mathscr{G}_0)$, Rem. 1.16]

(xi)   First of all, note that for $i = 1, 2$,

$$\forall A' \in \mathscr{A}'_X \ \forall \omega \in \Omega: \quad 1_{A'}[X_i(\omega)] = 1_{X_i^{-1}(A')}(\omega),$$

and

$$\{\omega \in \Omega : 1_{A'}[X_1(\omega)] \neq 1_{A'}[X_2(\omega)]\} \subset \{\omega \in \Omega : X_1(\omega) \neq X_2(\omega)\}. \tag{16.61}$$

Now, for all $A \in \sigma(X_2)$, let A' denote an element of $\mathscr{A}'_X$ for which $A = X_2^{-1}(A')$. Then,

$$\forall A \in \sigma(X_2): \quad 1_A = 1_{X_2^{-1}(A')}, \tag{16.62}$$

and, for all $A \in \sigma(X_2)$ and all $B \in \mathscr{E}$,

$$
\begin{aligned}
P(A \cap B \mid \mathscr{C}) &\underset{P}{=} E(1_A \cdot 1_B \mid \mathscr{C}) && [(10.2),\ (1.33)] \\
&\underset{P}{=} E(1_{X_2^{-1}(A')} \cdot 1_B \mid \mathscr{C}) && [(16.62)] \\
&\underset{P}{=} E(1_{X_1^{-1}(A')} \cdot 1_B \mid \mathscr{C}) && [X_1 \underset{P}{=} X_2,\ (16.61),\ \text{Box 10.1 (ix)}] \\
&\underset{P}{=} E(1_{X_1^{-1}(A')} \mid \mathscr{C}) \cdot E(1_B \mid \mathscr{C}) && [X_1 \underset{P}{\perp\!\!\!\perp} \mathscr{E} \mid \mathscr{C}] \\
&\underset{P}{=} E(1_{X_2^{-1}(A')} \mid \mathscr{C}) \cdot E(1_B \mid \mathscr{C}) && [X_1 \underset{P}{=} X_2,\ (16.61),\ \text{Box 10.1 (ix)}] \\
&\underset{P}{=} E(1_A \mid \mathscr{C}) \cdot E(1_B \mid \mathscr{C}) && [(16.62)] \\
&\underset{P}{=} P(A \mid \mathscr{C}) \cdot P(B \mid \mathscr{C}). && [(10.2)]
\end{aligned}
$$

(xii)   If $X = \alpha$, then $\sigma(X) = \{\Omega, \emptyset\}$ (see Example 2.10). Furthermore, Box 16.2 (v) implies $X \underset{P}{\perp\!\!\!\perp} \mathscr{E} \mid \mathscr{C}$. If $X \underset{P}{=} \alpha$, then applying Rule (xi) of Box 16.2 completes the proof.

16.4   Because $P(Y=1 \mid X, U) \underset{P}{=} 1 - P(Y=0 \mid X, U)$ and $P(Y=1 \mid U) \underset{P}{=} 1 - P(Y=0 \mid U)$ [see (10.2) and Box 10.2 (i) and (xv)],

$$
\begin{aligned}
& P(Y=1 \mid X, U) \underset{P}{=} P(Y=1 \mid U) \\
\Leftrightarrow\ & 1 - P(Y=0 \mid X, U) \underset{P}{=} 1 - P(Y=0 \mid U) \quad [\text{Rem. 2.73 (iii)}] \\
\Leftrightarrow\ & P(Y=0 \mid X, U) \underset{P}{=} P(Y=0 \mid U). \quad\quad\quad [(2.36)]
\end{aligned}
$$

16.5   Using $\sigma(\xi) \subset \sigma(U)$, the notation introduced in Equation (10.4), and Remark (10.7): For all $i = 1, \ldots, m$,

$$
\begin{aligned}
& P(Y_i = 1 \mid \xi, Y_1, \ldots, Y_{i-1}, Y_{i+1}, \ldots, Y_m) \\
&\underset{P}{=} E\big(P(Y_i=1 \mid U, Y_1, \ldots, Y_{i-1}, Y_{i+1}, \ldots, Y_m) \mid \xi, Y_1, \ldots, Y_{i-1}, Y_{i+1}, \ldots, Y_m\big) \\
& \hspace{8cm} [(13.37),\ \text{Box 10.2 (v)}] \\
&\underset{P}{=} E\big(P(Y_i=1 \mid U) \mid \xi, Y_1, \ldots, Y_{i-1}, Y_{i+1}, \ldots, Y_m\big) \quad\quad [(13.37),\ \text{Box 10.2 (ix)}] \\
&\underset{P}{=} E\left(\frac{\exp(\xi - \beta_i)}{1 + \exp(\xi - \beta_i)} \,\middle|\, \xi, Y_1, \ldots, Y_{i-1}, Y_{i+1}, \ldots, Y_m\right) \quad\quad [(13.40)] \\
&\underset{P}{=} \frac{\exp(\xi - \beta_i)}{1 + \exp(\xi - \beta_i)} \quad\quad\quad\quad\quad\quad\quad\quad\quad\quad [(13.41),\ \text{Box 10.2 (vii)}] \\
&\underset{P}{=} P(Y_i = 1 \mid U) \quad\quad\quad\quad\quad\quad\quad\quad\quad\quad\quad [(13.40)] \\
&\underset{P}{=} P(Y_i = 1 \mid \xi). \quad\quad\quad\quad\quad\quad\quad\quad\quad\quad\quad [(13.41)]
\end{aligned}
$$

16.6   According to Corollary 16.50 (i) and (ii), Equation (16.55) is equivalent to $\underset{P}{\perp\!\!\!\perp} Y_1, \ldots, Y_m \mid \xi$. Applying Corollary 16.47 yields that $\underset{P}{\perp\!\!\!\perp} Y_1, \ldots, Y_m \mid \xi$ is equivalent to (16.56).

17

Conditional distribution

In the previous chapters, we treated conditional probabilities and conditional expectations. In chapter 16, conditional expectations have been used to define conditional independence, and in this chapter we use them to introduce the concept of a *conditional distribution*. While a conditional expectation can be used to describe how the *expectation* of a numerical random variable depends on a σ-algebra or on a random variable, a conditional distribution can be used to describe how the *distribution* of a (not necessarily numerical) random variable depends on a σ-algebra or on a random variable.

17.1 Conditional distribution given a σ-algebra or a random variable

In section 5.1, we defined the *distribution* P_Y of a random variable $Y: (\Omega, \mathscr{A}, P) \to (\Omega'_Y, \mathscr{A}'_Y)$ by

$$P_Y(A') := P(Y \in A') = P[Y^{-1}(A')], \quad \forall A' \in \mathscr{A}'_Y.$$

There, we noted that $P_Y: \mathscr{A}'_Y \to [0, 1]$ is a probability measure on the measurable space $(\Omega'_Y, \mathscr{A}'_Y)$. According to Remark 5.33, P_Y is also the marginal distribution of Y with respect to the joint distribution of (X, Y), where X is any other random variable on $(\Omega, \mathscr{A}, P)$. Furthermore, in Equation (10.2) we defined the $\mathscr{C}$-conditional probability $P(A \mid \mathscr{C}) := E(1_A \mid \mathscr{C})$ of an event $A \in \mathscr{A}$ given the σ-algebra $\mathscr{C} \subset \mathscr{A}$.

In Definition 17.1, we consider the event $\{Y \in A'\} = Y^{-1}(A')$, where $A' \in \mathscr{A}'_Y$, and we use a $\mathscr{C}$-conditional probability $P(Y \in A' \mid \mathscr{C})$ of this event, which is an element of the set $\mathscr{P}(Y \in A' \mid \mathscr{C})$ of all versions of the $\mathscr{C}$-conditional probability of $\{Y \in A'\}$ (see Rem. 10.10).

We also consider functions $P_{Y\mid\mathscr{C}}: \Omega \times \mathscr{A}'_Y \to [0, 1]$ and the family $(P_{Y\mid\mathscr{C}}(\cdot, A'), A' \in \mathscr{A}'_Y)$ of functions $P_{Y\mid\mathscr{C}}(\cdot, A'): \Omega \to [0, 1]$, defined by

$$P_{Y\mid\mathscr{C}}(\cdot, A')(\omega) := P_{Y\mid\mathscr{C}}(\omega, A'), \quad \omega \in \Omega, \tag{17.1}$$

Probability and Conditional Expectation: Fundamentals for the Empirical Sciences, First Edition. Rolf Steyer and Werner Nagel.
© 2017 John Wiley & Sons, Ltd. Published 2017 by John Wiley & Sons, Ltd.
Companion website: http://www.probability-and-conditional-expectation.de

and the family $(P_{Y|\mathscr{C}}(\omega, \cdot), \omega \in \Omega)$ of functions $P_{Y|\mathscr{C}}(\omega, \cdot): \mathscr{A}'_Y \to [0, 1]$, defined by

$$P_{Y|\mathscr{C}}(\omega, \cdot)(A') := P_{Y|\mathscr{C}}(\omega, A'), \quad A' \in \mathscr{A}'_Y. \tag{17.2}$$

Definition 17.1 [Conditional distribution given a σ-algebra]

Let $Y: (\Omega, \mathscr{A}, P) \to (\Omega'_Y, \mathscr{A}'_Y)$ *be a random variable and* $\mathscr{C} \subset \mathscr{A}$ *be a σ-algebra. Furthermore, suppose that there is a function* $P_{Y|\mathscr{C}}: \Omega \times \mathscr{A}'_Y \to [0, 1]$ *satisfying the following two conditions:*

(a) *For all* $A' \in \mathscr{A}'_Y$,

$$P_{Y|\mathscr{C}}(\cdot, A') \in \mathscr{P}(Y \in A' \mid \mathscr{C}). \tag{17.3}$$

(b) *For all* $\omega \in \Omega$, *the function* $P_{Y|\mathscr{C}}(\omega, \cdot): \mathscr{A}'_Y \to [0, 1]$ *is a probability measure on* $(\Omega'_Y, \mathscr{A}'_Y)$.

Then $P_{Y|\mathscr{C}}$ *is called a version of the* $\mathscr{C}$ - *conditional distribution of Y.*

Remark 17.2 [X-conditional distribution]   Let $X: (\Omega, \mathscr{A}, P) \to (\Omega'_X, \mathscr{A}'_X)$ be a random variable. Using the σ-algebra $\sigma(X) = X^{-1}(\mathscr{A}'_X)$ generated by X, we define

$$P_{Y|X} := P_{Y|\sigma(X)} \tag{17.4}$$

and call it a version of the *X-conditional distribution of Y*, provided that it exists.   ◁

A version of a conditional distribution is also called a *stochastic kernel* or *Markov kernel*. Note that neither Y nor the random variable X have to be numerical; both might be nonnumerical random variables on the probability space $(\Omega, \mathscr{A}, P)$.

Remark 17.3 [The functions $P_{Y|\mathscr{C}}(\cdot, A')$]   Equation (17.3) implies that, for all $A' \in \mathscr{A}'_Y$, there is a version $P(Y \in A' \mid \mathscr{C}) \in \mathscr{P}(Y \in A' \mid \mathscr{C})$ such that

$$P_{Y|\mathscr{C}}(\cdot, A') = P(Y \in A' \mid \mathscr{C}). \tag{17.5}$$

This equation is equivalent to

$$P_{Y|\mathscr{C}}(\omega, A') = P(Y \in A' \mid \mathscr{C})(\omega), \quad \forall \omega \in \Omega. \tag{17.6}$$

Equation (17.5) is also equivalent to

$$P_{Y|\mathscr{C}}(\cdot, A') \underset{P}{=} E(1_{Y \in A'} \mid \mathscr{C}) \underset{P}{=} E[1_{A'}(Y) \mid \mathscr{C}], \tag{17.7}$$

where $1_{A'}(Y)$ denotes the composition of $Y: (\Omega, \mathscr{A}, P) \to (\Omega'_Y, \mathscr{A}'_Y)$ and the indicator function $1_{A'}: (\Omega'_Y, \mathscr{A}'_Y) \to (\mathbb{R}, \mathscr{B})$.

Note that, according to Equation (17.5), $P_{Y|\mathscr{C}}(\cdot, A'): \Omega \to [0, 1]$ is a $\mathscr{C}$-measurable random variable on $(\Omega, \mathscr{A}, P)$, because, for each $A' \in \mathscr{A}'_Y$, the function $P(Y \in A' \mid \mathscr{C})$ is a $\mathscr{C}$-conditional probability of the event $\{Y \in A'\}$.   ◁

Remark 17.4 [Existence of $P_{Y|\mathscr{C}}$]   Under the assumptions of Definition 17.1, a function $P_{Y|\mathscr{C}}$ satisfying condition (a) of Definition 17.1 always exists (see Th. 10.9). However, the function $P_{Y|\mathscr{C}}(\omega, \cdot): \mathscr{A}'_Y \to [0, 1]$ is not necessarily a measure. Therefore, the $\mathscr{C}$-conditional distribution of Y does not necessarily exist, and it is worthwhile studying sufficient conditions for its existence (see section 17.3.1). As we will see, one of the sufficient conditions of its existence is that Y is real-valued. Another condition under which it exists, is independence of Y and $\mathscr{C}$.   ◁

Remark 17.5 [The set $\mathscr{P}_{Y|\mathscr{C}}$]   Even if $P_{Y|\mathscr{C}}$ exists, this does not imply that it is uniquely defined. Therefore, we use $\mathscr{P}_{Y|\mathscr{C}}$ to denote the set of all functions satisfying (a) and (b) of Definition 17.1. Similarly, $\mathscr{P}_{Y|X}$ denotes the set of all functions satisfying these conditions with $\mathscr{C} = \sigma(X)$. If $\mathscr{P}_{Y|\mathscr{C}}$ is nonempty, then assuming $P_{Y|\mathscr{C}} \in \mathscr{P}_{Y|\mathscr{C}}$ means that $P_{Y|\mathscr{C}}$ is a version of the $\mathscr{C}$-conditional distribution of Y. Of course, $P_{Y|X} \in \mathscr{P}_{Y|X}$ has the same meaning for $\mathscr{C} = \sigma(X)$. Uniqueness of $P_{Y|\mathscr{C}}$ is treated in section 17.3.2.   ◁

Example 17.6 [Joe and Ann with self-selection – continued]   Table 17.1 shows an example that has already been introduced in chapter 11. However, now the table also contains four additional columns showing the conditional distribution $P_{Y|X}$. In Table 11.2, the values of the conditional expectation $E(Y \mid X)$ are displayed, which are repeated in the column headed $P(Y = 1 \mid X)$ of Table 17.1. According to Equations (17.4) and (17.7), $E(Y \mid X) = P_{Y|X}(\cdot, \{1\})$ (see the last but one column of Table 17.1). The other three of the last four columns can be computed using the fact that $P_{Y|X}(\omega, \cdot)$ is a probability measure on $(\Omega'_Y, \mathscr{A}'_Y)$ for each of the eight $\omega \in \Omega$ (see the rows of the table).

Table 17.1   Joe and Ann with self-selection: conditional distribution $P_{Y|X}$.

Outcomes ω			Observables			Conditional probabilities			$P_{Y|X}$			
Unit	Treatment	Success	$P(\{\omega\})$	Person variable U	Treatment variable X	Outcome variable Y	$P(Y=1 \mid X, U)$	$P(Y=1 \mid X)$	$P(X=1 \mid U)$	$P_{Y|X}(\cdot, \Omega'_Y)$	$P_{Y|X}(\cdot, \emptyset)$	$P_{Y|X}(\cdot, \{0\})$	$P_{Y|X}(\cdot, \{1\})$
(Joe, no, −)	.144	Joe	0	0	.7	.6	.04	1	0	.4	.6
(Joe, no, +)	.336	Joe	0	1	.7	.6	.04	1	0	.4	.6
(Joe, yes, −)	.004	Joe	1	0	.8	.42	.04	1	0	.58	.42
(Joe, yes, +)	.016	Joe	1	1	.8	.42	.04	1	0	.58	.42
(Ann, no, −)	.096	Ann	0	0	.2	.6	.76	1	0	.4	.6
(Ann, no, +)	.024	Ann	0	1	.2	.6	.76	1	0	.4	.6
(Ann, yes, −)	.228	Ann	1	0	.4	.42	.76	1	0	.58	.42
(Ann, yes, +)	.152	Ann	1	1	.4	.42	.76	1	0	.58	.42

For example, for $\omega_1 = (Joe, no, -)$ and Ω'_Y, $P_{Y|X}(\omega_1, \Omega'_Y) = 1$. Because $P_{Y|X}(\omega_1, \{1\}) = .6$, for $A' = \{0\} \in \mathcal{A}'_Y$, we receive

$$P_{Y|X}(\omega_1, \{0\}) = 1 - P_{Y|X}(\omega_1, \{1\}) = 1 - .6 = .4.$$

Furthermore, $P_{Y|X}(\omega_1, \emptyset) = 0$.   ◁

17.2 Conditional distribution given a value of a random variable

In the definition of $P_{Y|X}$, for each $A' \in \mathcal{A}'_Y$, we use a version $P(Y \in A' \mid X)$ of the X-conditional probability, which is measurable with respect to X. Therefore, Equation (17.4) and Definition 17.1 (a) imply

$$P_{Y|X}(\omega, A') = P(Y \in A' \mid X)(\omega) = P(Y \in A' \mid X=x), \quad \forall \, \omega \in \{X=x\} \quad (17.8)$$

(see Def. 10.33 and Rem. 10.37), which can also be written as

$$\begin{aligned} P_{Y|X}(\omega, A') &= E[1_{A'}(Y) \mid X](\omega) = E(1_{Y \in A'} \mid X=x) \\ &= E[1_{A'}(Y) \mid X=x], \quad \forall \, \omega \in \{X=x\}, \end{aligned} \quad (17.9)$$

where $E(1_{Y \in A'} \mid X=x) = g_{A'}(x)$, and $g_{A'}$ is a factorization of $E(1_{Y \in A'} \mid X)$. Equation (17.8) implies that, for a given $A' \in \mathcal{A}'_Y$, the function $P_{Y|X}(\omega, A')$ is constant for all $\omega \in \{X=x\}$, a fact to keep in mind while reading the following definition.

Definition 17.7 [Conditional distribution given $X=x$]
Let $X: (\Omega, \mathcal{A}, P) \to (\Omega'_X, \mathcal{A}'_X)$ and $Y: (\Omega, \mathcal{A}, P) \to (\Omega'_Y, \mathcal{A}'_Y)$ be random variables, and let $P_{Y|X} \in \mathscr{P}_{Y|X}$ and $(P_{Y|X=x}, x \in \Omega'_X)$ be the family of probability measures $P_{Y|X=x}: \mathcal{A}'_Y \to [0, 1]$, defined by

$$\forall \, x \in \Omega'_X \; \forall \, \omega \in \{X=x\} \; \forall \, A' \in \mathcal{A}'_Y: \quad P_{Y|X=x}(A') := P_{Y|X}(\omega, A'). \quad (17.10)$$

Then $P_{Y|X=x}, x \in \Omega'_X$, is called an $(X=x)$-conditional distribution of Y pertaining to $P_{Y|X}$.

Definitions 17.1 and 17.7 imply the following lemma.

Lemma 17.8 [A characterization of a family of $(X=x)$-conditional distributions]
Let $X: (\Omega, \mathcal{A}, P) \to (\Omega'_X, \mathcal{A}'_X)$ and $Y: (\Omega, \mathcal{A}, P) \to (\Omega'_Y, \mathcal{A}'_Y)$ be random variables. A family $(P_{Y|X=x}, x \in \Omega'_X)$ of functions $P_{Y|X=x}: \mathcal{A}'_Y \to [0, 1]$ is a family of $(X=x)$-conditional

distributions of Y pertaining to $P_{Y|X}$ if and only if the following two conditions hold:

(a) For all $A' \in \mathscr{A}'_Y$, there is a version $P(Y \in A' \mid X) \in \mathscr{P}(Y \in A' \mid X)$ with

$$\forall x \in \Omega'_X \colon P_{Y|X=x}(A') = P(Y \in A' \mid X=x). \tag{17.11}$$

(b) For all $x \in \Omega'_X$, the function $P_{Y|X=x}$ is a probability measure on $(\Omega'_Y, \mathscr{A}'_Y)$.

(Proof p. 534)

Lemma 17.8, P-uniqueness of the X-conditional probabilities $P(Y \in A' \mid X)$, and Corollary 10.39 (i) imply Corollary 17.9 that shows how the $(X=x)$-conditional distributions of Y are related to $(X=x)$-conditional probabilities $P^*(Y \in A' \mid X=x)$ pertaining to any version $P^*(Y \in A' \mid X) \in \mathscr{P}(Y \in A' \mid X)$. Reading this corollary, note that, for two functions $f_1, f_2 \colon \Omega'_X \to \mathbb{R}$,

$$f_1(x) \underset{P_X\text{-a.a.}}{=} f_2(x) \tag{17.12}$$

is a more convenient way to express

$$f_1(x) = f_2(x), \quad \text{for } P_X\text{-a.a. } x \in \Omega'_X. \tag{17.13}$$

Each of Equations (17.12) and (17.13) is equivalent to $f_1 \underset{P_X}{=} f_2$ (see Rem. 5.17).

Corollary 17.9 [Conditional distribution and conditional probabilities]
Let $X \colon (\Omega, \mathscr{A}, P) \to (\Omega'_X, \mathscr{A}'_X)$ and $Y \colon (\Omega, \mathscr{A}, P) \to (\Omega'_Y, \mathscr{A}'_Y)$ be random variables. If $P_{Y|X} \in \mathscr{P}_{Y|X}$ and $(P_{Y|X=x}, x \in \Omega'_X)$ is the family of $(X=x)$-conditional distributions pertaining to $P_{Y|X}$, then, for all $A' \in \mathscr{A}'_Y$ and all versions $P^*(Y \in A' \mid X) \in \mathscr{P}(Y \in A' \mid X)$,

$$P_{Y|X=x}(A') \underset{P_X\text{-a.a.}}{=} P^*(Y \in A' \mid X=x). \tag{17.14}$$

In Corollary 17.10, we consider the relationship between an $(X=x)$-conditional distribution $P_{Y|X=x}$ introduced in Definition 17.7 and the distribution $P_Y^{X=x}$ of Y with respect to the conditional-probability measure $P^{X=x}$ for an $x \in \Omega'_X$ with $P(X=x) > 0$ [see Eq. (9.4)].

Corollary 17.10 [Consistency of definitions of conditional distributions]
Let $X \colon (\Omega, \mathscr{A}, P) \to (\Omega'_X, \mathscr{A}'_X)$ and $Y \colon (\Omega, \mathscr{A}, P) \to (\Omega'_Y, \mathscr{A}'_Y)$ be random variables, assume that $P_{Y|X}$ exists and that $x \in \Omega'_X$ with $P(X=x) > 0$, and let $P_{Y|X=x}$ denote an $(X=x)$-conditional distribution of Y pertaining to $P_{Y|X}$. Then

$$P_Y^{X=x} = P_{Y|X=x}, \tag{17.15}$$

which is equivalent to

$$P_{Y|X=x}(A') = P_Y^{X=x}(A'), \quad \forall A' \in \mathcal{A}'_Y. \tag{17.16}$$

(Proof p. 535)

Example 17.11 [Joe and Ann with self-selection – continued] We continue the example presented in Table 17.1. Obviously, the information about the conditional distribution $P_{Y|X}$ is already contained in the two conditional distributions $P_{Y|X=0}$ and $P_{Y|X=1}$. According to Equation (17.10), for $A' = \{0\}$,

$$P_{Y|X=0}(\{0\}) = P_{Y|X}(\omega_1, \{0\}) = P_{Y|X}(\omega_2, \{0\}) = .4,$$

where $\omega_1 = (Joe, no, -)$ and $\omega_2 = (Joe, no, +)$. Hence, in this example, the $(X=0)$-conditional distribution of the outcome variable Y is

$$P_{Y|X=0}(\{0\}) = P(Y \in \{0\} \mid X=0) = P(Y=0 \mid X=0) = .4$$
$$P_{Y|X=0}(\{1\}) = P(Y \in \{1\} \mid X=0) = P(Y=1 \mid X=0) = .6$$
$$P_{Y|X=0}(\Omega'_Y) = P(Y \in \Omega'_Y \mid X=0) = 1$$
$$P_{Y|X=0}(\emptyset) = P(Y \in \emptyset \mid X=0) = 0,$$

and the $(X=1)$-conditional distribution of Y is

$$P_{Y|X=1}(\{0\}) = P(Y \in \{0\} \mid X=1) = P(Y=0 \mid X=1) = .58$$
$$P_{Y|X=1}(\{1\}) = P(Y \in \{1\} \mid X=1) = P(Y=1 \mid X=1) = .42$$
$$P_{Y|X=1}(\Omega'_Y) = P[Y^{-1}(\Omega'_Y) \mid X=1] = 1$$
$$P_{Y|X=1}(\emptyset) = P[Y^{-1}(\emptyset) \mid X=1] = 0.$$

Obviously, in this example, in which X takes on all its values x with $P(X=x) > 0$, the two conditional distributions $P_{Y|X=0}$ and $P_{Y|X=1}$ contain all information conveyed by $P_{Y|X}$, and both are probability measures on $(\Omega'_Y, \mathcal{A}'_Y)$, where $\Omega'_Y = \{0, 1\}$ and $\mathcal{A}'_Y = \{\Omega'_Y, \emptyset, \{0\}, \{1\}\}$. ◁

In the next corollary, we consider a random variable $X: (\Omega, \mathcal{A}, P) \to (\Omega'_X, \mathcal{A}'_X)$, a measurable function $h: (\Omega'_X, \mathcal{A}'_X) \to (\Omega', \mathcal{A}')$, and the composition $h(X)$. Reading this corollary, remember that $\delta_{h(x)}$ denotes the Dirac measure at $h(x)$ (see Example 1.52), which is a probability measure on $(\Omega', \mathcal{A}')$.

Corollary 17.12 [Conditional distribution of a composition]
If $X: (\Omega, \mathcal{A}, P) \to (\Omega'_X, \mathcal{A}'_X)$ is a random variable and $h: (\Omega'_X, \mathcal{A}'_X) \to (\Omega', \mathcal{A}')$ a measurable function, then there is a version $P_{h(X)|X} \in \mathcal{P}_{h(X)|X}$ such that

$$\forall x \in \Omega'_X: \quad P_{h(X)|X=x} = \delta_{h(x)}. \tag{17.17}$$

(Proof p. 535)

Remark 17.13 [Conditional distribution of a composition] Note that for the version $P_{h(X)|X}$, Equation (17.17) is equivalent to

$$\forall x \in \Omega'_X \ \forall A' \in \mathscr{A}': \quad P_{h(X)|X=x}(A') = \delta_{h(x)}(A') = 1_{A'}[h(x)] \qquad (17.18)$$

(see again Example 1.52). In other words,

$$\forall x \in \Omega'_X \ \forall A' \in \mathscr{A}': \quad P_{h(X)|X=x}(A') = \begin{cases} 1, & \text{if } h(x) \in A' \\ 0, & \text{otherwise.} \end{cases}$$

◁

17.3 Existence and uniqueness

In this section, we consider sufficient conditions for the existence and for uniqueness of the conditional distributions $P_{Y|\mathscr{C}}$, $P_{Y|X}$, and $P_{Y|X=x}$.

17.3.1 Existence

Reading the following lemma, remember that $Y \underset{P}{\perp\!\!\!\perp} \mathscr{C}$ denotes independence of $\sigma(Y)$ and $\mathscr{C}$ with respect to P (see Rem. 5.44).

Lemma 17.14 [$P_{Y|\mathscr{C}}$ if Y and $\mathscr{C}$ are independent]
Let $X: (\Omega, \mathscr{A}, P) \to (\Omega'_X, \mathscr{A}'_X)$, $Y: (\Omega, \mathscr{A}, P) \to (\Omega'_Y, \mathscr{A}'_Y)$ be random variables and $\mathscr{C} \subset \mathscr{A}$ a σ-algebra. If $Y \underset{P}{\perp\!\!\!\perp} \mathscr{C}$, then $P_{Y|\mathscr{C}}$ exists and

$$P_{Y|\mathscr{C}}(\omega, A') := P_Y(A'), \quad \forall (\omega, A') \in \Omega \times \mathscr{A}'_Y, \qquad (17.19)$$

defines a version of the $\mathscr{C}$-conditional distribution of Y. Correspondingly, if $Y \underset{P}{\perp\!\!\!\perp} X$, then $P_{Y|X}$ exists and

$$P_{Y|X}(\omega, A') := P_Y(A'), \quad \forall (\omega, A') \in \Omega \times \mathscr{A}'_Y, \qquad (17.20)$$

defines a version of the X-conditional distribution of Y.

(Proof p. 535)

Hence, if $\mathscr{C}$ and Y are independent, then the distribution of Y is a version of the $\mathscr{C}$-conditional distribution of Y. Correspondingly, if X and Y are independent, then the distribution of Y is a version of the X-conditional distribution of Y.

Remark 17.15 [Constant X] If X is P-a.s. constant (i.e., if $X \underset{P}{=} \alpha$, $\alpha \in \Omega'_X$), then, according to Lemma 5.51, $Y \underset{P}{\perp\!\!\!\perp} X$. Therefore, Equation (17.20) always yields a version $P_{Y|X}$ if $X \underset{P}{=} \alpha$.

◁

Another sufficient condition for the existence of $P_{Y|X}$ is that X is discrete (see Def. 5.56 and Cor. 17.10).

Theorem 17.16 [Existence and uniqueness if X is discrete]
Let $X: (\Omega, \mathcal{A}, P) \to (\Omega'_X, \mathcal{A}'_X)$ and $Y: (\Omega, \mathcal{A}, P) \to (\Omega'_Y, \mathcal{A}'_Y)$ be random variables, and let X be discrete. Then $P_{Y|X}$ exists, and for all $x \in \Omega'_X$ with $P(X=x) > 0$, the conditional distribution $P_{Y|X=x}$ is uniquely defined.

(Proof p. 536)

According to Theorem 17.17, the conditional distribution $P_{Y|X}$ also exists if Y is discrete.

Theorem 17.17 [Existence for a discrete Y]
Let $X: (\Omega, \mathcal{A}, P) \to (\Omega'_X, \mathcal{A}'_X)$ and $Y: (\Omega, \mathcal{A}, P) \to (\Omega'_Y, \mathcal{A}'_Y)$ be random variables, and assume that Y is discrete. Then $P_{Y|X}$ exists.

(Proof p. 536)

Other theorems provide sufficient conditions for the existence of $P_{Y|\mathcal{C}}$ (see, e.g., Bauer, 1996; Klenke, 2013). Some sufficient conditions that are important for our purposes are stated in Theorem 17.18.

Theorem 17.18 [Existence of a conditional distribution]
Let $X: (\Omega, \mathcal{A}, P) \to (\Omega'_X, \mathcal{A}'_X)$ and $Y: (\Omega, \mathcal{A}, P) \to (\Omega'_Y, \mathcal{A}'_Y)$ be random variables and $\mathcal{C} \subset \mathcal{A}$ a σ-algebra. If one of the following assumptions holds, then $P_{Y|\mathcal{C}}$ and $P_{Y|X}$ exist:

(a) $(\Omega'_Y, \mathcal{A}'_Y) = (\mathbb{R}^n, \mathcal{B}_n)$, $n \in \mathbb{N}$.

(b) $\Omega'_Y = \mathbb{N}^n$, $n \in \mathbb{N}$, and $\mathcal{A}'_Y = \mathcal{P}(\Omega'_Y)$.

(c) Ω'_Y is finite and $\mathcal{A}'_Y$ is a σ-algebra on Ω'_Y.

For a proof, see Klenke (2013, Th. 8.37).

17.3.2   Uniqueness of the functions $P_{Y|\mathcal{C}}(\cdot, A')$

Remark 17.19 [P-uniqueness of $P_{Y|\mathcal{C}}(\cdot, A')$]   If $P_{Y|\mathcal{C}}, P^*_{Y|\mathcal{C}} \in \mathcal{P}_{Y|\mathcal{C}}$, then according to Equations (10.2), (10.12), and (17.3),

(i) $\forall A' \in \mathcal{A}'_Y: \quad P_{Y|\mathcal{C}}(\cdot, A') \underset{P}{=} P^*_{Y|\mathcal{C}}(\cdot, A').$

(ii) $\forall A' \in \mathcal{A}'_Y \, \exists N_{A'} \in \mathcal{A}: \quad P(N_{A'}) = 0$ and $\forall \omega \in \Omega \setminus N_{A'}: P_{Y|\mathcal{C}}(\omega, A') = P^*_{Y|\mathcal{C}}(\omega, A').$

According to Remark 5.17, (i) is another notation for (ii).   ◁

Remark 17.20 [P-uniqueness of $P_{Y|X}(\cdot, A')$]   Similarly, if $P_{Y|X}, P^*_{Y|X} \in \mathscr{P}_{Y|X}$, then

(i) $\forall A' \in \mathscr{A}'_Y$: $P_{Y|X}(\cdot, A') \underset{P}{=} P^*_{Y|X}(\cdot, A')$.

(ii) $\forall A' \in \mathscr{A}'_Y \; \exists N_{A'} \in \mathscr{A}$: $P(N_{A'}) = 0$ and $\forall \, \omega \in \Omega \setminus N_{A'}$: $P_{Y|X}(\omega, A') = P^*_{Y|X}(\omega, A')$.

(iii) $\forall A' \in \mathscr{A}'_Y \; \exists N'_{A'} \in \mathscr{A}'_X$: $P_X(N'_{A'}) = 0$ and $\forall \, x \in \Omega'_X \setminus N'_{A'}$: $P_{Y|X=x}(A') = P^*_{Y|X=x}(A')$.

According to Equations (17.10) and (10.34), Propositions (ii) and (iii) are equivalent to each other. These propositions refer to null sets $N_{A'}$ and $N'_{A'}$. Note that for $A', B' \in \mathscr{A}'_Y$, $A' \neq B'$, the sets $N_{A'}$ and $N_{B'}$ may differ from each other, and the same applies to the sets $N'_{A'}$ and $N'_{B'}$. Finally, in order to express (iii), we write

$$\forall A' \in \mathscr{A}'_Y: \quad P_{Y|X=x}(A') \underset{P_X\text{-a.a.}}{=} P^*_{Y|X=x}(A'). \tag{17.21}$$

$\triangleleft$

17.3.3   Common null set uniqueness of a conditional distribution

Considering a family $(P_{Y|X=x}, x \in \Omega'_X)$ of probability measures, it is of interest if a uniqueness property holds that is stronger than Equation (17.21), that is, if there is a set N such that $N = N_{A'}$ for all $A' \in \mathscr{A}'_Y$. In other words, it is of interest if there is a *common null set* that does not depend on A'. For convenience, we introduce the following term for this kind of uniqueness.

Definition 17.21 [Common null set uniqueness]
Let $X: (\Omega, \mathscr{A}, P) \to (\Omega'_X, \mathscr{A}'_X)$ and $Y: (\Omega, \mathscr{A}, P) \to (\Omega'_Y, \mathscr{A}'_Y)$ be random variables, and let $\mathscr{C} \subset \mathscr{A}$ be a σ-algebra. If for all $P_{Y|\mathscr{C}}, P^*_{Y|\mathscr{C}} \in \mathscr{P}_{Y|\mathscr{C}}$,

$$\exists N \in \mathscr{A}: \quad P(N) = 0 \text{ and } \forall \, \omega \in \Omega \setminus N \; \forall A' \in \mathscr{A}'_Y: P_{Y|\mathscr{C}}(\omega, A') = P^*_{Y|\mathscr{C}}(\omega, A'),$$

then $P_{Y|\mathscr{C}}$ is called CNS-unique. Correspondingly, $P_{Y|X}$ is called CNS-unique if $P_{Y|\sigma(X)}$ is CNS-unique, and in this case we write

$$P_{Y|X=x} \underset{P_X\text{-a.a.}}{=} P^*_{Y|X=x}, \tag{17.22}$$

or, equivalently,

$$\text{for } P_X\text{-a.a. } x \in \Omega'_X, \forall A' \in \mathscr{A}'_Y: \quad P_{Y|X=x}(A') = P^*_{Y|X=x}(A'). \tag{17.23}$$

Remark 17.22 [Alternative formulations]   Equations (17.22) and (17.23) can equivalently be written in each of the following ways:

(i) $\exists N' \in \mathscr{A}'_X$: $P_X(N') = 0$ and $\forall \, x \in \Omega'_X \setminus N'$: $P_{Y|X=x} = P^*_{Y|X=x}$.

(ii) $\exists N' \in \mathscr{A}'_X$: $P_X(N') = 0$ and $\forall \, x \in \Omega'_X \setminus N' \; \forall A' \in \mathscr{A}'_Y: P_{Y|X=x}(A') = P^*_{Y|X=x}(A')$.

$\triangleleft$

Definition 17.23 [P_X-uniqueness of $(P_{Y|X=x}, x \in \Omega'_X)$]
Let $X: (\Omega, \mathscr{A}, P) \to (\Omega'_X, \mathscr{A}'_X)$ and $Y: (\Omega, \mathscr{A}, P) \to (\Omega'_Y, \mathscr{A}'_Y)$ be random variables. If

$$\forall P_{Y|X}, P^*_{Y|X} \in \mathscr{P}_{Y|X}: \quad P_{Y|X=x} \underset{P_X\text{-}a.a.}{=} P^*_{Y|X=x}, \tag{17.24}$$

then we say that the family $(P_{Y|X=x}, x \in \Omega'_X)$ of $(X=x)$-conditional distributions of Y is P_X-unique.

Remark 17.24 [P_X-uniqueness and CNS-uniqueness] An immediate implication of Definitions 17.21 and 17.23 is

$$(P_{Y|X=x}, x \in \Omega'_X) \text{ is } P_X\text{-unique} \quad \Leftrightarrow \quad P_{Y|X} \text{ is CNS-unique}. \tag{17.25}$$

◁

Remark 17.25 [CNS-uniqueness for discrete X] According to Theorem 17.16, if X is discrete and $P(X=x) > 0$ for all $x \in \Omega'_X$, then each $P_{Y|X=x}$ as well as the family $(P_{Y|X=x}, x \in \Omega'_X)$ of conditional distributions are uniquely defined. This implies that $(P_{Y|X=x}, x \in \Omega'_X)$ is P_X-unique, which in turn implies that $P_{Y|X}$ is CNS-unique.   ◁

In Remark 17.25, we provided a sufficient condition of CNS-uniqueness of $P_{Y|X}$ that refers to X. Now, we turn to conditions referring to Y that imply CNS-uniqueness of $P_{Y|\mathscr{C}}$ and $P_{Y|X}$. In Lemma 17.26, we choose a more general notation that proves useful in a number of proofs. Reading this lemma, remember: If $(\Omega, \mathscr{A})$ is a measurable space, then $\mathscr{A}$ is called *countably generated* if there is a finite or countable set $\mathscr{E} \subset \mathscr{A}$ such that $\sigma(\mathscr{E}) = \mathscr{A}$ (see Def. 1.24). Also remember that each of the σ-algebras $\mathscr{A}'_Y$ in (a) to (c) of Theorem 17.18 is countably generated (see Example 1.25 and Rem. 1.28).

Lemma 17.26 [CNS-uniqueness]
Let $(\Omega, \mathscr{A}, P)$ be a probability space, let $(\Omega', \mathscr{A}')$ be a measurable space, and let $K, K^*: \Omega \times \mathscr{A}' \to \mathbb{R}$ be functions such that the following three conditions hold:

(a) $\forall A' \in \mathscr{A}': \quad K(\cdot, A'), K^*(\cdot, A'): \Omega \to \mathbb{R}$ are $(\mathscr{A}, \mathscr{B})$-measurable.

(b) $\forall \omega \in \Omega: \quad K(\omega, \cdot), K^*(\omega, \cdot)$ are probability measures on $(\Omega', \mathscr{A}')$.

(c) $\forall A' \in \mathscr{A}' \exists N_{A'} \in \mathscr{A}: \quad P(N_{A'}) = 0 \text{ and } \forall \omega \in \Omega \setminus N_{A'}: K(\omega, A') = K^*(\omega, A').$

If $\mathscr{A}'$ is countably generated, then

$$\exists N \in \mathscr{A} \, \forall A' \in \mathscr{A}': \quad P(N) = 0 \text{ and } \forall \omega \in \Omega \setminus N: K(\omega, A') = K^*(\omega, A'). \tag{17.26}$$

(Proof p. 537)

Condition (c) of Lemma 17.26 may equivalently be written as

$$\forall A' \in \mathscr{A}': \quad K(\cdot, A') \underset{P}{=} K^*(\cdot, A'). \tag{17.27}$$

While this condition refers to null sets $N_{A'}$ that may depend on $A' \in \mathscr{A}'$, Equation (17.26) refers to a *common null set* N for all $A' \in \mathscr{A}'$.

The following corollary is a special case of Lemma 17.26.

Corollary 17.27 [Sufficient condition for CNS-uniqueness]
Let $X: (\Omega, \mathscr{A}, P) \to (\Omega'_X, \mathscr{A}'_X)$ *and* $Y: (\Omega, \mathscr{A}, P) \to (\Omega'_Y, \mathscr{A}'_Y)$ *be random variables, let* $\mathscr{C} \subset \mathscr{A}$ *be a σ-algebra, and suppose that the conditional distributions $P_{Y|\mathscr{C}}$ and $P_{Y|X}$ exist. If $\mathscr{A}'_Y$ is countably generated, then $P_{Y|\mathscr{C}}$ and $P_{Y|X}$ are CNS-unique.*

For many applications, Corollary 17.28 implies that conditional distributions are CNS-unique. This corollary immediately follows from Corollary 17.27, Example 1.25, and Remark 1.28.

Corollary 17.28 [Sufficient condition for CNS-uniqueness]
Let $X: (\Omega, \mathscr{A}, P) \to (\Omega'_X, \mathscr{A}'_X)$ *and* $Y: (\Omega, \mathscr{A}, P) \to (\Omega'_Y, \mathscr{A}'_Y)$ *be random variables and* $\mathscr{C} \subset \mathscr{A}$ *a σ-algebra. Then each of the conditions (a) to (c) of Theorem 17.18 implies that $P_{Y|\mathscr{C}}$ and $P_{Y|X}$ exist and are CNS-unique.*

Note that the σ-algebras appearing in Theorem 17.18 are countably generated. For simplicity, instead of CNS-uniqueness, we often assume that the σ-algebras are countably generated, which according to Corollary 17.27 implies that the conditional distributions of Y are CNS-unique.

Now we turn to X-conditional distributions of a discrete random variable Y.

Remark 17.29 [Discrete Y] If Y is discrete, then, according to Theorem 17.17, the conditional distribution $P_{Y|X}$ exists and, according to Definition 17.7, the conditional distributions $P_{Y|X=x}$ exist as well. Furthermore, the $(X=x)$-conditional distributions of Y are determined if the values $P_{Y|X=x}(\{y\})$ for the singletons $\{y\}$ are specified. More precisely, let $X: (\Omega, \mathscr{A}, P) \to (\Omega'_X, \mathscr{A}'_X)$ and $Y: (\Omega, \mathscr{A}, P) \to (\Omega'_Y, \mathscr{A}'_Y)$ be random variables, and assume that Y is discrete. Then, by Definition 17.1 (b) and Definition 17.7, the conditional distributions $P_{Y|X=x}$ are probability measures. Therefore, σ-additivity [see Def. 4.1 (c)] yields

$$\forall A' \in \mathscr{A}'_Y: \quad P_{Y|X=x}(A') \underset{P_X\text{-a.a.}}{=} \sum_{\substack{y \in \Omega'_Y \\ P(Y=y) > 0}} 1_{A'}(y) \cdot P_{Y|X=x}(\{y\}) \tag{17.28}$$

[see also Def. 5.56 and Eqs. (5.44) to (5.46)].

Note that, according to Theorem 3.19 (ii), any nonnegative measurable function $h: (\Omega'_Y, \mathscr{A}'_Y) \to (\mathbb{R}, \mathscr{B})$ can be described by a weighted sum of the indicators of a countable sequence $A'_1, A'_2, \dots \in \mathscr{A}'_Y$. Furthermore, a countable union of null sets $N_{A'_1}, N_{A'_2}, \dots$ [see

Rem. 17.20 (iii)] is a null set as well. Therefore, we can apply Remark 3.30, which yields: For all nonnegative measurable $h: (\Omega'_Y, \mathcal{A}'_Y) \to (\mathbb{R}, \mathcal{B})$,

$$\int h(y)\, P_{Y|X=x}(dy) \underset{P_X\text{-}a.a.}{=} \sum_{\substack{y \in \Omega'_Y \\ P(Y=y) > 0}} h(y) \cdot P_{Y|X=x}(\{y\}). \qquad (17.29)$$

$\triangleleft$

Remember, in Theorem 17.17 we showed that the conditional distribution $P_{Y|X}$ exists if Y is discrete. Now we prove that it is also CNS-unique.

Theorem 17.30 [CNS-uniqueness for a discrete Y]
Let $X: (\Omega, \mathcal{A}, P) \to (\Omega'_X, \mathcal{A}'_X)$ and $Y: (\Omega, \mathcal{A}, P) \to (\Omega'_Y, \mathcal{A}'_Y)$ be random variables, and assume that Y is discrete. Then $P_{Y|X}$ is CNS-unique.

(Proof p. 538)

Remark 17.31 [An implication of CNS-uniqueness]   Applying CNS-uniqueness of $P_{Y|X}$, we can strengthen Equation (17.28) as follows: If Y is discrete, then, for P_X-a.a. $x \in \Omega'_X$,

$$\forall A' \in \mathcal{A}'_Y: \quad P_{Y|X=x}(A') = \sum_{\substack{y \in \Omega'_Y \\ P(Y=y) > 0}} 1_{A'}(y) \cdot P_{Y|X=x}(\{y\}). \qquad (17.30)$$

This means that there is a common null set such that Equation (17.30) holds for all $A' \in \mathcal{A}'_Y$. In contrast, in (17.28), there can be different null sets for different $A' \in \mathcal{A}'_Y$.

$\triangleleft$

According to Corollary 5.24, the distributions of two P-equivalent random variables are identical. In the following corollary, we formulate a corresponding result for $(X=x)$-conditional distributions.

Corollary 17.32 [P-equivalence and conditional distributions]
Let $X: (\Omega, \mathcal{A}, P) \to (\Omega'_X, \mathcal{A}'_X)$ and $Y, Z: (\Omega, \mathcal{A}, P) \to (\Omega', \mathcal{A}')$ be random variables, and assume that $P_{Y|X}$ exists and $\mathcal{A}'$ is countably generated. If $Y \underset{P}{=} Z$, then $P_{Z|X}$ exists as well and

$$P_{Y|X=x} \underset{P_X\text{-}a.a.}{=} P_{Z|X=x}. \qquad (17.31)$$

(Proof p. 539)

17.4   Conditional-probability measure given a value of a random variable

Presuming $P(X=x) > 0$, we already introduced the $(X=x)$-conditional-probability measure $P^{X=x}: \mathcal{A} \to [0, 1]$ with

$$P^{X=x}(A) = P(A \mid X=x), \quad \forall A \in \mathcal{A}$$

[see Eq. (9.4)]. Now we extend this concept in such a way that the assumption $P(X=x) > 0$ is no longer required, utilizing the identity mapping introduced in Example 2.11. Furthermore, we consider the relationship between the $(X=x)$-conditional distributions $P_{Y|X=x}$ introduced in Definition 17.7 and the distributions of Y with respect to the conditional-probability measures $P^{X=x}$, $x \in \Omega'_X$.

Definition 17.33 [Conditional-probability measure given $X=x$]
Let $X: (\Omega, \mathcal{A}, P) \to (\Omega'_X, \mathcal{A}'_X)$ be a random variable, let id: $(\Omega, \mathcal{A}, P) \to (\Omega, \mathcal{A})$ denote the identity mapping, assume that $P_{id|X}$ exists, and let $(P_{id|X=x}, x \in \Omega'_X)$ be a family of $(X=x)$-conditional distributions of id pertaining to $P_{id|X}$. For all $x \in \Omega'_X$, the function $P^{X=x}: \mathcal{A} \to [0, 1]$ defined by

$$P^{X=x} := P_{id|X=x}, \qquad (17.32)$$

is called an $(X=x)$-conditional-probability measure on $(\Omega, \mathcal{A})$ pertaining to $P_{id|X}$.

Hence, a probability measure $P^{X=x}$ is a special $(X=x)$-conditional distribution, the $(X=x)$-conditional distribution of the identity mapping *id*. Existence and uniqueness of $(X=x)$-conditional distributions have been treated in section 17.3.

Definition 17.33 and Corollary 17.9 imply Corollary 17.34. Reading this corollary, remember that $P^*(A \mid X=x)$ denotes a value of a factorization of $P^*(A \mid X)$ (see Def. 10.33).

Corollary 17.34 [$(X=x)$-conditional probability]
Let $(P^{X=x}, x \in \Omega'_X)$ be a family of $(X=x)$-conditional-probability measures defined by Equation (17.32). Then, for all $A \in \mathcal{A}$ and all $P^*(A \mid X) \in \mathcal{P}(A \mid X)$,

$$P^{X=x}(A) \underset{P_X\text{-a.a.}}{=} P^*(A \mid X=x). \qquad (17.33)$$

(Proof p. 539)

Reading the following theorem, remember that according to Definition 5.3,

$$P_Y^{X=x}(A') = P^{X=x}[Y^{-1}(A')], \quad \forall A' \in \mathcal{A}'_Y,$$

defines the distribution of Y with respect to the measure $P^{X=x}$ defined by Equation (17.32).

Theorem 17.35 [Distribution of Y with respect to $P^{X=x}$]
Let $X: (\Omega, \mathcal{A}, P) \to (\Omega'_X, \mathcal{A}'_X)$, $Y: (\Omega, \mathcal{A}, P) \to (\Omega'_Y, \mathcal{A}'_Y)$ be random variables, and assume that $P_{Y|X}$ and $P_{id|X}$ exist. Furthermore, let $(P_{Y|X=x}, x \in \Omega'_X)$ denote a family of $(X=x)$-conditional distributions of Y pertaining to $P_{Y|X}$, and let $(P^{X=x}, x \in \Omega'_X)$ be a family of $(X=x)$-conditional-probability measures pertaining to $P_{id|X}$. Then

$$\forall A' \in \mathcal{A}'_Y: P_{Y|X=x}(A') \underset{P_X\text{-a.a.}}{=} P_Y^{X=x}(A'), \qquad (17.34)$$

and, if $\mathscr{A}'_Y$ is countably generated, then,

$$P_{Y|X=x} \underset{P_X\text{-}a.a.}{=} P_Y^{X=x}, \tag{17.35}$$

or, equivalently, for

$$P_X\text{-}a.a. \ x \in \Omega'_X, \forall A' \in \mathscr{A}'_Y: \quad P_{Y|X=x}(A') = P_Y^{X=x}(A'). \tag{17.36}$$

(Proof p. 540)

Hence, if $\mathscr{A}'_Y$ is countably generated, then the $(X=x)$-conditional distribution of Y and the distribution of Y with respect to an $(X=x)$-conditional-probability measure $P^{X=x}$ are identical for P_X-a.a. $x \in \Omega'_X$. This theorem extends the case $P(X=x) > 0$ already considered in Corollary 17.10.

Remark 17.36 [Uniqueness and consistency of definitions] Let $X: (\Omega, \mathscr{A}, P) \to (\Omega'_X, \mathscr{A}'_X)$ be a random variable, and let $(P^{X=x}, x \in \Omega'_X)$ be a family defined by Equation (17.32). If $x \in \Omega'_X$ with $P(X=x) > 0$, then $P^{X=x}$ defined by (17.32) is uniquely defined, and it is identical to $P^{X=x}$ defined in Equation (9.4) (see Exercise 17.1). ◁

17.5   Decomposing the joint distribution of random variables

In section 4.2.5, we treated the Theorem of Total Probability. If we consider the random variables $X: (\Omega, \mathscr{A}, P) \to (\Omega'_X, \mathscr{A}'_X)$ and $Y: (\Omega, \mathscr{A}, P) \to (\Omega'_Y, \mathscr{A}'_Y)$, assuming that X is discrete, and the event $\{Y \in B'\}$, $B' \in \mathscr{A}'_Y$, then this theorem yields

$$P(Y \in B') = \sum_{\substack{x \in \Omega'_X \\ P(X=x) > 0}} P(Y \in B' \mid X=x) \cdot P(X=x), \tag{17.37}$$

and, for the event $\{X \in A'\}$, $A' \in \mathscr{A}'_X$,

$$\begin{aligned}
P(X \in A', Y \in B') &= \sum_{\substack{x \in \Omega'_X \\ P(X=x) > 0}} P(X \in A', Y \in B' \mid X=x) \cdot P(X=x) \\
&= \sum_{\substack{x \in \Omega'_X \\ P(X=x) > 0}} 1_{A'}(x) \cdot P(Y \in B' \mid X=x) \cdot P(X=x)
\end{aligned} \tag{17.38}$$

(see Exercise 17.2). Furthermore, if Y is discrete as well, then

$$P(X \in A', Y \in B') = \sum_{\substack{x \in \Omega'_X \\ P(X=x) > 0}} \sum_{\substack{y \in \Omega'_Y \\ P(Y=y) > 0}} 1_{A'}(x) \cdot 1_{B'}(y) \cdot P(Y=y \mid X=x) \cdot P(X=x). \tag{17.39}$$

This equation is generalized in the following theorem:

Theorem 17.37 [Decomposition]
Let $X: (\Omega, \mathcal{A}, P) \to (\Omega'_X, \mathcal{A}'_X)$ and $Y: (\Omega, \mathcal{A}, P) \to (\Omega'_Y, \mathcal{A}'_Y)$ be random variables. If $P_{Y|X} \in \mathscr{P}_{Y|X}$, then for all $(A', B') \in \mathcal{A}'_X \times \mathcal{A}'_Y$,

$$P(X \in A', Y \in B') = \int 1_{A'}(x) \cdot P_{Y|X=x}(B') \, P_X(dx) \tag{17.40}$$

$$= \iint 1_{A'}(x) \cdot 1_{B'}(y) \, P_{Y|X=x}(dy) \, P_X(dx), \tag{17.41}$$

as well as for all $C' \in \mathcal{A}'_X \otimes \mathcal{A}'_Y$,

$$P_{X,Y}(C') = \iint 1_{C'}(x, y) \, P_{Y|X=x}(dy) \, P_X(dx). \tag{17.42}$$

In addition, for all $P_{X,Y}$-integrable or nonnegative measurable functions $h: \Omega'_X \times \Omega'_Y \to \overline{\mathbb{R}},$

$$\int h(x, y) \, P_{X,Y}[d(x, y)] = \iint h(x, y) \, P_{Y|X=x}(dy) \, P_X(dx). \tag{17.43}$$

(Proof p. 540)

Remark 17.38 [Marginal distribution] For $A' = \Omega'_X$, Equations (17.40), (5.2), and (5.3) yield

$$P_Y(B') = P(Y \in B') = \int P_{Y|X=x}(B') \, P_X(dx), \quad \forall \, B' \in \mathcal{A}'_Y. \tag{17.44}$$

This equation generalizes Equation (17.37); it also holds if X is continuous. If X is discrete, then Equation (6.15) or (6.16), respectively, yields

$$P_Y(B') = P(Y \in B') = \sum_{\substack{x \in \Omega'_X \\ P(X=x) > 0}} P_{Y|X=x}(B') \cdot P(X=x), \quad \forall \, B' \in \mathcal{A}'_Y, \tag{17.45}$$

which is equivalent to Equation (17.37) (see Lemma 17.8).   ◁

Remark 17.39 [Joint distribution for discrete X] If X is discrete, then applying Equations (5.45), (6.15), or (6.16) shows that Equation (17.40) is equivalent to

$$P(X \in A', Y \in B') = \sum_{\substack{x \in \Omega'_X \\ P(X=x) > 0}} 1_{A'}(x) \cdot P_{Y|X=x}(B') \cdot P(X=x) \tag{17.46}$$

$$= \sum_{\substack{x \in A' \\ P(X=x) > 0}} P_{Y|X=x}(B') \cdot P(X=x), \tag{17.47}$$

which is equivalent to Equation (17.38) (see Lemma 17.8).   ◁

Remark 17.40 [Mixture distributions] Reading the equations in Theorem 17.37 and Remark 17.38 from right to left yields formulas for mixture distributions. If $(X, Y): (\Omega, \mathcal{A}, P) \to (\Omega'_X \times \Omega'_Y, \mathcal{A}'_X \otimes \mathcal{A}'_Y)$ is a bivariate random variable, then the joint distribution $P_{X,Y}$ and the marginal distribution P_Y can be determined from a family $(P_{Y|X=x}, x \in \Omega'_X)$ of $(X=x)$-conditional distributions of Y and the distribution P_X. This is illustrated in Example 17.80 using densities. ◁

In Lemma 17.41, $P_{Y|Z}^{X=x}$ denotes a version of the Z-conditional distribution of Y with respect to the conditional-probability measure $P^{X=x}$ that has been defined by Equation (17.32). In other words, $P_{Y|Z}^{X=x}$ is specified by Definition 17.1 replacing P by the measure $P^{X=x}$. Furthermore, $\mathscr{P}_{Y|Z}^{X=x}$ denotes the set of all versions of the Z-conditional distribution of Y with respect to $P^{X=x}$ (see Rem. 17.5). Correspondingly, a $(Z=z)$-conditional distribution $P_{Y|Z=z}^{X=x}$ of Y with respect to $P^{X=x}$ is specified by Definition 17.7 replacing P by $P^{X=x}$.

Lemma 17.41 [$(Z=z)$-conditional distribution of Y with respect to $P^{X=x}$]
Let $X: (\Omega, \mathcal{A}, P) \to (\Omega'_X, \mathcal{A}'_X)$, $Y: (\Omega, \mathcal{A}, P) \to (\Omega'_Y, \mathcal{A}'_Y)$, and $Z: (\Omega, \mathcal{A}, P) \to (\Omega'_Z, \mathcal{A}'_Z)$ be random variables, and assume that the conditional distributions $P_{Y|Z}, P_{Y|X,Z}$, and $P_{Z|X}$ exist. Furthermore, for all $x \in \Omega'_X$, assume that the $(X=x)$-conditional-probability measures $P^{X=x}$ and $P_{Y|Z}^{X=x}$ exist. Then,

$$\forall\, B' \in \mathcal{A}'_Y: \quad P_{Y|Z=z}^{X=x}(B') \underset{P_{X,Z}\text{-a.a.}}{=} P_{Y|X=x, Z=z}(B'). \tag{17.48}$$

(Proof p. 541)

In other words, for each $B' \in \mathcal{A}'_Y$, the two probabilities $P_{Y|Z=z}^{X=x}(B')$ and $P_{Y|X=x, Z=z}(B')$ are equal for $P_{X,Z}$-almost all $(x, z) \in \Omega'_X \times \Omega'_Z$.

In Corollary 17.42, we assume $P(X=x) > 0$, which allows us to reformulate Proposition (17.48) in terms of the conditional distribution $P_Z^{X=x}$.

Corollary 17.42 [The special case $P(X=x) > 0$]
Let $X: (\Omega, \mathcal{A}, P) \to (\Omega'_X, \mathcal{A}'_X)$, $Y: (\Omega, \mathcal{A}, P) \to (\Omega'_Y, \mathcal{A}'_Y)$, and $Z: (\Omega, \mathcal{A}, P) \to (\Omega'_Z, \mathcal{A}'_Z)$ be random variables, and assume that $x \in \Omega'_X$ with $P(X=x) > 0$. Furthermore, assume that $P_{Y|Z}, P_{Y|X,Z}, P_{Z|X}$, and $P_{Y|Z}^{X=x}$ exist. Then,

$$\forall\, B' \in \mathcal{A}'_Y: \quad P_{Y|Z=z}^{X=x}(B') \underset{P_Z^{X=x}}{=} P_{Y|X=x, Z=z}(B'). \tag{17.49}$$

(Proof p. 542)

17.6   Conditional independence and conditional distributions

In Lemma 5.49, we showed that independence of random variables implies that their joint distribution is identical to the product measure of their distributions, and vice versa. Furthermore, in Theorem 16.32, we presented a product rule for conditional expectations. Now we turn to product rules for conditional distributions.

Theorem 17.43 [Product rule]

Let $X: (\Omega, \mathcal{A}, P) \to (\Omega_X', \mathcal{A}_X')$, $Y: (\Omega, \mathcal{A}, P) \to (\Omega_Y', \mathcal{A}_Y')$, and $Z: (\Omega, \mathcal{A}, P) \to (\Omega_Z', \mathcal{A}_Z')$ be random variables and assume that $P_{X,Y|Z}$, $P_{X|Z}$, and $P_{Y|Z}$ exist.

(i) *The following four propositions are equivalent to each other:*

(a) $X \underset{P}{\perp\!\!\!\perp} Y \mid Z$.

(b) *For all $(A', B') \in \mathcal{A}_X' \times \mathcal{A}_Y'$,*

$$P_{X,Y|Z=z}(A' \times B') \underset{P_Z\text{-a.a.}}{=} P_{X|Z=z}(A') \cdot P_{Y|Z=z}(B'). \qquad (17.50)$$

(c) *For all $D' \in \mathcal{A}_X' \otimes \mathcal{A}_Y'$,*

$$P_{X,Y|Z=z}(D') \underset{P_Z\text{-a.a.}}{=} P_{X|Z=z} \otimes P_{Y|Z=z}(D'). \qquad (17.51)$$

(d) *For all measurable nonnegative or $P_{X,Y,Z}$-integrable functions*
$h: (\Omega_X' \times \Omega_Y' \times \Omega_Z', \mathcal{A}_X' \otimes \mathcal{A}_Y' \otimes \mathcal{A}_Z') \to (\mathbb{R}, \mathcal{B})$,

$$\iint h(x, y, z) \, P_{X,Y|Z=z}[d(x,y)] \, P_Z(dz)$$

$$= \iiint h(x, y, z) \, P_{X|Z=z}(dx) \, P_{Y|Z=z}(dy) \, P_Z(dz). \qquad (17.52)$$

(ii) *If $\mathcal{A}_X'$ and $\mathcal{A}_Y'$ are countably generated, then each of (a) to (d) is equivalent to*

$$P_{X,Y|Z=z} \underset{P_Z\text{-a.a.}}{=} P_{X|Z=z} \otimes P_{Y|Z=z}. \qquad (17.53)$$

(Proof p. 542)

Now consider the special case $Z = X$. Because $X \underset{P}{\perp\!\!\!\perp} Y \mid X$ (see Rem. 16.18), Theorem 17.43 implies Corollary 17.44.

Corollary 17.44 [X-conditional distribution of (X, Y)]

Let $X: (\Omega, \mathcal{A}, P) \to (\Omega_X', \mathcal{A}_X')$ and $Y: (\Omega, \mathcal{A}, P) \to (\Omega_Y', \mathcal{A}_Y')$ be random variables. If the conditional distributions $P_{X,Y|X}$ and $P_{Y|X}$ exist, then,

(i) *For all measurable functions $h: (\Omega_X' \times \Omega_Y', \mathcal{A}_X' \otimes \mathcal{A}_Y') \to (\mathbb{R}, \mathcal{B})$ that are nonnegative,*

$$\int h(x^*, y) P_{X,Y|X=x}[d(x^*, y)] \underset{P_X\text{-a.a.}}{=} \int h(x, y) \, P_{Y|X=x}(dy). \qquad (17.54)$$

(ii) For all $D' \in \mathscr{A}'_X \otimes \mathscr{A}'_Y$,

$$P_{X,Y|X=x}(D') \underset{P_X\text{-}a.a.}{=} P_{Y|X=x}(D'_x), \qquad (17.55)$$

where $D'_x := \{y \in \Omega'_Y : (x, y) \in D'\}$.

(Proof p. 544)

Reading Theorem 17.45, remember that

$$P_{Y|Z}(\cdot, B') \underset{P}{=} E(1_{Y \in B'} \mid Z)$$

[see Def. 17.1 (a) and Eq. (17.4)].

Theorem 17.45 [Conditional distribution and conditional independence]
*Let $X: (\Omega, \mathscr{A}, P) \to (\Omega'_X, \mathscr{A}'_X)$, $Y: (\Omega, \mathscr{A}, P) \to (\Omega'_Y, \mathscr{A}'_Y)$, and $Z: (\Omega, \mathscr{A}, P) \to (\Omega'_Z, \mathscr{A}'_Z)$
be random variables and assume that the conditional distributions $P_{Y|Z}$ and $P_{Y|X,Z}$ exist.
Then, $X \underset{P}{\perp\!\!\!\perp} Y \mid Z$ if and only if*

$$\forall\, B' \in \mathscr{A}'_Y : \quad P_{Y|X,Z}(\cdot, B') \underset{P}{=} P_{Y|Z}(\cdot, B'). \qquad (17.56)$$

(Proof p. 544)

Remark 17.46 [Conditional distribution and conditional independence]   In other words,
under the assumptions of Theorem 17.45, $P_{Y|Z} \in \mathscr{P}_{Y|X,Z}$ if and only if $X \underset{P}{\perp\!\!\!\perp} Y \mid Z$ (see Exercise 17.3).   ◁

Remark 17.47 [Equivalent equations]   According to Equation (17.10) and Corollary 5.25
(i), Equation (17.56) is equivalent to

$$\forall\, B' \in \mathscr{A}'_Y : \quad P_{Y|X=x,Z=z}(B') \underset{P_{X,Z}\text{-}a.a.}{=} P_{Y|Z=z}(B'). \qquad (17.57)$$

If $\mathscr{A}'_Y$ is countably generated, then, according to Corollary 17.27, Equation (17.56) is also
equivalent to

$$P_{Y|X=x,Z=z} \underset{P_{X,Z}\text{-}a.a.}{=} P_{Y|Z=z}. \qquad (17.58)$$

For Z being a constant, this implies: If $\mathscr{A}'_Y$ is countably generated, then, according to Equation
(17.10) and Corollary 5.25 (i), Equation (17.58) is equivalent to

$$P_{Y|X=x} \underset{P_X\text{-}a.a.}{=} P_Y. \qquad (17.59)$$

◁

Replacing X by $1_{X=x}$, Theorem 17.45 and Remark 17.47 imply Corollary 17.48.

Corollary 17.48 [Conditional independence of $1_{X=x}$ and Y]
Let $X: (\Omega, \mathscr{A}, P) \to (\Omega'_X, \mathscr{A}'_X)$, $Y: (\Omega, \mathscr{A}, P) \to (\Omega'_Y, \mathscr{A}'_Y)$, and $Z: (\Omega, \mathscr{A}, P) \to (\Omega'_Z, \mathscr{A}'_Z)$ be random variables, and assume that $P_{Y|X,Z}$ and $P_{Y|Z}$ exist. If $x \in \Omega'_X$ with $P(X=x) > 0$ and $1_{X=x} \underset{P}{\perp\!\!\!\perp} Y \mid Z$, then

$$\forall B' \in \mathscr{A}'_Y: \quad P_{Y|X=x, Z=z}(B') = P_{Y|Z=z}(B'), \quad \text{for } P_{Z|X=x}\text{-a.a. } z \in \Omega'_Z. \tag{17.60}$$

(Proof p. 545)

Remark 17.49 [Countably generated] If, additionally to the assumptions of Corollary 17.48 $\mathscr{A}'_Y$ is countably generated, then

$$P_{Y|X=x, Z=z} = P_{Y|Z=z}, \quad \text{for } P_{Z|X=x}\text{-a.a. } z \in \Omega'_Z. \tag{17.61}$$

◁

The following corollary immediately follows from Remarks 17.47 and Theorems 17.45 and 16.34.

Corollary 17.50 [Conditional mean independence]
Let $X: (\Omega, \mathscr{A}, P) \to (\Omega'_X, \mathscr{A}'_X)$, $Y: (\Omega, \mathscr{A}, P) \to (\mathbb{R}, \mathscr{B})$, and $Z: (\Omega, \mathscr{A}, P) \to (\Omega'_Z, \mathscr{A}'_Z)$ be random variables, and assume that Y is nonnegative or with finite expectation. Then,

$$\left(\forall B' \in \mathscr{B}: \quad P_{Y|X=x, Z=z}(B') \underset{P_{X,Z}\text{-a.a.}}{=} P_{Y|Z=z}(B')\right) \Rightarrow E(Y \mid X, Z) \underset{P}{=} E(Y \mid Z).$$

The following corollary combines the decomposition theorem 17.37 with proposition (17.57).

Corollary 17.51 [Decomposition under conditional independence]
Let $X: (\Omega, \mathscr{A}, P) \to (\Omega'_X, \mathscr{A}'_X)$, $Y: (\Omega, \mathscr{A}, P) \to (\Omega'_Y, \mathscr{A}'_Y)$, and $Z: (\Omega, \mathscr{A}, P) \to (\Omega'_Z, \mathscr{A}'_Z)$ be random variables, and assume that the conditional distributions $P_{Y|X,Z}$, $P_{Y|Z}$, and $P_{X|Z}$ exist. If $X \underset{P}{\perp\!\!\!\perp} Y \mid Z$, then for all nonnegative measurable functions $h: (\Omega'_X \times \Omega'_Y \times \Omega'_Z, \mathscr{A}'_X \otimes \mathscr{A}'_Y \otimes \mathscr{A}'_Z) \to (\mathbb{R}, \mathscr{B})$,

$$\iiint h(x, y, z)\, P_{Y|Z=z}(dy)\, P_{X|Z=z}(dx)\, P_Z(dz)$$
$$= \iiint h(x, y, z)\, P_{Y|X=x, Z=z}(dy)\, P_{X|Z=z}(dx)\, P_Z(dz), \tag{17.62}$$

and for all $(A', B') \in \mathscr{A}'_X \times \mathscr{A}'_Y$,

$$P_{X|Z=z}(A') \cdot P_{Y|Z=z}(B') \underset{P_Z\text{-a.a.}}{=} \int 1_{A'}(x) \cdot P_{Y|X=x, Z=z}(B')\, P_{X|Z=z}(dx). \tag{17.63}$$

(Proof p. 546)

Now we consider the case $1_{X=x} \perp\!\!\!\perp_P Y$. In contrast to Lemma 17.14, the assumption of independence refers only to the event $\{X=x\}$ and not to the random variable X, unless $\sigma(X)$ and $\sigma(1_{X=x})$ are identical.

Corollary 17.52 [Independence of $1_{X=x}$ and Y]
Let $X\colon (\Omega, \mathcal{A}, P) \to (\Omega'_X, \mathcal{A}'_X)$ and $Y\colon (\Omega, \mathcal{A}, P) \to (\Omega'_Y, \mathcal{A}'_Y)$ be random variables, and assume that $P_{Y|X}$ exists. If $x \in \Omega'_X$ with $P(X=x) > 0$, then the following propositions are equivalent to each other:

(a) $1_{X=x} \perp\!\!\!\perp_P Y$.

(b) $P_{Y|X=x} = P_Y$.

(Proof p. 547)

In the following corollary, we introduce assumptions that are stronger than those made in Corollary 17.48. This yields a proposition for P_Z-a.a. $z \in \Omega'_Z$ instead of the corresponding proposition for $P_{Z|X=x}$-a.a. $z \in \Omega'_Z$.

Corollary 17.53 [Z-conditional independence of X and Y and $P_{Y|X=x,Z=z}$]
Let $X\colon (\Omega, \mathcal{A}, P) \to (\Omega'_X, \mathcal{A}'_X)$, $Y\colon (\Omega, \mathcal{A}, P) \to (\Omega'_Y, \mathcal{A}'_Y)$, and $Z\colon (\Omega, \mathcal{A}, P) \to (\Omega'_Z, \mathcal{A}'_Z)$ be random variables. Furthermore, assume that $P_{Y|X,Z}$ and $P_{Y|Z}$ exist and $X \perp\!\!\!\perp_P Y \mid Z$.

(i) Then, for all $x \in \Omega'_X$:

$$P(X=x \mid Z) \underset{P}{>} 0 \;\Rightarrow\; P_{Y|X=x,Z=z}(B') \underset{P_Z\text{-a.a.}}{=} P_{Y|Z=z}(B'), \quad \forall\, B' \in \mathcal{A}'_Y. \quad (17.64)$$

(ii) Furthermore, if $\mathcal{A}'_Y$ is countably generated, then,

$$\forall\, x \in \Omega'_X\colon \quad P(X=x \mid Z) \underset{P}{>} 0 \;\Rightarrow\; P_{Y|X=x,Z=z} \underset{P_Z\text{-a.a.}}{=} P_{Y|Z=z}. \quad (17.65)$$

(Proof p. 547)

Lemma 17.54 provides a proposition equivalent to $P(X=x \mid Z) \underset{P}{>} 0$ (see the premise of Cor. 17.53, and cf. Cor. 14.48).

Lemma 17.54 [A condition equivalent to $P(X=x \mid Z) \underset{P}{>} 0$]
Let $X\colon (\Omega, \mathcal{A}, P) \to (\Omega'_X, \mathcal{A}'_X)$ and $Z\colon (\Omega, \mathcal{A}, P) \to (\Omega'_Z, \mathcal{A}'_Z)$ be random variables. Then, for all $x \in \Omega'_X$, the following propositions are equivalent to each other:

(a) $P(X=x \mid Z) \underset{P}{>} 0$.

(b) $P(X=x) > 0$ and $P_Z \underset{\mathcal{A}'_Z}{\ll} P_Z^{X=x}$.

(Proof p. 548)

This lemma shows that $P(X=x \mid Z) \underset{P}{>} 0$ is stronger than $P(X=x) > 0$, assumed in Corollary 17.48. It is equivalent to assuming $P(X=x) > 0$ *and* absolute continuity of the measure P_Z with respect to $P_Z^{X=x}$ (see Def. 3.70).

17.7 Expectations with respect to a conditional distribution

The conditional expectation value $E(Y \mid X=x)$ has been defined in section 9.1 as the expectation of Y with respect to the $P^{X=x}$-conditional-probability measure, provided that $P(X=x) > 0$. In section 10.4.4, it has been defined more generally as the value $g(x)$ of a factorization of a version $g(X) = E(Y \mid X) \in \mathscr{E}(Y \mid X)$. In this section, we show how it can be computed from the $(X=x)$-conditional distribution $P_{Y\mid X=x}$. We start by considering $E[f(Y) \mid X=x]$.

Theorem 17.55 [Conditional expectation value of a composition]
Let $X: (\Omega, \mathscr{A}, P) \to (\Omega'_X, \mathscr{A}'_X)$ and $Y: (\Omega, \mathscr{A}, P) \to (\Omega'_Y, \mathscr{A}'_Y)$ be random variables. If $P_{Y\mid X} \in \mathscr{P}_{Y\mid X}, f: (\Omega'_Y, \mathscr{A}'_Y) \to (\overline{\mathbb{R}}, \overline{\mathscr{B}})$ is a measurable function that is nonnegative or such that $E[f(Y)]$ is finite, and $E[f(Y) \mid X] \in \mathscr{E}[f(Y) \mid X]$, then,

$$E[f(Y) \mid X=x] = \int f(y)\, P_{Y\mid X=x}(dy), \quad \text{for } P_X\text{-a.a. } x \in \Omega'_X. \tag{17.66}$$

(Proof p. 548)

Remark 17.56 [A special case]   Let $X: (\Omega, \mathscr{A}, P) \to (\Omega'_X, \mathscr{A}'_X)$, and $Y: (\Omega, \mathscr{A}, P) \to (\overline{\mathbb{R}}, \overline{\mathscr{B}})$ be random variables. If Y is nonnegative or with finite expectation $E(Y)$, then,

$$E(Y \mid X=x) = \int y\, P_{Y\mid X=x}(dy), \quad \text{for } P_X\text{-a.a. } x \in \Omega'_X. \tag{17.67}$$

This equation immediately follows from Theorem 17.55 if the identity function $id: (\overline{\mathbb{R}}, \overline{\mathscr{B}}) \to (\overline{\mathbb{R}}, \overline{\mathscr{B}})$ takes the role of f.   ◁

Corollary 17.57 [Transformation theorem for conditional distributions]
Let $X: (\Omega, \mathscr{A}, P) \to (\Omega'_X, \mathscr{A}'_X)$ and $Y: (\Omega, \mathscr{A}, P) \to (\Omega'_Y, \mathscr{A}'_Y)$ be random variables, let $f: (\Omega'_Y, \mathscr{A}'_Y) \to (\overline{\mathbb{R}}, \overline{\mathscr{B}})$ be a nonnegative measurable function or such that $E[f(Y)]$ is finite, and assume that $P_{Y\mid X} \in \mathscr{P}_{Y\mid X}$ and $P_{f(Y)\mid X} \in \mathscr{P}_{f(Y)\mid X}$. Then,

$$E[f(Y) \mid X=x] = \int_{\Omega'_Y} f(y)\, P_{Y\mid X=x}(dy)$$

$$= \int_{\overline{\mathbb{R}}} z\, P_{f(Y)\mid X=x}(dz), \quad \text{for } P_X\text{-a.a. } x \in \Omega'_X. \tag{17.68}$$

(Proof p. 549)

Remark 17.58 [Domains of integration] Note that $P_{Y|X=x}$ is a probability measure on $(\Omega'_Y, \mathscr{A}'_Y)$, whereas $P_{f(Y)|X=x}$ is a probability measure on $(\overline{\mathbb{R}}, \overline{\mathscr{B}})$. Hence, in the first line of (17.68), integration is over all $y \in \Omega'_Y$, whereas integration in the second line is over all $z \in \overline{\mathbb{R}}$.   ◁

Remark 17.59 [Discrete Y] If Y takes on only a finite number of values $y_1, \dots, y_n$, then Equation (17.66) can also be written as:

$$E[f(Y) \mid X=x] = \sum_{i=1}^{n} f(y_i) \cdot P_{Y|X=x}(\{y_i\}) \quad \text{for } P_X\text{-a.a. } x \in \Omega'_X, \tag{17.69}$$

and if, additionally, $P(X=x) > 0$, then,

$$E[f(Y) \mid X=x] = \sum_{i=1}^{n} f(y_i) \cdot P(Y=y_i \mid X=x), \tag{17.70}$$

with $P(Y=y_i \mid X=x) = P_Y^{X=x}(\{y_i\})$, where $P_Y^{X=x}$ denotes the distribution of Y with respect to the $(X=x)$-conditional-probability measure $P^{X=x}$ [see Eqs. (17.15) and (17.11)] (see Exercise 17.4).   ◁

According to Equation (17.67), the conditional expectation value $E(Y \mid X=x)$ is identical to the expectation of Y with respect to a conditional distribution $P_{Y|X=x}$. Now consider the equation

$$E(Y \mid X) \underset{P}{=} E[E(Y \mid X, Z) \mid X], \tag{17.71}$$

which is a special case of Rule (v) of Box 10.2. According to Theorem 17.60, this equation for conditional expectations can also be written in terms of expectations with respect to conditional distributions.

Theorem 17.60 [Marginalization]
Let $X: (\Omega, \mathscr{A}, P) \to (\Omega'_X, \mathscr{A}'_X)$, $Y: (\Omega, \mathscr{A}, P) \to (\overline{\mathbb{R}}, \overline{\mathscr{B}})$, and $Z: (\Omega, \mathscr{A}, P) \to (\Omega'_Z, \mathscr{A}'_Z)$ be random variables, and let Y be nonnegative or with finite expectation. If $P_{Z|X}$ exists, then,

$$E(Y \mid X=x) \underset{P_X\text{-a.a.}}{=} \int E(Y \mid X=x, Z=z) \, P_{Z|X=x}(dz). \tag{17.72}$$

(Proof p. 549)

Remark 17.61 [The case $P(X=x) > 0$] Suppose that $P(X=x) > 0$ for an $x \in \Omega'_X$. Then $P_{Z|X=x}$ is uniquely defined (see Cor. 17.10), and in this case Theorem 17.60 implies

$$E(Y \mid X=x) = \int E(Y \mid X=x, Z=z) \, P_{Z|X=x}(dz). \tag{17.73}$$

◁

Corollary 17.62 [Discrete Z]
If Z is discrete, then Equation (17.72) can be written as:

$$E(Y \mid X=x) \underset{P_X\text{-a.a.}}{=} \sum_{i=1}^{n} E(Y \mid X=x, Z=z_i) \cdot P(Z=z_i \mid X=x). \qquad (17.74)$$

If additionally $P(X=x) > 0$, then,

$$E(Y \mid X=x) = \sum_{i=1}^{n} E(Y \mid X=x, Z=z_i) \cdot P(Z=z_i \mid X=x). \qquad (17.75)$$

(Proof p. 549)

Theorem 17.60 and Equation (17.58) imply Corollary 17.63:

Corollary 17.63 [Conditional independence among regressors]
Let $X: (\Omega, \mathcal{A}, P) \to (\Omega'_X, \mathcal{A}'_X)$, $Y: (\Omega, \mathcal{A}, P) \to (\overline{\mathbb{R}}, \overline{\mathcal{B}})$, $Z: (\Omega, \mathcal{A}, P) \to (\Omega'_Z, \mathcal{A}'_Z)$, and $W: (\Omega, \mathcal{A}, P) \to (\Omega'_W, \mathcal{A}'_W)$ be random variables, and let Y be nonnegative or with finite expectation. If $P_{W|Z}$ exists and $X \underset{P}{\perp\!\!\!\perp} W \mid Z$, then,

$$E(Y \mid X=x, Z=z) \underset{P_{X,Z}\text{-a.a.}}{=} \int E(Y \mid X=x, Z=z, W=w)\, P_{W|Z=z}(dw). \qquad (17.76)$$

(Proof p. 549)

Example 17.64 [Joe and Ann with self-selection – continued]   Let us use the example displayed in Table 17.1 to illustrate Corollary 17.62. Applying Equation (17.75) to the conditional expectation values $E(Y \mid X=0)$ and $E(Y \mid X=0, U=u)$ yields,

$$E(Y \mid X=0) = \sum_{u} E(Y \mid X=0, U=u) \cdot P(U=u \mid X=0)$$

$$= E(Y \mid X=0, U=Joe) \cdot P(U=Joe \mid X=0)$$

$$+ E(Y \mid X=0, U=Ann) \cdot P(U=Ann \mid X=0)$$

$$= .7 \cdot \frac{.144 + .336}{.144 + .336 + .096 + .024} + .2 \cdot \frac{.096 + .024}{.144 + .336 + .096 + .024}$$

$$= .56 + .04 = .6.$$

This is exactly the same number as obtained in Example 9.22 using a different formula for its computation.   ◁

17.8   Conditional distribution function and probability density

Consider a bivariate real-valued random variable $(X, Y)\colon (\Omega, \mathscr{A}, P) \to (\mathbb{R}^2, \mathscr{B}_2)$ and its joint distribution function [see Eq. (5.56)] defined by:

$$F_{X,Y}(x, y) = P(X \leq x, Y \leq y), \quad \forall\, (x, y) \in \mathbb{R}^2. \tag{17.77}$$

If there is a nonnegative Riemann-integrable function $f_{X,Y}\colon \mathbb{R}^2 \to [0, \infty]$ with:

$$F_{X,Y}(x, y) = \int_{-\infty}^{y} \int_{-\infty}^{x} f_{X,Y}(u, v)\, du\, dv, \quad \forall\, (x, y) \in \mathbb{R}^2 \tag{17.78}$$

(see Remark 5.99), then $f_{X,Y}$ is a joint probability density of (X, Y). Often, it is useful to represent a conditional distribution $P_{Y|X=x}$ by its conditional distribution function and – if it exists – by a conditional probability density. These functions can be constructed as follows: If, for $x \in \mathbb{R}$ and $h > 0$, we assume $P(X \in\,]x-h, x]) = P(x-h < X \leq x) > 0$, then,

$$P(Y \leq y \mid x-h < X \leq x) = \frac{P(Y \leq y, x-h < X \leq x)}{P(x-h < X \leq x)}. \tag{17.79}$$

Furthermore, Theorem 1.68 (ii) yields

$$\lim_{h\downarrow 0} P(x-h < X \leq x) = P(X=x) \tag{17.80}$$

and

$$\lim_{h\downarrow 0} P(Y \leq y, x-h < X \leq x) = P(Y \leq y, X=x), \tag{17.81}$$

where $\lim_{h\downarrow 0}$ denotes the limit for $h \to 0$ from above. If X is a continuous random variable, these probabilities are 0, but nevertheless, the limit of the ratio in Equation (17.79) can have a value in the interval $[0, 1]$. Reading the following theorem, note the distinction between a continuous random variable (see Def. 5.94) and a continuous function in the sense of calculus (see ch. 2 of Ellis & Gulick, 2006).

> **Theorem 17.65 [Conditional distribution function and density]**
> Let $(X, Y)\colon (\Omega, \mathscr{A}, P) \to (\mathbb{R}^2, \mathscr{B}_2)$ be a continuous bivariate real-valued random variable with joint density $f_{X,Y}$ with respect to the Lebesgue measure. Furthermore, suppose $f_{X,Y}$ as well as the marginal density f_X are continuous functions. Then, for all $x \in \mathbb{R}$ with $f_X(x) > 0$:
>
> (i) The limit
>
> $$F_{Y|X=x}(y) := \lim_{h\downarrow 0} P(Y \leq y \mid x-h \leq X < x) \tag{17.82}$$

exists for all $y \in \mathbb{R}$, and it can be written as:

$$F_{Y|X=x}(y) = \frac{\int_{-\infty}^{y} f_{X,Y}(x, v)\, dv}{f_X(x)}, \quad \forall y \in \mathbb{R}. \tag{17.83}$$

(ii) *The function $F_{Y|X=x}: \mathbb{R} \to [0, 1]$ is the distribution function of a probability measure on $(\mathbb{R}, \mathscr{B})$, and it has a probability density $f_{Y|X=x}$ satisfying*

$$f_{Y|X=x}(y) = \frac{f_{X,Y}(x, y)}{f_X(x)}, \quad \forall y \in \mathbb{R}. \tag{17.84}$$

For a proof, see Fisz (1963).

Remark 17.66 [Joint density]   Equation (17.84) immediately yields

$$f_{X,Y}(x, y) = f_{Y|X=x}(y) \cdot f_X(x), \quad \forall y \in \mathbb{R}, \ \forall x \in \mathbb{R} \text{ with } f_X(x) > 0. \tag{17.85}$$

◁

Remark 17.67 [Convention]   For $f_X(x) = 0$, we define $F_{Y|X=x}(y) := 0$ as well as $f_{Y|X=x}(y) := 0$.   ◁

Lemma 17.68 [Consistency]
Under the assumptions of Theorem 17.65, let $(F_{Y|X=x}, x \in \mathbb{R})$ be the family of functions defined by Equation (17.82) and Remark 17.67. Then, there is a $P_{Y|X} \in \mathscr{P}_{Y|X}$ with a family $(P_{Y|X=x}, x \in \mathbb{R})$ of $(X=x)$-conditional distributions of Y pertaining to $P_{Y|X}$ such that, for P_X-a.a. $x \in \mathbb{R}$, $F_{Y|X=x}$ is the distribution function of $P_{Y|X=x}$.

(Proof p. 550)

This lemma justifies Definition 17.69.

Definition 17.69 [Conditional distribution function and density]
Let the assumptions of Theorem 17.65 be satisfied. Then,

(i) *$F_{Y|X=x}$ is called the $(X=x)$-conditional distribution function of Y.*

(ii) *$f_{Y|X=x}$ is called an $(X=x)$-conditional probability density of Y.*

In section 17.9, the definition of an $(X=x)$-conditional probability density is extended, dropping the assumption that X and Y are real valued.

Using the notation introduced in Definition 17.69, Equation (17.44) with $B = \,]-\infty, y]$ yields Lemma 17.70.

Lemma 17.70 [Marginal and conditional distribution functions]
Let the assumptions of Theorem 17.65 be satisfied. Then,

$$F_Y(y) = \int_{-\infty}^{\infty} F_{Y|X=x}(y) \cdot f_X(x)\, dx, \quad \forall\, y \in \mathbb{R}, \tag{17.86}$$

holds for the marginal distribution function of Y, and a marginal probability density of Y is specified by:

$$f_Y(y) = \int_{-\infty}^{\infty} f_{Y|X=x}(y) \cdot f_X(x)\, dx, \quad \forall\, y \in \mathbb{R}. \tag{17.87}$$

(Proof p. 551)

Remark 17.71 [Total probability]   In a sense, Equation (17.86) is a 'continuous version' of the formula of total probability (see Th. 4.25). This can be seen as follows: For a sequence $\ldots < x_{-1} < x_0 < x_1 < x_2 < \ldots$ with $P(x_i < X \leq x_{i+1}) > 0$,

$$
\begin{aligned}
F_Y(y) = P(Y \leq y) &= \sum_i P(Y \leq y \mid x_i < X \leq x_{i+1}) \cdot P(x_i < X \leq x_{i+1}) \\
&= \int F_{Y|X=x}(y) \cdot f_X(x)\, dx.
\end{aligned}
\tag{17.88}
$$

Hence, for a fixed $y \in \mathbb{R}$, the probability $F_Y(y) = P(Y \leq y)$ is represented as the integral with respect to x over the product $F_{Y|X=x}(y) \cdot f_X(x)$. ◁

Theorem 17.72 [Bayes' theorem]
Suppose that the assumptions of Theorem 17.65 hold. Then, for all $(x, y) \in \mathbb{R}^2$, with $f_X(x)$, $f_Y(y) > 0$,

$$f_{Y|X=x}(y) = \frac{f_{X|Y=y}(x) \cdot f_Y(y)}{\int_{-\infty}^{\infty} f_{X|Y=y^*}(x) \cdot f_Y(y^*)\, dy^*}, \quad \forall\, y \in \mathbb{R}, \tag{17.89}$$

specifies an $(X=x)$-conditional density of Y.

(Proof p. 551)

Analogously to Equation (6.10), Lemma 17.73 shows that replacing the density f_Y by the conditional density $f_{Y|X=x}$ yields a conditional expectation value $E(Y \mid X=x)$.

Lemma 17.73 [Conditional densities and conditional expectation]
Let the assumptions of Theorem 17.65 hold. Furthermore, assume that Y is nonnegative or with finite expectation. Then,

$$E(Y \mid X=x) = \int y \cdot f_{Y\mid X=x}(y)\, dy, \quad \forall x \in \mathbb{R}, \tag{17.90}$$

defines a version of the X-conditional expectation of Y.

(Proof p. 551)

17.9   Conditional distribution and Radon-Nikodym density

In section 17.8, we considered real-valued random variables X and Y and densities with respect to the Lebesgue measure. Now we generalize some of these results for random variables X and Y that are not necessarily real-valued. We consider the case in which $P_{X,Y}$ is absolutely continuous with respect to the product measure $\mu \otimes \nu$ of some σ-finite measures μ and ν on arbitrary measurable spaces (see Defs. 3.70 and 1.63 and Lemma 1.66).

Notation and assumptions 17.74
Let $(X, Y)\colon (\Omega, \mathcal{A}, P) \to (\Omega_X' \times \Omega_Y', \mathcal{A}_X' \otimes \mathcal{A}_Y')$ be a random variable, and let $P_{X,Y}$ denote its distribution. Furthermore, assume that μ and ν are σ-finite measures on $(\Omega_X', \mathcal{A}_X')$ and $(\Omega_Y', \mathcal{A}_Y')$, respectively, and

$$P_{X,Y} \underset{\mathcal{A}_X' \otimes \mathcal{A}_Y'}{\ll} \mu \otimes \nu.$$

In the following definition, we adapt Equation (17.84) and Definition 17.69 (ii).

Definition 17.75 [$(X=x)$-conditional density]
Let the assumptions 17.74 hold. For a fixed version $f_{X,Y} = \dfrac{d\,P_{X,Y}}{d\mu \otimes \nu}$ of the density of (X, Y) (see Th. 3.72) and its marginal density f_X, we define $f_{Y\mid X=x}$ by:

$$f_{Y\mid X=x}(y) := \begin{cases} \dfrac{f_{X,Y}(x, y)}{f_X(x)}, & \text{if } f_X(x) > 0 \\ 0, & \text{otherwise,} \end{cases} \quad \forall\, y \in \Omega_Y', \tag{17.91}$$

and call it an $(X=x)$-conditional density of Y (with respect to $\mu \otimes \nu$).

Now we show that this definition is consistent with the definition of conditional distributions.

Lemma 17.76 [Consistency]
Let the assumptions 17.74 hold. Furthermore, suppose that $P_{Y|X}$ exists and $\mathscr{A}'_Y$ is countably generated. Then, for P_X-almost all $x \in \Omega'_X$, the function $f_{Y|X=x}$ defined in Equation (17.91) is a density of $P_{Y|X=x}$ with respect to the measure ν.

(Proof p. 552)

As a generalization of Lemma 17.70 [Eq. (17.87)], using (5.48) we obtain Lemma 17.77.

Lemma 17.77 [Decomposition of the joint density]
If the assumptions 17.74 hold, then,

$$f_Y(y) = \int f_{Y|X=x}(y) \cdot f_X(x)\, \mu(dx), \quad \text{for } \nu\text{-a.a. } y \in \Omega'_Y. \tag{17.92}$$

(Proof p. 552)

Now we generalize Lemma 17.73. Now Y may be numerical, and its distribution P_Y does not necessarily have a density with respect to the Lebesgue measure. Furthermore, X can be any random variable on $(\Omega, \mathscr{A}, P)$. In particular, it may be multivariate.

Lemma 17.78 [Conditional densities and conditional expectation]
Let the assumptions 17.74 hold, and let $Y\colon (\Omega, \mathscr{A}, P) \to (\overline{\mathbb{R}}, \overline{\mathscr{B}})$ be nonnegative or with finite expectation. Then,

$$E(Y \mid X=x) = \int y \cdot f_{Y|X=x}(y)\, \nu(dy), \quad \forall\, x \in \Omega'_X \tag{17.93}$$

defines a version of the X-conditional expectation of Y.

(Proof p. 553)

Remark 17.79 [Existence of a linear logistic regression] In Remark 13.27, we already hinted at a sufficient condition for the existence of a linear logistic regression. Formulated in the notation and the concepts of the present chapter, that proposition reads as follows: Let $Y\colon (\Omega, \mathscr{A}, P) \to (\mathbb{R}, \mathscr{B})$ be a continuous random variable, let $A \in \mathscr{A}$ with $0 < P(A) < 1$, and let $X = 1_A$. Furthermore, assume that the $(X=x)$-conditional densities of Y are normal and $Var(Y \mid X=0) = Var(Y \mid X=1)$. Then, $P(X=1 \mid Y)$ has a linear logistic parameterization. This proposition is proved in Examples 17.80 to 17.82. ◁

Example 17.80 [A mixture of two normal distributions] Consider the random variable $(X, Y)\colon (\Omega, \mathscr{A}, P) \to (\mathbb{R}^2, \mathscr{B}_2)$, where X is an indicator variable with $P(X=0) = 1 - p$ and $P(X=1) = p$, that is,

$$P_X = P(X=0) \cdot \delta_0 + P(X=1) \cdot \delta_1, \tag{17.94}$$

where δ_0 and δ_1 are Dirac measures at 0 and 1 on $(\mathbb{R}, \mathscr{B})$, respectively (see Example 1.52). For the conditional distributions, assume $P_{Y|X=0} = \mathcal{N}(\mu_0, \sigma_0^2)$, and $P_{Y|X=1} = \mathcal{N}(\mu_1, \sigma_1^2)$, where $\mu_0, \mu_1 \in \mathbb{R}, \sigma_0^2, \sigma_1^2 > 0$ (see section 8.2.2). Then, according to Equation (17.45),

$$P_Y = P_{Y|X=0} \cdot P(X=0) + P_{Y|X=1} \cdot P(X=1). \tag{17.95}$$

According to Equations (17.94) and (17.46), for all $(A', B') \in \mathscr{B} \times \mathscr{B}$,

$$
\begin{aligned}
P_{X,Y}(A' \times B') &= P(X \in A', Y \in B') \\
&= \sum_{\substack{x \in \Omega'_X \\ P(X=x) > 0}} 1_{A'}(x) \cdot P_{Y|X=x}(B') \cdot P_X(\{x\}) \\
&= 1_{A'}(0) \cdot P_{Y|X=0}(B') \cdot P(X=0) + 1_{A'}(1) \cdot P_{Y|X=1}(B') \cdot P(X=1).
\end{aligned}
\tag{17.96}
$$

◁

Example 17.81 [A mixture of two normal distributions – densities] Define $\mu = \delta_0 + \delta_1$, and let λ be the Lebesgue measure on $(\mathbb{R}, \mathscr{B})$. Then, $P_{X,Y} \underset{\mathscr{B}_2}{\ll} \mu \otimes \lambda$ (see Exercise 17.5). This implies that the assumptions 17.74 are satisfied, and we can apply Definition 17.75. Hence, if $f_{Y|X=x}$ denotes the density (with respect to λ) of $P_{Y|X=x} = \mathcal{N}(\mu_x, \sigma_x^2), x = 0, 1$, then,

$$
f_{X,Y}(x, y) = \begin{cases} (1-p) \cdot f_{Y|X=0}(y), & \text{if } y \in \mathbb{R}, x = 0 \\ p \cdot f_{Y|X=1}(y), & \text{if } y \in \mathbb{R}, x = 1 \\ 0, & \text{otherwise,} \end{cases}
\tag{17.97}
$$

$$
f_X(x) = \begin{cases} 1-p, & \text{if } x = 0 \\ p, & \text{if } x = 1 \\ 0, & \text{otherwise,} \end{cases}
\tag{17.98}
$$

and

$$
f_Y(y) = (1-p) \cdot f_{Y|X=0}(y) + p \cdot f_{Y|X=1}(y)
\tag{17.99}
$$

[see Lemma 5.79 and Eq. (3.58)]. Furthermore, according to Definition 17.75 and Equation (17.97),

$$
f_{X|Y=y}(x) = \begin{cases} \dfrac{(1-p) \cdot f_{Y|X=0}(y)}{(1-p) \cdot f_{Y|X=0}(y) + p \cdot f_{Y|X=1}(y)}, & \text{if } x = 0 \\[3ex] \dfrac{p \cdot f_{Y|X=1}(y)}{(1-p) \cdot f_{Y|X=0}(y) + p \cdot f_{Y|X=1}(y)}, & \text{if } x = 1 \\[3ex] 0, & \text{otherwise.} \end{cases}
\tag{17.100}
$$

◁

Example 17.82 [A mixture of two normal distributions – logistic regression] If $0 < p < 1$, then Equation (17.100) for the conditional density $f_{X|Y=y}$ implies

$$P(X=1 \mid Y) \underset{P}{=} \frac{\exp(\beta_0 + \beta_1 Y + \beta_2 Y^2)}{1 + \exp(\beta_0 + \beta_1 Y + \beta_2 Y^2)} \tag{17.101}$$

with

$$\beta_0 = \ln\left(\frac{p\,\sigma_0}{(1-p)\,\sigma_1}\right) + \left(\frac{\mu_0^2}{2\sigma_0^2} - \frac{\mu_1^2}{2\sigma_1^2}\right), \tag{17.102}$$

$$\beta_1 = \left(\frac{\mu_1}{\sigma_1^2} - \frac{\mu_0}{\sigma_0^2}\right), \tag{17.103}$$

and

$$\beta_2 = \left(\frac{1}{2\sigma_0^2} - \frac{1}{2\sigma_1^2}\right) \tag{17.104}$$

(see Exercise 17.6). If $\sigma_0 = \sigma_1 =: \sigma$, then $\beta_2 = 0$ and Equation (17.101) simplifies to

$$P(X=1 \mid Y) \underset{P}{=} \frac{\exp(\beta_0 + \beta_1 Y)}{1 + \exp(\beta_0 + \beta_1 Y)} \tag{17.105}$$

with

$$\beta_0 = \ln\left(\frac{p}{(1-p)}\right) + \left(\frac{\mu_0^2 - \mu_1^2}{2\sigma^2}\right), \tag{17.106}$$

and

$$\beta_1 = \left(\frac{\mu_1 - \mu_0}{\sigma^2}\right). \tag{17.107}$$

Hence, under the specified assumptions, the conditional probability $P(X=1 \mid Y)$ has a linear logistic parameterization in Y [see Eq. (17.105) and Def. 13.10]. ◁

17.10   Proofs

Proof of Lemma 17.8

$\Rightarrow$ If $P_{Y|X} \in \mathscr{P}_{Y|X}$ and $(P_{Y|X=x}, x \in \Omega'_X)$ is the family of $(X=x)$-conditional distributions pertaining to $P_{Y|X}$, then,

$$\forall x \in \Omega'_X \; \forall \omega \in \{X=x\} \; \forall A' \in \mathscr{A}'_Y: \quad P_{Y|X=x}(A') = P_{Y|X}(\omega, A') \qquad [(17.10)]$$
$$= P(Y \in A' \mid X=x), \qquad [(17.8)]$$

which is condition (a). The definition of $\mathscr{P}_{Y|X}$ (see Rem. 17.5), Definition 17.1 (b), and Equation (17.8) imply that condition (b) is satisfied as well.

$\Leftarrow$ Let $(P_{Y|X=x}, x \in \Omega'_X)$ be a family of functions $P_{Y|X=x}: \mathscr{A}'_Y \to [0, 1]$ satisfying (a) and (b). Define the function $P_{Y|X}: \Omega \times \mathscr{A}'_Y \to [0, 1]$ by

$$P_{Y|X}(\omega, A') = P(Y \in A' \mid X)(\omega), \quad \forall \omega \in \Omega, \forall A' \in \mathscr{A}'_Y, \tag{17.108}$$

where $P(Y \in A' \mid X)$ is the version specified in (a). Then, for all $A' \in \mathcal{A}'_Y$, the function $P_{Y|X}(\cdot, A') = P(Y \in A' \mid X)$ is X-measurable [see Def. 10.2 (a)]. Furthermore, for all $\omega \in \Omega$, there is an $x \in \Omega'_X$ with $x = X(\omega)$ and

$$P_{Y|X}(\omega, \cdot) = P(Y \in A' \mid X = x) \quad [(17.108), (10.32), (10.28)]$$
$$= P_{Y|X=x}, \quad [(a)]$$

and, according to (b), this is a probability measure. Hence, $P_{Y|X}$ is a version of the X-conditional distribution of Y, and $(P_{Y|X=x}, x \in \Omega'_X)$ is the family of $(X=x)$-conditional distributions pertaining to $P_{Y|X}$.

Proof of Corollary 17.10

For all $A' \in \mathcal{A}'_Y$,

$$P_{Y|X=x}(A') = P(Y \in A' \mid X=x) \quad [(17.3), (17.10), (10.32)]$$
$$= P^{X=x}(Y \in A') \quad [\text{Rem. } 10.35]$$
$$= P_Y^{X=x}(A'). \quad [(14.39)]$$

Proof of Corollary 17.12

Let $1_{A'}(h(X))$ denote the composition of $h(X)$ and the indicator $1_{A'}$. Then, Box 10.2 (vii) and proposition (10.12) imply:

$$\forall\, A' \in \mathcal{A}': \quad 1_{A'}(h(X)) \in \mathscr{E}(1_{A'}(h(X)) \mid X) = \mathscr{P}(h(X) \in A' \mid X).$$

Therefore, defining

$$P_{h(X)|X}(\omega, A') := 1_{A'}(h(X))(\omega) = \delta_{h(X(\omega))}(A'), \quad \forall\, (\omega, A') \in \Omega \times \mathcal{A}',$$

yields a version of the X-conditional distribution of $h(X)$, because, for all $\omega \in \Omega$, the Dirac measure $\delta_{h(X(\omega))}$ is a probability measure on $(\Omega', \mathcal{A}')$. Then Equation (17.10) yields, for all $x \in \Omega'_X$, $\omega \in \{X=x\}$,

$$P_{h(X)|X=x}(A') = \delta_{h(x)}(A'), \quad \forall\, A' \in \mathcal{A}'.$$

Proof of Lemma 17.14

$$Y \underset{P}{\perp\!\!\!\perp} \mathscr{C}$$
$$\Rightarrow \forall\, A' \in \mathcal{A}'_Y: 1_{Y \in A'} \underset{P}{\perp\!\!\!\perp} \mathscr{C} \qquad\qquad [\text{Rem. } 5.44, \sigma(1_{Y \in A'}) \subset \sigma(Y)]$$
$$\Rightarrow \forall\, A' \in \mathcal{A}'_Y: E(1_{Y \in A'} \mid \mathscr{C}) \underset{P}{=} E(1_{Y \in A'}) = P_Y(A') \quad [\text{Box } 10.1 \text{ (vi)}, (6.4), (5.2), (5.3)]$$
$$\Rightarrow \forall\, A' \in \mathcal{A}'_Y: P_Y(A') \in \mathscr{P}(Y \in A' \mid \mathscr{C}). \qquad\qquad [(10.12), \text{Rem. } 10.10]$$

Therefore, if we define

$$P_{Y|\mathscr{C}}(\omega, A') := P_Y(A'), \quad \forall\, \omega \in \Omega,$$

then condition (a) of Definition 17.1 holds for $P_{Y|\mathscr{C}}$. Because, for all $\omega \in \Omega$, $P_{Y|\mathscr{C}}(\omega, \cdot) = P_Y$ is a probability measure, condition (b) of Definition 17.1 is satisfied as well.

Proof of Theorem 17.16

Let X be discrete, let $\Omega'_0 \subset \Omega'_X$ be finite or countable with $P_X(\Omega'_0) = 1$ (see Def. 5.56), and let $P(X=x) > 0$ for all $x \in \Omega'_0$. For all $A' \in \mathscr{A}'_Y$, we define a version $P(Y \in A' \mid X) \in \mathscr{P}(Y \in A' \mid X)$ by

$$\forall \omega \in \Omega: \quad P(Y \in A' \mid X)(\omega) := \begin{cases} P^{X=x}(Y \in A'), & \text{if } X(\omega) = x \text{ and } P(X=x) > 0 \\ P(Y \in A'), & \text{otherwise,} \end{cases}$$

where $P^{X=x}(Y \in A')$ is defined by (4.14) with $B = \{X=x\}$. According to Remarks 10.35 and 10.36, this defines a version $P(Y \in A' \mid X) \in \mathscr{P}(Y \in A' \mid X)$ for each $A' \in \mathscr{A}'_Y$. Therefore, condition (a) of Lemma 17.8 is satisfied if we define

$$P_{Y|X=x}(A') = P(Y \in A' \mid X=x), \quad \forall x \in \Omega'_X.$$

For all $x \in \Omega'_X$, the functions $P_{Y|X=x}$ are probability measures on $(\Omega'_Y, \mathscr{A}'_Y)$. [If $P(X=x) > 0$, see Th. 4.28. If $P(X=x) = 0$, then $P(Y \in A' \mid X)(\omega) = P(Y \in A')$ (see the last but one equation).] Therefore, condition (b) of Lemma 17.8 is satisfied as well. Now Lemma 17.8 implies that $(P_{Y|X=x}, x \in \Omega'_X)$ is the family of $(X=x)$-conditional distributions of Y pertaining to a conditional distribution $P_{Y|X}$. This proves the existence of $P_{Y|X}$. Uniqueness of $P_{Y|X}$ for all $x \in \Omega'_X$ with $P(X=x) > 0$ immediately follows from (17.14) and Remark 2.71.

Proof of Theorem 17.17

Define $\Omega'_> := \{y \in \Omega'_Y : P(Y=y) > 0\}$, which is finite or countable (see Def. 5.56). Fix a family $(P(Y=y \mid X), y \in \Omega'_{Y_>})$ of versions $P(Y=y \mid X) \in \mathscr{P}(Y=y \mid X)$ with $0 \le P(Y=y \mid X) \le 1$ [see Box 10.3 (v), (vii)]. Because $P(Y \in \Omega'_>) = 1$, there is a version such that $P(Y \in \Omega'_> \mid X) = 1$ [see (10.4) and Box 10.3 (vi)]. Hence, for this version,

$$\sum_{y \in \Omega'_>} P(Y=y \mid X) \underset{P}{=} \sum_{y \in \Omega'_>} E(1_{Y=y} \mid X) \qquad [(10.4)]$$

$$\underset{P}{=} E\left(\sum_{y \in \Omega'_>} 1_{Y=y} \,\middle|\, X \right) \qquad [\text{Box 10.2 (xvi), (10.20)}]$$

$$\underset{P}{=} E(1_{Y \in \Omega'_>} \mid X) \qquad [(1.36), (1.37)]$$
$$\underset{P}{=} P(Y \in \Omega'_> \mid X) \qquad [(10.4)]$$
$$= 1.$$

Therefore, $P(A) = 0$ holds for the set $A := \{\omega \in \Omega : \sum_{y \in \Omega'_>} P(Y=y \mid X)(\omega) \ne 1\}$ and $A \in \sigma(X)$ [see Rem. 2.67 (c)].

Now define the function $P_{Y|X}: \Omega \times \mathscr{A}'_Y \to [0, 1]$ by

$$P_{Y|X}(\omega, A') = \begin{cases} \displaystyle\sum_{y \in A' \cap \Omega'_>} P(Y=y \mid X)(\omega), & \text{if } \omega \in \Omega \setminus A \\ P_Y(A'), & \text{otherwise,} \end{cases}$$

with the convention $P_{Y|X}(\omega, A') = 0$ if $A' \cap \Omega'_> = \emptyset$. Now we show that $P_{Y|X}$ is in fact a version of the X-conditional distribution of Y. For all $\omega \in \Omega$, the function $P_{Y|X}(\omega, \cdot): \mathscr{A}'_Y \to [0, 1]$ is a probability measure because, for $\omega \in A$, the function $P_{Y|X}(\omega, \cdot)$ is a probability measure by definition, and for $\omega \in \Omega \setminus A$,

$$P_{Y|X}(\omega, \Omega'_Y) = \sum_{y \in \Omega'_>} P(Y=y \mid X)(\omega) = 1,$$

$$P_{Y|X}(\omega, A') \geq 0, \quad \forall A' \in \mathscr{A}'_Y,$$

and if $A'_1, A'_2, \ldots \in \mathscr{A}'_Y$ are pairwise disjoint, then

$$P_{Y|X}\left(\omega, \bigcup_{i=1}^{\infty} A'_i\right) = \sum_{y \in (\bigcup_{i=1}^{\infty} A'_i) \cap \Omega'_>} P(Y=y \mid X)(\omega) = \sum_{i=1}^{\infty} \sum_{y \in A'_i \cap \Omega'_>} P(Y=y \mid X)(\omega)$$

$$= \sum_{i=1}^{\infty} P_{Y|X}(\omega, A'_i).$$

This shows that conditions (a) to (c) of Definition 4.1 are satisfied. Therefore, condition (b) of Definition 17.1 holds for $P_{Y|X}$.

Now, for all $A' \in \mathscr{A}'_Y$,

$$P_{Y|X}(\cdot, A') \underset{P}{=} \sum_{y \in A' \cap \Omega'_>} P(Y=y \mid X) \qquad\qquad [P(A) = 0]$$

$$\underset{P}{=} P\big(Y \in (A' \cap \Omega'_>) \mid X\big) \qquad [(10.4), \text{Box } 10.2 \text{ (xvi)}, (10.20), (1.36), (1.37)]$$

$$\underset{P}{=} P(Y \in A' \mid X). \quad [P_Y(A') = P_Y(A' \cap \Omega'_>), 1_{Y \in A'} \underset{P}{=} 1_{Y \in (A' \cap \Omega'_>)}, \text{Box } 10.2 \text{ (vi)}]$$

Because $P_{Y|X}(\cdot, A')$ is X-measurable [see Th. 2.57 and Th. 1.92 of Klenke, 2013], proposition (10.12) implies $P_{Y|X}(\cdot, A') \in \mathscr{P}(Y \in A') \mid X)$ for all $A' \in \mathscr{A}'_Y$. Hence, condition (a) of Definition 17.1 holds for $P_{Y|X}$.

Proof of Lemma 17.26

This proof is analog to the proof of Theorem 44.2 (ii) of Bauer (1996). If $\mathscr{A}'$ is countably generated and $\mathscr{E}' = \{A'_i, i \in I\}$ is a finite or countable set system generating $\mathscr{A}'$, then define

$$\mathscr{C}' := \left\{ \bigcap_{i \in J} A'_i : J \subset I, J \text{ finite} \right\},$$

the set system of all intersections of finitely many elements of $\mathscr{E}'$. Then $\mathscr{C}'$ is also finite or countable, and moreover it is $\cap$-stable. Note that $\mathscr{E}' \subset \mathscr{C}'$, and $\sigma(\mathscr{E}')$ is closed with respect to finite intersection. Hence,

$$\mathscr{E}' \subset \mathscr{C}' \subset \sigma(\mathscr{E}') = \mathscr{A}'.$$

Monotonicity of generated σ-algebras (see Rem. 1.23) implies

$$\mathscr{A}' = \sigma(\mathscr{E}') \subset \sigma(\mathscr{C}') \subset \sigma(\mathscr{A}') = \mathscr{A}'.$$

Therefore, $\sigma(\mathscr{C}') = \mathscr{A}'$, that is, $\mathscr{C}'$ is a finite or countable set system generating $\mathscr{A}'$. Now define

$$N := \bigcup_{A' \in \mathscr{C}'} N_{A'},$$

using the sets $N_{A'}$ mentioned in (c). σ-Subadditivity of probability measures [see Box 4.1 (xi)] yields

$$P(N) \leq \sum_{A' \in \mathscr{C}'} P(N_{A'}) = 0.$$

Furthermore, (c) implies

$$\forall\, \omega \in \Omega \setminus N \,\forall\, A' \in \mathscr{C}': \quad K(\omega, A') = K^*(\omega, A').$$

Hence, for all $\omega \in \Omega \setminus N$, the probability measures $K(\omega, \cdot)$ and $K^*(\omega, \cdot)$ are identical on the $\cap$-stable generating system $\mathscr{C}'$. Therefore, Theorem 1.72 implies

$$\forall\, \omega \in \Omega \setminus N \,\forall\, A' \in \mathscr{A}': \quad K(\omega, A') = K^*(\omega, A').$$

Proof of Theorem 17.30

Define $\Omega'_> := \{y \in \Omega'_Y \colon P(Y = y) > 0\}$, which is finite or countable (see Def. 5.56). Furthermore, let $P_{Y|X}, P^*_{Y|X} \in \mathscr{P}_{Y|X}$ and, for all $y \in \Omega'_>$, define

$$N_y := \{\omega \in \Omega \colon P_{Y|X}(\omega, \{y\}) \neq P^*_{Y|X}(\omega, \{y\})\},$$

$$N_{\neq} := \{\omega \in \Omega \colon P_{Y|X}(\omega, \Omega' \setminus \Omega'_>) \neq 0\}, \quad \text{and} \quad N^*_{\neq} := \{\omega \in \Omega \colon P^*_{Y|X}(\omega, \Omega' \setminus \Omega'_>) \neq 0\}.$$

Condition (a) of Definition 17.1 and Proposition (10.12) imply $P(N_y) = 0$ for all $y \in \Omega'_>$. Furthermore, $P(Y \in (\Omega' \setminus \Omega'_>)) = 0$ implies $1_{Y \in (\Omega' \setminus \Omega'_>)} \underset{P}{=} 0$, and hence $P(Y \in (\Omega' \setminus \Omega'_>) \mid X) \underset{P}{=} 0$ [see Box 10.3 (iv)]. Condition (a) of Definition 17.1 and Proposition (10.12) imply $P_{Y|X}(\cdot, \Omega' \setminus \Omega'_>) \underset{P}{=} 0$, and therefore $P(N_{\neq}) = P(N^*_{\neq}) = 0$. Then, for

$$N := N_{\neq} \cup N^*_{\neq} \cup \bigcup_{y \in \Omega'_>} N_y,$$

σ-subadditivity of P [see Box 4.1 (xi)] yields

$$P(N) \leq P(N_{\neq}) + P\left(N_{\neq}^*\right) + \sum_{y \in \Omega_>'} P(N_y) = 0.$$

Now for all $\omega \in \Omega \setminus N$ and all $A' \in \mathscr{A}_Y'$,

$$
\begin{aligned}
P_{Y|X}(\omega, A') &= \sum_{y \in A' \cap \Omega_>'} P_{Y|X}(\omega, \{y\}) + P_{Y|X}(\omega, A' \cap (\Omega' \setminus \Omega_Y')) && \text{[17.28, Def. 17.1 (b)]} \\
&= \sum_{y \in A' \cap \Omega_>'} P_{Y|X}(\omega, \{y\}) && \text{[def. of } N_{\neq}, \text{ Def. 17.1 (b)]} \\
&= \sum_{y \in A' \cap \Omega_>'} P_{Y|X}^*(\omega, \{y\}) && \text{[def. of } N] \\
&= \sum_{y \in A' \cap \Omega_>'} P_{Y|X}^*(\omega, \{y\}) + P_{Y|X}^*(\omega, A' \cap (\Omega' \setminus \Omega_Y')) && \\
&&& \text{[def. of } N_{\neq}^*, \text{ Def. 17.1 (b)]} \\
&= P_{Y|X}^*(\omega, A'). && \text{[17.28, Def. 17.1 (b)]}
\end{aligned}
$$

This shows that $P_{Y|X}$ is CNS-unique.

Proof of Corollary 17.32

According to Remark 2.72 (ii), $Y \underset{P}{=} Z$ implies $1_{Y \in A'} \underset{P}{=} 1_{Z \in A'}$, for all $A' \in \mathscr{A}'$. Therefore, for all $A' \in \mathscr{A}'$,

$$
\begin{aligned}
P_{Y|X=x}(A') &\underset{P_x\text{-a.a.}}{=} E(1_{Y \in A'} \mid X=x) && \text{[(17.3), (17.10), (10.28)]} \\
&\underset{P_x\text{-a.a.}}{=} E(1_{Z \in A'} \mid X=x) && \text{[Box 10.2 (ix)]} \\
&\underset{P_x\text{-a.a.}}{=} P_{Z|X=x}(A'). && \text{[(10.28), (17.10), (17.3)]}
\end{aligned}
$$

Because $\mathscr{A}'$ is countably generated, Lemma 17.26 and Corollary 5.25 (i) imply

$$P_{Y|X=x} \underset{P_x\text{-a.a.}}{=} P_{Z|X=x}.$$

Proof of Corollary 17.34

For an $(X=x)$-conditional-probability measure $P^{X=x}$ defined by Equation (17.32) and all $A \in \mathscr{A}$,

$$
\begin{aligned}
P^{X=x}(A) &= P_{id|X=x}(A) && \text{[(17.32)]} \\
&\underset{P_x\text{-a.a.}}{=} P^*(id \in A \mid X=x) && \text{[(17.14)]} \\
&\underset{P_x\text{-a.a.}}{=} P^*(A \mid X=x). && \text{[}\{id \in A\} = A\text{]}
\end{aligned}
$$

Proof of Theorem 17.35

If $P^{X=x}$ exists for all $x \in \Omega'_X$, then, according to Definition 5.3,

$$\forall A' \in \mathscr{A}'_Y \; \forall \, x \in \Omega'_X: \quad P_Y^{X=x}(A') = P^{X=x}[Y^{-1}(A')].$$

This implies

$$\forall A' \in \mathscr{A}'_Y: \quad P_Y^{X=x}(A') \underset{P_X\text{-a.a.}}{=} P[Y^{-1}(A') \mid X=x] \qquad [(17.33)]$$

$$\underset{P_X\text{-a.a.}}{=} P(Y \in A' \mid X=x) \qquad [\{Y \in A'\} = Y^{-1}(A')]$$

$$\underset{P_X\text{-a.a.}}{=} P_{Y|X=x}(A'). \qquad [(17.11), (10.34)]$$

If $\mathscr{A}'_Y$ is countably generated, Lemma 17.26 implies $P_Y^{X=x} \underset{P_X\text{-a.a.}}{=} P_{Y|X=x}$.

Proof of Theorem 17.37

For $A' \in \mathscr{A}'_X$ and $B' \in \mathscr{A}'_Y$,

$$P(X \in A', Y \in B') = E(1_{X \in A'} \cdot 1_{Y \in B'}) \qquad [(1.33), (6.4)]$$

$$= E[E(1_{X \in A'} \cdot 1_{Y \in B'} \mid X)] \qquad [\text{Box } 10.2 \text{ (iv)}]$$

$$= E[1_{X \in A'} \cdot E(1_{Y \in B'} \mid X)] \qquad [\text{Box } 10.2 \text{ (xiv)}]$$

$$= \int 1_{A'}(x) \cdot E(1_{Y \in B'} \mid X=x) \, P_X(dx) \qquad [(6.13)]$$

$$= \int 1_{A'}(x) \cdot P_{Y|X=x}(B') \, P_X(dx) \qquad [(10.3), (17.11)]$$

$$= \int 1_{A'}(x) \int 1_{B'}(y) \, P_{Y|X=x}(dy) \, P_X(dx) \qquad [(3.9)]$$

$$= \iint 1_{A'}(x) \cdot 1_{B'}(y) \, P_{Y|X=x}(dy) \, P_X(dx), \qquad [(3.32)]$$

which proves Equations (17.40) and (17.41). Hence,

$$P(X \in A', Y \in B') = \int 1_{A'}(x) \cdot 1_{B'}(y) \, P_{X,Y}[d(x, y)] \qquad [\text{Th. } 3.57]$$

$$= \iint 1_{A'}(x) \cdot 1_{B'}(y) \, P_{Y|X=x}(dy) \, P_X(dx). \qquad [(17.41)]$$

Consider the set $\mathscr{C}' \subset \mathscr{A}'_X \otimes \mathscr{A}'_Y$ of all sets $C' \subset \Omega'_X \times \Omega'_Y$ satisfying

$$\int 1_{C'}(x, y) \, P_{X,Y}[d(x, y)] = \iint 1_{C'}(x, y) \, P_{Y|X=x}(dy) \, P_X(dx).$$

Because

$$P(A' \in \mathscr{A}'_X, B' \in \mathscr{A}'_Y) = P_{X,Y}(A' \times B') = \int 1_{A' \times B'}(x, y) \, P_{X,Y}[d(x, y)]$$

and $1_{A' \times B'}(x, y) = 1_{A'}(x) \cdot 1_{B'}(y)$ for all $(x, y) \in \Omega'_X \times \Omega'_Y$ [see Eq. (1.38)], Equation (17.41) can equivalently be written as:

$$\int 1_{A' \times B'}(x, y) \, P_{X,Y}[d(x, y)] = \iint 1_{A' \times B'}(x, y) \, P_{Y|X=x}(dy) \, P_X(dx).$$

Therefore, $\mathscr{E}' = \{A' \times B' : A' \in \mathscr{A}'_X, B' \in \mathscr{A}'_Y\} \subset \mathscr{C}'$. Because $\Omega'_X \times \Omega'_Y \in \mathscr{C}'$ and $\mathscr{C}'$ is closed with respect to complements [see Eqs. (1.35) and (3.36)] and countable unions of pairwise disjoint sets [see Eq. (3.65)], the set $\mathscr{C}'$ is a Dynkin system (see Def. 1.40). Hence,

$$\delta(\mathscr{E}') \subset \mathscr{C}' \subset \mathscr{A}'_X \otimes \mathscr{A}'_Y.$$

Because $\mathscr{E}'$ is ∩-stable (see Example 1.39), Theorem 1.41 (ii) yields

$$\delta(\mathscr{E}') = \sigma(\mathscr{E}') = \mathscr{A}'_X \otimes \mathscr{A}'_Y$$

(see Def. 1.31). Hence, we can conclude: $\mathscr{C}' = \mathscr{A}'_X \otimes \mathscr{A}'_Y$. This proves Equation (17.42). Now the proof of Equation (17.43) can be completed by standard methods of integration theory (see Rem. 3.30).

Proof of Lemma 17.41

For all $B' \in \mathscr{A}'_Y$, $D' \in \mathscr{A}'_X \otimes \mathscr{A}'_Z$, and $E' := B' \times D'$,

$$\int 1_{D'}(x, z) \cdot P^{X=x}_{Y|Z=z}(B') \, P_{X,Z}[d(x, z)]$$

$$\iint 1_{D'}(x, z) \cdot 1_{B'}(y) \, P^{X=x}_{Y|Z=z}(dy) \, P_{X,Z}[d(x, z)] \qquad [(3.9), (3.32)]$$

$$= \iint 1_{E'}(y, x, z) \, P^{X=x}_{Y|Z=z}(dy) \, P_{X,Z}[d(x, z)] \qquad [(1.38)]$$

$$= \iiint 1_{E'}(y, x, z) \, P^{X=x}_{Y|Z=z}(dy) \, P^{X=x}_Z(dz) \, P_X(dx) \qquad [(17.43), (17.35)]$$

$$= \iint 1_{E'}(y, x, z) \, P^{X=x}_{Y,Z}[d(y, z)] \, P_X(dx) \qquad [(17.43)]$$

$$= \iint 1_{E'}(y, x, z) \, P_{Y,Z|X=x}[d(y, z)] \, P_X(dx) \qquad [(17.35)]$$

$$= \int 1_{E'}(y, x, z) \, P_{Y,X,Z}[d(y, x, z)] \qquad [(17.43)]$$

$$= \iint 1_{E'}(y, x, z) \, P_{Y|X=x, Z=z}(dy) \, P_{X,Z}[d(x, z)] \qquad [(17.43)]$$

$$= \iint 1_{D'}(x, z) \cdot 1_{B'}(y) \, P_{Y|X=x, Z=z}(dy) \, P_{X,Z}[d(x, z)] \qquad [(1.38)]$$

$$= \int 1_{D'}(x, z) \cdot P_{Y|X=x, Z=z}(B') \, P_{X,Z}[d(x, z)]. \qquad [(3.9), (3.32)]$$

Applying Theorem 3.48 completes the proof.

Proof of Corollary 17.42

According to Equation (17.48), for all $B' \in \mathscr{A}_Y'$: $P_{X,Z}(N_B') = 0$, where

$$N_B' = \{(x, z) \in \Omega_X' \times \Omega_Z': P_{Y|Z=z}^{X=x}(B') \neq P_{Y|X=x, Z=z}(B')\}.$$

Defining $N_{B,x}' = \{z \in \Omega_Z': (x, z) \in N_B'\}$ for all $x \in \Omega_X'$,

$$P_{X,Z}(N_B') = 0 \quad \Leftrightarrow \quad \int\int 1_{N_{B,x}'}(z)\, P_{Z|X=x}(dz)\, P_X(dx) = 0. \qquad [(17.42)]$$

Hence,

$$\int P_{Z|X=x}(N_{B,x}')\, P_X(dx) = 0, \qquad [(3.9)]$$

which according to Theorem 3.43 implies

$$P_{Z|X=x}(N_{B,x}') \underset{P_X\text{-a.a.}}{=} 0.$$

If $x \in \Omega_X'$ and $P(X=x) > 0$, then this equation implies

$$P_Z^{X=x}(N_{B,x}') = P_{Z|X=x}(N_{B,x}') \qquad [(17.15)]$$
$$= 0. \qquad [\text{Rem. 2.71}]$$

Therefore, for $x \in \Omega_X'$ with $P(X=x) > 0$, proposition (17.48) implies (17.49).

Proof of Theorem 17.43

(i)

    (a) $\Leftrightarrow$ (b) Using (16.15) as well as Equations (17.5), (17.8), and (17.10) yields:

$$X \perp\!\!\!\perp_P Y \mid Z$$
$$\Leftrightarrow \forall\, (A', B') \in \mathscr{A}_X' \times \mathscr{A}_Y': \quad P(X \in A', Y \in B' \mid Z) \underset{P}{=} P(X \in A' \mid Z) \cdot P(Y \in B' \mid Z)$$
$$\Leftrightarrow \forall\, (A', B') \in \mathscr{A}_X' \times \mathscr{A}_Y': \quad P_{X,Y|Z}(\cdot, A' \times B') \underset{P}{=} P_{X|Z}(\cdot, A') \cdot P_{Y|Z}(\cdot, B')$$
$$\Leftrightarrow \forall\, (A', B') \in \mathscr{A}_X' \times \mathscr{A}_Y': \quad P_{X,Y|Z=z}(A' \times B') = P_{X|Z=z}(A') \cdot P_{Y|Z=z}(B'),$$
$$\text{for } P_Z\text{-a.a. } z \in \Omega_Z'.$$

    (b) $\Rightarrow$ (c) (17.50) and Theorem 3.48 imply

$$\forall\, (A', B', C') \in \mathscr{A}_X' \times \mathscr{A}_Y' \times \mathscr{A}_Z': \int\int 1_{A' \times B'}(x, y) \cdot 1_{C'}(z)\, P_{X,Y|Z=z}[d(x, y)]\, P_Z(dz)$$
$$= \int\int 1_{A' \times B'}(x, y) \cdot 1_{C'}(z)\, P_{X|Z=z} \otimes P_{Y|Z=z}[d(x, y)]\, P_Z(dz).$$

Now let $\mathscr{D}' \subset \mathscr{A}'_X \otimes \mathscr{A}'_Y$ denote the set of all sets $D' \in \mathscr{A}'_X \otimes \mathscr{A}'_Y$ satisfying

$$\forall\, C' \in \mathscr{A}'_Z: \quad \iint 1_{D'}(x, y) \cdot 1_{C'}(z)\, P_{X,Y|Z=z}[d(x, y)]\, P_Z(dz)$$

$$= \iint 1_{D'}(x, y) \cdot 1_{C'}(z)\, P_{X|Z=z} \otimes P_{Y|Z=z}[d(x, y)]\, P_Z(dz).$$

Then, just as in the proof of Equation (17.42) in Theorem 17.37, we can conclude $\mathscr{D}' = \mathscr{A}'_X \otimes \mathscr{A}'_Y$. Together with Theorem 3.48, this yields (c).

(c) $\Rightarrow$ (b) This implication is trivial because $A' \times B' \in \mathscr{A}'_X \otimes \mathscr{A}'_Y$ for all $(A', B') \in \mathscr{A}'_X \times \mathscr{A}'_Y$ (see Def. 1.31).

(d) $\Rightarrow$ (c) For $C' \in \mathscr{A}'_Z$ and $D' \in \mathscr{A}'_X \otimes \mathscr{A}'_Y$, choose

$$h(x, y, z) = 1_{D'}(x, y) \cdot 1_{C'}(z), \quad \forall\, (x, y, z) \in \Omega'_X \times \Omega'_Y \times \Omega'_Z.$$

Then,

$$\int 1_{C'}(z) \cdot P_{X,Y|Z=z}(D')\, P_Z(dz)$$

$$= \iint 1_{D'}(x, y) \cdot 1_{C'}(z)\, P_{X,Y|Z=z}[d(x, y)]\, P_Z(dz) \qquad [(6.4), (6.1)]$$

$$= \iiint 1_{D'}(x, y) \cdot 1_{C'}(z)\, P_{X|Z=z}(dx)\, P_{Y|Z=z}(dy)\, P_Z(dz) \qquad [(17.52)]$$

$$= \iint 1_{D'}(x, y) \cdot 1_{C'}(z)\, P_{X|Z=z} \otimes P_{Y|Z=z}[d(x, y)]\, P_Z(dz) \qquad [(3.80)]$$

$$= \int 1_{C'}(z) \cdot P_{X|Z=z} \otimes P_{Y|Z=z}(D')\, P_Z(dz). \qquad [(6.4), (6.1)]$$

Because these equations hold for all $C' \in \mathscr{A}'_Z$, Theorem 3.48 yields (c).

(c) $\Rightarrow$ (d) Applying (6.4) and (6.1), Equation (17.51) can equivalently be written as:

$$\forall\, D' \in \mathscr{A}'_X \otimes \mathscr{A}'_Y:$$

$$\int 1_{D'}(x, y)\, P_{X,Y|Z=z}[d(x, y)] \underset{P_Z\text{-}a.a.}{=} \int 1_{D'}(x, y)\, P_{X|Z=z} \otimes P_{Y|Z=z}[d(x, y)],$$

and Theorem 3.48 implies that this equation is equivalent to

$$\forall\, D' \in \mathscr{A}'_X \otimes \mathscr{A}'_Y\ \forall\, C' \in \mathscr{A}'_Z:$$

$$\iint 1_{D'}(x, y) \cdot 1_{C'}(z)\, P_{X,Y|Z=z}[d(x, y)]\, P_Z(dz)$$

$$= \iint 1_{D'}(x, y) \cdot 1_{C'}(z)\, P_{X|Z=z} \otimes P_{Y|Z=z}[d(x, y)]\, P_Z(dz).$$

Just like in the proof of Equation (17.42) in Theorem 17.37, we can conclude

$$\forall E' \in \mathscr{A}'_X \otimes \mathscr{A}'_Y \otimes \mathscr{A}'_Z:$$

$$\int\!\!\int 1_{E'}(x, y, z)\, P_{X,Y|Z=z}[d(x,y)]\, P_Z(dz)$$

$$= \int\!\!\int\!\!\int 1_{E'}(x, y, z)\, P_{X|Z=z}(dx)\, P_{Y|Z=z}(dy)\, P_Z(dz),$$

where, on the right-hand side, we applied Equation (3.80). Now the proof can be completed by standard methods of integration theory (see Rem. 3.30).

(ii)

(17.53) $\Rightarrow$ (b) This implication immediately follows from Lemma 1.66.

(b) $\Rightarrow$ (17.53) If $\mathscr{A}'_X$ and $\mathscr{A}'_Y$ are countably generated, then $\mathscr{A}'_X \otimes \mathscr{A}'_Y$ is countably generated as well (see Cor. 1.33). Therefore, (b), Lemmas 17.26 and 1.66, and (17.25) imply

$$P_{X,Y|Z=z} \underset{P_Z\text{-a.a.}}{=} P_{X|Z=z} \otimes P_{Y|Z=z}.$$

Proof of Corollary 17.44

(i) Note that $X \underset{P}{\perp\!\!\!\perp} Y \mid X$ [see Eq. (16.18)]. Therefore, for all nonnegative measurable functions $h: (\Omega'_X \times \Omega'_Y, \mathscr{A}'_X \otimes \mathscr{A}'_Y) \to (\mathbb{R}, \mathscr{B})$, proposition (d) of Theorem 17.43 with $Z = X$ yields

$$\forall A' \in \mathscr{A}'_X: \quad \int\!\!\int 1_{A'}(x) \cdot h(x^*, y)\, P_{X,Y|X=x}[d(x^*, y)]\, P_X(dx)$$

$$= \int\!\!\int\!\!\int 1_{A'}(x) \cdot h(x^*, y)\, P_{X|X=x}(dx^*)\, P_{Y|X=x}(dy)\, P_X(dx) \qquad [(17.52)]$$

$$= \int\!\!\int 1_{A'}(x) \cdot h(x, y)\, P_{Y|X=x}(dy)\, P_X(dx). \qquad [(17.17), (3.57)]$$

Because these equations hold for all $A' \in \mathscr{A}'_X$, Theorem 3.48 yields (17.54).

(ii) For $D' \in \mathscr{A}'_X \otimes \mathscr{A}'_Y$, apply (17.54) with $h = 1_{D'}$. Note that $1_{D'}(x, y) = 1_{D'_x}(y)$ for all $(x, y) \in \Omega'_X \times \Omega'_Y$. Therefore,

$$P_{X,Y|X=x}(D') = \int 1_{D'}(x^*, y)\, P_{X,Y|X=x}[d(x^*, y)] \qquad [(6.4), (6.1)]$$

$$\underset{P_X\text{-a.a.}}{=} \int 1_{D'}(x, y)\, P_{Y|X=x}(dy) \qquad [(17.54)]$$

$$= P_{Y|X=x}(D'_x). \qquad [(6.4), (6.1), \text{def. of } D'_x]$$

Proof of Theorem 17.45

$X \underset{P}{\perp\!\!\!\perp} Y \mid Z \Rightarrow$ (17.56) According to the definition of conditional independence:

$$X \underset{P}{\perp\!\!\!\perp} Y \mid Z \quad \Rightarrow \quad \forall B' \in \mathscr{A}'_Y: X \underset{P}{\perp\!\!\!\perp} 1_{Y \in B'} \mid Z. \qquad [\text{Box 16.3 (vi)}]$$

Hence, for all $B' \in \mathscr{A}'_Y$,

$$
\begin{aligned}
P_{Y|X,Z}(\cdot, B') &\underset{P}{=} E(1_{Y \in B'} \mid X, Z) && [(17.7)] \\
&\underset{P}{=} E(1_{Y \in B'} \mid Z) && [(16.37)] \\
&\underset{P}{=} P_{Y|Z}(\cdot, B'). && [(17.7)]
\end{aligned}
$$

$(17.56) \Rightarrow X \underset{P}{\perp\!\!\!\perp} Y \mid Z$ For all $(A', B', C') \in \mathscr{A}'_X \times \mathscr{A}'_Y \times \mathscr{A}'_Z$,

$$
\int 1_{C'}(z) \iint 1_{A'}(x) \cdot 1_{B'}(y) \, P_{X,Y|Z=z}[d(x,y)] \, P_Z(dz)
$$

$$
= \int 1_{A'}(x) \cdot 1_{B'}(y) \cdot 1_{C'}(z) \, P_{X,Y,Z}[d(x,y,z)] \qquad\qquad [(17.41)]
$$

$$
= \int 1_{A'}(x) \cdot 1_{C'}(z) \left[\int 1_{B'}(y) \, P_{Y|X=x, Z=z}(dy) \right] P_{X,Z}[d(x,z)] \qquad [(17.41)]
$$

$$
= \int 1_{A'}(x) \cdot 1_{C'}(z) \left[\int 1_{B'}(y) \, P_{Y|Z=z}(dy) \right] P_{X,Z}[d(x,z)] \qquad [(17.56), \text{Th. } 3.48]
$$

$$
= \int 1_{C'}(z) \iint 1_{A'}(x) \cdot 1_{B'}(y) \, P_{Y|Z=z}(dy) \, P_{X|Z=z}(dx) \, P_Z(dz). \qquad [(17.41)]
$$

Hence, Theorem 3.48 implies

$$
\forall \, (A', B') \in \mathscr{A}'_X \times \mathscr{A}'_Y \colon P_{X,Y|Z=z}(A', B') \underset{P_Z\text{-a.a.}}{=} P_{X|Z=z}(A') \cdot P_{Y|Z=z}(B'),
$$

which, according to Theorem 17.43, implies $X \underset{P}{\perp\!\!\!\perp} Y \mid Z$.

Proof of Corollary 17.48

Equation (17.57) implies

$$
\forall \, B' \in \mathscr{A}'_Y \colon P_{Y|1_{X=x}=v, Z=z}(B') = P_{Y|Z=z}(B'), \quad \text{for } P_{1_{X=x}, Z}\text{-a.a. } (v, z) \in \{0, 1\} \times \Omega'_Z.
$$

Hence, for all $B' \in \mathscr{A}'_Y$, there is a set $D' \subset \{0, 1\} \times \Omega'_Z$ with $P_{1_{X=x}, Z}(D') = 0$ and

$$
\forall \, (v, z) \in (\{0, 1\} \times \Omega'_Z) \setminus D' \colon P_{Y|1_{X=x}=v, Z=z}(B') = P_{Y|Z=z}(B').
$$

Now, if we define $D'_v := \{ z \in \Omega'_Z \colon (v, z) \in D' \}$, for $v = 0, 1$, then $D' = (\{0\} \times D'_0) \cup (\{1\} \times D'_1)$ and $(\{0, 1\} \times \Omega'_Z) \setminus D' = [\{0\} \times (\Omega'_Z \setminus D'_0)] \cup [\{1\} \times (\Omega'_Z \setminus D'_1)]$. This implies

$$
\forall \, z \in \Omega'_Z \setminus D'_1 \colon P_{Y|1_{X=x}=1, Z=z}(B') = P_{Y|Z=z}(B')
$$

and

$$P_{1_{X=x},Z}(D') = 0$$

$$\Rightarrow P_{1_{X=x},Z}(\{1\} \times D_1') = 0 \qquad [\{1\} \times D_1' \subset D']$$

$$\Rightarrow P(X=x, Z \in D_1') = 0 \qquad [(5.2)]$$

$$\Rightarrow \frac{P(X=x, Z \in D_1')}{P(X=x)} = 0 \qquad [P(X=x) > 0]$$

$$\Rightarrow P_{Z|X=x}(D_1') = 0. \qquad [(4.2), (9.4), \text{Cor. } 17.10]$$

Hence, $P_{Y|1_{X=x}=1, Z=z}(B') = P_{Y|Z=z}(B')$ for $P_{Z|X=x}$-a.a. $z \in \Omega_Z'$. Because $\{1_{X=x}=1\} = \{X=x\}$, the proof of (17.60) is complete.

Proof of Corollary 17.51

If $X \underset{P}{\perp\!\!\!\perp} Y \mid Z$, then, for all nonnegative measurable $h: \Omega_X' \times \Omega_Y' \times \Omega_Z' \to \mathbb{R}$,

$$\int\!\!\!\int\!\!\!\int h(x, y, z) \, P_{Y|Z=z}(dy) \, P_{X|Z=z}(dx) \, P_Z(dz)$$

$$= \int\!\!\!\int h(x, y, z) \, P_{X,Y|Z=z}[d(x, y)] \, P_Z(dz) \qquad [(17.52)]$$

$$= \int h(x, y, z) \, P_{X,Y,Z}[d(x, y, z)] \qquad [(17.43)]$$

$$= \int\!\!\!\int h(x, y, z) \, P_{Y|X=x, Z=z}(dy) \, P_{X,Z}[d(x, z)] \qquad [(17.43)]$$

$$= \int\!\!\!\int\!\!\!\int h(x, y, z) \, P_{Y|X=x, Z=z}(dy) \, P_{X|Z=z}(dx) \, P_Z(dz). \qquad [(17.43)]$$

This proves Equation (17.62).

Now, for

$$h = 1_{A'} \cdot 1_{B'} \cdot 1_{C'}, \quad (A', B', C') \in \mathscr{A}_X' \times \mathscr{A}_Y' \times \mathscr{A}_Z',$$

(17.62) yields

$$\int 1_{C'}(z) \cdot P_{Y|Z=z}(B') \cdot P_{X|Z=z}(A') \, P_Z(dz)$$

$$= \int 1_{C'}(z) \int 1_{A'}(x) \cdot P_{Y|X=x, Z=z}(B') \, P_{X|Z=z}(dx) \, P_Z(dz). \qquad [(6.4), (6.1)]$$

Therefore, (3.45) implies: For all $(A', B') \in \mathscr{A}_X' \times \mathscr{A}_Y'$,

$$P_{X|Z=z}(A') \cdot P_{Y|Z=z}(B') \underset{P_Z\text{-a.a.}}{=} \int 1_{A'}(x) \cdot P_{Y|X=x, Z=z}(B') \, P_{X|Z=z}(dx).$$

Proof of Corollary 17.52

$$P_{Y|X=x}(B') = P_Y(B'), \quad \forall\, B' \in \mathscr{A}'_Y,$$

$$\Leftrightarrow\ P_Y^{X=x}(B') = P_Y(B'), \quad \forall\, B' \in \mathscr{A}'_Y, \qquad\qquad [(17.15),\, P(X{=}x) > 0]$$

$$\Leftrightarrow\ \frac{1}{P(X{=}x)} \cdot P(X{=}x, Y{\in}B') = P(Y{\in}B'), \quad \forall\, B' \in \mathscr{A}'_Y, \qquad [\text{Defs. 4.12, 5.3}]$$

$$\Leftrightarrow\ P(X{=}x, Y{\in}B') = P(X{=}x) \cdot P(Y{\in}A'), \quad \forall\, B' \in \mathscr{A}'_Y,$$

$$\Leftrightarrow\ \{X{=}x\} \underset{P}{\perp\!\!\!\perp} Y \qquad\qquad\qquad\qquad\qquad [\text{Def. 4.40}]$$

$$\Leftrightarrow\ 1_{X=x} \underset{P}{\perp\!\!\!\perp} Y. \qquad\qquad\qquad\qquad\qquad [\text{Rem. 5.46}]$$

Proof of Corollary 17.53

Note that $\int 1_{\{x\}}(x')\, P_{X|Z=z}(dx') = P_{X|Z=z}(\{x\})$ [see Eqs. (6.4) and (6.1)]. Furthermore, $P(X{=}x \mid Z) \underset{P}{>} 0$ implies $P(X{=}x \mid Z{=}z) \underset{P_Z\text{-a.a.}}{>} 0$ [see Cor. 5.25 (ii)], which in turn implies $P_{X|Z=z}(\{x\}) \underset{P_Z\text{-a.a.}}{>} 0$ [see (17.11), (17.14), and (2.40)]. Therefore,

$$\int \frac{1}{P_{X|Z=z}(\{x\})} \cdot 1_{\{x\}}(x')\, P_{X|Z=z}(dx') = 1, \quad \text{for } P_Z\text{-a.a. } z \in \Omega'_Z. \qquad (17.109)$$

For all $B' \in \mathscr{A}'_Y$ and $x \in \Omega'_X$ with $P(X{=}x \mid Z) \underset{P}{>} 0$,

$$\int P_{Y|Z=z}(B')\, P_Z(dz)$$

$$= \iint 1_{B'}(y)\, P_{Y|Z=z}(dy)\, P_Z(dz) \qquad\qquad [(3.8)]$$

$$= \iint 1_{B'}(y)\, P_{Y|Z=z}(dy) \int \frac{1}{P_{X|Z=z}(\{x\})} \cdot 1_{\{x\}}(x')\, P_{X|Z=z}(dx')\, P_Z(dz) \qquad [(17.109)]$$

$$= \iiint 1_{B'}(y) \cdot \frac{1}{P_{X|Z=z}(\{x\})} \cdot 1_{\{x\}}(x')\, P_{Y|Z=z}(dy)\, P_{X|Z=z}(dx')\, P_Z(dz) \qquad [(3.32)]$$

$$= \iiint 1_{B'}(y) \cdot \frac{1}{P_{X|Z=z}(\{x\})} \cdot 1_{\{x\}}(x')\, P_{Y|X=x,\,Z=z}(dy)\, P_{X|Z=z}(dx')\, P_Z(dz) \qquad [(17.62)]$$

$$= \iint 1_{B'}(y)\, P_{Y|X=x,\,Z=z}(dy)\, P_Z(dz) \qquad\qquad [(17.109)]$$

$$= \int P_{Y|X=x,\,Z=z}(B')\, P_Z(dz). \qquad\qquad [(3.8)]$$

Applying Theorem 3.48 yields (17.64), and Lemma 17.26 completes the proof of (17.65).

Proof of Lemma 17.54

If $P(X=x \mid Z) \underset{P}{>} 0$, then,

$$
\begin{aligned}
P(X=x) &= E(1_{X=x}) & & [(6.4)] \\
&= E[E(1_{X=x} \mid Z)] & & [\text{Box } 10.2 \text{ (iv)}] \\
&= E[P(X=x \mid Z)] & & [(10.3)] \\
&> 0. & & [P(X=x \mid Z) \underset{P}{>} 0, (6.1), (3.51)]
\end{aligned}
$$

Now we can apply Corollary 14.48 with $B = \{X=x\}$, because $P(X=x) > 0$ in (a) as well as in (b). This implies

$$
\left(P(X=x) > 0 \text{ and } P \underset{\sigma(Z)}{\ll} P^{X=x} \right) \quad \Leftrightarrow \quad P(X=x \mid Z) \underset{P}{>} 0.
$$

Furthermore,

$$
\begin{aligned}
& P \underset{\sigma(Z)}{\ll} P^{X=x} \\
\Leftrightarrow \ & \forall\, C' \in \mathcal{A}'_Z\colon \quad P^{X=x}_Z(C') = P^{X=x}[Z^{-1}(C')] = 0 \Rightarrow P_Z(C') = P[Z^{-1}(C')] = 0 \\
& \hspace{9cm} [\text{Def. } 3.70, (5.2)] \\
\Leftrightarrow \ & P_Z \underset{\mathcal{A}'_Z}{\ll} P^{X=x}_Z.
\end{aligned}
$$

Proof of Theorem 17.55

Let $A' \in \mathcal{A}'_X$. Then, $A := X^{-1}(A') \in \sigma(X)$, and the definition of $E[f(Y) \mid X]$ (see Def. 10.2) implies

$$
\int 1_A \cdot E[f(Y) \mid X]\, dP = \int 1_A \cdot f(Y)\, dP,
$$

which, according to Theorem 3.57, is equivalent to

$$
\begin{aligned}
\int 1_{A'}(x) \cdot E[f(Y) \mid X=x]\, P_X(dx) &= \int 1_{A'}(x) \cdot f(y)\, P_{X,Y}[d(x,y)] \\
&= \iint 1_{A'}(x) \cdot f(y)\, P_{Y\mid X=x}(dy)\, P_X(dx) & & [(17.43)] \\
&= \int 1_{A'}(x) \left(\int f(y)\, P_{Y\mid X=x}(dy) \right) P_X(dx). & & [(3.32)]
\end{aligned}
$$

Because these equations hold for all $A' \in \mathcal{A}'_X$, Theorem 3.48 yields

$$
E[f(Y) \mid X=x] = \int f(y)\, P_{Y\mid X=x}(dy), \quad \text{for } P_X\text{-a.a. } x \in \Omega'_X.
$$

Proof of Corollary 17.57

For all nonnegative measurable functions $g: \Omega'_X \times \overline{\mathbb{R}} \to \overline{\mathbb{R}}$,

$$\iint g[x, f(y)]\, P_{Y|X=x}(dy)\, P_X(dx) = \int g[x, f(y)]\, P_{X,Y}[d(x, y)] \qquad [(17.43)]$$

$$= \int g(x, z)\, P_{X,f(Y)}[d(x, z)] \qquad [(3.59), (5.5)]$$

$$= \iint g(x, z)\, P_{f(Y)|X=x}(dz)\, P_X(dx). \qquad [(17.43)]$$

Choosing $g(x, z) = 1_{A'}(x) \cdot z$ for $A' \in \mathscr{A}'_X$, Theorem 3.48 implies (17.68).

Proof of Theorem 17.60

Consider Equation (17.66) and replace $f(Y)$ by $E(Y \mid X, Z)$, which is a function of (X, Z). Then, for P_X-a.a. $x \in \Omega'_X$,

$$E(Y \mid X=x) = E[E(Y \mid X, Z) \mid X=x] \qquad [\text{Box } 10.2 \text{ (v)}, (10.34)]$$

$$= \int E(Y \mid X=x', Z=z)\, P_{X,Z|X=x}[d(x', z)] \qquad [(17.66)]$$

$$= \int E(Y \mid X=x, Z=z)\, P_{Z|X=x}(dz). \qquad [\text{Cor. } 17.44 \text{ (i)}]$$

Proof of Corollary 17.62

If Z is discrete, then for P_X-a.a. $x \in \Omega'_X$,

$$E(Y \mid X=x) = \int E(Y \mid X=x, Z=z)\, P_{Z|X=x}(dz) \qquad [(17.72)]$$

$$= \sum_{\substack{z \in \Omega'_Z \\ P(Z=z) > 0}} E(Y \mid X=x, Z=z) \cdot P_{Z|X=x}(\{z\}) \qquad [(17.29)]$$

$$= \sum_{\substack{z \in \Omega'_Z \\ P(Z=z) > 0}} E(Y \mid X=x, Z=z) \cdot P(Z=z \mid X=x). \qquad [(17.8), (17.10)]$$

Proof of Corollary 17.63

Replacing in Theorem 17.60 W by Z and X by (X, Z): $(\Omega, \mathscr{A}, P) \to (\Omega'_X \times \Omega'_Z, \mathscr{A}'_X \otimes \mathscr{A}'_Z)$ yields, for $P_{X,Z}$-a.a. $(x, z) \in \Omega'_X \times \Omega'_Z$,

$$E(Y \mid X=x, Z=z) = \int E(Y \mid X=x, Z=z, W=w)\, P_{W|X=x,Z=z}(dw)$$

$$= \int E(Y \mid X=x, Z=z, W=w)\, P_{W|Z=z}(dw). \qquad [X \underset{P}{\perp\!\!\!\perp} W \mid Z, (17.58)]$$

Proof of Lemma 17.68

For all $x \in \mathbb{R}$ with $f(x) > 0$, the distribution function $F_{Y|X=x}$ defines a probability measure on $(\mathbb{R}, \mathscr{B})$ (see Rem. 5.83), which in this proof is denoted by $P_{(x)}$. Furthermore, define

$$C' := \{x \in \mathbb{R}: f_X(x) > 0\}.$$

If $f_X(x) = \int f_{X,Y}(x, y)\, dy = 0$, then $f_{X,Y}(x, \cdot) \underset{P_Y}{=} 0$ (see Th. 3.43). Hence, for all measurable nonnegative functions $h: \mathbb{R}^2 \to \mathbb{R}$,

$$\int_{\mathbb{R} \setminus C'} \int h(x, y) \cdot f_{X,Y}(x, y)\, dy\, dx = 0. \tag{17.110}$$

Note that, according to Theorem 17.18 (a), $P_{Y|X} \in \mathscr{P}_{Y|X}$ exists. Furthermore, for all $P_{Y|X} \in \mathscr{P}_{Y|X}$,

$$\int\int h(x, y)\, P_{Y|X=x}(dy)\, P_X(dx)$$

$$= \int h(x, y) \cdot P_{X,Y}[d(x, y)] \qquad\qquad [(17.43)]$$

$$= \int\int h(x, y) \cdot f_{X,Y}(x, y)\, dy\, dx \qquad\qquad [(3.72)]$$

$$= \int_{C'} \int h(x, y) \frac{f_{X,Y}(x, y)}{f_X(x)}\, dy\, f_X(x)\, dx + \int_{\mathbb{R} \setminus C'} \int h(x, y) f_{X,Y}(x, y)\, dy\, dx \qquad [\text{def. of } C']$$

$$= \int_{C'} \int h(x, y)\, P_{(x)}(dy)\, P_X(dx). \qquad\qquad [(17.84), (17.110)]$$

This implies that for all $A' \in \mathscr{A}'_X$, $B' \in \mathscr{A}'_Y$, and $h(x, y) = 1_{A'}(x) \cdot 1_{B'}(y)$,

$$\int 1_{A'}(x) \cdot P_{Y|X=x}(B')\, P_X(dx) = \int 1_{A' \cap C'}(x) \cdot P_{(x)}(B')\, P_X(dx).$$

Because $P_X(\mathbb{R} \setminus C') = \int_{\mathbb{R} \setminus C'} f_X(x)\, P_X(dx) = 0$, we can conclude $P_X(C') = 1$. Therefore, according to Box 4.1 (viii), $P_X(A' \cap C') = P_X(A')$. Hence, $1_{A'} \underset{P_X}{=} 1_{A' \cap C'}$, and we obtain

$$\int 1_{A'}(x) \cdot P_{Y|X=x}(B')\, P_X(dx) = \int 1_{A' \cap C'}(x) \cdot P_{Y|X=x}(B')\, P_X(dx)$$

$$= \int 1_{A' \cap C'}(x) \cdot P_{(x)}(B')\, P_X(dx)$$

$$= \int 1_{A'}(x) \cdot P_{(x)}(B')\, P_X(dx).$$

Now Theorem 3.48 and Corollary 17.28 imply

$$P_{Y|X=x} \underset{P_X\text{-a.a.}}{=} P_{(x)},$$

where we define $P_{(x)} = 0$ for $x \in \mathbb{R} \setminus C'$.

Proof of Lemma 17.70

Equation (17.83) implies, for all $y \in \mathbb{R}$,

$$F_{Y|X=x}(y) \cdot f_X(x) = \int_{-\infty}^{y} f_{X,Y}(x, v) \, dv,$$

which also holds if $f_X(x) = 0$ [see Rem. 17.67 and (17.110)]. Taking the Riemann integral with respect to x and applying Theorem 3.62 yield, for all $y \in \mathbb{R}$,

$$
\begin{aligned}
\int_{-\infty}^{\infty} F_{Y|X=x}(y) \cdot f_X(x) \, dx &= \int_{-\infty}^{\infty} \int_{-\infty}^{y} f_{X,Y}(x, v) \, dv \, dx \\
&= \int_{-\infty}^{y} \int_{-\infty}^{\infty} f_{X,Y}(x, v) \, dx \, dv \quad \text{[Th. 3.76]} \\
&= \int_{-\infty}^{y} f_Y(v) \, dv \qquad\qquad\qquad \text{[(5.50)]} \\
&= F_Y(y). \qquad\qquad\qquad\qquad \text{[(5.64)]}
\end{aligned}
$$

Furthermore, for all $y \in \mathbb{R}$,

$$
\begin{aligned}
\int_{-\infty}^{\infty} f_{Y|X=x}(y) \cdot f_X(x) \, dx &= \int_{-\infty}^{\infty} f_{X,Y}(x, y) \, dx \qquad \text{[(17.85)]} \\
&= f_Y(y), \qquad\qquad\qquad \text{[(5.50)]}
\end{aligned}
$$

where f_Y is the density of the marginal distribution of Y.

Proof of Theorem 17.72

We assume $f_X(x) > 0$ and $f_Y(y) > 0$. Therefore,

$$
\begin{aligned}
f_{Y|X=x}(y) &= \frac{f_{X,Y}(x, y)}{f_X(x)} && \text{[(17.84)]} \\
&= \frac{f_{X|Y=y}(x) \cdot f_Y(y)}{f_X(x)} && \text{[(17.85)]} && (17.111) \\
&= \frac{f_{X|Y=y}(x) \cdot f_Y(y)}{\int_{-\infty}^{\infty} f_{X|Y=y^*}(x) \cdot f_Y(y^*) \, dy^*}. && \text{[(17.87)]}
\end{aligned}
$$

Proof of Lemma 17.73

Define the function $g: \mathbb{R} \to \overline{\mathbb{R}}$ by

$$
\begin{aligned}
\forall x \in \mathbb{R}: g(x) &= \int y \cdot f_{Y|X=x}(y) \, dy \\
&= \begin{cases} \frac{1}{f_X(x)} \int y \cdot f_{X,Y}(x, y) \, dy, & \text{if } f_X(x) > 0 \text{ and } \int y \cdot f_{X,Y}(x, y) \, dy \text{ exists} \\ 0, & \text{otherwise.} \end{cases}
\end{aligned}
$$

The function $g: \mathbb{R} \to \overline{\mathbb{R}}$ is $(\mathcal{B}, \overline{\mathcal{B}})$-measurable (see Th. 3.76). Therefore, according to Theorem 2.49, $g(X): \Omega \to \overline{\mathbb{R}}$ is $(\sigma(X), \overline{\mathcal{B}})$-measurable. If $A \in \sigma(X)$, then there is an $A' \in \mathcal{A}'_X$ with $A = X^{-1}(A')$ and

$$\int 1_A \cdot g(X) \, dP = \int 1_{A'}(x) \cdot g(x) \, P_X(dx) \qquad [(6.13)]$$

$$= \int 1_{A'}(x) \cdot \int y \cdot f_{Y|X=x}(y) \, dy \, P_X(dx) \qquad [\text{def. of } g]$$

$$= \iint 1_{A'}(x) \cdot y \cdot f_{Y|X=x}(y) f_X(x) \, dy \, dx \qquad [(3.72)]$$

$$= \iint 1_{A'}(x) \cdot y \cdot f_{X,Y}(x, y) \, dy \, dx \qquad [(17.85)]$$

$$= \int 1_{A'}(x) \cdot y \, P_{X,Y}[d(x, y)] \qquad [(3.72)]$$

$$= \int 1_A \cdot Y \, dP. \qquad [(6.13)]$$

Hence, according to Definition 10.2, $g(X) \in \mathcal{E}(Y \mid X)$.

Proof of Lemma 17.76

In the proof of Lemma 17.68, let the $P_{(x)}$ be defined by the densities $f_{Y|X=x}, x \in \Omega'_X$, and replace dx by $\mu(dx)$ and dy by $\nu(dy)$ (i.e., replace the integration with respect to the Lebesgue measure by the integration with respect to μ and ν, respectively).

Proof of Lemma 17.77

Let $f_{X,Y}$ denote a density of $P_{X,Y}$ with respect to $\mu \otimes \nu$. Note that, for $A' := \{x \in \Omega'_X : f_X(x) = 0\}$ and all $B' \in \mathcal{A}'_Y$,

$$\int 1_{A'}(x) \cdot 1_{B'}(y) \cdot f_{X,Y}(x, y) \, (\mu \otimes \nu)[d(x, y)]$$

$$= \int 1_{A'}(x) \cdot 1_{B'}(y) \, P_{X,Y}[d(x, y)] \qquad [(5.40)]$$

$$= P_{X,Y}(A' \times B') \qquad [(3.8)]$$

$$\leq P_{X,Y}(A' \times \Omega'_Y) \qquad [\text{Box 4.1 (v)}]$$

$$= P_X(A') \qquad [(5.21)]$$

$$= \int 1_{A'}(x) \, P_X(dx) \qquad [(3.8)]$$

$$= \int 1_{A'}(x) \cdot f_X(x) \, \mu(dx) = 0. \qquad [(5.40), \text{Lemma 5.79}, (3.40)]$$

This implies: for all $B' \in \mathcal{A}'_Y$,

$$\int 1_{B'}(y) \cdot \left(\int f_{Y|X=x}(y) \cdot f_X(x) \, \mu(dx) \right) \nu(dy)$$

$$= \int 1_{B'}(y) \int f_{X,Y}(x, y) \, \mu(dx) \, \nu(dy) \qquad \text{[Def. 17.75, Th. 3.76]}$$

$$= \int 1_{B'}(y) \cdot f_Y(y) \, \nu(dy). \qquad \text{[(5.48)]}$$

Theorem 3.48, then, implies the proposition of the lemma.

Proof of Lemma 17.78

The proof is analogous to the proof of Lemma 17.73 using the function $g \colon \Omega'_X \to \overline{\mathbb{R}}$ with

$$g(x) = \int y \cdot f_{Y|X=x}(y) \, \nu(dy), \quad \forall \, x \in \Omega'_X,$$

and replacing dy by $\nu(dy)$ and dx by $\mu(dx)$, respectively.

Exercises

17.1 Prove the proposition of Remark 17.36.

17.2 Suppose that $X \colon (\Omega, \mathcal{A}, P) \to (\Omega'_X, \mathcal{A}'_X)$ and $Y \colon (\Omega, \mathcal{A}, P) \to (\Omega'_Y, \mathcal{A}'_Y)$ are random variables, $A' \in \mathcal{A}'_X$, $B' \in \mathcal{A}'_Y$, and $P(X=x) > 0$. Show that

$$P(X \in A', Y \in B' \mid X=x) = 1_{A'}(x) \cdot P(Y \in B' \mid X=x).$$

17.3 Prove the proposition of Remark 17.46.

17.4 Prove the propositions of Remark 17.59.

17.5 Consider Example 17.81 and prove $P_{X,Y} \underset{\mathcal{B}_2}{\ll} \mu \otimes \lambda$.

17.6 Prove Equations (17.101) to (17.104).

Solutions

17.1 For an $(X=x)$-conditional-probability measure $P^{X=x}$ defined by Equation (17.32) and for all $A \in \mathcal{A}$, there is a version $P(id \in A \mid X)$ such that

$$\begin{aligned}
P^{X=x}(A) &= P_{id|X=x}(A) & \text{[(17.32)]} \\
&= P(id \in A \mid X=x) & \text{[(17.10)]} \\
&= P^{X=x}(id \in A) & \text{[Rem. 10.35, (9.4)]} \\
&= P^{X=x}(A). & [\{id \in A\} = A]
\end{aligned}$$

According to (9.4), this value is unique for all $A \in \mathcal{A}$.

17.2   For $A' \in \mathcal{A}'_X$, consider the events

$$\{X \in A'\} = \{\omega \in \Omega: X(\omega) \in A'\} = \{\omega \in \Omega: 1_{A'}(X)(\omega) = 1\}$$

and, for $x \in \Omega'_X$,

$$\begin{aligned}
\{X \in A'\} \cap \{X = x\} &= \{\omega \in \Omega: X(\omega) \in A'\} \cap \{\omega \in \Omega: X(\omega) = x\} \\
&= \{\omega \in \Omega: 1_{A'}(X)(\omega) = 1, X(\omega) = x\} \\
&= \begin{cases} \{X = x\}, & \text{if } x \in A' \\ \varnothing, & \text{if } x \notin A'. \end{cases}
\end{aligned} \qquad (17.112)$$

Hence, if $x \in A'$ (i.e., if $1_{A'}(x) = 1$), then

$$\begin{aligned}
P(X \in A', Y \in B' \mid X = x) &= \frac{P(\{X \in A'\} \cap \{Y \in B'\} \cap \{X = x\})}{P(\{X = x\})} && [(4.2)] \\
&= \frac{P(\{Y \in B'\} \cap \{X = x\})}{P(\{X = x\})} && [(17.112)] \\
&= P(Y \in B' \mid X = x) && [(4.2)] \\
&= 1_{A'}(x) \cdot P(Y \in B' \mid X = x).
\end{aligned}$$

If $x \notin A'$ (i.e., if $1_{A'}(x) = 0$), then

$$\begin{aligned}
P(X \in A', Y \in B' \mid X = x) &= \frac{P(\{X \in A'\} \cap \{Y \in B'\} \cap \{X = x\})}{P(\{X = x\})} && [(4.2)] \\
&= \frac{P(\varnothing)}{P(\{X = x\})} && [(17.112)] \\
&= 0 \\
&= 1_{A'}(x) \cdot P(Y \in B' \mid X = x).
\end{aligned}$$

17.3   We show that (17.56) is equivalent to $P_{Y|Z} \in \mathcal{P}_{Y|X, Z}$. We assume that $P_{Y|Z}$ and $P_{Y|X,Z}$ exist. Hence, according to (17.3) and (17.4), for all $B' \in \mathcal{A}'_Y$ there is a $P(Y \in B' \mid X, Z) \in \mathcal{P}(Y \in B' \mid X, Z)$ and a $P(Y \in B' \mid Z) \in \mathcal{P}(Y \in B' \mid Z)$ with

$$P(Y \in B' \mid X, Z)(\omega) = P_{Y|X,Z}(\omega, B'), \quad \forall \, \omega \in \Omega,$$

and

$$P(Y \in B' \mid Z)(\omega) = P_{Y|Z}(\omega, B'), \quad \forall \, \omega \in \Omega.$$

Hence, (17.56) is equivalent to

$$\forall\, B' \in \mathscr{A}'_Y: \quad P(Y \in B' \mid X, Z) \underset{P}{=} P(Y \in B' \mid Z) \tag{17.113}$$

Because $\sigma(Z) \subset \sigma(X, Z)$, the random variable $P(Y \in B' \mid Z)$ is (X, Z)-measurable, and proposition (10.12) implies that, for all $B' \in \mathscr{A}'_Y$, $P(Y \in B' \mid Z) \in \mathscr{P}(Y \in B' \mid X, Z)$ is equivalent to (17.113). However, according to Definition 17.1, $P(Y \in B' \mid Z) \in \mathscr{P}(Y \in B' \mid X, Z)$ is equivalent to $P_{Y|Z} \in \mathscr{P}_{Y|X, Z}$.

17.4   If Y takes on only a finite number of values $y_1, \ldots, y_n$, then,

$$f(y) = \sum_{i=1}^{n} 1_{\{y_i\}}(y) \cdot f(y_i).$$

Therefore, for P_X-a.a. $x \in \Omega'_X$,

$$E[f(Y) \mid X = x] = \int f(y)\, P_{Y|X=x}(dy) \qquad [(17.66)]$$

$$= \sum_{i=1}^{n} f(y_i) \cdot P_{Y|X=x}(\{y_i\}). \qquad [(17.29)]$$

If $P(X = x) > 0$, then Equation (17.70) follows from Equation (17.11) for $A' = \{y_i\}$.

17.5   According to Theorem 5.77, the probability function p_X of X (see Def. 5.56) is a density of P_X with respect to $\mu = \delta_0 + \delta_1$. If $f_{Y|X=x}$ denotes the density (with respect to λ) of $P_{Y|X=x} = \mathscr{N}(\mu_x, \sigma_x^2)$, $x = 0, 1$, then, for all nonnegative measurable functions $h: \mathbb{R}^2 \to \mathbb{R}$,

$$\int h(x, y)\, P_{X,Y}[d(x, y)] = \int\!\!\int h(x, y)\, P_{Y|X=x}(dy)\, P_X(dx) \qquad [(17.43)]$$

$$= \int\!\!\int h(x, y) \cdot f_{Y|X=x}(y)\, \lambda(dy)\, p_X(x)\, \mu(dx) \qquad [(3.79)]$$

$$= \int\!\!\int h(x, y) \cdot f_{Y|X=x}(y) \cdot p_X(x)\, \lambda(dy)\, \mu(dx)$$

$$= \int h(x, y) \cdot f_{Y|X=x}(y) \cdot p_X(x)\, \mu \otimes \lambda[d(x, y)]. \qquad [(3.80)]$$

Hence, choosing $h = 1_{D'}$ for $D' \in \mathscr{B}_2$, Theorem 3.65 implies that $f_{Y|X=x}(y)\, p_X(x)$, where $(x, y) \in \mathbb{R}^2$, describes a Radon-Nikodym density of $P_{X,Y}$ with respect to $\mu \otimes \lambda$, which implies $P_{X,Y} \underset{\mathscr{B}_2}{\ll} \mu \otimes \lambda$.

17.6 According to Lemma 17.78, for all $y \in \Omega_Y'$:

$$E(X \mid Y=y)$$

$$= \int x \cdot f_{X|Y=y}(x)\,\mu(dx)$$

$$= f_{X|Y=y}(1) \qquad\qquad [\mu = \delta_0 + \delta_1, (3.57)]$$

$$= \frac{p \cdot f_{Y|X=1}(y)}{(1-p) \cdot f_{Y|X=0}(y) + p \cdot f_{Y|X=1}(y)} \qquad\qquad [(17.100)]$$

$$= \frac{p \cdot \dfrac{1}{\sigma_1} \cdot \exp\left[-\dfrac{1}{2}\left(\dfrac{y-\mu_1}{\sigma_1}\right)^2\right]}{(1-p)\cdot\dfrac{1}{\sigma_0}\cdot\exp\left[-\dfrac{1}{2}\left(\dfrac{y-\mu_0}{\sigma_0}\right)^2\right] + p\cdot\dfrac{1}{\sigma_1}\cdot\exp\left[-\dfrac{1}{2}\left(\dfrac{y-\mu_1}{\sigma_1}\right)^2\right]}$$

$$[(8.23)]$$

defines a version of $E(X \mid Y)$. Note that the term $\dfrac{1}{\sqrt{2\pi}}$ in the densities $f_{Y|X=x}$, $x=0,1$, of the normal distributions cancels out. If we define

$$q_0 := \frac{1-p}{\sigma_0}, \qquad q_1 := \frac{p}{\sigma_1}, \qquad c := \frac{q_1}{q_0},$$

and

$$a := -\frac{1}{2}\left(\frac{y-\mu_0}{\sigma_0}\right)^2, \qquad b := -\frac{1}{2}\left(\frac{y-\mu_1}{\sigma_1}\right)^2,$$

then,

$$E(X \mid Y=y) = \frac{q_1 \cdot e^b}{q_0 \cdot e^a + q_1 \cdot e^b} \qquad\qquad [\text{cancel by } q_0 \cdot e^a]$$

$$= \frac{c \cdot e^{b-a}}{1 + c \cdot e^{b-a}} \qquad\qquad [\text{replace } c,\, c = e^{\ln c}]$$

$$= \frac{e^{b-a+\ln c}}{1 + e^{b-a+\ln c}}.$$

Now consider

$$b - a = -\frac{1}{2}\left[\left(\frac{y-\mu_1}{\sigma_1}\right)^2 - \left(\frac{y-\mu_0}{\sigma_0}\right)^2\right]$$

$$= -\frac{1}{2}\left[\frac{1}{\sigma_1^2}(y^2 - 2\mu_1 y + \mu_1^2) - \frac{1}{\sigma_0^2}(y^2 - 2\mu_0 y + \mu_0^2)\right]$$

$$= \left(\frac{\mu_0^2}{2\sigma_0^2} - \frac{\mu_1^2}{2\sigma_1^2}\right) + \left(\frac{\mu_1}{\sigma_1^2} - \frac{\mu_0}{\sigma_0^2}\right)\cdot y + \left(\frac{1}{2\sigma_0^2} - \frac{1}{2\sigma_1^2}\right)\cdot y^2.$$

If $\sigma_0 = \sigma_1 =: \sigma$, then this equation simplifies to

$$b - a = \left(\frac{\mu_0^2 - \mu_1^2}{2\sigma^2}\right) + \left(\frac{\mu_1 - \mu_0}{\sigma^2}\right)\cdot y.$$

References

Agresti, A. (2015). *Foundations of linear and generalized linear models*. Hoboken, NJ: John Wiley & Sons.

Bauer, H. (1996). *Probability theory*. Berlin, Germany and New York, NY: de Gruyter.

Bauer, H. (2001). *Measure and integration theory*. Berlin, Germany and New York, NY: de Gruyter.

Billingsley, P. (1995). *Probability and measure* (3rd ed.). New York, NY: Wiley-Interscience.

Bronshtein, I. N., Semendyayev, K. A., Musiol, G., & Mühlig, H. (2015). *Handbook of mathematics* (6th ed.). New York, NY: Springer.

Ellis, R., & Gulick, D. (2006). *Calculus* (6th ed.). Mason, OH: Thomson.

Elstrodt, J. (2007). *Maß- und Integrationstheorie [Measure and integration theory]* (4th ed.). Berlin, Germany: Springer.

Fisz, M. (1963). *Probability theory and mathematical statistics*. New York, NY: Wiley.

Fristedt, B., & Gray, L. (1997). *A modern approach to probability theory*. Boston, MA: Birkhäuser.

Georgii, H.-O. (2008). *Stochastics – Introduction to probability and statistics*. Berlin, Germany: de Gruyter.

Harris, J. W., & Stocker, H. (1998). *Handbook of mathematics and computational science*. New York, NY: Springer.

Hoffmann-Jørgensen, J. (1994). *Probability with a view toward statistics* (Vol. 1). New York, NY: Chapman & Hall.

Horn, R. A., & Johnson, C. R. (1991). *Matrix analysis*. Cambridge, UK: Cambridge University Press.

Johnson, N. L., Kemp, A. W., & Kotz, S. (2005). *Univariate discrete distributions* (3rd ed.). New York, NY: Wiley.

Johnson, N. L., Kotz, S., & Balakrishnan, N. (1995). *Continuous univariate distributions* (2nd ed., Vol. 2). New York, NY: Wiley.

Kheyfits, A. (2010). *A primer in combinatorics*. Berlin, Germany and New York, NY: De Gruyter.

Klenke, A. (2013). *Probability theory – A comprehensive course* (2nd ed.). London, UK: Springer.

Kolmogoroff, A. N. (1933/1977). *Grundbegriffe der Wahrscheinlichkeitsrechnung* (reprinted ed.). Berlin, Germany: Springer.

Kolmogorov, A. N. (1956). *Foundations of the theory of probability* (2nd ed.; N. Morrison, Trans.). New York, NY: Chelsea.

Mayer, A., Thoemmes, F., Rose, N., Steyer, R., & West, S. G. (2014). Theory and analysis of total, direct, and indirect causal effects. *Multivariate Behavioral Research, 49*(5), 425–442.

McCullagh, P., & Nelder, J. A. (1989). *Monographs on statistics and applied probability: Vol. 37. Generalized linear models* (2nd ed.; D. R. Cox, D. V. Hinkley, N. Reid, D. B. Rubin, & D. V. Silverman, Eds.). New York, NY: Chapman & Hall.

Michel, H. (1978). *Maβ- und Integrationstheorie I. [Measure and integration theory I]*. Berlin, Germany: VEB Deutscher Verlag der Wissenschaften.

Rao, C. R. (1973). *Linear statistical inference and its applications* (2nd ed.). New York, NY: Wiley.

Rasch, G. (1960/1980). *Probabilistic models for some intelligence and attainment tests* (expanded ed.). Chicago, IL: University of Chicago. (Originally published by Nissen & Lydicke, 1960. Kopenhagen.)

Rosen, K. (2012). *Discrete mathematics and its applications* (7th ed.). New York, NY: McGraw-Hill.

Steyer, R. (2003). *Wahrscheinlichkeit und Regression [Probability and regression]*. Berlin, Germany: Springer.

Steyer, R., Mayer, A., & Fiege, C. (2014). Causal inference on total, direct, and indirect effects. In Michalos, A. C. (Ed.), *Encyclopedia of Quality of Life and Well-Being Research* (pp. 606–631). Dordrecht, the Netherlands: Springer.

Süli, E., & Mayers, D. F. (2003). *An introduction to numerical analysis*. Cambridge, UK: Cambridge University Press.

Tong, Y. L. (1990). *The multivariate normal distribution*. New York, NY: Springer.

Wise, G. L., & Hall, E. B. (1993). *Counterexamples in probability and real analysis*. New York, NY: Oxford University Press.

List of Symbols

¬	not
∧	and
∨	or
⇒	implies
⇔	equivalent to
∃	there is (synonymously, there exists)
∀	for all
∈	element of
1_A	indicator (function) of the set A
∪	union of sets
∩	intersection of sets
\	set difference
A^c	complement of a set A with respect to a set Ω, that is, $A^c := \Omega \setminus A$
⊂	subset or equal
×	Cartesian product or product set
⊗	product σ-algebra
⊗	product measure
∘	composition of two mappings
⊙	measure with density
$=_\mu$	μ-equivalence of mappings
$<_\mu$	smaller than except for a μ-null set
$>_\mu$	greater than except for a μ-null set
$\leq_\mu$	smaller than or equal except for a μ-null set
$\geq_\mu$	greater than or equal except for a μ-null set
$=_P$	P-equivalence of random variables
$\underset{\mu\text{-}a.a.}{=}$	equal for μ-almost all $\omega \in \Omega$
$\underset{\mathscr{A}}{\ll}$	absolute continuity of a measure with respect to another measure
$\underset{\mathscr{A}}{\approx}$	null-set equivalence of two measures
$\underset{P}{\perp\!\!\!\perp}$	independence with respect to the probability measure P
$[a, b]$	closed interval between real numbers a and b
$]a, b]$	half-open interval including b but not a

Probability and Conditional Expectation: Fundamentals for the Empirical Sciences, First Edition. Rolf Steyer and Werner Nagel.
© 2017 John Wiley & Sons, Ltd. Published 2017 by John Wiley & Sons, Ltd.
Companion website: http://www.probability-and-conditional-expectation.de

$\{x\}$	singleton (i.e., the set that contains x as the only element)
$\emptyset$	empty set
$(A_i, i \in I)$	family of sets A_i, $i \in I$, where the index set I may be finite, countable, or uncountable
$\bigcup_{i \in I} A_i$	union of the sets A_i, $i \in I$
$\bigcap_{i \in I} A_i$	intersection of the sets A_i, $i \in I$
$\bigcup_{i=1}^{n} A_i$	union of finitely many sets $A_1, \dots, A_n$
$\bigcup_{i=1}^{\infty} A_i$	union of countably many sets $A_1, A_2, \dots$
$\bigcap_{i=1}^{n} A_i$	intersection of finitely many sets $A_1, \dots, A_n$
$\bigcap_{i=1}^{\infty} A_i$	intersection of countably many sets $A_1, A_2, \dots$
$\times_{i=1}^{n} A_i$	Cartesian product or product set of n sets A_i
$\lim_{n \to \infty} a_n$	limit of a sequence $a_1, a_2, \dots$ of real numbers
$\sum_{i=1}^{n} a_i$	sum of the numbers $a_1, \dots, a_n$
$\sum_{i=1}^{\infty} a_i$	$\lim_{n \to \infty} \sum_{i=1}^{n} a_i$
$\prod_{i=1}^{n} a_i$	product of the real numbers $a_1, \dots, a_n$
$\mathscr{A}_1 \otimes \dots \otimes \mathscr{A}_n$	product σ-algebra of the $\mathscr{A}_1, \dots, \mathscr{A}_n$
$\bigotimes_{i=1}^{n} \mathscr{A}_i$	product σ-algebra of the $\mathscr{A}_1, \dots, \mathscr{A}_n$
$\int f \, d\mu$	integral of a measurable function $f : (\Omega, \mathscr{A}, \mu) \to (\overline{\mathbb{R}}, \overline{\mathscr{B}})$ with respect to the measure μ
$\int_A f \, d\mu$	integral of a measurable function $f (\Omega, \mathscr{A}, \mu) \to (\mathbb{R}, \mathscr{B})$ with respect to the measure μ over a subset A of Ω
$\int_a^b f(x) \, dx$	Riemann integral of the function f from a to b
$\mathscr{A} \vert_{\Omega_0}$	trace of the set system $\mathscr{A}$ in the set Ω_0
$(\Omega, \mathscr{A})$	measurable space
$(\Omega, \mathscr{A}, \mu)$	measure space
$(\Omega, \mathscr{A}, P)$	probability space
$f : A \to B$	mapping f assigning to each $a \in A$ one and only one element $b \in B$
$f(A)$	image of the set A under f
$f^{-1}(A')$	inverse image of the set A' under f
$\{f \in A'\}$	$:= f^{-1}(A')$
$\{f = \omega'\}$	$:= f^{-1}(\{\omega'\})$
$(f_i, i \in I)$	family of mappings
$f^{-1}(\mathscr{E}')$	set of the inverse images of all sets $A' \in \mathscr{E}'$. If $\mathscr{E}'$ is a σ-algebra, then $f^{-1}(\mathscr{E}')$ is the σ-algebra generated by f
$g \circ f, g(f)$	composition of f and g
$\vert f \vert$	absolute value function of f
f^+	positive part of the function f
f^-	negative part of the function f
$f_n \uparrow f$	the sequence $f_1, f_2, \dots$ of functions converges pointwise and monotonically from below to f
$f : (\Omega, \mathscr{A}) \to (\Omega', \mathscr{A}')$	$(\Omega, \mathscr{A}), (\Omega', \mathscr{A}')$ are measurable spaces and the mapping $f : \Omega \to \Omega'$ is $(\mathscr{A}, \mathscr{A}')$-measurable
$f : (\Omega, \mathscr{A}, \mu) \to \Omega'$	$(\Omega, \mathscr{A}, \mu)$ is a measure space and $f : \Omega \to \Omega'$ is a mapping

$f: (\Omega, \mathscr{A}, \mu) \to (\Omega', \mathscr{A}')$	$f: \Omega \to \Omega'$ is an $(\mathscr{A}, \mathscr{A}')$-measurable mapping and μ is a measure on $(\Omega, \mathscr{A})$
$f \underset{\mu}{=} g$	the mappings f and g are μ-equivalent
$f \underset{\mu}{<} g$	f is smaller than g except for a set A of arguments with $\mu(A) = 0$
$f \underset{\mu}{>} g$	f is greater than g except for a set A of arguments with $\mu(A) = 0$
$f \underset{\mu}{\leq} g$	f is smaller than or equal to g except for a set A of arguments with $\mu(A) = 0$
$f \underset{\mu}{\geq} g$	f is greater than or equal to g except for a set A of arguments with $\mu(A) = 0$
$f(\omega) \underset{\mu\text{-}a.a.}{=} g(\omega)$	$f(\omega) = g(\omega)$, for μ-almost all $\omega \in \Omega$
μ_f	image measure of μ under f
$\nu \underset{\mathscr{A}}{\ll} \mu$	the measure ν is absolutely continuous with respect to the measure μ
$\nu \underset{\mathscr{A}}{\approx} \mu$	the measures μ and ν are null-set equivalent (i.e., they are absolutely continuous with respect to each other)
$A \underset{P}{\perp\!\!\!\perp} B$	the events A and B are independent with respect to the probability measure P
$\underset{P}{\perp\!\!\!\perp} (A_i, i \in I)$	a family of events A_i that are independent with respect to the probability measure P
$A \underset{P}{\perp\!\!\!\perp} C \mid B$	the events A and C are B-conditionally independent with respect to the probability measure P
$\mathscr{E}_1 \underset{P}{\perp\!\!\!\perp} \mathscr{E}_2$	the set systems $\mathscr{E}_1$ and $\mathscr{E}_2$ are independent with respect to the probability measure P
$\underset{P}{\perp\!\!\!\perp} (\mathscr{E}_i, i \in I)$	a family of set systems $\mathscr{E}_i$ that are independent with respect to the probability measure P
$A \underset{P}{\perp\!\!\!\perp} \mathscr{E}$	the event A and the set system $\mathscr{E}_1$ are independent with respect to the probability measure P
$\underset{P}{\perp\!\!\!\perp} (X_i, i \in I)$	a family of random variables X_i that are independent with respect to the probability measure P
$\underset{P}{\perp\!\!\!\perp} X_1, \dots, X_n$	the random variables $X_1, \dots, X_n$ are independent with respect to the probability measure P
$\mathscr{E} \underset{P}{\perp\!\!\!\perp} X$	the set system $\mathscr{E}$ and the random variable X are independent with respect to the probability measure P
$X \underset{P}{\perp\!\!\!\perp} (Y_i, i \in I)$	the random variable X and the (σ-algebra generated by the) family of random variables $Y_i, i \in I$, are independent with respect to the probability measure P
$\underset{P}{\perp\!\!\!\perp} (\mathscr{E}_i, i \in I) \mid B$	a family of set systems $\mathscr{E}_i$ that are B-conditionally independent with respect to the probability measure P
$\underset{P}{\perp\!\!\!\perp} A_1, A_2, A_3 \mid \mathscr{C}$	the events A_1, A_2, A_3 are $\mathscr{C}$-conditionally independent with respect to P
$\underset{P}{\perp\!\!\!\perp} (A_i, i \in I) \mid \mathscr{C}$	a family of events A_i that are $\mathscr{C}$-conditionally independent with respect to P
$\underset{P}{\perp\!\!\!\perp} (A_i, i \in I) \mid Z$	a family of events A_i that are Z-conditionally independent with respect to P
$\underset{P}{\perp\!\!\!\perp} (\mathscr{E}_i, i \in I) \mid \mathscr{C}$	a family of set systems $\mathscr{E}_i$ that are $\mathscr{C}$-conditionally independent with respect to P

$\underset{P}{\perp\!\!\!\perp}\,(\mathcal{E}_i, i \in I) \mid Z$	a family of set systems $\mathcal{E}_i$ that are Z-conditionally independent with respect to P
$\underset{P}{\perp\!\!\!\perp}\,(X_i, i \in I) \mid \mathcal{C}$	a family of random variables X_i that are $\mathcal{C}$-conditionally independent with respect to P
$\underset{P}{\perp\!\!\!\perp}\,(X_i, i \in I) \mid Z$	a family of random variables X_i that are Z-conditionally independent with respect to P
$A \underset{P}{\perp\!\!\!\perp} B \mid \mathcal{C}$	the events A and B are $\mathcal{C}$-conditionally independent with respect to the probability measure P
$A \underset{P}{\perp\!\!\!\perp} B \mid Z$	the events A and B are Z-conditionally independent with respect to the probability measure P
$\mathcal{D} \underset{P}{\perp\!\!\!\perp} \mathcal{E} \mid \mathcal{C}$	the set systems $\mathcal{D}$ and $\mathcal{E}$ are $\mathcal{C}$-conditionally independent with respect to the probability measure P
$\mathcal{D} \underset{P}{\perp\!\!\!\perp} \mathcal{E} \mid Z$	the set systems $\mathcal{D}$ and $\mathcal{E}$ are Z-conditionally independent with respect to the probability measure P
$X \underset{P}{\perp\!\!\!\perp} Y \mid \mathcal{C}$	the random variables X and Y are $\mathcal{C}$-conditionally independent with respect to the probability measure P
$X \underset{P}{\perp\!\!\!\perp} Y \mid Z$	the random variables X and Y are Z-conditionally independent with respect to the probability measure P
$\overline{Y}$	arithmetic mean (sample mean) of the real-valued random variables $Y_1, \ldots, Y_n$
$\mathcal{A}, \mathcal{B}, \mathcal{C}, \mathcal{D}, \mathcal{E}$	set systems, sometimes σ-algebras
$\mathcal{A}_1 \otimes \ldots \otimes \mathcal{A}_n$	product σ-algebra of the σ-algebras $\mathcal{A}_1, \ldots, \mathcal{A}_n$
$\mathcal{A}\vert_{\Omega_0}$	trace of the set system $\mathcal{A}$ in the set Ω_0
$\mathcal{B}$	Borel σ-algebra on $\mathbb{R}$
$\mathcal{B}_n$	Borel σ-algebra on $\mathbb{R}^n$
$\overline{\mathcal{B}}$	Borel σ-algebra on $\overline{\mathbb{R}}$
$\overline{\mathcal{B}}_n$	Borel σ-algebra on $\overline{\mathbb{R}}^n$
$B_{n,p}$	binomial distribution with parameters n and p
$b_{n,p}$	probability function of the binomial distribution with parameters n and p
$\mathcal{C}'_f$	final σ-algebra of $\mathcal{C}$ under f
$Cov\,(X, Y)$	covariance of the random variables X and Y
$Cov\,(Y_1, Y_2 \mid X=x)$	$(X=x)$-conditional covariance of Y_1 and Y_2
$Cov\,(Y_1, Y_2 \mid \mathcal{C})$	a version of the $\mathcal{C}$-conditional covariance of Y_1 and Y_2
$Corr\,(X, Y)$	correlation of the random variables X and Y
$Cov\,(Y_1, Y_2 \mid X)$	a version of the X-conditional covariance of Y_1 and Y_2
$Corr\,(Y_1, Y_2 \mid X=x)$	$(X=x)$-conditional correlation of Y_1 and Y_2
$Corr\,(Y_1, Y_2; \mathcal{C})$	partial correlation of Y_1 and Y_2 given $\mathcal{C}$
$Corr\,(Y_1, Y_2; X)$	partial correlation of Y_1 and Y_2 given X
χ_n^2	central χ^2-distribution with n degrees of freedom
$\dfrac{d\nu}{d\mu}$	Radon-Nikodym density (also Radon-Nikodym derivative) of ν with respect to μ
δ_ω	Dirac measure at (point) ω
$E(Y)$	expectation of the random variable Y
$E(x)$	column vector of expectations

$E^B(Y)$	expectation of the random variable Y with respect to the probability measure P^B
$E(Y \mid B)$	conditional expectation value of Y given the event B
$E(Y \mid X=x)$	conditional expectation value of Y given the event $\{X=x\}$, also denoted by $E(Y \mid \{X=x\})$
$E(Y \mid X=x)$	$(X=x)$-conditional expectation value of Y
$E_Y(g)$	expectation of the random variable g with respect to the distribution of the random variable Y
$E(X)$	matrix of expectations
$E_Y^{X=x}(g)$	expectation of g with respect to the distribution $P_Y^{X=x}$
$E^{X=x}(Y)$	expectation of Y with respect to the conditional-probability measure $P^{X=x}$
$E^{\{X=x\}}(Y)$	expectation of Y with respect to the conditional-probability measure $P^{X=x}$
ε	residual of a random variable Y with respect to its $\mathscr{C}$-conditional expectation
ϵ	residual with respect to a (multiple) linear quasi-regression
$E(Y \mid X)$	a version of the X-conditional expectation of Y
$\mathscr{E}(Y \mid X)$	set of all versions of the X-conditional expectation of Y
$E(Y \mid X_1, \dots, X_n)$	a version of the conditional expectation of Y given the multivariate regressor $X_1, \dots, X_n$
$E(Y \mid \mathscr{C})$	a version of the $\mathscr{C}$-conditional expectation of Y
$E(Y \mid \mathscr{C}, \mathscr{D})$	a version of the $\sigma(\mathscr{C} \cup \mathscr{D})$-conditional expectation of Y
$E(Y \mid \mathscr{C}, Z)$	a version of the $\sigma[\mathscr{C} \cup \sigma(Z)]$-conditional expectation of Y
$\mathscr{E}(Y \mid \mathscr{C})$	set of all versions of the $\mathscr{C}$-conditional expectation of Y
$E^B(Y \mid \mathscr{C})$	a version of the $\mathscr{C}$-conditional expectation of Y with respect to the measure P^B
$\mathscr{E}^B(Y \mid \mathscr{C})$	the set of all versions of the $\mathscr{C}$-conditional expectation of Y with respect to the measure P^B
$E^B(Y \mid X)$	a version of the X-conditional expectation of Y with respect to the measure P^B
$E^B(Y \mid X=x)$	an $(X=x)$-conditional expectation value of Y with respect to the measure P^B
$\mathscr{E}^B(Y \mid X)$	the set of all versions of the X-conditional expectation of Y with respect to the measure P^B
$E^{Z=z}(Y \mid \mathscr{C})$	a version of the $\mathscr{C}$-conditional expectation of Y with respect to the measure $P^{Z=z}$
$\mathscr{E}^{Z=z}(Y \mid \mathscr{C})$	the set of all versions of the $\mathscr{C}$-conditional expectations of Y with respect to the measure $P^{Z=z}$
$E^{Z=z}(Y \mid X=x)$	an $(X=x)$-conditional expectation value of Y with respect to the measure $P^{Z=z}$
$E(Y \mid X, Z=z)$	a version of the partial $(X, Z=z)$-conditional expectation of Y (with respect to the measure P)
$E^{Z=z}(Y \mid X)$	a version of the X-conditional expectation of Y with respect to the measure $P^{Z=z}$
$\mathscr{E}^{Z=z}(Y \mid X)$	the family of all versions of the X-conditional expectation of Y with respect to the measure $P^{Z=z}$

$E^{Z=z}(Y \mid X = x)$	an $(X = x)$-conditional expectation value of Y with respect to the measure $P^{Z=z}$
$E^{Z=z}(Y \mid X, W)$	a version of the (X, W)-conditional expectation of Y with respect to the measure $P^{Z=z}$
$E^{Z=z}(Y \mid \mathcal{C}, \mathcal{D})$	a version of the $\sigma(\mathcal{C} \cup \mathcal{D})$-conditional expectation of Y with respect to the measure $P^{Z=z}$
$E^{Z=z}(Y \mid \mathcal{C}, Z)$	a version of the $\sigma[\mathcal{C} \cup \sigma(Z)]$-conditional expectation of Y with respect to the measure $P^{Z=z}$
F_X	distribution function of a real-valued random variable X
$F_{X_1,\dots,X_n}$	joint distribution function of $X_1, \dots, X_n$
$F(x)\vert_a^b$	$F(b) - F(a)$
$F_{Y\mid X=x}$	$(X = x)$-conditional distribution function of Y
$f_{Y\mid X=x}$	$(X = x)$-conditional-probability density of Y
f_X	probability density of a continuous real-valued random variable X
$(f_i, i \in I)$	family of mappings
$f \odot \mu$	measure with density f with respect to μ
$F_{m,n}$	F-distribution with m and n degrees of freedom
Γ	gamma function
$\mathcal{G}_p$	geometric distribution with parameter p
$\mathbb{1}_A$	indicator (function) of the set A
$\mathbb{1}_{X \in A'}$	indicator of the event $\{X \in A'\}$, that is, $\mathbb{1}_{X \in A'} = \mathbb{1}_{X^{-1}(A')}$
$\mathbb{1}_{X=x}, \mathbb{1}_{\{X=x\}}$	indicator of the event $\{X = x\}$
$\mathcal{J}_1$	set system of all half-open intervals in $\mathbb{R}$
id	identity mapping
$\mathcal{J}_n$	set system of all half-open cuboids in $\mathbb{R}^n$
logit	logit transformation
$\mathrm{logit}\,[P(Y = 1 \mid \mathcal{C})]$	logit of $P(Y = 1 \mid \mathcal{C})$
λ, λ_n	Lebesgue measure on $(\mathbb{R}^n, \mathcal{B}_n)$, where $\lambda := \lambda_1$
$\mu_\#$	counting measure
μ, ν	general symbols for measures
μ_f	image measure of μ under f
$\mathbb{N}$	set of all positive integers without zero (i.e., $\mathbb{N} = \{1, 2, \dots, \})$
$\mathbb{N}_0$	set of all nonnegative integers including zero (i.e., $\mathbb{N}_0 = \{0, 1, 2, \dots, \})$
$\mathcal{N}_{\mu,\sigma^2}$	univariate normal distribution with parameters μ and σ^2
$\mathcal{N}_{\boldsymbol{\mu},\boldsymbol{\Sigma}}$	multivariate normal distribution with parameters $\boldsymbol{\mu}$ and $\boldsymbol{\Sigma}$
$(\Omega, \mathcal{A}, P)$	probability space
$\mathcal{P}(\Omega)$	power set of Ω
P	probability measure
$P(A)$	probability of the event A
$P(X = x)$	probability of the event $\{X = x\} = X^{-1}(\{x\})$
$P(X \in A')$	probability of the event $\{X \in A'\} = X^{-1}(A')$
$P(A \mid B)$	conditional probability of A given B (with respect to the probability measure P)
$P(X_1 \in A', X_2 \in B')$	probability of the event $\{X_1 \in A'\} \cap \{X_2 \in B'\}$
P^B	B-conditional-probability measure

$P(A \mid X=x)$ conditional probability of the event A given the event $\{X=x\}$, also called $(X=x)$-conditional probability of A

$P(Y=y \mid \{X=x\})$ conditional probability of the event $\{Y=y\}$ given the event $\{X=x\}$ with $P(\{X=x\}) > 0$, also denoted by $P(Y=y \mid X=x)$

$P(Y=y \mid X=x)$ conditional probability of the event $\{Y=y\}$ given the event $\{X=x\}$, also called $(X=x)$-conditional probability of $\{Y=y\}$ and also denoted by $P(\{Y=y\} \mid X=x)$

$P(A \mid \mathscr{C})$ a version of the $\mathscr{C}$-conditional probability of the event A

$P(A \mid \mathscr{C}, \mathscr{D})$ a version of the $\sigma(\mathscr{C} \cup \mathscr{D})$-conditional probability of the event A

$P(A \mid \mathscr{C}, Z)$ a version of the $\sigma[\mathscr{C} \cup \sigma(Z)]$-conditional probability of the event A

$\mathscr{P}(A \mid \mathscr{C})$ set of all versions of the $\mathscr{C}$-conditional expectation of the event A

$P(A \mid X)$ a version of the X-conditional probability of the event A

$\mathscr{P}(A \mid X)$ set of all versions of the X-conditional expectation of the event A

$P(Y=y \mid X)$ a version of the X-conditional probability of the event $\{Y=y\}$

$P^{X=x}$ the $(X=x)$-conditional-probability measure on $(\Omega, \mathscr{A})$

P_X distribution of the random variable X (with respect to P)

$P_{X_1,\ldots,X_n}$ joint distribution of the random variables $X_1, \ldots, X_n$, the distribution of the multivariate random variable $X = (X_1, \ldots, X_n)$

$(P_X)_g$ image measure of P_X under g

Φ distribution function of the standard normal distribution

P_X^B distribution of X with respect to the conditional-probability measure P^B

p_X probability function of a discrete random variable X

p_{X_1, X_2} probability function of the bivariate random variable $X = (X_1, X_2)$

π_j jth projection mapping

$\mathcal{P}_\lambda$ Poisson distribution with parameter λ

$P^B(A \mid \mathscr{C})$ a version of the $\mathscr{C}$-conditional probability of the event A with respect to the measure P^B

$\mathscr{P}^B(A \mid \mathscr{C})$ the set of all versions of the $\mathscr{C}$-conditional probability of the event A with respect to the measure P^B

$P^B(A \mid X)$ a version of the X-conditional probability of the event A with respect to the measure P^B

$P^B(A \mid X=x)$ an $(X=x)$-conditional probability of A with respect to the measure P^B

$P^{Z=z}(A \mid X=x)$ an $(X=x)$-conditional probability of A with respect to the measure $P^{Z=z}$

$\mathscr{P}^B(A \mid X)$ the set of all versions of the X-conditional probability of the event A with respect to the measure P^B

$P^{Z=z}(A \mid \mathscr{C})$ a version of the $\mathscr{C}$-conditional probability of the event A with respect to the measure $P^{Z=z}$

$\mathscr{P}^{Z=z}(A \mid \mathscr{C})$ the family of all versions of the $\mathscr{C}$-conditional probability of the event A with respect to the measure $P^{Z=z}$

$P^{Z=z}(A \mid X)$ a version of the X-conditional probability of the event A with respect to the measure $P^{Z=z}$

$\mathscr{P}^{Z=z}(A \mid X)$ the family of all versions of the X-conditional probability of the event A with respect to the measure $P^{Z=z}$

$P_{Y\mid\mathscr{C}}$	a version of the conditional distribution of Y given the σ-algebra $\mathscr{C}$
$\mathscr{P}_{Y\mid\mathscr{C}}$	the set of all versions of the conditional distribution of Y given $\mathscr{C}$
$P_{Y\mid X=x}$	$(X=x)$-conditional distribution of Y
$P_{Y\mid X}$	a version of the conditional distribution of Y given the random variable X
$P_{Y\mid Z}^{X=x}$	a version of the Z-conditional distribution of Y with respect to the probability measure $P^{X=x}$
$P_{Y\mid Z=z}^{X=x}$	$(Z=z)$-conditional distribution of Y with respect to the probability measure $P^{X=x}$
Q_X	quantile function of a real-valued random variable X
$\mathbb{Q}$	set of all rational numbers
$Q_{lin}(Y\mid X)$	the composition of X and the linear quasi-regression (or linear least-squares regression of Y on X)
$Q_{lin}(Y\mid X_1,\dots,X_n)$	linear quasi-regression of Y on $X_1,\dots,X_n$
$R_{Y\mid X}$	multiple correlation of Y and X
$R_{Y\mid X_1,\dots,X_n}$	multiple correlation of Y and $(X_1,\dots,X_n)$
$R_{Y\mid\mathscr{C}}^2$	coefficient of determination of $E(Y\mid\mathscr{C})$
$R_{Y\mid X}^2$	coefficient of determination of $E(Y\mid X)$
$R_{Y\mid X_1,\dots,X_n}^2$	coefficient of determination of $E(Y\mid X_1,\dots,X_n)$
$\mathbb{R}$	set of all real numbers
$\mathbb{R}^2$	Cartesian product $\mathbb{R}\times\mathbb{R}$
$\mathbb{R}^n$	n-fold Cartesian product $\mathbb{R}\times\dots\times\mathbb{R}$
$\overline{\mathbb{R}}$	extended set of all real numbers (i.e., $\overline{\mathbb{R}}=\mathbb{R}\cup\{\infty,-\infty\}$)
$SD(Y)$	standard deviation of the random variable Y
$SE(\overline{Y})$	standard error of the sample mean of the random variables $Y_1,\dots,Y_n$
$\mathrm{sgn}(f)$	sign function of f
$SD(Y\mid X=x)$	$(X=x)$-conditional standard deviation of Y
$SD(Y\mid\mathscr{C})$	a version of the $\mathscr{C}$-conditional standard deviation of Y
$SD(Y\mid X)$	a version of the X-conditional standard deviation of Y
$\sigma(\mathscr{E})$	σ-algebra generated by the set system $\mathscr{E}$
$\sigma(f)$	σ-algebra generated by the mapping f
$\sigma(f_1,\dots,f_n)$	σ-algebra generated by the mappings $f_1,\dots,f_n$
$\sigma(X)$	σ-algebra generated by the random variable X
$\sigma(f,\mathscr{A})$	the σ-algebra generated by the union of $\sigma(f)$ and $\mathscr{A}$
Σ_{xy}	covariance matrix of x and y
Σ_{xx}	variance–covariance matrix of x
t_n	t-distribution with n degrees of freedom
$\mathcal{U}_B$	continuous uniform distribution on the set B
$Var(Y)$	variance of the random variable Y
$Var(Y\mid X=x)$	$(X=x)$-conditional variance of Y
$Var(Y\mid X)$	a version of the X-conditional variance of Y
$Var(Y\mid\mathscr{C})$	a version of the $\mathscr{C}$-conditional variance of Y
x	column vector of numerical random variables
X	matrix of numerical random variables

$X^{-1}(\mathcal{A}'_X)$	σ-algebra generated by the random variable X
$X \sim \mathcal{L}$	the random variable X has the distribution $\mathcal{L}$. Examples for $\mathcal{L}$ are $\mathcal{B}_{n,p}, \mathcal{P}_\lambda, \mathcal{N}_{0,1}$, or $F_{m,n}$
$X \underset{P}{=} Y$	X and Y are P-equivalent
$X(\omega) \underset{P\text{-}a.a.}{=} Y(\omega)$	X and Y are identical for P-almost all $\omega \in \Omega$
Z_Y	Z-transformation of the random variable Y

Author index

Probability and Conditional Expectation: Fundamentals for the Empirical Sciences, First Edition. Rolf Steyer and Werner Nagel.
© 2017 John Wiley & Sons, Ltd. Published 2017 by John Wiley & Sons, Ltd.
Companion website: http://www.probability-and-conditional-expectation.de

Subject index

Probability and Conditional Expectation: Fundamentals for the Empirical Sciences, First Edition. Rolf Steyer and Werner Nagel.
© 2017 John Wiley & Sons, Ltd. Published 2017 by John Wiley & Sons, Ltd.
Companion website: http://www.probability-and-conditional-expectation.de